Essentials of
Microbiology

Dr. S. Rajan, M.Sc., Ph.D.,
Assistant Professor
Department of Microbiology
M.R. Government Arts College
Mannargudi - 614 001

Mrs. R. Selvi Christy, M.Sc., M.Phil.
Microbiologist

CBSPD

CBS Publishers & Distributors Pvt Ltd

New Delhi • Bengaluru • Chennai • Kochi • Kolkata • Lucknow • Mumbai
Hyderabad • Jharkhand • Nagpur • Patna • Pune • Uttarakhand

Essentials of
Microbiology

ISBN 978-93-86827-67-8

CBS Reprint: 2018
 Reprint: 2020, 2023, **2025**
First Edition: September 2015
Revised Print: August 2016

Published by **Satish Kumar Jain** and produced by **Varun Jain** for

CBS Publishers & Distributors Pvt Ltd
4819/XI Prahlad Street, 24 Ansari Road, Daryaganj, New Delhi 110 002, India.
Ph: 011-23266838, 23289259   Website: www.cbspd.com
                                e-mail: delhi@cbspd.com

Corporate Office: 204 FIE, Industrial Area, Patparganj, Delhi 110 092
Ph: 011-4934 4934   Fax: 011-4934 4935
                                e-mail: publishing@cbspd.com; publicity@cbspd.com

Branches

Bengaluru: Seema House 2975, 17th Cross, KR Road, Banasankari 2nd Stage, Bengaluru 560 070, Karnataka, India
Ph: +91-80-26771678/79   Fax: +91-80-26771680   e-mail: bangalore@cbspd.com
Chennai: 18/8B, Subbarayan Street, Shenoy Nagar, Chennai 600 030, Tamil Nadu, India
Ph: +91-44-42032115, 26681266   e-mail: chennai@cbspd.com
Kochi: 42/1325, 1326, Power House Road, Opp KSEB, Power House, Ernakulum Kochi 682 018, Kerala, India
Ph: +91-484-4059061-65,67   Fax: +91-484-4059065   e-mail: kochi@cbspd.com
Kolkata: 147, Hind Ceramics Compound, 1st Floor, Nilgunj Road, Belghoria, Kolkata-700056, West Bengal, India
Ph: +033-25633055, 033-25633056   e-mail: kolkata@cbspd.com
Lucknow: Basement, Khushnuma Complex, 7 Meerabai Marg (Behind Jawahar Bhawan), Lucknow-226001, UP, India
Ph: +0522-4000032   e-mail: tiwari.lucknow@cbspd.com
Mumbai: PWD Shed, Gala no 25/26, Ramchandra Bhatt Marg, Next to JJ Hospital Gate no. 2, Opp. Union Bank of India, Noorbaug, Mumbai-400009, Maharashtra, India
Ph: 022-66661880/89   e-mail: mumbai@cbspd.com

Representatives

| Hyderabad | 0-9885175004 | Jharkhand | 0-9811541605 | Nagpur | 0-8692091830 |
| Patna | 0-9334159340 | Pune | 0-9664372571 | Uttar akhand | 0-9716462459 |

Printed at Rashtriya Printers, Dilshad Garden, Delhi, India

Dr. N. THAJUDDIN
Professor & Head
Department of Microbiology
Dean, Faculty of Science, Engineering & Technology

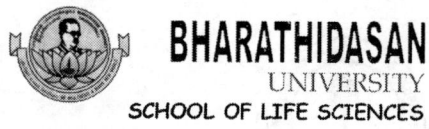

BHARATHIDASAN
UNIVERSITY
SCHOOL OF LIFE SCIENCES

July 25, 2015

Foreword

"*Essentials of Microbiology*", book written by Dr. S. Rajan, an experienced teacher and researcher, provides a comprehensive account of microbiology in its holistic perspective to introduce students to microbiology. The subject matter has been dealt in 17 chapters covering morphological diversity of microbes, their ecobiology, ultrastructure, biochemistry, physiology, genetics, immunology and biotechnology in a very lucid manner to enable the student community to prepare for their highly competitive future. The way in which the author has introduced the subject in a step-by-step approach is admirable. He has made a focused effort to provide both the fundamentals and the complexities related to microbiology in a very simple perspective. Though several books on microbiology are available, the present book covers basic information on microbiology in a systematic manner providing not only the accumulated information, but also the recent developments in the subject, profusely illustrated with simple line drawings readily reproducible by students. Certainly, the content of the book would be useful to the students pursuing courses both at undergraduate and post-graduate levels in all biological disciplines as well as teachers in colleges and universities. I am confident that this book will prove to be a useful compendium and cater to the needs of students at different levels and lead to the transition from student to peer and novice to authority in microbiology. I congratulate Dr. S. Rajan for bringing out this book, which would fill long-felt need of both teachers and students.

(N. THAJUDDIN)

M. R. Government Arts College

VOC Road
Mannargudi – 614 001

Dr. S. Karthikeyan
Principal

     Microbiology is pioneering science which deals with organisms invisible to naked eye and ubiquitous in nature. This branch of Science came to the limelight in the 19th century even though these are first evolved organisms in the world. This branch of science actually took a shape after the inventions and discoveries of Louis Pasteur, the father of modern Microbiology. This fascinating branch of Science includes the study of bacteria, viruses, fungi, Algae, protozoa, Actinomycetes, Chlamydia etc and their interaction with biotic and abiotic factors. Microorganisms are both beneficial and harmful and all living organisms interact with the microorganisms either positively or negatively man has started exploiting micro organisms for making various products from time immemorial even without the knowing the nature and type of microorganisms some microorganisms are really harmful Contagious and result in morbidity and mortality of human beings and animals. Even Plants and small tiny Creatures are not spared by these microbes

     The author of this book has taken meticulous effort to compile the various facet of microbiology right from the fundamentals to Molecular level. The author has taken strenuous effort to incorporate all essential features in the field of microbiology. He has penned not only the conventional method of identification of microbes but also the recent 16S & 23S rRNA sequencing techniques to identify the microbes. He has dealt in length the classification of all microorganisms. He has also included the important equipments used in micro biology laboratory and their applications. The author has included in the book the microbial physiology and microbial interactions in the soil and their ability to transform various nutrients in soil for the benefit of mankind. The author has shown much to interest to include fermentation techniques and microbial products and has also included microbial diseases, their etiology and the ways to control them.

     I Sincerely appreciate the effort of the author to write a comprehensive book on microbiology which will be of immense help to UG & PG students of all biological sciences.

Dr. S. Karthikeyan

PREFACE

Essentials of Microbiology is a text book for all the students of Life Science. This book is prepared based on a decade experience on the field of Microbiology teaching and research. This book provides ample informations regarding *basic and applied concepts of Microbiology.*

This book is divided into 17 prime chapters which provides basic information related to respective topics. This book covers the concepts of general microbiology, microbial physiology, immunology, medical microbiology, soil and agricultural microbiology, environmental microbiology, microbial genetics, molecular biology, microbial biotechnology, industrial microbiology, food microbiology and general virology. This book has been structured in such a way, without sacrificing the completeness or accuracy of information, that students can access information in small, well identifiable steps. Illustrations make the subject more interesting and easy to understand. We hope this book will promote self learning, creates interests and will help the students for quick revision of the therotic aspects for their university as well as competitive examinations.

Our First and foremost, duty is to thank our microbiology guru, Dr. T. Thirunalasundari, Professor, Department of Industrial Biotechnology, Bharathidasan University, Tiruchirapalli who has been providing constant encouragement to sustain in this field. We are highly thankful to Dr. N. Thajuddin, Head, Department of Microbiology, Bharathidasan University for his encouragement and support. We should not forget the support of Principal, M.R. Govt. Arts college, Mannargudi, Head and Faculty members of MRGAC, Mannargudi.

We owe our gratitude to all the scientists in the field of Microbiology, without their contribution, genesis of this book would not have been possible.

We are highly thankful to the Publisher M/s. Anjanaa Book House for their suggestions & encouragements without which this book would not have come in this form. We are also delightful when thanking Ms. N. Thilagavathi for her typesetting and Cover design.

We are highly indebted to our family members for their constant support, cooperation, encouragement throughout our life.

Last but not least, we thank god almighty for making my attempt to reality.

We earnestly invite comments and suggestions from the readers to enable us to improve the book in the subsequent editions, for any suggestions please feel free to contact us through our **E-mail Id : ksrajan99@gmail.com**

- Authors

CONTENTS

hydrocarbon-Polychlorinated   biphenyls (PCB) biodegradation;
U. Biodeterioration- Paper, Textile, Metal, Leather and wood

AA. Hepatitis, AB. AIDS, AC. Chicken Pox; , AD. Dengue, AE. Yellow fever; AF. Cold Sore; AG. Superficial Mycosis (Pityriasis versicolor, Tinea Nigra, Piedra, Dermatophytosis), AH. Subcutaneous Mycoses (Sporotrichosis, Chromoblastomycoses, Mycetoma); AI. Systemic Mycosis (Coccidioidomycosis, Histoplasmosis, Blastomycosis, Candidiasis, Cryptococcosis); AJ. Amoebiosis, AK. Giardiasis, AL. Malaria, AM. Ascariasis, AN. Filariosis; AO. Zoonotic infections; AP. Hospital Borne Infection.

INTRODUCTION

Microbiology is the branch of science that deals with microorganisms, which are seen under the microscopes. Microbiology covers several disciplines, including virology (study of viruses), bacteriology (study of bacteria), mycology (study of fungi), and parasitology (study of parasites). Specialization within microbiology may include microbial physiology (i.e., microbial growth, metabolism, structure), microbial genetics and evolution, environmental microbiology (i.e., microbial ecology), industrial microbiology (i.e., industrial fermentation, wastewater treatment), and food microbiology (i.e., use of microbes for food production, fermentation). Major milestones in the field of biology is only due to microorganisms. Micro organisms are a tiny life exist in all places whereever life is possible.

A. HISTORY OF MICROBIOLOGY

Microbiology is a relatively new discipline and its development is described below.

"Today's News Tomorrows History"

During the last century people gained some knowledge about microbiology. Now microbial importance is well known to mankind. Ancient people regarded diseases as supernatural in origin and sent by the God as punishment for the sins of human beings. Varro recorded the diseases by living creaters in the 2nd century BC. Fracastro (1546) suggested that invisible organisms cause diseases. The years between 1590 – 1609, Jansen developed first useful compound microscope. However, the first person to observe the microbes and explained microorganisms was by amateur microscopist Antony van Leeuwenhoek (1676). He also given the name 'little animalcules' to the microorganisms.

General Microbiology

1590 - **Jensen & Hans** designed the first compound microscope.

1786 - **Mueller** first classified the bacteria.

1805 - **Nicholas Appert** developed Appertization technique.

1854 - **John Snow** showed the reason for cholera outbreak. He was called the **father of epidemiology.**

1859 - **Darwin** postulated about origin of species.

1865 - **Joseph Lister** introduced aseptic technique.

1870 - **His** developed microtome for cutting sections of tissue cells.

1878	-	**Ernst Karl Abbe** introduced oil immersion lens.
1884	-	**Walter Hesse** used agar agar as a solidifying agent.
1884	-	**Christian Gram** - Gram staining technique.
1886	-	**Ernst Karl Abbe** - Abbe condenser.
1886	-	**Mac Munn** discovered cytochromes.
1887	-	**Richard petri** devised petriplate

Charles chamberland discovered bacterial filter and autoclave.
Ferdin and Colin discovered the multiplication of bacteria.

1890	-	**Sergei Winogradsky** observed Auxotrophic growth of chemolithotrophs.
1894	-	**Ehrlich** articulated the principle of selective toxicity.
1897	-	**Buchner** discovered cell free alcohol fermentation.
1901	-	**Martinous Beijerinck** -Enrichment culture methods.
1903	-	**Buchner** discovered enzymes.
1906	-	**Tswell** - developed Chromatography.
1907	-	**Harrison** developed tissue culture techniques.
1912	-	**Carel** deviced the technique for tissue culture
1926	-	**Sved Berg**-Ultra centrifugation
1937	-	**Krebs** - TCA cycle
1932	-	**Knoll&Ruska** discovered electron microscope.
1953	-	**Zernike** discovered Phase contrast microscope.
1954	-	**Salk** introduced inactive polio virus vaccine.
1955	-	**Sabin** introduced live attenuated polio vaccine.
1977	-	**Carl Woese** discovered the Archae.
1978	-	**Carl Woese** deviced classification of microorganisms.
1980	-	WHO proclaimed small pox as an extinct disease.
1986	-	First Hepatitis B vaccine produced by genetic engineering and approved for human use.
1995	-	*Haemophillus influenzae* gene sequenced.
1996	-	Yeast gene sequenced.
1997	-	Discovery of *Thiomargarita namibienis,* the largest bacterium in the world.
1997	-	*E.coli* gene sequenced.

2000 - Scientists discovered that *V.cholerae* has two separate chromosomes.

2009 - *Janibacter hoylei , Bacillus isrealeansis* and *Bacillus aryabhattai* are identified from the space.

2011 - *Halomonas titanicae,* a new bacteria found on the rusting hull of the *Titanic.*

The period between 1860 and 1910 is considered as the **Golden Age of Microbiology** because during this time various types of observations were made by Louis Pasteur and Robert Koch

Chemotherapy

1910 - **Paul Ehrlich** developed Salvarson a magic bullet for syphilis

1929 - **Alexander Fleming** discovered Penicillin.

1935 - **Domag** discovered sulpha drugs.

1944 - **Waksman** discovered Streptomycin.

1962 - Nalidixic Acid was discovered.

1990 - **Murry & Johnson** used immuno suppressive agents to perform successful transplantation.

Immunology

1798 - **Edward Jenner** developed smallpox vaccine

1885 - **Louis Pasteur** produced Rabies vaccine

1884 - **Metchnikoff** discovered phagocytosis

Jules Bordet - Complement Fixation test (CF).

Almorth Wright - role of opsonin in phagocytosis.

1903 - **Wright** discovered antibodies in blood.

1913 - **Richet** worked on anaphylaxis

1921 - **Fleming** discovered lysozyme.

1930 - **Landsteiner-**ABO blood grouping

1931 - **Lewis** discovered Pinocytosis

1953 - **Medavar** discovered immune tolerance

1955 - **Jerne and Burnet** proposed the clonal selection theory.

1957 - **A.Isaacs & J.Lindermann** discovered interferon.

1959 - **Yalow -** Radial Immuno Assay technique

1963   -   **Porter** proposed the structure of IgG

1972   -   **G.Edelman, R.Porter-** structure of antibodies.

1975   -   **Kohler and Milstein** -Monoclonal antibody synthesis.

1980   -   **Cerraft, Snell & Daussel-** histo compatibility antigen.

1987   -   **Susumu Tonegawa desenbed-**The genetic principle for generation of antibody diversity and got nobel prize.

1996   -   **P.Doherty & R.Zinkerngel-**Discovery of the mechanism by which T lymphocytes to recognize virus-infected cell.

Medical Microbiology

1871   -   **Gerhard Hansen** - *Mycobacterium leprae*

1873   -   **Otto H.F.Obermier** - bacteria for Relapsing fever.

1884   -   **George Gaffley** cultivated typhoid bacilli.

1884   -   **Loeffler** cultivated Diphtheria bacilli.

            Kitasato discovered tetanus bacilli.

            Albert Calmette-Nonvirulent strain of TB bacilli.

            Rose discovered malaria causing agent

1892   -   **William Welch** identified gas gangrene bacilli.

1887   -   **Richard Pfeiffer** identified cause of meningitis.

1894   -   **Kitasato &Yersin** identified Plaque causing agent

1898   -   **Kiyoshi Shiga-***Shigella dysentriae.*

1876   -   **Koch-** *Bacillus anthracis*

1879   -   **Neisser-** *Neisseria gonorrhoeae*

1885   -   **Escherich-***E.coli*

1882   -   **Koch-***Mycobacterium tuberculosis*

1883   -   **Koch-***Vibrio cholerae*

1886   -   **Fraenkel-***Streptococcus pneumoniae.*

1887   -   **Bruce-**Brucella

1884   -   **Loeffler –** Soluble toxin of *C. diphtheriae.*

1888   -   **Roux and Yersin –** Antitoxins of *C. diphtheriae.*

1888   -   **Van Ermengem-***Clostridium botulinum.*

1905   -   **Schaudinn & Hoffmann -***Treponema pallidum*

1906 - **Bordet & Gongou**-*Bordetella pertusis*

1909 - **Ricketts**-*Rickettsia rickettsi.*

1912 - **McCooy Chapin**-*Francisella tularensis.*

1913 - **Schick** -skin test for diphtheria in humans.

1960 - *Kingella kingae* discovered by **Elizabeth King**

1975 - Lyme disease discovered

1997 - Legionellaris –*Legionella pneumophila.*

1977 - Enteric disease by *Campylobacter jejuni.*

1981 - Toxic shock syndrome diagnosed

1982 - Hemorrhagic colitis caused by *E.coli* O157:H7

1982 - **Barry Marshall and Robin Warren**-Peptic ulcer–*Helicobacter pylori* - Got Nobel prize in 2005.

1982 - **Willy Burgdorfer** discovered *Borrelia burgdorferi*

2003 - **Dr. David E. Greenberg** isolated *Granulobacter bethesdensis* from lymph node of human.

Virology – Details Refer Virology section Page No. 364-367

1892 - **Ivanovsky**-Discovered TMV

1900 - **W.Reed**-Yellow fever causative agent was discovered.

1903 - **P.Remlinger**-Discovered of Rabies virus

1903 - **A.Negri**-Demonstrated inclusion bodies of Rabies virus

1908 - **V.Ellermann &O.Bang**–Avian leukemia virus

1911 - **P.Rous**- Avian sarcoma causative virus was identified.

1932 - **Furth**-Used mice as a host for virus

1935 - **Stanley** – crystallized viruses

1933 - **R.E.Shope**-A virus causing mammalian cancer identified

1939 - **G.A.Kausche**-Virus visualized under electron microscope.

1949 - **J.F.Enders**-Human cell cultures for the growth of Poliovirus

1953 - **W.P.Rowe**-Adenovirus discovered.

1955 - **F.L.Schaffer**-Polio virus crystallized

1959 - Parvovirus discovered

1963 - Occurrence of double stranded RNA virus is discovered.

1981   -   **Luc Montagnier**-Discovered HIV

1997   -   **S.Prusiner**-Discovery of prions.

2003   -   SARS(Severe Acute Respiratory Syndrome) identified at China-Corona virus causes SARS.

2014   -   MERS- Maddle East Respiratory Syndrome-A viral disease recognized.

Parasitology

1976   -   Cryptosporidiasis

1985   -   Picrosporidiasis

1991   -   Bebiasis

1998   -   Myositis-Brachiola

B.   CONCEPTS OF MICROBIOLOGY

Spontaneous Generation Theory

Generation of living matter from nonliving matter is known as spontaneous generation. **In 384-322BC Aristotle** said that animals could originate from soil. In 1665, **Francesco Redi** (1665) disproved the spontaneous generation theory, where as later (1728-1799) **Spallanzani** introduced the sterile culture media and disproved spontaneous generation theory. **John Needham** (1749) proved the spontaneous generation. Doubts of Spallanzani was cleared by the following scientists. **Franz Schulze** (1815-1873) – passed air through concentrated sulphuric acid to disprove spontaneous generation. **Theoder Schwann** (1810-1882)-passed air through red-hot tube to disprove spontaneous generation. In 1859, a controversy of **Pouchet** experiment was observed. During this time **Pasteur** prepared boiled broth in flask with long, narrow and goose neck tubes that are open to air. There is no growth because microbes settle in the goose neck region. Finally **John Tyndall** (1820-1893) proved dust carried germs and the spontaneous generation story is completed.

Germ Theory of Disease

Casual nature of infectious disease was first established only in the later half of the 19th century. In early 1800s, **Agastino Bassi,**proved that a fungus causes a disease in silkworm called Muscardine in France. In 1840s, **Oliver Wendell Holms** wrote about contagiousness of Puerperal fever. **Ignaz Semmelweis** (1847) postulated that blood victims of puerperal fever contained the causative agent of disease. In 1861, **Ignaz Semmelweis** introduced antiseptic technique, and it was not fully realized until the late 1870s, when Joseph Lister demonstrated the value of spraying operating room with phenol.

Theory of Fermentation

In 1880s, **Pasteur** demonstrated about diseases of wine. In 1850, **Pollander, Rayer** and **Davine** observed rod shaped bacteria in the blood of animals (Anthrax). 1n 1857 Pasteur demonstrated Lactic acid fermentation. In 1929, Flemming demonstrated penicillin production. In 1982 - Rabies vaccine was developed.

C.   MAJOR CONTRIBUTIONS OF MICROBIOLOGY

Antony Van LeeuwenHoek (1676)

He is the father of Microbiology

He was a merchant, Haber dasher, and owner of a dry goods shop in Delf, Holland.

He was a qualified surveyor and the town's official wine taster.

In his spare time, he ground pieces of glass in to fine lenses.

He was not an educationalist but had keen mind.

Magnification of Antonys microscope was between 100 and 200.

Figure 1   Leeuwenhoek

He observed minute objects through his lens and named as Very little *Animalcules*.

He observed microbes from rainwater, pepper infusions, saliva, tooth scrapings and excreta.

He communicated his findings to the Royal Society of London and the observations were published in 1677 in the proceedings of the Royal Society as a series of letter.

In 1680, he was elected as a *Fellow of the society*.

Robert Koch (1843 – 1910)

He was a German physician.

He is a father of **Medical Microbiology.**

In 1876, he proposed germ theory of diseases & demonstrated the causative nature of anthrax.

In 1881, he cultured bacteria on Gelatin and also used agar to demonstrates pure culture technique. He also discovered Tubercle Bacilli in 1882, Discoverd Cholera Bacilli in 1883 and published Koch postulates in 1884.

Figure 2   Robert koch

Koch Postulates

The microbes must be present in every case of disease.

The suspected microorganism must be isolated and grown in a pure culture.

The same disease must result when the isolated microbe is inoculated in to the healthy host.

The same organism must be isolated again from the infected host.

Exceptions of Koch postulates are

Some diseases are caused by Opportunistic pathogens (Comedo).

If the animals are immunized with the pathogen, the disease does not arise (Chickenpox).

Some disease requires cooperation between the pathogens (Impetigo).

Some organisms cannot culture as pure culture *(Treponema pallidum)*.

Usefulness of Koch's Postulates

1. It is useful in determining pathogenic organisms.
2. To differentiate the pathogenic and nonpathogenic microorganism.
3. For the classification of organisms.
4. To detect the susceptibility, resistance of the laboratory animals.

Joseph Lister (1827 – 1912)

He was a pioneer of **antiseptic surgery**. Joseph Lister developed antiseptic method for preventing infection using carbolic acid to treat wounds in 1867. He developed "serial dilution technique" in liquid media. He identified the bacteria, *Bacterium lactis* from milk sample. Lister used bandages soaked in carbolic acid to dress wounds caused by compound fractures. His discovery of chemicals which prevent infections greatly increased survival rates of the wounded patients. His antiseptic principles guide today's modern surgical procedures.

Figure 3   Joseph Lister

Martinus W. Beijerinck (1851 – 1931)

Beijerinck isolated root nodule causing bacteria. He published the results on tobacco mosaic disease in 1898 and 1900. He proposed that the TMV disease was caused by an entity that is entirely different from bacteria. He called viruses are a filterable agents. Beijerinck observed that the virus would multiply only in living plant cells. Beijerinck showed that the viruses could survive for long periods in a dried state. He made fundamental contributions to microbial ecology. He isolated the aerobic nitrogen fixing bacterium *Azotobacter* and sulfate reducing bacterium. He developed enrichment culture technique and proposed the uses of selective media along with Winogradsky

Figure 4   Beijerinck

Elie Metchnikoff (1845 – 1916)

He found out the concept of Phagocytosis. His work on antitoxin, provided evidence that, immunity could result from soluble substances in the blood, now known as antibodies. He described that blood cells are important in immunity. He discovered that some blood leukocytes could engulf disease causing bacteria. He called these cells as phagocytes, which is an important process in immunology.

Fannie Eilshemius and Walther Hesse (1850 – 1934)

Fannie and Hesse developed agar agar as better alternate to gelatin. They are the assistants of Koch. Hesse suggested the use of agar as a solidifying agent. She showed that agar was not attacked by most bacteria and did not melt until reaching a temperature of above 100°C. This development made possible the isolation of pure cultures. The discovery of agar directly stimulated progress in all areas of bacteriology.

Sergei. N. Winogradsky (1856 – 1953)

The Russian microbiologist Sergei N. Winogradsky made many contributions to soil Microbiology. He discovered that soil bacteria would oxidize iron, sulfur and ammonia to obtain energy. He also showed that many bacteria could incorporate CO_2 into organic matter much like photosynthetic organisms do. Winogradsky also isolated anaerobic nitrogen – fixing soil bacteria and studied the decomposition of cellulose. He developed the enrichment – culture technique and the use of selective media, which have been of great importance in microbiology along with Beijerinck

Figure 5   Fannie & Hesse

Figure 6   Winogradsky

Paul Ehrlich (1854 – 1915)

Paul Ehrlich in 1910, worked on Chemotherapy. He used an arsenic based drug called salvarsan to treat syphilis, a sexually transmitted disease, caused by *Treponema pallidum*. In 1898, he proposed that cells possess a wide variety of side chains on their surfaces. He won the Nobel Prize in 1908 for his work on immunity.

Figure 7   Paul Ehrlich

Karl Landsteiner (1868 – 1943)

Landsteiner discovered blood group antigens and their corresponding agglutinins in 1900. Landsteiner was a dominant figure in immunology for 40 years, developing the concept of the antigenic determinant and demonstration. The discovery of blood groups leads to the blood transfusion without provoking reactions. In 1900, he won the Nobel Prize for his discovery of blood group antigens and antibodies.

Figure 8   Karl Landsteiner

James Watson and Francis Crick

James Watson and Francis crick proposed the double helical structure of DNA in 1953. They rely on the simple laws of structural chemistry. Watson and crick shared the Nobel Prize in 1953 for Medicine with Maurice Wilkins. They explained the model of DNA & how they can transmit hereditary information. The discovery of DNA lead to the development of fields like Genetic engineering, Biotechnology etc., Rapid amplification of DNA by PCR technique is possible nowadays only because of their discovery.

Figure 9   Watson and Crick

Sir Alexander Fleming (1881- 1955)

Sir Alexander Fleming discovered the antibiotic penicillin in 1929. He observed that the mould *Penicillium notatum* killed his cultures of bacterium *Staphylococeus aureus*. After separating the fluid from the cell, Fleming discovered that the cell free liquid was an inhibitor for many bacterial species. He was the first person to demonstrate that a substance produced by microorganisms would inhibit or kill other microorganisms. His discovery made the modern era of drug therapy, and lead to the discovery of therapeutic value of penicillin.

Figure 10   Fleming

Louis Pasteur (1822 – 1895)

Figure 11   Louis Pasteur

He was a Professor of Chemistry at the University of Lille, France. He is the father of **fermentation technology.** He conducted Swan Neck flask experiment to disprove the spontaneous generation theory (1859). In 1865 he demonstrated that the microscopic germs are responsible for silkworm disease. In 1881- Introduced vaccination against anthrax. In 1885, Pasteur announced to the French Academy of Sciences that he had developed a vaccine for preventing a dread disease, rabies. The term vaccine was given by Pasteur to honour **Edward Jenner.** Pasteur introduced the concept of **pasteurization** (1860) in which the microorganisms can be removed without affecting the flavour and taste of grape juice. He discovered causative agent of alcohol fermentation. Life without air, Animals cannot live in the absence of microorganisms were the statements given by him.

Edward Jenner

He is the **Father of Immunization.** The significance of immunization against smallpox was come to the light by the work of Edward Jenner. He observed that individual who attended cow with cowpox, a disease of cattle caused by a similar virus, rarely develope small pox. In May 14, 1796, Jenner extracted the contents of a pustule from the arm of cowpox infected milkmaid, and injected into another person. As he expected, no symptoms were developed when the person was inoculated with smallpox virus. In 1798 Jenner reported to the Royal Society of London on the value of vaccination with cowpox as a means of protecting against smallpox, which is the basis for the immunological prevention of disease.

Dmitri Iwanowski (1864-1920)

Figure 12   Jenner

He provided evidence for virus causation of Mosaic Disease of Tobacco (TMV). On 12th February 1892, **Dmitri Iwanowski**, a Russian botanist, presented a paper to the St. Petersburg Academy of Science which showed that extracts from diseased tobacco plants could transmit disease to other plants after passage through ceramic filters fine enough to retain the smallest known bacteria. This is generally recognised as the beginning of Virology. Unfortunately, Iwanowski did not fully realize the significance of these results.

D.   THE SCOPE OF MICROBIOLOGY

Bacteria were the first living organisms on our planet. They live virtually everywhere life is possible; hence bacteria are called omnipresent life. They are more numerous than any other kind of organism. Bacteria are considered as the largest component of the earth's biomass. We live in the Age of Bacteria. The whole ecosystem depends on bacterial activities. They influence human society in countless ways. Thus modern microbiology is a large discipline with many different specialties. Microbiology has a great impact on fields such as medicine, agricultural, food sciences, ecology, genetics, biochemistry, and molecular biology. In the 1970s new discoveries in microbiology led to the development of recombinant DNA technology and genetic engineering.

Microbiology has both basic and applied aspects. Depends on the pattern of work, microbiologists are called virologists (viruses), bacteriologists (bacteria), phycologists or algologists (algae), mycologists (fungi), or protozoologists (protozoa). Some microbiologists are interested in microbial morphology and work in fields such as microbial cytology, microbial physiology, microbial ecology, microbial genetics, molecular biology and microbial taxonomy. Many microbiologists have a more applied orientation and work on practical problems in fields such as medical microbiology, food and dairy microbiology, and public health microbiology.

Medical microbiology - Medical microbiologists are involved in identifying the agents involved in the infectious disease. They also involved in the controlling of diseases. They are frequently involved in tracking down new, unidentified pathogens. They are also study the ways in which microorganisms cause disease.

Public health microbiology-Main work of public health microbiologists are to control the spread of communicable diseases. They often monitor community food establishments and water supplies in an attempt to keep them safe and free from infectious disease agents.

Immunology -Those people work on immune system related work are called immunologists. They are concerned with how the immune system protects the body from pathogens and the response of infectious agents. It is one of the fastest growing areas in science.

Agricultural microbiology-It is concerned with the impact of microorganisms on agriculture. Agricultural microbiologists try to combat plant diseases that attack important food crops, work on methods to increase soil fertility and crop yields.

Microbial Ecology-The field of microbial ecology is concerned with the relationships between microorganisms and their living and nonliving habitats. Microbial ecologists study the contributions of microorganisms to the cycling of various nutrients or elements. Microbial ecologists are employing microorganisms in bioremediation to reduce pollution effects.

Food and dairy microbiology - Main work of the food microbiologists is to prevent microbial spoilage of food and the transmission of foodborne diseases. They also use microorganisms to make foods such as cheeses, yogurts, pickles and beer.

Industrial microbiology - Microorganisms are used to make products such as antibiotics, vaccines, steroids, alcohols, vitamins, amino acids, and enzymes.

Biomining - Microorganisms can even leach valuable minerals from low-grade ores.

Microbial genetics and molecular biology - It focuses on the nature of genetic information and how it regulates the development and function of cells and organisms. The use of microorganisms has been very helpful in understanding gene function. Microbial geneticists play an important role in applied microbiology by producing new microbial strains that are more efficient in synthesizing useful products. Genetic techniques are used to test substances for their ability to cause cancer.

Genetic engineering – It has arisen from work in microbial genetics and molecular biology and will contribute substantially to microbiology. Engineered microorganisms are used to make hormones, antibiotics, vaccines, and other products. New genes can be inserted into plants and animals.

E.   THE FUTURE OF MICROBIOLOGY

Microbiology has a clearer mission than other scientific disciplines. It is confident of its value because of its practical significance. Microbiology is required both to face the threat of new and reemerging human infectious diseases and to develop industrial technologies that are more efficient and environmentally friendly.

The following brief list should give some idea of what the future may hold:

1.   Every day microbes changing its nature and new diseases are arising. Hence microbiologists need to work on new infections and infectious agents.

2.   Some of the microbes becoming widespread and destructive. Microbiologists will have to respond to these threats, many of them presently unknown.

3. Microbiologists must find ways to stop the spread of established infectious diseases.

4. Microbiologists along with pharmacologist's and chemists have to create new drugs and find ways to slow or prevent the spread of drug resistance microorganisms.

5. New vaccines must be developed to protect against diseases such as AIDS. It will be necessary to use techniques in molecular biology and recombinant DNA technology.

6. Microbiologists should assess the association between infectious agents and chronic diseases such as autoimmune and cardiovascular diseases.

7. We are only now beginning to understand how pathogens interact with host cells and the ways in which diseases arise. There also is much to learn about how the host resists pathogen invasions.

8. Microbial diversity is another area requiring considerable research.

9. It is estimated that less than 1% of the earth's microbial population has been cultured. We must develop new isolation techniques for the isolation of microorganisms.

10. Much work needs to be done on microorganisms living in extreme environments. The discovery of new microorganisms may well lead to further advances in industrial processes and enhanced environmental control.

11. Microbial communities often live in biofilms, and these biofilms are of profound importance in both medicine and microbial ecology. Research on biofilms is in its infancy and it needs complete study.

12. The genomes of many microorganisms already have been sequenced, and many more will be determined in the coming years.

13. Microorganisms are essential partners with higher organisms in symbiotic relationships. Greater knowledge of symbiotic relationships can help improve our appreciation of the living world. It also will lead to improvements in the health of plants, livestock, and humans.

The future of microbiology is bright. Microbiology is one of the most rewarding of professions because it gives its practitioners the opportunity to be in contact with all the other natural sciences and thus to contribute in many different ways to the betterment of human life.

F.   PROKARYOTES AND EUKARYOTES

Microorganisms and all other living organisms are classified as prokaryotes or eukaryotes. Prokaryotes and eukaryotes are distinguished on the basis of their cellular characteristics. Prokaryotic cells lack a nucleus and other membrane-bound structures known as organelles. Eukaryotic cells have both a nucleus and organelles.

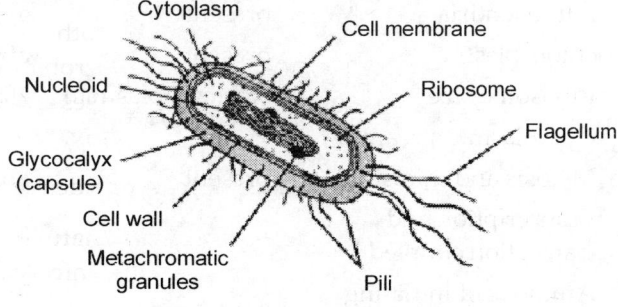

Figure 13   Prokaryotic Cell

Prokaryotic and eukaryotic cells are similar in several ways. Both types of cells are enclosed by cell membranes (plasma membranes), and both use DNA for their genetic information.

Prokaryotes include several kinds of microorganisms, such as bacteria and cyanobacteria. Eukaryotes include fungi, protozoa, algae, plant and animals.

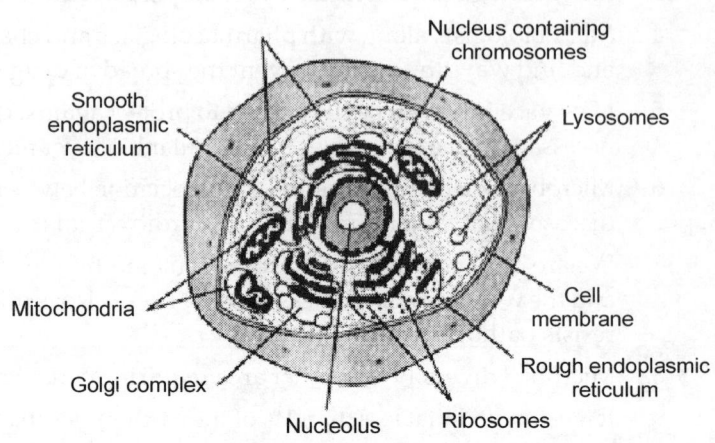

Figure 14   Eukaryotic cell

Table 1.1   Differential Properties of Prokaryotes (Bacteria, Archaea) and Eukaryotes

Property	Biological Domain		
	Eukarya	**Bacteria**	**Archaea**
Cell configuration	eukaryotic	prokaryotic	prokaryotic
Nuclear membrane	Present	absent	absent
Number of chromosomes	>1	1	1
Chromosome topology	linear	circular	circular
Murein in cell wall	-	+	-
Cell membrane lipids	ester-linked glycerides; unbranched; polyunsaturated	ester-linked glycerides; unbranched; saturated or monounsaturated	ether-linked branched; saturated
Cell membrane sterols	present	absent	absent
mitochondria	present	absent	absent
chloroplasts	present	absent	absent
Ribosome size	80S (cytoplasmic)	70S	70S
Cytoplasmic streaming	+	-	-
Meiosis and mitosis	present	absent	absent
Transcription and translation coupled	-	+	+
Amino acid initiating protein synthesis	methionine	N-formyl methionine	methionine

2

CLASSIFICATIONS OF MICROORGANISMS

A. TAXONOMY

Taxonomy is defined as the science of biological classification. It has three interrelated parts. They are classification, nomenclature and identification. **Classification** is the arrangement of organisms into groups or taxa. **Nomenclature** is concerned with the assignment of names to taxonomic groups. **Identification** is the process of determining that a particular isolate belongs to a recognized taxon. The systematics is the scientific study of organisms with the ultimate object of characterizing and arranging them in an orderly manner. The most desirable classification system is a **natural classification.** It arranges organisms into groups based on its common characteristics. Linnaeus developed the first natural classification, based on anatomical characteristics. Organisms can be grouped together based on overall similarity to form a phenetic system. The development of computers has made possible the quantitative approach known as **numerical taxonomy.** Peter H. A. Sneath and Robert Sokal have defined numerical taxonomy as "the grouping by numerical methods of taxonomic units into taxa on the basis of their character states." The results of numerical taxonomic analysis are often summarized with a treelike diagram called a **dendrogram**

Taxonomic Ranks

The basic taxonomic group in microbial taxonomy is the species. A prokaryotic species is the collection of strains that share many stable properties and differ significantly from other groups of strains. A **strain** is a population of organisms that is distinguishable from atleast some other population within a particular taxonomic category. **Biovars** are varient strains based on biochemical or physiological characters. **Morphovars** differ morphologically. Serovars have distinctive antigenic properties. Microbiologists name microorganisms by using the binomial system. This system of nameology provides two parts to each name. The first part first letter is capitalized and the second part is uncapitalized but both parts should be italized. Eg. *Staphylococcus aureus.* Approved list of microbial names were first published in 1980 in the International Journal of Systematic Bacteriology and new valid names are published periodically. Bergeys manual systematic bacteriology contains the currently accepted system of classification system of prokaryotic taxonomy.

Taxonomic Hierarchy

This is the complete list of taxa (classification groups) which are used to classify living organisms:

Domain (this is the relatively new one)

Kingdom; Phylum; Class; Order; Family; Genus and Species

Major characteristics used in taxonomy

Characteristics used in taxonomy are categorized in to two groups. They are classical characteristics and molecular characteristics. Classical characteristics used in taxonomy are morphological, physiological, biochemical, ecological and genetic characteristics. Molecular characteristics includes the study of proteins and nucleic acids.

B.   CLASSIFICATION

Classification is the arrangement of organisms into groups or taxa. Various scientists from different countries classified living things in a different way. Few are described below.

Classification based of Major kingdom

Aristatle classified organisms as either plants or animal kingdoms. This is called two kingdom systems. Algae, fungi, bacteria and plants are included under plants and all animals and protozoa are belongs to animals.

Classification based on the nature of nucleic acid

Based on the nature of nucleus and other cellular structures of microbial cells are classified as either prokaryotic or eukaryotic. Eubacteria, actinomycetes, archaebacteria and cyanobacteria are categorized as prokaryotes. Algae, fungi, protozoa, plants and animals are belongs to eucaryotes.

Classification based on type of cell

Based on the nature of cells, all microorganisms are classified as unicellular, multicellular and acellular organisms. Bacteria, some protozoa, some algae, some fungi belongs to the category unicellular cell type. Viruses are acellular type and fungi, algae, plants and animals are available within multicellular category.

Three Kingdom System (1866)

Haeckel (1866), a Swiss naturalist, was the first to create a natural kingdom for the microbes, which had been discovered nearly two centuries before by Antony van Leeuwenhoek. Haeckel placed all unicellular organisms in a new kingdom, **"Protista"**, on the level with the existing kingdoms for plants (**Plantae**) and animals (**Animalia**), which are multicellular (macroscopic) organisms.

Five Kingdom concept

Most bacteriologists favour the three-domain system, many protozoologists, botanists, and zoologists still think in terms of five or more kingdoms. The first classification system to have gained popularity in the last few decades is the five-kingdom system. It was first suggested by Robert H. Whittaker in the 1969. Organisms are placed into five kingdoms based on at least three major criteria:

1. Cell type-prokaryotic or eukaryotic,
2. Level of organization-solitary and colonial unicellular organization or multicellular
3. Nutritional type.

According to this system organisms are classified as monera, protista, fungi, plantae and animalia.

Kingdom Monera - It is also called as Prokaryotae- it includes all prokaryotes, which means all bacteria.

Kingdom Protista-it includes unicellular eukaryocytes. Eg., protozoa, unicellular algae, slime molds mode of nutrition is autotrophic, holozoic or heterotrophs.

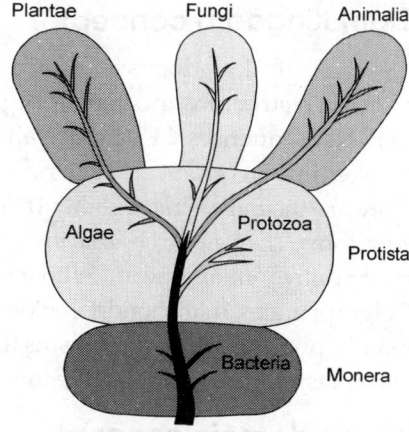

Figure 15 Five kingdom

Kingdom Fungi-unicellular yeasts, multicellular (but still microscopic) molds, macroscopic mushrooms are included in this group. They showed absorptive mode of nutrition.

Kingdom Plantae-plants. All multicellular and all carry on photosynthesis.

Kingdom Animalia- it includes all animals. Mode of nutrition is ingestive type.

The five-kingdom system is not accepted by many biologists. A major problem is its lack of distinction between archaea and bacteria.

Six kingdom concept

The six-kingdom system is the simplest concept. It is a modified version of five kingdom concept. It divides the kingdom *Monera* or *Prokaryotae* into two kingdoms, the *Eubacteria* and *Archaeobacteria*. Other kingdoms included in this concept are Protista, plantae, fungi and animalia.

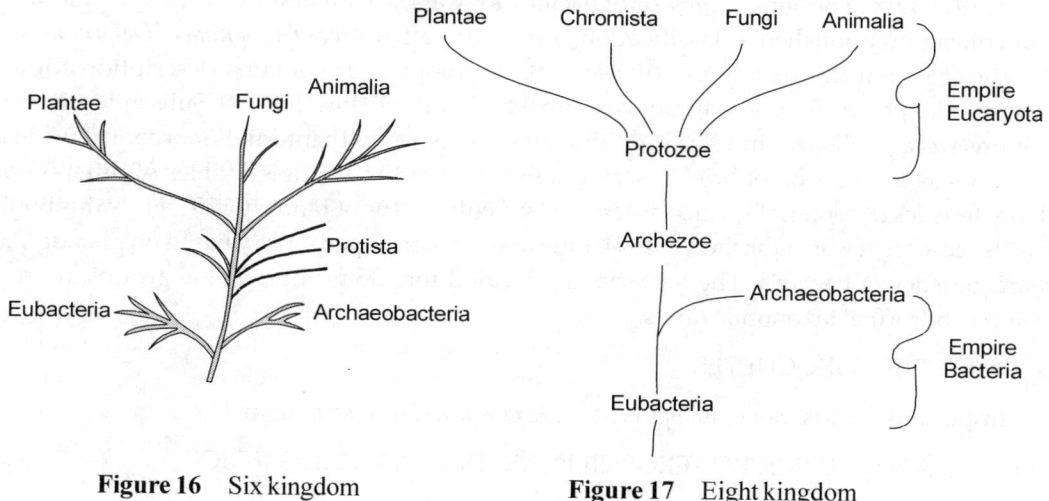

Figure 16 Six kingdom   **Figure 17** Eight kingdom

Eight Kingdom concept

The eight-kingdom system is proposed by Cavalier-Smith in 1987. This concept is based on cellular structure and genetic organization. He used ultrastructural characteristics as well as rRNA sequences. He divides all organisms into two empires and eight kingdoms. Two empires in this concept are bacteria and eukaryota. The empire *Bacteria* contains two kingdoms. They are the *Eubacteria* and the *Archaeobacteria*. The second empire, the *Eucaryota*, contains six kingdoms. They are Archezoa, protozoa, Plantae, Chromista, fungi and Animalia. The *Archezoa* are primitive eukaryotic unicellular organisms such as *Giardia* that have 70S ribosomes and lack Golgi apparates, mitochondria, chloroplasts, and peroxisomes. The kingdom *Chromista* contains mainly photosynthetic organisms that have their chloroplasts within the lumen of the rough endoplasmic reticulum. Eg. Diatoms, brown algae, cryptomonads, and oomycetes.

Three domain concept

Carl woese and his collegues in 1990 used rRNA studies to group all living organisms. They grouped the organisms into three domains. They are archae, bacteria and eucarya. It is based on the phylogenetic system of classification. Archae and bacteria are fall under prokaryotic category. Archae are differ from bacteria and eucaryotes in some ways. The 3 domains are **Eukarya**-this includes all animals, plants, fungi, and eukaryotic microbes; **Bacteria**-these are the common, everyday bacteria; **Archaea**-unusual bacteria.

C.   CLASSIFICATION OF BACTERIA

Many schemes were there for identification before 1923. During 1916-1918 -Robert Buchanan was the first to prepare a comprehensive scheme for the classification of bacteria. In 1920 – American society for microbiology submitted a report on various schemes which was the beginning of new outline for bacterial classification.

Bergeys manual

In 1923, David Bergey, Professor of bacteriology at the University of Pennsylvania, and four colleagues published a classification of bacteria called *Bergey's Manual of Determinative Bacteriology* from the society of American Bacteriologists. It contains descriptions of all procaryotic species. Second edition was published in 1925, third in 1930. Subsequently five editions were published. In 1974 8th edition was published with international contributions. This book is a collection of brief descriptions of bacteria and detailed tables of differential characteristics of bacterial species described and cultured as of January 1991. Now this book is in 9th edition. It was published in 1994. Information is arranged strictly based on phenotypic characteristics of bacteria. The bacteria are divided into 35 groups. These groups are not meant to be formal taxonomic ranks.

GROUP 1: THE SPIROCHETES

Important Genus: Borrelia, *Spirochaeta, Leptospira, Treponema*

GROUP 2: AEROBIC/MICROAEROPHILIC, MOTILE, HELICAL/VIBRIOID

Important Genus: *Campylobacter and Helicobacter*

GROUP 3: NON MOTILE (OR RARELY MOTILE) GRAM NEGATIVE CURVED BACTERIA

Genus: Spirosoma, Ancyclbacter

Group 4: GRAM-NEGATIVE AEROBIC/MICROAEROPHILIC RODS AND COCCI

Alcaligenes, Legionella, Bordetella Neisseria, *Brucella Pseudomonas*

Group 5: FACULTATIVELY ANAEROBIC GRAM-NEGATIVE RODS

Citrobacter, Aeromonas, Escherichia, Plesiomonas

Group 6: GRAM-NEGATIVE, ANAEROBIC, STRAIGHT, CURVED AND HELICAL RODS

Important Genus: *Bacteroides, Fusobacterium, Porphyromonas*

Group 7: DISSIMILATORY SULFATE OR SULPHUR REDUCING BACTERIA

Important Genus: *Desulfotomaculum, Desulfovibrio, Desulfococcus*

Group 8: ANAEROBIC GRAM-NEGATIVE COCCI

Important Genus: *Veillonella*

Group 9: THE RICKETTSIAS AND CHLAMYDIAS

Important Genus: *Rickettsia, Chlamydia*

Group 10: ANOXIGENIC PHOTOTROPHIC BACTERIA

Important Genus: *Chlorobium*

Group 11: OXYGENIC PHOTOTROPHIC BACTERIA

Important Genus: *Chrococcus, Oscillatoria, Anaebena, Nostoc.*

Group 12: AEROBIC CHEMOLITHOTROPHIC BACTERIA AND ASSOCIATED ORGANISMS

Important Genus: *Thiobacillus, Sulfobacillus*

Prosthecate Bacteria – Caulobacter

Planctomycetales – Planctomyces

Budding or appendaged bacteria – Gallonella

Group 14: SHEATHED BACTERIA

Leptothrix

Group 15: NON PHOTOSYNTHETIC, NON FRUITING AND GLIDING BACTERIA

Cytophaga, Thiothrix

Group 16: THE FRUITING, GLIDING BACTERIA : MYXOBACTERIA

Cytobacter

Group 17: GRAM POSITIVE COCCI

Important Genus: *Enterococcus , Micrococcus, Peptococcus, Staphylococcus, Streptococcus*

Group 18: ENDOSPORE-FORMING GRAM-POSITIVE RODS AND COCCI

Important Genus: *Bacillus, Clostridium*

Group 19: REGULAR, NONSPORING GRAM-POSITIVE RODS

Important Genus: *Lactobacillus, Erysipelothrix, Listeria*

Group 20: IRREGULAR, NON-SPORING GRAM-POSITIVE RODS

Important Genus: *Acetobacterium, Bifidobacterium, Brevibacterium, Corynebacterium*

Group 21 : THE MYCOBACTERIA

Important genus: *Mycobacterium*

Group 22-29 THE ACTINOMYCETES

Gram-positive bacteria that form branching filaments or hyphae. Some are motile

Group 22: NOCARDIFORM ACTINOMYCETES - *Nocardia.*

Group 23: GENERA WITH MULTILOCULAR SPORANGIA – Frankia

Group 24: ACTINOPLANETS - Micromonospora

Group 25: STREPTOMYCES AND RELATED GENUS - Streptomyces

Group 26: MADURAMYCETES - Actinomadura

Group 27: THERMOMONOSPORA AND RELATED GENUS - Thermomonospora

Group 28: THERMOACTINOMYCETES - Thermoactinomyces

Group 29: OTHER GENERA - Glucomyces, Saccharothrix

Group 30 MYCOPLASMAS

Important genus: *Mycoplasma, Spiroplasma, Ureoplasma*

Group 31: THE METHANOGENS

Important Genus: Methanobacterium, Methanococcus

Group 32 : ARCHEAL SULFATE REDUCERS

Important Genus: Archaeoglobus

Group 33: HALOBACTERIA

Important Genus: Haloarcula, Halococcus.

Group 34 :CELLWALL LESS ARCHAE

Important Genus: Thermoplasma

Group 35: EXTREMLY THERMOPHILIC, HYPER THERMOPHILIC S⁰ METABOLITES

   Important Genus: Desulfurolobus, Sulfolobus

The first edition of bergey's manual of systematic bacteriology

The system given in the first edition of *Bergey's Manual* is primarily phenetic. It contains 33 sections in the four volumes.

Volume 1 (1984) - It describes Gram-negative *Bacteria* of general, medical or industrial importance.

Volume 2 (1986) - It illustrates Gram-positive *Bacteria* other than *Actinomycetes*.

Volume 3 (1989) - It describes *Archaeobacteria, Cyanobacteria,* and remaining Gram-negative *Bacteria*.

Volume 4 (1989) – It illustrates Actinomycete.

The second edition of bergey's manual of systematic bacteriology

There has been enormous progress in prokaryotic taxonomy since 1984. In particular, the sequencing of rRNA, DNA and proteins has made phylogenetic analysis of prokaryotes feasible. As a consequence, the second edition of *Bergey's Manual was published based on phylogenetic characters*. The second edition is made of 5 volumes.

Volume 1 (2001) - It explains the *Archaea* and the deeply branching and phototrophic *Bacteria*.

Volume 2 (2005) - It describes the characters of *Proteobacteria*.

Volume 3 (2009) - It clearly depicts the *Firmicutes*.

Volume 4 (2011) - The nature of the *Bacteroidetes, Spirochaetes, Mollicutes, Acidobacteria, Fibrobacteres, Fusobacteria, Dictyoglomi, Gemmatimonadetes, Lentisphaerae, Verrucomicrobia, Chlamydiae,* and *Planctomycetes*.

Volume 5 (2012) - It describes The *Actinobacteria*

Summary of Bergey's Manual of Systematic Bacteriology 2nd edition

Volume 1. The Archaea and the Deeply Branching and Phototrophic Bacteria

Domain Archaea

Phylum Crenarchaeota Thermoproteus, Pyrodictium, Sulfolobus

Phylum Euryarchaeota

   Class I   - Methanobacteria - Methanobacterium
   Class II   - Methanococci - Methanococcus
   Class III - Methanomicrobia
   Class IV   - Halobacteria - Halobacterium, Halococcus
   Class V   - Thermoplasmata - Thermoplasma, Picrophilus

Class VI - Thermococci -Thermococcus, Pyrococcus

Class VII - Archaeoglobi- Archaeoglobus

Class VIII- Methanopyri -Methanopyrus

Domain Bacteria

Phylum Aquificae -Aquifex, Hydrogenobacter

Phylum Thermotogae -Thermotoga, Geotoga

Phylum Thermodesulfobacteria- Thermodesulfobacterium

Phylum "Deinococcus-Thermus" Deinococcus, Thermus

Phylum Chrysiogenetes- Chrysogenes

Phylum Chloroflexi- Chloroflexus, Herpetosiphon

Phylum Thermomicrobia- Thermomicrobium

Phylum Nitrospira- Nitrospira

Phylum Deferribacteres- Geovibrio

Phylum Cyanobacteria- Prochloron, Synechococcus, Pleurocapsa, Oscillatoria, Anabaena, Nostoc, Stigonema

Phylum Chlorobi- Chlorobium, Pelodictyon

Volume 2. The Proteobacteria

Phylum Proteobacteria

Class I   –Alphaproteobacteria- Rhodospirillum, Rickettsia, Caulobacter, Rhizobium, Brucella, Nitrobacter, Methylobacterium, Beijerinckia, Hyphomicrobium

Class II - Betaproteobacteria -Neisseria, Burkholderia, Alcaligenes, Comamonas, Nitrosomonas, Methylophilus, Thiobacillus

Class III - Gammaproteobacteria- Chromatium, Leucothrix, Legionella, Pseudomonas, Azotobacter, Vibrio, Escherichia, Klebsiella, Proteus, Salmonella, Shigella, Yersinia, Haemophilus

Class IV - Deltaproteobacteria- Desulfovibrio, Bdellovibrio, Myxococcus, Polyangium

Class V - Epsilonproteobacteria- Campylobacter, Helicobacter

Volume 3. The Low G - C Gram-Positive Bacteria

Phylum Firmicutes

Class I   - Clostridia- Clostridium, Peptostreptococcus, Eubacterium, Desulfotomaculum, Heliobacterium, Veillonella

Class II - Mollicutes- Mycoplasma, ureaplasma, Spiroplasma, Acholeplasma

Class III - Bacilli- Bacillus, Caryophanon, Paenibacillus, Thermoactinomyces, Lactobacillus, Streptococcus, Enterococcus, Listeria, Leuconostoc, Staphylococcus

Volume 4. The Planctomycetes, Spirochaetes, Fibrobacteres, Bacteriodetes, and Fusobacteria

Phylum Planctomycetes -Planctomyces, Gemmata

Phylum Chlamydiae-Chlamydia

Phylum Spirochaetes-Spirochaeta, Borrelia, Treponema, Leptospira

Phylum Fibrobacteres-Fibrobacter

Phylum Acidobacteria-Acidobacterium

Phylum Bacteroides-Bacteroides, Porphyromonas, Prevotella, Flavobacterium, Sphingobacterium, Flexibacter, Cytophaga

Phylum Fusobacteria-Fusobacterium, Streptobacillus

Phylum Verrucomicrobia-Verrucomicrobium

Phylum Dictyoglomi-Dictyoglomus

Volume 5. The High G- C Gram-Positive Bacteria

Phylum Actinobacteria

Class Actinobacteria- Actinomyces, Micrococcus, Arthrobacter, Corynebacterium, Mycobacterium, Nocardia, Actinoplanes, Propionibacterium, Streptomyces, Thermomonospora, Frankia, Actinomadura, Bifidobacterium.

D. CLASSIFICATION OF ALGAE

According to Whittaker's five kingdom concept, algae has seven divisions. All the seven divisions are available within two kingdoms namely Protista and plantae. The divisions chrysophyta, euglenophyta, pyrrhophyta, charophyta, chlorophyta belongs to Protista and phaeophyta and rhodophyta belongs to plantae. Some of the important properties includes in the classical classification are as follows.

Cell wall – chemistry and morphology. Form in which food or products of photosynthesis are stored. Chlorophyll and accessory pigments. Flagella number and location. Morphology of the cell or body. Habitat. Reproductive structures. Life history pattern.

Chlorophyta

- ᝄ It is also called as green algae.
- ᝄ They grow in fresh and salt water and also in soil.
- ᝄ They have chlorophyll a and b along with α carotene, β carotene and xanthophyll.
- ᝄ They exhibit wide variety of body forms ranging from unicellular to colonial, filamentous and tubular forms.
- ᝄ Some species have holdfast that anchors them to the substratum.
- ᝄ Reproduction is by both sexual and asexual methods.
- ᝄ Starch and fructose are the reserved food material.
- ᝄ Cell wall contains cellulose, mannan, protein and calcium carbonate.

➤   *A green algae Prototheca moriforms* causes protothecosis in human and animals.

➤   Important species are Chlamydomonas, Chlorella, Volvox.

Euglenophyta

➤   It is also called as Euglenoids.

➤   They share the characters of charophyta and chlorophyta.

➤   They grow in fresh &brackish water and also in soil.

➤   They have chlorophyll a and b along with α carotene, β carotene and xanthophyll.

➤   They exhibit unicellular to colonial & filamentous forms.

➤   Reproduction is by both sexual and asexual methods.

➤   Reserved food material is paramylon (a polysaccharide).

➤   Important species are Euglena.

➤   They often form water bloom

Chrysophyta

➤   Organism present in this class shows different nature with respect to pigment, cell wall and type of flagella.

➤   Common names are golden brown, yellow green algae and diatoms.

➤   The division is divided into three classes namely golden brown, yellow green algae and diatoms.

➤   They grow in fresh & salt water, brackish water and also in soil.

➤   They have chlorophyll a and c1/c2 along with β carotene, fucoxanthin and xanthophyll.

➤   Reproduction is by asexual methods.

➤   Chrysolaminarin is the reserved food material.

➤   The major chemical constituents of the cell wall are cellulose, silica, chitin, and calcium carbonate.

➤   Important species are Cyclotella.

➤   Some species produce unpleasesnt odour.

Phaeophyta

➤   It is also called as Brown algae.

➤   It consists of multicellular organisms.

➤   They grow exclusively in sea water.

➤   They have chlorophyll a and c along with β carotene, fucoxanthin and xanthophyll.

➤   Reproduction is by asexual methods.

➤   Reserved food material is laminarin.

➤   Important species are Sargassum.

➤   Cellwall contains cellulose, alginic acid and fucoidan.

Rhodophyta

➤   It is also called as Red algae.

➤   It includes most of the seaweeds that are filamentous & multicellular in nature. Some species are unicellular.

- They grows in fresh, brackish and salt water.
- Some red algae grows upto 1meter long
- They have chlorophyll a. They also contains phycoerythrin, phycobilins and phycocyanins as pigment.
- Reproduction is by asexual methods.
- Reserved food material is Floridean Starch.
- Important species is Gelidium.
- Cellwall contains cellulose, xylans, galactans and calcium carbonate.
- The matrix contains sulfated polymer of galactose called agar, funori, porphysan and carrageenan.

Pyrrhophyta

- It is also called as dinoflagellates.
- They are unicellular, photosynthetic alveolate algae.
- Most dinoflagellates are marine, but some live in fresh water.
- They have chlorophyll a, c1 and c2 along with β carotene, fucoxanthin, peridinin and dinoxanthonin.
- Reproduction is by asexual methods.
- Reserved food material is Starch and glucan.
- Important species are Nictiluca, Pyrodinium.
- Cellwall contains cellulose, alginic acid and fucoidan.

Charophyta

- They are also called as stoneworts or brittleworts.
- It contains chlorophyll a, b. It also contains α, β, ω carotene and xanthophylls.
- Reserved food material is starch.
- They survive in fresh and brackish water.
- Cellwall contains cellulose and calcium carbonate.

Fritsch classified algae in to eleven classes. They are Chlorophyceae (Green algae), Xanthophyceae (Yellow green algae), Chrysophyceae, Bacillariophyceae, Cryptophyceae, Dinophyceae, chloromonadineae, Euglenophyceae, Phaeophyceae, Rhodophyceae, myxophyceae.

E.   CLASSIFICATION OF PROTOZOA

In 1985 the Society of Protozoologists divides protozoa into seven phyla. They are Sarcomastigophora, Labyrinthomorpha, Apicomplexa, Microspora, Acestospora, Myxozoa and Ciliophora. Sarcomastigophora and Apicomplexa contains human disease causing species. This scheme is based on morphology as revealed by light, electron and scanning microscopy.

Phylum Sarcomastigophora

Locomotion by flagella, pseudopodia or both'

Sexual reproduction by syngamy

Single type of nucleus is present

There are two sub phylum. They are mastigophora and sarcodina.

Mastigophora Locomotion is by flagella. Cell division by longitudinal binary fission. Chromatophores absent. Amoeboid forms present. Sexuality is present in some groups. Most of the genus are parasitic. Eg. Trypanosoma, Giardia, Trichomonas, Leishmania.

Sarcodina locomotion by pseudopodia. Shells often present, reproductive stage forms contains flagella. Asexual reproduction by fission. Mostly free living. Some are parasitic. Eg. Amoeba, Elphidium.

Phylum: Labyrinthomorpha

Spindle shaped cells. Capable of producing mucous tracks. Non amoeboid cells. Saprozoic and parasitic on algae and sea grass.

Eg. Labyrinthula.

Phylum: Apicomplexa

All members have spore forming stage in their life cycle.

It is also called as sporozoa.

Cell contains an apical complex. Sexual reproduction by syngamy.

All species are parasitic in nature. Cysts are present. Cilia is absent.

Eg. Plasmodium, Toxoplasma, Eimeria, Cryptosproridium.

Phylum: Microspora

It is a small microsporans. It is a obligatory intracellular parasite. There is no mitochondria in the cells. Unicellular spores are produced, which is transmitted from one host to other.

Eg. *Nosema bombycis* causes pebrine on silk worm and *Nozema apis* causes serious dysentery in honey bees. *Nosema locustae* is one of the biocontrol agents for certain insects.

Phylum : Acestospora

It is a very small phylum. It is characterized by spores lacking polar filaments. Eg. Haplosporidium are parasitic to the cells, tissues and body cavities of mollusks.

Phylum: Myxozoa

All Myxozoans are parasitic. They have a resistant spore with one to six coiled polar filaments. Eg. *Myxosoma cerebralis* infects the nervous system of salmon fish.

Phylum: Ciliophora

It is the largest phylum of the seven protozoa. These are unicellular, heterotrophic protists. Cilia is available as a locomotary organ. Two types of nuclei are present. They are micro nucleus and macronucleus. Contractile vacuole present. Asexual reproduction is by transverse binary fission. Sexual reproduction by conjugation. Most species are free living commensal, some are parasitic. Eg. Balantidium, Paramecium

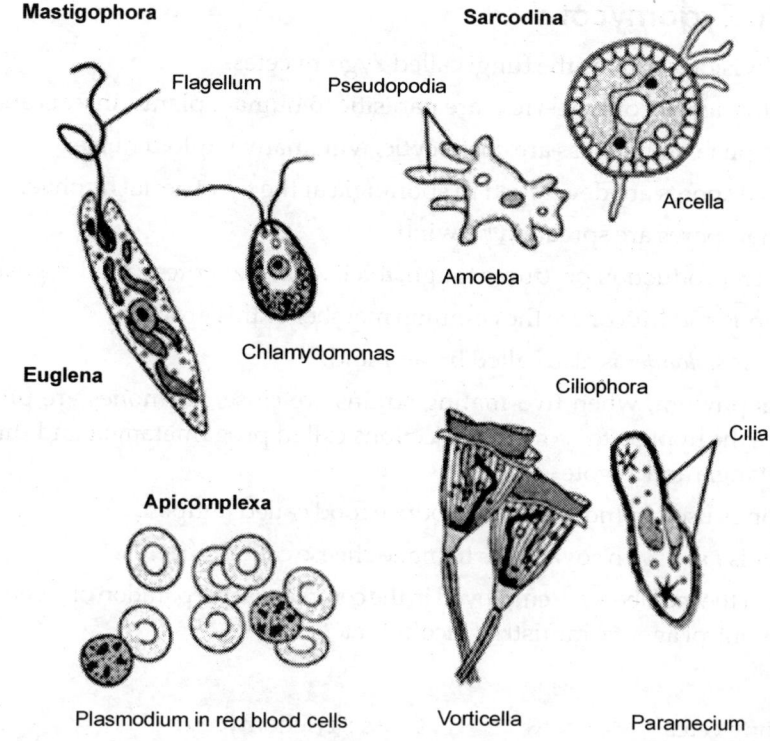

Figure 18   Some protozoans

F.   CLASSIFICATION OF FUNGI

The five kingdom system places the fungi in the kingdom fungi. According to the universal phylogenetic tree, fungi are the members of the domain Eukarya. Fungi are believed to constitute a monophyletic group referred to as a true fungi or eumycota. Eumycota contains five divisions namely Zygomyota, Ascomycota, Basidiomycota, duteromycota and Chytridiomycota. According to Alexopoulous, the fungi are considered in three different groups. They are kingdom fungi (true fungi), kingdom stramenophila and protists. The kingdom fungi includes four phyla namely chytridiomycota, Zygomycota, Ascomycota and Basidiomycota. Stramenophila include the phyla oomycota, Hyphochytriomycota and Labyrinthulomycota. Third group protists includes Plasmodiophoromycota, Dictyoteliomycota, Acrasiomycota and Myxomycota. Protists are also called Slime mold.

I.   KINGDOM FUNGI (TRUE FUNGI)

1.   Phylum Chytridiomycota

- ⅄   The simplest of true fungi belongs to this group.
- ⅄   These are simple terrestrial and aquatic fungi.
- ⅄   Reproduce by asexual method by forming motile zoospores with single, posterior whiplash flagella. Eg., Allomyces, Rozella, chytriomyces

2.   Phylum Zygomycota

⅄   This division contains the fungi called zygomycetes.

⅄   Most live as saprophytes; few are parasitic to human, plants, insects and animals.

⅄   The hyphae of this class are coenocytic, with many haploid nuclei.

⅄   Asexual spores are developed in sporangia at the tip of aerial hyphae.

⅄   Asexual spores are spreaded by wind.

⅄   Sexual reproduction produces tough, thick walled zygotes called zygospores.

⅄   Rhizopus and Mucor are the common member of this group.

⅄   *Rhizopus stolonifer* is also called bread mold.

⅄   In this phylum, when two mating strains are close, hormones are produced that induce the hyphae to   form projections called progametangia and this matures a gametangia and zygote.

⅄   Rhizopus used in Indonesia to produce food called tempeh.

⅄   Mucor is used with soyabeans to make cheese called sufer.

⅄   Some of the members are employed in the commercial preparation of some anaesthetics, birth control agents, industrial alcohols etc.,

Class: Zygomycetes	Class : Trichomycetes
Orders Mucorales, Dimargaritales, Kickxellales, Endogonales, Glomales, Entamopthorales, Zopagales.	Orders Asellariales, Harpellales, Eccrioales, Amoebidiales

3.   Phylum Ascomycota

⅄   Members of this class are called ascomycetes.

⅄   They are commonly called as sac fungi.

⅄   They produce characteristic reproductive structure, the sac like ascus.

Asexual Ascomycete	Sexual reproducer
Duteromycetes	Orders
	Taphrinales, Schizosaccharomycetales, saccharomycetales, Plectomycetes members, Pyrenomycetes members (Endophytes), Ophistomatales (Animal pathogens), Hypocreales (Arthropod associates), filamentous ascomycetes with apotheca, filamentous ascomycetes with ascostroma,

- Mycelium is composed of septate hyphae.
- Ascospores are involved in sexual reproduction.
- Sexual reproduction involves + and - strains. Fusion of these strains produces asci.
- Thousands of asci packed together in a cup or flask shaped ascocarp.
- *Neurospora crassa* is a pink bread mold. Yeast is one of the members of this division.
  Examples: Saccharomyces, Aspergillus, Penicillium, Fusarium, Mucor.

4. Phylum Basidiomycota

- This phylum contains basidiomycetes.
- They are commonly called club fungi.
- This phylum is so named for their characteristic structure, the basidium.
- The basidium is produced at the tip of hyphae and normally is club shaped.
- Most of the members are saprophytes, some are parasitic.
  Eg. Agaricus , Cryptococcus, Puccinia, Polyphorus

Orders and groups: Agariales, Gastreomycetes group (6 orders), Aphyllophorales, Uredinales, Ustilaginales.

II. STRAMENOPILA

1. Phylum Oomycota

- They are Unicellular, coenocytic and filamentous forms.
- They are Saprophytic and parasitic in nature.
- Food reserve is glycogen.
- Habitats are marine, freshwater, terrestrial.
  Examples- Saprolegnia and Phytophthora.

  - The group is arranged into six orders.
  - The Lagenidiales are the most primitive; some are filamentous, others unicellular; they are generally parasitic.
  - The Leptomitales have wall thickenings that give their continuous cell body the appearance of septation. They bear chitin and often reproduce asexually.
  - The Rhipidiales use rhizoids to attach their thallus to the bed of stagnant or polluted water bodies.
  - The Saprolegniales are the most widespread. Many break down decaying matter; others are parasites.
  - The Peronosporales are mainly saprophytic or parasitic on plants, and have an aseptate, branching form. Many of the most damaging agricultural parasites belong to this order.

⋏   The Albuginales are considered by some authors to be a family (Albuginaceae) within the Peronosporales, although it has been shown that they are phylogenetically distinct from this order.

2.   Phylum Hyphochytriomycota

⋏   Hyphochytriomycota, a phylum of mostly aquatic fungi that contains approximately 23 species.

⋏   The phylum is distinguished by the asexual production of motile cells (zoospores) with a single, anterior, feathery, whiplike structure (flagellum).

⋏   Sexual reproduction has not been found among these fungi.

Order : Hypochytriales

3.   Phylum Labyrinthulomycota

⋏   They are commonly called net slime molds.

⋏   The members show the presence of an ectoplasmic network of branched, anastomosing, wall-less filaments produced by spindle shaped or spherical cells that move by gliding within the network. The branched, anastomosing wall-less filaments are produced by cells with a specialized cell surface organelle, called a bothrosome or sagenogen or sagenogenetosome. This specialized cell surface organelle is an invaginated organelle at the cell surface.

⋏   Some members produce zoospores each provided with two flagella, the longer directed anteriorly provided with mastigonemes (i.e. tinsel flagellum) and the shorter is whiplash directed posteriorly.

This phylum has two orders and two family

⋏   Labyrinthulales – Labyrinthulaceae - Labyrinthula
⋏   Thraustochytriales – Thraustochytriaceae - Thraustochytrium

III.   PROTISTS

1. Phylum Plasmodiophoromycota

Members of Plasmodiophoromycota are plant parasites.

⋏   They are commonly referred to as endoparasitic slime molds.

⋏   The most important genera are *Plasmodiophora* and *Spongospora*.

⋏   All members are obligate parasites of algae, fungi, or plants, causing cell enlargement, especially of the roots.

⋏   They are distinguished by the production of motile cells (zoospores) with two unequal anterior whiplike threads (flagella).

⋏   *Plasmodiophora brassicae* causes clubroot of cabbage and related plants. *Spongospora subterranea* causes powdery scab of potato.

2.   Phylum Dictyoteliomycota

- ▲ Dictyosteliomycota are also called cellular slime molds
- ▲ They are Amoeba-like cells that form slugs.
- ▲ They Engulf bacteria.
- ▲ Food reserve is glycogen.
- ▲ There is no Flagella.
- ▲ Habitat of this phylum is terrestrial.
- ▲ Examples- Dictyostelium.

3.   Phylum Acrasiomycota

They have both fungus and protozoa characteristics. Spores germinate into amoebas that feed on bacteria, When food is depleted, cyclic AMP is released, causing amoebas to aggregate into one unit, the crowd of amoebas form a slug that migrates cells from the slug form a stalk with a capsule, which makes spores. Spores are released, and the cycle repeats. Eg. Fonticula

4.   Phylum Myxomycota

- ▲ Myxomycota comprises the plasmodial slime molds.
- ▲ Grow as a single, spreading mass or plasmodium
- ▲ They Feed on decaying vegetation
- ▲ When the environment becomes unfavorable (no food or water), they form stalks and spore-producing capsules
- ▲ Haploid spores germinate into amoeboid or flagellated cells
- ▲ These cells fuse to form a new diploid plasmodium
- ▲ They thrive in moist environments with bacteria, usually on decaying organic matter. There are two main group of slime molds: the cellular slime molds and the plasmodial slime molds.
- ▲ The plasmodial slime molds can exist as solitary amoeboid cells.
- ▲ This phylum contains 5orders, 14 families, 62 genera and 888 species. Orders are Echinosteliales, Liceales, Physarales, Stemonitales and Trichiales

G.   CLASSIFICATION OF VIRUSES

Viruses are acellular, subviral particles, which infects bacteria, plants and animals. Viruses are classified differently by different peoples.

Viral properties used in taxonomy are

Virion morphology (size, shape, capsid symmetry, envelope availability), Genome structure, Sensitivity to physical and chemical agents, Viral constituents, Antigenic properties, Replication strategy, Host range, Mode of transmission, Pathogenicity and Nomenclature.

Holmes system of Classification

Holmes (1948) followed the Linnaen system of binomial nomenclature. Viruses were grouped under the order Virales, which was divided into three suborders:

Phaginae - infecting bacteria, **Phytophaginae** - infecting plants, **Zoophaginae** - infecting animals.

LHT system of Classification of Viruses

In 1962 Lowff R.W.Horne & P.Tournier classified all viruses based on Linnaean hierarchical system. It is also called LHT classification system. Classification based on **phylum, class, order, family, genus, and species.** Viruses grouped according to the nature of nucleic acid, Symmetry of capsid, Presence or absence of envelope, Dimension of the virion.

Based on virion shape

Some capsids are icosahedral in shape. Eg: Papovavirus, Adenovirus, Parvovirus, Reovirus, Picornovirus, Calcivirus. Some viruses are helical in shape. Eg: Baculovirus.

Other viral shapes are **Bullet shaped** - Rhabdovirus, **Brick shaped** - Small Poxvirus, **Complex shaped** - T4 Phage, **Cylindrical shaped** - Baculovirus.

Based on the presence of envelope

Many viruses have an envelope, an outer membranous layer surounding the nucleocapsid.Enveloped viruses are sensitive to disinfectants, solvents etc;

Eg. Enveloped viruses–Rhabdoviruses, Influenza viruses.

Non enveloped viruses –Adenoviruses, Reoviruses.

Capsid symmetry and structure

Capsids are the protective layer of nucleic acid. They are made up of proteins, carbohydrates& lipids. It aids in transfering between host cells. Capsid is made up of capsomeres, which is made up of protomer.

1. Icosahedral – Herpes virus.
2. Helical – Tobacco mosaic virus.

Based on the genome nature

Type of nucleic acid

The genome of viruses contains either DNA or RNA. Based on the genome of viruses they are classified as DNA viruses and RNA viruses.

DNA virus-Herpes virus;

RNA virus-Orthomyxo virus.

Strandedness of nucleic acid

Based on the nature of genome strand the viruses are classified as

SS DNA Virus-Parvo virus

DS DNA virus-Pox virus

SS RNA virus-Corona virus

DS RNA virus-Reo virus.

Classification of Virus by Casjens and King (1975)

The Major groups of virus according to the type of nucleic acid, symmetry presence or absence of an envelope and site of assembly. Ds DNA viruses

SSDNA viruses

RNA and DNA reverse transcribing viruses

DS RNA viruses

Negative strand SS RNA viruses

Positive strand SS RNA viruses

Subviral agents

ICTV System of Classification

In the current (2011) ICTV (International Committee on Taxonamy of Viruses) taxonomy, Six orders have been established, the Caudovirales, Herpesvirales, Mononegavirales, Nidovirales, and Picornavirales and Tymovirales. A seventh order Ligamenvirales has also been proposed. The committee does not formally distinguish between subspecies, strains and isolates. In total there are 6 orders, 94 families, 22 subfamilies, 395 genera, 2,475 species and about 3,000 types yet unclassified. ICTV system also includes prions, viroids and satellites.

The International Committee on Taxonomy of Viruses (ICTV) developed the current classification system and put a greater certain virus properties to maintain family uniformity. The general taxonomic structure is as follows:

Order (-virales)

Family (-viridae)

Subfamily (-virinae)

Genus (-virus)

Species (-virus)

Baltimore Classification

David Baltimore proposed a classification system for animal viruses based on genetic system. Main properties adopted for this classification are nucleic acid and its polarity. mRNA is defined as positive strand because it contain immediately translatable information's. DNA strand of equalant polarity is designated as + strand. The DNA & RNA complement of + strand are designated as – strand. Numbering system is adapted in Baltimore classification system.

Nucleic acid and its polarity   Baltimore classification

+/-DNA   -I

+DNA		-II
+/-RNA		-III
+RNA	-RNA	-IV
-RNA		-V
-RNA	+RNA	-VI

Baltimore system	Animal viral group
I	Papovaviridae
	Adenoviridae
	Hepadnaviridae
	Herpesviridae
	Iridoviridae
	Baculoviridae
	Poxviridae
II	Parvoviridae
III	Reoviridae
	Birnaviridae
IV	Caliciviridae
	Picornaviridae
	Flaviviridae
	Togaviridae
	Coronaviridae
V	Filoviridae
	Rhabdoviridae
	Bunyaviridae
	Orthomyxoviridae
	Paramyxoviridae
	Arenaviridae
VI	Retroviridae

3

GENERAL CHARACTERS AND ULTRA STRUCTURE

A. GENERAL CHARACTERS OF BACTERIA

Definition: A very large group of microorganisms comprising one of the three domains of living organisms. They are prokaryotic, unicellular, and either free-living in soil or water or parasites of plants or animals.

- ᴧ They are microscopic in nature
- ᴧ The genus bacterium is first established by Eherenberg in 1829.
- ᴧ In 1786 Muller established first classification of bacteria.
- ᴧ They are omnipresent in nature.
- ᴧ They are a major group of microorganisms.
- ᴧ They exist as single cell or colonies.
- ᴧ They are in the form of rod, cocci or spiral in shape.
- ᴧ Cell structure is enclosed by plasma membrane and cell wall.
- ᴧ They show absorptive type of nutrition.
- ᴧ Some of the cells produce spores during unfavourable condition.
- ᴧ A nuclear material of the bacteria is represented by nucleoid.
- ᴧ They multiplied by budding, binary fission, or fragmentation.
- ᴧ They transfer its nucleic acid from one cell to other cell by conjugation, transformation and transduction.
- ᴧ Some of the cells having capsule, glycocalyx or slime.
- ᴧ They are often called as gram positive or gram negative based on the chemical nature of the cell wall.
- ᴧ Size of the bacteria ranges from $0.5\mu m$ to $600\mu m$.
- ᴧ Some of the bacteria have plasmid, an extra chromosomal DNA.
- ᴧ Pili and flagella are present in bacteria as extracellular appendage.
- ᴧ Golgi apparatus, endoplasmic reticulum, mitochondria, lysosomes, nucleus, nucleolus are absent in bacteria.

B. BENEFICIAL ASPECTS OF BACTERIA

⋏ Bacteria are important in every phase of life.

⋏ Some of the bacteria are used as a food, feed etc.,

⋏ Bacteria are able to produce antibiotics, vitamins, enzymes, organic acids etc.,

⋏ Some bacteria are able to fix nitrogen, phosphorus solubilization , mineral mobilization etc.,

⋏ Bacteria are predominately involved in recycling of waste.

⋏ Bacteria like *Pseudomonas* involved in heavy metal, hydrocarbon and pesticide degradation.

⋏ Some of the bacteria are considered as a vaccine.

⋏ Probiotics are also made from bacteria.

⋏ Bacteria are an ideal material for all kinds of genetic research.

⋏ Bacteria are enhancing soil fertility.

⋏ Bacteria are able to improve overall immune status of an individual.

⋏ Bacteria are the part of biogeochemical cycling.

⋏ Cellulose degradation in nature as well as in ruminants are mediated by the group of bacteria.

⋏ Methanogenic bacteria are involved in energy generation.

⋏ *Thiobacillus* group of bacteria are involved in mining process.

C. ULTRA STRUCTURE OF BACTERIA

Bacteria are unicellular, prokaryotic and microscopic cells. They exist in various size and shapes. Morphological features like size, shape, arrangement of cell are important to study the nature of bacteria. They are single, discrete, self multiplying and physiologically complete organisms.

Size

Bacteria show great variations in size. A spherical form is measures by its diameter. A rod or spiral form is measured by its length and width. Bacterial cell is measured by microns (m). One micron is equal to 10^{-3} (1/1000mm). Size of spherical bacteria ranges from 0.5-1.25m in diameter. Rod shaped bacterial size ranges from 0.5 -1m width and 2-5m in diameter. Smaller size of the bacteria confers many advantages. That are, more nutrients can enter into the cell, More wastes are removed easily, Rate of growth is high and Rate of metabolism also high. More recently larger bacterium *Thiomargarita namibiensis* has been discovered in ocean sediment. It is a spherical bacterium, between 100 and 750μm in diameter. They are larger than normal eukaryotic cell. Smallest bacterium is *Bdellovibrio*.

Shape

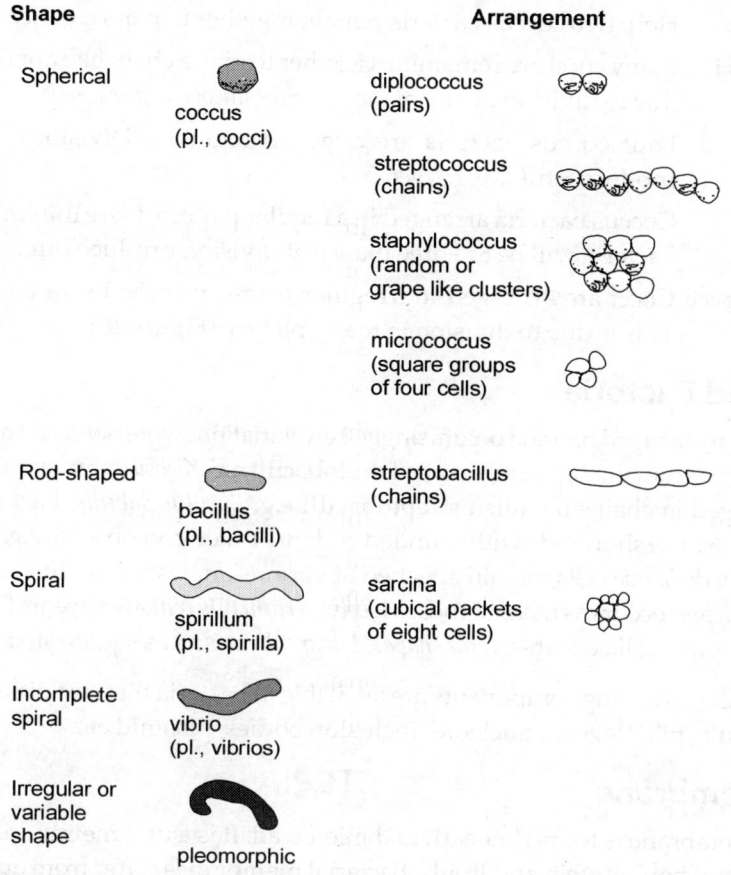

Figure 19 Shape and Arrangements of bacteria

The shape of the bacteria is governed by its rigid cellwall. Bacteria falls into four morphological shapes namely spheres, rods, spirals and pleomorphs.

Spheres: Bacteria showing spherical shapes are generally termed as cocci. E.g. *Staphylococcus aureus*

Rods: Rod shaped bacteria is generally termed as bacilli.e.g. *Bacillus subtilus*

Spirals: They are helically coiled or curved e.g. *Treponema pallidum*

Pleomorphic: Some bacteria exhibit varieties of shapes and they are said to be pleomorphic E.g. *Mycoplasma*. (Figure 19)

Arrangements of Cells

Bacterial cells are usually arranged in a characteristic manner. These arrangements are useful for their identification. The bacteria arrange themselves in a definite pattern soon after cell division. The following patterns of arrangements were observed.

Spherical Bacteria

Diplococci: Here two coccus bacteria remain together in pairs e.g. *Neisseria.*

Streptococci: Many cocci are remaining together to give a chain like appearance. This is due to division in one plane e.g. *Streptococcus faecalis*

Tetrads: Four coccus bacteria are attached together. Divisions in two planes produce tetrads e.g. *Micrococci*

Sarcina: Coccus bacteria arranged in a regular pattern. Here minimum number of bacteria will be 8. Three planes of division produce sarcina.

Staphylococci: Cocci are arranged in irregular forms, it looks like a bunch of grapes. This is due to division in many planes. (Figure 19)

Rod Shaped Bacteria

Most of the rod shaped bacteria occurs singly. Few variations were observed microscopically. They are, Two rods remain together are called **diplobacilli** e.g. *Klebsiella pneumoniae*. Some of the rods are arranged in chains be called **streptobacilli** e.g. *Bacillus subtilis*. Lactobacillus forms branching chains. The short rods with rounded ends are called coccobacilli. e.g. *Yersinia spp*. In *Corynebacterium dipheriae* cells remain attached at various angles resembling chineese letters. *Mycobacterium leprae* occurs as mass of rods. *Mycobacterium tuberculosis* appears like a tree. *Vibrio* shows curved rods. Stella occurs as *star shape*. *Haloarcula* occurs as square shaped one.

Structure the following components are available in bacteria they are plasma membrane, cell wall, capsule, pili, flagella, nucleoid, inclusion bodies, plasmid etc.

Plasma Membrane

Plasma membrane is found beneath to the cell wall. It is a unit membrane. It is a lipid bi layer. It consist of both protein and lipids. Bacterial membranes differ from eukaryotic cell in lacking sterols such as cholesterol. It is a very thin structure, about 5 to 10μm in thick. Most widely accepted current model for plasma membrane structure is the fluid mosaic model of S. Jonathan Singer and Garth Nicholson. (Figure 20)

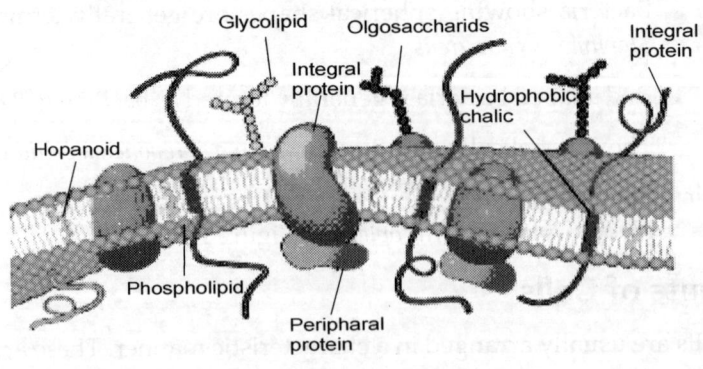

Figure 20   Plasma membrane

Functions

- ➤ It retains cytoplasm, particularly in cells without cellwall.
- ➤ It serves as a selectively permeable barrier.
- ➤ It allows the entry or release of ions either into or out of the cell.
- ➤ It prevents leakage of ions and molecules.
- ➤ It caries the functions like nutrient uptake, waste excretion and protein secretion.
- ➤ It is the location of a variety of crucial metabolic process such as respiration, photosynthesis, the synthesis of lipids and cell wall constituents.
- ➤ It is essential for the survival of the microorganisms.

Cytoplasmic Matrix

- ➤ It is the substance lying between the plasma membrane and the nucleoid.
- ➤ 70% of bacterial matrix is made up of water.
- ➤ Cytoplasmic matrix is a major part of the protoplast.
- ➤ It contains higher quantities of ribosomes.
- ➤ It also contains inclusions bodies.

Inclusion Bodies

- ➤ It is present in cytoplasmic matrix.
- ➤ They are used for storage.
- ➤ It also reduces osmotic pressure by tying up molecules in particulate form.
- ➤ Some inclusions are available freely in the cytoplasm (poly phosphate, cyanophysin, some glycogen) and some are available with simple membrane (PHB, sulphur granules).

Glycogen

- ➤ It is a carbon storage material.
- ➤ It is a organic inclusion body.
- ➤ It is a polymer of glucose units composed of long chains formed by α (1–4) glycosidic bonds, branched chains are connected by β (1–6) glycosidic bonds.
- ➤ It is stained with iodine and turns reddish brown.
- ➤ It is dispersed more evenly throughout the matrix.

PHB

- ➤ It is expanded as Poly β Hydroxyl Butyrate.
- ➤ It contains β hydroxyl butyrate molecules linked together by ester bonds between carboxyl and hydroxyl groups of adjacent molecules.
- ➤ Size of PHB are 0.2 to 0.7µm in diameter.

⅄ It is stained with sudan black.

⅄ It is a carbon storage reservoir providing materials for energy.

Cyanophycin Granules

⅄ These are found in cyanobacteria.

⅄ It is one of the organic inclusion body.

⅄ It is made up of large polypeptide with equal amount of arginine and aspartic acid.

⅄ It stores extra nitrogen for bacterial survival.

Gas Vacuoles

⅄ It is present in cyanobacteria, purple and green photosynthetic bacteria, halobacteria, thiothrix etc.,

⅄ Gasvacuoles give buoyancy to bacteria to float on the water body.

⅄ Small, hollow, cylindrical gas vesicles are aggregated and produced gas vacuoles.

⅄ It also contains proteins.

Carboxysomes

Carboxysomes are bacterial microcompartments that contain enzymes involved in carbon fixation. Carboxysomes are made of polyhedral protein shells. It is about 80 to 140 nanometres in diameter. Carboxysomes are found in all cyanobacteria, some nitrifying bacteria, and *Thiobacilli*. The carboxysome is involved in the carbon-concentrating mechanism and enhances CO_2 fixation by co-localizing the two enzymes ribulose-1, 5-bisphosphate carboxylase/oxygenase (RuBisCO) and carbonic anhydrase (CA) inside a thin shell that is assembled from thousands of protein subunits.

Polyphosphate / Volutin granules

⅄ It is an inorganic inclusion body.

⅄ It is a phosphate storage granule.

⅄ It is a linear polymer of orthophosphate joined by ester bond.

⅄ It is also called metachromatic granules because it shows the metachromatic effect.

⅄ It appears as red or different shades of blue when stained with methylene blue.

Nucleoid

⅄ It is a place of prokaryotic nucleic acid.

⅄ In prokaryotic cell, nucleic acid is packed and irregularly arranged and named as nucleoid.

⅄ It is also called nuclear body, chromatin body, nuclear region.

⅄ Usually prokaryotic cell contains only one circular Double stranded DNA but some have linear DNA.

⅄ *Vibrio cholerae* have two chromosomes.

- Nucleoid is stained with fuelgen stain.
- In actively growing cells, the nucleoid has projections that extend into the cytoplasmic matrix.
- Chemical analysis showed that nucleoid contains 60% DNA, 30% RNA and 10% Protein.

Plasmids

Plasmid is a circular, double-stranded unit of DNA that replicates within a cell independently of the chromosomal DNA. Plasmids are most often found in bacteria and are used in recombinant DNA research to transfer genes between cells.

Structure

Plasmids are usually made up of circular double stranded non-chromosomal DNA. They make their structure circular by combining the two ends of the double stranded DNA together. These ends are combined through covalent bonds.

Function

- The main function of plasmids is to carry antibiotic resistant genes.
- The other function of plasmids is to carry those genes which are involved in metabolic activities and are helpful in digesting the pollutants from the environment.
- They are also capable of producing antibacterial proteins.
- Plasmids are also able to carry the genes which are concerned with increasing the pathogenicity of bacteria.
- Plasmids have a significant role in gene therapy. They are mostly used for the insertion of therapeutic genes in the human body to fight against diseases.

Types

- There are five types of plasmids which are used for different purposes.
- **Fertility F-plasmids** - it contain *tra* genes. They are involved of conjugation.
- **Resistance (R) plasmids** - it contain genes that can build a resistance against antibiotics or poisons and help bacteria produce pili.
- **Col plasmids** - it contain genes that code for bacteriocins, proteins that can kill other bacteria.
- **Degradative plasmids** - they enable the digestion of unusual substances, e.g. toluene and salicylic acid.
- **Virulence plasmids** - it converts the bacterium into a pathogen.

Cell Wall of Bacteria

- It is a rigid layer formed just outside to the plasma membrane.
- It provides the shape to the bacteria.

⅄   It protects bacteria from osmotic lysis.

⅄   In 1884, Christian Gram developed gram staining technique, which differentiates the bacteria into gram positive and gram negative based on its cell wall chemical nature. Those bacteria that retain grams dye (grams crystal violet and grams iodine) are called gram positive and the bacteria that loses grams dye are called gram negative.

⅄   Gram positive cell wall consists of a single 20-80nm in thick homogenous peptidoglycan (PG) or murein layer along with lipoteichoic acid.

⅄   Gram negative bacterial cell wall is complex and has 2-7nm thick peptidoglycan layer surrounded by 7-8nm thick outer membrane (LPS).

Peptidoglycan (PG)

⅄   It is also called as murein.

⅄   It is a polymer composed of many identical subunits.

⅄   Polymer of PG consists of N Acetyl Glucosamine (NAG) and N Acetyl Muramic Acid (NAM) and L Alanine, D Alanine, D Glutamic acid and Meso Diaminopimelic acid (DAP).

⅄   The presence of DAP protects the cell wall against the attack by most of the peptidases.

⅄   Back bone of PG layer composed of alternating NAG and NAM residues. A peptide chain of four alternating D and L aminoacids is connected to the carboxyl group of a NAM.

⅄   Chain linked PG subunits are joined by cross links between the peptide.

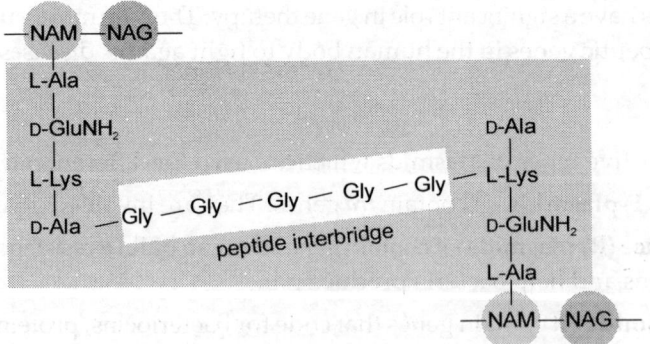

Figure 21   PG layer

⅄   The carboxyl group of the terminal D Alanine is connected directly to the amino group of DAP with peptide interbridge. (Figure 21)

Gram Positive Cell Wall

⅄   Gram positive cell wall is composed of thick PG layer and small quantity of Lipo teichoic acid. Teichoic acid is a polymer of glycerol or ribbitol joined by phosphate group.

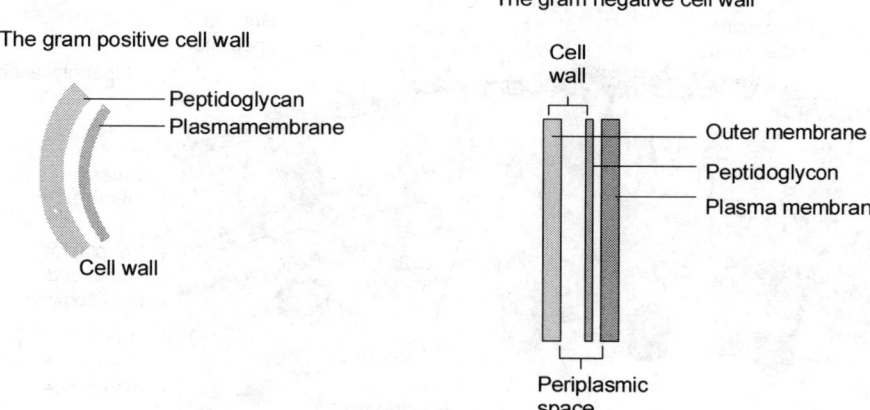

Figure 22 Cell wall of bacteria

⋏ Aminoacids and sugars are attached to the glycerol or ribbitol group.

⋏ Teichoic acids are connected to the PG layer covalent bond with the six hydroxyl of NAM. They are negatively charged. Teichoic acid is used to maintain the structure of the wall. (Figure 23)

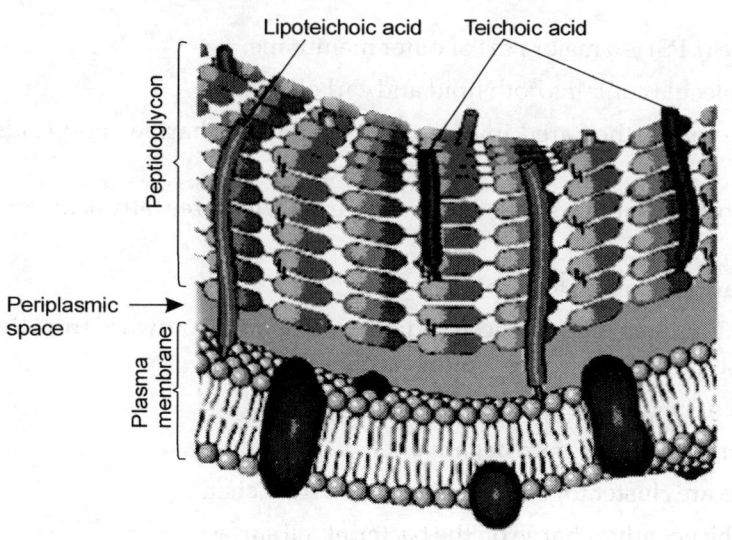

Figure 23 Gram positive bacteria cell wall

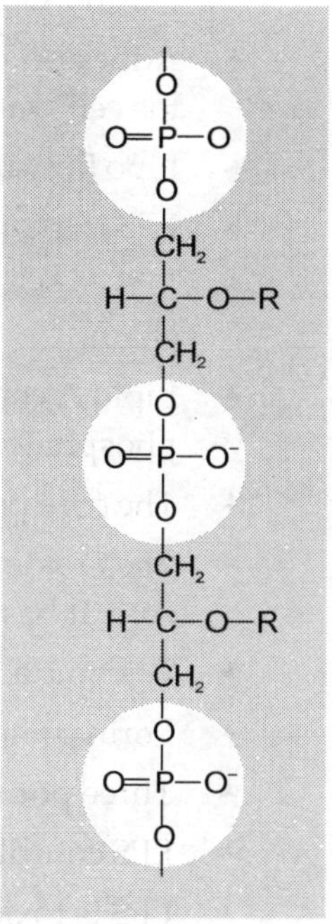

Gram Negative Cell Wall

⋏ It is composed of the PG layer followed by outer membrane.

⋏ Outer membrane is found outside the thin PG layer.

Figure 24 Chemical nature of Teichoic acid

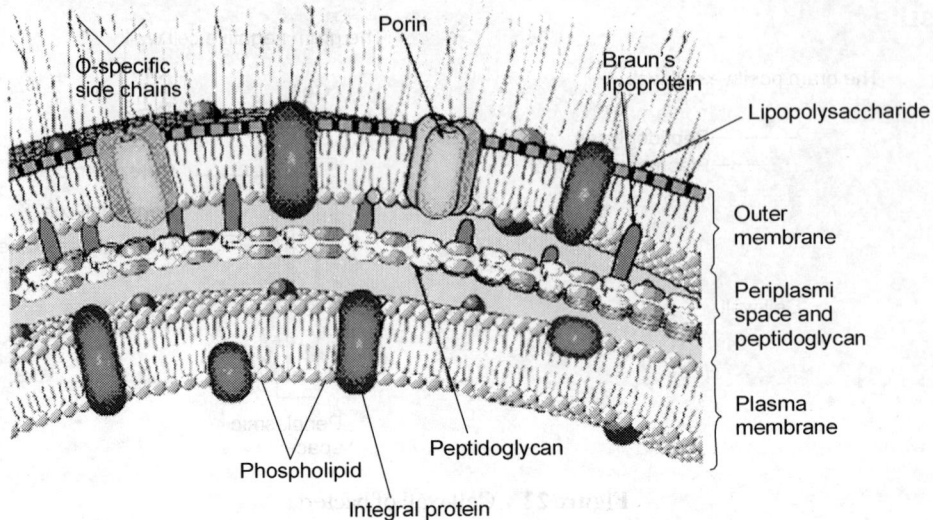

Figure 25   The Gram negative bacterial cell wall

⅄   The most aboundant membrane protein of bacteria is Braun's lipoprotein. It is present on the outer membrane. Brauns lipoproteins are covalently joined to the underlying PG and Embedded in the outermembrane by its hydrophobic end.

⅄   Adhesion site is present in outer membrane which is responsible for strengthening of the cell wall.

⅄   Lipo Poly Saccharide (LPS) is a major part of outer membrane.

⅄   LPS is a complex molecule contains both lipid and carbohydrate.

⅄   LPS consists of three parts, they are Lipid A, the core poly saccharide and O side chain.

⅄   Lipid A contains two glucosamine sugar derivatives, each with three fatty acids and phosphate.

⅄   The core polysaccharide is joined to a lipid A.

⅄   The O side chain or O antigen is a polysaccharide chain extending outward from the core. It has several sugars and its composition varies with strains.

⅄   Teichoic acid is absent in gram negative cell wall PG layer.

⅄   Porin proteins are present in the gram negative cell wall.

⅄   Three porin proteins are cluster together and form a narrow channel.

⅄   LPS contributes to the negative charge on the bacterial cell surface.

⅄   LPS is toxic, as a result the LPS can acts as an endotoxin.

⅄   It is a protective barrier.

⅄   It prevents / slows the entry of bile salt, antibiotics etc.,

⅄   It protects bacteria against osmatic pressure.

Capsule

- ▲ Capsule is a layer lying outside of the cellwall.
- ▲ It is composed of polysaccharides.
- ▲ It is observed under light microscope when the cells are stained by negative stain.
- ▲ Capsule resists phagocytosis.
- ▲ It can protect bacteria against dessiccation.
- ▲ It protects cell against phagocytosis, complement attack etc.,
- ▲ Capsulated strains produce smooth and mucoid colonies Eg. *Klebsiella pneumoniae.*
- ▲ Noncapsulated bacteria produce rough colonies.
- ▲ There are two types of capsule based on the thickness of the capsule. They are microcapsule and macrocapsule. Macrocapsule are morethan 20nm in thick.
- ▲ Capsule is water soluble and non ionic in nature.
- ▲ It is one of the most important virulent factor of bacteria.

Slime Layer

- ▲ It is a zone of diffuse, unorganized material present in some bacteria.
- ▲ It is removed easily.
- ▲ It is composed of polysaccharides.
- ▲ Gliding bacteria often produce slime which presumably aids in their motility.

S Layer

- ▲ Most of the gram positive and gram negative bacteria have regularly structured layer called S layer on their surface.
- ▲ They are common in Archaebacteria.
- ▲ It is made up of protein or glycoprotein.
- ▲ It protects cell against ion and pH fluctuations, osmotic stress, enzymes etc.,
- ▲ It helps maintains the shape and rigidity of the cell.
- ▲ It can promote cell adhesions.
- ▲ It protect cell against complement attack and phagocytosis.
- ▲ It is one of the virulence factor of the cell.

Pili

- ▲ They are fine hair like appendages.
- ▲ One to ten pili are present in every cell.
- ▲ Pili are larger than fimbriae.
- ▲ It is determined by sex factor or conjugative plasmid.
- ▲ It is required for conjugation. If the pili is involved in conjugation are called sex pili.

Fimbriae

  ▲ It is a short fine, hairlike appendage present in gram negative bacteria.

  ▲ It is thinner than flagella and pili. They are not involved in motility.

  ▲ Each cell may be covered with upto 1000 fimbriae.

  ▲ It is made up of protein subunits.

Flagella

  ▲ Flagella is one of the locomotary organ.

  ▲ It is found in motile bacteria.

  ▲ It is a thread like appendage extending outward from the plasma membrane and cell wall.

  ▲ They are slender, rigid structure.

  ▲ It is about 15 or 20μm long.

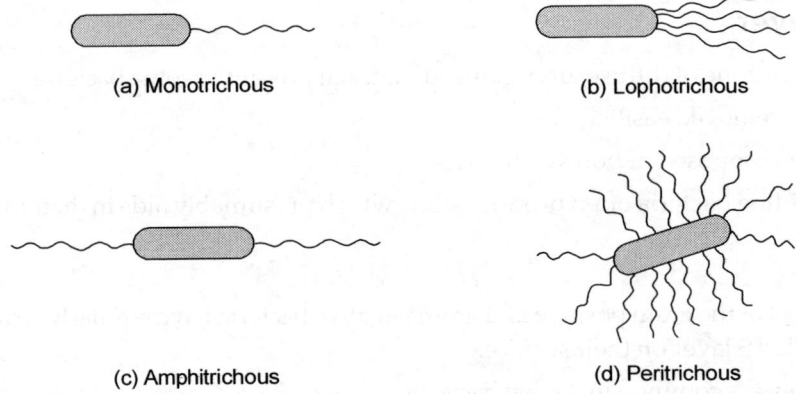

(a) Monotrichous   (b) Lophotrichous

(c) Amphitrichous   (d) Peritrichous

Figure 26   Flagella types

  ▲ Different types of flagella are found in the cells.

  ▲ Flagella maybe classified into following types

Monotrichous	-flagella at one end e.g., *Vibrio cholerae*
Amphitrichous	-flagella at both ends e.g., *Pseudomonas*
Lophotrichous	- tuft of flagella present at one or both ends e.g., *Spirillum*
Peritrichous	-flagella found around the surface of the cell e.g., *Salmonella typhi*

Ultrastructure of Flagella

  ▲ Flagella is made up of three parts. They are filament, basal body and hook. The longest and most obvious portion is the filament. It extends from the cell surface to the tip.

  ▲ A **basal body** is embedded in the cell.

⅄ A short curved segment is the **hook**. It links the filament and acts as a coupling.

⅄ Filament is made up of single protein called flagellin.

⅄ Hook is made up of different protein subunits.

⅄ The basal body is the most complex part of a flagellum.

⅄ In gram negative bacteria basal body has 4 rings connected to a central rod. Outer L and P rings are associated with LPS and PG layer respectively. The M ring contacts the PG layer.

⅄ Gram positive bacterial flagella have only two basal body rings. The inner ring connected to the plasma membrane and the outer one attached to the PG layer.

⅄ At least 20–30 genes are responsible for the synthesis of flagella.

⅄ The direction of flagellar rotation determines the nature of bacterial movement.

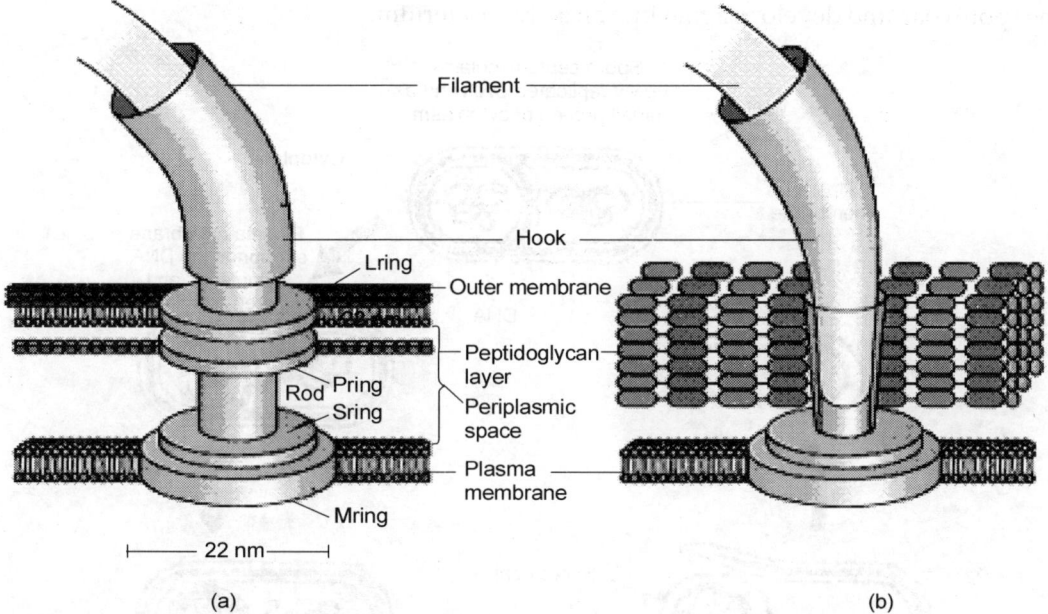

(a)   (b)

Figure 27   The Ultrastructure of bacterial flagella. (a) gram negative and (b) gram-positive bacteria.

The Bacterial Endospore

A number of gram-positive bacteria can form a special resistant, dormant structure called an **endospore**. Endospores are present in the species of *Bacillus* and *Clostridium* (rods), *Sporosarcina* (cocci). These structures are extraordinarily resistant to environmental stresses such as heat. Endospores are of great practical importance in food, industrial, and medical microbiology. Endospores often survive when boiling for an hour or more.

Spore formation normally commences when growth ceases due to lack of nutrients. It is a complex process and may be divided into seven stages (**Figure** 28). An axial filament of nuclear material is formed (stage I). It is followed by an inward folding of the cell membrane to enclose part of the DNA and produce the forespore septum (stage II). The membrane continues to grow and engulfs the immature spore in a second membrane (stage III). Next, cortex is laid down in the space between the two membranes, and both calcium and dipicolinic acid are accumulated (stage IV). Protein coats then are formed around the cortex (stage V), and maturation of the spore occurs (stage VI). Finally, lytic enzymes destroy the sporangium releasing the spore (stage VII). Sporulation requires only about 10 hours in *Bacillus megaterium.*

The transformation of dormant spores into active vegetative cells seems almost as complex a process as sporogenesis. It occurs in three stages: (1) activation, (2) germination, and (3) outgrowth. This process is characterized by spore swelling, rupture or absorption of the spore coat, loss of resistance to heat and other stresses, loss of refractility, release of spore components, and increase in metabolic activity. Many normal metabolites or nutrients (e.g., amino acids and sugars) can trigger germination after activation. Germination is followed by the third stage, outgrowth. The spore protoplast makes new components, emerges from the remains of the spore coat, and develops again into an active bacterium.

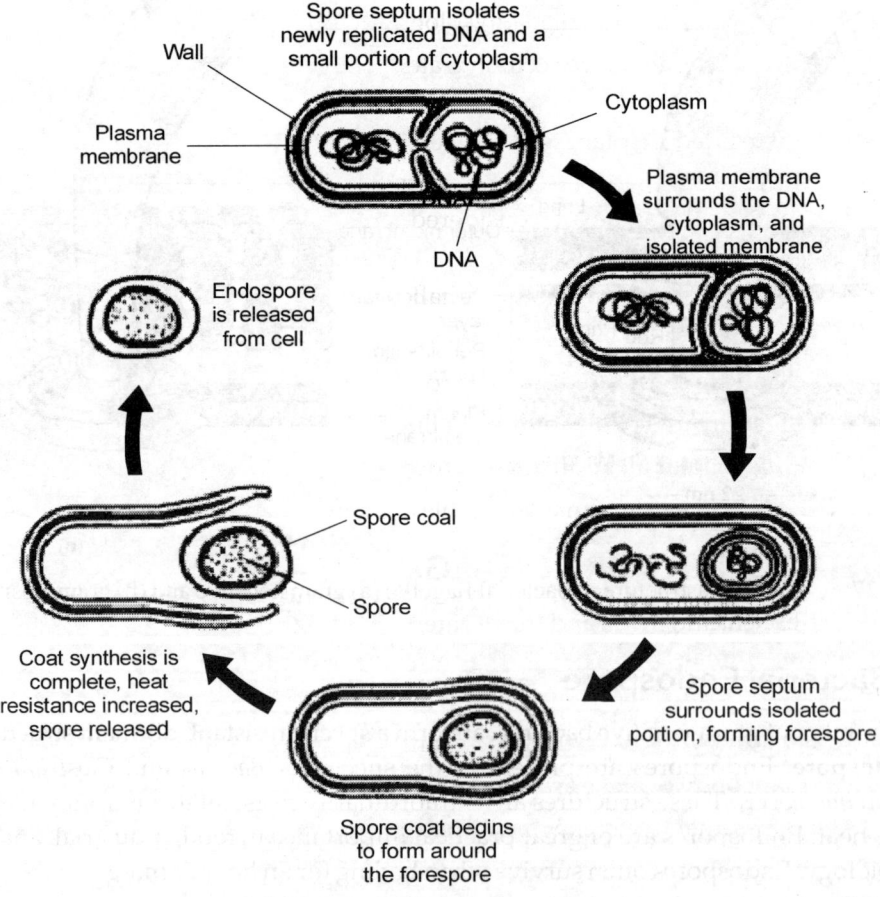

Figure 28 Endospore formation

D. CELL WALL-LESS FORMS

A few bacteria are able to live or exist without a cell wall. The mycoplasmas are a group of bacteria that lack a cell wall. Mycoplasmas have sterol-like molecules incorporated into their membranes and they are usually inhabitants of osmotically-protected environments. *Mycoplasma pneumoniae* is the cause of primary atypical bacterial pneumonia, known in the vernacular as "walking pneumonia". For obvious reasons, penicillin is ineffective in treatment of this type of pneumonia. Sometimes, under the pressure of antibiotic therapy, pathogenic bacteria can revert to cell wall-less forms (called **spheroplasts** or **protoplasts**) and persist or survive in osmotically-protected tissues. When the antibiotic is withdrawn from therapy the organisms may regrow their cell walls and reinfect unprotected tissues.

E. GENERAL CHARACTERS OF ALGAE

Definition: Algae are a group of eukaryotic cells, which are generally called aquatic plants.

- Study of algae are called phycology or algology.
- Most of the algae are occur in water. It may be suspended or attached and living on the bottom.
- Suspended or planktonic algae consists of free floating, mostly microscopic organisms.
- Some of the algae are considered as a endosymbionts of protozoa, mollusks, worms and corals.
- Symbiotic algae and fungi are called lichens.
- Size of the algae varies from microscopic to several meter in length.
- Algae are considered as a feed for aquatic animals.
- Algae are unicellular or multicellular.
- Multicellular algae are filamentous or colonial form.
- Filamentous algae may be simple or branched.

F. ULTRA STRUCTURE OF ALGAE

- Algae are found in sea and fresh water, damp soil, rocks, bark of trees, on plants and also on animals.
- Algae needs sufficient light for growth.
- Euglena and Volvox are the good examples for algae.
- The plant body of algae are called thallus.
- Algae are either free living or attached forms.
- Algae are unicellular (*Chlamydomonas*) or multicellular (*Spirogyra*).
- Multicellular algae forms filaments (*Spirogyra*) or colonies (*Volvox*).

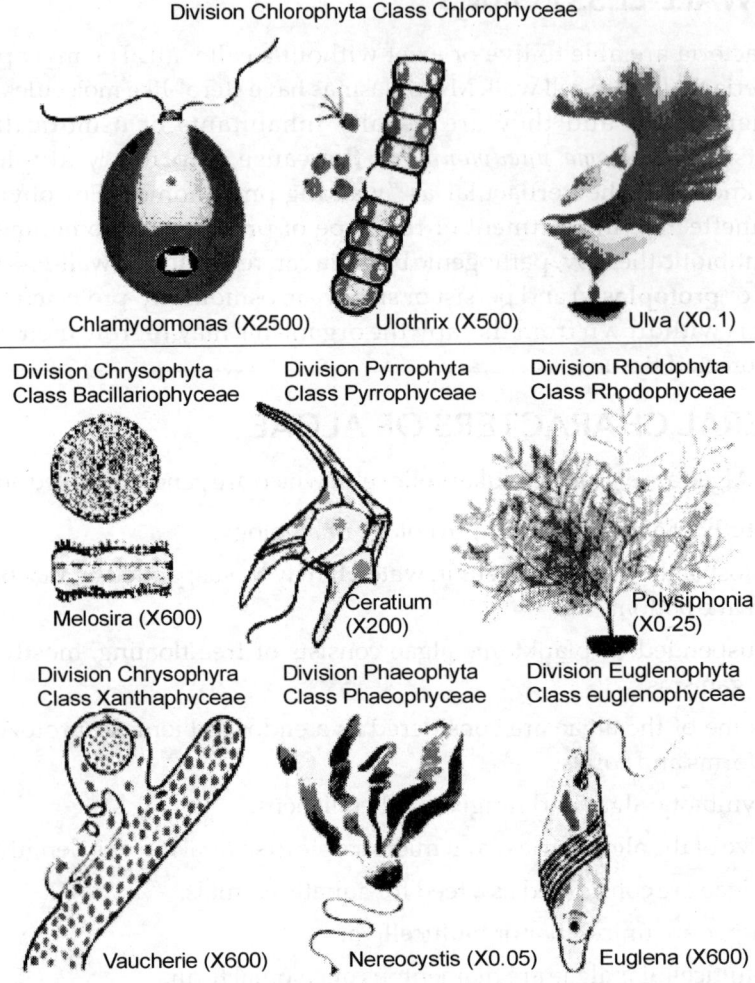

Division Chlorophyta Class Chlorophyceae

Chlamydomonas (X2500)   Ulothrix (X500)   Ulva (X0.1)

Division Chrysophyta
Class Bacillariophyceae

Division Pyrrophyta
Class Pyrrophyceae

Division Rhodophyta
Class Rhodophyceae

Melosira (X600)   Ceratium (X200)   Polysiphonia (X0.25)

Division Chrysophyra
Class Xanthaphyceae

Division Phaeophyta
Class Phaeophyceae

Division Euglenophyta
Class euglenophyceae

Vaucherie (X600)   Nereocystis (X0.05)   Euglena (X600)

Figure 29   Algae

⅄ Filamentous algae are simple or branched. Simple filamentous algae consists of single row or cells (eg. *Spirogyra*).

⅄ Some of the filamentous algae having trichomes. Some filamentous algae attached to the surface through holdfast.

⅄ Algal cell is surrounded by a thin, rigid cell wall.

⅄ Some algae have outer flexible, gelatinous matrix, that found outside of the cellwall.

⅄ Flagella are the locomotary organ in the unicellular algae.

⅄ The nucleus has typical nuclear envelope with pores. Nucleolus, chromatin and karyolymph are present inside the nucleus.

⅄ The chloroplast have membrane bound sacs called thylakoids that carryout light reactions of photosynthesis. Pyrenoid is associated with the chloroplasts.

➤ Variable nature of mitochondria is present in different classes of algae. Euglenoids have discoid cristae, green and red algae contains discoid cristae and golden brown, yellow green, brown, diatoms have tubular cristae.

➤ Three types of pigments are present in the algae. They are chlorophylls, carotenoids and biliproteins.

➤ Chlorophyll a, b, c, d, e are the chlorophyll (green) pigments found in different classes of algae. Chlorophyll a is found in all the classes.

➤ α, β, λ, δ carotene and xanthophylls are present in different algal groups.

➤ Phycocyanin (blue) and phycoerythrin (red) are the biliproteins found in algae.

➤ The algae showed autotrophic or heterotrophic nutrition.

➤ There are three types of asexual algal reproduction. They are fragmentation, spores and binary fission. Motile spores are called zoospores and nonmotile spores are called aplanospore. In sexual reproductions gametes fuse to produce zygote.

➤ **Akinetes -** An **akinete** is a thick-walled dormant cell derived from the enlargement of a vegetative cell. It serves as a survival structure. It is a resting cell of cyanobacteria and unicellular and filamentous green algae. Under magnification, akinetes appear thick walled with granular-looking cytoplasms.

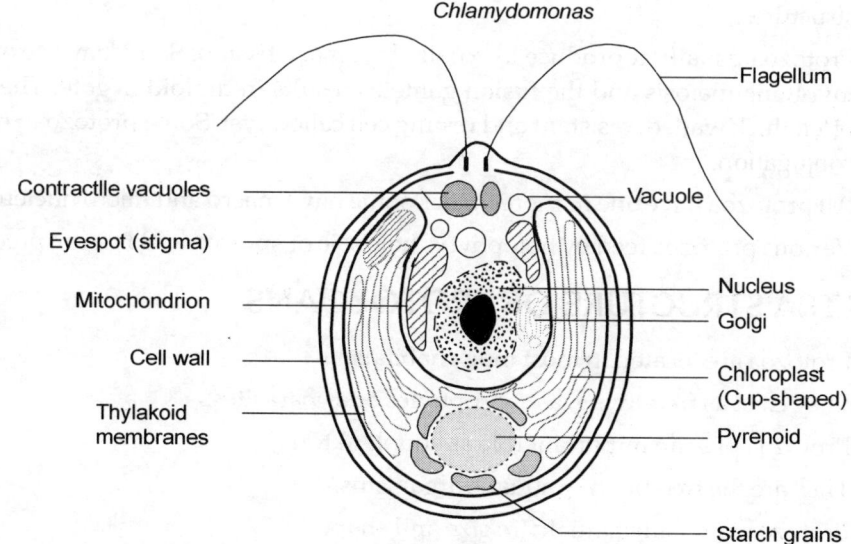

Figure 30

➤ **Heterocyst - Heterocysts** are specialized nitrogen-fixing cells formed by some filamentous cyanobacteria, such as *Nostoc punctiforme*, *Cylindrospermum stagnale* and *Anabaena sphaerica*, during nitrogen starvation. They fix nitrogen from dinitrogen (N_2) in the air using the enzyme nitrogenase, in order to provide the cells in the filament with nitrogen for biosynthesis. Nitrogenase is inactivated by oxygen, so the heterocyst must create a microanaerobic environment. The heterocysts' unique structure and physiology require a global change in gene expression.

G.   ECONOMIC IMPORTANCE OF ALGAE

- ∧ Algae are considered as an autotrophic plants. It servs as a food for aquatic animals.
- ∧ Industrial / commercial products are obtained form algae. Agar agar is obtained from seaweed Gelidium.
- ∧ An alga improves the fertility of the soil.
- ∧ Some algae are considered as a food for man.
- ∧ Some algae are considered as a medicine (Chlorellin).

H.   GENERAL CHARACTERS OF PROTOZOA

- ∧ Protozoa are unicellular eukaryotic microorganism.
- ∧ Protozoa are the first animal.
- ∧ Study of protozoa are called protozoology.
- ∧ Protozoa exhibits heterotrophic nutrition.
- ∧ They showed various types of locomotion.
- ∧ They occupy a vast array of habitats and niches.
- ∧ They contains organelles similar to other eukaryotic cells and also contains specialized structures.
- ∧ Protozoa usually reproduce asexually by binary fission. Some have sexual cycles, involving meiosis and the fusion gametes results in diploid zygote. The zygote is often thick walled, resistant and resting cell called cyst. Some protozoan undergoes conjugation.
- ∧ All protozoa have one or more nuclei, some have macro and micro nucleus.
- ∧ Various protozoa feed by holophytic, holozoic or sporozoic. Some are predatory.

I.   ULTRA STRUCTURE OF PROTOZOANS

- ∧ Protozoa are located in most moist habitats.
- ∧ They exist as free-living, as saprophytes or as parasites.
- ∧ Protozoa play an important role as zooplankton.
- ∧ They are the free-floating aquatic organisms.
- ∧ Protozoa vary substantially in size and shape.
- ∧ Protozoal cells have no cell walls and therefore can assume an infinite variety of shapes.
- ∧ Some genera have cells surrounded by hard shells, while the cells of other genera are enclosed only in a cell membrane.
- ∧ Many protozoa alternate between a free-living vegetative form known as a trophozoite and a resting form called a cyst.
- ∧ Most protozoa have a single nucleus, but some have both a macronucleus and one or more micronuclei.

▲ Contractile vacuoles may be present in protozoa to remove excess water, and food vacuoles are often observed.

▲ Protozoa are heterotrophic microorganisms.

▲ Most species obtain large food particles by phagocytosis.

▲ Many protozoal species move independently by one of three types of locomotor organelles: flagella, cilia, and pseudopodia.

▲ Respiration is accomplished by the diffusion of dissolved gases through the cell membrane.

▲ Carbon dioxide and water, the waste products of this oxidation, diffuse out of the cell.

▲ Reproduction is usually asexual, occurring mostly by cell division, or binary fission. Some forms reproduce asexually by budding or by the formation of spores. In certain groups sexual reproduction sometimes also occurs. In these instances, cell division is preceded by the fusion of two individuals or, in ciliates, by conjugation and exchange of nuclear material.

▲ Waste materials in protozoans are removed from the cell by diffusion through the cell. They are transported out of the cell by food vacuoles that come in contact with the surface. This is known as exocytosis.

Entamoeba histolytica

▲ It is a unicellular organism.

▲ It lives inside of the large intestine.

▲ It is an endoparasite.

▲ Fully grown and mature Entamoeba is called as trophozoite.

▲ Body is covered by plasmalemma.

▲ The plasmalemma encloses the cytoplasm.

▲ The cytoplasm has outer ectoplasm and inner endoplasm.

▲ Endoplasm contains nucleus and food vacuole.

▲ The food vacuole contains RBC, WBC, Epithelial cells etc.,

▲ There is no contractile vacuole.

▲ Nucleus is spherical in shape.

▲ Locomotion is by pseudopodia.

Figure – Refer Plate 8, Page No. 141.

It is a protozoan. Commonly available in tropical subtropical, temporate regions.It is a Endoparasite. Present in man and other mammals. Alive in mucous layer of colon. Feeds dissolved tissues, bacteria and RBCs.Causes fatal and serious disease. Infected individual discharge mucous and blood in their stool. *E histolytica* has a relatively simple life cycle that alternates between trophozoite and cyst stages.

Trophozoite - The trophozoite is the actively metabolizing, mobile stage, and the cyst is dormant and environmentally resistant. When they are alive they may be actively motile (Unidirectional motility). Amoebas are anaerobic organisms and do not have mitochondria. The finely granular endoplasm contains the nucleus and food vacuoles, which in turn may contain bacteria or red blood cells. Nuclear morphology is best seen in permanent stained preparations. The nucleus has a distinctive central karyosome and a rim of finely beaded chromatin lining the nuclear membrane. Finger like pseudopodia is available.

Cyst - The cyst is a spherical structure, 10-20 μm in diameter, with a thin transparent wall. Fully mature cysts contain four nuclei with the characteristic amebic morphology. Rod-like structures (chromatoidal bars) are present variably, but are more common in immature cysts. Inclusions in the form of glycogen masses also may be present. A number of non-pathogenic amoebae can parasitize the human gastrointestinal tract and may cause diagnostic confusion

Giardia lamblia

It is a flagellated protozoan. It was first seen by Leewen hoek in 1681 while examining his own stool. It is world wide in distribution. It mostly found in Duodenum and upper part of ileum. The Giardia life cycle involves two stages. The trophozoite and the cyst. The *G. lamblia* trophozoite is easily recognized under a microscope. It is about 12 to 15 μm long, shaped like tennis racket.The dorsal surface is convex and the ventral surface is concave with a sucking disc, and has two nuclei that resemble eyes, structures called median bodies that resemble a mouth, and four pairs of flagella that look like hair; these combine to give the stained trophozoite the eerie appearance of a face. The flagella help these organisms to migrate to a given area of the small intestine, where they attach by means of an adhesive disk to epithelial cells and thus maintain their position despite peristalsis. It is bilaterally symmetrical. Anterior end is broad and the posterior end is tapers to a sharp point.

The Giardia cyst - the form usually seen in the feces - is ovoid, 6 to 12 μm long, and can often be seen to contain two to four nuclei at one end and prominent diagonal fibrils. Flagella and sucking disc are seen inside of cytoplasm.

Figure – Refer Plate 9, Page No. 142

J.   ECONOMIC IMPORTANCE OF PROTOZOA

⅄   The skeletons of *Foramininfera* make up much of the limestone and chalk on the Earth.

⅄   Flagellates are the primary component in the marine food chain.

⅄   Some protozoa are uses in sewage treatment. Zooglea is used in trickling filter.

⅄   Protozoa is used as a biocontrol agent. Nosema used to control insect vector.

⅄   Protozoa is used in oil exploration.

K.   GENERAL CHARACTERS OF FUNGI

Definition :   A fungus is a eukaryotic, spore forming organism that has absorptive nutrition and lacks chlorophyll. They reproduces asexually, sexually or both methods. Cells of fungi are surrounded by cell walls, which usually contains chitin.

- Fungi are widely distributed (omnipresent) and are found wherever moisture is present.

- The body or vegetative structure of a fungi are called a thallus.

- Fungi are grouped into moulds or yeast based on the development of the thallus.

- Yeasts are unicellular fungi that have a single nucleus. They reproduce either asexually by budding and transverse division and sexually through spore formation.

- Molds are long branched, filamentous thread like hyphae, which form mycelium.

- Hyphae may be either septate or coenocytic.

- Some fungi are dimorphic in nature.

- They are the great importance to human in both beneficial and harmful ways.

- Fungi exist primarily as filamentous hyphae. A mass of hyphae is called mycelium.

- Like some bacteria, fungi digest insoluble organic matter by secreting exoenzymes, then absorbing solubilized nutrients.

- Two reproductive structures occur in the fungi that are sporangia from asexual spores and gametangia from sexual gametes.

- The zygomycetes are characterized by resting structures called zygospores, cells in which zygotes are formed.

- The ascomycetes form zygote within a characteristic saclike structure, the ascus. The ascus ciontains two or more ascospores.

- Yeasts are unicellular fungi mainly ascomycetes.

- Basidiomycetes possess dikaryotic hyphae with two nuclei, one of each mating type. The hyphae divide uniquely, forming basidiocarps within which club shaped basidia can be found. The basidia bear two or more basidiospores.

- The duteromycetes (Fungi imperfecti) have either lost the capacity for sexual reproduction or it has never been observed.

- The chytrids are a group of terrestrial and aquatic fungi that reproduced by motile zoospores with single, posterior, whiplash flagella and represent a link between fungi and protists.

- The slime and water moulds resemble the fungi only in appearance and lifestyle. In their cellular organization, reproduction and lifecycles, they are phylogenetically distinct.

L. ULTRA STRUCTURE OF FUNGI

A fungus is a eukaryotic, spore forming organism that has absorptive nutrition and lacks chlorophyll. The study of fungi is called mycology. The body or vegetative structure of fungi is called a thallus. There is no root, stem and leaves as like plants. Most of the fungus grows best at terrestrial habitat, few are aquatic growers. Some fungi form beneficial relationship with other group of plants / algae (VAM, Lichens). Size of the fungal cell ranging from unicellular (yeast) or multicellular (mold). Molds are long branched, filamentous structure called mycelium.

Each filament of mycelium is called hyphae. Hyphae may be either septate or coenocytic. Some hyphae have crosswalls septa and are called sptate hyphae. Some hyphae are uninterupted by cross walls are called coenocytic hyphae. Septa having single or multiple pores that permit cytoplasmic streaming.

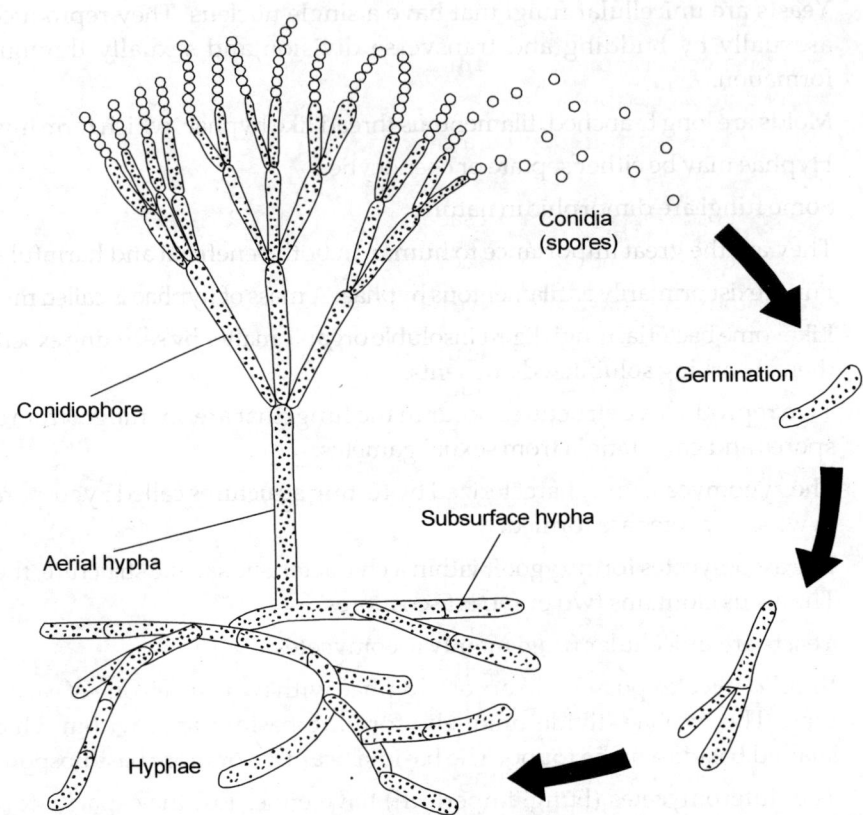

Figure 31 Simple fungal life

Each cell is composed of outer cellwall and inner lumen. Cytoplasm, cell inclusions and organelles are the major content of the inner lumen. Plasma membrane surrounds the cytoplasm and is present next to the cell wall. Cell wall is made up of chitin. Cytoplasm contains nucleus. Septate hyphae may be uninucleate or multinucleate. The organelles include golgi apparatus, mitochondria, endoplasmic reticulum, vacuoles etc., The inclusions include glycogen, oil droplets, pigments etc., Some fungi are dimorphic in nature. Dimorphic fungi can change from the yeast form in the animal to the mold form in the external environment. This form change is called YM shift.

Fungi grow best in dark, moist places. Most fungi are saprophytes and some fungi excrete extracellular enzymes, which digest complex materials and utilize the soluble products. They are chemoorganoheterotrops and use organic compounds as carbon, electron and energy.

Glycogen is the primary storage polysaccharide.

Fungi are usually aerobic. Obligatory anaerobic fungi are found in rumen of cattle.

Reproduction in fungi is usually asexual or sexual. The most common method of asexual reproduction is a spore formation. The following are the different types of asexual spore production observed in different classes of fungi.

They are

1. A hyphae can fragment to form a cell that act as a spore. These are called arthroconidia or arthrospores.

2. If the cells are surrounded by a thick wall before separation, they are called chlamydospores.

3. If the spore developed within a sac at the hyphal tip are called sprangiospore.

4. If the spores are not enclosed in a sac but produced at the tip are called conidiospores.

5. Spores produced from a vegetative mother cell by budding are called blastospores.

Sexual reproduction in fungi involves the union of nucleus. Gametangia are produced by sexual reproduction. Spores of sexual reproduction are named differently in different classes. Zygomycetes produce zygospore, ascomycetes produce ascospore and basidiomycetes produce basidiospore.

Aspergillus

It is a multicellular, eukaryotic fungi. It is a saprophytic fungus. These organisms cause opportunistic infections. Some species are used in the industry for the production of some enzymes. About 167 species are identified , of these 16 species are the etiological agent of Aspergillosis. The organism is found worldwide in soil, air, on mouldy storage grains and on decaying vegetables. It is a haploid fungi. Hyphae are septate in nature. The hyphal cells are multinucleate (Coenocytic). Most of the hyphae are horizontal and vertical in nature. Vertical hyphae produce conidiophores, which contains vesicle. Vesicle bear bottle shaped cells called sterigmata. Sterigmata contains chains of conidia. It reproduces by vegetative, asexual and sexual methods. Important species are *A.niger, A.flavus, A.fumigatus and A.terreus*. Identifying features of various agents are,

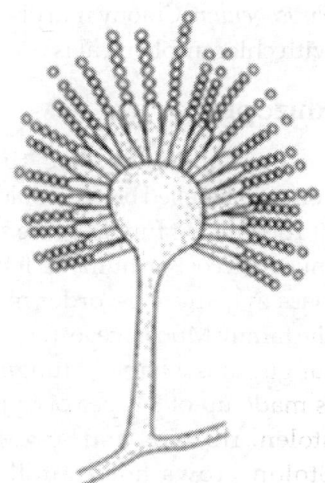

Figure 32   Aspergillus

A.niger - Coarse black granules against the creamy colony.Globose vesicles with biseriate phialids. Large echinulate jet-black conidia in chains.

A.flavus - Yellow to yellow green colonies. Globose to sub globose vesicles with uniseriate or biseriate phialides.Conidial heads that radiate and are loosely formed.

A.fumigatus - Bluish green to grey colonies. Flask shaped vesicles.Uniseriate arrangment of phialides.

A.terreus - Smallest Aspergilli.Cinnamon to buff brown colonies.Dome shaped vesicles with biseriate phialides. Long and compact conidial heads.Submerged hyphae that may form globose conidia.

Penicillium

Penicillium sp are ubiquitous and omnipresent throughout the world. They are found in soil and decaying vegetation. It belongs to the class eurotiomycetes; order moniliales and family moniliaceae. It is a multicellular fungi. It is commonly called green mould. The body of penicillium is called thallus. Thallus is filamentous in nature. The filamentous body is called mycelium. Filaments shows septate hyphae. There are two types of hyphae. They are vertical and horizontal. Vertical hyphae bear conidia. Phialides are formed as blunt tips. Chains of conidia from the phialides. Phialides may be arranged in whorls.

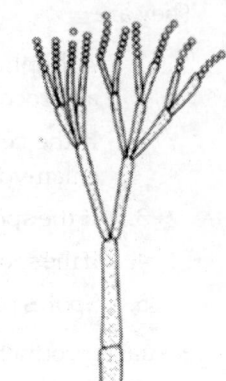

It reproduces vegetatively by fragmentation, conidia by asexually and sexually by plasmogamy. It causes opportunistic infections like pulmonary infection, keratomycetitis, onchomychosis, cutaneous lesions, bladder infection. *Penicillium*

Figure 33   Penicillium

is one of the most common laboratory contaminants. Penicillin is produced by *Penicillium chrysogenum*. Colony morphology is flat granular that are typically blue green. Modified SDA with chloramphenical is used for cultivation .

Rhizopus

It is a eukaryotic fungus. It is commonly called bread mould. It is a genus of rot-causing fungi. Some species cause mucormycosis in humans. It belongs to the class zygomycetes; order mucorales and the family Mucoraceae. It is a saprophytic fungus. It is a haploid fungus. Mycelium is made up of 3 types of hyphae namely stolen, rhizoids and sporangiophores. Stolon grows horizontally above the substratum. Rhizoids grow downwards. It reproduces by vegetative, sexual and asexual methods. *Rhizopus* produce cottony to woolly olive gray colonies that rapidly fill the SDA plate. The surface becomes covered with dark spots when sporangia appear, so *Rhizopus* is described as salt and pepper appearance. On microscopical examination broad

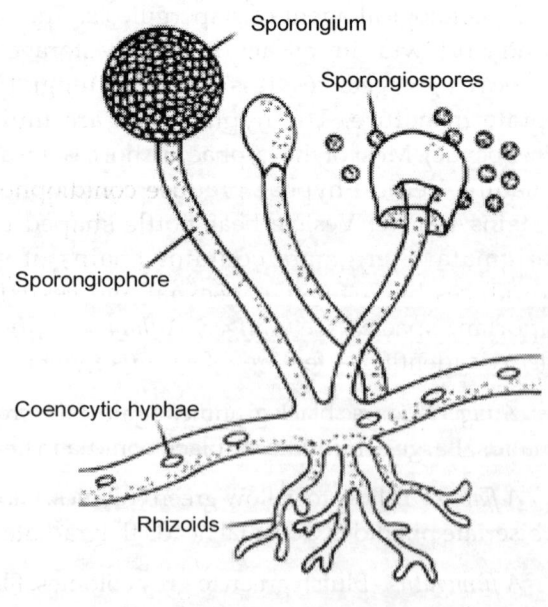

Figure 34   Rhizopus

irregular hyphae that are aseptate or sparsely septate. They have Well-developed rhizoids. Branching sporangiophore with hemispherical columellae arising from nodes adjacent to rhizoids. Grows best at 25°C and 37°C sometimes up to 50°C. It involves in alcoholic and lactic acid fermentation. Eg. *R. stolonifer.*

Mucor

It is a eukaryotic fungi. Mucor is a genus of about 40 species of molds commonly found in soil and on plant surfaces, as well as in rotten vegetable matter. It belongs to division Zygomycota and family Mucoraceae. It is also commonly called bread mould. It is a saprophytic fungi. The vegetative body is called mycelium. Mycelium contains branched filaments called hyphae. Hyphae are long and slender. The hyphae is aseptate and coenocytic. . It reproduces by vegetative, sexual and asexual methods. On modified SDA it grows at 25°C after 2-4 days, Colonies of

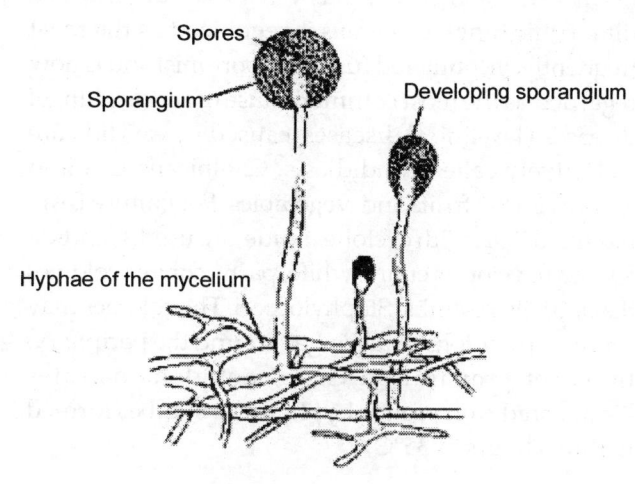

Figure 35 Mucor

Mucor are *wooly* and rapidly fill the entire petriplate with an abundant matted mycelium. The colony is white at first and becomes grey or yellow. Broad irregular hyphae that are aseptate, septate or sparsely septate. Branching sporangiophores with collumellae supporting sporangia filled with sporangiospores. Absence of rhizoids.

Saccharomyces cerevisiae (*Yeast*)

It is a unicellular eukaryotic fungi. It is commonly called bakers yeast and brewers yeast. It survives saprophytically in sugar solutions, ripe fruits etc., it s also found in soil. It is a working yeast. Yeasts are oval in shape. Cells are covered with outer cell wall and inner plasma membrane. The plasma membrane encloses protoplast. The protoplast is made up of cytoplasm and nucleus. The cytoplasm contains mitochondria, golgi apparatus, endoplasmic reticulum, ribosomes. Various strains are used in industry to make bread, beer, wine and industrial alcohol. PDA is used for the

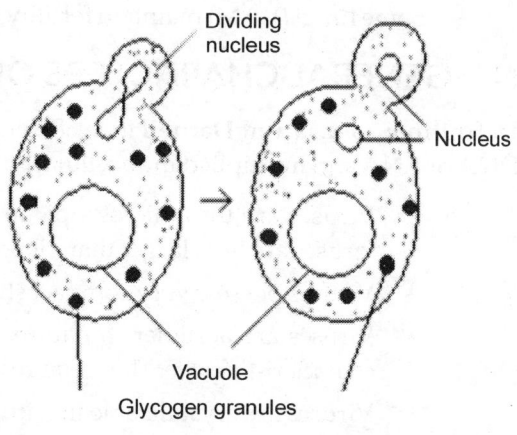

Figure 36 Yeast

isolation of *Saccharomyces cerevisiae*. On Rose Bengal agar, milky white colonies are formed. It is sensitive to Cycloheximide and produce blastoconidia, but neither germ tube nor chlamydospores. Pseudohyphae may be formed. It produced single multi lateral budding yeast. It is facultative anaerobic one.

Candida albicans

It is a small unicellular yeast like fungi. It is a dimorphic fungi. *C.clbicans* is regonized as the most frequently encountered fungal opportunist and is now regarded as the most common cause of serious fungal disease. The clinical diseases caused by candida are collectively called candidiasis. C. albicans is found worldwide on fruits and vegetables. For culture BAP, modified SDA with cycloheximide are used. Candida species develop as entire, white, pasty, convex colonies that initially resemble Staphylococci. The colonies may produce pseudohyphal fringes around the periphery. In a wet preparation C.albicans demonstrates blastoconidia on pseudohyphye. Germ tubes formed within 3 hours at 35°C

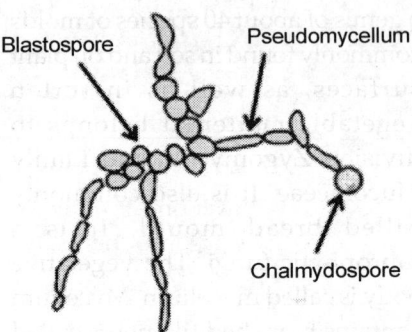

Figure 37   Candida

M. ECONOMIC IMPORTANCE OF FUNGI

- ⋏ Antibiotics like penicillin are obtained from fungi.
- ⋏ Mushroom fungus *Agaricus* is used as a food.
- ⋏ Yeast is used in alcoholic fermentation.
- ⋏ Yeasts are also used in baking industry.
- ⋏ Various enzymes used in food and other industry were made from fungal fermentations.
- ⋏ Some fungi like *Gibberella* are able to produce plant growth hormone gibberillin.
- ⋏ Some fungi (VAM) maintain fertility of the soil.

N. GENERAL CHARACTERS OF VIRUSES

Definition: **Luria and Darnell** in 1968 defined viruses as entities whose genome is either DNA or RNA and multiplied intracellularly. Complete virus particle is called as **virion.**

- ⋏ Viruses are ultramicroscopic in size ranging from 10 nm to 450 nm. Smallest viruses are little larger than ribosomes. Largest viruses are like smallest bacteria.
- ⋏ Viruses possess very compact structure.
- ⋏ Viruses are acellular in nature. Because they do not independently fulfill the characteristic of life. They require host cell for their replication and protein synthesis.
- ⋏ Viruses are crystalizable in nature.

- Basic structure of virus consists of protein capsid and nucleic acid.
- Capsid is made up of repeating unit called **capsomer**. It protects nucleic acid. Nucleic acid can be either DNA or RNA but not both.
- Nucleic acid types include single stranded RNA, double standard RNA, double standard DNA and single stranded DNA.
- Molecules on cell surface impart high specificity for host cell. Spike proteins of enveloped virus and capsid protein of non-enveloped virus are responsible for antigenicity and pathogenicity of viruses.
- Viruses lack machinery for synthesizing proteins. They are obligate intracellular parasites of bacteria, protozoa, fungi, algae, plants and animals.
- All the viruses are infectious in nature.
- Viral genome is replicated within an appropriate host cell and directs the synthesis of viral components by host cellular system.
- Progeny viruses are formed by assembly.
- Viruses do not have any cell organelles like chloroplast, ribosomes, mitochondria etc.
- A virus contains single genome only (either DNA or RNA).

Viruses differ from living cells by

- Their simple, acellular organization.
- Absence of both nucleic acid (RNA and DNA) in same virion.
- Their inability reproduces independently.
- There is no cell division as like prokaryotes and eukaryotes.

O. ULTRA STRUCTURE OF VIRUSES

Viruses are a unique group of infectious agents. The complete viral particles is called virion. It consist of one or more molecules of DNA or RNA. It is enclosed in a coat protein and envelope. Size of the virion ranges from about 10-300 or 400nm in diameter.

Capsid

Nucleic acid core of the viruses are surrounded by protein coat called capsid. Capsid protects the viral genetic material and aids in its transfer between host cells. The capsid is composed of a number of subunits known as capsomeres.

Capsomeres are assembled and give rise to viral symmetry. There are 3 morphological types of capsid:

1. Polyhedral/Icosahedral/Spherical symmetry.
2. Helical/Cylindrical/Rod like symmetry.
3. Complex/Binal symmetry.

Icosahedral

It is a regular polyhedron with 20 equilateral triangular faces and 12 intersecting points. Watson and Crick shown the polyhedral capsids exhibit 3 types of symmetry.

Eg: Tetrahedral, Octahedral and Icosahedral. Viruses employ the Icosahedral shape because it is a most efficient way to enclose a space.

Capsids are large macromolecular structures constructed from many copies of one or more or few types of protein subunits called protomers. Five or 6 protomers combined form a structure, which was represented as capsomers.

Capsid is made up of pentamers and hexamers. .

Helical Symmetry

It consists of monomers arranged in a helix around a single rotational axis.

Eg. TMV. Helical capsids are shaped much like hollow tubes with protein wall size of the helical capsid is influenced by its protomers and Nucleic acid enclosed within the capsid.

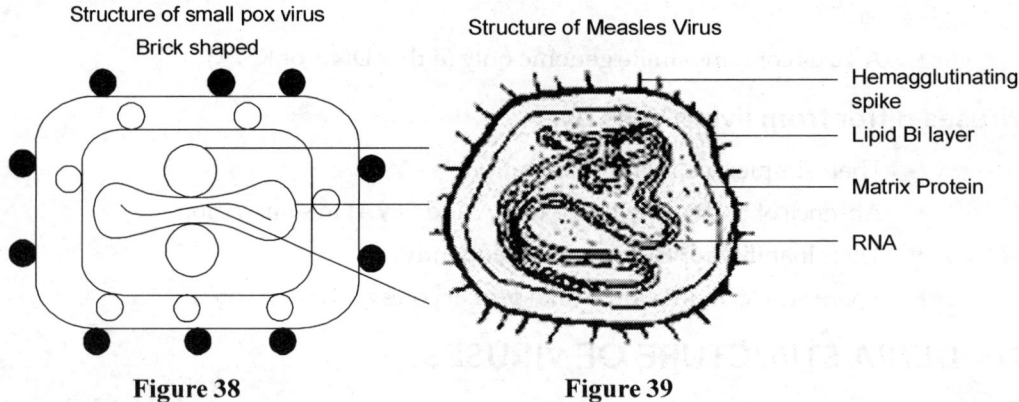

Structure of small pox virus
Brick shaped

Structure of Measles Virus

Hemagglutinating spike

Lipid Bi layer

Matrix Protein

RNA

Figure 38

Figure 39

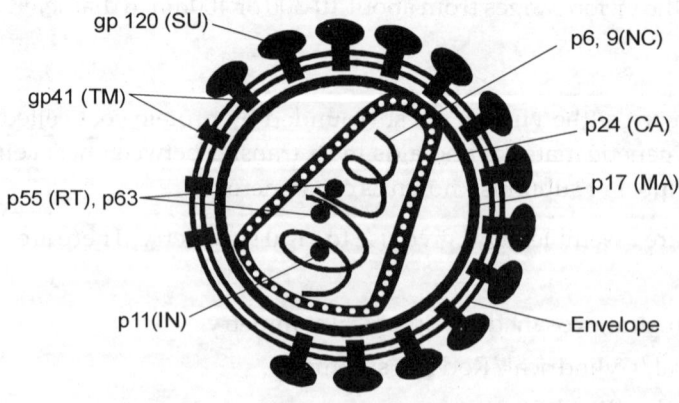

gp 120 (SU)

gp41 (TM)

p55 (RT), p63

p11(IN)

p6, 9(NC)

p24 (CA)

p17 (MA)

Envelope

Figure 40   HIV

Complex symmetry

Complex viruses have capsid symmetry that is neither purely icosahedral nor helical. They may have fail and other structure or have multilayered wall surrounding Nucleic acid.

Complex viruses with both heads and tails are said to have binal symmetry because they possess a combination of icosahedral head and helical tail.

Viral Envelope and its Constituents

Many animal viruses and some plant viruses are bounded by an outer membranous layer, in addition to capsid are called an envelope. Chemically envelope is made up of carbohydrate protein and lipids. Lipids and proteins are associated to form lipoproteins. Carbohydrates and proteins associated and forms glycoproteins. It is made from host cell plasma membrane.

Viral glycoproteins are embedded in the lipid bi-layer by a short membrane-spanning domain. It is a major antigenic determinant and mediate fusion during entry.

Envelope is a flexible, membranous structure so enveloped viruses can have various shaped and are pleomorphic.

- It gives structure to the virus. It protects Nucleic Acid and Capsid
- It helps in adherence of virus on to receptors of host cell.

Spikes

- Some of the proteins projects from the envelope and are called spikes or peplomers. It is composed of glycoproteins.
- They adhere to the receptors of host cell and play an important role in infection.
- Spikes mediate host viral inferaction.
  Eg. Influenza virus, Rhabdhovirus, HIV

Viral Enzymes

Enzymes catalyze a variety of biological reactions and are called biocatalysts.

DNA polymerase:

- It is also known as DNA dependentt DNA polymerase.
- Used in polymerization of Nucleic acid.
  Eg. Involved in T4 DNA replication

RNA polymerase:

RNA viruses code for RNA dependent RNA polymerase or DNA dependent RNA polymerase.

Viral Nucleic acid

Viruses are exceptionally flexible with respect to the nature of their genetic materials. They have all 4 Possible Nucleic acid types.

DS DNA	DS RNA
SS DNA	SS RNA

Q. RICKETTSIA

General characters

The name Rickettsia is given by Howard Taylor Ricketts (1871–1910). He identified this organisms as the causative agent of typhus and rocky mountain spotted fever. Rickettsia are non-motile, Gram-negative, non-spore forming, highly pleomorphic bacteria. They are present as cocci , rods or thread-like. It is a obligate intracellular parasites. Rickettsia cannot live in artificial nutrient environments and are grown either in tissue or embryo cultures. They grow in the cytoplasm of infected cells. They lack flagella. Arthropods are the primary host. They grow mostly in endothelial cells of mammals. Rickettsial organisms possess true cell walls similar to other gram-negative bacteria. The majority of Rickettsia bacteria are susceptible to antibiotics of the tetracycline group. Rickettsia species are carried by many ticks, fleas, and lice, and cause diseases in humans such as typhus, rickettsialpox, Boutonneuse fever, African tick bite fever, Rocky Mountain spotted fever, Flinders Island spotted fever and Queensland tick typhus. They have also been associated with a range of plant diseases. The method of growing Rickettsia in chicken embryos was invented by Ernest William Goodpasture.

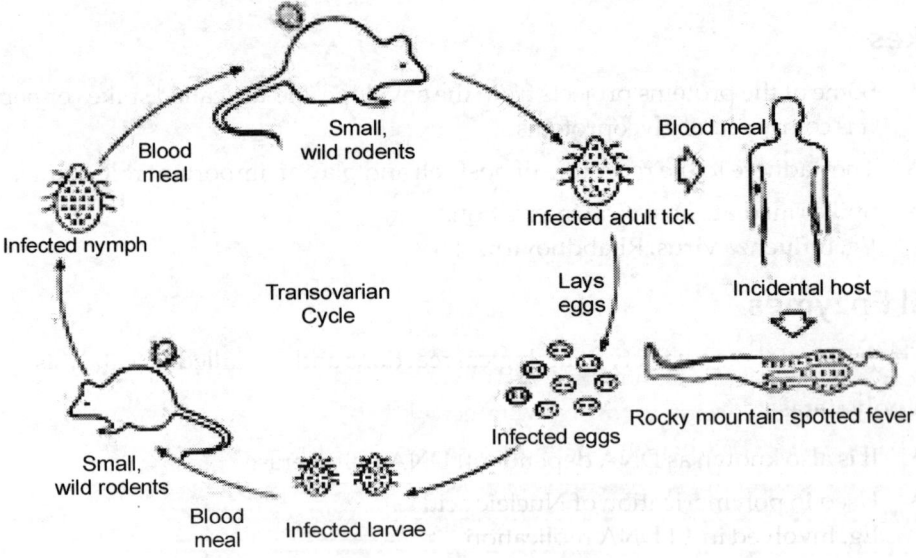

Figure 41 Rickettsia life cycle

Classification

The genus Rickettsia are classified into three groups. They are spotted fever, typhus and scrub typhus. The scrub typhus group has been reclassified as a new genus – Orientia – but many medical textbooks still list this group under the rickettsial diseases.

Spotted fever group

Rickettsia rickettsii - Rocky Mountain spotted fever

Rickettsia akari - Rickettsial pox

Typhus group

Rickettsia prowazekii - Epidemic typhus, recrudescent typhus and sporadic typhus

Rickettsia typhi -Murine typhus (endemic typhus)

Scrub typhus group

The causative agent of scrub typhus formerly known as *R. tsutsugamushi* has been reclassified into the genus Orientia.

Scientific classification

Domain: Bacteria, Phylum: Proteobacteria, Class: Alphaproteobacteria, Order: Rickettsiales, **Family:** Rickettsiaceae, Genus: Rickettsia

R. CHLAMYDIA

Chlamydia belongs to the domain bacteria.Chlamydia is part of the class chlamydiae, order Chlamydiales and the family Chlamydiaceae. Chlamydia is an obligate intracellular parasites.

They are non motile. It is a coccoid bacteria. The important Chlamydia species include *Chlamydia trachomatis* (a human pathogen), *Chlamydia suis* (affects only swine), *Chlamydia pecorum* (infects ruminants, koalas and pigs) and *Chlamydia muridarum* (affects only mice and hamsters). They multiply within membrane bound vacuoles in the cytoplasm of the host cell.

Cell wall is devoid of muramic acid. Glucose is not oxidized with ATP generation, hence parasite depends on host cell ATP. There are two type of cell formed during lifecycle. They are Reticulate body (RB) and Elementary body (EB). The reticulate body divides by binary fission to form new infectious elementary body. EB bodies are converted to RB Body. The incubation period may be up to 21 days. the reticulate body transforms back to the elementary form and is released by the cell by exocytosis. Studies on the growth cycle of *C. trachomatis* and *C. psittaci* in cell cultures *in vitro* revealed that the infectious elementary body develops into a noninfectious reticulate body (RB) within a cytoplasmic vacuole in the infected cell. It is multiplied by a process by which EB body is attached to the host cell. It is followed by phagocytosis, conversion, reproduction, condensation and release of EB.

S. THE SPIROCHAETES

General characters

Spirochaetes are long, slender, flexible spiral-shaped organisms; most appear as helical coils. They are tightly coiled, and so look like telephone cords. Spirochaetes are chemoheterotrophic in nature. Lengths of spirochete ranges between 5 and 250 µm and diameter are around 0.1–0.6 µm.

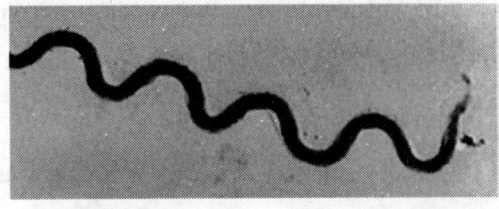

Figure 42   Treponema-a spirochaete

They are gram negative bacteria, but Gram stain not useful because the width of many spirochetes is at or just below the resolving power of the light microscope. They are observed by dark-field microscopy or by staining with special reagents. They are aerobic, microaerobic or anaerobic in nature. They are free-living and parasitic in nature. Spirochaetes are distinguished from other bacterial phyla by the location of their Flagella or periplasmic flagella. It is also called axial filaments. It is located between inner membrane and outer membrane in periplasmic space. A spirochaete will undergo asexual transverse binary fission. Many spirochaetes have not yet been cultivated *in vitro*.

Basal body of the flagella is responsible for swimming motility.

Classification

Spirochetes are placed within a single order Spirochaetales. This order contains two important families namely spirochaetaceae and Leptospiraceae.

Spirochaetaceae

They are stringent anaerobes, facultative anaerobes and microaerophiles. Carbohydrates or aminoacid used as a carbon and energy source. This family contains 4 genus namely Spirochaeta, Cristispira, Treponema and Borrelia. *Borrelia burgdorferi, B. garinii* and *B. afzelii*, which cause Lyme disease. *Borrelia recurrentis* causes relapsing fever. *Treponema pallidum* subspecies which cause treponematoses such as syphilis and yaws. Nonpathogenic treponemes can be found in the oral cavity, GI tract, and urogenital tract of animals. One species, in the genus *Cristispira*, has only been found growing on the crystalline style in the digestive tract of certain bivalve mollusks. Some species of *Treponema* live in the rumen of a cow's stomach, where they break down cellulose and other difficult to digest plant polysaccharides for their host.

Leptospiraceace

Some species are harmless inhabitants of fresh water. Leptospira is able to cause leptospirosis. They are aerobic bacteria. Long chain Fatty acid is used as a carbon and energy source. These bacteria are found throughout the world, except in Antarctica. *Leptospira* require iron for growth. The genome of pathogenic *Leptospira* consists of two chromosomes. Eg, *Leptospira interrogans*.

T. ACTINOMYCETES

Actinomycetes are a group of Gram-positive bacteria. It is also called actinobacteria or filamentous bacteria. Shape of the actinomycetes varied from rod shaped to filamentous in nature.Most of the members are aerobic in nature, but a few, such as *Actinomyces israelii*, can grow under anaerobic conditions. Actinobacteria are the good source of secondary metabolites like antibiotics. In 1942 Selman Waksman discovered actinomycin from the genus *Streptomyces* called Streptomycin.They also produce anticancer, antihelminthic and immunosuppressive drugs.The actinobacteria undergoes complex life cycle. It includes development of filamentous cells called hyphae and spores.They were all believed to have high guanine and cytosine content in their DNA. However, The G+C content of freshwater Actinobacteria can be as low as 42%. They can be terrestrial or aquatic. Actinobacteria is one of the dominant bacterial phyla and contains one of the largest of bacterial genera, *Streptomyces*. The phylum is large and very complex; it contains one class. They play an important role in the decomposition of organic materials, such as cellulose and chitin, and thereby playing a vital part in organic matter turnover and the carbon cycle. They also play an important role in the formation of humus.

Actinobacteria inhabit plants and animals. Some actinobacteria are pathogens. They are *Mycobacterium, Corynebacterium, Nocardia* and *Rhodococcus*. Some types of Actinomycetes are responsible for the peculiar odour emanating from the soil after rain (Petrichor), mainly in warmer climates. The chemical that produces this odour is known as Geosmin. The *Actinomycetes* are also known to form intracellular inclusions of polyhydroxyalkanoates under certain environmental conditions (with an excessive supply of carbon sources).

Most Actinobacteria of medical or economic significance are in subclass Actinobacteridae order Actinomycetales.

Scientific classification

Domain:	Bacteria
Phylum:	Actinobacteria
Class:	Actinobacteria
Subclass:	Actinobacteridae
Order:	Actinomycetales

Suborders

Actinomycineae	-Actinomyces
Corynebacterineae	-Corynebacterium, Mycobacterium, Nocardia
Frankineae	-Frankia
Micrococcineae	-Micrococcus
Micromonosporineae	-Actinoplanes
Propionibacterineae	-Propionibacterium

Pseudonocardineae   -Pseudonocardia

Streptomycineae   -Streptomyces

Streptosporangineae -Thermomonospora

Glycomycineae   -Glycomyces

Pseudonocardia found living mutualistically in the metapleural glands of the leaf cutter ants. Second edition of bergeys manual classified actinomycetes as actinobacteria with high G+C content containing group of bacteria using 16S rRNA Data. Actinomyces are irregularly shaped nonsporing rods that can cause disease in cattles and human. Arthrobacter has an unusual rod coccus growth cycle and carriesout snapping division. Mycobacterium forms either rod or filaments that readily fragment. The cellwall contains high lipid content & mycolic acid. These are acid fast bacteria. Nocardiofarm actinomycetes have hyphae that readily fragment into rods and coccoid elements and often form aerial mycelium with spores.

Actinoplanes have an extensive substrate mycelium and form special aerid sporangium. *Propionibacterium* are common skin and intestinal inhabitants and important in cheese manufacture and development of acnevulgaris.

U. ARCHAEBACTERIA

It is one of the bacteria. It contains prokaryotic microorganisms. Many archaea are also called as extremophiles that flourish at high temperature, low or high pH, or high salt. Some archaea are fundamental components of the biogeochemical cycles on earth or dominate special ecosystems that are of great interest (such as the methanogens). Two phyla are recognized in the Archaea: the Crenarchaeota and the Euryarchaeota. The cultured Crenarchaota are represented largely by hyperthermophilic aerobes and anaerobes. The aerobes are most acidophiles. The anaerobes are frequently neutrophilic. *Pyrodictium abyssi* is a typical organism. Optimum temperature for growth is 97-105°C. A facultative autotroph, it use either H_2 or organic compounds as electron donors. The Euryarchaeota includes the methanogens, the extreme halophiles, the 'wall-less' archaea such as Thermoplasma, sulfate-reducing organisms and some hyperthermophilic heterotrophs. *Nanoarchaeum equitans*: a small obligate symbiont, isolated in coculture with *Ignicoccus* as a very small coccus found attached to *Ignicoccus* cells or free in suspension cell diameter about 400 µm, stained with DAPI (which is specific for DNA). During exponential growth of the *Ignicoccus*, *Nanoarchaeum* grows attached to the *Ignicoccus* cell. At the end of the exponential growth phase, about 80 % of the *Nanoarchaeum* cells detach and become free in the medium infection has no obvious affect on growth of *Ignicoccus*

Nanoarchaeum is a new phylum within the archaea or a highly evolved species of the Pyrococcus/Thermococcus species.

Archaebacteria, commonly called ancient bacteria lives and thrives in environments that are normally not conducive to life and often do not contain oxygen. They can be found in swamps and habitats with high salt contents such as the Dead Sea. They also live in habitats with high temperatures and high acidity such as near underwater thermal vents and sulfur springs.

The three main groups of archaebacteria are methanogens, extreme thermophiles, and extreme halophiles. The methanogens are bacteria that produce methane gas. The thermophiles live in environments with extremely hot temperatures (up to 110°C). The halophiles live in environments that have a high salt content.

Isoprenoid glycerol diether or diglycerol tetraether lipid is available in the membrane.

They can stain either gram positive or gram negative.

Cells may be spherical, rod, spiral, cuboidal or pleomorphic in nature.

Archaea reproduce asexually by binary fission, fragmentation, or budding

In general archae have no PG layer. Methanobacteriales do have cell walls containing pseudopeptidoglycan, which resembles eubacterial peptidoglycan in morphology, function, and physical structure, but pseudopeptidoglycan is distinct in chemical structure; it lacks D-amino acids and N-acetylmuramic acid.

Methanogens

Methanobacteriales

Methanobacteriaceae	Methanobacterium
Methanothermaceae	Methanothermus

Methanococcales

Methanococcaceae	Methanococcus

Methanomicrobiales

Methanomicrobiaceae	Methanomicrobium
Methanocorpusculaceae	Methanocorpuculum
Methanospirillaceae	Methanospirillum

Methanosarcinales

Methanosarcinaceae	Methanosarcina
Methanosetaceae	Methanosaeta

Domain:	Archaea
Kingdom:	Euryarchaeota
Phylum:	Euryarchaeota
Class:	Methanopyri
Order:	Methanopyrales
Family:	Methanopyraceae
Genus:	*Methanopyrus*

Methanopyrales

    Methanopyraceae   Methanopyrus

Extremely Halophilic Bacteria

Halobacteriales

    Halobacteriaceae   Halobacterium, Haloarcula, Halococcus

Thermophilic Sulfur metabolizing bacteria

Sulfolobales

    Sulfolobaceae   Sulfolobus

Thermoproteales

    Thermoproteaceae   Thermoproteus

    Desulfurococcaceae   Desulfurococcus

    Staphylothermaceae   Staphylothermus

    Pyrodictiaceae   Pyridictium

    Thermodiscaceae   Thermodiscus

Intermediate group

    Thermoplasmales

        Thermoplasmaceae   Thermoplasma

Archaeoglobales

    Archaeoglobaceae   Archchaeoglobus

Thermococcales

    Thermococcaceae   Thermococcus

V. CYANOBACTERIA

Cyanobacteria belongs to the domain bacteria, kingdom eubacteria and phylum cyanobacteria. They show great diversity in form and shape. Some are spherical; some are rod-shaped, while few of them are unicellular or multicellular. Vacuoles are formed in these bacteria. Some forms are covered with sheath. Unicellular forms are non-motile but trichome forms have capability of gliding movement. Flagella are absent. They are found in soil, clean water, and marine environment. Some forms can live symbiotically. Some have the capacity to fix atmospheric nitrogen. Some of them live symbiotically with protozoans and such forms of bacteria are called as Cyanellae and disassociation is described as Syncyanoses. Nucleus is incipient like prokaryotes and is devoid of nuclear membrane and other membrane. They contain chlorophyll 'a' and not bacterio chlorophyll, 'a' absorbs red light. Along with chlorophyll, 'a' water-insoluble caroterioids and water-soluble phycobilins are also present. These two are the main light absorbing pigments. In some forms of bacteria red phycobilin (phycoerythrin) is found which can absorb light of short wave length (470 to 600 /mm). Cell division incorporates cytoplasmic division as well as cell plate formation. Stored food is Cyanobacteria starch. In trichomes of Cyanobacteria heterocysts are also present which help in fixation of free nitrogen. Cell wall is much thicker in comparison with the Qthei cells of the trichome. In some forms thick-walled akinetes are also found which help them to survive in unfavorable conditions. Like bacteria they also lack sexual reproduction.

Cyanobacteria are able to prepare their own food through photosynthesis but they are different from photosynthetic bacteria. Cyanobacteria are a critical component of the Earth's biosphere and are largely responsible for life as we know it. Cyanobacteria gets its common name from the blue green pigment, phycocyanin and chlorophyll a. Cyanobacteria may be single celled or colonial. The single celled forms are coccoid and are among the most abundant phytoplankton in the middle of tropical oceans where nutrient levels are extremely low. Cyanobacteria reproduce asexually by fission. The colonial and filamentous cyanobacteria reproduce by fragmentation. In fragmentation segments of the parents break off and float away. These fragments then grow into new cells.

Anabaena

Group : Cyanobacteria

Order : Nostocales

Family: Nostocaceae

Genus: Anabaena

They are oxygenic photoautotrophic bacteria in which akinetes are usually formed. They cylindrical, spherical or ovoid forms generally having 2-10 um size. The plant body consists of vegetative cells

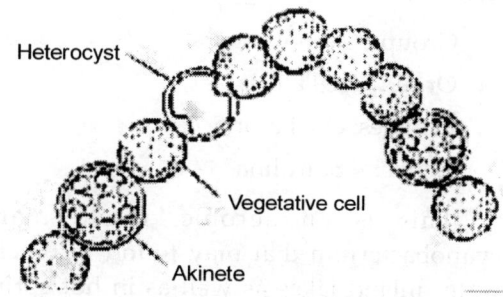

Figure 43   Anabaena

as well as heterocysts and akinetes. The heterocysts are present either in intercalary or in terminal or both positions. The cells contain slime covering and a distinct individual sheath is absent the trichomes are normally motile and colonies are not formed. Their species establish symbiotic associations with fungi (lichens), bryophytes (Anthoceros), pteridophytes (Azolla) and gymnosperms (cycas). Examples are *A. azollae*, *A. cycadae*, etc., *Anabaena* is also filamentous, but it contains heterocysts. Some species of Anabaena form symbiotic relationships with plants, while other produce very potent neurotoxins.

Oscillatoria

Group : Cyanobacteria

Order : Oscillatoriales

Family: Oscillatoriaceae

Genes : Oscillatoria

It is gram-negative, filamentous cyanobacterium which is found in fresh water ponds, pools, lakes and sub-aerial habitats. It is an unbranched filamentous alga. The filaments (trich omes) occur singly or matted together to form thin or thick sheets. The cells exhibit a typical prokaryotic structure and its protoplast is differentiated in to the peripheral pigmented chromoplasm and the central centroplasm. The photosynthetic pigments are found in the sufgace of the thylakoids. *Oscillatoria* exhibits intercalary growth. It has G+C% 40 – 50. The filaments of *Oscillatoria* move on the left and right side o f the axis similar to pendulum of a wall clock due to the presence of oscillatory movement. The main photosynthetic pigment is chlorophyll a and chlophyll – c. The accessory pigments are found in phycobilisomes possessing phycobiulins i.e phycocyanin –c, allophycocyanin and phycoerythrin – c. The reproduction occurs with the help of hormogonia and fragmentation of parental trichomes . Due to the property of oscillation , the terms *Oscillatoria* has been coined Examples are *Oscillatoria formosa, O.prolifica*, etc.,

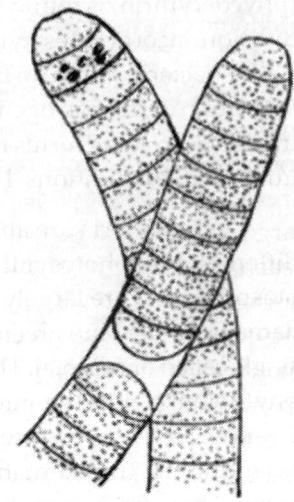

Figure 44   Oscillatoria

Spirulina

    Group : Cyanobacteria

    Order: Oscillatoriales

    Families: Oscillatoriaceae

    Genus: Spirulina

This is an aerobic, fresh, marine cyanobacterium that may found in brackish water inland lakes as well as in hot springs. Generally they grow in closed right-handed or left-handed helix. The cross walls are thin and are invisible or nearly so with light microscopy.

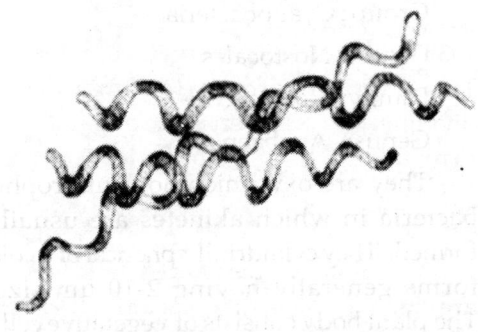

Figure 45   Spirulina

They are self pH adjusters that grow between 8.5 and 11. They have gliding motility, consists of a 'turning of the screw' to form continuous helical coil with thin cross walls. These are significant due to their industrial importance in the form of rich protein value (62%). Their colour is variable from blue green to red and mol% G +C contents is 54. *Spirulina* is a filamentous cyanobacteria. This organism is often packaged as a human nutritional supplement due to its high quality and quantity of vitamins, proteins, and minerals.

Nostoc

    Group : Cyanobacteria

    Order: Nostocales

    Families: Nostocacaceae

    Genus: Nostoc

They are oxygenic phototrophic bacteria. The trichomes with conspicuous constructions at cross walls are present giving the typical contorted appearance. The cells are cylindrical, spherical or ovoid in shape. The heterocysts are intercalary and trichomes are present in confluent gel. Some colonies are in ball shaped while few form flattened discs or large sheats. The size of the colonies sometimes reached to 20cm in diameter. Vegetative trichomes are not capable of gliding motility. The hormogonoia are often filled with gas and their width is lessthan that trichomes. The DNA base composition ranges from 39-45 mol% G+C. *Nostoc* is a filamentous cyanobacteria that forms heterocysts and is covered with a gelatinous sheath.

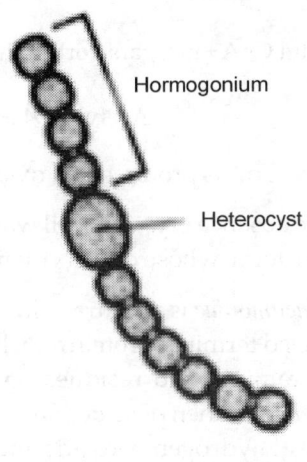

Figure 46   Nostoc

W. L-FORM BACTERIA

L-form bacteria is also known as L-phase bacteria, L-phase variants, and cell wall-deficient (CWD) bacteria. They are strains of bacteria that lackcell walls. They were first isolated in 1935 by Emmy Klieneberger. He named them "L-forms" after the Lister Institute in London. Two types of L-forms are distinguished: unstable L-forms, spheroplasts that are capable of dividing, but can revert to the original morphology and stable L-forms, L-forms that are unable to revert to the original bacteria.

X. MYCOPLASMA

The mycoplasmas are bacteria lacking a rigid cell wall during their entire life cycle. They are much smaller than bacteria. It is called pleuropneumonia organism (PPO). These are commonly referred to as pleuropneumonia-like organisms or PPLO. A certain group of mycoplasmas produce extremely tiny colonies on agar plates, and are called the T-strains. These organisms are the smallest known free-living organisms. They do not stain with the Gram stain. They are more pleomorphic in nature. They appear as tiny pleomorphic cocci, short rods, short spirals, and sometimes as hollow ring forms when stained with giemsa stain. Most mycoplasmas require a rich medium containing a sterol and serum proteins for growth. On solid media, they form minute, transparent colonies. The colony looks like a fried egg. The cytoplasm contains ribosomes, but lacks mesosomes. The parasitic mycoplasmas have truncated respiratory systems, lacking quinones and cytochromes. The major source for ATP is the arginine dihydrolase pathway.

$$\text{Arginine} + H_2O \xrightarrow{\text{arginine deaminase}} \text{citrulline} + NH_3$$

$$\text{Citrulline} + \text{inorganic orthophosphate} \xrightarrow{\text{ornithine carbamoyl transferase}} \text{ornithine} + \text{carbamoyl PO}$$

$$\text{Carbamoyl PO4} + \text{ADP} \xrightarrow{\text{carbamate kinase}} \text{ATP} + CO_2 + NH_3$$

Another mechanism for ATP generation is:

$$\text{Acetyl CoA} + \text{inorganic orthophosphate} \xrightarrow{\text{phosphate acetyl transferase}} \text{acetyl PO4} + \text{CoA}$$

$$\text{Acetyl PO4} + \text{ADP} \xrightarrow{\text{acetate kinase}} \text{acetate} + \text{ATP}$$

Acetyl CoA is produced by oxidative decarboxylation of pyruvate.

In the absence of a rigid cell wall, the pattern of replication is quite different from that of typical bacteria, whose division starts with the formation of a well-defined septum.

M. pneumoniae is an extracellular pathogen that adheres to the respiratory epithelium by a specialized terminal protein attachment factor. This adherence protein interacts specifically with neuraminic acid residues on the epithelial cell surface. Ciliastasis occurs following attachment and then destruction of the superficial layer of epithelial cells. Destruction is due to release of hydrogen peroxide and superoxide anion.

The human diseases caused by mycoplasmas are Primary atypical pneumonia, Non-gonococcal urethritis (NGU). Primary atypical pneumonia is usually selflimiting and does not require antibiotic treatment. However, if antibiotics are needed, the drug of choice is one of the macrolide antibiotics like Azithromycin, Clarithromycin, Dirithromycin and Erythromycin. Urogenital diseases may be treated with: Metronidazole and Clindamycin.

P.   GENERAL CHARACTERS AND ULTRASTRUCTURE OF LICHENS

Lichens are composite organisms composed of algae and fungi. Lichens are a good example for symbiotic relationship between two group of microorganisms. It consists of two partners. One partner is called the **mycobiont, the fungal partner** and the phycobiont or photobiont, a photosynthetic partner. Phycobiont is usually either a green algae or cyanobacterium. The usual algal partner of lichen is Trebouxia and the cyanobacterium is *Nostoc*. In this association, two partners are mutually benefitted. The algae supplies carbohydrates to the fungus and the fungus supplies nitrogen and other nutrients to the algae. The fungal also protects algae from the external environment.

Ecologists have shown that many species of lichens are very sensitive to air pollutants, such as **sulfur dioxide**. Thus, they are often used as **indicator species** for **air pollution**.

Many lichens can inhabit harsh environments and withstand prolonged periods of desiccation. In the temperate region of **North America**, lichens often grow on tree trunks and bare **rocks** and **soil**. In the arctic and antarctic regions, lichens constitute a large proportion of the **ecosystem biomass**. Many lichens are even found growing upon and within rocks in **Antarctica**.

Taxonomy and Classification

Lichens are named based on the fungal component. Fungal component plays the primary role in determining the lichen's form. Major part of the lichen is made up of fungus.

On the basis of the type of fungal partner lichens are classified into two. They are ascolichens and basidiolichens.

Ascolichens: Fungal partner of this group are a member of the Ascomycota. Eg. Corapavonia

Basidiolichens: Fungal partner belong to the class Basidiomycota and termed **basidiolichens**. Eg. *Geosiphon pyriforme*. Algal component of the lichen belongs to Myxophycophyta. Eg. *Nostoc, Rivularia*.

Lichens are informally classified into seven by the growth of thallus. They are

Crustose lichens: It grows like a paint. Thallus of this group lichen is flat. Growth is with leathery texture. E.g., *Caloplaca flavescens*

Filamentous: It grows like hair-like, e.g., *Ephebe lanata*

Foliose: Thallus of this group are like leaf, e.g., *Hypogymnia physodes*

Fruticose: Thallus may be branched, cylindrical. Thallus may be erect, e.g., *Cladonia evansii, C. subtenuis* and *Usnea australis*

Leprose: Thallus shows powdery appearance. e.g., *Lepraria incana*

Squamulose: It consisting of small scale-like structures. It lack a lower cortex. E.g., *Normandina pulchella*

Gelatinous lichens: In this group the cyanobacteria produce a polysaccharide that absorbs and retains water.

The Lichen Thallus

In the traditional sense of lichens, their thallus can be artificially divided into three forms: foliose, crustose and fruticose.

Foliose Lichens

Lichen thallus is generally "leaf-like", in appearance and attached to the substrate at various points by root-like structures called rhizines. Because of their loose attachment, they can easily be removed. These are the lichens which can generally be mistaken for bryophytes, specifically liverworts. It is possible, or even probable, that herbaria still contain lichens that have been mistakenly identified as liverworts. If we look at these a foliose lichen in longitudinal section, from top to bottom, we would be able to distinguished the following layers:

Upper Cortex: Often composed of tightly interwoven mycelium, which gives it a cellular appearance. This cellular appearance is referred to as pseudoparenchymatous.

Algal Layer: Composed of interwoven hyphae with the host algal cells. This is the ideal location for the algal cells. Beneath the upper cortex so that it receives the optimal amount of solar radiation, for photosynthesis, but not direct solar radiation which would be harmful.

Medullary Layer: Composed of loosely interwoven mycelium. Layer is entirely fungal.

Lower Cortex: Usually same composition as the upper cortex and attached to the substrate by root-like structures called rhizines. The rhizines are entirely fungal, in origin, and serve to anchor it to the substrate. Thus, the foliose lichens also have what is referred to as a dorsiventral thallus, i.e. a distinct upper and lower surface.

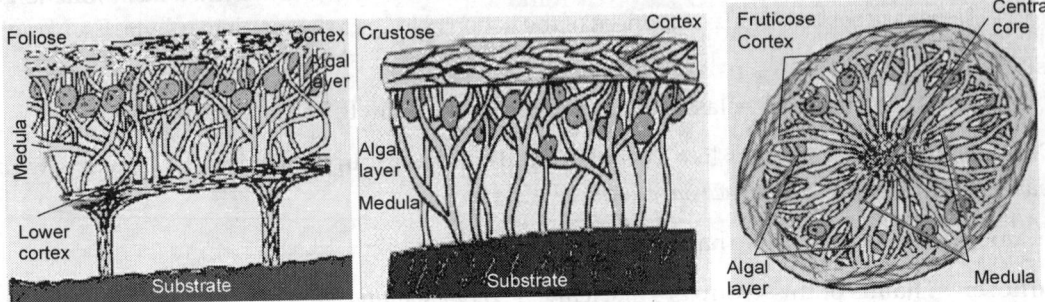

Figure 47   Lichen structure

Crustose Lichens

Lichen thallus which is very thin and flattened against the substrate. The entire lower surface is attached to the substrate. These lichens are so thin that they often appear to be part of the substrate on which they are growing. Crustose species that are brightly coloured often give the substrate a "spray-painted" appearance. The thallus has the upper cortex, algal and medullary layers in common with the foliose lichens, but does not have a lower cortex. The medullary layer attached directly to the substrate and the margins are attached by the upper cortex.

Fruticose Lichens

The thallus is often composed of pendulous ("hair-like) or less commonly upright branches (finger-like). The thallus is attached at a single point by a holdfast. In cross section, the thallus can usually be seen to be radially symmetrical, i.e. does not have a top and bottom. The layers that can be recognized are the cortex, algal layer, medullary layer, and in some species the center has a "cord" which is composed of tightly interwoven mycelium. Other species have a hollow center that lack this central cord.

Economic Relevance

Economically, lichens have little significance. Perhaps this is why there is so little interest in this group of organisms. One way that they have been utilized is in the extraction of blue, red, brown or yellow dyes in the garment industry. Also, the indicator pigments used in litmus paper was also derived from lichens. Previously, we briefly mentioned lichens as a source of pharmaceutical compounds. You can include some "folk" remedies in this category as well. They are also used in the cosmetic industry, in the making of perfumes and essential. Finally, some species have been used as food. One species, *Lecanora esculenta*, is a species that grows in the mountains near Israel and are typically blown free from their substrate. Desert tribes grind up the lichen, dry it and mix it with dry meal to form a flour. It is postulated that this is the species lichen that is referred to as "Manna from Heaven" when Moses led the Hebrews across the desert during biblical time. One species, *Cladonia rangiferina* (reindeer moss), is fed upon by reindeers and cattle. This has led to the discovery that lichens readily absorb radioactive elements. After open-air, atomic testing, both Alaskan Eskimos and Scandinavian Laplanders were found to have high levels of radioactive contamination, which they had absorbed from eating reindeer, which in turn ate lichens.

Lichens are conspicuously absent in and surrounding cities because many species are sensitive to pollution, especially to sulfur dioxide and flourine, which are common pollutants. For this reason, they have been commonly used as indicators of pollutants. In urban areas, where lichen surveys have been carried out, the absence of certain indicator species is used as early warnings of decrease in air quality.

Lichens also play a very significant role in nature. They are the pioneers in rocky substrates, where there is no soil. Lichens break down the rocky substrate into soil and their decomposing thallus fertilize the newly produced soil, making it possible for the plant habitation.

Reproduction

Reproduction of the lichen is entirely asexual. It may occur by soredia (sing.: soredium), spore-like structures, composed of alga cells and hyphae that are formed from the algal layer and become exposed when the cortex ruptures. This is best seen in a sectioned lichen. The other means of asexual reproduction is by isidia (sing.: isidium), columnar to swollen structures that are part of the lichen thallus that are likely to break off to form new lichens. Ascospores and conidia also form, but these will only reproduce the fungus. It is assumed that these structures will come in contact with a suitable algal host and resynthesis the lichen thallus. However, the latter are not thought to be significant in lichen reproduction.

Many lichens reproduce asexually, either by vegetative reproduction or through the dispersal of diaspores containing algal and fungal cells. *Soredia* small groups of algal cells surrounded by fungal filaments that form in structures called *soralia*, from which the soredia can be dispersed by wind. Many lichen fungi appear to reproduce sexually in a manner typical of fungi, producing spores that are presumably the result of sexual fusion and meiosis.

Economic importance of Lichens

Lichens are eaten by many different cultures across the world. Iceland moss (*Cetraria islandica*) was an important human food in northern Europe, and was cooked as a bread, porridge, pudding, soup, or salad. Wila (*Bryoria fremontii*) was an important food in parts of North America, where it was usually pitcooked. Northern peoples in North America and Siberia traditionally eat the partially digested reindeer lichen (*Cladina* spp.). Rock tripe (*Umbilicaria* spp. and *Lasalia* spp.) is a lichen that has frequently been used as an emergency food in North America. *Umbilicaria esculenta*, is used in a variety of traditional Korean and

Japanese foods. Many lichens produce secondary compounds, including pigments that reduce harmful amounts of sunlight and powerful toxins that reduce herbivory or kill bacteria. Lichens being used to extract purple and red colours. The pH indicator litmus is a dye extracted from the lichen genus *Rocella tinctoria* by boiling. Extracts from many *Usnea* species were used to treat wounds. *Lobaria pulmonaria* was collected in large quantities as "Lungwort" and, due to its lung-like appearance, was sold as a cure for lung diseases. *Peltigera leucophlebia* was used as a supposed cure for thrush.

4

MICROSCOPY AND STAINING

A. MICROSCOPY

The study of Microbiology requires appropriate methods for observing microbes. Microscopy is the use of a microscope to view objects too small to be visible with the naked eye. Microorganisms are the tiny particles seen through microscope.

In 1676, Antony Von Leeuwenhoek observed minute objects and named as animalcules through his ground pieces of glasses. His microscope magnification is around 50 – 300 times. The common initial microscope may be invented by Zacharias Janseen from Netherlands or Galielo Galielei of Italy. Advanced compound microscope was invented by Robert Hook in 18th century.

Two key characteristics of a reliable microscope are **magnification** or the ability to enlarge objects and **resolving power** or the ability to show detail. Lenses act like a collection of prisms operating as a unit. When a light source is distant so that the parallel rays of light strikes the lens, a convex lens will focus these rays at a specific point, the focal point. The distance between the centre of the lens and the focal point is called focal length. Many different types of microscopes have been developed over the past two centuries. Each has its own characteristics features that provide it with a specific value in microscopy. Basically there are two kinds of microscopes. They are light and electron.

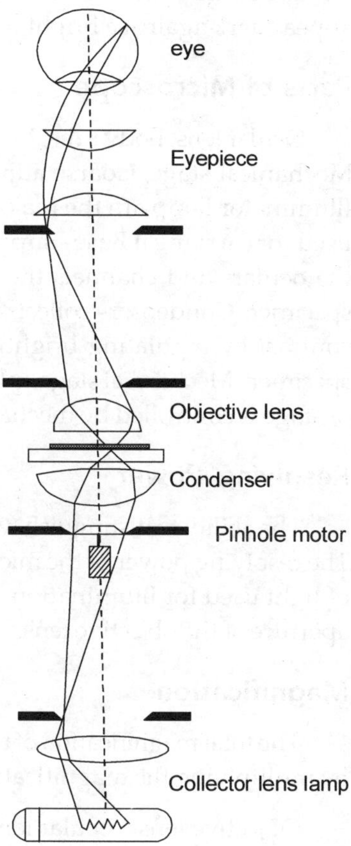

Figure 48 Light path of Bright field microscope

Light microscope uses light as its source of illumination and has four types. They are (1) Bright field Microscope, (2) Dark field Microscope, (3)Fluorescent Microscope and (4)Phase contrast Microscope. Electron microscopes uses a beam of electrons instead of light waves to produce an image.There are two types of electron microscopes. They are (1)Transmission Electron Microscopes and (2)Scanning Electron Microscope.

Compound Microscope (Bright Field) (Figure 48)

The standard instrument used in the laboratory to observe microorganism is the compound microscope. This is called compound because it contains two or more sets of lenses. It is also called bright field microscope. Modern compound microscope has a condenser lens, which focuses light on the objective lenses that are close to the specimen and magnify it, and ocular lenses that are close to eye and further magnify the image.

Light rays pass through the specimen in a bright field microscope. Since objects are seen against the light background and staining of the specimen enhances its contrast against the background. Specimens that are stained and view with the bright field (light field) microscope appear dark against a bright background.

Parts of Microscope

Ocular lens, Body tube, Movable arm, Nose piece, Objective lens, Body of the microscope, Mechanical stage, Coarse adjustment, Fine adjustment, Condenser, Iris diaphragm, Base, Illuminator lamp are the major parts of microscope. Arm - supports the microscope and is used for carrying it Base - support structure; used in carrying microscope. Body tube - holds the oculars and channels the light rays. Coarse adjustment knob- used to initially focus specimen. Condenser - collects the light rays into a single cone of light. Diaphraghm - provides contrast by regulating brightness of light. Fine adjustment knob- used for final focus of specimen. Mechanical stage - platform for holding slide; contains clips to hold slide; movement of stage is controlled by mechanical stage knobs. Revolving nosepiece - holds the objectives

Resolving Power

The ability to distinguish two objects as separate and distinct entities also called resolution. The resolving power of the microscope is determined by three factors i.e., (a). The wavelength of light used for illumination, (b). Numerical aperture of the condenser and (c). Numerical aperture of the objective lens.

Magnification

The total magnification of a specimen seen through a compound microscope is determined by multiplying the magnifications of objective lens and ocular lens.

Objective lens × ocular lens = magnification

$$010 \times 10 = 0100$$

$$040 \times 10 = 0400$$

$$100 \times 10 = 1000$$

The **refractive index** varies with the medium used between the lens and the specimen. Air is generally used as the standard surrounding medium. The refractive index of air is assumed to be 1.00. Since the value of sin θ remain constant for any given lens, the numerical aperture of an objective lens can be increased only by inserting a medium with a refractive index higher than 1.00 between the specimen and lenses. Water has a refractive index of 1.33

where as immersion oil- has refractive index of 1.52 by changing the medium from air to oil, we will increase the numerical aperture of objective lens. Resolving power also depends on condensers numerical aperture. **Abbe condenser** used in student microscopes. This consists of two or more lenses that are not corrected for spherical or chromatic aberration. Other type condenser has 6 elements, which correct aberration and provide neat perfect images of the light source. An adjustable multi leaf **iris diaphragm** is located in the condenser. It controls the diameter of light leaving the condenser and striking the specimen. Adjusting voltage mode of **light source** controlled light intensity in microscopy.

Table 4.1   Differential properties of microscopic objectives

Character	Low power	High power	Oil immersion
Magnification	100-150	400-450	1000-1500
Numerical aperture	0.25	0.55-0.65	1.25-1.4
Focal length	16mm	4mm	1.8-2mm
Working distance	4-8mm	0.5-0.7mm	0.1mm
Resolving power	0.9m	0.35m	0.18m

Dark Field Microscopy (Figure 49)

It is designed to eliminate the need for staining to achieve contrast between the specimen and background. Dark field microscope uses dark field condenser. It contains an opaque disc. The disc blocks light that enters the objective directly. Only light that is reflected off the specimen enters the objective lens. The dark field condenser directs the rays of light into the specimen field at such an angle that only the rays strictly an object in the field are bent, or refracted in to the objective.

Light enters the microscope for illumination of the sample. A specially sized disc, the *patch stop* (see figure 49) blocks some light from the light source, leaving an outer ring of illumination. A wide phase annulus can also be reasonably substituted at low magnification.

The condenser lens focuses the light towards the sample. The light enters the sample. Most is directly transmitted, while some is scattered from the sample. The **scattered light** enters the objective lens,

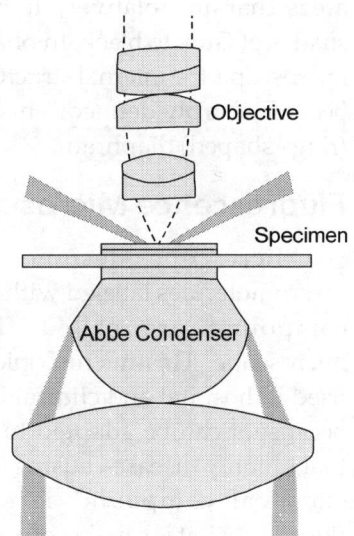

Figre 49   Principle of dard field microscopy

while the **directly transmitted light** simply misses the lens and is not collected due to a *direct illumination block*. Only the scattered light goes on to produce the image, while the directly transmitted light is omitted.

Phase Contrast Microscope (Figure 50)

It is useful because it permits detailed examination of internal structures in living microorganisms. It is not necessary to fix or stain the specimen. The principle of phase contrast microscopy is based on the wave nature of light rays, and the fact that light rays can be in phase or out of phase. If the wave peak of light rays from one source coincides with the wave peak of light rays from another source, rays interact to produce reinforcement (relative brightness). One set of light rays directly from the light source, another set is from light that is reflected or diffracted from the particular structure in the specimen. When the two sets of light rays -direct rays and reflected or diffracted rays are brought together, they form an image of the specimen on the ocular lens, containing areas that are relatively light, through shades of Grey to black. In phase contrast microscope, the internal structures of a cell become sharply defined. This microscope uses special condenser that contains an annular (ring-shaped) diaphram.

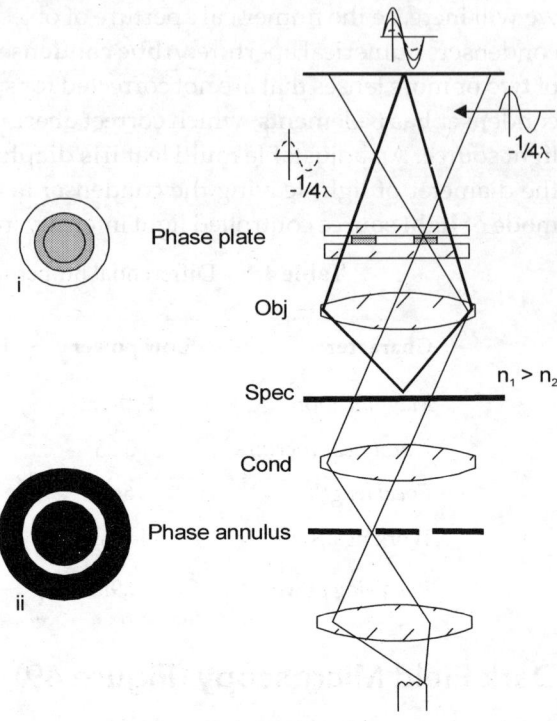

Figure 50 Principle of phase contrst microscope

Fluorescence Microscope (Figure 51)

Microscope designed to view macromolecules labeled with fluorescent compound are called fluorescent microscope. This microscopic technique used in hospital and clinical laboratories because it can be adapted to rapid tests that identify disease-causing microbes. A chemical compound is said to be fluorescent if it is capable of absorbing UV light and remitting the energy as visible light. Some fluorescent dyes are used for fluorescent microscopy.

Typical components of a fluorescence microscope are a light source (xenon arc lamp or mercury-vapor lamp), the excitation filter, the dichroic mirror (or

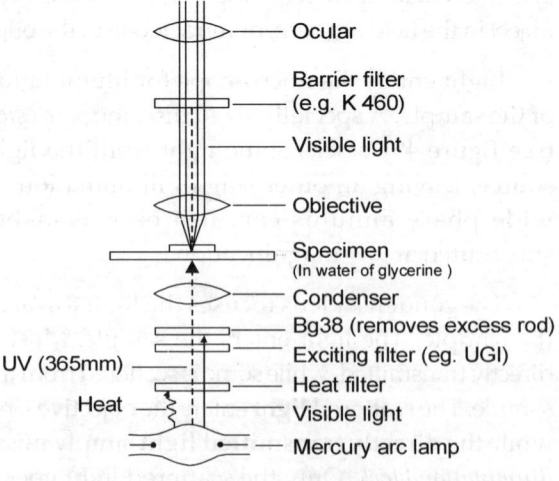

Figure 51 Principle of fluorescent microscopy

dichroic beamsplitter), and the emission filter. The filters and the dichroic are chosen to match the spectral excitation and emission characteristics of the fluorophore used to label the specimen

Electron Microscope

Electron microscopes are scientific instruments that use a beam of highly energetic electrons to examine objects on a very fine scale. It permits much greater resolution and thus obtains higher useful magnification than the light microscope. Electron microscopes functions exactly as their optical counterparts except that they use a focused beam of electrons instead of light to 'image' the specimen and gain information as to its structure and composition.

The basic steps involved in all electron microscopes are

A stream of electrons is formed (by the electron source) and accelerated toward the specimen using a positive electrical potential.

The stream is confined and focused using metal apertures and magnetic lenses into a thin, focused, monochromatic beam.

- This beam is focused onto the sample using magnetic lens.
- Interactions occur inside the irradiated sample, affecting the electron beam.
- These interactions and effects are detected and transformed into an image.
- Total magnification of electron microscope is 10,000 X plus

Two main types of electron microscopes are

1. Transmission Electron Microscope (TEM)
2. Scanning Electron Microscope (SEM)

Transmission Electron Microscope (TEM) (Figure 52)

- It was the first type of electron microscope.
- Max Knoll and Ernst Ruska in Germany developed it in 1931.
- TEM works much like a slide projector.
- Specimen for TEM is prepared by *shadow casting technique*. This will give three-dimension structure.
- Structures appear after these types of microscopy are called artifacts.
- Electron gun produces a stream of monochromatic electrons.
- This is focused to a small, thin, coherent beam by the use of condenser lenses 1 and 2.
- The beam strikes the specimen and parts of it are transmitted
- This transmitted portion is focused by the objective lens into an image.

Optional objective and selected area metal apertures can restrict the beam; the objective aperture enhancing contrast by blocking out high angle diffracted electrons, the selected area aperture enabling the user to examine the periodic diffraction of electrons by ordered arrangements of atoms in the sample.

The image is passed down the column through the intermediate and projector lenses, being enlarged all the way.

The image strikes the phosphor image screens and light is generated, allowing the user to see the image. This image represents two areas of the specimen. They are thicker area and thinner area. In thinner area electrons are passed easily.

Specimens are prepared by sectioned into extremely thin slices (20-100nm) and stained or soaked in metals that will increase image contrast.

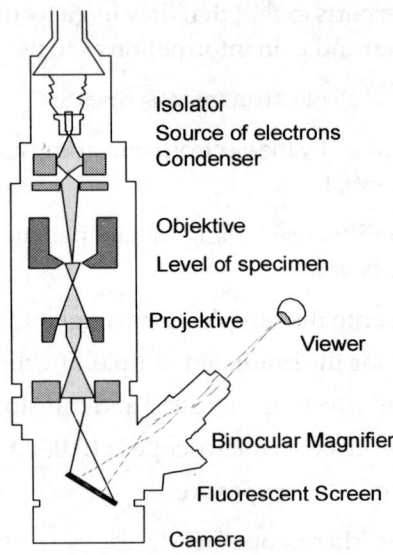

Isolator
Source of electrons
Condenser

Objektive

Level of specimen

Projektive

Viewer

Binocular Magnifier

Fluorescent Screen

Camera

Figre 52 Principle of TEM

Scanning Electron Microscope (SEM) (Figure 53)

The first scanning electron microscope debuted in 1942 with the first commercial instrument around 1965. Its later development is due to the electronics involved in scanning the beam of electrons across the sample. Electron gun produces a stream of monochromatic electrons. First condenser lens condenses the stream. The condenser aperture, eliminating some high angle electrons then constricts the beam. The second condenser lens forms the electrons into a thin, tight, coherent beam and is usually controlled by the ' fine probe current knob'. A user selectable objective aperture further eliminates high angle electrons from the beam. A set of coins then scan or sweep the beam in the grid fashion, dwelling on points for a period of time determined by the scan speed. The final lens, the objetive, focuses the scanning beam onto the part of the specimen desired. When the beam strikes the sample interactions occur inside the sample and are detected with various instruments.

Advantages of electron microscope over light microscope. It is useful for the detection of fastidious gastroenteritis viruses such as rota, adeno, astro, Norwalk and calici viruses. It is also used for the detection of herpes viruses. All viral inclusions were detected with electron microscopes. Internal structures of the viruses are also detected with the help of electron microscopes.

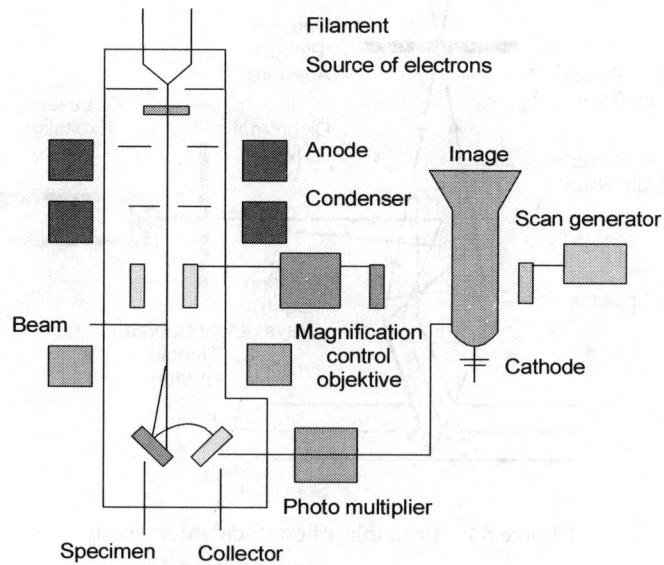

Figure 53 Principle mechanism of SEM

Confocal microscope (Figure 54)

Confocal microscopy is an optical imaging technique used to increase micrograph contrast and/or to reconstruct three-dimensional images by using a spatial pinhole to eliminate out-of-focus light or flare in specimens that are thicker than the focal plane. This technique has been gaining popularity in the scientific and industrial communities.

The principle of confocal imaging was patented by Marvin Minsky in 1957. In a conventional (i.e., wide-field) fluorescence microscope, the entire specimen is flooded in light from a light source. Due to the conservation of light intensity transportation, all parts of the specimen throughout the optical path will be excited and the fluorescence detected by a photodetector or a camera. In contrast, a confocal microscope uses point illumination and a pinhole in an optically conjugate plane in front of the detector to eliminate out-of-focus information. Only the light within the focal plane can be detected, so the image quality is much better than that of wide-field images. As only one point is illuminated at a time in confocal microscopy, 2D or 3D imaging requires scanning over a regular raster (i.e. a rectangular pattern of parallel scanning lines) in the specimen. The thickness of the focal plane is defined mostly by the square of the numerical aperture of the objective lens, and also by the optical properties of the specimen and the ambient index of refraction.

Three types of confocal microscopes are commercially available: Confocal laser scanning microscopes, spinning-disk (Nipkow disk) confocal microscopes and Programmable Array Microscopes (PAM).

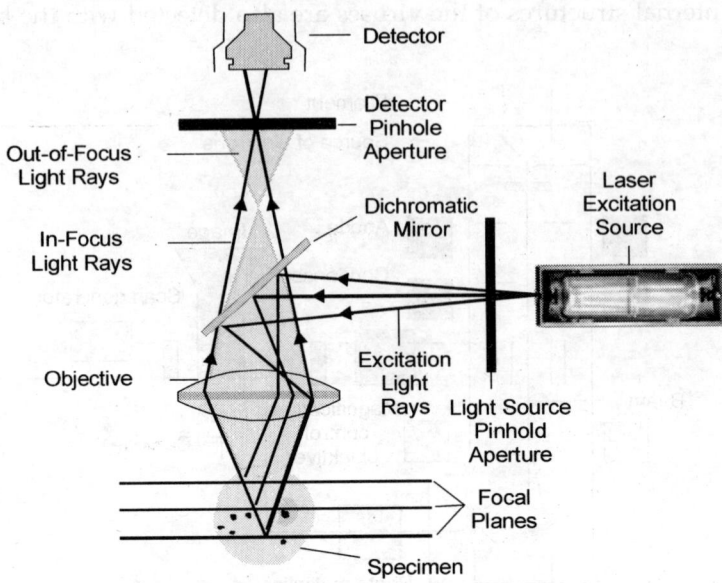

Figure 54   Principle of confocal microscopy

Table 4.2   Difference between light and electron microscopes

Character	Light microscope	Electron microscope
Highest practical magnification	About 1000-15000	Over 100,000
Best resolution	0.2m	0.5m
Radiation source	Visible light	Electron beam
Medium of travel	Air	High vaccum
Types of lens	Glass	Electromagnet
Source of contrast	Differential light absorption	Scattering of electrons
Focusing mechanism	Adjust lens portion mechanically	Adjust current to the magnetic lens
Method of changing magnification	Switch the objective lens	Adjust current to the magnetic lens

B.   STAINING

Staining is a process of colouring the cells using special treatments and methods. Visualization of microbial population in living state is more difficult because they are small transparent and colourless. If it is stained with any one of the dyes, it increases its visibility.

Today various staining techniques are available to study the properties of various microorganisms and differentiation into specific groups / genera/ species. The chemical substances commonly used to stain bacteria are known as dyes. Dyes are classified as natural or synthetic. Chemically a dye is defined as organic compound containing a benzene ring plus a chromophore and auxochrome group. Such dyes are acidic, basic or neutral. Many types of dyes used to stain microorganisms have two features in common. (1) They have **chromophore groups,** groups with conjugated double bonds that give the dye its colour. (2) They can bind with cells by ionic, covalent, or hydrophobic bonding. For example, a positively charged dye binds to negatively charged structures on the cell. Ionizable dyes may be divided into two general classes based on the nature of their charged group. **Basic dyes** – methylene blue, basic fuchsin, crystal violet, safranin, malachite green – have positively charged groups. Basic dyes bind to negatively charged molecules like nucleic acids and many proteins. Because the surfaces of bacterial cells also are negatively charged, basic dyes are most often used in bacteriology. **Acid dyes** – eosin, rose bengal, and acid fuchsin – possess negatively charged groups such as carboxyls ($-COOH$) and phenolic hydroxyls ($-OH$). Acid dyes, because of their negative charge, bind to positively charged cell structures.

Neutral dyes are prepared by mixing both basic and acidic components.

Dyes are crude colouring agents whereas **stains** are prepared from purified dye. The stain is prepared by dissolving particular dye in distilled water or alcohol.

Categories of Staining

Staining are broadly categorized into two based on the process of staining and nature of staining principle. They are simple and differential staining. **Simple staining** - Methylene blue staining and Negative staining

Differential Staining

Separation into group - Gram staining and   Acid - fast staining

Visualization of structures - Capsule staining; Spore staining and Flagellar staining

Staining Procedure

It involves smear preparation, drying of smear, fixing of smear, use of stain and washing off excess stain.

Smear Preparation, Drying and Fixation (Figure 55)

Proper staining of microorganisms need good fixed smear. Microscopical examination of stained preparations enables the morphology, relative sizing and arrangement of microorganism to be seen clearly. Better observation of microorganisms needs better smearing and fixation. Every slide should be labeled clearly with the date and the number.

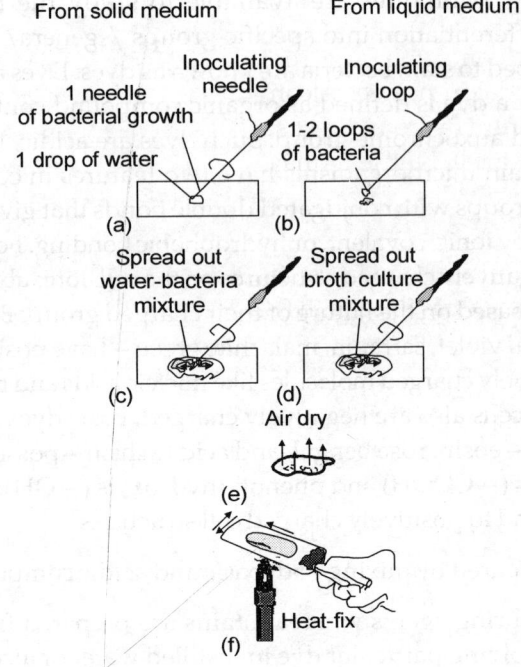

From solid medium   From liquid medium

Inoculating   Inoculating
1 needle   needle   loop
of bacterial growth   1-2 loops
of bacteria
1 drop of water

(a)   (b)

Spread out   Spread out
water-bacteria   broth culture
mixture   mixture

(c)   (d)

Air dry

(e)

Heat-fix

(f)

Figure 55   Smear preparation

Smearing

Smears should be spreaded evenly covering an area of about 15-20mm diameter on a slide. Purulent materials are smeared thinly using a sterile wire loop. The flame-sterilized loop must be allowed to cool before it is used. Emulsify colonies in sterile distilled water before the preparation of thin smear. Non-purulent materials are centrifuged before smearing. A drop of well-mixed specimen is used for smearing. Swabs are smeared by rolling procedure. Sticks are used to smear sputum and faeces.

Drying

After the preparation of smears, the slides should be kept in a safe place to air dry, protected from flies and dust. If the smears cannot be stained immediately, they should be fixed and placed in a covered container.

Fixation

The purpose of fixation is to preserve microorganisms and to prevent the smear being washed from slides, during staining. Smears should be fixed by heat, alcohol or other chemicals before staining.

Heat Fixation

It is a widely used method but can damage organisms. It may alter staining reactions especially if excessive heat is employed. Heat fixation is performed if the smear is prepared from solid media.

Alcohol Fixation

This fixation is far less damaging to microorganisms than heat. Cells are well preserved. Alcohol is more bactericidal in nature. Alcohol fixation is performed when the smear is prepared from liquid culture or from direct samples (Clinical samples). Alcohol removes excess nutrients from the smear. Nutrients may interferewith staining reactions.

Other Chemical Fixation

Anthrax Bacilli containing smear is fixed with 4% potassium permanganate. *Mycobacterium* smear is sometimes fixed with Formaldehyde vapour.

Application of stain

After proper fixation, depends on the technique apply proper stain over the fixed smear.

Washing of excess stain

In general, dry preparation should be visualized under the microscope. Excess stain can be removed by pouring small quantity of water to the stained preparations.

Simple Staining

The process of visualizing the morphology of microbial population by using single basic stain is called simple staining. In this staining, bacterial smear is stained with a single stain. Basic stains are usually preferred to determine the shapes and arrangements of bacteria. Bacterial nucleic acids, cell wall components are negatively charged and strongly attract positive dye and imparts same colour to the bacteria. Most commonly used basic dyes are Methylene blue, Crystal violet and Carbol fuschin. Simple staining is performed by making use of the following method. Smear is made on a clean slide. Smear is dried and fixed with heat. Methylene blue is applied and allowed for one minute. Excess stain is removed by pouring gentle stream of water after proper staining. Slide is allowed to dry and observed under microscope.

Negative Staining Technique

Staining method that stains the background and not the bacteria are called negative staining. This technique is more advantageous than others are because, heat fixation is not required and the cells are not subjected to the distorting effect of chemicals and heat, their natural shape and size can be stained. Secondly, it is possible to observe bacteria that are difficult to stain e.g. Some spirilli. Negative staining requires the use of an acidic stain such as eosin, india ink or nigrosin. The acidic stain with its negatively charged chromogen, will not penetrate the cell because of the negative charge on the surface of the bacteria.

It is performed by mixing a drop of dye and the bacterial cells on a clean glass slide. Coverslip is placed over the preparation and observed under the microscope.

Modifications of Gram staining

Jensens modification–Alcohol as decolourizer, weak neutral red as counter stain. Used for the examinations of *Neisseria* sp.,

Kopeloff and Beermans modifications–Uses acetone as decolourizer

Preston and Morrells modifications–Iodine acetone as decolourizer, gives good result, no need for carefull timing.

Gram staining

This is one of the most important staining techniques in bacteriology. It is also called differential staining. This staining is called gram staining because it was discovered by Dr. Christian Gram. It divides bacterial cells into two major groups, gram positive and gram negative, which makes it an essential tool for classification and differentiation of microorganisms. Those bacteria accepts and retain grams dye like grams crystal violet are called gram positive bacterium. If the bacteria losses grams dye are called as gram negative.

The gram stain uses four different reagents and their mechanism of actions are as follows. **Crystal violet** (primary stain) is a cationic stain attracted by anionic cell wall and stains all the cells purple. **Gram's Iodine** (mordant) forms an insoluble complex by binding to the primary stain. The resultant CV – I complex serves to intensify the colour of the stain and all the cells will appear purple black. Only in gram positive cells, this CV – I complex binds to the magnesium-ribonucleic acid component of the cell wall. The resultant Mg-RNA-CV-I complex is more difficult to remove than the smaller CV-I complex. **Ethyl alcohol 95%** serves as a protein-dehydrating agent. Its action is determined by the lipid concentration of the microbial cell wall. In gram positive cells, the low lipid concentration is important for the retention of Mg-RNA-CV-I complex. Therefore small amount of lipid content is readily dissolved by the action of the alcohol, causing formation of minute cell wall pores. These are then closed by alcohol's dehydrating effect. As a consequence, the tightly bound primary stain is difficult to remove and the cells remain purple. In gram negative cells, the high lipid concentration found in the outer layers of the cell wall is dissolved by the alcohol, creating large pores in the cell wall that do not close appreciably on dehydration of cell wall proteins. This facilitates release of the unbound CV-I complex, leaving these cells as colourless or unstained. **Safranin** is the final stain (Counter stain), which stains decolourized cells only. Gram negative cells appear pink colour and Gram positive cells appear as purple colour.

It is performed by using the following simple technique.

Smear is made on the clean slide and fixed using heat. Stain the smear with grams crystal violet for one minute. Stain is washed completely using gentle stream of water. Stain is treated with Grams iodine and allowed for one minute. It is washed with water. Stained cells are decolourized with Grams decolourizer. Smear is counter stained with safranin and wait for 30 seconds. Excess stain on the smear is washed off with water and blot dry. Stained smear is observed under bright field microscope and looked for pink coloured (Gram negative) or violet coloured (gram positive) cells

Capsule Staining

Some bacteria secrete chemical substances that accumulate on the outer surface of the cell walls are called **capsule**. Capsule is water-soluble and non-ionic in nature. They have distinct chemical structure that can be clearly differentiated from the cell wall. There are two types of capsules, they are microcapsule and macro capsule. Macro capsules are 20 nm or more in size easily seen through light microscope. Microcapsule is less than 20 nm in diameter in size and seen under electron microscope. Capsules may be seen in stained or unstained preparations as a clear zone around the bacteria. Two types of staining procedures are employed to demonstrate capsules. They are positive staining and negative staining. In the positive staining procedure, the capsule is stained and coloured where as in the negative staining procedure the background is stained the capsule is seen as unstained hallow around the organisms. Two stains are used to distinguish capsule from cellwall. Cell wall of bacteria is initially stained with any positive dye and non ionic capsule is partially stained with neutral solutions like copper sulphate.

Prepare a smear from a 12 to 18 hour culture of slant or plate. Allow the smear to air dry. DO NOT HEAT FIX. Cover the slide with 1% crystal violet for 2 minutes. Rinse gently with a 20% solution of copper sulfate. Air dry the slide. DO NOT BLOT. Examine the slide under an oil immersion lens. Bacterial cells and the proteinaceous background will appear purplish while the capsules will appear transparent.

Spore Staining

It is performed by Schaeffer- Fulton Method. Bacteria belonging to the genera *Bacillus* and *Clostridium* possess resting and resisting structures called endospores. They are present intracellularly or as free spores. The position of the spore may be central, sub terminal or terminal. The heat resistant property of spores has been linked to their high content of calcium and dipicolinic acid. Spore is formed by a process called sporulation. Endospore is surrounded by impermeable layers called spore coats. Endospores are completely resistant to heat, radiation, chemicals and agents that are lethal to microbial growth.

Endospores strongly resist the application of simple dyes but once stained, are quite resistant to decolourization. Unlike other cells, the spore will not accept the primary stain easily. Heat is applied to increase penetration. After heating vegetative cells and spores appear greenish. Once the spore accepts the stain, it cannot be decolourized by tap water, which removes only excess stain. The spore will remain green. On the other hand, the vegetative cells do not have strong affinity for stain. Hence, water removes it and vegetative cells look colourless. To make the distinction between the spore and vegetative portion of the cell, a contrasting counter stain is usually applied in the ordinary fashion the resulting picture shows the initial stain taken up by the spore and stain appear in the cytoplasm.

Prepare smears of organisms and heat fix. Cut absorbent paper to fit the slide leaving one end for handling. Place the slide on wire gauze on a ring stand. Saturate the paper with Malachite green . Heat the slides with a hand-held bunsen burner until steam can be seen rising from the surface. Alternately remove the burner and reheat the slide to maintain steaming for 3-5 minutes. As the paper begins to dry during the staining process add a drop or two of malachite green to keep the slide moist. Adding too much stain will cool the slide. Overheating the slide or letting it dry will distort the cells. The process is steaming and not baking. Remove the paper with tweezers and rinse the slide thoroughly with tap water. Drain the slide and counterstain 45 seconds with 0.5% safranin. Wash, blot, and examine. The vegetative cells will appear red and the spores will appear green.

Acid Fast Staining

It is a differential staining. Paul Ehrlich developed it in 1882, which was later on modified by Ziehl-Neelsen and is being used by present day microbiologists. Certain species of bacteria, particularly the organisms of the genus *Mycobacterium* and some strains of *Nocardia*, once stained with dyes like carbol fuchsin, resist decolourization by strong mineral acid solution. This feature of acid fastness is associated with intact cellwall and the presence of large quantities of unsaponifiable wax fraction (mycolic acid), which makes penetration by stains extremely difficult.

Bacteria are classified as acid-fast if they retain the primary stain after washing with strong acid and appear red or as non acid-fast of they lose their colour on washing with acid. Heat is applied to increase the penetration of dye during primary staining. Once stained, the stain cannot be removed when it is treated with strong acid. Mycolic acid content of bacterium confers acid fastness to the bacterium. Non acid fast bacterium decolourizes easily and readily accepts counter stain. Acid fast bacteria appears as red colour. Degree of acid fastness vary depends on the species.

Prepare smears of organisms and Heat fix. Cut or tear absorbent paper (bibulous paper) to fit the slide leaving one end for handling. Do not allow the paper to protrude beyond the slide, but the smears must be covered. Place the slide on wire gauze on a ring stand. Saturate the paper with carbolfuschin. Heat the slides with a hand-held bunsen burner until steam can be seen rising from the surface. Alternately remove the burner and reheat the slide to maintain steaming for 3-5 minutes. As the paper begins to dry during the staining process add a drop or two of carbolfuschin to keep the slide moist. At the end of staining remove the paper with tweezers and wash the slide thoroughly. Drain the slide. Decolourize with acid-alcohol for 30 seconds. Rinse, drain, and counterstain with methylene blue for 45 seconds. Rinse, blot, and examine. First observe each organism on its separate smear. Then examine the mixed smear. Acid-fast organisms will appear red and non-acid-fast organisms will be blue.

Alberts Staining

The Albert technique is used to stain the volutin or metachromatic granules of *Corynebacterium diphtheriae*. The granules are most numerous after the organism has been cultured on a protein rich medium such as Dorset egg or loeffler serum. Metachromatic granules can also be found in Corynebacterium species and occasionally in some Bacillus species. These granules are madeup of poly meta phosphate and are seen in unstained preparations as round, refractile bodies within cytoplasm. With basic dyes, granules tend to stain more strongly than the rest of the bacterium. Neisser stain, Pouch stain, Alberts stains are used to stain these granules. Granules of Diphtheria bacilli exhibit metachromasia when it is stained with alberts staining.

5

MICROBIAL GROWTH

A. NUTRIENT AND NUTRITION

Nutrient is defined as a substance which enhances the growth of cell. **Nutrition** is the process of taking nutrients from the surrounding environment. Nutrients are classified into three types. They are macronutrients, micronutrients and trace elements. Macronutrients are required in gram quantities, micronutrients are required in milligram quantities where as trace elements are required in microgram quantity. Analysis of microbial cell shows that over 95% of cell dry weight is made up of carbon, oxygen, hydrogen, nitrogen, sulfur, phosphorus, potassium, calcium, magnesium, and iron. C, O, H, N, S, and P are macroelements or macronutrients. They are the components of carbohydrates, lipids, proteins, and nucleic acids. The elements like potassium, calcium, magnecium and iron were existing in the cell as cations and play a variety of roles. Potassium is required for activity of enzymes. Calcium contributes to the heat resistance of bacterial endospores. Magnesium serves as a cofactor for many enzymes. Iron is a part of cytochromes and a cofactor for enzymes and electron-carrying proteins. The micronutrients like manganese, zinc, cobalt, molybdenum, nickel and copper are needed by most cells for synthesis and repair. Micronutrients are normally a part of enzymes and cofactors. They aid in the catalysis of reactions and maintenance of protein structure. Zinc is present at the active site of some enzymes. Manganese aids many enzymes catalyzing the transfer of phosphate groups. Molybdenum is required for nitrogen fixation and cobalt is a component of vitamin B_{12}. Trace elements may have particular requirements that reflect the special nature of their morphology or environment. Diatoms need **silicic acid** (H_4SiO_4) to construct their beautiful cell walls of silica. K^+ is required for the maintenance of ionic strength; cofactor for certain enzymes. Ca^{++} acts as a Cofactor for certain enzymes and Fe^{++} present in cytochromes and other metalloenzymes. Cobalt, Copper, Molybdenum and nickel are the trace elements present in certain metalloenzymes.

Requirements for Carbon, Hydrogen, and Oxygen

Carbon is a backbone of all organic molecules. Carbon frequently serve as energy sources. One important carbon source that does not supply hydrogen or energy is carbon dioxide (CO_2). **Autotrophs** can use CO_2 as their sole or principal source of carbon. Some autotrophs oxidize inorganic molecules and derive energy from electron transfers. The reduction of CO_2 is a very energy-expensive process. Organisms that use reduced, preformed organic molecules as carbon sources are **heterotrophs**. For example, the glycolytic pathway produces carbon skeletons for use in biosynthesis and also releases energy as ATP and NADH.

Actinomycetes will degrade amyl alcohol, paraffin, and even rubber. Some bacteria use almost anything as a carbon source; for example, *Burkholderia cepacia* can use over 100 different carbon compounds. Cultures of methylotrophic bacteria metabolize methane, methanol, carbon monoxide, formic acid, and related one-carbon molecules. Parasitic members of the genus *Leptospira* use only long-chain fatty acids as their major source of carbon and energy.

Requirements for Nitrogen, Phosphorus and Sulfur

Nitrogen is needed for the synthesis of amino acids, purines, pyrimidines, some carbohydrates, lipids, enzyme cofactors, and other substances. Many microorganisms can use the nitrogen in amino acids, and ammonia. Most phototrophs and many nonphotosynthetic microorganisms reduce nitrate to ammonia and incorporate the ammonia in assimilatory nitrate reduction. Phosphorus is present in nucleic acids, phospholipids, nucleotides like ATP, several cofactors, some proteins, and other cell components. Sulfur is needed for the synthesis of substances like the amino acids cysteine and methionine, some carbohydrates, biotin, and thiamine. Most microorganisms use sulfate as a source of sulfur and reduce it by assimilatory sulfate reduction; a few require a reduced form of sulfur such as cysteine.

Requirements of Growth Factors

Organic compounds required because they are essential cell components or precursors of such components and cannot be synthesized by the organism are called growth factors. There are three major classes of growth factors: (1) amino acids, (2) purines and pyrimidines, and (3) vitamins. Amino acids are needed for protein synthesis, purines and pyrimidines for nucleic acid synthesis. Vitamins are small organic molecules that usually make up all or part of enzyme cofactors, and only very small amounts sustain growth. Some microorganisms require many vitamins. *Enterococcus faecalis* needs eight different vitamins for growth. *Haemophilus influenza* requires X and V factor for growth.

Biotin is needed for Carboxylation (CO_2 fixation) process. Molecular rearrangement process needed Cyanocobalamin (B_{12}). Folic acid is useful for one-carbon metabolism. Pantothenic acid is a precursor of coenzyme A. It carries acyl groups. Pyridoxine (B6) is responsible for amino acid metabolism. Niacin (nicotinic acid) is a precursor of NAD and NADP. Riboflavin (B_2) acts as a precursor of FAD and FMN. Thiamine (B_1) is helpful in Aldehyde group transfer during pyruvate decarboxylation.

B.   NUTRITIONAL TYPES OF MICROORGANISMS

Microorganisms are classified differently based on the utilization of elements, energy, electron and others. Commonly microorganisms are classified as **autotrophs, heterotrophs** based on carbon utilization. There are only two sources of energy available to organisms. They are light energy and the energy derived from oxidizing organic or inorganic molecules. **Phototrophs** use light as their energy source; **chemotrophs** obtain energy from the oxidation of chemical compounds. Microorganisms also have only two sources for electrons. **Lithotrophs** use reduced inorganic substances as their electron source, whereas **organotrophs** extract electrons from organic compounds.

Photolithotrophic autotrophs (photoautotrophs or photolithoautotrophs) use light energy and have CO_2 as their carbon source. Eucaryotic algae and cyanobacteria employ water as the electron donor and release oxygen. Purple and green sulfur bacteria cannot oxidize water but extract electrons from inorganic donors like hydrogen, hydrogen sulfide, and elemental sulfur. Chemoorganotrophic heterotrophs (often called chemoheterotrophs, chemoorganoheterotrophs, or even heterotrophs) use organic compounds as sources of energy, hydrogen, electrons, and carbon. All pathogenic microorganisms are chemoheterotrophs. The other two nutritional classes have fewer microorganisms but often are very important ecologically. Some purple and green bacteria are photosynthetic and use organic matter as their electron donor and carbon source. These photoorganotrophic heterotrophs (photoorganoheterotrophs) are common inhabitants of polluted lakes and streams. Some of these bacteria also can grow as photoautotrophs with molecular hydrogen as an electron donor. The fourth group, the chemolithotrophic autotrophs (chemolithoautotrophs), oxidizes reduced inorganic compounds such as iron, nitrogen, or sulfur molecules to derive both energy and electrons for biosynthesis. Chemolithotrophs contribute greatly to the chemical transformations of elements that continually occur in the ecosystem.

Some of the microbes are sometimes called mixotrophic, because they combine chemolithoautotrophic and heterotrophic metabolic processes.

Table 5.1   Major Nutritional Types of Prokaryotes

Nutritional Type	Energy Source	Carbon Source	Examples
Photoautotrophs	Light	CO_2	Cyanobacteria, some Purple and Green Bacteria
Photoheterotrophs	Light	Organic compounds	Some Purple and Green Bacteria
Chemoautotrophs or Lithotrophs (Lithoautotrophs)	Inorganic compounds, e.g. H_2, NH_3, NO_2, H_2S	CO_2	A few Bacteria and many Archaea
Chemoheterotrophs or Heterotrophs	Organic compounds	Organic compounds	Most Bacteria, some Archaea

C. CULTURE MEDIA

Culture media means any materials that support the growth of organisms. It must contain many nutrients. Micronutrients and macronutrients are required for the growth of microorganisms. Microbial culture medium was basically classified into two types. They are **Defined medium** and **Complex medium**. Medium, which contains known chemical constituents, are called defined medium. Eg., Minimal Medium. Those medium which contains unknown chemical constituents are called complex medium Eg., Nutrient Agar. Most essential culture media are available commercially in readymade dehydrated form.

Types of Media

Media are classified based on its usage. They are Simple medium, Enriched and enrichment medium, Selective medium, Differential medium and Transport medium .

Simple Medium

These are simple nutrient medium that will support the growth of microorganisms that do not require special nutrition. They are often used to prepare enriched media, storing of stock cultures and sub culturing. Eg., Nutrient Agar.

Enriched and Enrichment Medium

These are the medium that enriched with whole blood, lysed blood serum extra peptones special extracts or vitamins to support the growth of fastidious organisms. Eg.,*Haemophillus influenzae* require X and V factor for growth. Chocolate agar provides X and V factor. The term enrichment is used to describe a fluid medium that increases the numbers of pathogens by enhancing the growth and discouraging the multiplication of unwanted pathogens. Eg., GN broth discourage the growth of Enterobacteriaceae members other than *Salmonella, Shigella* and *Escherichia*.

Selective Medium

These are the media, which contains substances that prevent or slowdown the growth of microbes other than pathogens for which the media are intended. Example XLD medium selects *Salmonella* and *Shigella*. Now a day antimicrobial agents have became increasingly used as selective agents. Example New York City Agar medium used to select *Neisseria gonorrhoea* which contains Colistin, Nalidixic acid, Nystatin and Trimethoprim Sulphate. These antibiotics inhibit the growth of all gram positive, gram negative bacteria except *Neisseria*.

Differential Medium

This type of the medium is used to differentiate various pathogens. Main differential part of the medium is indicators and dyes. Eg., TCBS medium contain bromo thymol blue which differentiate sucrose fermenting *Vibrio* from others.

Transport Medium

These are mostly semisolid media that contain ingredients to prevent the over growth of commensals and ensure the growth of aerobic and anaerobic pathogens. Example Cary Blair medium used for the preserving enteric pathogens.

Classification of Media Based on Solidification

Culture media used in three forms they are Solid, Semisolid and Fluid medium.

Solid Medium

Medium with agar or gelatin are called solid medium. On solid media culture is grown as colonies. Based on colony morphology on solid medium microbiologist preliminarily identify the organisms. These mediums are used for slant preparation, deep preparation and also for subculturing and stock culture preparation.

Semisolid Medium

This media is prepared by adding lower quantities of agar (0.4-0.5%) to fluid medium. They are used as transport medium and for motility testing.

Fluid Medium

This is prepared without agar. In this, the growh and multiplicaton is described into four stages Lag, Log, Stationary and decline. Growth was shown by turbidity and used as enrichment medium, biochemical tests and blood culture medium.

Microbial Growth Requirements

Approximately 80% of the living weight of microbial cell is made up of water and of dry weight 2-5% phosphorus remaining part is made up of minerals and combination oxygen, hydrogen and nitrogen containing organic compounds.

D.   COMMON INGREDIENTS OF CULTURE MEDIA

Peptone

This is a general term for the water-soluble products obtained from the break down of animal or plant protein. The proteins are commonly from meat, milk and soyabean meal. They are hydrolyzed by acids or by enzymes such as pepsin, trypsin and papain. These products are free from aminoacids, peptides and proteases. All forms of peptones are not coagulated by heat. Peptone provides nitrogen for growing microorganisms. Plant protein such as Soya peptone provides carbohydrates and most peptones contain nucleic acid fractions, minerals and vitamins. Peptone powder should be light in colour, dry and have neutral in pH. The concentration and forms of peptone used depends on the use of individual culture media. Various types of peptones are described on the basis of its nitrogen and sodium chloride concentration.

Mycological peptone	-9.5% nitrogen and 1.1% sodium chloride
Bacteriological peptone	-14% nitrogen and 1.6% sodium chloride
Proteose peptone	-12.7% nitrogen and 8% sodium chloride
Soya peptone	-8.7% nitrogen and 0.4% sodium chloride
Special peptone	-11.7% nitrogen and 3.5% sodium chloride

Meat Extracts

Beef extracts supply amino acids, vitamins and minerals such as phosphorus and sulfates to the organisms. It is an essential ingredient in most general-purpose culture media.

Trypsin digested meat extract is used for hartley's broth

Mineral Salts

Sulphates are required as the sources of sulphur and phosphates are the sources of phosphorus for cell growth. Culture media should also contain traces of magnesium, potassium, iron, calcium and other minerals, which are required for bacterial enzyme activity. Sodium chloride is an essential ingredient in most culture media.

Agar

This is an inert polysaccharide and mucopolysaccharide obtained from red- purple seaweed called Gelidium. It is one of the agarophyte group of marine algae. It consists of two main polysaccharides, agarose (70-75%) and agaropectin (20-25%). For use in culture media, agar must be free from pigments and substances toxic to bacteria.

Agar is used as a solidifying agent in all culture media because of gelling strength and its setting temperature of 32-39°C and melting temperature of 90-95°C. Normal concentration of agar in culture media is 1.5 %. If it is below 1% it act as a semi solid media. It act as a nutrient medium to some marine microorganisms like *Pseudomonas*, *Cytophaga* etc. It has two important properties called gelation and solation.

Water

This is highly essential for the growth of all organisms. It must be free from any chemicals, which inhibit the bacterial growth. Deionized water or distilled water must be used for the preparation of culture media.

E.   STERILIZATION OF MEDIA

Sterilization means complete destruction of microbes from an article or freeing of a bacteria and spores from article. Moist heat, Filtration and Radiations are the some of the methods available for sterilization. Mostly microbiological media are sterilized by moist heat sterilization at 121°C for 15 minutes. Heat sensitive media are sterilized by filteration.

F.   CULTURAL VARIATIONS OF BACTERIA

The cultural charecteristics of bacteria refers to their macroscopic appearance in various media. The abundance of growth, size and colour of colonies provide certain useful clues for the identification. Bacterial appearance can vary substansially depending on the organism, the medium and growth conditions. Some differentiation of bacteria can be well elucidated commonly. Agar plates, agar slants, gelatin stabs and broth media in tubes are used for studying cultural characteristics of bacteria.

The macroscopically visible growth of microorganisms on agar plate commonly arrising from a single isolated cell and are called a colony. The main features of growth are summarized below.

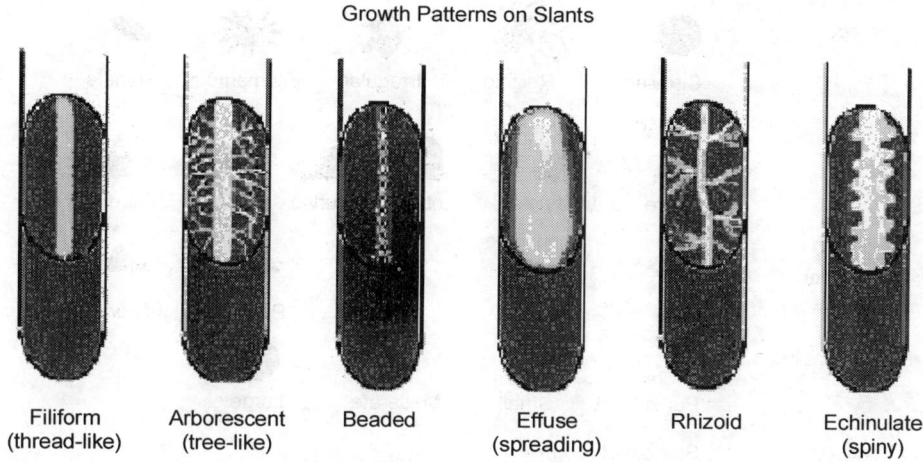

Figure 56 Growth pattern on slants

I.Growth on Nutrient Agar Slants

Nutrient agar in tubes are prepared and sterilized at 121 ° C and made it as a slant. Test organism is streaked on the surface as straight single line streak. Incubate tubes at 37°C for 24 hours. After incubation observe tubes for various morphological features. One organism has one type of morphology at any one condition.

a) Abundance of growth : Amount of growth is described as none, slightly, moderate or large.

b) Consistency: This may be evaluated based on the amount of light transmitted through growth. They may be opaque (no light transmission), translucent (partial transmission) or transparent (full transmission).

C) Form : The appearance of the single line streak growth on the agar surface maybe as follows. (Figure 56)

    i) Filiform : Uniform growth along the line of inoculation

    ii) Echinulate : Margin of growth exhibit tooth like appearance

    iii) Beaded : Separate or semiconfluent column is found along the line of inoculation

    iv) Effuse : Growth is thin and vein like appearance

    v) Arborescent : Branched tree like appearance

    vi) Rhizoid : Root like appearance

II.Growth on nutrient agar plates

Nutrient agar plates are prepared and inoculated the plates by pour plate, spread plate or streak plate method.

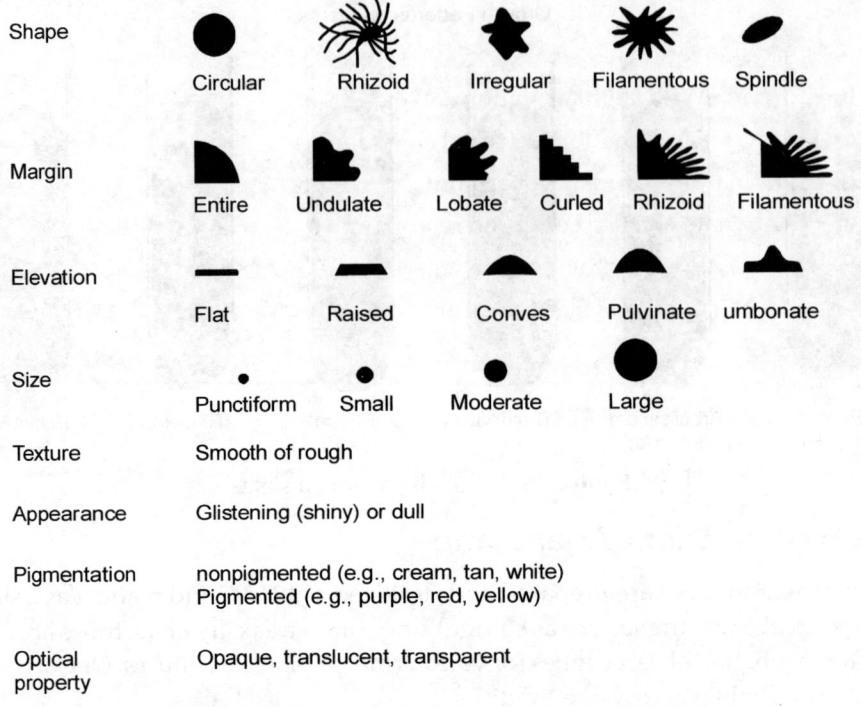

Shape

Circular   Rhizoid   Irregular   Filamentous   Spindle

Margin

Entire   Undulate   Lobate   Curled   Rhizoid   Filamentous

Elevation

Flat   Raised   Conves   Pulvinate   umbonate

Size

Punctiform   Small   Moderate   Large

Texture   Smooth of rough

Appearance   Glistening (shiny) or dull

Pigmentation   nonpigmented (e.g., cream, tan, white)
Pigmented (e.g., purple, red, yellow)

Optical
property   Opaque, translucent, transparent

Figure 57   Growth pattern on Agar Plate

A) **Size** - Pinpoint, small, moderates, large.

B) **Pigmentation** - Pale yellow, colourless, brown, yellow, black, pink, red and green

C) **Form** - This represents the shape of the colony

    a)   Punctiform - Small circular pin headed colonies

    b)   Irregular - peripheral edged colonies

    c)   Circular - unbroken peripheral edge

    d)   Filamentous - threadlike spreading branch

    e)   Rhizoid - root like spreading growth with branches

    f)   Spindle - looks like a spindle with a central bulged area at both sides, tapering ends.

D) **Margin** - Here the appearance of outer edge is taken into consideration

    a)   Entire - Shortly defined smooth even margin

    b)   Undulate - Wavy indundations

    c)   Lobate - Marked edge

    d)   Erose

    e)   Filamentous - Thread like spreading edge

    f)   Curled - Intermediate between wavy and smooth resulting in concentric folding

E) **Elevation**- Degree to which the colony growth appears raised on the agar surface
  a) Flat - Elevation not clear
  b) Raised - With slight elevation
  c) Convex - A slight dome shaped elevation
  d) Pulvinate - A drop like deep convexed region
  e) Umbonate - Raised convexed region with uplevel
  f) Umblicate - A double convex region with a central depression

F) **Opacity** - Opaque, transleucent, transparent.

III.Growth on Nutrient Broth

These are evaluated based on turbidity and appearance of growth on the broth surface.

a) Uniform turbidity - Uniform growth through out the medium
b) Flocculent - Aggregates of growth dispersed in flocs throughout the medium
c) Pellicle - Thick pads like parallel growth on the surface.
d) Sediment - Concentration of growth in the bud region of the broth culture is high
e) Membranous - A thin membrane on the surface with evenly dispersed growth
f) Ring - A slightly thicker membrane with a central vacant hole

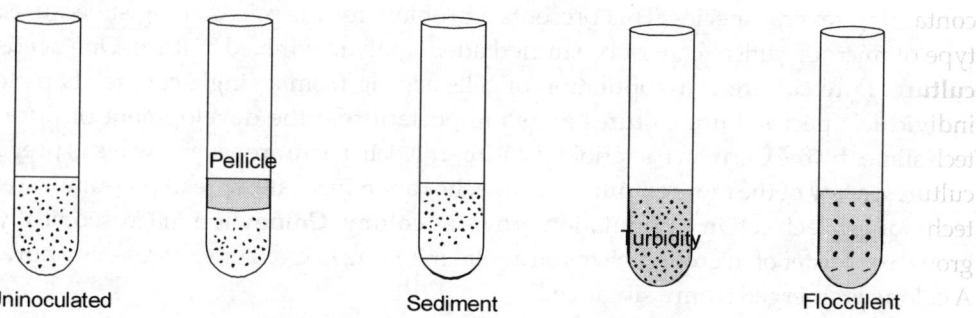

Growth Patterns in Broths

Figure 58   Growth pattern on Broth

IV.Growth on Nutrient Gelatin

Liquefaction of the gelatin is taken into consideration

a) Crateriform - Liquefied surface in saucer form
b) Napiform - Bulbous liquefaction of the surface
c) Infundibuliform - Funnel shaped
d) Startiform - Complete liquefaction of the upper half of the medium
e) Scalate - Elongated tubular liquefaction

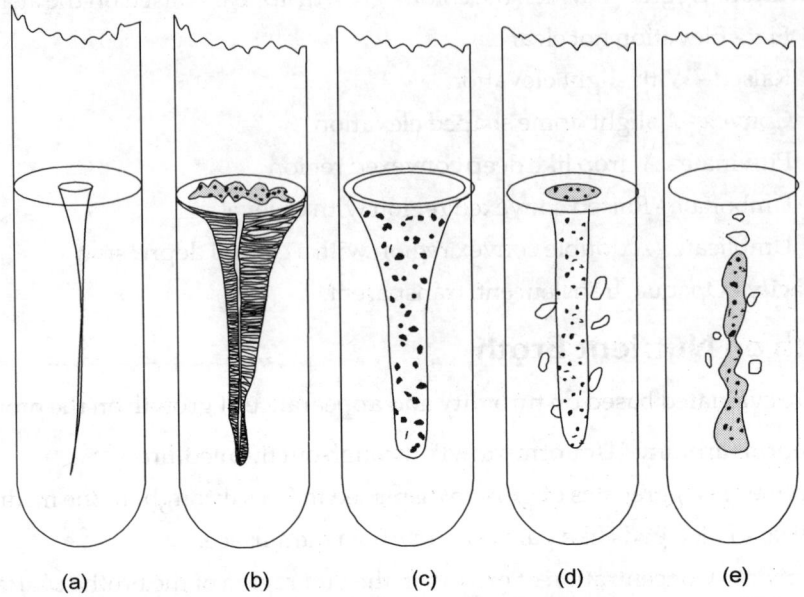

Figure 59   Growth pattern on nutrient gelatin

G.   ISOLATION OF PURE CULTURES

In natural habitats, microorganisms usually grow as complex, mixed populations containing several species. This presents a problem for the microbiologist because a single type of microorganism cannot be studied adequately in a mixed culture. One needs a **pure culture.** Pure culture is a population of cells arising from a single cell, to characterize an individual species. Pure cultures are so important that the development of pure culture techniques by the German bacteriologist **Robert Koch.**There are several ways to prepare pure cultures; a few of the more common approaches are pour plate, spread plate and streak plate techniques. Each cell in a population grows as **colony. Colony** is a macroscopically visible growth or cluster of microorganisms on a solid medium. Each colony represents a pure culture. A colony is emerged from a single cell.

Serial Dilution

To perform viable cell counts on agar plates, it is often necessary to dilute the original sample to make counting easier. Countable plates are typically considered to hold 30-300 colonies on the surface of the agar. Serial dilution is the stepwise dilution of a substance in solution. Usually the dilution factor at each step is constant, resulting in a geometric progression of the concentration in a logarithmic fashion. Serial dilution may also be used to reduce the concentration of microscopic organisms or cells in a sample. A tenfold dilution for each step is called a logarithmic dilution.

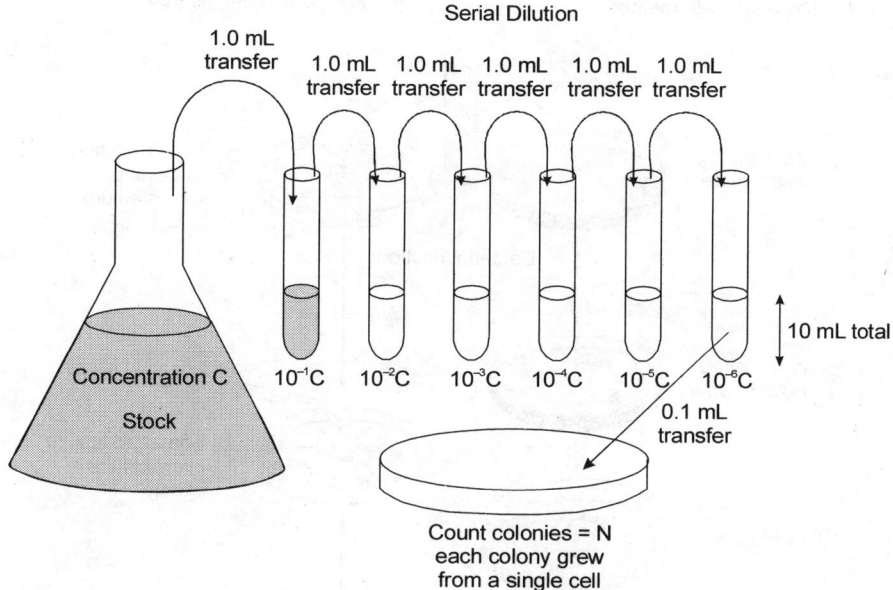

Figure 60 Serial dilution

It is performed by transferring 1ml of stock sample into the first sterile water blank. Label this tube as 1:10. Mix the tube contents. Using another transfer pipette, transfer 1ml of the 1:10 tube to the next sterile blank tube. Label this tube 1:100. Continue making transfers until 1:100 and 1:10000 dilutions have been made. After all dilution have been completed, transfer 0.1ml/1ml sample into the appropriately labeled plates by pour or spread or streak plate method.

1. The Pour Plate

It is one of the pure culture techniques. It is extensively used for the isolation of bacteria and fungi. This technique can yield isolated colonies. In this method, the original sample is diluted several times to reduce the microbial population sufficiently to obtain separate colonies when plating (**Figure** 60). Small volumes of several diluted samples are added to the sterilized petriplates. Then liquid agar medium that has been cooled to about 45°C and poured immediately into sterile culture dishes. Most bacteria and fungi are not killed by a brief exposure to the warm agar. After the agar has hardened, each cell is fixed in place and forms an individual colony. Plates containing between 30 and 300 colonies are counted after proper incubation. The total number of colonies equals the number of viable microorganisms in the diluted sample. Colonies growing on the surface also can be used to inoculate fresh medium and prepare pure cultures.

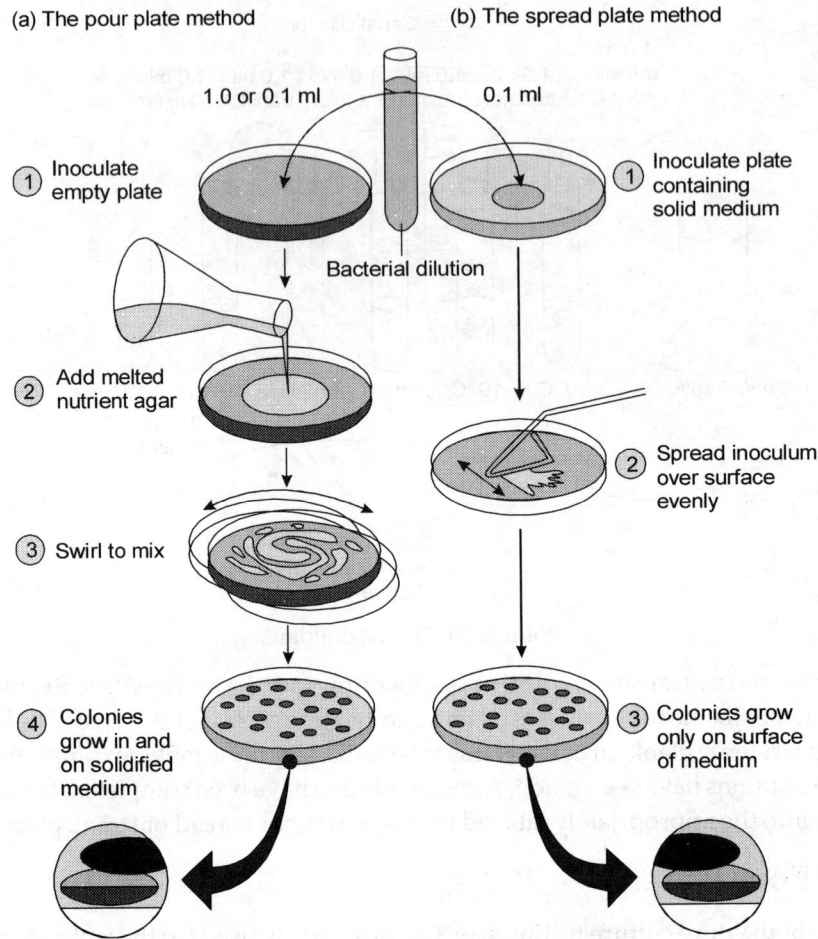

(a) The pour plate method   (b) The spread plate method

1.0 or 0.1 ml   0.1 ml

① Inoculate empty plate   ① Inoculate plate containing solid medium

Bacterial dilution

② Add melted nutrient agar

② Spread inoculum over surface evenly

③ Swirl to mix

④ Colonies grow in and on solidified medium   ③ Colonies grow only on surface of medium

Figure 61   Pour and spread plate method

2. Spread Plate

The **spread plate** is an easy, direct way of achieving pure culture. A small volume (0.1) of diluted microbial mixture is transferred to the center of an agar plate and spread evenly over the surface with a sterile bent-glass rod (**Figure 61**). The dispersed cells develop into isolated colonies. Because the number of colonies should equal the number of viable organisms in the sample, spread plates can be used to count the microbial population.

3. The Streak Plate

Pure colonies also can be obtained from **streak plates.** The microbial mixture is transferred to the edge of an agar plate with an inoculating loop or swab and then streaked out over the surface of the medium T streak (Figure 62) or Quadrant streat (Figure 63). At some point in the process, single cells drop from the loop as it is rubbed along the agar surface and develop into separate colonies.

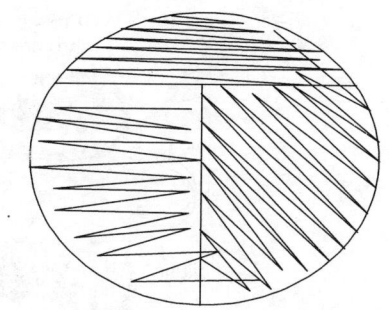

Figure 62   T Streak

Steps in the Streak Plate Method

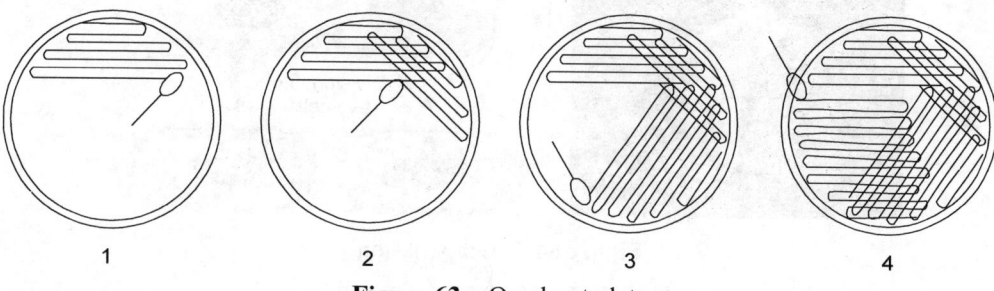

1   2   3   4

Figure 63   Quadrant plate

H. GROWTH

Growth is defined as an increase in cellular constituents and may results in an increase in a microorganism's size, population number or both. The microorganisms are reproduced by binary fission, fragmentation, budding, transverse fission.

Mechanisms of Bacterial Multiplication

Binary fission is an asexual reproduction by a separation of the body into two new bodies. In the process of binary fission, an organism duplicates its genetic material, or deoxyribonucleic acid (DNA), and then divides into two parts (cytokinesis), with each new organism receiving one copy of DNA. Binary fission is the primary method of reproduction of prokaryotic organisms.

Types of Binary Fission

Based on the plane of cytoplasmic division, Binary fission is of Three types

   a)   Simple binary fission
   b)   Transverse binary fission
   c)   Longitudinal binary fission

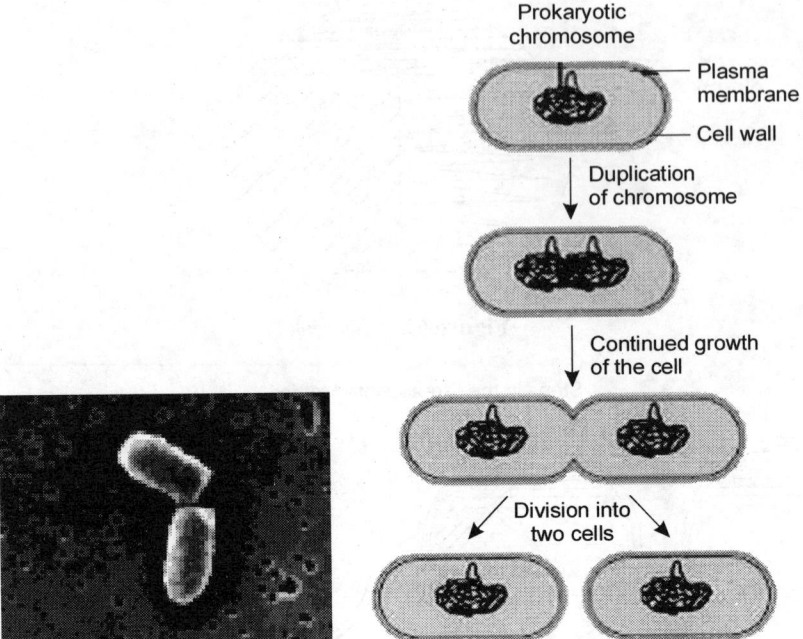

Figure 64   Binary fission

It depends on the axis of cell separation. They are as follows.

a) **Simple binary fission**   When the cytoplasmic division passes through any plane, the fission is called simple binary fission. Example: *Amoeba*

b) **Transverse binary fission**   When the plane of cytoplasmic division coincides with the transverse axis of the individual, the fission is called transverse binary fission. Example: *Paramoecium, Planaria* and Bacteria (Figure 64).

c) **Longitudinal binary fission**   When the plane of cytoplasmic division coincides with the longitudinal axis of the individual, the fission is called longitudinal binary fission. Example: *Euglena*

Multiple Fission

In some organismsthe nucleus of the parent divides into many daughter nuclei by repeated divisions (amitosis). This is followed by the division of the cytoplasm into several parts with each part enclosing one nucleus. So a multiple number of daughter cells are formed from a single parent at the same time. This kind of fission is known as multiple fission.

Budding

Some cells split via budding resulting in a 'mother' and 'daughter' cell. The offspring organism is smaller than the parent. Budding is also known on a multicellular level;

Eg. *Sacharomyces cerevistee.*

Fragmentation

In some organisms the body of the organism breaks into several parts. Each part then develops into a complete organism. Eg. *Amoeba*

I. CULTURE OF MICROORGANISMS

A microbiological culture, or microbial culture, is a method of multiplying microbial organisms by letting them reproduce in predetermined culture media under controlled laboratory conditions. Microbial cultures are used to determine the type of organism, its abundance in the sample being tested, or both. It is one of the primary diagnostic methods of microbiology and used as a tool to determine the cause of infectious disease by letting the agent multiply in a predetermined medium.

Solid medium and liquid medium are used for the cultivation of microorganisms. There are batch culture, continueous culture, synchromous growth culture, diauxic growth culture and fed bactch culture method.

Batch Culture

In batch culture, growth of microbes occur in a limited volume of liquid medium. In liquid medium, microbes are usually grown as batch culture. When a bacterium grown under batch system of culture. It may under goes four phases of growth and consdered as growth phases.

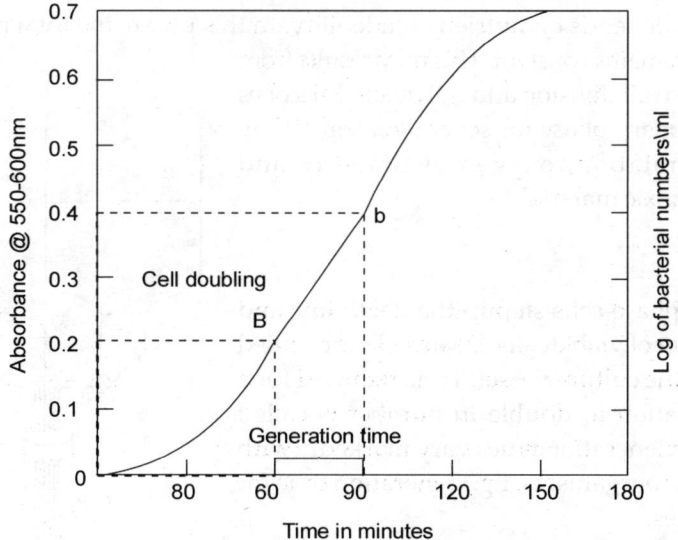

Figure 65   Growth curve

Growth Phases

Population growth is studied by analyzing the growth curve of a microbial culture. In a closed system, population growth remains exponential for only a few generations and then enters a stationary phase due to factors such as nutrient. The growth of microbes can be plotted as the logarithm of cell number versus the incubation time. The resulting curve exhibits four distinct phases. They are Lag phase, Log phase, Stationary phase and Decline phase. (Figure 65)

Lag Phase

When microorganisms are introduced into fresh culture medium, usually no immediate increase in cell numbers or mass occurs and therefore this period is called the lag phase. In this phase, all the microbes adopt themselves and ready to prepare ATP, essential cofactors and ribosomes. Lag phase varies considerably in length with the condition of the microorganisms and nature of the medium. Inoculation of culture into chemically different medium results in long lag phase. When young cultures are inoculated to the same sterile medium, the lag phase will be short or absent.

Log Phase

It is also called Exponential phase. In this phase microbes grow and divide at higher rate. The rate of growth is constant. The population is most uniform in terms of chemical and physiological properties.

Stationary Phase

In this phase, growth is attained by bacteria at a population level of around 10^9 cells/mL. Final population depends on nutrient availability. In this phase, the total number of viable microorganisms remains constant. This may results from a balance between cell division and cell death. Microbes enter into a stationary phase for several reasons. They are nutrient limitation, oxygen availability and accumulation of toxic material.

Death Phase

During this phase cells stepup the death line and reduce the number of viable cells. Death cells are settled in the bottom of the culture vessel. Time required for a microbial population to double in number is called generation time. Generation times vary markedly with the species of microorganisms. Eg. Generation time for *E.coli* is 20 minutes.

Continuous Culture of Bacteria

Bacterial cultures can be maintained in a state of exponential growth over long periods of time using a system of **continuous culture**. Medium is contiuously added to the vessel to maintain the cells at logphase. It is done by using chemostat or turbidostat (Figure 66 and 67).

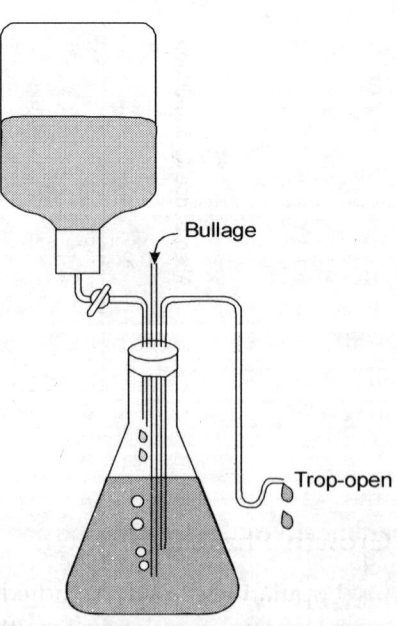

Figure 66 Chemostat

Chemostat

Chemostat can be used to maintain a bacterial population at a constant density. In a chemostat, the growth chamber is connected to a reservoir of sterile medium. Once growth is initiated, fresh medium is continuously supplied from the reservoir. The volume of fluid in the growth chamber is maintained at a constant level by some sort of overflow drain. Fresh medium is allowed to enter into the growth chamber at a rate that limits the growth of the bacteria. The bacteria grow (cells are formed) at the same rate that bacterial cells (and spent medium) are removed by the overflow. The bacterial culture can be grown and maintained at relatively constant conditions, depending on the flow rate of the nutrients.

Turbidostat

A turbidostat is a continuous culturing method where the turbidity of the culture is held constant by manipulating the rate at which medium is fed. If the turbidity tends to increase, the feed rate is increased to dilute the turbidity back to its setpoint. When the turbidity tends to fall, the feed rate is lowered so that growth can restore the turbidity to its setpoint. Turbidostat uses photo electric cell for the detection of turbidity.

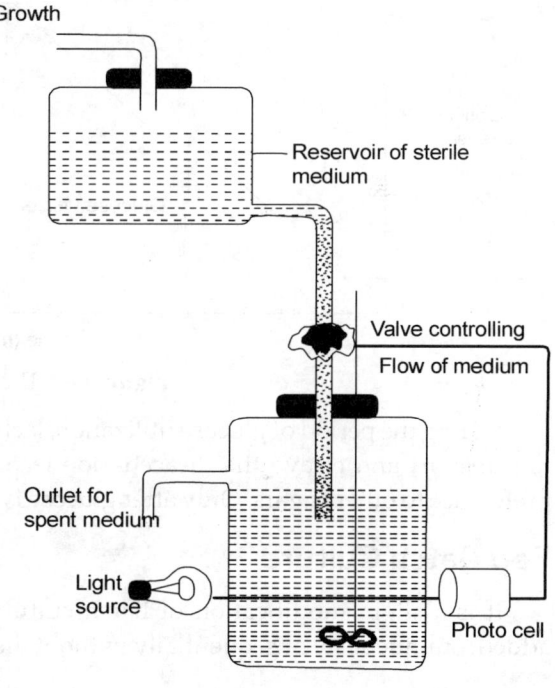

Figure 67   Turidostat

Synchronous Growth of Bacteria

The cells in a culture divide at the same time called synchronous culture. Growth behavior of individual bacteria can be obtained by the study of synchronous cultures. Synchronized cultures must be composed of cells which are all at the same stage of the bacterial cell cycle. Measurements made on synchronized cultures are equivalent to measurements made on individual cells.

A number of clever techniques have been devised to obtain bacterial populations at the same stage in the cell cycle. Some techniques involve manipulation of environmental parameters which induces the population to start or stop growth at the same point in the cell cycle. Synchronous cultures rapidly lose synchrony because not all cells in the population divide at exactly the same size, age or time.

Diauxic Growth

Jacques Monod discovered diauxic growth in 1941 during his experiments with *Escherichia coli* and *Bacillus subtilis*. Diauxic growth curve which showed two distinct phases of active growth (Figure 6.68). During the first phase of exponential growth, the bacteria utilize glucose as a source of energy until all the glucose is exhausted. Then, after a secondary lag phase, the lactose is utilized during a second stage of exponential growth.

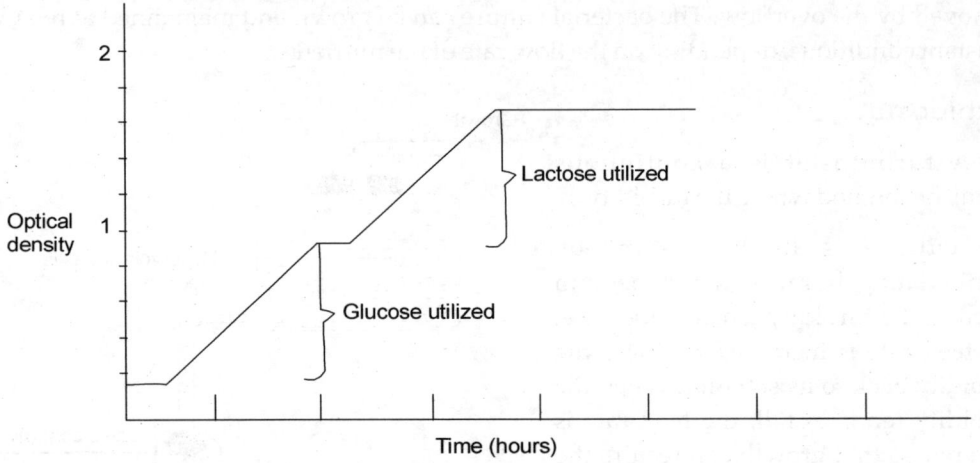

Figure 68 Diauxic growth

During the period of glucose utilization, lactose is not utilized because the cells are unable to transport and cleave the disaccharide lactose. Glucose is always metabolized first in preference to other sugars. Only after glucose is completely utilized is lactose degraded.

Fed Batch Culture

It is a slight modification of batch culture. In the growth environment, medium is addedcontinueously or sequentially without the removal of culture fluid.

J.   MEASUREMENT OF GROWTH

Growth can be measured in terms of two different parameters. They are measurement of **cell mass** and measurement of **cell numbers**.

Methods for Measurement of Cell Mass

Methods for measurement of the cell mass involve both direct and indirect techniques.

1.   Direct **physical measurement** of dry weight, wet weight, or volume of cells after centrifugation.

2.   Direct **chemical measurement** of some chemical component of the cells such as total N, total protein, or total DNA content.

3.   **Turbidity measurements**

Methods for Measurement of Cell Numbers

Measuring techniques involve direct counts, visually or instrumentally, and indirect viable cell counts.

1. **Direct microscopic counts**
2. **Electronic counting chambers**
3. **Indirect viable cell counts**, also called **plate counts**
4. Membrane filter technique

Measurement of Cell Numbers

1. Direct Microscopic count /Breed Method

A known volume of microbial cell suspension (0.01 ml) is spread uniformly over a glass slide covering a specific area (1 sq. cm). The smear is then fixed by heating, stained and examined under oil immersion lens, and the cells are counted. Customarily, cells in a few microscopic fields are counted because it is not possible to scan the entire area of smear. The counting of total number of cells is determined by calculating the total number of microscopic fields per one square cm. area of the smear. The total number of cells can be counted with the help of following calculations:

a) Area of microscopic field = πr^2

r (oil immersion lens) = 0.08 mm.

Area of the microscopic field under the oil immersion lens = πr^2 = 3.14 × (0.08 mm)^2 = 0.02 sq. mm.

b) Area of the smear one sq. cm. = 100 sq. mm. Then, the no. of microscopic fields = 100 / 0.02 = 5000

c) No. of cells 1 sq. cm. (or per 0.01 ml microbial cell suspension) = Average no. of microbes per microscopic field × 5000

Counting Chamber Technique

Neubauer counting chamber and ordinary measured microscopic slide are the most obvious way to determine microbial numbers. The use of a counting chamber is easy, inexpensive and relatively quick. It also gives information about the size and morphology of microorganisms. It is performed by placing the drop of bacterial suspension on the center of the platform using RBC pipette. Count the cells in RBC counting areas of the haemocytometer platform and calculate number of cells per mL by the following calculations

Calculation

25 squares cover an area of 1mm × 1mm area.

Chamber depth is about 0.02mm

Volume of the chamber $\quad$ = 1mm × 1mm × 0.02mm

$\qquad\qquad\qquad\qquad$ = 1/10cm × 1/10 cm × 1/200 cm^3

$\qquad\qquad\qquad\qquad$ = 0.00005 cm^3

0.00005 cm³ area hold 0.00005 mL of fluid

That is 5×10^{-4}

5×10^{-4} holds 5×10^{-4} mL fluid

Bacteria per cm³ is = number of bacteria/ square ×25

Number of bacteria per mL is = Number of cells / square × 25 × 5 × 10^{-4}

Viable Count

The bacterial culture need not contain all living cells. There might be few dead as well. Only living cells will form colony when grown in proper solid medium and under standard set or growth conditions. This fact is used to estimate number of living or dead bacterial cells (viable count) in the given culture. Estimates thus obtained are expressed as a colony forming unit (CFU).

Viable count technique is very much useful in the dairy industry and the food industry for quantitative analysis of milk and spoilage of food products. For convenience, to obtain a colony count for bacteria in milk, 1 ml of well mixed milk is placed in 99 ml of sterile dilute solution (may be water or nutrient broth or saline solution).

This results in s dilution of 1: 100 or 1×10^{-2} To the petri dish containing pre solidified medium 1 ml of 1: 100 dilution is transferred and incubated at desired is repeated for the preparation of further dilution as 1 : 1000 or 1 : 10, 0000 of bacteria per ml in original sample can be found by multiplying bacterial colony count by the reciprocal of he dilution and of the volume used.

For Example, CFU = 50 for 1: 10, 000 if volume used is 1 ml then, CFU = 50 × 10, 000 × 1

CFU = 5×10^{5}

Electronic Coulter Counter

Coulter counter is an electronic used to count number of microbes preferably protozoa microalgae and yeasts. In This method, the sample of microbes is forced through a small orifice (small hole). On the both sides of the orifice, electrodes are present measure the electric resistance or conductivity when electric current is passed through the orifice. Every time a microorganism passes through the orifice, electrical resistance increases or the conductivity drops and the cell is counted. The Coulter counter gives accurate results with larger cells. The precaution to be taken in this method is that the suspension of samples should be free of any cell debris or other extraneous matter.

Membrane-Filter Technique

Microbial cell numbers are frequently determined using special membrane filters possessing millipores small enough to trap bacteria. In this technique a water sample containing microbial cells passed through the filter. The filter is then placed on solid agar medium or on a pad soaked with nutrient broth (liquid medium) and incubated until each cell develops into a separate colony. Membranes with different pore sizes are used to trap different microorganisms. Incubation times for membranes also vary with medium and the microorganism. A colony count gives the number of microorganisms in the filtered sample, and specific media can be used to select for specific microorganisms. This technique is especially useful in analyzing aquatic samples (Figure 69).

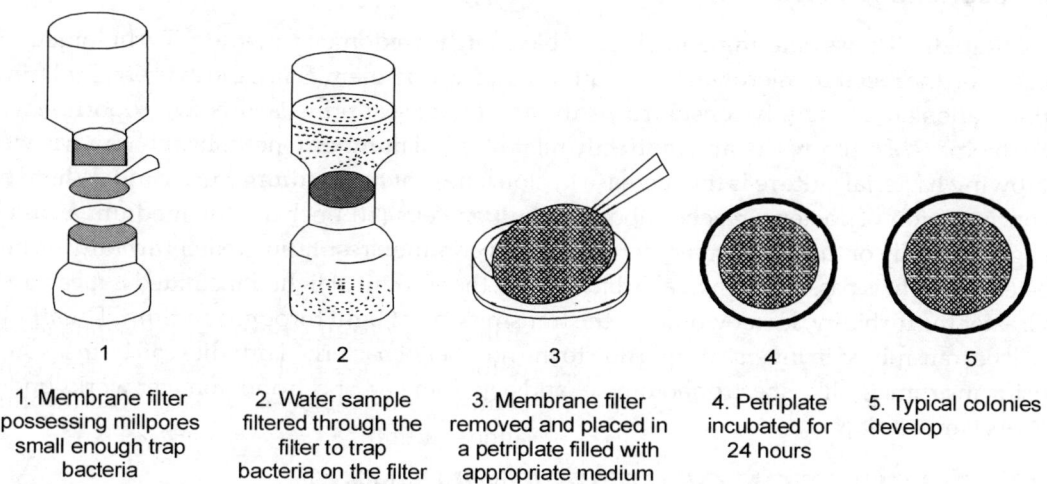

1	2	3	4	5
1. Membrane filter possessing millpores small enough trap bacteria	2. Water sample filtered through the filter to trap bacteria on the filter	3. Membrane filter removed and placed in a petriplate filled with appropriate medium	4. Petriplate incubated for 24 hours	5. Typical colonies develop

Figure 69   Membrane filter technique

Measurement of Cell Mass

Increase in the total cell mass as wells as in cell numbers accompany population growth. Some people performed measurement of cell mass by measuring dry weight. It is a time consuming and not very sensitive method. Spectrophotometric method is the more rapid and sensitive technique depends upon the fact that microbial cells scatter light striking them. Because microbial cells in a population are of roughly constant size, the amount of scattering is proportional to the concentration of cells present. This type of measurement is also called Turbidity measurement. It is performed with the help of calorimeter or spectrophotometer.

1. Dry Weight Technique

The cell mass of a very dense cell suspension can be determined by this technique. In this technique, the microorganisms are removed from the medium by filtration and the microorganisms on filters are washed to remove all extraneous matter, and dried in dessicator by putting in weighing bottle (previously weighted). The dried microbial content is then weighted accurately. This technique is especially useful for measuring the growth of microfungi. It is time consuming and not very sensitive. Since bacteria weigh so little, it becomes necessary to centrifuge several hundred millions of culture to find out a sufficient quantity to weigh.

Measurement of nitrogen content

As the microbes (bacteria) grow, there is an increase in the protein concentration (i.e. nitrogen concentration) in the cell. Thus, cell mass can be subjected to quantitative chemical analysis methods to determine total nitrogen that can be correlated with growth. This method is useful in determining the effect of nutrients or antimetabolites upon the protein synthesis of growing culture

Measurement of Turbidity (Turbidometry)

Rapid cell mass determination is possible using turbidometry method. Turbidometry is based on the fact that microbial cells scatter light striking them. Since the microbial cells in a population are of roughly constant size, the amount of scattering is directly proportional to the biomass of cells present and indirectly related to cell number. One visible characteristic of growing bacterial culture is the increase in cloudiness of the medium (turbidity). When the concentration of bacteria reaches about 10 million cells (10^7) per ml, the medium appears slightly cloudy or turbid. Further increase in concentration results in greater turbidity. When a beam of light is passed through a turbid culture, the amount of light transmitted is measured, Greater the turbidity, lesser would be the transmission of light through medium. Thus, light will be transmitted in inverse proportion to the number of bacteria. Turbidity can be measured using instruments like spectrophotometer and nephlometer. Spectrophotometer works under Beers lamberts law.

K. CALCULATION OF GENERATION TIME

When bacteria grows exponentially by binary fission, the increase in a bacterial population is by geometric progression. If we start with one cell, when it divides, there are 2 cells in the first generation, 4 cells in the second generation, 8 cells in the third generation, and so on. The generation time is the time interval required for the cells (or population) to divide.

G (generation time) = (time, in minutes or hours)/n(number of generations)

G = t/n

t   = time interval in hours or minutes

B = number of bacteria at the beginning of a time interval

b = number of bacteria at the end of the time interval

n = number of generations (number of times the cell population doubles during the time interval)

b = B x 2^n (This equation is an expression of growth by binary fission)

Solve for n:

logb   = logB + nlog2

$$n = \frac{logb - logB}{log2}$$

$$n = \frac{logb - logB}{.301}$$

n = 3.3 logb/B

G = t/n

Solve for G

$$G = \frac{t}{3.3 \log b/B}$$

Example   What is the generation time of a bacterial population that increases from 10,000 cells to 10,000,000 cells in four hours of growth?

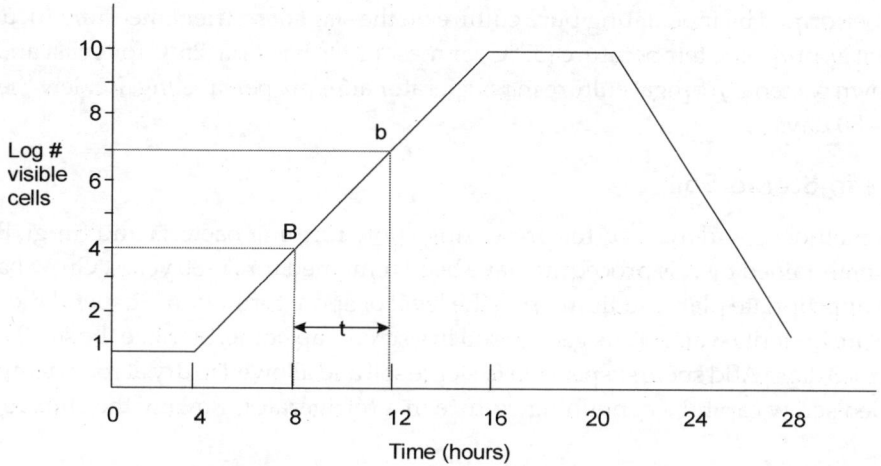

Figure 70   Growth curve for generation time calculation

$$G = \frac{t}{3.3 \log b/B}$$

$$G = \frac{240 \text{ minutes}}{3.3 \log 10^7/10^4}$$

$$G = \frac{240 \text{ minutes}}{3.3 \times 3}$$

$$G = 24 \text{ minutes}$$

L.   PRESERVATION OF MICRO ORGANISMS

Pure cultures of microorganisms are required for research as well as for other purposes. The methods used for maintenance and preservation should conserve all the characteristics of the organisms. Long-term use of microorganisms needs to be stored in proper environmental conditions. Most of the cultures are preserved by culture collection centers. Main objectives of culture collection centers are depositing cultures from different sources, supplying authentic cultures to researchers and also for teaching. Authentic cultures are helpful in identification of unknown cultures, production of industrially economically viable compounds. Different methods are available to preserve microorganisms. They are as follows

Maintanence in Fresh Media

Microbial cultures can be maintained by periodic transfer on fresh, sterile media in tubes. The frequency of transfer however varies with the organisms. To keep the cultures viable, it is necessary to use an appropriate growth medium and a proper storage temperature. Many heterotrophic bacteria can be maintained on a medium such as nutrient agar with transfers to fresh medium after every 20-30 days.

It is performed by inoculating pure culture on the slant of nutrient medium. Incubate the cultures at appropriate temperature (37°C for mesophilic bacteria; 25°C for yeast and fungi). Store grown bacterial / fungal culture in refrigerator after proper labeling. Review the culture after 15 – 30 days.

Storage in Sterile Soil

This method is widely used for preserving spore forming bacteria and fungi. Bacterial cultures maintained by this procedure have been remained for 70-80 years. Grow bacteria / fungi on appropriate plate medium up to the level of spore formation. Suspend the spore of pure culture in sterile water. Take garden soil in a screw cap bottle. Sterilize the soil 2 – 3 hours at 110°C for 3 days. Add spore suspension to sterile soil and allowed to dry at room temperature. Store dried screw cap tube containing culture in a refrigerator. Review the culture after 50 days.

Storage in Mineral Oil

This method also recommended for the preservation of bacteria and fungi. Mineral oil / liquid paraffin of Specific gravity 0.83 – 0.89 are used. It prevents dehydration, maintain slow metabolic rate and also reduce oxygen tension. Mineral oil covered cultures are stored at 0-5°C. Some microorganisms have been preserved satisfactorily for more than 15-20 years by this method. Inoculate test organism on slant as zig zag streak. Incubate at appropriate temperature for 24-48 hours. Sterilize liquid paraffin / mineral oil at 120°C for 40 minutes twice. Pour sterile mineral oil on the surface of the growth to a level of 1cm above the highest point of agar slant. Store cultures at 0-5°C. Review cultures after 90 – 100 days.

Storage at Low Temperature / liquid nitrogen

In this method, cultures are frozen in the presence of a protective agent such as glycerol or dimethylsulfoxide in liquid nitrogen (-196°C). This procedure has been successful with many organisms which cannot be preserved by lyophilization.

Collect 0.5mL of cell suspension. Transfer the cell suspension in storage based cryo tube and Store the culture in liquid nitrogen containing ultra low temperature cabinet after proper labeling.

Lyophilization

Lyophilization is a process in which the cell suspensions are placed in small vials, which are then frozen by immersing in a mixture of dry ice and acetone or liquid nitrogen. The vials are then evacuated and dried under vacuum, sealed and stored at a low temperature. This method provides long-term culture survival without a change in any characteristic. Also, lyophilized cultures take little space for storing. Although cultures can be stored for long periods by this method, viability depends on the quality of glass vials used. It has been found that loss of vacuum during storage leads to inactivation of cultures. Lyophilization needs high cost equipment.

Storage in Silica Gel

Both bacteria and yeast can be **stored in silica gel powder** at low temperature for a period of 1-2 years. In this method, finely powdered, heat sterilized and cooled silica powder is mixed with a thick suspension (paste) of cells, mixed and stored at a low temperature. The basic principle in this technique is quick desiccation at low temperature. Which allows the cells to remain viable for a long period.

6

IDENTIFICATION OF MICROORGANISM

A. IDENTIFICATION OF BACTERIA

Definitions

Identification is the procers of characterizing the isolate. Identification of bacteria is the most vital topic in microbiology. The methods of bacterial identification fall into **three categories**. They are phenotypic methods, immunological methods and genotyping methods. **Strain** is a population of organisms that decends from a single organisms. Serovar is a strain differentiated by serological means. It is also called serotype. Biorans are strains that are differentiated by biochemical test. It is also called biotype. Morphovar or morphotype is a strain which is differented n the basis of morphological distinctions. An Isolate is a pure culture derived from a heterogenous, wild population of microorganisms.

Steps in Basic Identification

Bacterials isolates are identified by making use of the following methods.

1. Phenotypic characterization
2. Genotypic methods
3. Immunological methods.

Phenotypic Methods

It includes Morphological features of colony (Macroscopy) and bacterial cells (Microscopy) i.e., cultural variation of bacterial colony & morphological variation of bacterial cells. It also includes physiological and biochemical character of the bacteria.

Microscopic morphology include a combination of cell shape, size, Gram stain, acid fast reactions, special structures e.g. endospores, granule and capsule. **Macroscopic morphology** are traits that can be accessed with the naked eye (e.g.) appearance of colony including texture, shape, pigment, speed of growth and growth pattern in broth.

Physiology/Biochemical characteristic are traditional mainstay of bacterial identification. These include enzymes (catalase, oxidase, decarboxylase), fermentation of sugars, capacity to digest or metabolize complex polymers and sensitivity to drugs can be used in identification.

The successful identification of microbe depends on:

Using the proper **aseptic techniques**.

Correctly obtaining the specimen.

Correctly **handling the specimen**

Quickly transporting the specimen to the lab.

Once the specimen reaches the lab, it is **cultured and identified**.

Phenotypic Methods of Identification

Microbiologists use 5 basic techniques to **grow, examine and characterize microorganisms** in the lab.

They are called the **5 'I's: inoculation, incubation, isolation, inspection and identification.**

Inoculation - To culture microorganisms a tiny sample (inoculum) is introduced into medium (inoculation).

Isolation – it involves the **separating one species from another**.

Incubation: once the media is inoculated it is incubated which means putting the culture in a controlled environment (incubation) to **allow for multiplication**.

After incubation the organisms are **inspected** and **identified** phenotypically, immunologically or genetically.

Specimen Collection

Successful identification depends on how the specimen is collected, handled and stored. It is important that general aseptic procedures be used including sterile sample containers and sampling methods to prevent contamination of the specimen.

E.g. Throat and nasopharyngeal swabs should not touch the cheek, tongue or salvia.

After collection the specimen must be transported to the laboratory and stored appropriately (e.g. refrigeration) and processed.

Processing of the Specimen

Processing of specimen include microscopic examination of specimen and prepare the media for the inoculation of specimen. During throat specimen processing using direct microscopic Examination of Gram stained smear. We may note the presence of polymorphonuclear cells and organisms. Similary wet mount preparation (KOH, Saline, Iodine), giemsa staining, lactophenol cotton blue staing may provide preliminary idea about type of organism found (bacteri, fungi, protozoa, helminthes), nature of contamination, grams nature of bacteria, whether the pathogen is intracellular or extracellular. Direct fluorescent antibody test (DFA) used to highlight the presences of microorganisms in a specimen. DFA test are available for *Staphylococcus aureus, Streptococcus pyogenes, Neisseria gonorhoeae* and *Haemophilus influenza*.

Based on the results, technician submit preliminary report to the physician for treatment, microbiologist select the media and incubation condition for bacterial cultivation.

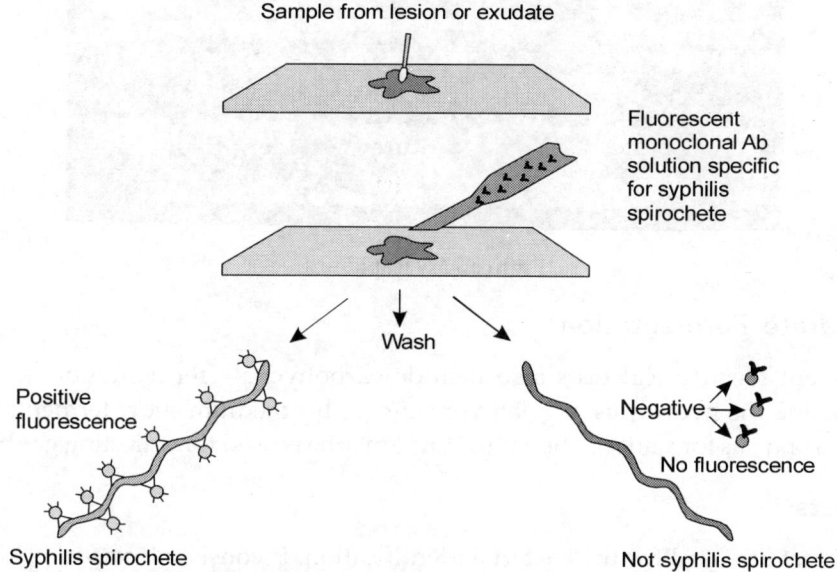

Figure 71   Direct fluorescent antibody test

Media Inoculation

Sample is to be inoculated on any one type of the following medium. It will depend on the type of organism tobe isolated. Inoculate the media using asepting technique.

Incubation

After inoculation, media must be incubated appropriately. Aerobic environment or micro aerophilic environment or anaerobic envirment are selected for incubation.

Inspection

After 24 to 48 hours of incubation, media is looked for the growth and note the colony morphology and microscopic nature. After preliminary observation, pure cuture of the isolate will be made by making use of storage media on tube as slant or deep.

Identification

Generic level confirmed identification will be performed by using various physiological and biochemical tests.

Biochemical Tests

Biochemical tests like Methyl Red Test, Voges Proskauer test, Nitrate Reduction, Starch Hydrolysis, Catalase Test, H_2S production, Indole test, Oxidase test, Oxidation fermentation, Phenylalanine deaminase test are performed to identify bacteria. Details Refer Experimental Procedures in Life Sciences, Experiments in Microbiology by S. Rajan & R. Selvi Christy.

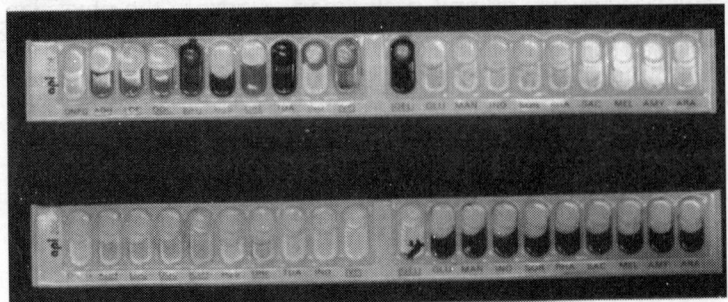

Figure 72   Rapid test

Carbohydrate Fermentation

Prominent biochemical tests also include carbohydrate fermentation, acid or gas production and the hydrolysis of gelatin or starch. This medium show fermentation (acid production) and gas formation. The small Durham tube is used for collecting gas bubbles.

Rapid Tests

Rapid test are used for the bacterial identification. It consist of plastic strips with 20 different dehydrated biochemical substrates used to detect biochemical characteristics.

The biochemical substrates are inoculated with pure cultures and suspended in physiological saline. After 5 hrs-overnight the 20 tests are converted to **digital profile.** E.g. Enterotubes, API methods (Figure 72).

Bacteriophage Typing

Phage typing is a method used for detecting single strain of bacteria. This method can be used for identification and differentiating of bacterial pathogens. It is used to trace the source of outbreaks of infections. The viruses that infect bacteria are called bacteriophages and some of these can only infect a single strain of bacteria. These phages are used to identify different strains of bacteria within a single species. A culture of the strain is grown in the agar and dried. A grid is drawn on the base of the petri dish to mark out different regions. Inoculation of each square of the grid is done by a different phage. The phage drops are allowed to dry and are incubated: The susceptible phage regions will show a circular clearing where the bacteria have been lysed, and this is used in differentiation.

Identification of Unculturable Organisms

Environmental researchers estimate that **< 1%** of microorganisms **are culturable** and therefore it is not possible to use phenotypic method of identification.

Flow cytometry is used for counting microbes in environmental samples.

Flow Cytometry

Flow cytometry allows single or multiple microorganism detection. It is an easy, reliable method. In flow cytometry, microorganisms are identified on the basis of the **cytometry parameters** or by means of certain dyes called **fluorochromes** that can be used independently or bound to specific antibodies. The cytometer forces a suspension of cells through a **laser beam** and measures the light they scatter or the fluorescence the cell emits as they pass through the beam. The cytometer also can measure the cell's shape, size and the content of the DNA or RNA.

Immunological Methods

The study of antibody(Ab) - antigen(Ag) reactions in *in vitro* is called **serology**. Serological reactions are the basic of **immunological identification** and diagnostic methods.

Precipitation Reactions

Precipitation is the interaction of a soluble Ag with a soluble Ab to form an **insoluble complex.** The complex formed is an aggregate of Ag and Ab. Precipitation reactions occurs maximally only when the **optimal proportions** of Ag and Ab are present. Precipitation can also be done in agar referred to as **immunodiffusion**. Precipitation test uses antibodies to detect for streptococcal group antigens.

Agglutination Reactions

Agglutination is the **visible clumping** of an Ag when mixed with a specific Ab. Agglutination tests are widely used because they are simple to perform, highly specific, inexpensive and rapid. Standardized tests are available for the determination of **blood groups** and **identification of pathogens and their products**.

E.g. Blood typing and detection of *Mycoplasma pneumoniae*.

Indirect (passive) agglutination. Ab/Ag is **adsorbed or chemically coupled** to the cell, latex beads or charcoal particles which serves as an inert carrier. **Latex Agglutination Tests** are available for the detection of *Staphylococcus aureus, Streptococcus pyogenes, Haemophilus influenza* and *Camplyobacter* spp.

Fluorescent Antibodies

Antibodies can be chemically modified with **fluorescent dyes** such as rhodamine B, fluorescent red, fluorescien isothiocynate and fluoresces yellow or green. Cells with bound fluorescent antibody emit a **bright red, orange, yellow or green light** depending on the dye used.

There are two distinct fluorescent Antibody procedure. They are **direct and indirect**.

In the **direct method** the fluorescent Ab is directed to surface Ag of the organism.

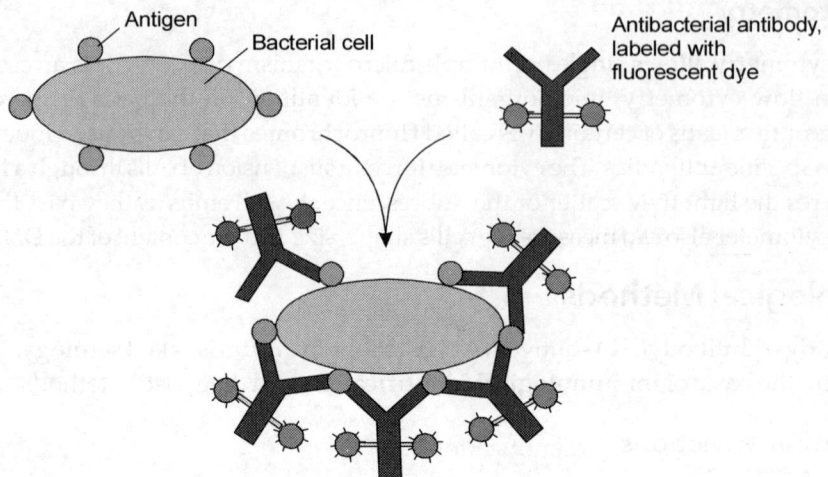

Figure 73 Fluorescent staining

In the **indirect method** a non-fluorescent Ab reacts with the organism's Ag and a fluorescent Ab reacts with the non-fluorescent Ag. **Fluorescent Ab** can be used to detect microorganisms **directly in tissue**, long before a primary isolation technique yield the suspected pathogen. Fluorescent Ab has been used for the detection of *Bacillus anthracis* and HIV virus.

ELISA

Engval and Perlmann developed it in 1970. **Enzyme-linked immunosorbent assay (ELISA)**. It is a popular test to detect the presence an antigen or antibody in a sample. The ELISA has been used as a diagnostic tool in medicine. It is a useful tool for determining serum antibody concentrations (such as with the HIV test). It has also found applications in the food industry in detecting potential food allergens. There are three important types of ELISA. They are Competitive ELISA, Sandwich ELISA and Indirect ELISA.

Immunoblot/Western Blot

It involves separation of viral proteins by SDS PAGE (Sodium Dodecylsulphate – Polyacrylamide Gel Electrophoresis), transferring the resolved antigen to nitrocellulose membrane and identifying specific antigen through antigen and antibody reactions (Immuno blotting). It is used for the confirmation of HIV detected in ELISA.

Genotypic Methods

Nucleic acid probes, PCR (RT-PCR, RAPD-PCR), Nucleic acid sequence analysis, rRNA analysis, RFLP,Plasmid fingerprinting are the Genotypic methods used for the microbe identification.

Nucleic Acid Hybridization

Nucleic acid hybridization is one of the most powerful tools available for microbe identification. Hybridization detects for a **specific DNA sequence** associated with an organism. The process uses a **nucleic acid probe** which is specific for that particular organism. The target DNA (from the organism) is **attached to a solid matrix** such as a nylon or nitrocellulose membrane. A single stranded probe is added and if there is **sequence complementality** between the target and the probe a positive hybridization signal will be detected. Hybridization is detected by a **reporter molecule** (radioactive, fluorescent, chemiluminescent) which is attached to the probe. Nucleic acid probes have been marketed for the identification of many pathogens such as *N. gonorrhoeae.*

Polymerase Chain Reaction (PCR)

PCR is widely used for the identification of microorganisms.Sequence **specific primers** are used with PCR in the amplification of DNA or RNA of specific pathogens.PCR allows for the detection even if **only a few cells are present** and can also be used on **viable nonculturables**. The presence of the **appropriate amplified PCR product** confirms the presence of the organisms. **Primers** are available for the identification of *Niesseria gonorrhoeae,* and to monitor food for the presence of *Salmonella* and *Staphylococcus.*

Real Time PCR and RT-PCR

Currently many PCR tests employ **real time PCR**. This involves the use of **fluorescent primers**. The PCR machine **monitors the incorporation of the primers** and display an **amplification plot** which can be viewed continuously through the PCR cycle. Real time PCR **yields immediate results**. Another application of PCR is **RT-PCR** (reverse trancriptase PCR). During RT-PCR an RNA template is used to **generate cDNA and from this dsDNA** is generated. The enzyme used is **reverse transciptase**. RT-PCR is used to detect for HIV and to monitor the progress of the disease.

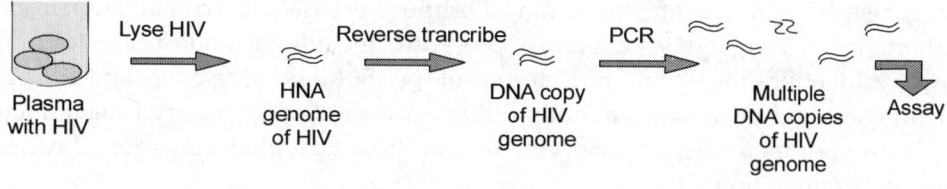

Figure 74 RT PCR

RAPD-PCR

Random amplified polymorphic DNA PCR uses **a random primer** (10-mer) to generate a DNA profile. **The primer** anneals to several places on the DNA template and generate a DNA profile which is used for microbe identification.

RAPD has many advantages, they are, Pure DNA is not needed, Less labour intensive than RFLP. There is not need for prior DNA sequence data.RAPD has been used to fingerprint the outbreak of *Listeria monocytogenes* from milk.

DNA Sequencing

The determination of a small amount of DNA sequence can be used for microbial identification. The most common sequence used for microbe identification is DNA sequence of the **16S rRNA gene**. PCR is used to amplify the 16S rRNA gene and the sequence determined. rRNA is a major component for ribosome and **ribosome have the same function** in protein synthesis in all cells. Computer analysis of 16S rRNA sequence has revealed the presence of **signature sequences**, short oligonucleotides **unique to certain groups of organisms** and useful in their identification. rRNA sequence can be used to fine tune identity at the species level e.g differentiating between *Mycobacterium* and *Legionella*. 16s rRNA sequence can also be used **to identify microorganisms from a community**.

Restriction Fragment Length Polymorphism (RFLP)

RFLP involves **digestion** of the genomic DNA of the organism **with restriction enzymes.** The restricted fragments are **separated** by agarose gel electrophoresis. The DNA fragments **are transferred to a membrane** and **probed** with probes specific for the desired organisms. A DNA profile emerges which can be used for microbe identification.

Plasmid Fingerprinting

Plasmid fingerprinting identifies **microbial species or similar strains** as related strains often contain the same number of plasmids with the same molecular weight. Plasmid of many strains and species of *E. coli, Salmonella, Camylobacter* and *Psseudomonas* has demonstrated that this methods is **more accurate** than phenotypic methods such as biotyping, antibiotic resistance patterns, phage typing and serotyping.

B. ISOLATION OF FUNGI FROM CLINICAL SAMPLES

Fungi are significant, sometimes overlooked, human pathogens. Infection caused by the fungus ranges from mild to life threatening. Diagnosis is based on a combination of clinical and laboratory investigations. Laboratory procedure includes, Demonstration of fungi by microscopy, Identification by culture, Detection of specific humoral response and Detection of fungal antigens and metabolites in body fluids. Successful of laboratory fungal diagnosis depends on specimen selection, Specimen collection, Specimen transport, Processing, Microscopic examination, Culturing.

Specimen Collection

Specimen collection for the detection of an etiologic agent of mycoses is very similar to specimen collection for bacteria. But fungal detection requires large quantity of specimen than bacterial identification. Fingernails and toenails suspected of Dermatophytosis should be cleaned extensively with 70% ethanol. Swabs are not optimal specimen collection for fungal identification. Depends on the site of infection, sample collection method and specimen type may vary.

For dermatophytic infection : Skin scrapings, Pus, Nail clippings, Hair plug not cut,

Subcutaneuous infection: Pus, biopsy tissues, Aspirated fluid, Skin scrapings

Systemic mycoses : Sputum, Bronchial washings, Exudates from cutaneous lesion, CSF, Urine, Tissue biopsy, Vaginal swab, Blood.

Specimen Processing

Liquid specimens may have low concentration of fungus and will require centrifugation to increase fungal cell concentration. Hard specimens should be mined before inoculation.

Concentration

Large volume of fluid should be concentrated by centrifugation (2000 rpm for 10 min). Use the resulting pellet for culture and KOH examination. Body fluids, Urine, CSF, Sputum are used after digestion with N-acetyl L-cysteine.

Mincing or homogenization

Biopsy samples, tissues and nails must be processed to increase the recovery. With sterile scalpels, mince specimens into small pieces in a petriplate with a few drops of sterile distilled water. If *Histoplasma* is suspected, specimen is homogenize by using tissue grinder after mincing.

Media selection

Media selection depends on body site, nature of infection and nature of contamination. To prevent bacterial pathogens antibiotics should be used or reduce the pH of the medium Brain heart infusion agar with antibiotics, Potato Dextrose Agar, Mold agar, Sabouraud Glucose Agar, Niger seed agar, Yeast extract phosphate medium. , Buffered charcoal yeast extract agar, Rose Bengal chloramphenical agar. From the above list of mycotic media, any one or two media selected for cultivation.

Inoculation

Concentrated aspirated body fluids are directly inoculated on to the medium by making use of inoculation loop. Swabs are vortexed in distilled water before inoculation. Vortexed specimens are inoculated on medium after concentration. After inoculation hair and skin scrapings are pressed firmly on the surface of the medium. Nail should be pulverized using scalpels and press fragments firmly onto medium surface.

Incubation

All fungal media were incubated at 28 to 30°C for 4 weeks, because most fungi will produce colonies by the end of the 3rd Week.. Vaginal cultures may be incubated for only 7 days. Observe the media once in two days. *Histoplasma capsulatum* and *Blastomyces dermatidis* may require 4-8 weeks. Aerobic non clinical fungus may produce fruiting bodies within 72 hours.

Examination of fungal growth on primary media

Read primary plates daily for the first week, every other day for the second week and twice weekly for the remaining two weeks. When growth appears, differentiate yeast from mold by microscopic examination. Once colonies are formed on the media, the organism should be viewed microscopically. Zygomycetes are observed under dissecting microscope. Yeast is observed by Wet preparation, Lacto phenol cotton blue staining and Indian ink preparation. Mold is observed by Wet preparation, Scotch tape , Tease preparation , Slide cultures, Lactophenol cotton blue staining , Ascospore.

KOH Mount

KOH may be used to examine hair, nails, skin scrapings, fluids, exudates or biopsies. The fungal structures such as hyphae, large yeast (*Blastomyces*), spherules, and sporangia may be distinguished. Examine slides with reduced light (narrow the iris diaphragm) and examine negative smears on several consecutive days. The fungal structures may be enhanced by using a phase-contrast microscope. Specimens placed in a drop of 15% KOH will dissolve at a greater rate than fungi because fungi have chitinous cell walls. The clearing effect throughout the clinical specimen can be accelerated by gently heating the KOH preparation. Visualization of fungi can be further enhanced by the addition of Parker Superquink permanent black ink to the preparation.

Mycelium and spore were observed within tissue. KOH clears hard tissue and facilitates easy observation. KOH mount provides first hand information about fungal infection in tissues.

KOH-DMSO-INK mount/stain

Specimens with tissue placed in a drop of 15% KOH will dissolve at a greater rate than fungi because fungi have chitinous cell walls. The clearing effect throughout the clinical specimen can be accelerated by gently heating the KOH preparation. Addition of Ink to KOH enhances contrast. This method is most useful for detecting *Malassezia furfur* in skin scrapings. DMSO present in staining reagent eliminates the need of heat during staining

Mycelial elements clearly demonstrated, which indicates the availability of fungal infection in particular tissue. Spore along with conidium describes mould species.

KOH-Calcoflour fluorescent-stain

Calcoflour white stain may be used for direct examination of most specimens using fluorescent microscopy. The cell walls of the fungi bind the stain and fluoresce blue-white or apple-green depending on the filter combination used. The use of calcofluor white (CFW), a fluorescent brightener used in the textile industry, with the addition of potassium hydroxide (KOH) will enhance the visualization of fungal elements in specimens for microscopic examination. The CFW nonspecifically binds to the chitin and cellulose in the fungal cell wall and fluoresces a bright green to blue. A substantial amount of non-specific fluorescence from human cellular materials and natural and synthetic fibers should be expected. The CFW highlights suspicious structures but the interpretation of the structures relies on traditional fungal morphologic features.

India ink Preparations

An India ink preparation can be used for the rapid detection of the encapsulated yeasts based on negative staining. The capsule repels the carbon particles of the Indian ink, giving a clear well-demonstrated halo around each encapsulated stain.

Giemsa Stain for *Histoplasma capsulatum*

Giemsa stain is used for examining intracellular structures and is applied to primary specimens of bone marrow tissue and WBC's in which *H.capsulatum* is suspected. Necrotic cells in the specimen will have pink cytoplasm. Normal cell- light blue - violet lavender cytoplasm. Phagocytised yeast cells will stain light to dark blue, and each will have a clear halo around it.

Look for purple pseudoencapsulated yeast forms of *H.capsulatum* inside PMN cells and monocytes.

C. IDENTIFICATION OF YEAST

Yeasts are a heterogenous group of fungi that superficially appear to be homogeneous. Yeasts grow in a conspicuous unicellular form that reproduces by fission, budding, or a combination of both. True yeasts reproduce sexually, developing ascospores or basidiospores under favorable conditions. Yeast-like fungi (imperfect yeasts) reproduce only by asexual means. The identification of these fungi is based upon a combination of morphological and biochemical criteria. Morphology is primarily used to establish the genera, whereas biochemical assimilations are used to differentiate the various species.

Principal Criteria and Tests for Identifying Yeasts

1. Culture characteristics - Colony colour, shape, texture
2. Asexual structures: a. Shape and size of cells; b. Bipolar, fission, multipolar or unipolar "budding"; c. Absence or presence of arthroconidia, ballistoconidia, blastoconidia, clamp connections, endoconidia, germ tubes, hyphae, pseudohyphae, or sporangia and sporgangiospores.
3. Sexual structures - Arrangement, cell wall ornamentation, number, shape and size of ascospores or basidiospores
4. Physiological studies: a. Assimilation; c. Fermentation d. Nitrogen utilization e. Urea hydrolysis

Cultural characters

Clinical sample is streaked on Sabourad dextrose agar or Rose Bengal chloramphenicol agar plates. Plates were incubated at 25-30°C for 48 hours and observed for colony morphology, colour, shape and texture. Record the result and interpretate the results.

CHROM agar test

CHROM agar contains enzymatic substrates that are linked to chromogenic compounds. When specific enzymes cleave the substrates, the chromogenic substrates produce colour.

The action of different enzymes produced by yeast species results in colour variations useful for the presumptive identification of yeasts. The test provides only presumptive identification of yeast eg: *C.kruses, C.tropicalis, C.albicans.*

Cultures are streaked on the medium and incubate at 35°C in humidified dark chamber for 48-72 hours. After 72 hours observe colour change and interpretate the results.

C.albicans: A medium sized, green, smooth matte colony with a very slight green halo ion the surrounding medium.

C.tropicalois : Smooth medium sized matte colony, which is blue to blue gray with a pale pink edge. The colony may have a dark brown to purple halo, which diffuses into the agar.

C.krusei : A large, spreading, rough pink colony with a pale pink to white edge.

Germ tube test

Candida are the members of the normal flora of skin. More than 100 species are available in candida. Among these *Candida albicans* cause most of the human infections. The germ tube test provides a simple, reliable and economical procedure for the presumptive identification of *Candida albicans.* About 95% of the clinical isolates produce germ tubes when incubated in serum at 35°C for 2.5-3 hours. A germ tube represent the initiation of a hypha directly from the yeast cell. They have parallel walls at their point of origin. Germ tube formation is influenced by the medium, inoculum size and temperature of incubation. Fresh normal pooled human sera or a commercially available germ tube solutionare to be used as the medium for the test. The inoculum should result in a very faintly turbid serum suspension. Over-inoculation will inhibit the development of germ tubes. Incubate in at 35°C-37°C for 2.5-3 hours.

India ink preparations

India ink can be added to specimens such as spinal fluids or exudates to provide a dark background that will highlight hyaline yeast cells and capsular material (halo effect). Hence, it should be used to examine specimens suspected of containing *Cryptococcus neoformans.* White blood cells may be distinguished from *Cryptococcus neoformans* because of the irregular edge of the halo and the pale cell wash

Giemsa staining

Giemsa stain is used for examining intracellular structures and is applied to primary specimens of22 bone marrow tissue and WBC's in which *H.capsulatum* is suspected.

Urease test

The rapid urea hydrolysis test is used to screen isolates for *Cryptococcus neoformans.* Under these conditions, *C. neoformans* will rapidly hydrolyze urea, which results in a pink to red colour.

Nitrate test

Yeasts have the ability to use ammonium sulfate, asparagine, peptone, and urea aerobically as sole sources of nitrogen if adequate vitamins are provided. In contrast, aliphatic amines, potassium nitrate, sodium nitrate, and some amino acids are utilized selectively by different yeasts.

Ascospore Induction and Detection

One step in indentifying a yeast involves determining whether or not the isolate has the ability to form ascospores. Some yeasts will readily form ascospores on primary isolation medium, whereas others require special media. The ability to form ascospores varies from isolate to isolate and may be lost in old laboratory strains. If only one mating type of a heterothallic yeast is present, no ascospores will be formed. Ascospore media contain small amounts of carbohydrates; this restricts vegetative growth while enhancing ascospore formation.

Modified kinyons Acid fast staining

Modified acid-fast stains are recommended for demonstrating ascosspores. Unlike the Ziehl-Neelsen modified acid-fast stain, the modified Kinyoun acid-fast stain does not require heating the reagents used for staining

Carbohydrate fermentation

It is performed as like bacterial carbohydrate fermentation test.

Carbohydrate assimilation

This is performed using the medium phenol red agar base. On the medium spread 0.1mL of yeast culture and place carbohydrate disc. Incubate plates at 37°C for 24 to 48 hours and observed for colour change around disc.

Table 6.1 Identification and differentiation features of Different Yeasts

Organism	Pseudo hyphae	True hyphae	Urease	Blasto conidia	Ascospores	Nitrate	Germ tube
Candida albicans	Present	Present	Negative	Present	Absent	Positive	Positive
Trichosporan	Present	Present	Positive	Present	Absent	Negative	Negative
Saccharomyœs	Present	Absent	Negative	Present	Present	Negative	Negative
Crytococcus neoformans	Present	Absent	Positive	Present	Absent	Negative	Negative
Torulopsis	Absent	Absent	Negative	Present	Absent	Negative	Negative
Rhodotorulla	Absent	Absent	Positive	Present	Absent	Negative	Negative
Prototheca	Absent	Absent	Negative	Present	Absent	Negative	Negative
Geotrichum	Absent	Present	Negative	Absent	Present	Negative	Negative
Other Candida	Absent	Absent	Positive	Absent	Present	Negative	Negative
Hanseula	Absent	Absent	Negative	Absent	Present	Negative	Negative

Table 6.2   Differentiation features of Different Yeasts based on carbohydrate assimilation and fermentation

Assimilation

Yeast/Sugar	Glucose	Maltose	Sucrose	Lactose	Galactose	Melibiose	Inositol	Xylose	Trehalose
Candida albicans	Positive	Positive	Positive	Negative	Positive	Negative	Negative	Negative	Positive
Candida lipolytica	Positive	Negative	Negative	Negative	Negative	Negative	Negative	Positive	Negative
Candida krusei	Positive	Negative	Negative	Negative	Negative	Negative	Negative	Negative	Negative
Cryptococcus neoformans	Positive	Negative	Negative	Negative	Negative	Negative	Negative	Negative	Negative

Fermentation

Yeast / Sugar	Glucose	Maltose	Sucrose	Lactose	Galactose	Melibiose	Inositol	Xylose	Trehalose
Candida albicans	Fermentative	Fermentative	Negative	Negative	Fermentative	Negative	Negative	Negative	Fermentative
Candida lipolytica	Negative	Negative	Negative	Negative	Negative	Negative	Negative	Negative	Negative
Candida krusei	Fermentative	Negative	Negative	Negative	Negative	Negative	Negative	Negative	Negative
Cryptococcus neoformans	Fermentative	Fermentative	Fermentative	Negative	Fermentative	Negative	Negative	Negative	Negative

Result

Based on the above-mentioned test and microscopic morphology specific type of yeasts are identified.

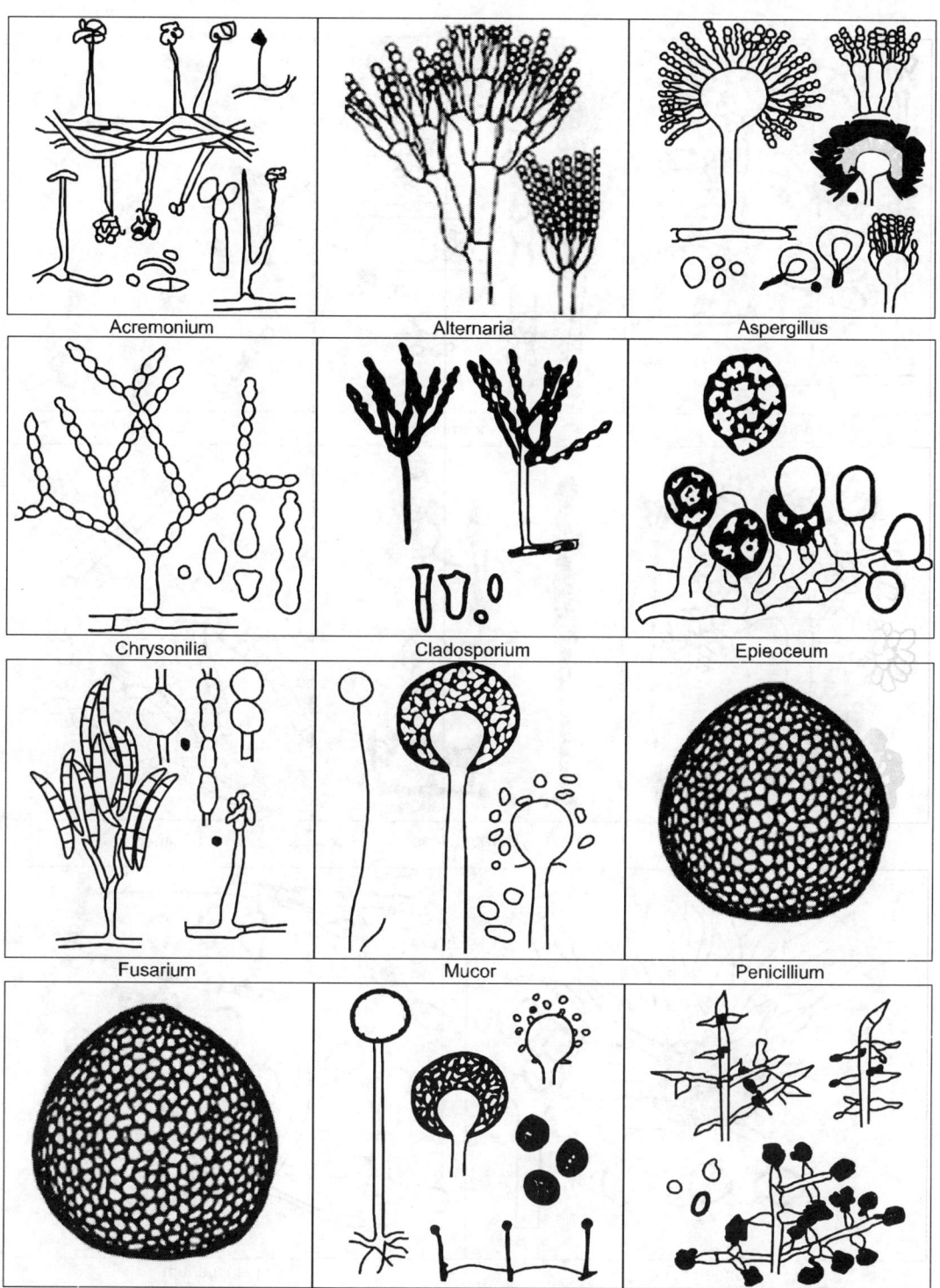

Plate 1 - Microscopic morphology of Fungus.

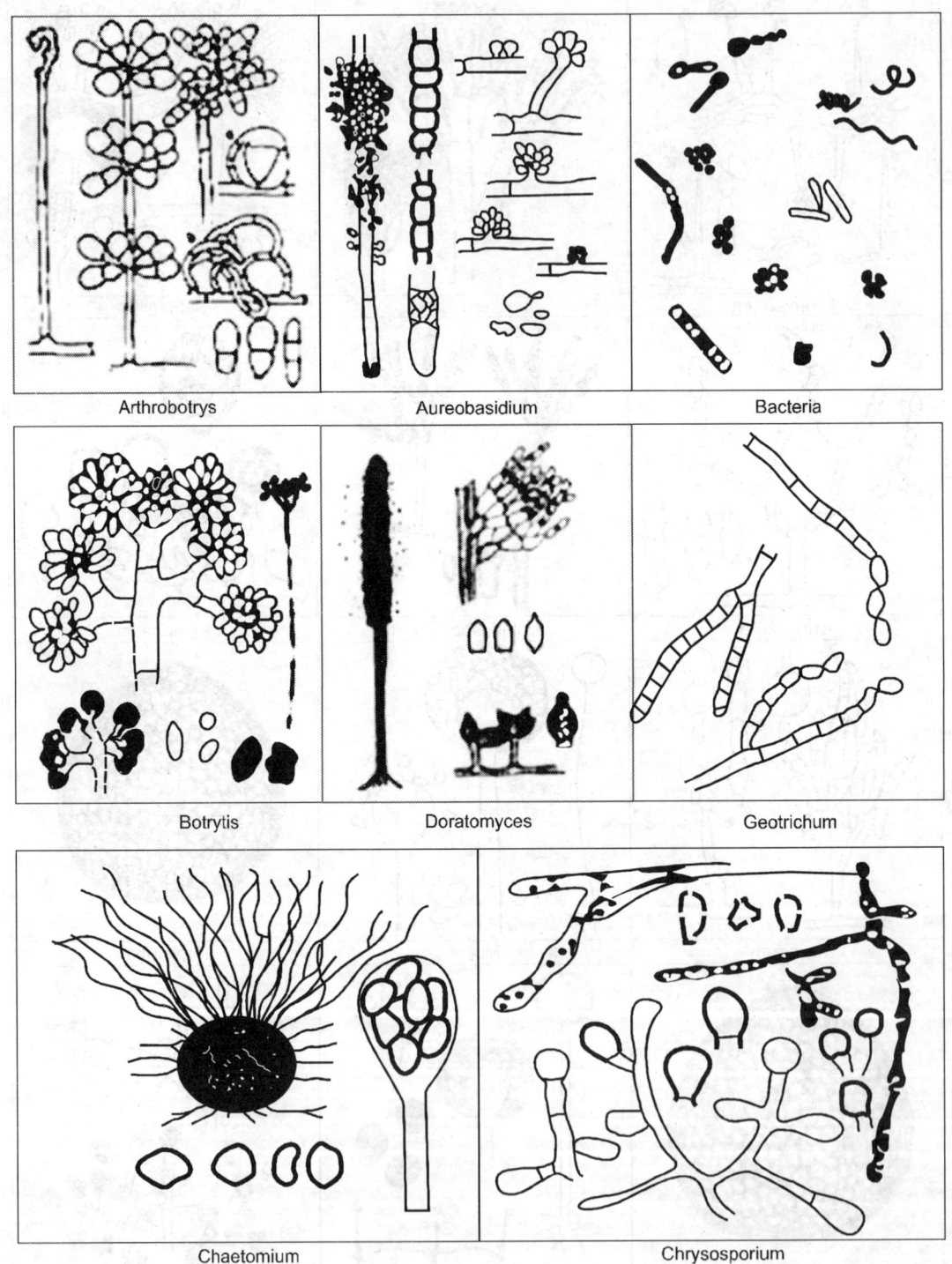

Arthrobotrys

Aureobasidium

Bacteria

Botrytis

Doratomyces

Geotrichum

Chaetomium

Chrysosporium

Plate 2 - Microscopic morphology of fungus

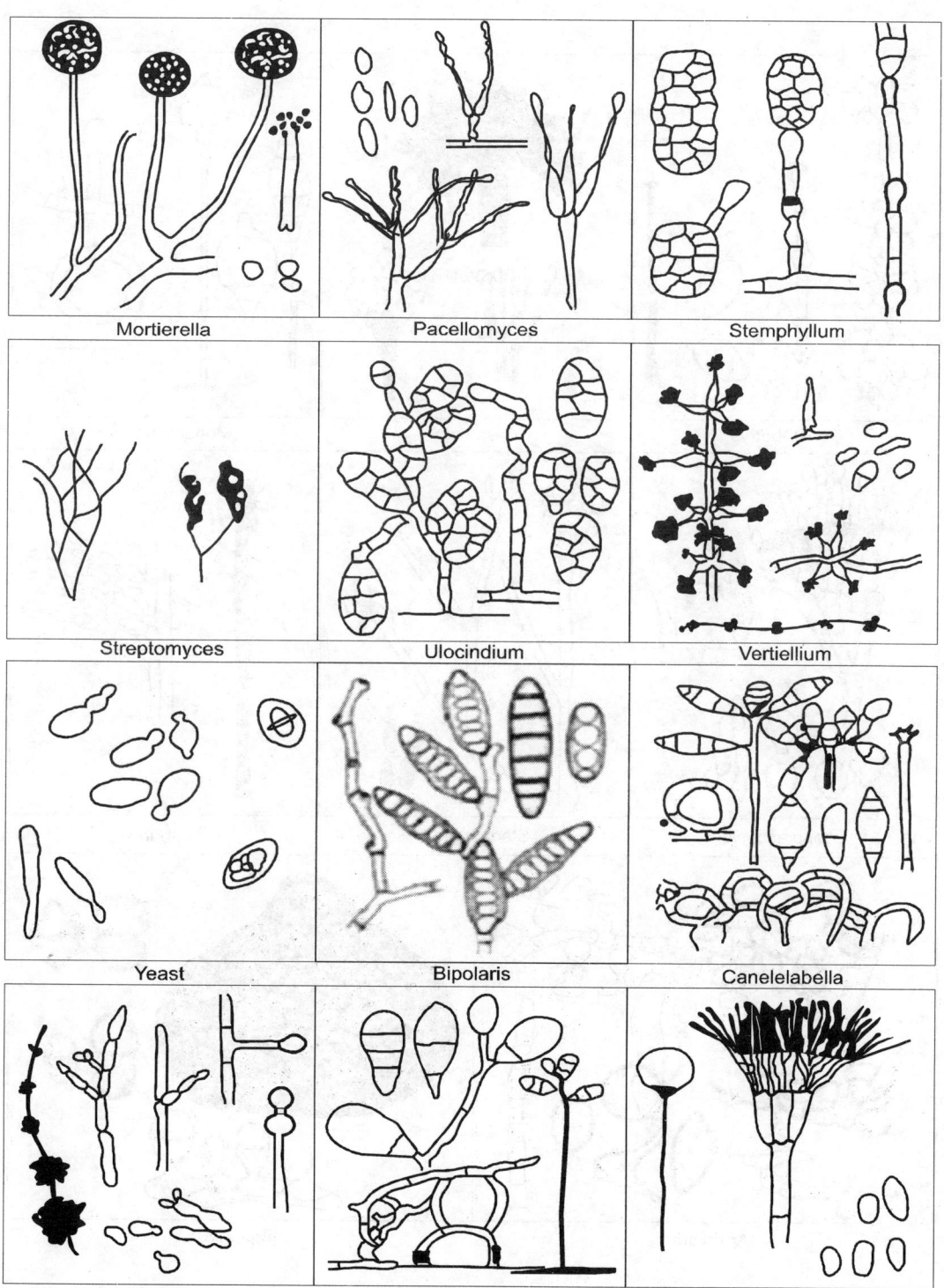

Plate 3 - Microscopic morphology of fungus

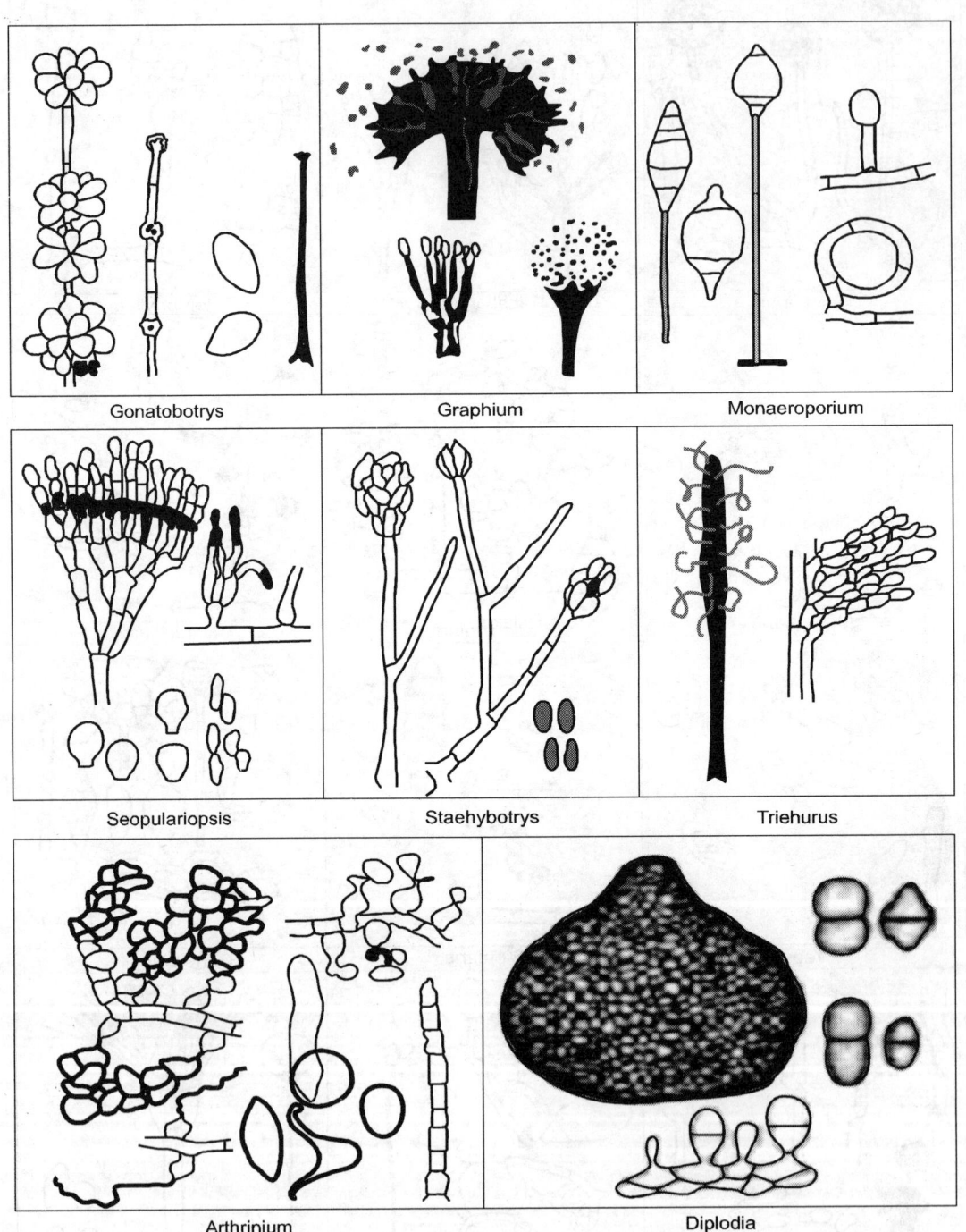

Gonatobotrys

Graphium

Monaeroporium

Seopulariopsis

Staehybotrys

Triehurus

Arthrinium

Diplodia

Plate 4 - Microscopic morphology of fungus

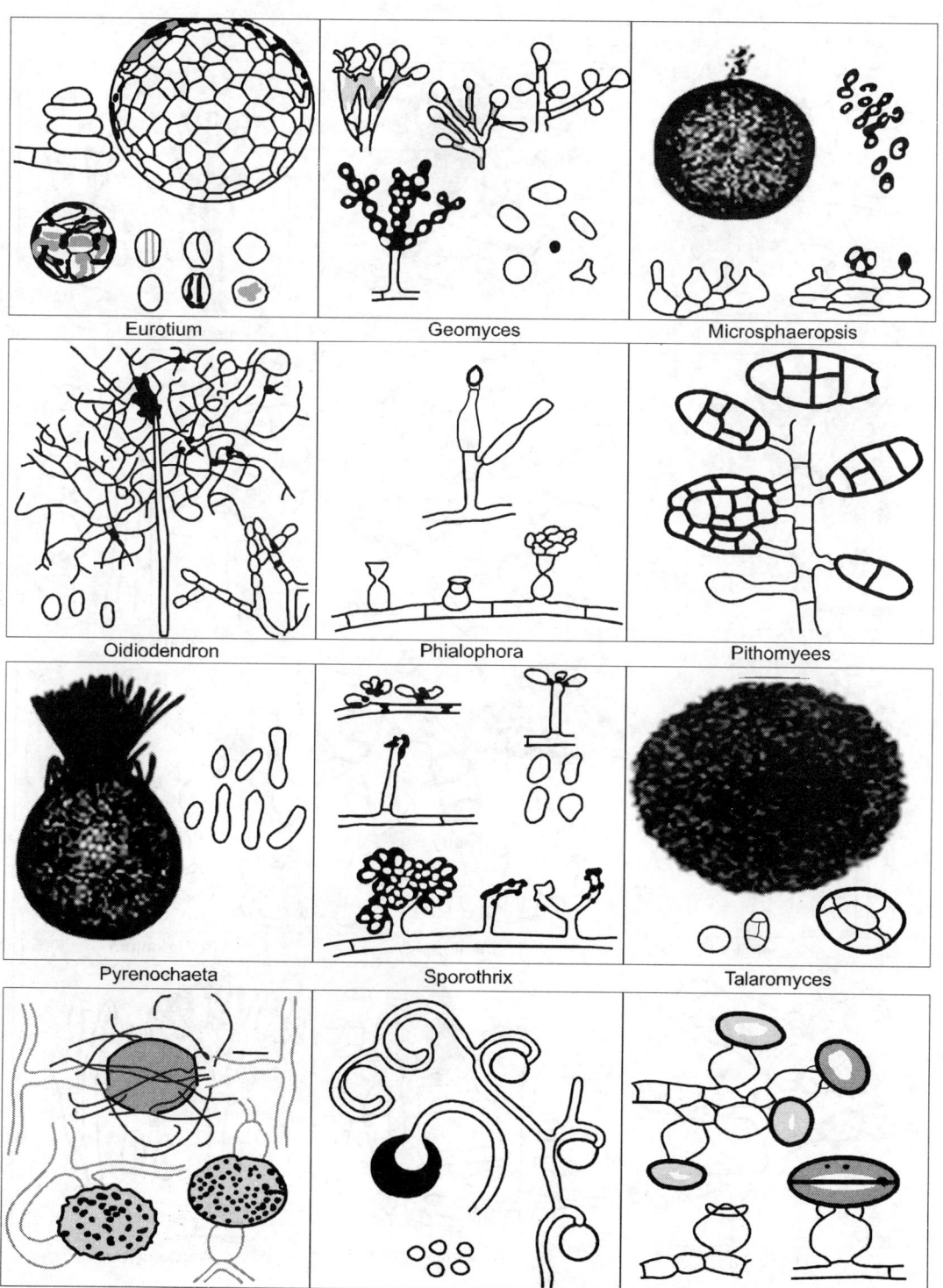

Plate 5 - Microscopic morphology of fungus

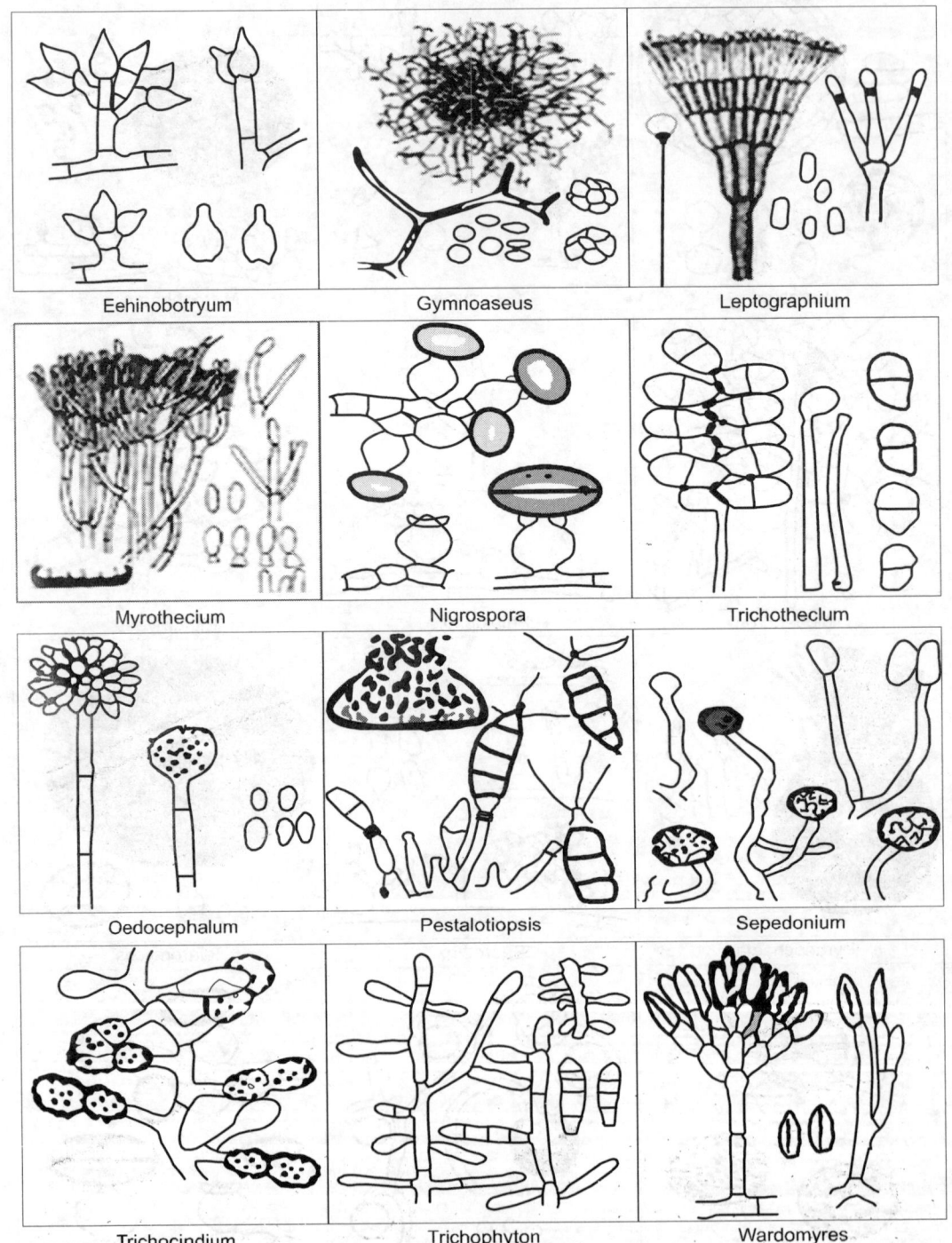

Plate 6 - Microscopic morphology of fungus

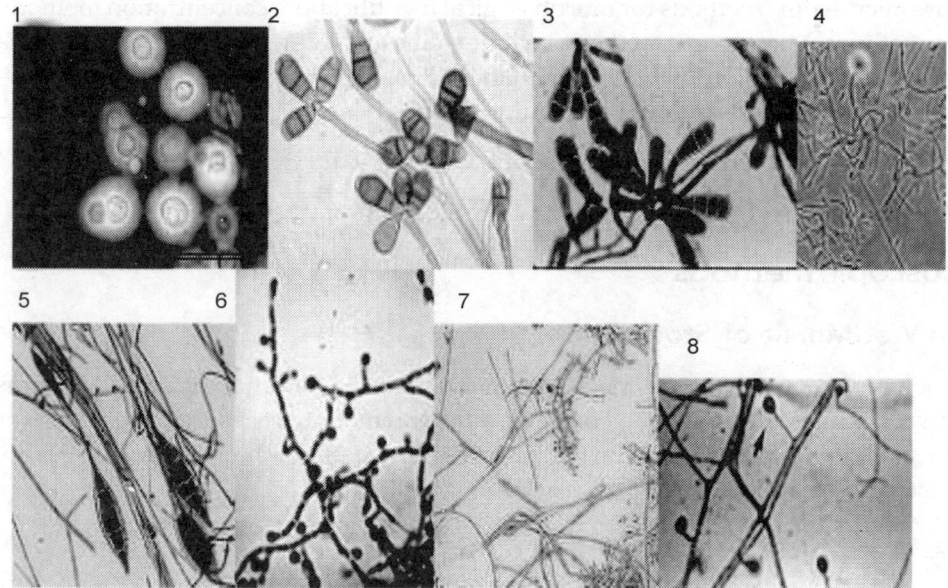

1. Cryptococcus neoformans
2. Curvularia lunata
3. Epidermophyton flocossum
4. Fusarium
5. Microsporum canis
6. Paracoccidioides brasiliensis
7. Trichophyton mentagrophytes
8. Blastomyces dermatidis

Plate 7 - Microscopic morphology of fungus

D. IDENTIFICATION OF INTESTINAL PARASITES

Introduction

Protozoan and helminthes are two major groups of organisms. They are responsible for varieties of human intestinal infection. Majority of intestinal parasites are detected using stool as specimen. All parasites have distinctive morphology and is studied by microscopic examination of faecal materials. Faecal specimens shows cyst and trophozoites of protozoans, adult worm, segments, larva and eggs of helminthes.

The following parasites may be detected by making use of microscopic techniques. They are *Entamoeba histolytica, Giardia lamblia, Trichomonas hominis, Balantidium coli, Isospora bellei, Taenia solium, Fasciola hepatica, Schistosoma sp, Trichostrongylus,Capilaria, Ascaris lumbricoides, Paragonimus, Enterobius vermicularis, Faciola buski, T. trichura*

Fresh faecal specimen is important for the detection of parasitic infections. Faecal specimen should be examined within a day (formed stool -1 day, semiformed stool-1 hours, liquid stool-30 min). Faecal specimen should not incubated for parasitic examinations.

Colour, consistency, presence of blood, mucus and segments are to be noted during faecal specimen examinations. Following methods are adopted to diagnose intestinal parasites.

They are microscopic methods for morphological identification, concentration methods and culture methods. Microscopic methods include saline and iodine wet mount. Concentration methods are sedimentation method (formal ether, formal detergents) and floatation techniques (Sodium chloride, $ZnSO_4$ and sugar floatation method)

Microscopic methods - Saline wet mount, Iodine wet mount

Concentration technique – Sedimentation, Floatation

Microscopic methods

Saline Wet Mount of Stool

Stool microscopy is a rapid method employed for the detection of intestinal parasites. Saline wet mount is more advantageous than other wet mount procedure adopted to examine parasites. Saline wet mount maintain viability / motility status of the parasite. It also facilitates demonstration of chromotoidal bodies in the cyst.

Iodine Wet Mount

Iodine wet mount is performed by making use of Dobell's, O'Connor's, lugols and Antoine's Iodine solutions. It is mainly used for the detection of protozoan parasites. This method clearly demonstrates the presence of nuclei as brown dots. It also demonstrates the yellowish cytoplasm and brown glycogen mass present in the cyst. Disadvantages of this technique includes trophozoites are killed by iodine; chromatoidal bars in protozoan cysts are not clearly demonstrable.

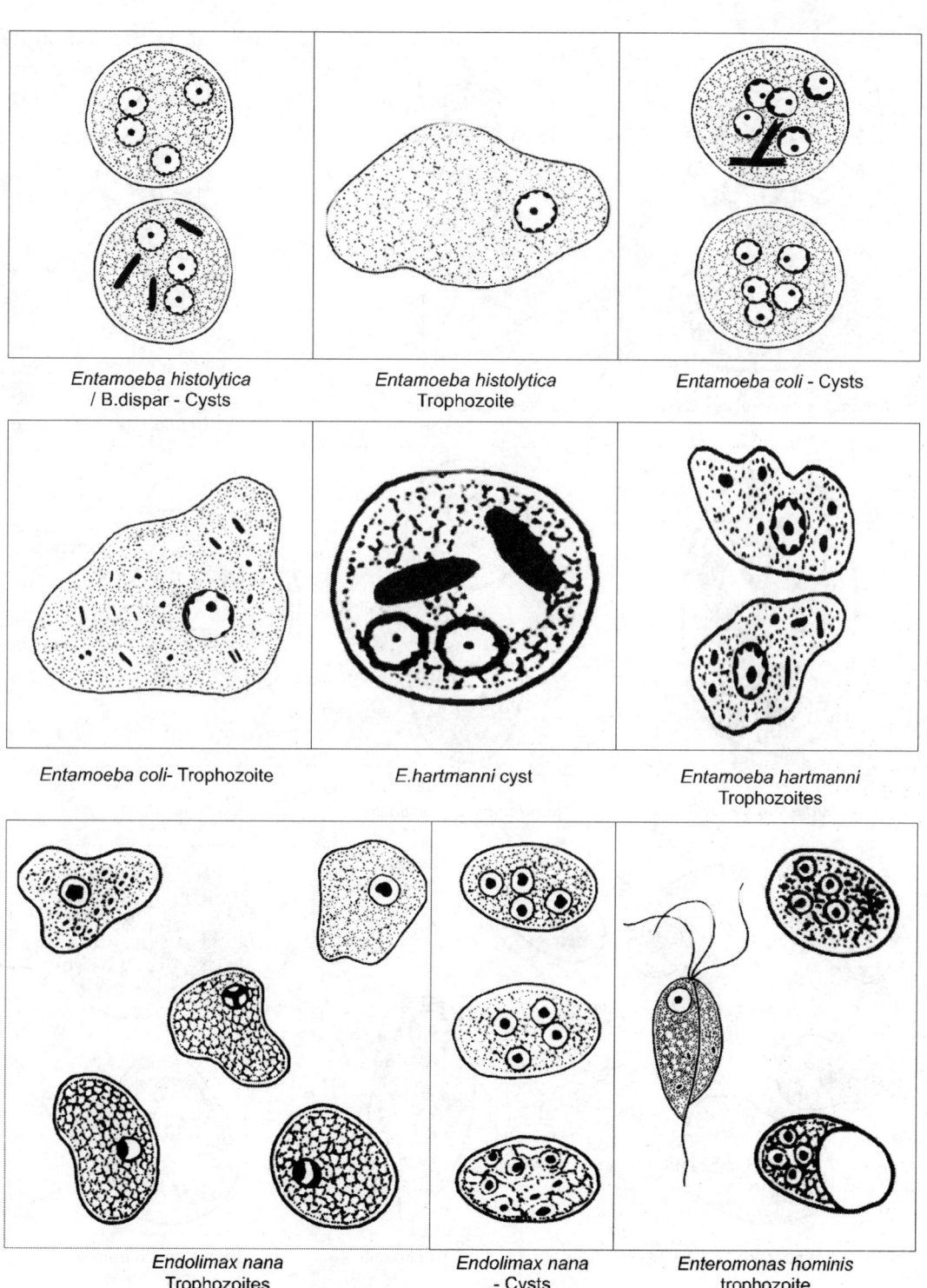

Entamoeba histolytica
/ B.dispar - Cysts

Entamoeba histolytica
Trophozoite

Entamoeba coli - Cysts

Entamoeba coli- Trophozoite

E.hartmanni cyst

Entamoeba hartmanni
Trophozoites

Endolimax nana
Trophozoites

Endolimax nana
- Cysts

Enteromonas hominis
trophozoite

Plate 8 - Intestinal protozoans

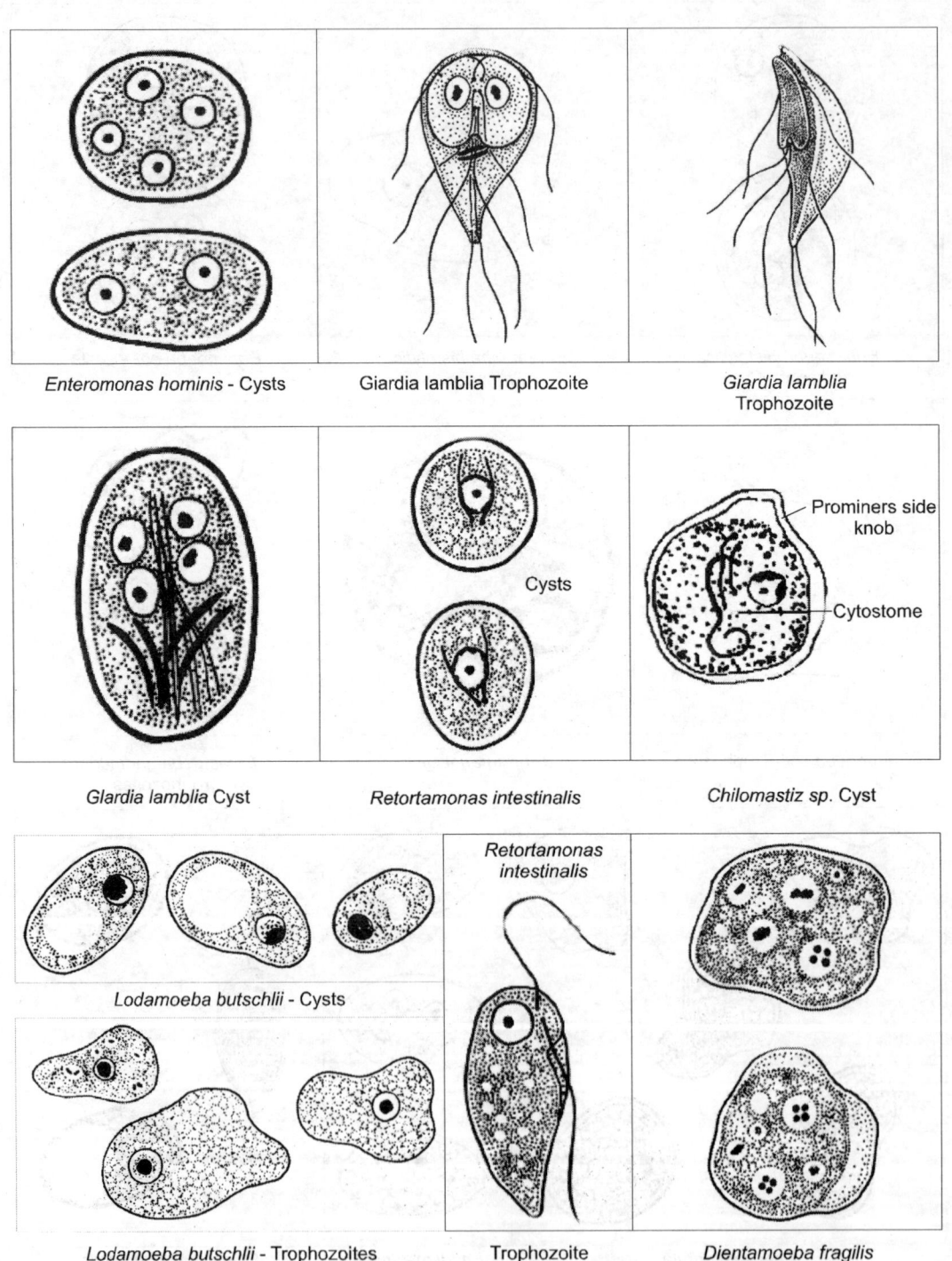

Enteromonas hominis - Cysts   Giardia lamblia Trophozoite   *Giardia lamblia* Trophozoite

Glardia lamblia Cyst   *Retortamonas intestinalis*   *Chilomastiz sp.* Cyst

Prominers side knob

Cytostome

Cysts

Lodamoeba butschlii - Cysts

Lodamoeba butschlii - Trophozoites

Retortamonas intestinalis

Trophozoite

Dientamoeba fragilis - Trophozoites

Plate 9 - Intestinal Protozoans

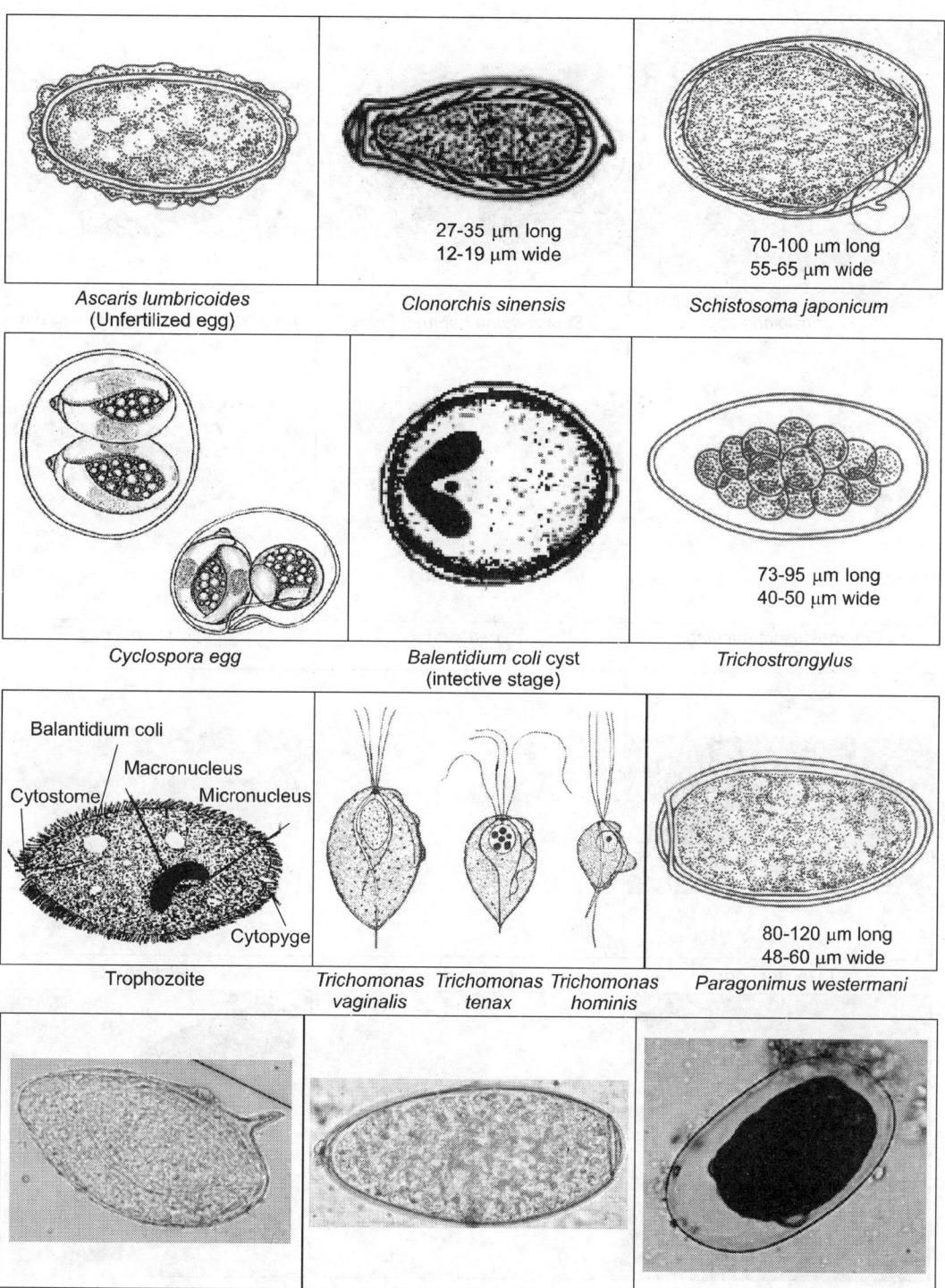

	27-35 µm long 12-19 µm wide	70-100 µm long 55-65 µm wide
Ascaris lumbricoides (Unfertilized egg)	*Clonorchis sinensis*	*Schistosoma japonicum*

		73-95 µm long 40-50 µm wide
Cyclospora egg	*Balentidium coli* cyst (intective stage)	*Trichostrongylus*

Balantidium coli
Macronucleus
Cytostome
Micronucleus
Cytopyge

			80-120 µm long 48-60 µm wide
Trophozoite	*Trichomonas* *vaginalis*	*Trichomonas* *tenax* *Trichomonas* *hominis*	*Paragonimus westermani*

S.mansoni	*Fasciola hepatica*	*Hookworm egg*

Plate 10 - Intestinal Protozoans

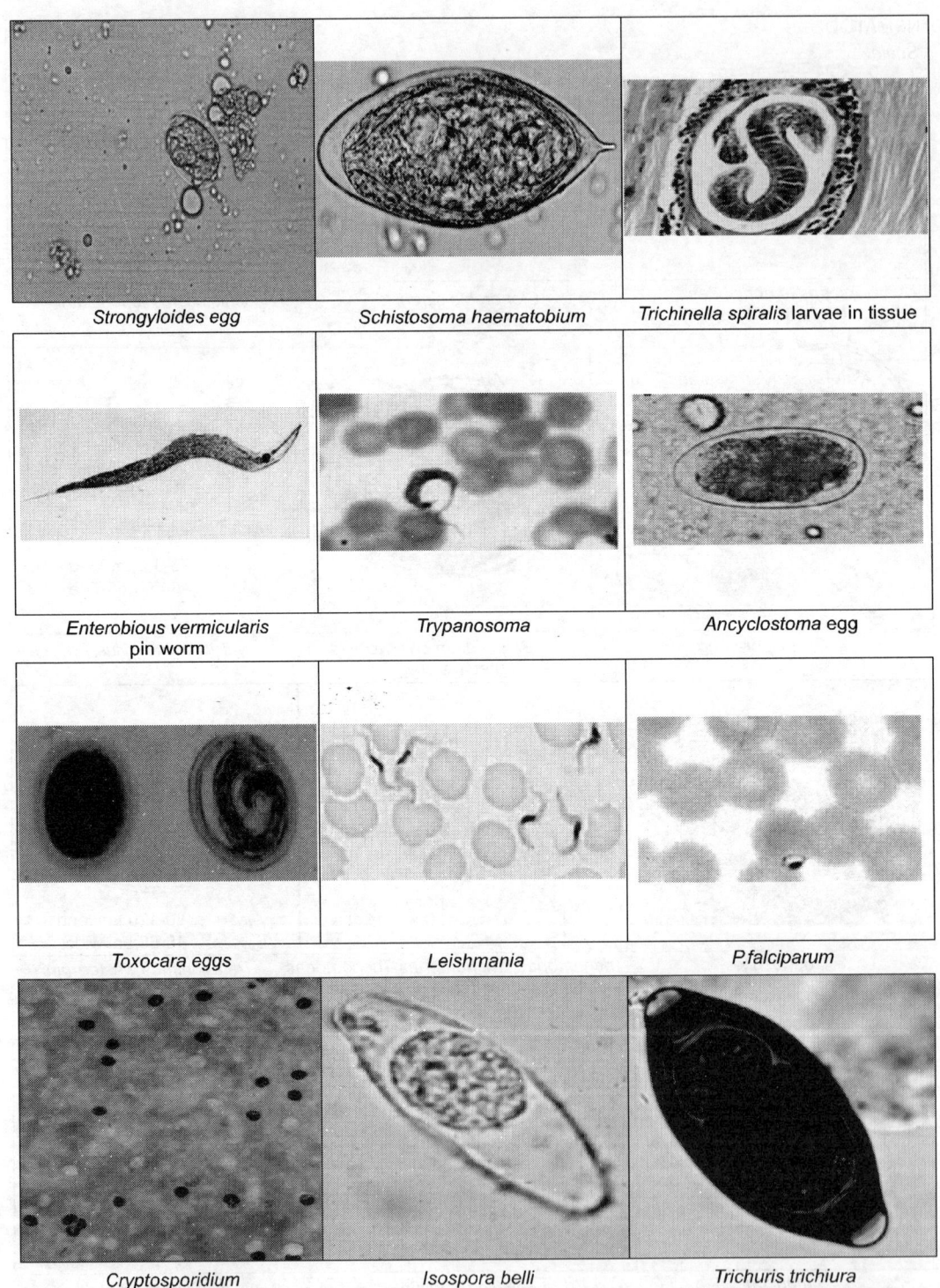

Strongyloides egg   Schistosoma haematobium   Trichinella spiralis larvae in tissue

Enterobious vermicularis   Trypanosoma   Ancyclostoma egg
     pin worm

Toxocara eggs   Leishmania   P.falciparum

   Cryptosporidium   Isospora belli   Trichuris trichiura
   acid fast staining

Plate 11 - Intestinal, tissue protozoans and helminthes

NEMATODES
Scale:

0   24   48µm

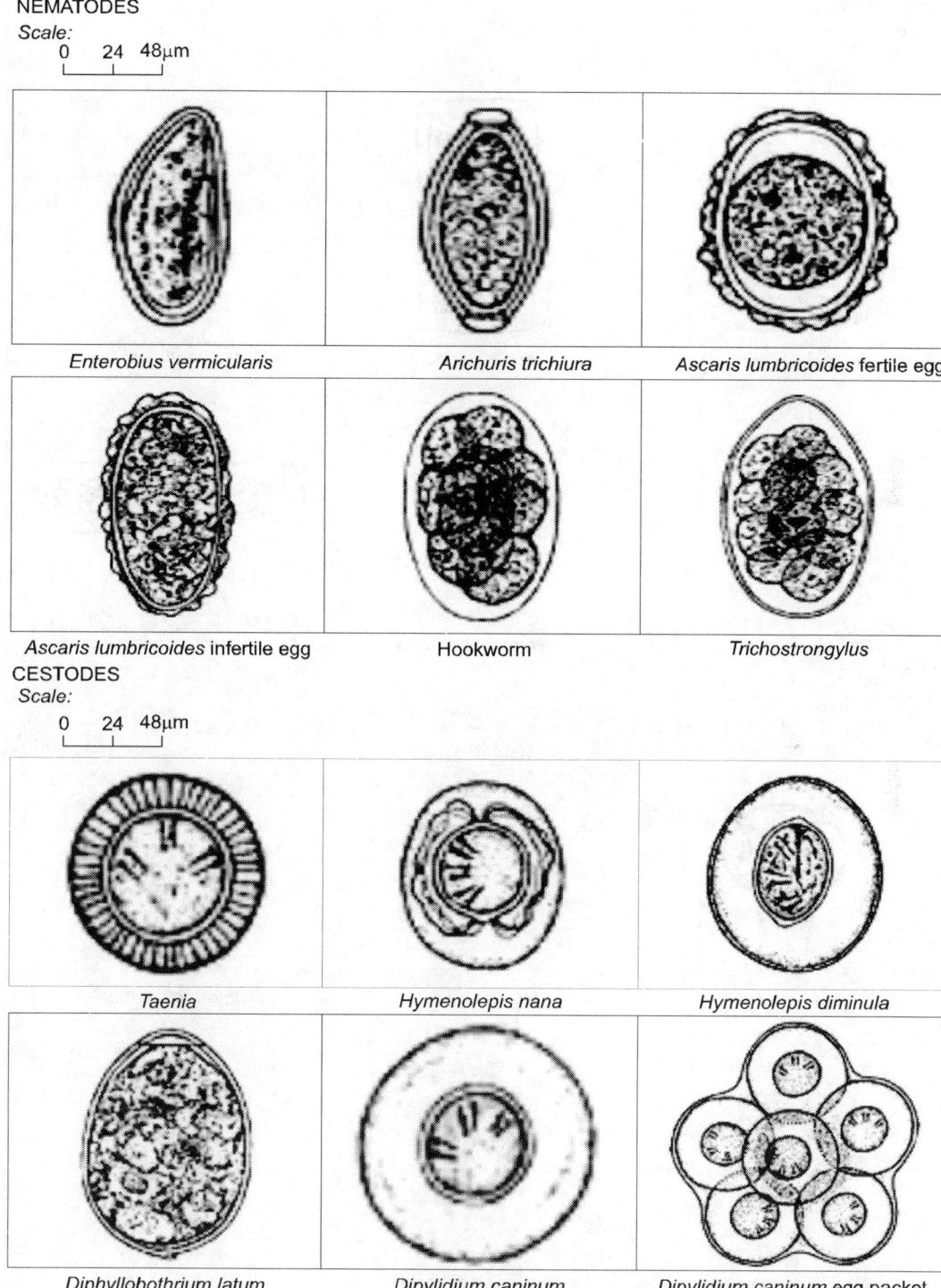

Enterobius vermicularis   Arichuris trichiura   *Ascaris lumbricoides* fertile egg

Ascaris lumbricoides infertile egg   Hookworm   Trichostrongylus

CESTODES
Scale:

0   24   48µm

Taenia   Hymenolepis nana   Hymenolepis diminula

Diphyllobothrium latum   Dipylidium caninum   Dipylidium caninum egg packet

Plate 12 - Intestinal helminthes eggs

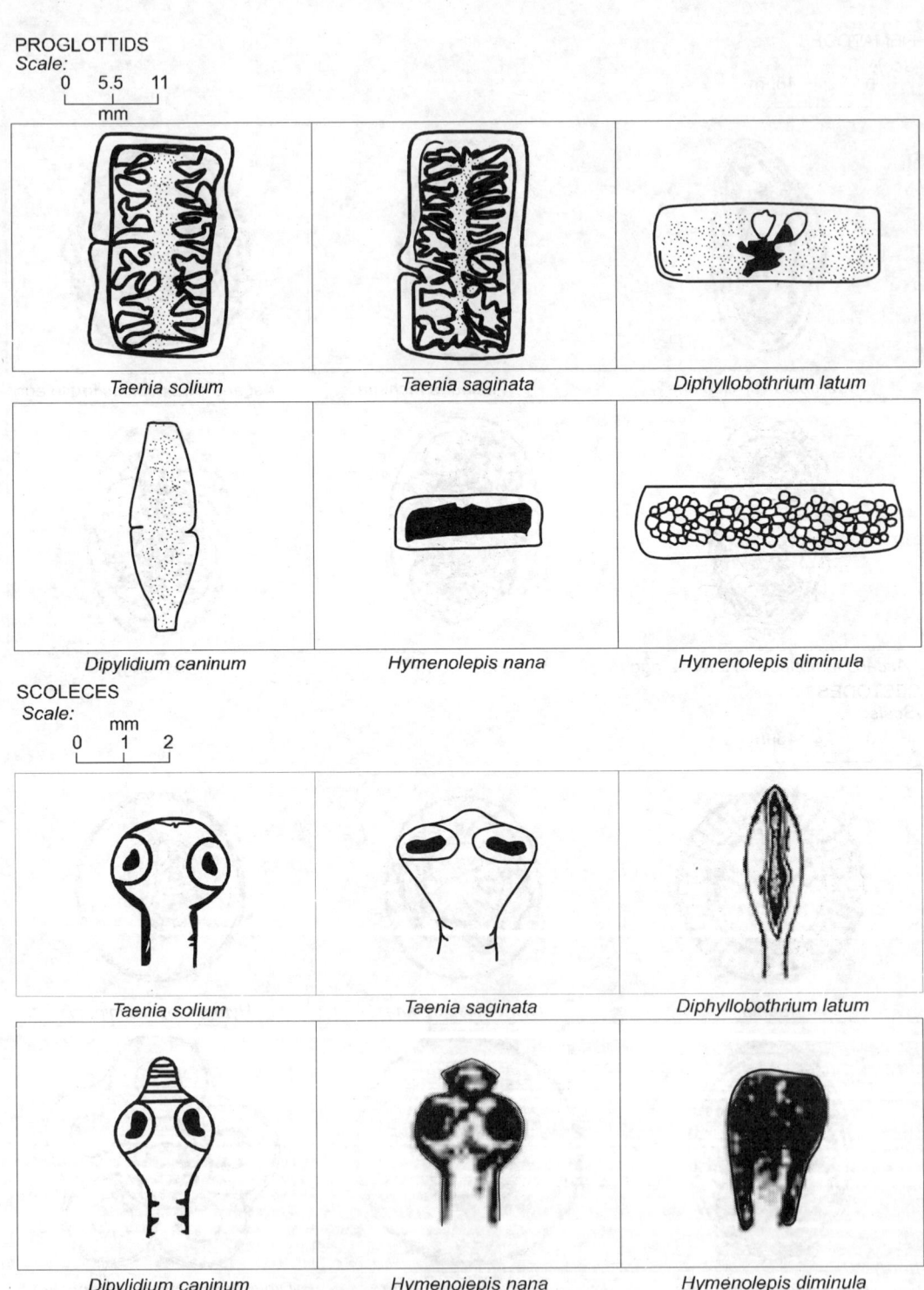

PROGLOTTIDS
Scale:
0   5.5   11
mm

Taenia solium

Taenia saginata

Diphyllobothrium latum

Dipylidium caninum

Hymenolepis nana

Hymenolepis diminula

SCOLECES
Scale:
mm
0   1   2

Taenia solium

Taenia saginata

Diphyllobothrium latum

Dipylidium caninum

Hymenolepis nana

Hymenolepis diminula

Plate 13 - Segments of helminthes present in human stool

E. IDENTIFICATION OF BLOOD AND TISSUE PARASITES

Introduction

Different types of protozoans and lesser number of helminthes played a vital role in blood and tissue infections. *Plasmodium sp, Tryphanosoma sp,Leishmania sp, T. vaginalis, W. bancrofti, Loa loa, Schistosoma, Paragonimus, Enterobius vermicularis, Taenia sp*, are some of the parasites usually detected using blood, urine, sputum, CSF, bone marrow, spleen, skin tissue, deodinal aspirate, corneal scrapes samples. Thick and thin blood smears from peripheral blood is used to screen blood parasites. Microscopic methods are the best methods to screen. Stained blood films are the most reliable and efficient means for definitive diagnosis.

Blood parasite can be detected by means of giemsa and wrights staining. It is used to differentiate nuclear or cytoplasmic morphology of platelets, RBCs, WBCs and parasites. Blood film should be prepared to observe clear morphology of blood parasites. There are two types of blood films , they are thin blood film and thick blood film. Stained blood films are the most reliable and efficient means for definitive diagnosis of nearly all blood parasites. For some parasites it is recommended to prepare a thin film in one slide, a thick film on other slide and combination on third slide.

Thin Blood Film Preparation

This is identical to a differential blood cell count film. It provides a good area for examining the morphology of parasites and RBCs and is used to confirm identity of parasites. This is less sensitive than thick film.

Thick Blood Film Preparation

The thick film essentially condenses into an area suitable for examination above 20 times more blood than thin film. The RBCs are lysed during the staining process so that only parasites, platelets and WBCs remain visible. It is used to differentiate a low parasitemia.

Giemsa Stain

In 1891, Giemsa stain was discovered by Ramonowsky the Russian protozoologist. Giemsa stain is also called Ramonowsky stain. Until 1960s, Giemsa stain is used to describe chromosomes of cells. Giemsa stain is used to differentiate nuclear & cytoplasmic morphology of platelets (pink) RBC(pink), nuclei and granules of PMN (purple), parasite (blue), flagella of parasites (red). Giemsa stain is most useful to stain thick blood film. Giemsa stain must be diluted with buffered water ph 6.5 or 7 or 7.2 before using.

Report any parasite, including the stage(s) seen (do not use abbreviations). Examples: *Plasmodium falciparum* rings and gametocytes, rings only. *Plasmodium vivax* rings, trophozoites, schizonts, and gametocytes. *Wuchereria bancrofti* microfilariae. *Trypanosoma brucei gambiense/ rhodesiense* trypomastigotes. *Trypanosoma cruzi* trypomastigotes. *Leishmania donovani* amastigotes

Wrights stain

Wright stain is the combination of acid dye (Eosin) and a basic dye (Methylene blue) for use of staining the blood smear. It highlights the differences among different types blood leukocytes for easier recognition of eosinophils and basophils.Wrights stain can be used to stain thin blood films. Wrights stain also called ramanovsky dye.

Leishman's Stain

It is also called Romanowskys stain. Leishman stain is a combination of eosin and methylene blue. It also contain oxidation product of methylene blue called azures. These azures provide further contrast in smears of peripheral blood.

F.   IDENTIFICATION OF VIRUSES FROM HUMAN

Diagnosis of human virus infections is broadly classified into clinical diagnosis and laboratory diagnosis. Clinical diagnosis is performed by the physician but laboratory diagnosis is performed by the medical microbiologists those who are specialists in virology. In this chapter we will describe about human viral infection diagnosis process

Viral laboratory diagnosis is broadly classified into 4 sections. They are

- ⋏   Direct examination
- ⋏   Indirect examination
- ⋏   Serology and
- ⋏   Molecular diagnosis

Direct Examination

Clinical specimens are directly examined for the presence of virus particles, viral antigens or viral nucleic acid.

Methods for direct examination:

- ⋏   Electron Microscopy
- ⋏   Light Microscopy for inclusion bodies
- ⋏   Antigen detection by ImmunoFluorescence
- ⋏   Immunoelectron microscopy

Direct examination methods are often called as Rapid Diagnostic Method because they can usually give results with in the same or next day.

Indirect Examination

In this method viruses are cultured and their antigens are detected. Eg: Cell Culture.

Serology

Detetion of rising titers of antibody between acute and convolescent stages of infection or the detection of IgM in primary infection is called Serology. Serum is used in the serology.

Classical techniques of Serology are

- ⅄ Complement fixation test
- ⅄ Hemagglutination inhibition test
- ⅄ Immunofluorescence technique.
- ⅄ Neutralization technique
- ⅄ Single radial hemolysis.

New techniques are,

- ⅄ Radioimmunoassay
- ⅄ ELISA
- ⅄ Particle hemagglutination
- ⅄ Western blot
- ⅄ Recombinant immuno blot assay

Viral Genome Detection or Molecular Methods

Molecular methods like PCR & Nucleic acid based amplification are used for the detection of viral genome.

- ⅄ The Classical techniques includes Dot blot and Southern blot, Northern blot and Insitu hybridization.
- ⅄ New molecular techniques are Polymerase Chain Reaction, Ligase chain reaction, Nucleic acid based Amplification and Branched DNA.

Direct Examination

a. Antigen Detection

Immunofluorescence is one of the methods adapted to detect antigens from clinical specimens

Eg. Influenza A, B & Adenovirus detection from respiratory specimens and Detection of Rotavirus antigen in faces. It is widely used for the rapid diagnosis of viral antigens as well as the detection of viral specific antibodies. This test makes use of a fluorescence labelled antibody to stain specimens containing specific viral antigens, so that the stained cells fluoresce under UV illumination.

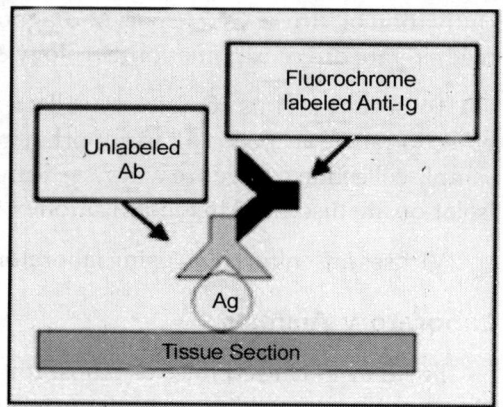

Figure 75 Immunofluoresence test

There are two types of Immuno Fluoresence; they are Direct Immuno Fluoresence and Indirect Immuno Fluoresence . Direct Immuno Fluoresence detects antigens where as indirect Immuno Fluoresence test detects antibodies. Fluorescein , Rhodamine and Phycoerythrine are the dyes used for labelling antibodies.

b. Electron microscopy

Viral particles are detected and identified on the basis of morphology. A magnification of around 50,000 is normally used.

Eg. Detection of Rotavirus, Calicivirus from feces.

Electron microscopes use a beam of highly energetic electrons to examine objects on a very fine scale.

c. Light Microscopy

Replicating virus often produce histological changes in infected cells. These changes may be characteristic or non-specific. Virus inclusion bodies are basically a collection of replicating virus particles either in nucleus or cytoplasm.

(e.g.) Negri bodies examination in Rabies virus infection.Cow dry type A inclusion bodies of Herpes virus.

Indirect Examination

Viruses are cultured and their antigens are detected by using cell cultures, embryonated eggs or laboratory animals. Viruses are obligatory intracellular parasite, which multiplies only inside of the host cell. Host for viruses may be animals, plants, human, bacteria, fungus, protozoan and algae. Hence virologists use any one of the host systems for viral cultivation.

Cultivation of Animal Viruses

Viruses are considered as the primary infection causing agents in human. Viruses cause varieties of human infections. It causes mild common cold infection to life threatening AIDS infections to human beings. Better treatment of infections needs proper diagnosis. Isolation or cultivation of viruses is one part of diagnosis. It is also called **indirect examination**. Other methods are direct examination, Serology & Molecular diagnosis

To isolate the virus, samples are collected from the infected sites in a proper way. Specimen collection and transport play an important role in viral diagnosis. Based on the site of infection sample collection procedure will vary. Sample collection procedures for various human viral isolation are discussed in identification of human virus section

Viruses are cultured by using laboratory animals, embryonated eggs and cell cultures.

Laboratory Animals

Furth in 1932 used mice as a host for viral cultivation. Here after animals like Rabbit, Guinea pig, rat, Suckling mice, hamsters and monkeys are used for the viral cultivation. Inoculation site may vary depends up on the type of virus and its target site of infection.

Eg. Intra cerebral inoculation is performed on mice to cultivate Arboviruses.

After inoculation observe morphological and physiological changes of animals.

Advantages

All types of animal viruses are cultivated. Used to understand pathogenesis. Used to understand the immune response of the animal. Used to study the efficiency of vaccines. Used to develop vaccines. Used to study interactions of drugs

Figure 76 Laboratory mouse

Disadvantages

Expensive. Difficult to handle. Show biological diversity. May leads to latent infections. Require efficient maintanence. Objections from blue cross members.

Embryonated Egg

Woodruff and Good Pasteur (1931) used fertilized chicken egg for viral cultivation. This is a simpler technique than animal inoculation, are inexpensive and easily available. Eggs usually not interfere with virus multiplication due to the absence of immune response. Suitable cells for the growth of viruses are available in embryo and its membrane, which may facilitate the growth of viruses. 8-11 day old chick embryo is used for cultivation. Incubate for 2-9 days after inoculation. Duration of incubation is also depends on the type of virus and the route of inoculation.

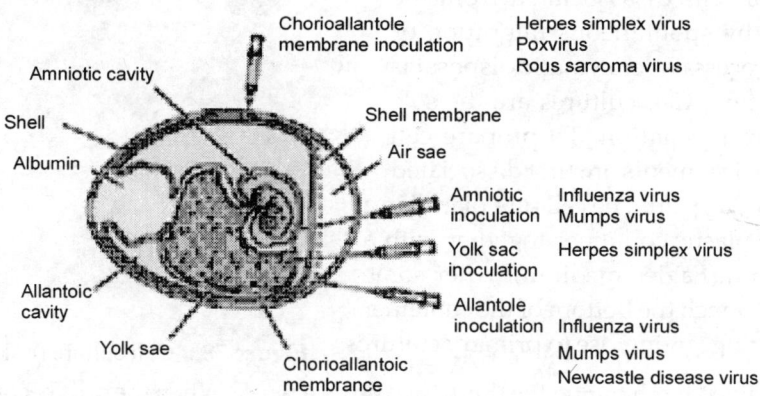

Figure 77 Embryonated egg

Viruses are inoculated various sites of embryonated egg (Figure 78) . To prepare the egg for viral cultivation, the shell surface is disinfected with Iodine. Make hole with the help of drill. Inoculate viruses in any one of the method mentioned in figure. After inoculation the hole is sealed with gelatin and the egg incubated.

Viruses reproduce only certain parts of the egg. Table shows the site of respecific virus for cultivation.

Site of inoculation	Virus
Chorio Allontoic Membrane inoculation (CAM)	Herpes simplex virusPox virusRous sarcoma virus
Amniotic inoculation	Influenza virusMumps virus.
Allontoic inoculation	Influenza virusMumps virus.New castle disease virus
Yolk sac inoculation	Herpes simplex virus

Viral suspension or virus containing tissue is injected into the fluid of the egg. Virus is injected into the proper location in the egg (viruses are able to reproduce only in certain parts of the embryo).   Death of the embryo or formation of typical pocks (Pock assay), Heamagglutination or lesions on the membranes of the egg that result from viral growth. After growth viruses are identified with the help of serological techniques. Eg Pox viruses produce Pock or lesions. Influenza virus produce haemagglutination

Cell Culture /Tissue Culture

Now a days cell cultures are mainly employed in cultivation/ isolation of viruses. Cell cultures are classified into different types based on their performances.

Cell Culture

Growth of cells disassociated from the parent tissue by spontaneous migration or mechanical processs or enzymatic disposal is called cell culture. Cell cultures are the sole system for virus isolation. To prepare cell cultures, tissue fragments are first dissociated with the help of trypsin. The cells are placed in a flat bottomed plastic container together with a suitable medium(Eagles medium). After some time, cells will attach the bottom of the container and start dividing, giving rise to primary cultures.

Figure 78   Cell culture flask with media

It is maintained by changing the fluid 2 or 3 times a week. When the cells become crowded, the cells are detached from the vessel wall by trypsin, and portions are used to initiate secondary cultures. Cells tend to adhere to the glass or plastic container and reproduce to form a monolayer.

Media Used For Mammalian Cell Culture

The basic component of cell culture medium is a **Balanced Salt Solution**(BSS). It provides essential inorganic ions, correct osmolarity, and correct pH. Two common BSS are **Hanks and Eagles Balanced Salt Solutions**. Some of the additional nutrients also required for the proliferation of cells. They are 13 essential amino acids, eight vitamins, antibiotics to prevent bacterial contamination, glucose, phenol red and 5 % Fetal Calf Serum. BSS contain all the necessary nutrients for cell growth. Eg. Eagles medium.

Types of Cell Cultures

Primary cells Prepared directly from animals and can be subcultured only once or twice. These are widely used as the best culture system because they support the widest range of viruses. E.g. Chick embryo fibroblast cells, Monkey Kidney cells. Once the primary culture is subcultured it becomes known as **cell lines**. A characterized cell line derived by selection or cloning are called **cell strains.**

Semi-continuous cells Derived from human fatal tissue and can be subcultured 20-50 times. E.g. Human embryonic kidney and skin fibroblasts.

Continuous cells Derived from tumours of human. E.g. HeLa (derived from Langerhan's of Cancer patient Henritta) and Vero cell lines.

The following viruses may be isolated by making use of cell culture technique

Readily isolated viruses	Less frequently isolated viruses
Herpes simplex virusCytomegalovirus Adenovirus Poliovirus Coxsackie B virus Echovirus Influenza virus Parainfluenza virus Mumps virus Respiratory Syncytial virus	Varicella – Zoster virusMeasles Rubella Rhinovirus Coxsackie A virus

Vero, Hep2, Human diploid cells (HEK&HEL), human amnion, human diploid fibroblasts, MK, BGM, LLC-MK2, Rhadomyosarcoma, MDCK, RK13 cell lines are used for the cultivation of viruses.

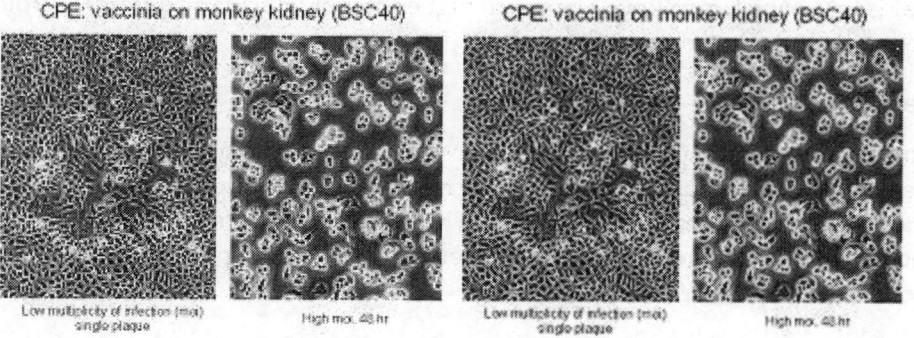

Figure 79 Cell culture cytopathic effect

Cultured viruses are identified by Cytopathic effect, Haemadsorption, Interference and ImmunoFluorescence tests.

Cytopathic effect

Viruses cause morphological changes in the cell culture and are observed under microscopic examination. Morphological changes are called cytopathic effect.

Hemadsorption

If hemagglutinating viruses are multiplying in the cell culture, the erythrocytes will adsorb onto the surface of cells. This is known as hemadsorption.

Interference

The growth of first virus will inhibit second virus infection is called interference.

Hemagglutination

It can be used to detect a virus carrying hemagglutinin in its envelope. The virus is mixed with RBCs, the virus cross-link the RBC and cause agglutination.

Advantages

Exact glycosylation. Can be transformed to continuous cell lines. Some (e.g., CHO cells) grow in suspension culture. Some carcinoma-derived cells (e.g., HeLa cells) have rapid growth rates. Extracellular expression.

Disadvantages

Some grow as adherent layer (anchorage dependent). Fastidious growth requirements. Very slow doubling time (15-25h). Very susceptible to contamination. Very sensitive to shear stress.

Serological Examination

Serology forms the mainstay of viral diagnosis. During viral exposure, the first antibody to appear is IgM, which is followed by much higher titer of IgG. Many different serological tests are available to diagnose viral infections.

Neutralization test

Neutralization of a virus is defined as the loss of infectivity through reaction of the virus with specific antibody. Virus and the serum are mixed under appropriate condition and then inoculated into cell culture, eggs or animals. The presence of un-neutralized virus may be detected by reactions such as cytopathic effect, haemadsorption, interference and immunofluorescence tests.

Complement Fixation Test

Complement fixation test (CFT) was first introduced by Wasserman in 1909 for the diagnosis of Syphilis. Now this technique is also used for the diagnosis of viral infections. CFT is convenient and rapid to perform, demand for equipment and reagent is small and varieties of test antigens are readily available.

It is a simple test and has two antigens –antibody reactions, one of which is the indicator system.2

Two test tubes are taken. Add antigen to both tubes. Test serum is added to one tube. If antibody present Ag-Ab complex will form. When complement is added, if complement is present it fix complement and consume it. Indicator cells and antierythrocytic antibodies are added. If complement is present the indicator cell will lyses (negative result). If the complement is consumed, no lysis of cells (positive result).

First reaction taken place between a known virus and a specific antibody in the presence of complement. Complement is fixed by antigen antibody complex

Second antigen –antibody reaction consists of reacting sheep RBC with haemolysin. When this indicator system is added to the reactants, the sensitized RBCs will only lyse in the presence of free complement.

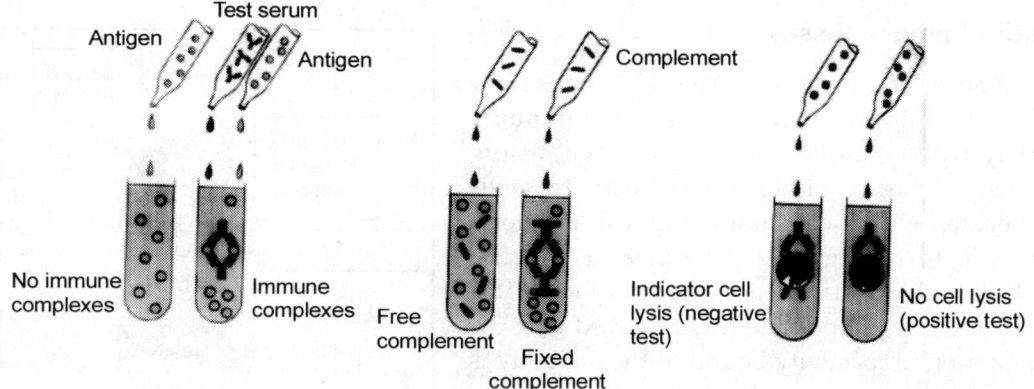

Figure 80 CF test

Hemagglutinin Inhibition Test (HAI)

Some viruses have the ability to agglutinate the erythrocytes of mammalian or avian species. Eg. Influenz virus,Para influenza virus, Adenovirus, Rubella virus, Alphaviruses, Bunya viruses, Flaviviruses, and some strains of Picorna viruses. Antibodies against hemagglutinin prevent haemagglutination, which is the main principle of this test. HAI test is simple to perform and requires inexpensive equipments and reagents.

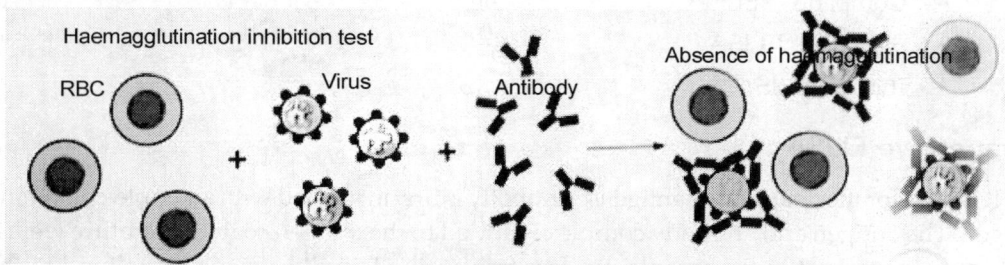

Figure 81 HAI test

Procedure

Dilute the patients sera from 1:8 to 1:1024 with the help of Bovine Albumin Vernol Buffer (BAVB). Non-specific inhibitors of viral Hemagglutination is removed by Kaolin, potassium periodate or by heat treatment. Add the serum to the fixed dose of viral agglutinin containing microtitre plate. Add agglutinable erythrocytes and incubate. Control erythrocytes only shows button at the bottom of the well. Absence of agglutination indicates infection.

Single Radial Hemolysis

It is routinely used for the detection of Rubella with its specific IgG and also for the diagnosis of Mumps. Test sera are placed in wells on a plate containing Rubella antigen-coated RBC and complement. The presence of Rubella specific IgG is detected by the lysis of rubella antigen –coated RBC. The zone of lysis around the well is depend on the level of specific antibody present

Radio Immuno Assay

One of the most sensitive techniques for detecting antigen or antibody is Radio Immuno Assay (RIA). S.A.Berson and Rosalyn Yalow developed this technique in 1960. Microtitre wells are coated with a constant amount of antibody specific for antigen. A serum and specific radiolabelled antigen are then added. After incubation, the supernatant is removed and the amount of radioactivity bound to the antibody is determined. If the sample is infected, the amount of label bound will be less than in control with un infected serum.

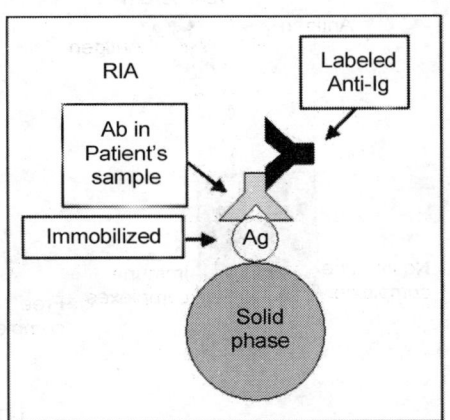

Figure 82 RIA test

Gamma emitting isotope ^{125}I and beta emitting isotope ^{3}H are commonly used for radio immuno assay.

Enzyme Linked Immuno Sorbent Assay (ELISA)

Engval and Perlmann developed it in 1970. There are three important types of ELISA

- Competitive ELISA
- Sandwich ELISA
- Indirect ELISA

Competitive ELISA

It is used for the detection of antigens. Antibody is first incubated with a sample-containing antigen. The antigen and antibody complex is added to the antigen coated microtitre well. If more antigen present in the sample, the less free antibody will be available to bind to the antigen coated well. Addition of an enzyme conjugated secondary antibody specific to the primary antibody can be used to determine the amount of primary antibody bound to the well. It is a quantitative test for the Antigen detection.

Sandwich ELISA

Antigen can be detected by this method. Antibody is immobilized on a microtitre plate. A sample-containing antigen is added and allowed to react with antibody. After washing enzyme linked second antibody is added and allowed to react with bound antigen. Substrate is added to measure colour reaction.

Indirect ELISA

Antibody can be detected with an indirect ELISA. Serum containing antibody can be detected by adding antigen coated microtitre well and allowed to react with antigen. After washing the non bound primary antibody enzyme conjugated secondary antibody is added, which binds to the primary antibody. Free secondary antibodies are washed and a substrate for the enzyme is added. The amount of colour is directly proportional to the quantity of antibody present in the serum.

Southern Blotting

This method has been used to distinguish between closely related viruses. Eg. HSV1 and HSV2, HPV6 and HPV16. There are two methods are applied, they are electrophoretic separation of DNA and hybridization. The sample can be PCR product or restriction enzyme digest. (Details Refer Chapter 16).

Northern Blotting

This is similar to southern blotting but the difference is that the sample is RNA. (Details Refer Chapter 16).

Polymerase Chain Reaction (PCR)

PCR is an extremely powerful research tool, which has become the part of the routine clinical laboratory tests, especially in the field of Microbiology, Virology and diagnostic genetics. PCR is the technique used to detect the presence of a specific segment or sequence of Nucleic acid, which could be diagnostic for infectious agents such as bacteria, virus and fungus. This technique is based on nucleic acid amplifiDenaturation of DNA duplex at high temperature. (Details Refer Chapter 16).

Western blotting

It involves the primary separation of viral proteins by SDS PAGE (Sodium Dodecylsulphate – Polyacrylamide Gel Electrophoresis). The second phase involves adsorption of protein from gel into nitrocellulose membrane. Viral antigens are identified through antigen and antibody reactions (Immuno blotting). This test is used for the confirmation of ELISA. Figure shows various proteins of HIV viruses detected based on MOLECULAR weights in western blotting technique.(gp120, gp41 etc.)

7 CONTROL OF MICROORGANISMS

Most of the microorganisms are useful to human. Some microbial activities may have undesirable consequences, such as food spoilage and disease. Therefore it is essential to be able to kill a wide variety of microorganisms or inhibit their growth to minimize their destructive effects. The goal is two fold: (1) to destroy pathogens and prevent their transmission, and (2) to reduce or eliminate microorganisms responsible for the contamination of water, food, and other substances.

A. DEFINITIONS AND TERMS RELATED TO THE CONTROL OF MICROORGANISMS

Sterilization is the destruction or removal of all viable organisms from an object or environment. **Disinfection** means killing, inhibition, or removal of pathogenic microorganisms mainly from inanimate objects. A **disinfectant** is a chemical capable of killing microbial cells. **Antisepsis** is the prevention of microbial infection in living tissue. **Asepsis** refers to those practices that keep microorganism from entering the sterile tissues. The application of these practices is referred to as aseptic technique. **Antiseptics** are those chemicals that can be applied to tissue surfaces to kill or inhibit the growth of microorganisms. **Sanitization** means reduction of microbial populations to a safe level in accord with public health standards. **Microbicidal** agents are chemicals that will kill or destroy microorganisms. **Fungicidal** agents which are designed to kill fungi; **bactericidal** agents which are designed to kill bacteria; **sporicidal** agents which are designed to destroy endospores; **viricidal** agents which are designed to destroy viruses. **Microbiostasis** refers to the inhibition of growth of microorganisms. This does not mean that the organisms are killed simply that they are unable to grow. Refrigeration and many antimicrobial drugs exert a microbistatic effect. **Bacteriostatic** agents are chemicals that inhibit the growth of bacteria. **Fungistatic** agents are chemicals that inhibit growth of fungi.

B. FACTORS AFFECTING GROWTH
The Use of Physical Methods in Control

The physical method of sterilization includes high temperature, low temperature, dessication, radiation, filtration and osmotic pressure etc.

Heat

Heating is one of the most popular way to destroy microorganisms. Either moist or dry heat may be applied. Moist heat readily kills viruses, bacteria, and fungi. Exposure to boiling water for10 minutes is sufficient to destroy vegetative cells and eucaryotic spores. Boiling water (100°C) is not high enough to destroy bacterial endospores. Therefore boiling can be used for disinfection of drinking water and objects not harmed by heat, but boiling does not sterilize. Effectiveness of heat was expressed in terms of thermal death point (TDP), the lowest temperature at which a microbial suspension is killed in 10 minutes. Thermal death time (TDT) is now more commonly used. This is the shortest time needed to kill all organisms in a microbial suspension at a specific temperature and under defined conditions.

Moist Heat Sterilization

Moist heat destroys microorganisms and its spores very effectively than dry heat. Autoclaving, tyndallization and pasteurization are the examples for moist heat sterilization. Moist heat sterilization was first performed by Robert koch.

Autoclaving

Water is boiled to produce steam, which is released through the jacket and into the autoclave's chamber. The air initially present in the chamber is forced out until the chamber is filled with saturated steam and the outlets are closed. Hot, saturated steam continues to enter until the chamber reaches the desired temperature and pressure, usually 121°C and 15 pounds

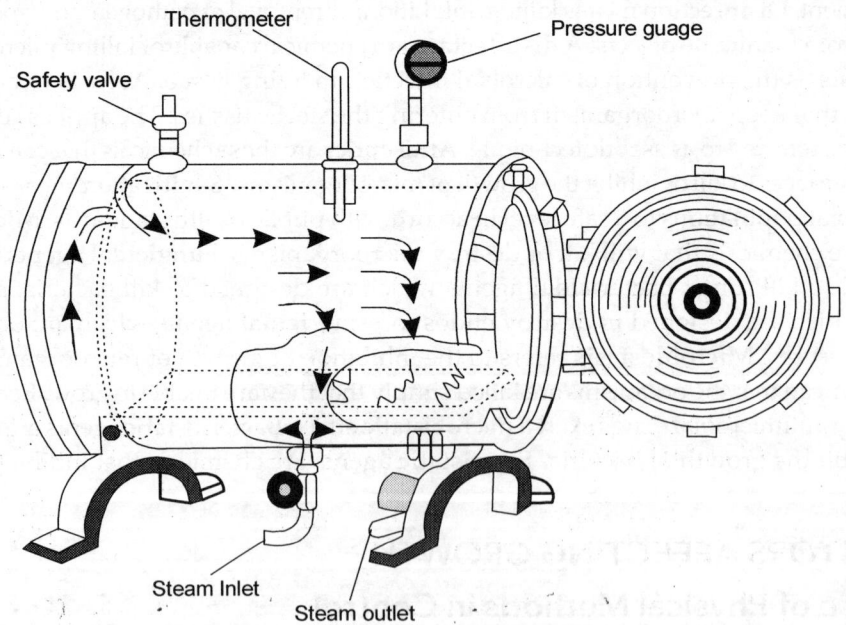

Figure 83 Autoclave

of pressure. At this temperature saturated steam destroys all vegetative cells and endospores in a small volume of liquid within 10 to 12 minutes. Treatment is continued for about 15 minutes to provide a margin of safety. Moist heat is thought to kill so effectively by degrading nucleic acids and by denaturing enzymes and other essential proteins. It also may disrupt cell membranes. The chamber should not be packed too tightly because the steam needs to circulate freely and contact everything in the autoclave. Bacterial endospores will be killed only if they are kept at 121°C for 10 to 12 minutes. Autoclaving is monitered by biological indicator. This indicator commonly consists of a culture tube containing a sterile ampule of medium and a paper strip covered with spores of *Bacillus stearothermophilus*. After autoclaving, the ampule is observed and its indicator strip that changes colour upon sufficient heating is autoclaved with a load of material. If the colour changes after autoclaving, the material is supposed to be sterile. These approaches are convenient and save time but are not as reliable as the use of bacterial endospores.

Pressure Cooker

It works under the principle, moist heat sterilization. It is used in the laboratory instead of autoclave. It is easy to use and require minimum time.

Tyndalization or fractional sterilization

It is used to sterilize media. It consists of steaming on three successive days. On the first day the medium is steamed at 100°C for 30 minutes to kill the vegetative cells. It is then left overnight at room temperature to encourage germination of endospores. The second and third day the medium is steamed for a further 30 minutes to ensure the destruction of the germinated spores.

Figure 84   Pressure Cooker

Pasteurization

Sterilization of heat labile fluids such as milk, beverages may be done by heating for 1 hour at 56°C or for 10 minutes at 65°C - 75°C. This treatment is sufficient to kill mesophilic vegetative bacteria. Vaccines prepared from culture of nonsporing bacteria maybe sterilized at 60°C for 1 hour.

Boiling

Boiling is performed by heating the materials at 100°C. This is a best example for moist heat sterilization. This method is very effective in destroying vegetative forms within 10 minutes where as spores is not destroyed.

Dry Heat Sterilization

Many objects are best sterilized in the absence of water by dry heat sterilization. The items to be sterilized are placed in an oven at 160 to 170°C for 30minutes to 1hour. Microbial death apparently results from the oxidation of cell constituents and denaturation of proteins. Although dry air heat is less effective than moist heat. Dry heat does not corrode glassware and metal instruments as moist heat does, and it can be used to sterilize powders, oils, and similar items. Most laboratories sterilize glass petri dishes and pipettes with dry heat.

Flaming

It is used to destroy all forms of vegetative cells and also spores by slow passage through Bunsen burner. Bunsen burner (named after R.W. Bunsen) is a type of gas burner. It produces non-luminous flame. Bunsen burner and spirit lamp is used to sterilize inoculation loops/needles. It is also used for flaming the mouths of test tubes, media containing flasks. Heating of inoculation wires and loops by holding them almost vertically in Bunsen flame until red hot along their whole length, almost to the tip of the metal holder. Flaming of inoculation loop reaches 1860°C at its hottest point. This process is called red heat.

Incineration

Incineration in which all contaminated materials is burned to ashes. Eg., waste media, dressing materials and solid wastes.

Hot Air Oven

An oven is based on the principle where dry heat or hot air accomplishes sterilization. The sterilization process in an oven is longer than autoclaving. Dry heat removes water from microorganisms while moist heat adds water to them. In addition, moist heat has greater penetrating power than dry heat. It is used for sterilizing glasswares like petriplates, test tubes pipettes, metal instruments, oils, powers, waxes etc... Commonly used temperature for hot air oven sterilization is 20 minutes at 180°C, 40 minutes at 170°C, 60 minutes at 160°C, 150 minutes at 150°C,180 minutes at 140°C,480 minutes at 120°C. An oven consists of an insulated cabinet, which is held at a constant temperature by means of an electric heating mechanism and thermostat. It is fitted with a fan to keep the hot air circulating at a constant temperature.

Figure 85   Hot Air Oven

Filtration

Filtration, the process of passing liquid or gases through a filter that retains many microorganisms. It is used to sterilize solutions, which are thermolabile, and to remove particulate matter from the materials. Vaccum or pressure is required for passing solutions through the filters. Basically there are two types of filters. They are Depth filters and Membrane filters.

Depth Filters

It consists of fibrous or granular material that are pressed, wound, fired or bound into a maze of flow channels. Various types of depth filters are designated on the basis of its basic components. They are describesd as follows.

Seitz Filters

These consist of an asbestos disk inserted into a metal holder, which is connected to a flask. The disks have various degrees of porosity and are known as claryfing, normal and special. The last two donot allow bacteria pass through it.

Sintered Glass Filters

It is similar to seitz filter, but the disks are made of finely ground glass.

Barkefeld Filters or Earthenware Filters

These are made from a fossil diatomaceous earth called kiesselguhr. According to the size of the granules forming the substance of filters, there are three grades of porosity, that are V (coarse), W (fine) and N (intermediate). After use, these filters should be brushed with a stiff mail brush and then boiled in distilled water to regenerate for further use.

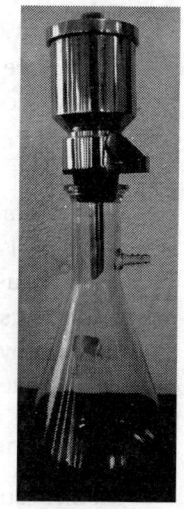

Figure 86 Filtration Unit

Chamberland Filters

These are made from unglazed porcelin and have various degrees of porocity, the finest type of particles pass only through the membrane.

Membrane Filters

The filter membrane are composed of cellulose nitrate or cellulose acetate. They can be obtained in various diameters (13-293mm) and in porocities from 0.015-12 μm.

- ⊿ 0.22 μm retains *Pseudomonas diminuta*
- ⊿ 0.45 μm retains coliform bacilli.
- ⊿ 0.8 μm retains air borne bio particles.
- ⊿ 0.1 μm filters retains viruses.

Filters are disposable materials. The unit can be sterilized by autoclaving. The filtration is performed by adding the fluid to be filtered into the filter holder of the unit above the filter and apply pressure, then fluid can pass through the filter. Generally negative pressure (vaccum) is applied.

Syringe Filters

Membranes of 13mm and 25mm diameter can be fitted in syringe like holders of stainless steel or polycarbonate. This is used to sterilize small volume of heat labile fluid. The fluid is forced through the filters by pressing down the piston.

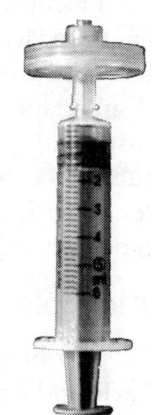

Figure 87 Syringe Filter

Air Filters

Air may be filtered through HEPA (High Efficiency Particle Arrester) filters. HEPA filters are used to decontaminate the air input of the laminar airflow chamber.

Radiation

Ionizing radiations like gamma rays provides an reliable means of sterilizing plastic and other materials that are liable to be damaged by heat. UV rays effective against nonsporing bacteria. It is used in the microbiology laboratory to disinfect the internal surfaces of safety cabinets. It is used in operation theaters and sterile room in some factories. Infra red radiations may be employed to sterilize metal instruments and glass syringes in hospitals. The infra red rays are directed from electrically heated elements onto the objects to be sterilized.

Gaseous Chemicals

Medical surgical articles that cannot withstand even heating at 73°C can be treated with ethylene oxide. This lethal gas is an alkylating agent and kill microorganism including viruses. Ethylene oxide is toxic and highly explosive

The Use of Chemical Agents in Control

A variety of non-volatile chemicals are generally used in the laboratory to sterilize discarded glassware, disk, hand gloves, etc. The primary objectives of the use of such chemicals are to kill potentially dangerous microorganisms present on such articles, and also to reduce the laboratory atmosphere from fungal spores. These are wide varieties of disinfectants; some important ones are as follows:

Phenolics

Phenol was the first widely used antiseptic and disinfectant. In 1867 Joseph Lister employed it to reduce the risk of infection during operations. Phenol such as cresols, xylenols, and orthophenylphenol are used as disinfectants in laboratories and hospitals. The commercial disinfectant Lysol is made of a mixture of phenolics. Phenolics act by denaturing proteins and disrupting cell membranes. They have some real advantages as disinfectants: phenolics are tuberculocidal, effective in the presence of organic material, and remain active on surfaces long after application. Hexachlorophene has been one of the most popular antiseptics because it persists on the skin once applied and reduces skin bacteria for long periods.

Alcohols

Alcohols are among the most widely used disinfectants and antiseptics. They are bactericidal and fungicidal but not sporicidal; it destroy lipid-containing viruses. The two most popular alcohol germicides are ethanol and isopropanol. They act by denaturing proteins andpossibly by dissolving membrane lipids. A 10 to 15 minute soaking is sufficient to disinfect thermometers and small instruments.

Halogens

Elements like fluorine, chlorine, bromine, iodine, and astatine are called as halogens. They exist as diatomic molecules in the free state and form salt like compounds with sodium and most other metals. The halogens iodine and chlorine are important antimicrobial agents. Iodine is used as a skin antiseptic and kills by oxidizing cell constituents and iodinating cell proteins. At higher concentrations, it may even kill some spores. Iodine often has been applied as tincture of iodine. Chlorine is the usual disinfectant for municipal water supplies and swimming pools and is also employed in the dairy and food industries. It may be applied as chlorine gas, sodium hypochlorite, or calcium hypochlorite, all of which yield hypochlorous acid (HClO) and then atomic oxygen. The result is oxidation of cellular materials and destruction of vegetative bacteria and fungi, although not spores.

Table 7.1 Mode of action and uses of chemical compound

Chemical	Action	Uses
Ethanol (50-70%)	Denatures proteins and solubilizes lipids	Antiseptic used on skin
Isopropanol (50-70%)	Denatures proteins and solubilizes lipids	Antiseptic used on skin
Formaldehyde (8%)	Reacts with NH_2, SH and COOH groups	Disinfectant, kills endospores
Tincture of Iodine (2% I_2 in 70% alcohol)	Inactivates proteins	Antiseptic used on skin Disinfection of drinking water
Chlorine (Cl_2) gas	Forms hypochlorous acid (HClO), a strong oxidizing agent	Disinfect drinking water; general disinfectant
Silver nitrate ($AgNO_3$)	Precipitates proteins	General antiseptic and used in the eyes of newborns
Mercuric chloride	Inactivates proteins by reacting with sulfide groups	Disinfectant, although occasionally used as an antiseptic on skin
Detergents (e.g. quaternary ammonium compounds)	Disrupts cell membranes	Skin antiseptics and disinfectants
Phenolic compounds (e.g. carbolic acid, lysol, hexylresorcinol, hexachlorophene)	Denature proteins and disrupt cell membranes	Antiseptics at low concentrations; disinfectants at high concentrations
Ethylene oxide gas	Alkylating agent	Disinfectant used to sterilize heat-sensitive objects such as rubber and plastics
Ozone	Generates lethal oxygen radicals	Purification of water, sewage

Heavy Metals

For many years the ions of heavy metals such as mercury, silver, arsenic, zinc, and copper were used as germicides. More recently these have been superseded by other less toxic and more effective germicides. 1% solution of silver nitrate is often added to the eyes of infants to prevent ophthalmic gonorrhoea. Silver sulfadiazine is used on burns. Copper sulfate is an effective algicide in lakes and swimming pools. Heavy metals combine with proteins, often with their sulfhydrylgroups, and inactivate them. They may also precipitate cell proteins.

Quaternary Ammonium Compounds

Detergents organic molecules that serve as wetting agents and emulsifiers because they have both polar hydrophilic and nonpolar hydrophobic ends. Quaternary ammonium compounds characterized by a positively charged quaternary nitrogen and a long hydrophobic aliphatic chain. They disrupt microbial membranes and may also denature proteins. Cationic detergents like benzalkonium chloride and cetylpyridiniumchloride kill most bacteria. Cationic detergents are often used as disinfectants for food utensils and small instruments and as skin antiseptics. Several brands are on themarket. Zephiran contains benzalkonium chloride and Ceepryn, cetylpyridinium chloride.

Aldehydes

Formaldehyde and glutaraldehyde are the common aldehyde. They are highly reactive molecules that combine with nucleicacids and proteins and inactivate them. They are sporicidal and can be used as chemical sterilants. Formaldehyde is usually dissolved in water or alcohol before use. 2% buffered solution of glutaraldehyde is an effective disinfectant.

Sterilizing Gases

Many heat-sensitive items such as disposable plastic petri dishes and syringes, heart-lung machine components, sutures, and catheters are now sterilized with ethylene oxide gas. Ethylene oxide (EtO) is both microbicidal and sporicidal and kills by combining with cell proteins. It is a particularly effective sterilizing agent because it rapidly penetrates packing materials, even plastic wraps.

Betapropiolactone (BPL) is occasionally employed as a sterilizing gas. In the liquid form it has been used to sterilize vaccines and sera. BPL decomposes to an inactive form after several hours and is therefore not as difficult to eliminate as EtO. It also destroys microorganisms more readily than ethylene oxide but doesnot penetrate materials well and may be carcinogenic. Recently vapor-phase hydrogen peroxide has been used to decontaminate biological safety cabinets.

C. ANTIMICROBIAL CHEMOTHERAPY

Paul Ehrlich is called father of Chemotherapy. He discovered p-rosaniline, which has antitrypanosomal effects. Ehrlich described the term magic bullet, which means chemicals that were selectively toxic for parasites but not toxic to humans.

Biochemical Action of Antimicrobial Agents

Antimicrobial agents interfere with specific processes that are essential for growth and/or division. Based on the mode of action antimicrobials are classified into

Inhibitors of bacterial and fungal cell walls,

Inhibitors of cytoplasmic membranes,

Inhibitors of nucleic acid synthesis, and

Inhibitors of ribosome function.

Antimicrobial agents may be either bactericidal i.e., killing the target bacterium or bacteriostatic, i.e., inhibiting its growth.

Inhibition of bacterial cell wall synthesis

Gram-positive and gram negative bacterial cell wall contains peptidoglycan. This layer is essential for the survival of bacteria in hypotonic environments; loss or damage of this layer destroys the rigidity of the bacterial cell wall, resulting in death. The critical attack site of cell-wall inhibiting agents is the peptidoglycan layer. Penicillins and cephalosporins/cephamycins are widely used to inhibit both Gram-positive and Gram-negative bacilli. Vancomycin interrupts cell wall synthesis.

Antibiotics that affect the function of cytoplasmic membranes

Bacterial Membranes

Biological membranes are composed basically of lipid, protein, and lipoprotein which act as a diffusion barrier for water, ions, nutrients, and transport systems. Antimicrobial agents can cause disorganization of the membrane. The best-known compounds are polymyxin B and polymyxin E. Polymyxins disorganize membrane permeability so that nucleic acids and cations leak out and the cell dies.

Fungal Membranes

The polyene antibiotics act on sterol membrane by binding to rigid hydrophobic center and a flexible hydrophilic section. Polyene interact with fungal cells to produce a membrane-polyene complex that alters the membrane permeability, resulting in internal acidification of the fungus with exchange of K+ and sugars; loss of phosphate esters, organic acids, nucleotides; and eventual leakage of cell protein. Amphotericin B is used systemically. Nystatin is used as a topical agent and Primaricin as an ophthalmic preparation.

Other agents interfere with the synthesis of fungal lipid membranes are referred to as miconazole, ketoconazole, clotrimazole and fluconazole.

Antibiotics that Inhibit Nucleic Acid Synthesis

Antimicrobial agents can interfere with nucleic acid synthesis at several different levels. They can inhibit nucleotide synthesis and they can interfere with the polymerases involved in the replication and transcription of DNA.

Interference with Nucleotide Synthesis

A large number of agents interfere with purine and pyrimidine synthesis. Flucytosine (5-fluorocytosine) inhibits yeast species. Adenosine arabinoside inhibits viruses. Acyclovir is a nucleoside analog that inhibits herpes viruses. Zidovudine (AZT) inhibits Human Immunodeficiency Virus.

Inhibition of DNA-Directed RNA Polymerase

Rifamycins are a class of antibiotics that inhibit DNA-directed RNA polymerase.

Inhibition of DNA Replication

The newer fluoroquinolones such as ciprofloxacin, norfloxacin, and ofloxacin interact with DNA gyrase and possess a broad spectrum of antimicrobial activity. Metronidazole binds to DNA, and cause DNA breakage.

Antimicrobial Inhibitors of Ribosome Function

A number of antibacterial agents act by inhibiting ribosome function. Aminoglycosides act by binding to specific ribosomal subunits. Chloramphenicol is a bacteriostatic agent that inhibits peptide bond formation. Erythromycin inhibits *Haemophilus, Mycoplasma, Chlamydia,* and *Legionella* by peptidyl transferase reaction

Antibacterial Agents that Affect Mycobacteria

Isoniazid is a mycocidal nicotinamide derivative that inhibits *Mycobacterium*. It affects the synthesis of lipids. Ethambutol is mycostatic.

Table 7.2   Some antimicrobials and its mode of action

Examples	Biological source	Spectrum (effective against)	Mode of action
Penicillin G, Cephalothin	*Penicillium notatum* and *Cephalosporium* species	Gram-positive bacteria	Inhibits steps in cell wall (peptidoglycan) synthesis and murein assembly
Ampicillin, Amoxycillin		Gram-positive and Gram-negative bacteria	Inhibits steps in cell wall (peptidoglycan) synthesis and murein assembly
Clavamox is clavulanic acid plus amoxycillin	*Streptomyces clavuligerus*	Gram-positive and Gram-negative bacteria	Suicide inhibitor of beta-lactamases
Aztreonam	*Chromobacter violaceum*	Gram-positive and Gram-negative bacteria	Inhibits steps in cell wall (peptidoglycan) synthesis and murein assembly
Imipenem	*Streptomyces cattleya*	Gram-positive and Gram-negative bacteria	Inhibits steps in cell wall (peptidoglycan) synthesis and murein assembly
Streptomycin	*Streptomyces griseus*	Gram-positive and Gram-negative bacteria	Inhibit translation (protein synthesis)
Gentamicin	*Micromonospora* species	Gram-positive and Gram-negative bacteria esp. Pseudomonas	Inhibit translation (protein synthesis)
Vancomycin	*Streptomyces orientales*	Gram-positive bacteria, esp. Staphylococcus aureus	Inhibits steps in murein (peptidoglycan) biosynthesis and assembly
Clindamycin	*Streptomyces lincolnensis*	Gram-positive and Gram-negative bacteria esp. anaerobic Bacteroides	Inhibits translation (protein synthesis)
Erythromycin	*Streptomyces erythreus*	Gram-positive bacteria, Gram-negative bacteria not enterics, Neisseria, Legionella, Mycoplasma	Inhibits translation (protein synthesis)
Polymyxin	*Bacillus polymyxa*	Gram-negative bacteria	Damages cytoplasmic membranes
Bacitracin	*Bacillus subtilis*	Gram-positive bacteria	Inhibits steps in murein (peptidoglycan) biosynthesis and assembly

Contd.

Amphotericin	*Streptomyces nodosus*	Fungi	Inactivate membranes containing sterols
Nystatin	*Streptomyces noursei*	Fungi (Candida)	Inactivate membranes containing sterols
Rifampicin	*Streptomyces mediterranei*	Gram-positive and Gram-negative bacteria, Mycobacterium tuberculosis	Inhibits transcription (eubacterial RNA polymerase)
Tetracycline	*Streptomyces species*	Gram-positive and Gram-negative bacteria, Rickettsias	Inhibit translation (protein synthesis)
Doxycycline		Gram-positive and Gram-negative bacteria, Rickettsias Ehrlichia, Borrelia	Inhibit translation (protein synthesis)
Chloramphenicol	*Streptomyces venezuelae*	Gram-positive and Gram-negative bacteria	Inhibits translation (protein synthesis)

D. BACTERIAL RESISTANCE

There are a number of ways in which bacteria can become resistant. Most of the early studies of bacterial resistance focused on single-step mutational events of chromosomal origin. This resistance was due not to a chromosomal change, but rather to the presence of extrachromosomal DNA that was transmissible. This type of resistance is called plasmid-mediated resistance. Bacteria also contain transposons, which can insert into plasmids and also into the chromosome. Transposon-mediated resistance to most of the major antibiotics has been found in the past few years.

Mechanisms of Resistance

The basic mechanisms by which a microorganism can resist an antimicrobial agent are

1. To alter the receptor for the drug (the molecule on which it exerts its effect);
2. To decrease the amount of drug that reaches the receptor by altering entry or increasing removal of the drug;
3. To destroy or inactivate the drug; and
4. To develop resistant metabolic pathways. Bacteria can possess one or all of these mechanisms simultaneously.

Antibiotics are frequently used in combination for the following reasons:

1. To treat a life-threatening infection;
2. To prevent emergence of bacterial resistance;
3. To treat mixed infections of aerobic and anaerobic bacteria;

4. To enhance antibacterial activity (synergy); and

5. To use lower doses of a toxic drug.

Mechanism to Reduce Bacterial Resistance

Proper selection of new antibiotics will be a major force in slowing the development of antimicrobial resistance.

- ▲ Control, reduce or cycle antibiotic usage
- ▲ Improve hygiene in hospitals, hospital personnel
- ▲ Discover and develop new antibiotics
- ▲ Modify existing antibiotics
- ▲ Develop inhibitors of antibiotic modifying enzymes
- ▲ Define agents that would 'cure' resistance plasmids
- ▲ Determination of the level of antimicrobial activity

E. DETERMINATION OF THE LEVEL OF ANTIMICROBIAL ACTIVITY

Dilution Susceptibility Test

This test can be used to determine Minimum Inhibitory Concentration (MIC). In the broth dilution test, a series of broth tubes containing antibiotics concentration in the range 0.1-128μg/ml is prepared and inoculated with standard test organisms. Lowest concentration of antibiotics resulting in no growth after 16-20 hours of incubation is the MIC.

Disc Diffusion Test

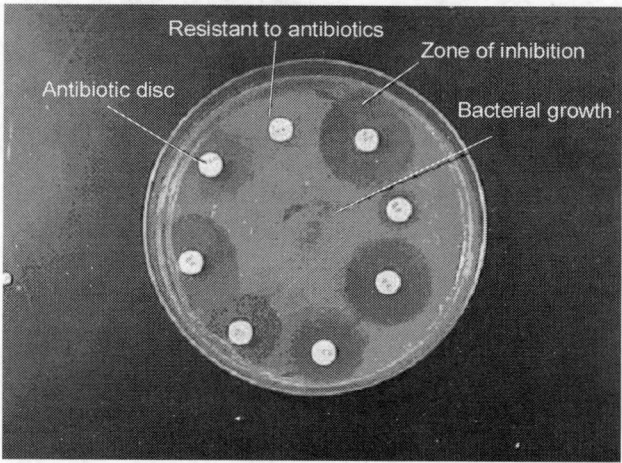

Figure 88   Disc Diffusion Test

It is also called Kirby & Bauer test. This is one of the more commonly used methods of antimicrobial susceptibility testing. In this test, small filter paper discs (6 mm) impregnated with a standard amount of antibiotic are placed onto an agar plate to which bacteria have been swabbed. The plates are incubated overnight, and the zone of inhibition of bacterial growth is used as a measure of susceptibility. Large zones of inhibition indicate that the organism is susceptible, while small or no zone of inhibition indicates resistance. An interpretation of intermediate is given for zones which fall between the accepted cutoffs for the other interpretations.

Etest

It is also called as episilon test. It is a well-established method for antimicrobial resistance testing in microbiology laboratories around the world. Etest consists of a predefined gradient of antibiotic concentrations on a plastic strip and is used to determine the Minimum Inhibitory Concentration (MIC) of antibiotics, antifungal agents and antimycobacterial agents. It Provides Minimum Inhibitory Concentrations (MICs) for slow-growing and fastidious organisms that have unique growth requirements and cannot be testing by automated methods.

Figure 89   Etest

8

MICROBIAL METABOLISM

A. METABOLISM

Metabolism is the sum of the chemical reactions that take place within each cell of a living organism and that provide energy for vital processes and for synthesizing new organic material. There are two phases of metabolism. They are Anabolism and catabolism. The study of synthesis (Anabolism) and degradation (Catabolism) of biomolecules is biochemically termed as metabolism.

$$\underset{\text{(Synthesis)}}{\text{Anabolism}} + \underset{\text{(Degradation)}}{\text{Catabolism}} = \text{Metabolism}$$

Anabolism is a constructive metabolism. It is also called as the synthesis reaction. Anabolism is the set of constructive metabolic processes where the energy released by catabolism is used to synthesize complex molecules. In general, the complex molecules that make up cellular structures are constructed step-by-step from small and simple precursors. It involves three basic stages. They are the production of precursors such as aminoacids, monosaccharides, isoprenoids and nucleotides, their activation into reactive forms using energy from ATP and the assembly of these precursors into complex molecules such as proteins, polysaccharides, lipids and nucleic acids.

Catabolism is a destructive metabolism and also called as the decomposition reactions. Catabolism is the set of metabolic processes that break down large molecules. These include breaking down and oxidizing food molecules. The purpose of the catabolic reactions is to provide the energy and components needed by anabolic reactions.

B. CARBOHYDRATE METABOLISM

Introduction

Carbohydrates are widely distributed in plants in which they are formed from carbon dioxide of the atmosphere and water by photosynthesis. Plants use carbohydrates as the precursor for the synthesis of proteins, lipids and other organic compounds. Animals obtain their carbohydrates from plants.

Functions of Carbohydrates

Carbohydrates have a variety of functions in the animal and human body.

1.   They supply energy for body functions and for doing work.

2. They are structural components of many organisms.

3. They exert a sparing action on proteins.

4. They provide the carbon skeleton for the synthesis of some nonessential amino acids and fats.

5. Some carbohydrates are present as tissue constituents.

6. Starch forms main source of carbohydrates in the diet.

Classification of Carbohydrates

Carbohydrates are classified as follows: They are

i. Monosaccharides

ii. Oligo saccharides and

iii. Polysaccharides

Monosaccharides

Monosaccharides are defined as polyhydroxy aldehydes or ketones, which cannot be further hydrolysed to simple sugars. Monosaccharides are divided into two groups according to their functional groups.

Disaccharides

Disaccharides are sugars containing two molecules of monosaccharides. Disaccharides are formed by the condensation of two molecules of monosaccharides with the elimination of one molecule of water.

Oligosaccharides

These are carbohydrates that yield 2-10 monosaccharide units on hydrolysis.eg. Maltotriose.

Polysaccharides

Polysaccharides, which are also known as glycans composed of number of monosccharide units. They represent condensation products of several molecules of simple sugars or monosaccharides.

According to this nature polysaccharides are classified into two groups, homopolysaccharides and heteropolysaccharides. Homopolysaccharides are composed of only one type of monosaccharides. On hydrolysis they yield only one type of monosaccharides Eg. starch, glycogen, cellulose etc. which yield only glucose on hydrolysis. Heteropolysaccharides are composed of a mixture of monosaccharides. On hydrolysis, they yield a mixture of monosaccharides. Eg. Hyaluronic acid, Heparin, Mucopolysaccharides.

Carbohydrate as a source of energy

The major function of carbohydrate in metabolism is to serve as fuel and get oxidised to provide energy for other metabolic processes. The metabolic intermediates are used for various biosynthetic reactions.

Glycolysis

Oxidation of glucose to pyruvate is called glycolysis. It was first described by Embden-Meyerhof and Parnas. Hence it is also called as Embden-Meyerhof pathway.

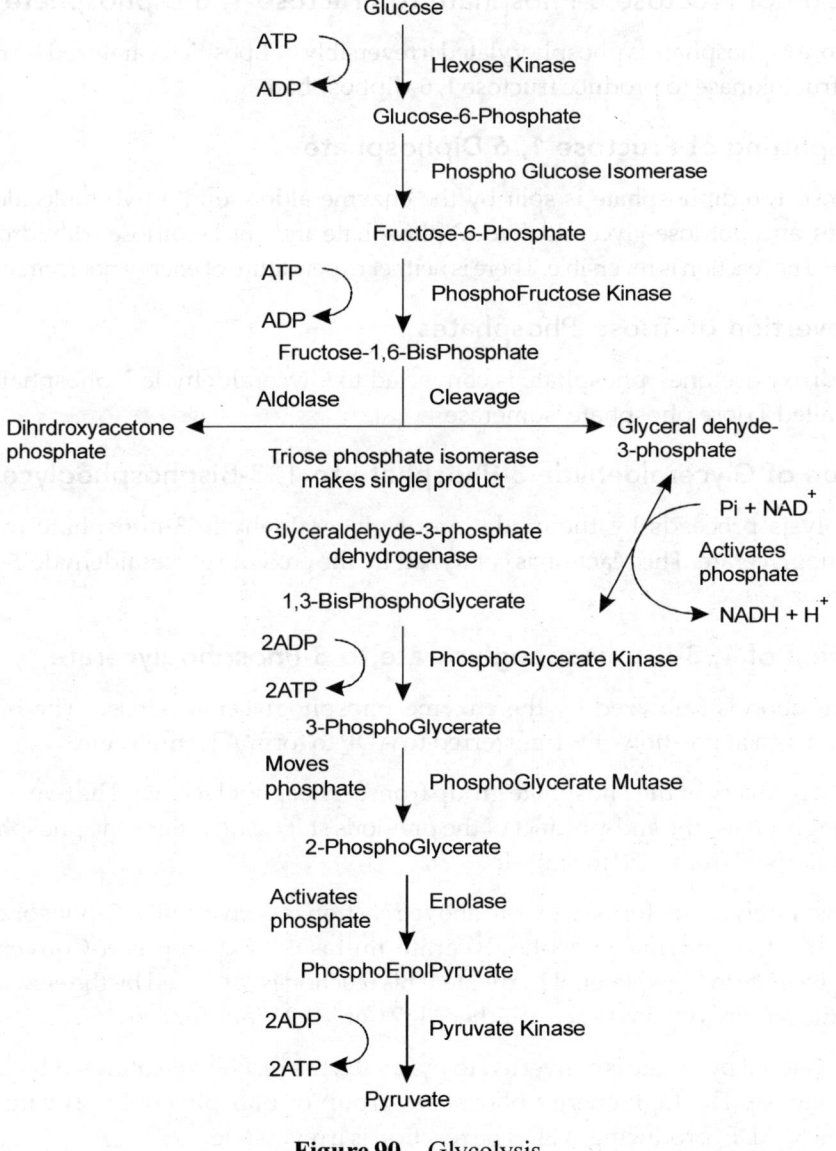

Figure 90   Glycolysis

Uptake of Glucose by Cells and its Phosphorylation

Glucose is freely permeable to cells. Glucose enters through active transport. Glucose is phosphorylated to form glucose 6-phosphate. The enzyme involved in this reaction is glucokinase. This reaction is irreversible.

Conversion of Glucose 6-Phosphate to Fructose 6-Phosphate

Glucose 6-phosphate is converted to fructose 6-phosphate by the enzyme phosphogluco isomerase.

Conversion of Fructose 6-Phosphate to Fructose 1, 6 Diphosphate

Fructose 6-phosphate is phosphorylated irreversibly at 1 position catalyzed by the enzyme phosphofructokinase to produce fructose 1, 6-diphosphate.

Actual Splitting of Fructose 1, 6 Diphosphate

Fructose 1, 6 diphosphate is split by the enzyme aldolase into two molecules of triose phosphates, an aldotriose-glyceraldehyde 3-phosphate and one ketotriose - dihydroxy acetone phosphate. The reaction is reversible. There is neither expenditure of energy nor formation of ATP.

Interconvertion of Triose Phosphates

Dihydroxy acetone   phosphate is converted to Glyceraldehyde 3-phosphate using an enzyme called Triose phosphate isomerase

Oxidation of Glyceraldehyde 3-Phosphate to 1, 3-Bisphosphoglycerate

Glycolysis proceeds by the oxidation of glyceraldehyde 3-phosphate to form 1,3-bisphosphoglycerate. The reaction is catalyzed by the enzyme glyceraldehyde 3-phosphate dehydrogenase.

Conversion of 1, 3-Bisphosphoglycerate to 3-Phosphoglycerate

The reaction is catalyzed by the enzyme phosphoglycerate kinase. The high energy phosphate bond at position - 1 is transferred to ADP to form ATP molecule.

It is the recovery of the phosphate group from 3-phosphoglycerate. The two molecules of 3-phosphoglycerate, the end-product of the previous stage, still retains the phosphate group, originally derived from ATP in Stage I.

3-phosphoglycerate formed by the above reaction is converted to 2-phosphoglycerate, catalyzed by the enzyme phosphoglycerate mutase. Next step is a Conversion of 2-Phosphoglycerate to Phosphoenol Pyruvate. This reaction is catalyzed by the enzyme enolase. The enzyme requires the presence of either Mg^{2+} or Mn^{2+} ions for activity.

Phosphoenol pyruvate is converted to pyruvate, the reaction is catalysed by the enzyme pyruvate kinase. The high energy phosphate group of phosphoenol pyruvate is directly transferred to ADP, producing ATP. The reaction is irreversible.

Summary of glycolysis

Net gain = 8 ATP

Reactions Catalyzed	ATP used	ATP formed
Stage I		
1. Glucokinase (for phosphorylation)	1	
2. Phosphofructokinase I (for phosphorylation)	1	
Stage II		
3. Glyceraldehyde 3-phosphate dehydrogenase (oxidation of 2 NADH in respiratory chain)		6
4. Phosphoglycerate kinase (substrate level phosphorylation)		2
Stage IV		
5. Pyruvate kinase (substrate level phosphorylation)		2
Total	2	10

Tricarboxylic acid cycle (TCA cycle)

This cycle is the aerobic phase of carbohydrate metabolism and follows the anaerobic pathway from the stage of pyruvate and is called as citric acid cycle or TCA cycle. The name citric acid cycle stems from citric acid which is formed in the first step of this cycle. This cycle is also named "Kerbs cycle" after H.A. Krebs, an English biochemist who worked on the process of this cycle.

Under aerobic conditions, pyruvate is oxidatively decarboxylated to acetyl coenzyme A (active acetate) before entering the citric acid cycle.

1. Formation of Citrate

The first reaction of the cycle is the condensation of acetyl CoA with oxaloacetate to form citrate, catalyzed by citrate synthase. This is an irreversible reaction.

2. Formation of Isocitrate Via Cis Aconitate

The enzyme aconitase catalyzes the reversible transformation of citrate to isocitrate, through the intermediary formation of cis aconitate.

3. Oxidation of Isocitrate to A-ketoglutarate and CO2

In the next step, isocitrate dehydrogenase catalyzes oxidative decarboxylation of isocitrate to form α-ketoglutarate.

4. Oxidation of A-keto Glutarate to Succinyl CoA and CO_2

The next step is another oxidative decarboxylation, in which a-ketoglutarate is converted to succinyl CoA and CO_2 by the action of the a-ketoglutarate dehydrogenase complex. The reaction is irreversible.

5. Conversion of Succinyl CoA to Succinate

The product of the preceding step, succinyl CoA is converted to succinate to continue the cycle. GTP is formed in this step (substrate level phosphorylation). The enzyme that catalyzes this reversible reaction is called succinyl CoA synthetase or succinic thiokinase.

6. Oxidation of Succinate to Fumarate

The succinate formed from succinyl CoA is oxidized to fumarate by the enzyme succinate dehydrogenase

7. Hydration of Fumarate to Malate

The reversible hydration of fumarate to malate is catalyzed by fumarase.

8. Oxidation of Malate to Oxaloacetate

The last reaction of the citric acid cycle is, NAD linked malate - dehydrogenase which catalyses the oxidation of malate to oxaloacetate.

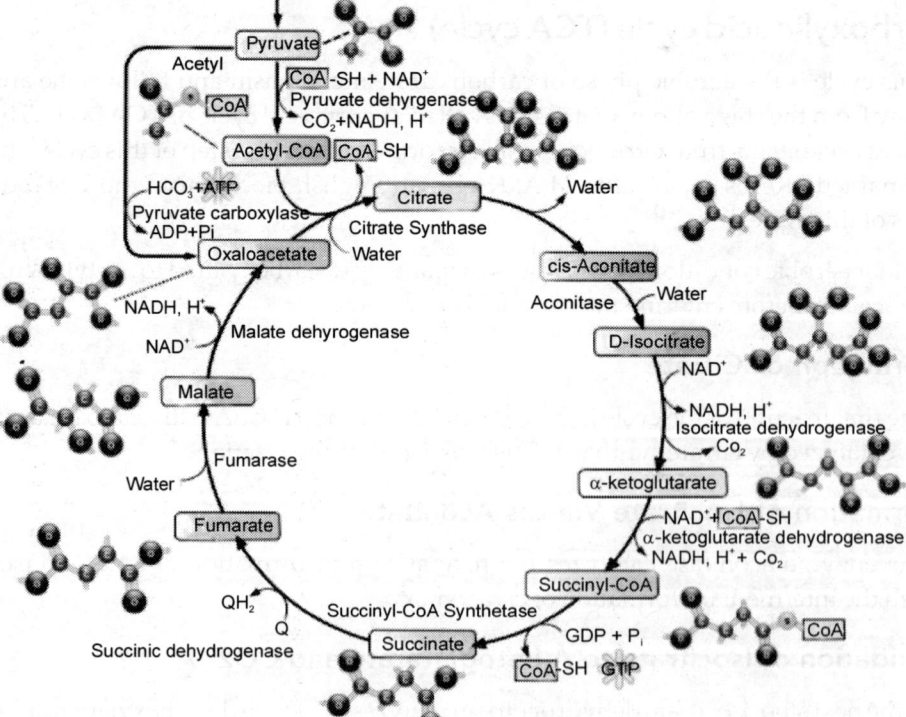

Figure 91 TCA cycle

Energy yield from TCA cycle

If one molecule of the substrate is oxidized through NADH in the electron transport chain three molecules of ATP will be formed and through FADH2, two ATP molecules will be generated. As one molecule of glucose gives rise to two molecules of pyruvate by glycolysis, intermediates of citric acid cycle also result as two molecules.

Reactions	No. of ATP formed
1. 2 isocitrate 2 a-ketoglutarate	
(2 NADH + 2H +) (2 × 3)	6
2. 2 a-ketoglutarate 2 succinyl CoA	
(2 NADH + 2H+) (2´ 3)	6
3. 2 succinyl CoA 2 succinate	
(2 GTP = 2ATP)	2
4. 2 succinate 2 Fumarate	
(2 FADH2) (2´ 2)	4
5. 2 malate 2 oxaloacetate	
(2 NADH + 2H+) (2´ 3)	6
Total No.of ATP formed 24	

HMP shunt pathway

A number of alternative pathways are also discovered other than Glycolysis and TCA cycle. The most important one is Hexose Monophosphate Shunt Pathway (HMP shunt). The pathway occurs in the extra mitochondrial soluble portion of the cells. It is also called Pentose Phosphate Pathway (PPP). HMP shunt generates a different type of metabolic energy - the reducing power. Some of the electrons and hydrogen atoms of fuel molecules are conserved for biosynthetic purposes rather than ATP formation. This reducing power of cells is NADPH (reduced nicotinamide adenine dinucleotide phosphate).

The fundamental difference between NADPH and NADH (reduced nicotinamide adenine dinucleotide) is that NADH is oxidised by the respiratory chain to generate ATP whereas NADPH serves as a hydrogen and electron donor in reductive biosynthesis, for example in the biosynthesis of fatty acids and steroids. The first reaction of the pentose phosphate pathway is the dehydrogenation of glucose 6-phosphate by glucose 6-phosphate dehydrogenase to form 6-phosphoglucono 1, 5-lactone.

Step 1: Glucose 6-phosphate in the presence of NADP and the enzyme glucose 6-phosphate dehydrogenase, forms 6-phospho glucono-1, 5-lactone. The first molecule of NADPH is produced in this step.

Step 2: The 6-phospho glucono 1, 5-lactone is unstable and the ester spontaneously hydrolyses to 6-phosphogluconate. The enzyme that catalyses the reaction is lactonase

Step 3: 6-phospho gluconate further undergoes dehydrogenation and decarboxylation by 6-phosphogluconate dehydrogenase to form the ketopentose, D-ribulose 5-phosphate. This reaction generates the second molecule of NADPH.

Step 4: The enzyme phosphopentose isomerase converts ribulose 5-phosphate to its aldose isomer, D-ribose 5-phosphate. The net result is the production of NADPH, a reductant for biosynthetic reactions, and ribose 5-phosphate, a precursor for nucleotide synthesis.

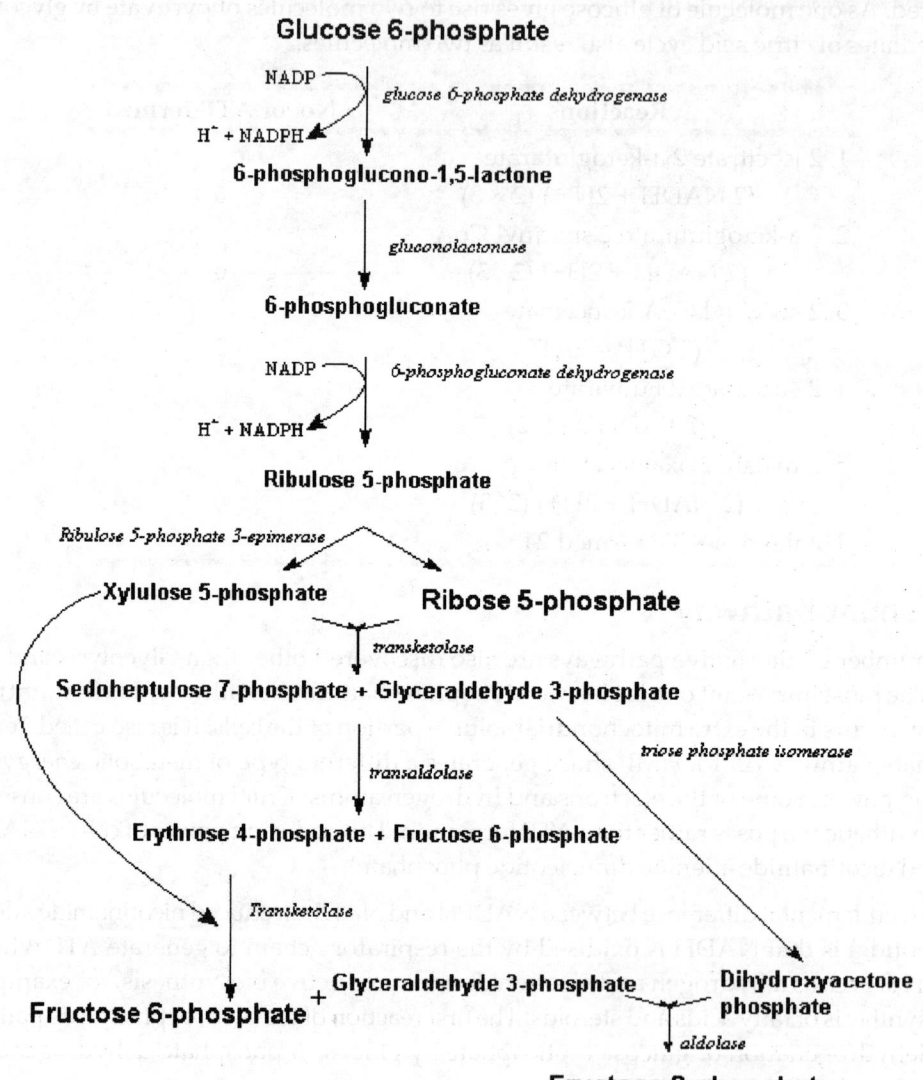

Figure 92 HMP pathway

ED Pathway

The **Entner-Doudoroff pathway** begins with the formation of glucose 6-phosphate and 6-phosphogluconate. Instead of being further oxidized, 6-phosphogluconate is dehydrated to form 2-keto-3-deoxy-6-phosphogluconate or KDPG, the key intermediate in this pathway. KDPG is then cleaved by KDPG aldolase to pyruvate and glyceraldehydes 3-phosphate. The glyceraldehyde 3-phosphate is converted to pyruvate in the bottom portion of the glycolytic pathway.

If the Entner-Doudoroff pathway degrades glucose to pyruvate in this way, it yields one ATP, one NADPH, and one NADH per glucose metabolized.

Most bacteria have the glycolytic and pentose phosphate pathways, but some substitute the Entner-Doudoroff pathway for glycolysis. The Entner-Doudoroff pathway is generally found in *Pseudomonas, Rhizobium, Azotobacter, Agrobacterium,* and a few other gram-negative genera. Very few gram-positive bacteria have this pathway, with *Enterococcus faecalis* being a rare exception.

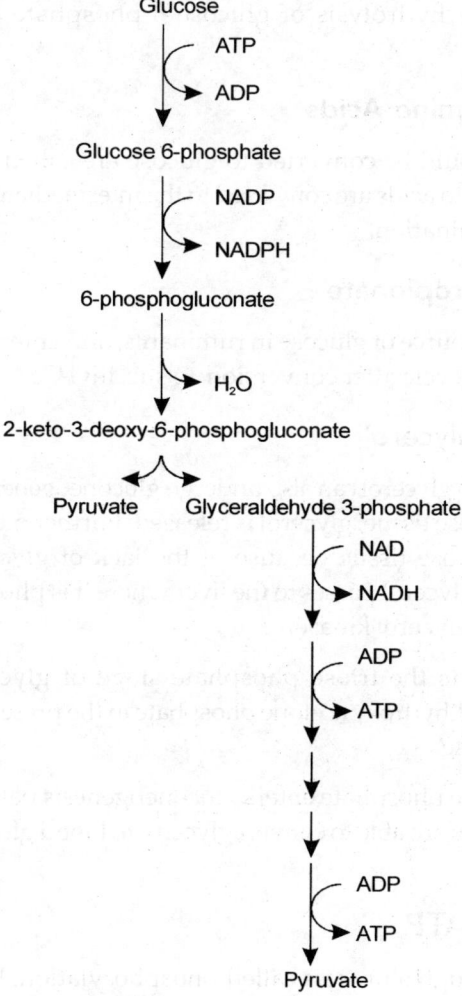

Figure 93   ED pathway

Gluconeogenesis

The synthesis of glucose from non-carbohydrate precursors is known as gluconeogenesis. The major site of gluconeogenesis is liver. It usually occurs when the carbohydrate in the diet is insufficient to meet the demand in the body, with the intake of protein rich diet and at the time of starvation, when tissue proteins are broken down to amino acids.

Reactions of gluconeogenesis

1. The formation of phosphoenol pyruvate begins with the carboxylation of pyruvate at the expense of ATP to form oxalo acetate. Oxaloacetate is converted to phosphoenolpyruvate by phosphorylation with GTP, accompanied by a simultaneous decarboxylation.

2. Fructose 6-phosphate is formed from fructose 1,6-diphosphate by hydrolysis and the enzyme fructose 1,6-diphosphatase catalyses this reaction.

3. Glucose is formed by hydrolysis of glucose 6-phosphate catalysed by glucose 6-phosphatase.

Gluconeogenesis of Amino Acids

Amino acids which could be converted to glucose are called glucogenic amino acids. Most of the glucogenic amino acids are converted to the intermediates of citric acid cycle either by transamination or deamination.

Gluconeogenesis of Propionate

Propionate is a major source of glucose in ruminants, and enters the main gluconeogenic pathway via the citric acid cycle after conversion to succinyl CoA.

Gluconeogenesis of Glycerol

At the time of starvation glycerol can also undergo gluconeogenesis. When the triglycerides are hydrolysed in the adipose tissue, glycerol is released. Further metabolism of glycerol does not take place in the adipose tissue because of the lack of glycerol kinase necessary to phosphorylate it. Instead, glycerol passes to the liver where it is phosphorylated to glycerol 3-phosphate by the enzyme glycerol kinase.

This pathway connects the triose phosphate stage of glycolysis, because glycerol 3-phosphate is oxidized to dihydroxy acetone phosphate in the presence of NAD+ and glycerol 3-phosphate dehydrogenase.

This dihydroxy acetone phosphate enters gluconeogenesis pathway and gets converted to glucose. Liver and kidney are able to convert glycerol to blood glucose by making use of the above enzymes.

The Generation of ATP

ATP is generated through a process called phosphorylation. It is a chemical process in which a phosphate group is added to an organic molecule (ADP). In living cells phosphorylation is associated with respiration. The energy released during metabolic or photosynthetic processes is captured in the energy-rich phosphate bonds of certain molecules, most commonly in the high-energy bonds of adenosine triphosphate (ATP). Microorganisms use three mechanisms of phosphorylation to generate ATP from ADP. They are *Substrate-Level Phosphorylation, Oxidative Phosphorylation and Photo Phosphorylation*

Substrate-Level Phosphorylation

In substrate-level phosphorylation, ATP is usually generated when a high-energy is directly transferred from a phosphorylated compound (a substrate) to ADP.

Oxidative Phosphorylation

In **oxidative phosphorylation,** electrons are transferred from organic compounds to one group of electron carriers (usually to NAD$^+$ and FAD). Then, the electrons are passed through a series of different electron carriers to molecules of oxygen (O_2) or other oxidized inorganic and organic molecules. This process occurs in the plasma membrane of prokaryotes and in the inner mitochondrial membrane of eukaryotes. The sequence of electron carriers used in oxidative phosphorylation is called an **electron transport chain (system).** The transfer of electrons from one electron carrier to the next releases energy, some of which is used to generate ATP from ADP through a process called *chemiosmosis.*

Photophosphorylation

This occurs only in photosynthetic cells, which contain light-trapping pigments such as chlorophylls. In photosynthesis, organic molecules, especially sugars, are synthesized with the energy of light from the energy-poor building blocks carbon dioxide and water. Photophosphorylation starts this process by converting light energy to the chemical energy of ATP and NADPH, which, in turn, are used to synthesize organic molecules. As in oxidative phosphorylation, an electron transport chain is involved.

Electron Transport Chain

An **electron transport chain (system)** consists of a sequence of carrier molecules that are capable of oxidation and reduction. As electrons are passed through the chain, there occurs a stepwise release of energy, which is used to drive the chemiosmotic generation of ATP. In eukaryotic cells, mitochondria is the site for electron transport chain; in prokaryotic cells, it is found in the plasma membrane.

There are three classes of carrier molecules in electron transport chains. The first are **flavoproteins.** These proteins contain flavin, a coenzyme derived from riboflavin (vitamin B2), and are capable of performing alternating oxidations and reductions. One important flavin coenzyme is flavin mononucleotide (FMN). The second class of carrier molecules are **cytochromes,** proteins with an iron-containing group (heme) capable of existing alternately as a reduced form (Fe^{2+}) and an oxidized form (Fe^{3+}). The cytochromes involved in electron transport chains include cytochrome *b* (cyt *b*), cytochrome *c1* (cyt *c*1), cytochrome *c* (cyt *c*), cytochrome *a* (cyt *a*), and cytochrome *a3* (cyt *a*3). The third class is known as **ubiquinones,** or **coenzyme Q,** these are small nonprotein carriers.

The first step in the mitochondrial electron transport chain involves the transfer of high-energy electrons from NADH to FMN.

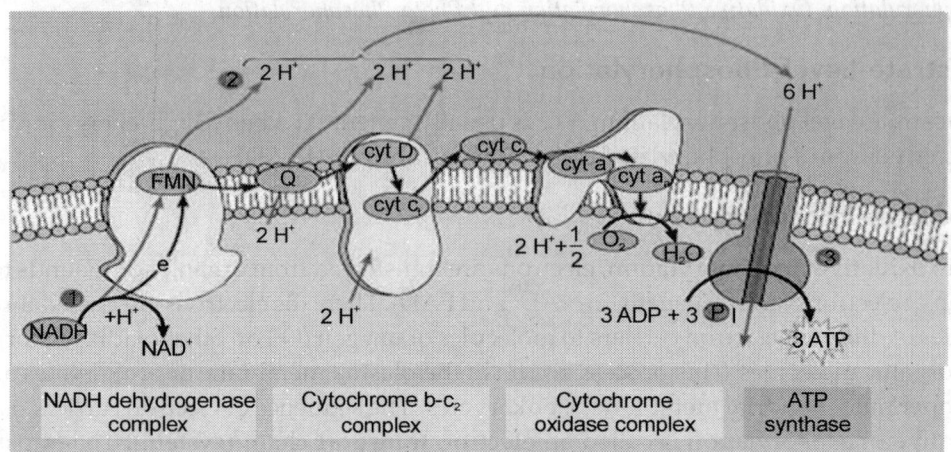

Figure 94 Electron transport

This transfer actually involves the passage of a hydrogen atom with two electrons to FMN, which then picks up an additional H^+ from the surrounding aqueous medium. As a result of the first transfer, NADH is oxidized to NAD^+, and FMN is reduced to $FMNH_2$. In the second step in the electron transport chain, $FMNH_2$ passes $2H^+$ to the other side of the mitochondrial membrane and passes two electrons to Q. As a result, $FMNH_2$ is oxidized to FMN. Q also picks up an additional $2H^+$ from the surrounding aqueous medium and releases it on the other side of the membrane.

The next part of the electron transport chain involves the cytochromes. Electrons are passed successively from Q to cyt b, cyt $c1$, cyt c, cyt a, and cyt $a3$. Each cytochrome in the chain is reduced as it picks up electrons and is oxidized as it gives up electrons. The last cytochrome, cyt $a3$, passes its electrons to molecular oxygen (O_2), which becomes negatively charged and then picks up protons from the surrounding medium to form H_2O.

$FADH_2$ is derived from the Krebs cycle. $FADH_2$ adds its electrons to the electron transport chain at a lower level than NADH. Because of this, $FADH_2$ produces 2 ATP whereas NADH generats three ATP.

Electron flow down the chain is accompanied at several points by the active transport (pumping) of protons from the matrix side of the inner mitochondrial membrane to the opposite side of the membrane. The result is a buildup of protons on one side of the membrane. Just as water behind a dam stores energy that can be used to generate electricity, this buildup of protons provides energy for the generation of ATP by the chemiosmotic mechanism.

The Chemiosmotic Mechanism of ATP Generation

The mechanism of ATP synthesis using the electron transport chain is called **chemiosmosis**.

As energetic electrons from NADH (or chlorophyll) pass down the electron transport chain, some of the carriers in the chain pump—actively transport—protons across the membrane. Such carrier molecules are called *proton pumps.* The phospholipid membrane is normally impermeable to protons, so this one-directional pumping establishes a proton gradient (a difference in the concentrations of protons on the two sides of the membrane). In addition to a concentration gradient, there is an electrical charge gradient. The excess H+ on one side of the membrane makes that side positively charged compared with the other side. The resulting electrochemical gradient has potential energy, called the *proton motive force.*

The protons on the side of the membrane with the higher proton concentration can diffuse across the membrane only through special protein channels that contain an enzyme called *ATP synthase.*When this flow occurs, energy is released and is used by the enzyme to synthesize ATP from ADP.

C. PROTEIN METABOLISM

Protein Biosynthesis

The biosynthesis of protein molecules in the cell by sequential addition of various amino acids using peptide bond is called protein synthesis. The amino acids are linked together in succession to produce a linear polypeptide chain.   The polypeptide chain is a unit of a protein molecule.

In a protein molecule amino acids are joined together by peptide bonds. In the process of protein synthesis also known as translation of mRNA, the amino acids are added sequentially in a specific number. The protein synthesizing mechanism involves the following steps:

Transcription

The formation of RNA complementary to a DNA strand is called transcription. In this process, the RNAs required for protein synthesis are synthesized on DNA strands. This reaction is catalyzed by the enzyme RNA polymerase.

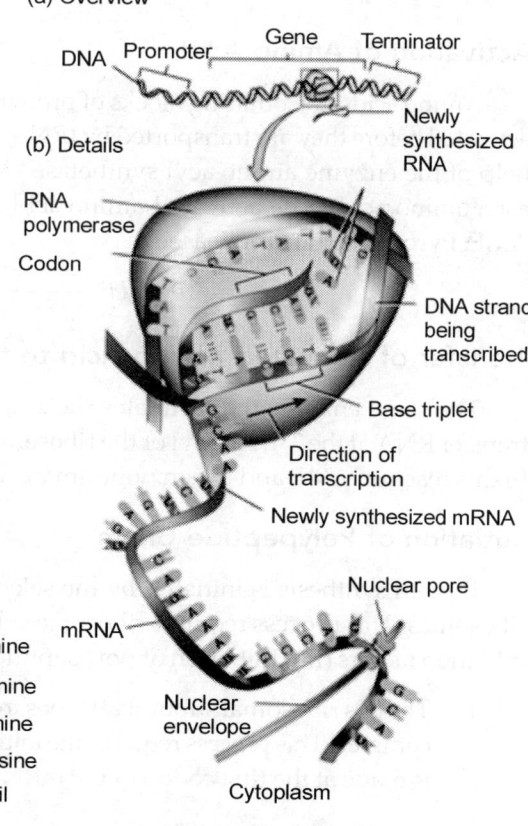

(a) Overview

DNA Promoter Gene Terminator

Newly synthesized RNA

(b) Details

RNA polymerase

Codon

DNA strand being transcribed

Base triplet

Direction of transcription

Newly synthesized mRNA

Nuclear pore

Key:
A = Adenine
G = Guanine
T = Thymine
C = Cytosine
U = Uracil

mRNA

Nuclear envelope

Cytoplasm

Figure 95   Transcription

The enzyme, RNA polymerase I, II, III are involved in the synthesis of rRNA (ribosomal RNA) mRNA (messenger RNA) and tRNA (transfer RNA) respectively in the eukaryotes. In prokaryotes only one type of RNA polymerase is present to synthesize all the three classes of RNA. In the DNA double helix, one of the strands serves as a template to produce RNA. The RNA produced by transcription is inactive and is called pre-RNA. They become active after further processing. All these RNA are processed through chemical reactions and structural modifications.

Translation

Translation is a process by which the base sequence of DNA transcribed to the mRNA is interpreted into amino acid sequence of a polypeptide chain.

Translation involves the following steps:

1. Activation of amino acid
2. Transfer of activated amino acid to tRNA
3. Initiation of polypeptide chain
4. Elongation of polypeptide chain
5. Termination of polypeptide chain

Activation of Amino Acid

Amino acids, the building blocks of proteins. They are present in the cytoplasm. They are activated before they are transported by tRNA. The amino acids are activated by ATP with the help of the enzyme amino acyl synthetase. Amino acyl synthetase is specific in activating each amino acid. The activated amino acid is called amino acyl adenylate or amino acyl AMP. Pyrophosphate is released.

$$AA + ATP \longrightarrow \text{aminoacyl } AMP$$

Transfer of Activated Amino Acid to tRNA

The same enzyme that activates the amino acid catalyses its transfer to a molecule of transfer RNA at the 3' hydroxyl of the ribose, an ester with a high potential for group transfer. In this reaction AMP and the enzyme amino acyl synthetase are released.

Initiation of Polypeptide Chain

Protein synthesis is initiated by the selection and transfer of the first amino acid into ribosomes. This process requires ribosome subunits, amino acyl tRNA complex, mRNA and initiation factors (IF). Initiation of polypeptide chain involves the following steps.

1. The 30s ribosomal subunit attaches to the 5' end of the mRNA to form an mRNA 30s complex. This process requires the initiation factor IF-3 and Mg2+ ions. The attachment is made at the first codon of the mRNA.

2. The first codon of mRNA will be always AUG. This codon specifies the amino acid methionine. So the first amino acid in the synthesis of any polypeptide chain is methionine.

3. The tRNA having the anticodon UAC (complementary to AUG) transports methionine to the 30s ribosome and attaches itself to the initiation codon on mRNA. The tRNA, mRNA and 30s ribosome subunit form a complex called 30s - pre initiation complex. This process requires initiation factors and GTP.

4. 30s - pre initiation complex joins with 50s ribosomal subunit to form initiation complex. The initiation complex is formed of 70s ribosome, mRNA and met -RNA (methionine RNA).

5. The 70s ribosome has two slots for the entry of amino acyl tRNA, namely P site (peptidyl site) and A site (amino acid site). The first tRNA i.e. met RNA is attached to the P site of 70s ribosome.

Elongation of Polypeptide Chain

Elongation refers to sequential addition of amino acids to methionine, as per the sequence of codon in the mRNA. It involves the following steps:

1. The second codon in the mRNA is recognised and as per the recognition, the amino acyl tRNA containing the corresponding anticodon moves to the 70s ribosome and fits into the A-site. Here the anticodon of tRNA base pairs with the second codon of mRNA.

2. A peptide bond is formed between the carboxyl group (-COOH) of first amino acid of site P and the amino group (-NH2) of second amino acid of A-site. The peptide bond links two amino acids to form a dipeptide. The bonding is catalysed by the enzyme peptidyl transferase which is present in 50s ribosomal subunit.

3. After the formation of peptide bond, the methionine and tRNA are separated by an enzyme called tRNA deacylase.

4. The dissociated tRNA is then released from P-site into the cytoplasm for further aminoacylation.

5. Now the ribosome moves on the mRNA in the 5′ to 3′ direction so that the first codon goes out of ribosome, the second codon comes to lie in the P-site from A-site and the third codon comes to lie in the A-site. Simultaneously, the second tRNA is shifted from A-site to P-site. All these events, the movement of ribosome, the release of first tRNA from P-site and shifting of second tRNA from A-site to P-site constitute *translocation*. Translocation is catalyzed by the enzyme *translocase*.

6. The third codon is recognised and the amino acyl tRNA containing the corresponding anticodon moves to the 70s ribosome and fits into the A-site. The anticodon base pairs with the codon. A peptide bond is formed between the third amino acid of site-A and the second amino acid of the dipeptide present in the P-site. Thus a tripeptide is formed.

7.   The amino acids are added one by one as per the codon in the mRNA and hence the tripeptide is converted into polypeptide chain. The polypeptide chain elongates by the addition of more and more amino acids.

8.   The elongation of polypeptide chain is brought about by a number of protein factors called elongation factors.

Termination of Polypeptide Chain

Termination is the completion of polypeptide chain. By termination, a polypeptide chain is finished and released. The polypeptide chain is completed, when the ribosome reaches the 3′ end of mRNA. The 3′ end contains stop codons or termination codons. They are UAG or UAA or UGA. Termination is helped by the terminating protein factors. The terminated polypeptide chain is released from the ribosome.

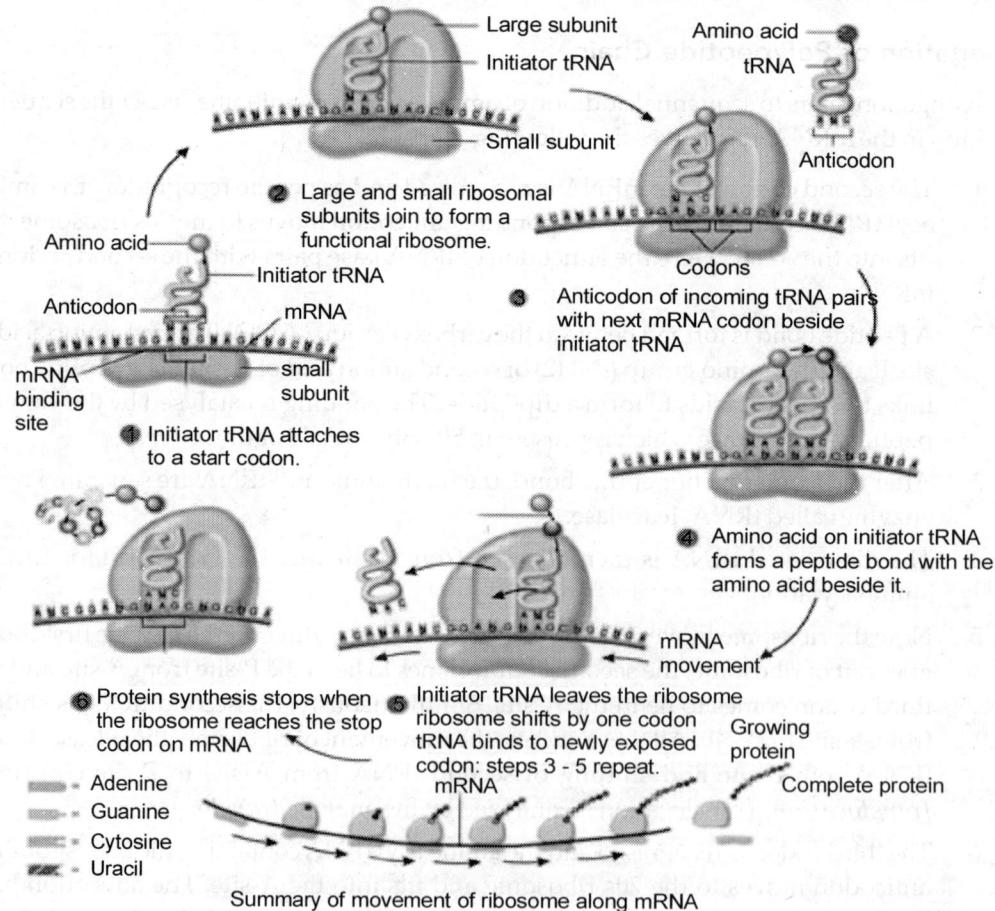

Summary of movement of ribosome along mRNA

Figure 96   Translation mechanism

After the release of polypeptide chain, the 70s unit dissociates into 50s and 30s sub-units. These subunits are again used in the formation of another initiation complex.

The polypeptide chain released after translation is inactive. It is processed to make it active. In the processing the initiating amino acid methionine is removed. Along with methionine a few more amino acids are removed from the N-terminal of the polypeptide. The processing is carried out by deformylase and amino peptidase. This processing is called as post translational modifications.

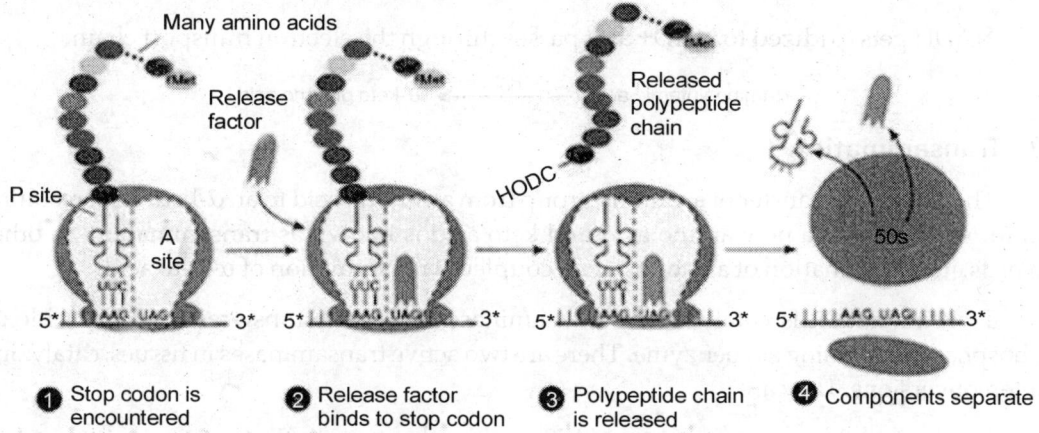

| ① Stop codon is encountered | ② Release factor binds to stop codon | ③ Polypeptide chain is released | ④ Components separate |

Figure 97   Termination of protein synthesis

Catabolism of amino acids

Amino acid follows its own specific metabolic pathway, a few general reactions are found to be common in the catabolism of nearly all the amino acids. Most of the amino acids are converted to α-keto acids by the removal of nitrogen in the form of ammonia which is quickly transformed into urea or it gets incorporated into some other amino acids.

1. Oxidative deamination

Deamination means removal of the amino groups from amino acids. This is the mechanism here in the amino acids lose two hydrogen atoms (dehydrogenation) to form keto acids and ammonia. Oxidative deamination is accompanied by oxidation and is catalysed by specific amino acid oxidases or more appropriately, dehydrogenases present in liver and kidneys. The process of oxidative deamination takes place in two steps. The first step is oxidation (dehydrogenation) of amino acid resulting in the formation of imino acid. The imino acid then undergoes the second step, namely hydrolysis which results in a keto acid and ammonia.

Amino acid ⟶ Imino acid ⟶ Keto acid ⟶ Ammonia

The first reaction is catalyzed by amino acid oxidase (also called dehydrogenase) and the coenzyme FAD or FMN takes up the hydrogen. There are two types of amino acid oxidases depending upon the substrate on which they act, namely,

1. L-amino acid oxidases which act on L-amino acids (FMN acts as coenzyme).

2. D-amino acid oxidases which act on D-amino acids (FAD acts as coenzyme).

The oxidative deamination of L-glutamic acid is an exceptional case where the deamination needs not only the zinc-containing enzyme L-glutamic acid dehydrogenase but also NAD+ or NADP+ as coenzymes.

$$\text{L-Glutamic acid + NAD}^+ \xrightarrow{\substack{\text{Glutamic acid}\\ \text{dehydrogenase}}} \alpha\text{-Iminoglutamic acid + NADH+H}^+$$

NADH gets oxidized to NAD+ as it passes through the electron transport chain.

$$\alpha\text{-Iminoglutamic acid} \xrightarrow{H_2O} \alpha\text{-keto glutaric acid}$$

2. Transamination

The process of transfer of an amino group from an amino acid to an α-keto acid, resulting in the formation of a new amino acid and keto acid is known as transamination. In other words, it is deamination of an amino acid, coupled with amination of α-keto acid.

Transamination is catalyzed by transaminases or aminotransferases with pyridoxal phosphate functioning as coenzyme. There are two active transaminases in tissues, catalyzing interconversions. They are

1. Aspartate aminotransferase (AST) is also known as Glutamate - oxalo acetate transaminase (GOT)

2. Alanine aminotransferase (ALT) is also known as Glutamate- pyruvate transaminase (GPT)

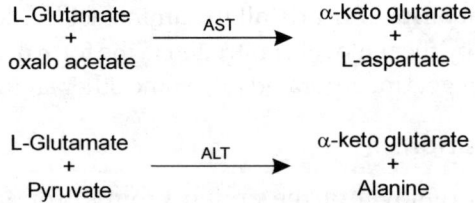

It catalyses the transfer of NH_2 group from glutamate to pyruvate, resulting in the formation of α-ketoglutaric acid and alanine.

3. Decarboxylation

This refers to the removal of CO_2 from the carboxyl group of amino acids. The removal of CO_2 needs the catalytic action of enzymes decarboxylases and the pyridoxal phosphate coenzyme. The enzymes act on amino acids resulting in the formation of the corresponding amines with the liberation of CO_2.

$$\text{Amino acid} \xrightarrow{\substack{\text{Amino acid}\\ \text{decarboxylase}}} \text{Amine} + CO_2$$

4. Transmethylation

The transfer of methyl group from one compound to another is called transmethylation and the enzymes involved in the transfer are known as transmethylases.

Biosynthesis of Aminoacids

Amino acid synthesis is the set of biochemical processes by which the various amino acids are produced from other compounds. The substrates for these processes are various compounds in the organism's diet or growth media. Not all organisms are able to synthesise all amino acids. All amino acids are derived from intermediates in glycolysis, the citric acid cycle, or the pentose phosphate pathway. Nitrogen enters these pathways by way of glutamate and glutamine. Different organisms vary greatly in their ability to synthesize the 20 amino acids. A useful way to organize the amino acid biosynthetic pathways is to group them into 6 families corresponding to the metabolic precursor of each amino acid.

α-Ketoglutarate	Oxaloacetate	Phosphoenolpyruvate and erythrose-4-phosphate
Glutamate	Aspartate	Tryptophan*
Glutamine	Asparagine	Phenylalanine*
Proline	Methionine*	Tyrosine
Argine	Threonine*	
	Lysine	
	Isoleucine	
3-Phosphoglycerate	**Pyruvate**	**Ribose-5-phosphate**
Serine	Alanine	Histidine*
Glyceine	Valine*	
Cysteine	Leucine	

α-Ketoglutarates

The α-ketoglutarate family of amino acid synthesis (synthesis of glutamate, glutamine, proline and arginine) begins with α-ketoglutarate, an intermediate in the Citric Acid Cycle.

Erythrose 4-phosphate and phosphoenolpyruvate

Phenylalanine, tyrosine, and tryptophan are known as the aromatic amino acids. The synthesis of all three share a common beginning to their pathways; the formation of chorismate from phosphoenolpyruvate (PEP) and erythrose 4- phosphate (E4P).

Oxaloacetate/aspartate

The oxaloacetate/aspartate family of amino acids is composed of lysine, asparagine, methionine, threonine and isoleucine. Aspartate can be converted into lysine, asparagine, methionine and threonine. Threonine also gives rise to isoleucine.

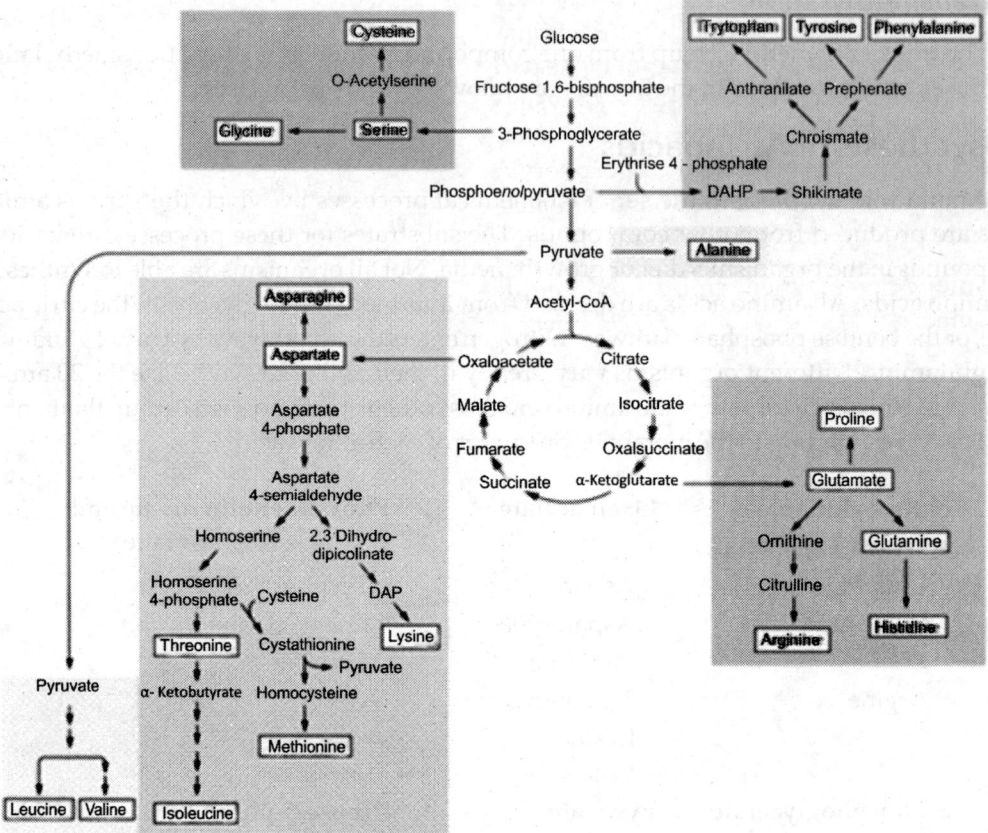

Figure 98 Aminoacid biosynthesis

Aspartate: The enzyme aspartokinase, which catalyzes the phosphorylation of aspartate and initiates its conversion into other amino acids, can be broken up into 3 isozymes, AK-I, II and III. AK-I is feed-back inhibited by threonine, while AK-II and III are inhibited by lysine. As a sidenote, AK-III catalyzes the phosphorylation of aspartic acid that is the commitment step in this biosynthetic pathway. The higher the concentration of threonine or lysine, the more aspartate kinase becomes downregulated. (AK-Aspartate kinase)

Lysine: Lysine is synthesized from aspartate via the diaminopimelate (DAP) pathway. The initial two stages of the DAP pathway are catalyzed by aspartokinase and aspartate semialdehyde dehydrogenase and play a key role in the biosynthesis of lysine, threonine and methionine.

Asparagine: There are two different asparagine synthetases found in bacterial species. These two synthetases, which are both referred to as the AsnC protein, are coded for by two genes: AsnA and AsnB.

Methionine: Methionine synthesis is under tight regulation. The repressor protein MetJ, in cooperation with the corepressor protein S-adenosyl-methionine, mediates the repression of methionine's biosynthetic pathway.

Threonine: The biosynthesis of threonine is regulated via allosteric regulation of its precursor, homoserine, by structurally altering the enzyme homoserine dehydrogenase. This reaction occurs at a key branch point in the pathway, with the substrate homoserine serving as the precursor for the biosynthesis of lysine, methionine, threonin and isoleucine. High levels of threonine result in low levels of homoserine synthesis. The synthesis of aspartate kinase (AK), which catalyzes the phosphorylation of aspartate and initiates its conversion into other amino acids, is feed-back inhibited by lysine, isoleucine, and threonine, which prevents the synthesis of the amino acids derived from aspartate.

Isoleucine: The enzymes threonine deaminase, dihydroxy acid dehydrase and transaminase are controlled by end-product regulation.

Ribose 5-phosphates

The synthesis of histidine in *E. coli* is a complex pathway involving 10 reactions and 10 enzymes. Synthesis begins with 5-phosphoribosyl-pyrophosphate (PRPP) and finishes with histidine and occurs through the reactions of the following enzymes.

HisG-> HisE/HisI-> HisA-> HisH-> HisF-> HisB-> HisC-> HisB-> HisD (HisE/I and HisB are both bifunctional enzymes)

3-Phosphoglycerates

Serine: Serine is the first amino acid in this family to be produced; it is then modified to produce both glycine and cysteine (and many other biologically important molecules). Serine is formed from 3-phosphoglycerate in the following pathway:

3-phosphoglycerate-> phosphohydroxyl-pyruvate-> phosphoserine-> serine

The conversion from 3-phosphoglycerate to phosphohydroxyl-pyruvate is achieved by the enzyme phosphoglycerate dehydrogenase.

Glycine: Glycine is synthesized from serine using the enzyme serine hydromethyltransferase (SHMT), which is coded by the gene glyA. The enzyme effectively removes a hydroxyl group from serine and replaces it with a methyl group to yield glycine. This reaction is the only way *E. coli* can produce glycine.

Cysteine: Cysteine is a very important molecule for a bacterium's survival. This amino acid harbors a sulfur atom and can actively participate in disulfide bond formation.

Pyruvates

Pyruvate is the end result of glycolysis and can feed into both the TCA cycle and fermentation processes. Reactions beginning with either one or two molecules of pyruvate cause the synthesis of alanine, valine, and leucine.

Alanine: Alanine is produced by the transamination of one molecule of pyruvate using two alternate steps: 1) conversion of glutamate to α-ketoglutarate using a glutamate-alanine transaminase, and 2) conversion of valine to α-ketoisovalerate via Transaminase C.

Valine: Valine is produced by a four-enzyme pathway. It begins with the reaction of two pyruvate molecules catalyzed by Acetohydroxy acid synthase yielding á-acetolactate. Step two is the NADPH+ + H+ - dependent reduction of α-acetolactate and migration of the methane groups to produce α, β-dihydroxyisovalerate. This is catalyzed by Acetohydroxy isomeroreductase. The third reaction is the dehydration reaction of α, β-dihydroxyisovalerate catalyzed by Dihydroxy acid dehydrase resulting in α-ketoisovalerate. Finally, a transamination catalyzed either by an alanine-valine transaminase or a glutamate-valine transaminase results in valine.

Leucine: The leucine synthesis pathway diverges from the valine pathway beginning with α-ketoisovalerate. α-Isopropylmalate synthase reacts with this substrate and Acetyl CoA to produce α-isopropylmalate. An isomerase then isomerizes α-isopropylmalate to β-isopropylmalate. The third step is the NAD$^+$ dependent oxidation of â-isopropylmalate via the action of a dehydrogenase to yield α-ketoisocaproate. Finally is the transamination via the action of a glutamate-leucine transaminase to result in leucine.

Urea Cycle

Living organisms excrete the excess nitrogen resulting from the metabolic breakdown of amino acids in one of three ways. Many aquatic animals simply excrete ammonia. Where water is less, plentiful processes have evolved that convert ammonia to less toxic waste products

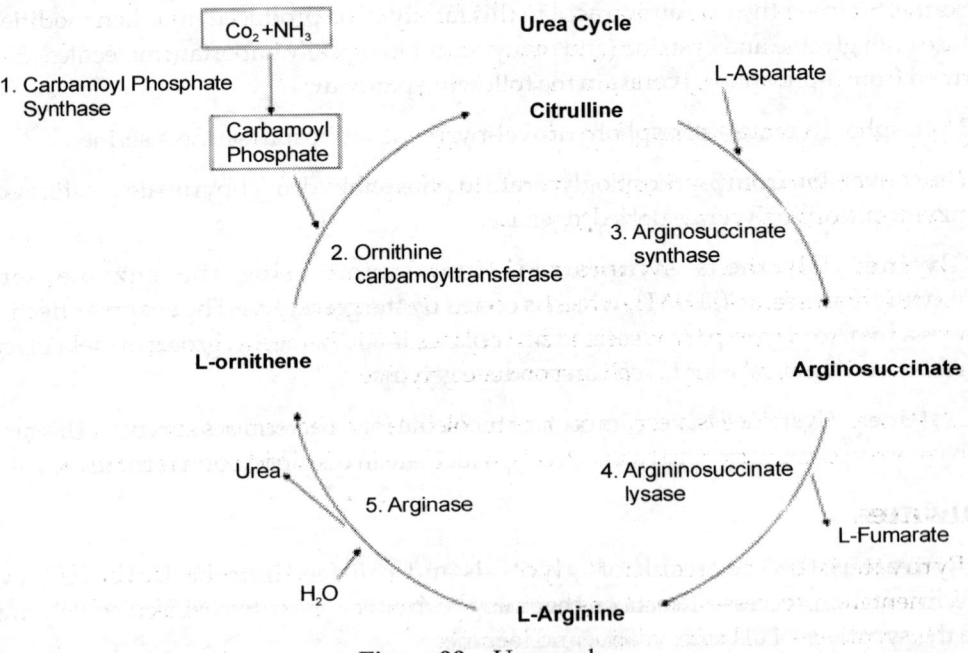

Figure 99   Urea cycle

which require less water for excretion. One such product is urea, which is excreted by most terrestrial vertebrates; another is uric acid, which is excreted by birds and terrestrial reptiles.

Accordingly, living organisms are classified as being either ammonotelic (ammonia excreting), urotelic (urea excreting) and uricotelic (uric acid excreting). Some animals can shift from ammonotelism to urotelism or uricotelism if their water supply becomes restricted. Urea is synthesised in the liver by the enzymes of the urea cycle. It is then secreted into the blood stream and sequestered by the kidneys for excretion in the urine.

The urea cycle reactions were elucidated by Hans Krebs and Kurt Henseleit. This cycle starts with the amino acid ornithine. The cycle is confined only to the mitochondria and cytoplasm of the cells of liver and it is found that the enzyme, arginase which is required in the final step of urea formation is present only in the liver and absent in all the other tissues.

Urea cycle occurs partially in the mitochondria and partially in the cytosol with ornithine and citrulline being transported across the mitochondrial membrane by specific membrane systems. The following are the various reactions in the process of urea formation.

1. Carbamoyl phosphate formation

Carbamoyl phosphate synthetase catalyses the condensation and activation of $NH4 +$ and HCO_3 to form carbamoyl phosphate.

2. Citrulline formation from ornithine

Ornithine transcarbamylase transfers the carbamoyl group of carbamoyl phosphate to ornithine, yielding citrulline. The reaction occurs in the mitochondria so that ornithine, which is produced in the cytosol, must enter the mitochondria via a specific transport system. Like wise, since the remaining urea cycle reactions occur in the cytosol, citrulline must be transported from the mitochondria.

3. Argininosuccinate formation

Citrulline undergoes condensation with amino group of aspartate to form arigininosuccinate this reaction requires ATP, Mg_2^+ and the enzyme argininosuccinate synthetase.

4. Formation of arginine and fumarate

The enzyme argininosucccinase catalyses the elimination of arginine from the aspartate carbon skeleton forming fumarate.

5. Formation of urea

The fifth and the final reaction in the urea cycle is the hydrolysis of arginine by the enzyme arginase to yield urea and ornithine. Ornithine is then returned to the mitochondria for another round of the cycle.

D.   LIPID METABOLISM

Introduction

Lipids are a group of water insoluble greasy organic compound . They are soluble in organic solvents. They are widely distributed in living organisms. They are called esters of fatty acids.

The term lipid was introduced by Bloor in 1943. Fatty acids and alcohol are the major components of lipid.

Lipids are classified into three types they are simple lipids, compound lipids and derived lipids. Triglycerides and waxes are the example for simple lipid. Phospholipid and glycolipids are the example for compound lipid and steroid. Terpenes andcaroternoids are the examples for derived lipids. Simple lipids are made up of glycerol and three fatty acids. Triglyceride is a simple lipid. It is also called neutral fat.

Lipids plays an important role in metabolism as a fuel for the production of ATP. The first step in lipid metabolism is hydrolysis of lipid in to fatty acids and glycerol through the action of lipolytic enzymes.

$$\text{Fat} \xrightarrow[\text{3H}_2\text{O}]{\text{Lipase}} \text{Glycerol + 3 Fatty acids}$$

Oxidation of Triglycerides

Triglycerides are generally used by chemoorganotrophic microorganisms. Lipase enzyme hydrolyses triglyceride in three steps of reactions with the uptake of 3 molecule of water. This hydrolysis releases glycerol and fatty acids

$$\text{Fat} \xrightarrow[\text{3H}_2\text{O}]{\text{Lipase}} \text{Glycerol + 3 Fatty acids}$$

Oxidation of glycerol

Glycerol is phosphorylated by making use of enzyme glycerokinase and converted to glycerol 3 phosphate in the presence of ATP. Glycerol 3 phosphate is dehydrogenated and form dihydroxy acetone phosphate in the presence of NAD. Dihydroxy acetane phosphate undergone isomerization process with the aid of Triose phosphate isomerases and converted to glyceroldehyde 3 phosphate. Glyceroldehyde 3 phosphate undergoes forward or reverse reactions of glycolysis. Forward reactions leads to the formation of pyruvic acid. Reverse reactions leads to the formation of glucose. Condensation of dihydroxy acetone phosphate and glyceroldehyde 3 phosphate form fructose 1, 6 di phosphate, which enters reverse glycolysis and form glucose.

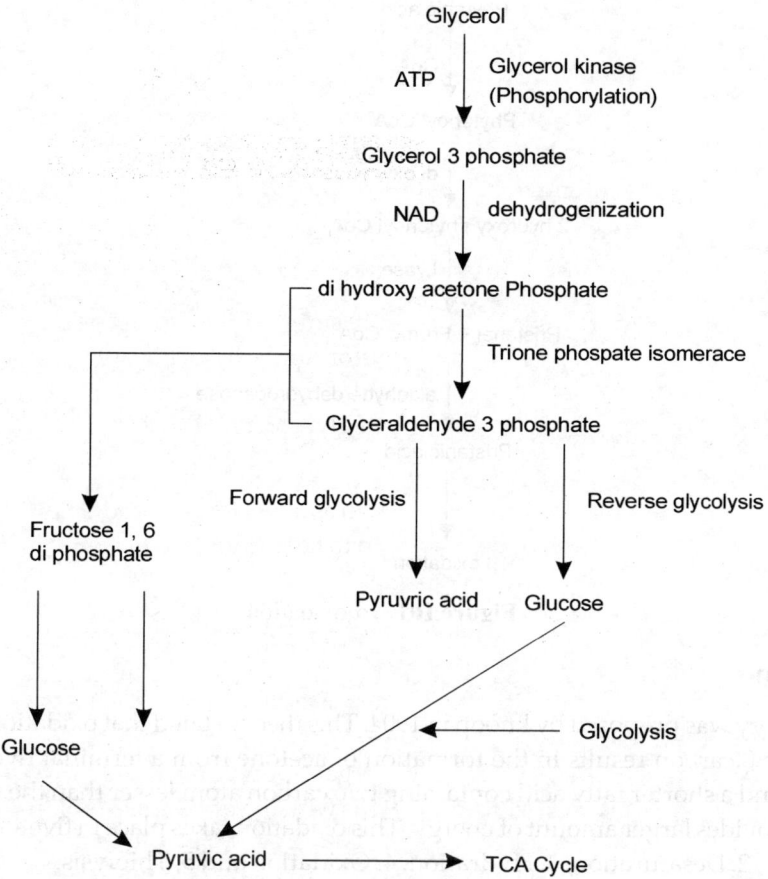

Figure 100 Oxidation of glycerol

Oxidation of fatty acids

Fatty Acids are the major source of energy. They are oxidized to CO_2 and water with the liberation of energy. Fatty acid oxidation takes place in mitochondria in eukaryotes and cytoplasm in prokaryotes. Three major theories explain the process of fatty acid oxidation. They are α oxidation, β oxidation and ω oxidation.

α oxidation

It is a process of fattyacids breakdown in the carboxyl end with the removal of a single carbon from α position. This theory was proposed by P. K. Stump. α oxidation is takes place in peroxisomes to break down dietary phytanic acid. Phytanic acid cannot undergoes α oxidation due to its α methyl branch. Phytanic acid is converted to pristanic acid then acquire acetyl CoA and subsequently become α oxidized. Phytanic acid is attached to Co A to form phytanoyl CoA. Phytanoyl CoA is oxidized by phytanoyl CoA dioxygenase and form hydroxy phytanoyl CoA. Hydroxy phytanoyl CoA is cleaved by lyase enzyme and form pristanal and formyl CoA. Pristanol is oxidized by aldehyde dehydrogenase to form pristanic acid. Pristanic acid may involve in α oxidation.

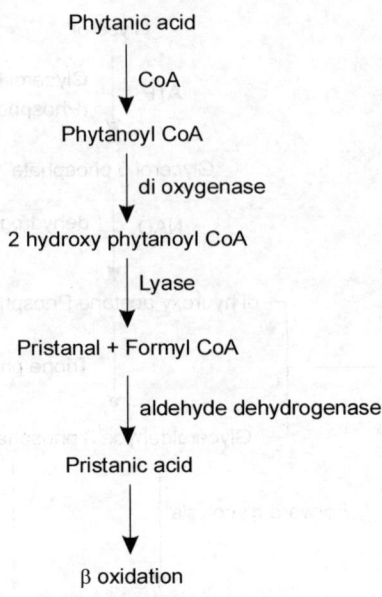

Figure 101 α oxidation

β oxidation

This theory was proposed by knoop in 1904. This theory stated that oxidation of fatty acid occurs at the β carbon results in the formation of acetone from a terminal two carbon and leaving behind a shorter fatty acid containing two carbon atom lesser than the original. This oxidation provides larger amount of energy. This oxidation takes place in five steps. They are: 1. Activation, 2. Desaturation, 3. Hydration, 4. Oxidation and 5. Thiolysis.

1. Activation

Fattyacids are activated by the addition of CoA, ATP and Magnesium ions with the help of thiokinase. Activated fatty acid is known as fatty acyl CoA.

$$\text{Fatty acid} + \text{CoA} + \text{ATP} \longrightarrow \text{Acetyl CoA derivative}$$

2. Desaturation

Activated fatty acid dehydrogenated by acetyl; CoA dehydrogenase in to α, β factty acid coA derivative.

$$\text{Acetyl CoA} \longrightarrow \alpha, \beta \text{ fatty acid CoA}$$

3. Hydration

α, β factty acid coA undergoes hydration and combined with a molecule of water using an enzyme hydratase and form α hydroxyl acyl CoA

$$\alpha, \beta \text{ fatty acid CoA} + \text{Water} \longrightarrow \alpha, \beta \text{ hydroxyl acyl CoA}$$

4. Oxidation

β hydroxyl acyl CoA derivative undergoes oxidation to form keto fatty acid in the presence of dehydragenase and NAD

$$\alpha \text{ hydroxyl acyl CoA + NAD} \longrightarrow \alpha\beta \text{ keto fatty acid CoA}$$

5. Thiolysis

β keto fatty acid CoA is split into acetyl coA and active fatty acid with the help of thiolase.

$$\alpha \text{ keto fatty acid CoA} \longrightarrow \alpha \text{ Acetyl CoA + active fatty acid}$$

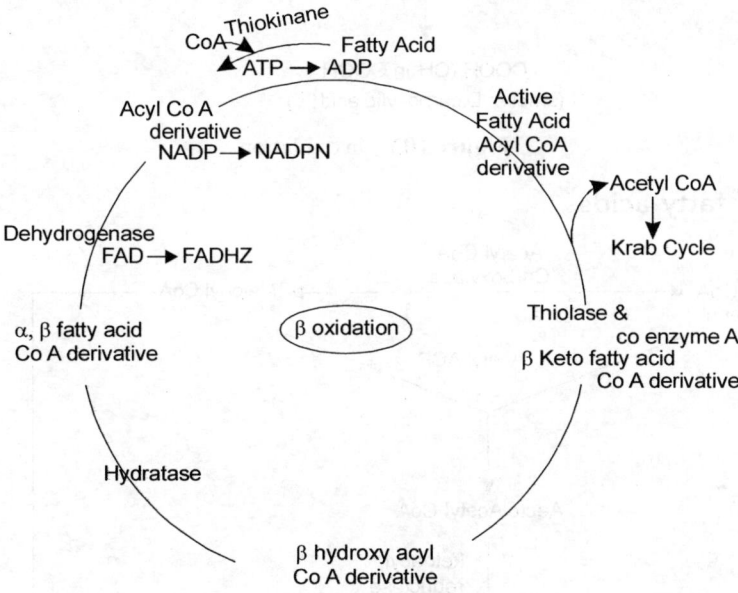

Figure 102 β oxidation

The newly formed acetyl CoA may enter in TCA cycle. Active fatty acid acyl CoA reenters the cycle and repeated till the fatty acid is completely splitup into acetyl units.

ω oxidation

This theory was proposed by Verkade. This theory stated that oxidation occurs at the carbon atom situated farthest from the carboxyl group [ω carbon]. Omega oxidation results in the formation of Dicarboxylic acid. After this process, ω oxidation starts at both ends successively. ω oxidation is quicker than regular ω oxidation process. It occurs in three steps

1. First step introduces hydroxyl group in to the ω carbon. It is done by mixed function oxidase.

2. Oxidation of the hydroxyl group to an aldehyde by NAD. It is by alcohol dehydrogenase.

3. Oxidation of the aldehyde group by aldehyde dehydrogenase to a carboxylic acid by NAD.

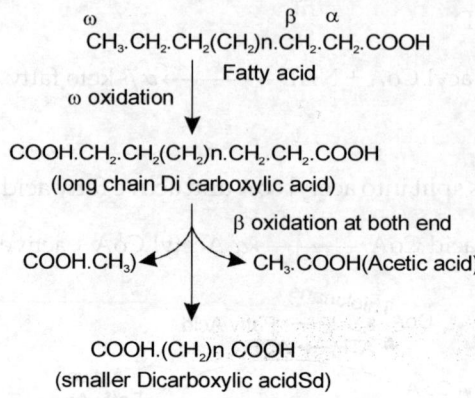

$$\overset{\omega}{CH_3}.CH_2.CH_2(CH_2)n.CH_2.\overset{\beta}{CH_2}.\overset{\alpha}{COOH}$$

Fatty acid

ω oxidation

$COOH.CH_2.CH_2(CH_2)n.CH_2.CH_2.COOH$
(long chain Di carboxylic acid)

β oxidation at both end

$COOH.CH_3$ ◄──── ────► $CH_3.COOH$(Acetic acid)

$COOH.(CH_2)n\ COOH$
(smaller Dicarboxylic acidSd)

Figure 103   ω oxidation

Synthesis of fatty acids

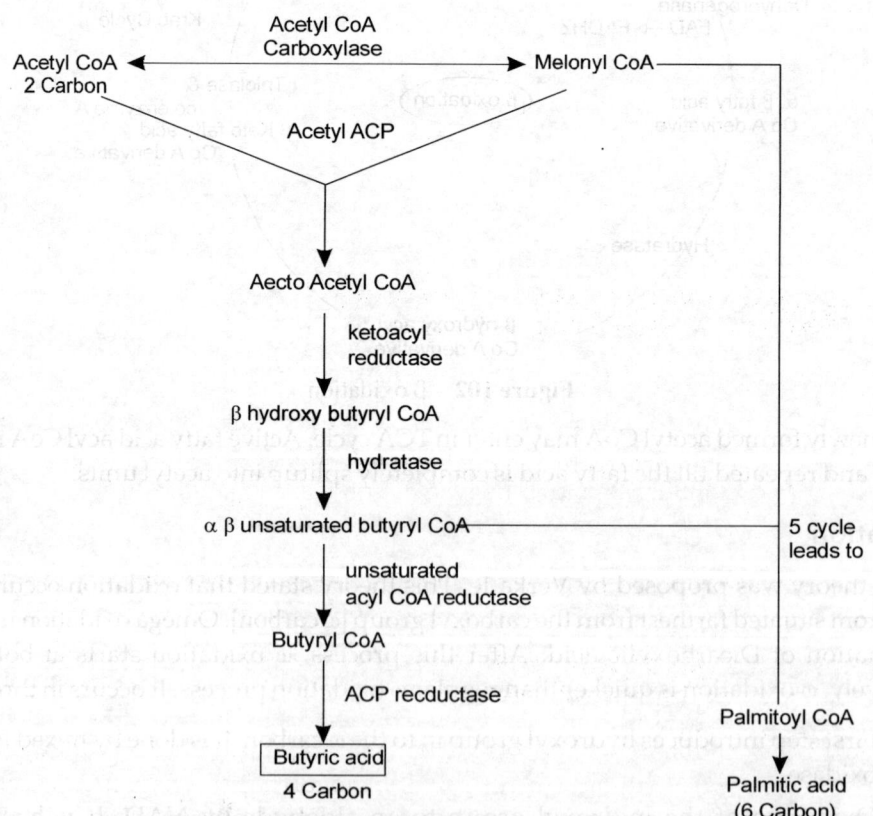

Figure 104   Fatty acid synthesis

The process of synthesizing fatty acids are called lipogenesis. It takes place in the cytoplasm. Fatty acid synthesis occurs in all organisms. Multienzyme complex called fattyacid synthetase is responsible for fatty acid synthesis.

⮞ Melonyl CoA reacts with acetyl CoA and gets decarboxylated in the presence of condensing enzyme to form aceto acetyl CoA.

⮞ Aceto acetyl CoA is reduced to â hydroxy butyryl CoA in the presence of an enzyme ketoacyl CoA reductase.

⮞ β hydroxy butyryl CoA, by the removal of water molecule, converted to a, b unsaturated butyryl CoA in the presence of hydratase.

⮞ α, β unsaturated butyryl CoA is reduced to butyryl CoA by the action of acyl CoA reductase.

⮞ Final step of first cycle is the formation of butyric acid (4 Carbon) by the action of enoyl ACP reductase.

In the second step butyryl ACP condenses with melonyl ACP with the similar sequence of reactions results in 6 Carbon ACP. Similar cycle follows until 16 carbon ACP is formed. 16 carbon fatty acid is palmitic acid.

Biosynthesis of triglycerides

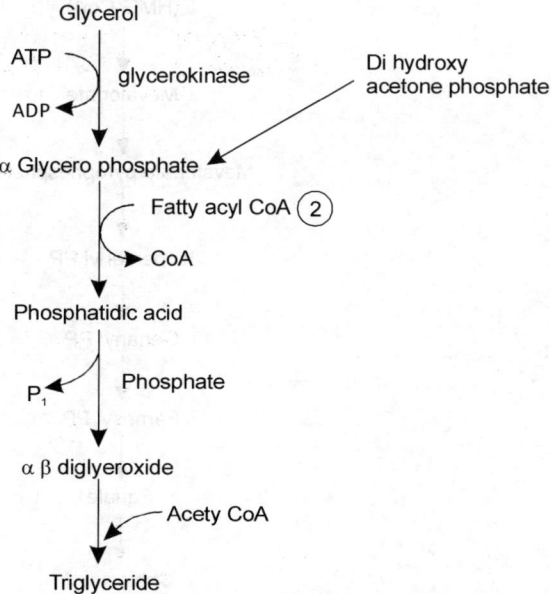

Figure 105 Triglycerides synthesis

Fatty acids combine with glycerol to form triglycerides. The glycerol can be derived from dihydroxy acetone phosphate. It is an energy yielding process. The first step in triglyceride synthesis is the activation of glycerol and fatty acids by ATP. Glycerol is activated by ATP in the presence of glycerokinase to form α glycerol phosphate. Fatty acid is activated to form acyl

Co A by the enzyme thiokinase in the presence of ATP, CoA and SH. The α glycerol phosphate is esterified with 2 molecules of aceyl CoA to form α β diglyceride phosphate (phosphatidic acid) in the presence of glycerol phosphate acyl transferase. The phosphatidic acid is dephosphorylated by the presence of phosphatase to form α, β diglyceride. Another one molecule of acyl CoA is esterified with the diglyceride, to form triglyceride.

Biosynthesis of cholesterol

Cholesterol is synthesized from 2 carbon units in the form of acetyl CoA. Two molecules of acetyl Co A condense to form aceto acetyl coA. Aceto acetyl coA reacts with actyl coA to form α hydroxyl α methyl glutaryl CoA (HMG CoA), which inturn gives rise to mevalonic acid. mevalonic acid is phosphorylated three times and forms mevalonate pyro phosphate, which is converted to isopentenyl pyrophosphate, genanyl pyrophosphate and farnesyl phyrophosphate. The two molecule of and farnesyl phyrophosphate condense to form a hydrocarbon squalene. It undergoes ring closure and loss of methyl group and converted to cholesterol.

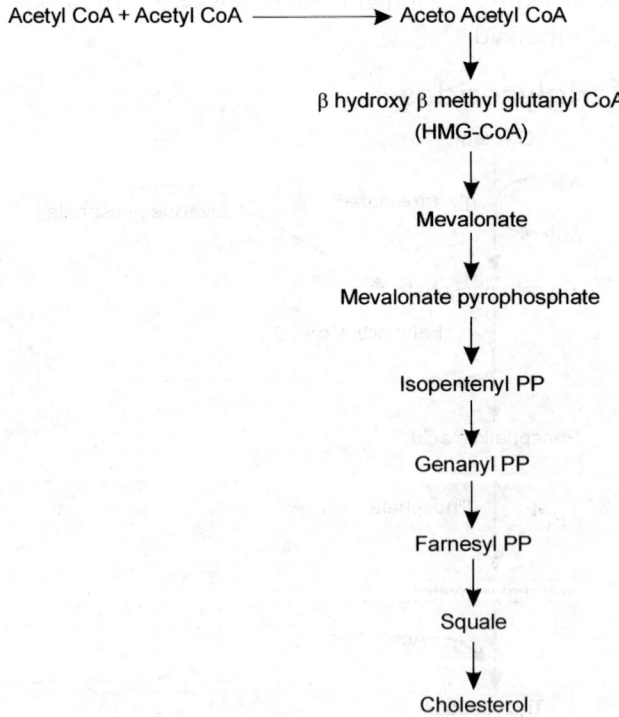

Figure 106   Biosynthesis of cholesterol

Biosynthesis of lecithin/phospholipid

Phospholipids are synthesized from fatty acids and its precursors. Triose phosphate is reduced by the dihydroxy acetone phosphateto 3 glycerol phosphate, which is subsequently esterified by 2 fatty acid residues. The resulting diglyceride phosphatic acid is then activated by CTP to form CDP di glyceride. It undergoes transfer reactions with serine and a glycerophosphate releasing CMP. The reaction between CDP diglyceride and glycerol phosphate leads to the forformation of phospholipid classes like phosphotidyl glycerol, lecithin and cardiolipin.

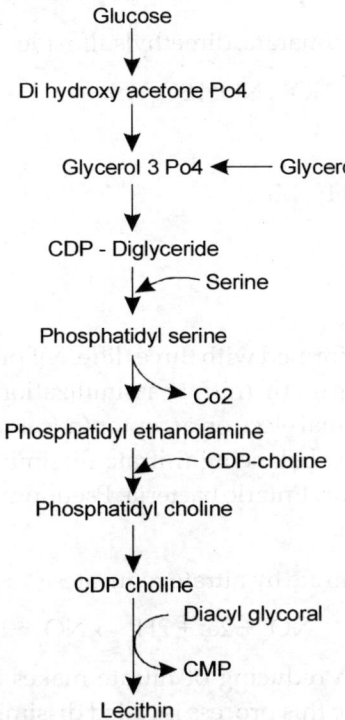

Figure 107   Biosynthesis of phospholipid

E.   ANAEROBIC RESPIRATION

It is a type of respiration using electron acceptors other then oxygen this respiration is mainly used by procaryotes. Many cells are capable of generating ATP using pathways that either rely on oxygen or do not rely on oxygen. Aerobic pathways produce more ATP per gram than do the anaerobic pathways. The most common terminal electron acceptors of anaerobic respiration are nitrate, sulfate & Co_2. Facultative anaerobes, microaerophils, obligate anaerobe are able to perform anaerobic respiration. Facultative anaerobes shift aerobic respiration to anaerobic process in the absence of oxygen only. Final electron acceptor is an inorganic substance other than O_2. Some bacteria such as *Pseudomonas* and *Bacillus* can use a nitrate ion (NO_{3}^{-}), in the presence of an enzyme called nitrate reductase, as a final electron acceptor, the nitrate ion is reduced to nitrite ion (NO_2^-). Nitrite ion can be converted to nitrous oxide (N_2O), or nitrogen gas (N_2) (denitrification process) which helps in recycling of nitrogen. Other bacteria like *Desulfovibrio* use sulfate (SO_4^{2-}) as the final electron acceptor and forms hydrogen sulfide (H_2S). Some bacteria use carbonate (CO_3^{2-}) to form methane (CH_4). Anaerobic respiration by bacteria using nitrate and sulfate as final electron acceptors is essential for the nitrogen and sulfur cycles that occur in nature. Amount of ATP generated varies with the organisms and the pathway. Because only a part of the Krebs cycle operates and since not all the carriers in the electron transport chain participate, ATP yield is less and accordingly anaerobes tend to grow more slowly than aerobes.

Diversity of electron acceptors for respiration

Organic compounds: Eg. Fumarate, dimethylsulfoxide

Inorganic compounds: Eg. $NO^{3-}, NO^{2-}, SO_4^{2-}, S^0$

Metals: Eg. $Fe^{3+}, Mn^{4+}, Cr^{6+}$

Minerals/solids: Eg. $Fe(OH)_3, MnO_2$

Gasses: Eg. NO, N_2O, CO_2

Nitrate respiration

Nitrate reduction can be performed with three different purposes. They are utilization of nitrate as a nitrogen source for growth (nitrate assimilation), the generation of metabolic energy by using nitrate as a terminal electron acceptor (nitrate respiration) and dissipation of excess reducing power for redox balancing (nitrate dissimilation). Some bacteria can use nitrate as a final electron acceptor. Enteric bacteria, Pseudomonas, Bacillus, Paracoccus are example for denitrifying bacteria.

Nitrate may be reduced to nitrate by nitrate reducates.

$$NO_3 + 2e^- + 2H^+ \rightarrow NO_2 + N_2O$$

In the anaerobic respiration reducing of nitrate makes it unavailable to the cells for assimilation or uptake. Therefore this process is called dissimilatery nitrate reduction.

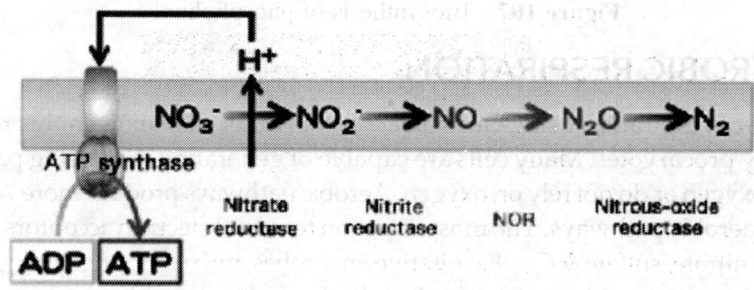

Figure 108 Nitrate respiration

Reducing of NO_3 to NO_2 is not an efficient way of making ATP because large amount of NO_3 is required for growth. Nitrite formed is quite toxic therefore nitrite after is further reduced to nitrogen gas. This process is called de nitrification.

$$NO_3 + 10e^- + 12H^+ \rightarrow N_2 + 6H_2O$$

De Nitrification is a multistep process involves 4 enzymes namely nitrate reductase, Nitrite reductase, nitric oxide reductase, nitrous oxide reductase.

$$NO_3 \rightarrow NO_2 \rightarrow NO \rightarrow NO_2 \rightarrow N_2$$

Electron transport chain in nitrate reduction are highly branched and cylochrome a complex is replaced with cytochrome b and c celectrons are passed from coenzyme to cyst b for the reduction of nitrate to nitrite. Electrons then flow through cyt e for the sequential oxidation of nitrite to N_2. Not much protons are pumped across the membrane, but nonetheless a PMP is established which results in ATP production two types of bacterial Nitrite reductase catalyze the formation of No in bacteria. Nitrite reductase is present in periplasmic space in or bacteria. Denitrification in anoxic soil results in the loss of soil N_2 and advaersely affects soil fertility. Four types of nitrate reductases catalyse the two electron reduction of nitrate to nitrite.

Methane Respiration

Methanogens are involved in anaerobic respiration, they are obligate anaerobic bacteria. Methanogenic bacteria use CO_2 or carbonate as terminal electron acceptor. These organisms are called methanogens because the electron acceptor is reduced to methane. Eg. Methanobacter.

Biochemistry of Methanogenesis

Methanogenesis in microbes is a form of anaerobic respiration. Methanogens do not use oxygen to respire; in fact, oxygen inhibits the growth of methanogens. The terminal electron acceptor in methanogenesis is not oxygen, but carbon. The carbon can occur in a small number of organic compounds, all with low molecular weights. The two best described pathways involve the use of carbon dioxide and acetic acid as terminal electron acceptors:

$$CO_2 + 4H_2 \rightarrow CH_4 + 2H_2O$$
$$CH_3COOH \rightarrow CH_4 + CO_2$$

However, depending on pH and temperature, methanogenesis has been shown to use carbon from other small organic compounds, such as formic acid (formate), methanol, methylamines, dimethyl sulfide and methanethiol.

Sulphate respiration

Sulfate reduction is a type of anaerobic respiration that utilizes sulfate as a terminal electron acceptor in the electron transport chain. Sulfate reduction is a relatively energetically poor process. It is a vital mechanism for bacteria and archaea living in oxygen-depleted, sulfate-rich environments.

Many sulfate reducers are organotrophic, using carbon compounds, such as lactate and pyruvate (among many others) as electron donors, while others are lithotrophic, and use hydrogen gas (H_2) as an electron donor. Some unusual autotrophic sulfate-reducing bacteria (e.g., Desulfotignum phosphitoxidans) can use phosphite (HPO_3^{-}) as an electron donor, whereas others (e.g., Desulfovibrio sulfodismutans, Desulfocapsa thiozymogenes and Desulfocapsa sulfoexigens) are capable of sulfur disproportionation (splitting one compound into two different compounds, in this case an electron donor and an electron acceptor) using elemental sulfur, sulfite, and thiosulfate to produce both hydrogen sulfide and sulfate.

Before sulfate can be used as an electron acceptor, it must be activated. This is done by the enzyme ATP-sulfurylase, which uses ATP and sulfate to create adenosine 5'-

phosphosulfate (APS). APS is subsequently reduced to sulfite and AMP. Sulfite is then further reduced to sulfide, while AMP is turned into ADP using another molecule of ATP. The overall process, thus, involves an investment of two molecules of the energy carrier ATP, which must to be regained from the reduction.

All sulfate-reducing organisms are strict anaerobes. Because sulfate is energetically stable, it must be activated by adenylation to form APS (adenosine 5-phosphosulfate) to form APS before it can be metabolized, thereby consuming ATP. The APS is then reduced by the enzyme APS reductase to form sulfite and AMP. In organisms that use carbon compounds as electron donors, the ATP consumed is accounted for by fermentation of the carbon substrate. The hydrogen produced during fermentation is actually what drives respiration during sulfate reduction.

$$So_4^{2+} 8e^- + 8h^+ \rightarrow S^{2-} + 4H_2O$$

Sulfate-reducing bacteria can be traced back to 3.5 billion years ago and are considered to be among the oldest forms of microorganisms, having contributed to the sulfur cycle soon after life emerged on Earth. Sulfate-reducing bacteria are common in anaerobic environments (such as seawater, sediment, and water rich in decaying organic material) where they aid in the degradation of organic materials. In these anaerobic environments, fermenting bacteria extract energy from large organic molecules; the resulting smaller compounds (such as organic acids and alcohols) are further oxidized by acetogens, methanogens, and the competing sulfate-reducing bacteria.

Many bacteria reduce small amounts of sulfates in order to synthesize sulfur-containing cell components; this is known as assimilatory sulfate reduction. By contrast, sulfate-reducing bacteria reduce sulfate in large amounts to obtain energy and expel the resulting sulfide as waste; this is known as "dissimilatory sulfate reduction. " Most sulfate-reducing bacteria can also

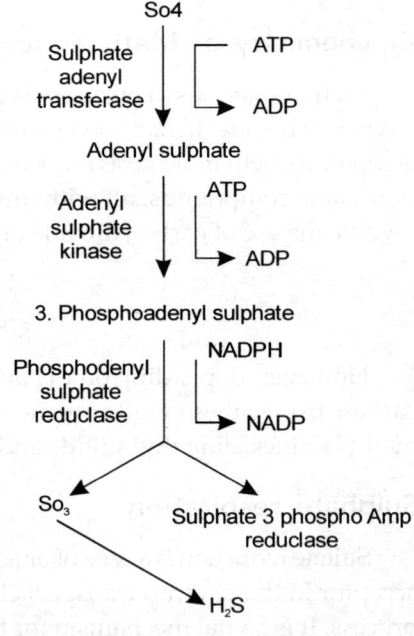

Figure 109   Sulphur respiration

reduce other oxidized inorganic sulfur compounds, such as sulfite, thiosulfate, or elemental sulfur (which is reduced to sulfide as hydrogen sulfide).

Toxic hydrogen sulfide is one waste product of sulfate-reducing bacteria; its rotten egg odor is often a marker for the presence of sulfate-reducing bacteria in nature. Sulfate-reducing bacteria are responsible for the sulfurous odors of salt marshes and mud flats. Much of the hydrogen sulfide will react with metal ions in the water to produce metal sulfides. These metal sulfides, such as ferrous sulfide (FeS), are insoluble and often black or brown, leading to the dark color of sludge. Thus, the black color of sludge on a pond is due to metal sulfides that result from the action of sulfate-reducing bacteria .

F. FERMENTATION

Fermentation is an energy yielding metabolism that involves a sequence of oxidation - reduction reaction. Quantum of ATP generation in fermentation is lower than oxidation. According to Louis Pasteur Fermentation is a anaerobic process. He stated that fermentation means life without air. In fermentation more substrates are utilized than in respiration. Depends on the substrates used for formation & product yielded, the fermentation is of following types. They are lactic acid fermentation, ethanolic fermentation, propionic acid fermentation, mixed acid fermentation, Butanediol fermentation, butyric acid fermentation and aminoacid fermentation.

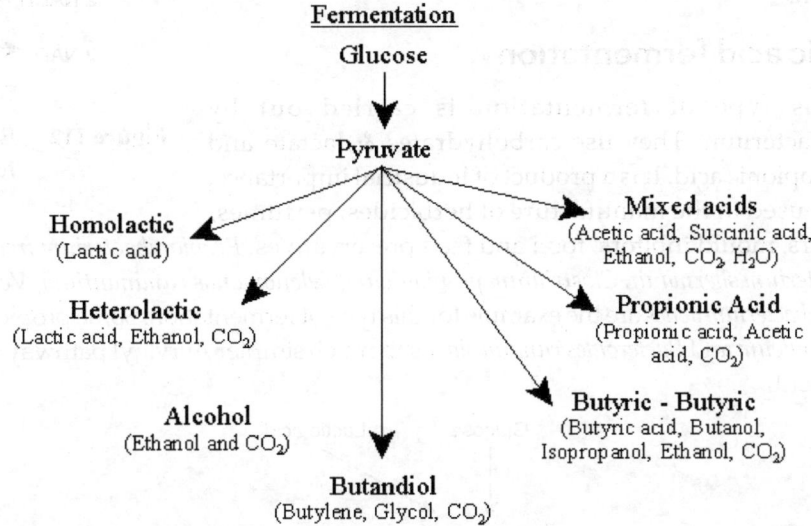

Figure 110   Fate of fermentation

Lactic acid Fermentation

The reduction of pyruvate to lactic acid is called lactic acid fermentation. Lactobacillus and Bacillus are involved in lactic acid fermentation. There are two types of lactic acid fermentation. They are homolactic & heterolactic fermentation. In Homolactic fermentation the only end product is lactate. *Streptococcus, Lactococcus, Lactobacillus, Enterococcus, Pediococcus* are involved in this type of fermentation and they are said to be homolactic fermenters. Homolactic fermenters use EMP pathway and directly reduces almost all puruvate to lactate with the help of enzyme lactate dehydrogenase. This type of fermentation is necessary for industrial production purposes, particularly in dairy industry for the preparation of sour milk & the production of cheese,

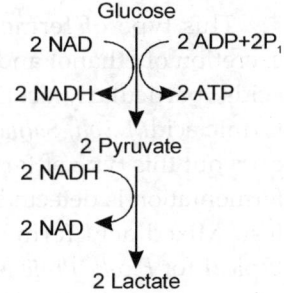

Figure 111   Lactic acid fermentation

yoghurt. Hetero lactic fermenters form substantial amounts of products other than lactate. They produce ethanol and Co_2 other than lactic acid. Enterobacterial members are example for herterolactic fermenters.

Glucose + ADP + Pi → Lactic acid + ethanol + CO_2 + ATP.

Ethanolic fermentation

Those fermentation that results in alcohol production are called ethanolic fermentation. Alcohol fermentation produces alcoholic beverages like beer, wine. Similarly CO2 from this fermentation causes bread to rise.

$$\text{Glucose} + 2ADP + 2\,Pi \rightarrow 2\,\text{ethanol} + 2\,CO_2 + 2\,ATP.$$

Yeast & *Zymomonas mobilis* are actively involved in Ethanolic fermentation.

Propionic acid fermentation

This type of fermentation is carried out by Propionicbacterium. They use carbohydrates & lactate and produce propionic acid. It is a product of industrial importance. It is widely used in the manufacture of herbicides, perfumes, fruit flavours, mould inbitors, food and feed preservatives. *Propionibacterium freudenreichii*, *Propionibacterium shermanii*, *Clostridium propionicum*, *Selenomonas ruminantium*, *Veillonella sp.* and *Propionibacterium acnes* are the example for this type of fermentation. Some propionibacteria, e.g. *C. propionicum* and *Bacteroides ruminicola*, use a much simpler Acryloyl pathway to produce propionic acid.

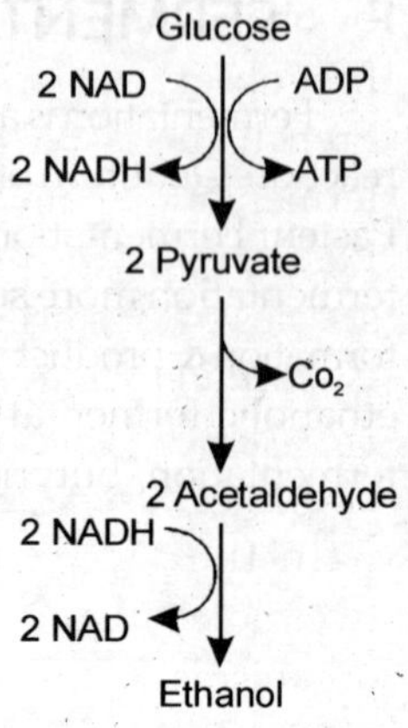

Figure 112   Ethanol fermentation

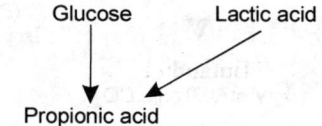

Figure 113   Propionic fermentation

Mixed acid fermentation

This type of fermentation results in the excretion of ethanol and a complex mixture of acids, particularly acetic, lactic, succinic and formic acid. *E.coli, Salmonella* are the microbes carryout this type of fermentation. Mixed acid fermentation is detected by making use of MR test. Mixed acid fermentation pathways are typical for *E.coli, Proteus, Salmonella* and many other gas-producing bacterial species which express the formate hydrogen lyase (FHL) enzyme system.

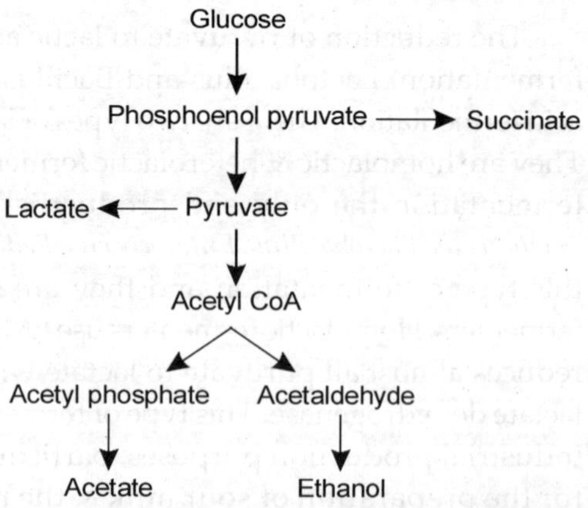

Figure 114   Mixed acid fermentation

Butyric acid Fermentation

This fermentation is also called as Butanol fermentation. It is carried by *Clostridium acetobutylicum*. Butyrate is produced as a end product of a fermentation process solely performed by obligate anaerobic bacteria. Butyric acid was first observed (in impure form) in 1814 by the French chemist Michel Eugène Chevreul. This fermentation pathway was discovered by Louis Pasteur. Examples of butyrate-producing species of bacteria are *Clostridium butyricum, Clostridium kluyveri, Clostridium pasteurianum, Fusobacterium nucleatum, Butyrivibrio fibrisolvens* and *Eubacterium limosum*. The pathway starts with the glycolytic cleavage of glucose to two molecules of pyruvate, as happens in most organisms. Pyruvate is then oxidized into acetyl coenzyme A using a unique mechanism that involves an enzyme system called pyruvate-ferredoxin oxidoreductase. Two molecules of carbon dioxide (CO_2) and two molecules of elemental hydrogen (H2) are formed as waste products from the cell.

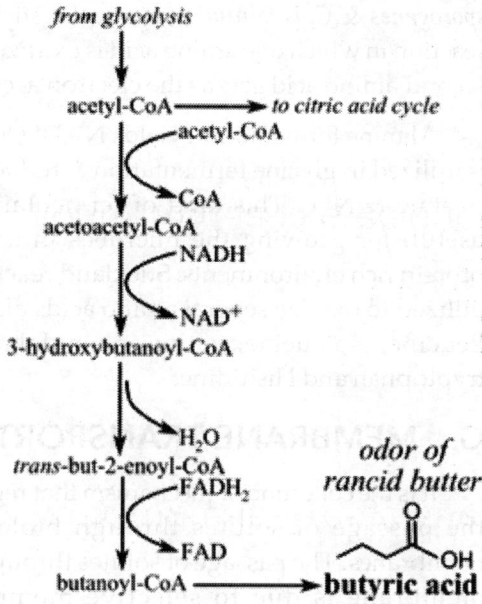

Figure 115 Butyric acid fermentation

Butanediol fermentation

It is a characteristic feature of Enterobacker, Serratia, Erwina, Klebsiella. Butanediol is a natural end product of carbohydrate fermentation. Acetoin / Acetyl methyl carbinol is an intermediate product of use of voges proskauer (VP) test. In this fermentation, pyruvate is converted to acetoin, which is then reduced to 2, 3 butanediol with NADH.

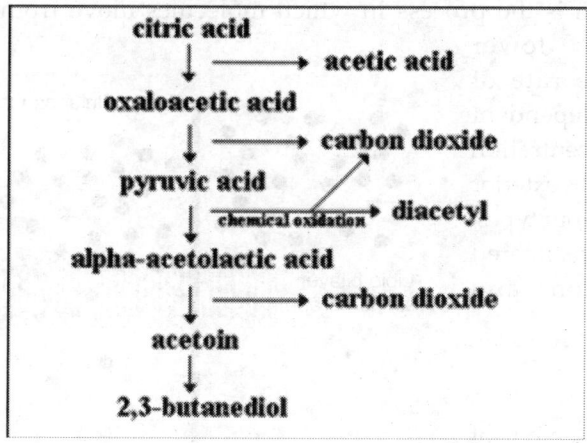

Figure 116 Butanediol fermentation

Aminoacid Fermentation

Most of the microorganisms carryout carbohydrate fermentation. Some members of the genus Clostridium carryout amino acid fermentation. Proteolytic clostridium such *Clostridium sporogenes* & *C. botalium* carryout the stickland reaction in which one amino acid is oxidized & a second amino acid acts as the electron accepter.

Alanine fermentation yields NADH, which is utilized in glycine fermentation & reduced to Acetate & NH_3. This kind of fermentation is usefull for growing the microbes in anoxic, protein rich environments. Stickland reaction is utilized to oxidize several amino acids alanine, Leucine, isoleucine, valine, phenylalamine, tryptophan and Hisitidine.

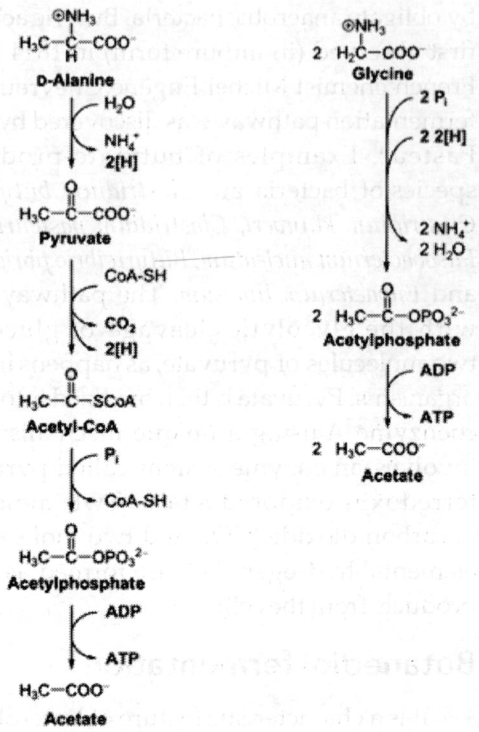

G. MEMBRANE TRANSPORT

It is the collection of mechanism that regulate the passage of solutes through biological membranes. The passage of solutes through the membrane is due to selective membrane permeability.

Membrane transport is of three types. They are passive transport, active transport and group translocation.

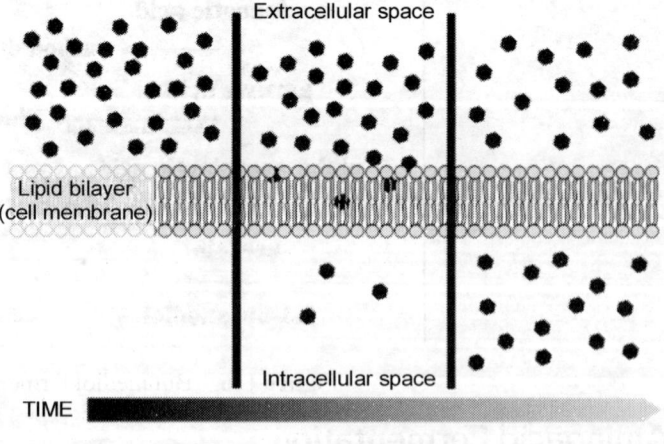

Figure 117 Amino acid fermentation

Passive Transport

Passive diffusion is the process in which molecules move from a region of higher concentration to lower concentration. The rate of passive diffusion is dependent on the size of the concentration gradient between a cells exterior and its interior. It is of four types. They are diffusion, facilitated diffusion, filtration and osmosis.

Diffusion

It is the net movement of material from an area of higher concentration to an area with lower concentration. It is depends on concentration gradient.

Figure 118 Diffusion

Facilitated Diffusion

Small molecules such as H_2O, O_2 and CO_2 often move across membranes by diffusion. Large molecules, ions and polar substances do not cross membranes by diffusion. In such condition, the rate of diffusion across selectively permeable membrane is greatly increased by using carrier proteins, called permeases. It is present in the (carrier) Plasma membrane.

Because carrier protein aids in the diffusion process, it is called facilitated diffusion. The rate of facilitated diffusion increases with the concentration gradient.

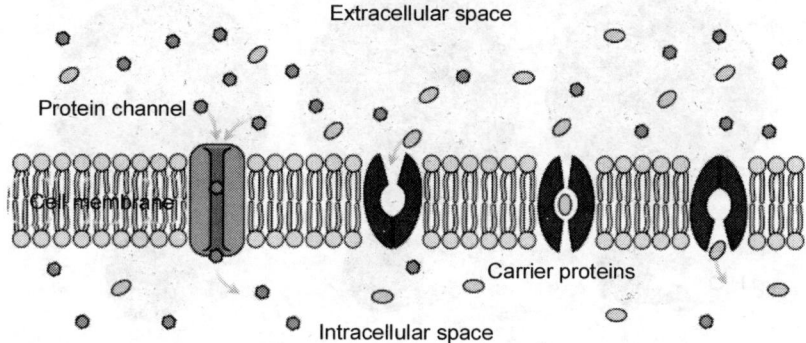

Figure 119   Facilitated diffusion

Filteration

It is movement of water and solute across the cell membrane due to hydrostatic pressure. Eg. Urine Filtration.

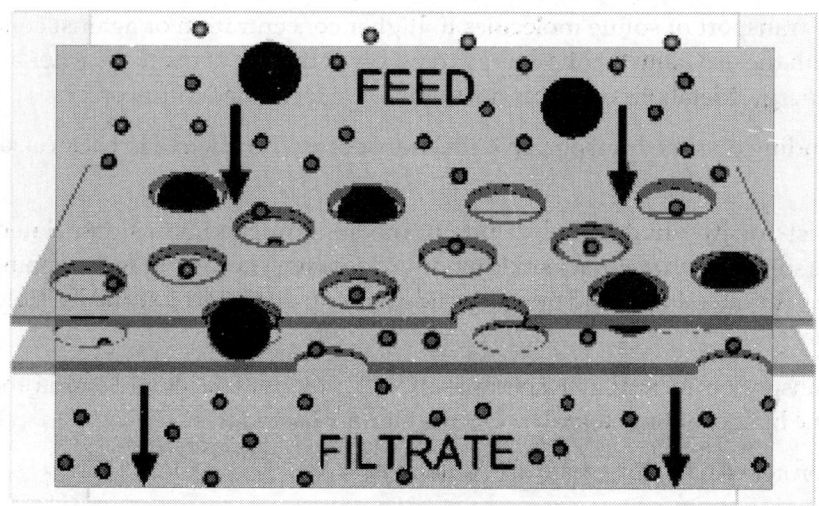

Figure 120   Filteraion

Osmosis

It is the diffusion of water molecules across the selectively permeable membrane. Depends on the concentration. There are three types of solution. They are Hypertonic, Isotonic and Hypotonic. A cell with a less negative water potential will draw in water but this depends on other factors as well such as solute potential (pressure in the cell) and pressure potential (External pressure).

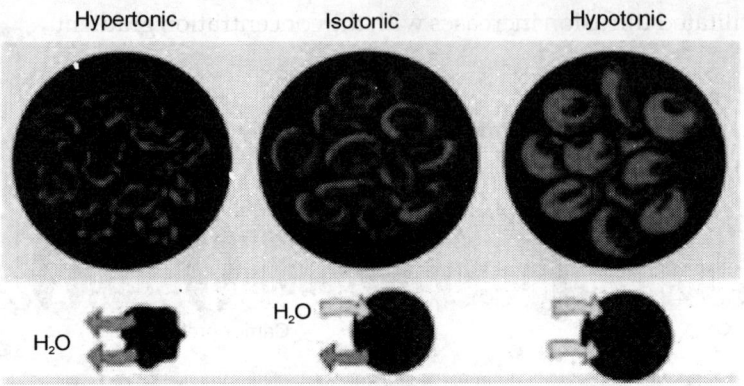

Figure 121   Osmosis

If the cell is paced is Hypertonic solution, exosmosis taken place and leads to shrinkage of cells.

If it is placed in isotonic solution, there is no change in cell. If the cell is placed in Hypotonic solution, endosmosis takes place, lead to enlargement of cell and finally bursting of cell.

Active Transport

It is the transport of solute molecules to higher concentration or against concentration gradient with the use of metabolic energy. It involves the use of protein carrier activity and metabolic energy. Metabolic energy is not involved in facilitated diffusion.

ATP binding cassette transported (ABC transporters) are active in bacteria, archae and eukaryotes.

It consists of two hydrophobic membrane spanning domains associated on their cytoplasmic surface with two nucleotide binding domains. The membrane spanning domains form a pore in the membrane and the nucleotide binding domains bind and hydrolyze ATP to drive uptake.

ABC transporters located in the periplasmic space of gram negative bacteria and attached to membrane lipids on the external face of the Gram + bacteria.

These proteins may participate in chemotaxis.

There are two types of active transport. They are primary active transport and secondary active transport.

Primary Active Transport

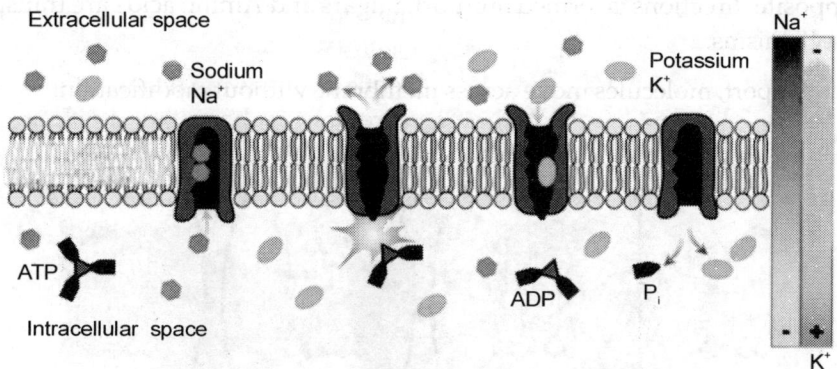

Figure 122   Primary active transport

The action of the sodium potassium pump is an example for primary active transport. Transmembrane ATPase helps to maintain cell potential, which uses ATP for the use of membrane transport.

Other energy used for primary active transport are redox energy and photon energy.

Secondary Active Transport

It is also called to transport. In this, there is no direct coupling of ATP, instead , the electrochemical potential difference created by pumping ions out of the cell is used. The two main forms of this are antiport and symport.

The lactose permease enzyme transport lactose molecule inward as a proton simultaneously enters the cell. Such linked transport of two substances in the same direction is called symport. These energy stored as a proton gradient drives solute transport..

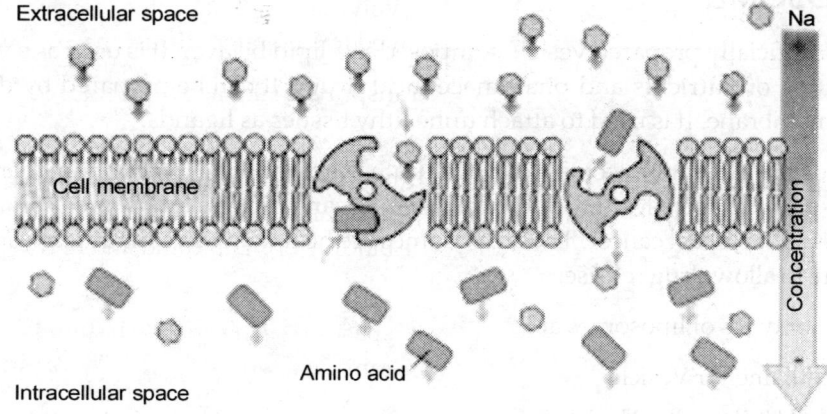

Figure 123   Secondary active transport

A proton gradient also can power transport indirectly, often through the formation of sodium ion gradient, Ex. *E.coli* sodium transport system pumps sodium outward in response to the inward movement of protons. Such linked transport in which the transported substances move in opposite directions is termed anitport. Sugars and Amino acids are transported by antiport mechanisms.

In this transport, molecules move across membrane without modification.

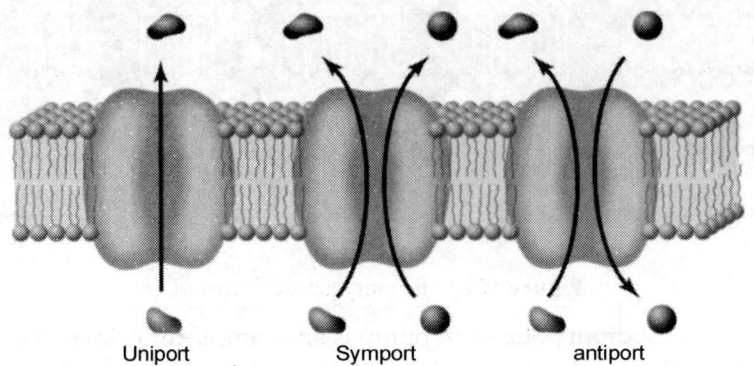

Uniport   Symport   antiport

Figure 124   Function of symport and antiport

Group translocation

Many prokaryotes takeup molecule by group translocation. It is process in which a molecule is transported into the cell after the modification. The best known examples is the phosphoenol pyruvate; Sugar phosphor transferase system (PTs).

It transports variety of sugars into cell after phosphorylation using phosphor enol pyruvate as the phosphate donor.

$$PEP-Sugar \xrightarrow{\hspace{1.5cm}} Pyruvate-Sugar\text{-}P$$
$$\text{(Outside)} \hspace{3.5cm} \text{(Inside)}$$

H.   LIPOSOME

It is an artificially prepared vesicle composed of a lipid bi layer. It is used as a vehicle for administration of nutrients and pharamaceutical drugs. It can be prepared by disrupting biological membrane. It is used to attach unhealthy tissues as ligands.

They're basically fat molecules (carbohydrate head and long hydrophobic carbon tails) that form spherical micelles which can be used to carry drugs. Liposomes can also carry drugs targetted to breast cancer. These liposomes are heat-activated, made leaky at a certain temperature to allow drug release.

The major types of liposomes are

Multilamellor Vesicle

Small Unilamellar Vesicle

Large Unilamellar Vesicle

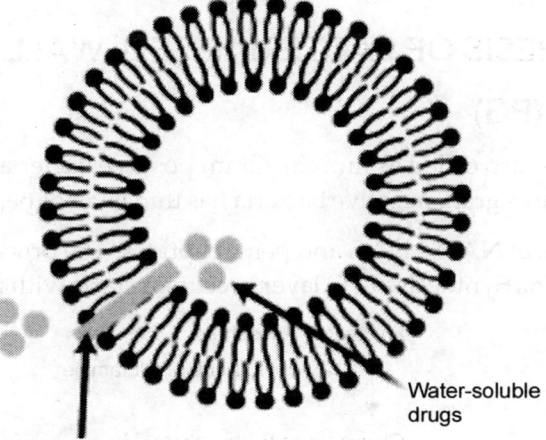

Liposomes can entrap drugs inside for subsequent delivery to a disease site.

Lipid-soluble drugs in membrane

Water-soluble drugs

Figure 125 Liposome

I. BIOLUMINESCENCE

Bioluminescence is the production and emission of light by living organisms. It is done by the organisms through the oxidation of organic compound luciferin mediated by an enzyme luciferase.

Luminescent organism include bacteria, fungi, fish, insects, algae. Luminescent organisms found in marine, freshwater and terrestrial habitats. They are free living, symbiotic, saprophytic or parasitic in nature. *Photobacterium, Vibrio* and *Photorhabdus* are the important genus which had bioluminouscent bacteria.

Many luminous bacteria are parasitic, they infecting marine crustacean, Vibrio harveiji infects prawn sp. The bacterial luminesceance reaction involves the oxidation of a long chain aliphatic aldehyde and reduced flavin mononucleotide [$FMNH_2$] with the liberation of excess free energy in the form of a blue green light at 490nm.

$$FNM\ H_2 + RCHO + O_2 \rightarrow FMN + RCOOH + H_2O + Light$$

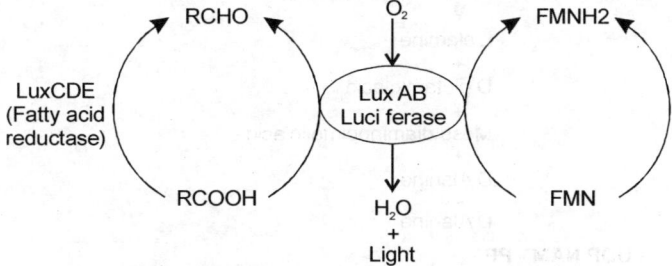

Figure 126 Bioluminescent effect

Gene coding for luciferase is Lux. Lux gone operon cenfan LUXABCDE. Lux AB for bacterial luciferase LUXC - Reductase, LUXD – Synthetase, LUX E – Transferase. Light is only produced when the organisms are present at high cell densities.

J. BIOSYNTHESIS OF BACTERIAL CELL WALL

Peptidoglycon (PG)

Peptidoglycon is also called as murein. Gram positive bacteria consist of thick layer of peptidoglycan where as gram negative bacteria has thin layer of peptidoglycan.

PG layer consists of NAM – NAG and pentapeptides. The process of PG layer synthesis occur in the cytoplasm. Synthesis of PG layer precursor starts with fructose 6 phosphate.

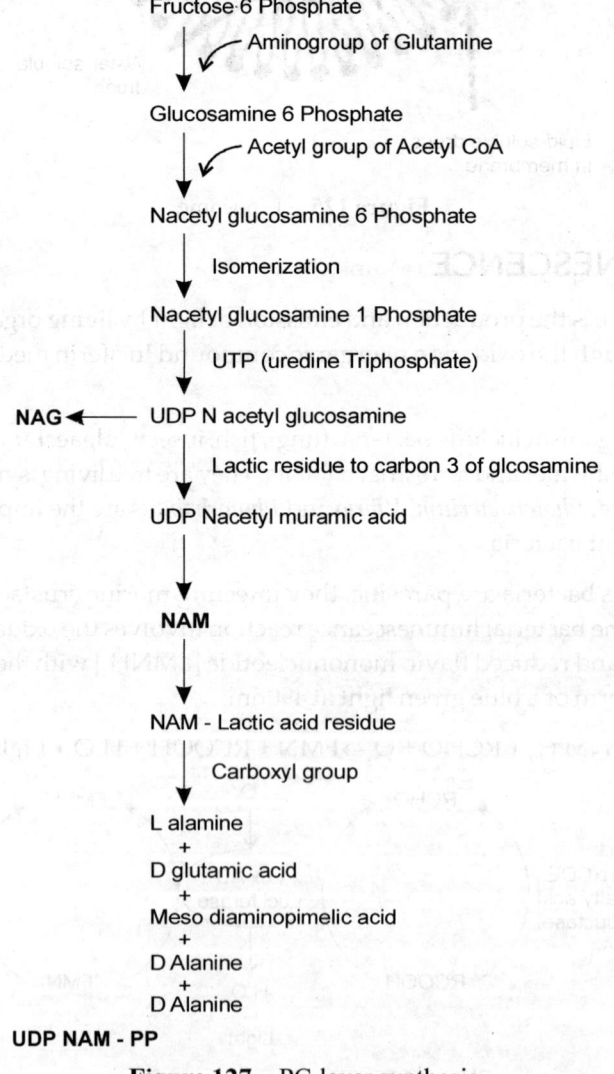

Figure 127 PG layer synthesis

Polymerization

NAM, NAG, are synthesized in the cytoplasm and transported to the cell membrane. It is mediated by a 55 carbon isoprenoid lipid carrier, known as bactoprenol.

The process of PG layer formation occurs in the cytoplasm – cytoplasmic membrane junction and is mediated by bactoprenol. Bactoprenol is also called as undecaprenold (UDPRP)

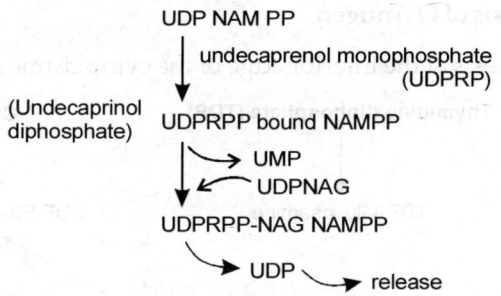

Figure 128 Polymerization of PG layer

Teichoic acid

It is a second major component of the G + cell. They are composed of polymers of ribitol or glycerol connected by phosphor diester linkage.

Polymer of ribitol are associated with the wall through covalent linkages. Polymers of glycerol are attached to a glycolipid in the membrane.

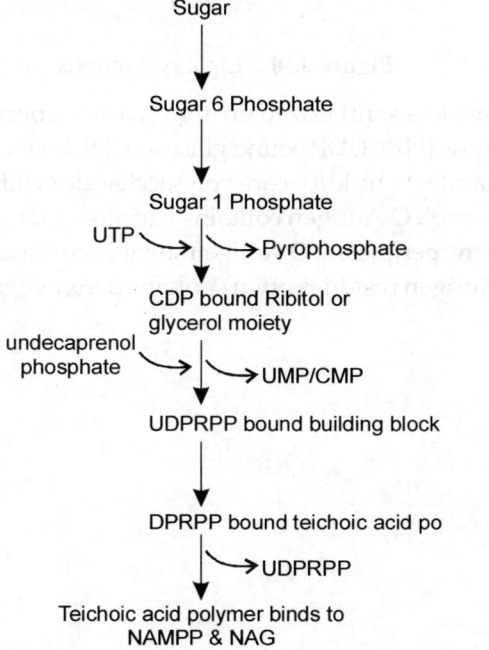

Figure 129 Teichoic acid synthesis and assembly

Synthesis of Lipopolysaccharide

Lipopolysaccharide is present only in gram negative bacteria. The process of LPS biosynthesis is divided into 4 major process. They are

1. Lipid A synthesis
2. Synthesis of core polysaccharide
3. Synthesis of O Antigen

Lipid A is synthesis at the interior edge of the cytoplasmic membrane.

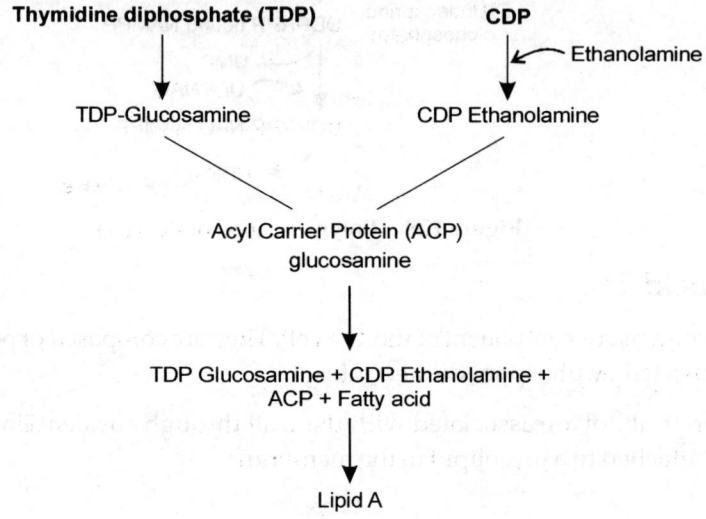

Figure 130   Lipid A synthesis

The core polysaccharide is synthesized on lipid A. CMP bound ketodeoxyoctulonic acid (KDO). ATP bound Heptose (HP). UDP bound glucose. CDP bound cthanolamine are combine to form a core polysaccharide. Vehicle for core polysaccharide synthesis are UDP carbohydrate precursor and undecaprenol. O Antigen contains various sugar residues. Addition of the O Antigen sugar occurs in the periplasm. O Antigen sugar occurs as repeating tri, tetra, or penta sugar residues. The O Antigen residues often contains deoxy sugars.

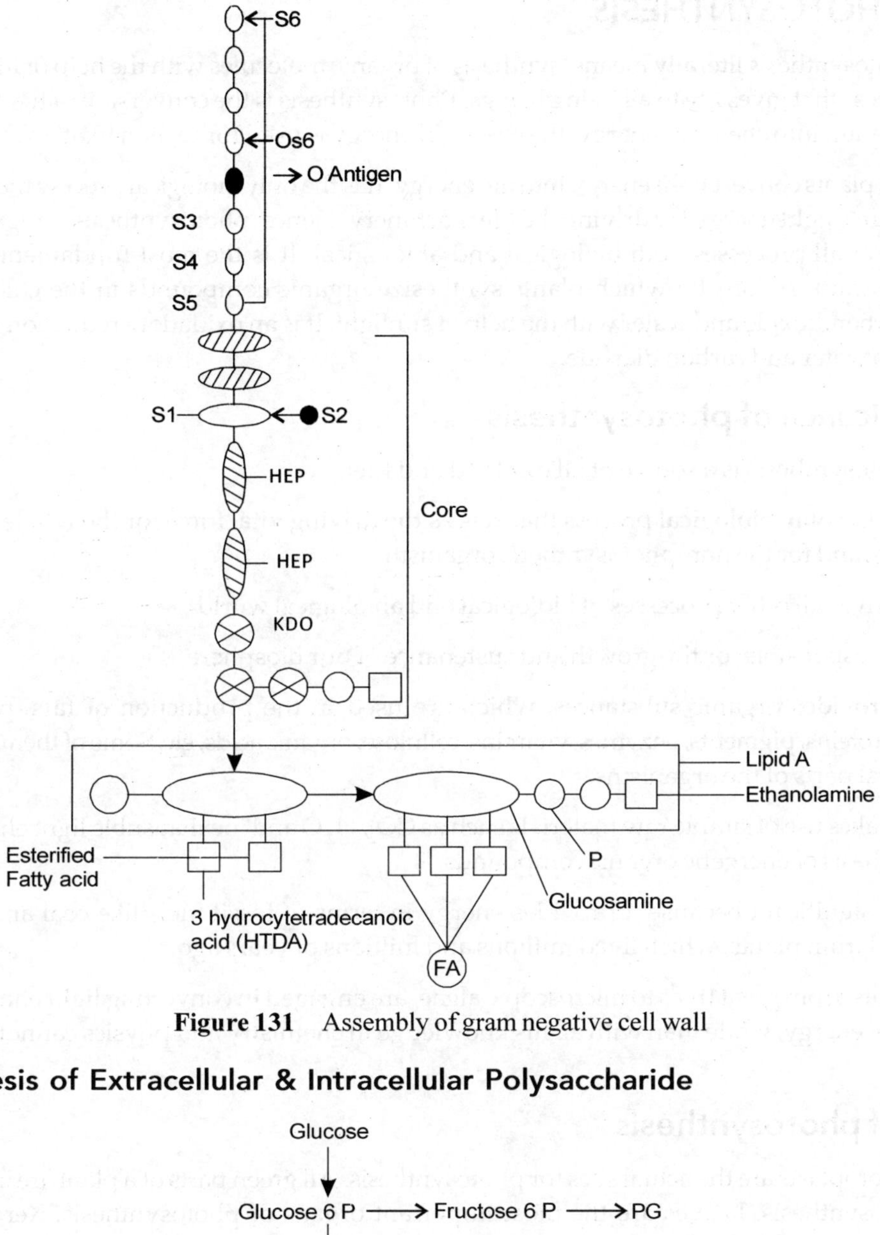

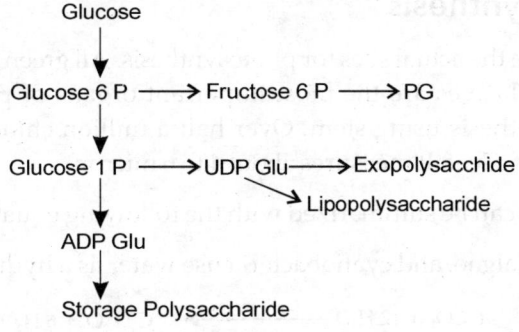

Figure 131   Assembly of gram negative cell wall

Synthesis of Extracellular & Intracellular Polysaccharide

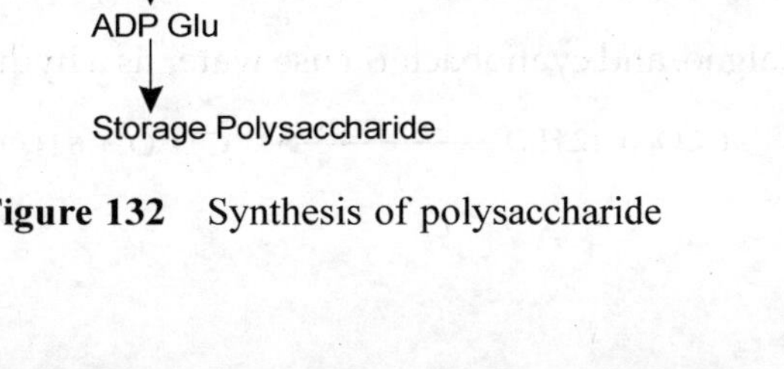

Glucose

Glucose 6 P ⟶ Fructose 6 P ⟶ PG

Glucose 1 P ⟶ UDP Glu ⟶ Exopolysacchide
                    ↘ Lipopolysaccharide

ADP Glu

Storage Polysaccharide

Figure 132   Synthesis of polysaccharide

K. PHOTOSYNTHESIS

Photosynthesis literally means 'synthesis of organic molecules with the help of light'. It is the process that gives life to all living beings. Photosynthesis is the conversion of light energy from the sun into chemical energy; the chemical energy is used for carbon fixation.

The plants convert light energy into life energy. It is the only biological process that makes use of sun's light energy for driving the life machinery. Hence, photosynthesis is regarded as 'leader' of all processes both biological and abiological. It is the most fundamental of all biochemical reactions by which plants synthesize organic compounds in the chloroplast from carbondioxide and water with the help of sunlight. It is an oxidation–reduction reaction between water and carbon di oxide.

Significance of photosynthesis

Photosynthesis is a source of all our food and fuel.

It is the only biological process that acts as the driving vital force for the whole animal kingdom and for the non-photosynthetic organism.

It drives all other processes of biological and abiological world.

It is responsible for the growth and sustenance of our biosphere.

It provides organic substances, which are used in the production of fats, proteins, nucleoproteins, pigments, enzymes, vitamins, cellulose, organic acids, etc. Some of them become structural parts of the organisms.

It makes use of simple raw materials such as CO_2, H_2O and inexhaustible light energy for the synthesis of energetic organic compounds.

It is significant because it provides energy in terms of fossil fuels like coal and petrol obtained from plants, which lived millions and millions of years ago.

Plants, from great trees to microscopic algae, are engaged in converting light energy into chemical energy, while man with all his knowledge in chemistry and physics cannot imitate them.

Site of photosynthesis

Chloroplasts are the actual sites for photosynthesis. All green parts of a plant are involved in photosynthesis. Leaves are the most important organs of photosynthesis. Xerophytes performs photosynthesis using stem. Over half a million chloroplasts are present in one square millimetre of a leaf. It measures about 4 to 6 micron.

Photosynthesis can be summarized with the following equations:

1. Plants, algae, and cyanobacteria use water as a hydrogen donor, releasing O_2.

$$6\,CO_2 + 12\,H_2O \xrightarrow{\text{Light energy}} C_6H_{12}O_6 + 6\,H_2O + 6O_2$$

2. Purple sulfur and green sulfur bacteria use H2S as a hydrogen donor, producing sulfur granules.

$$6\ CO_2 + 12\ H_2S \xrightarrow{\text{Light energy}} C_6H_{12}O_6 + 6\ H_2O + 12S$$

In the course of photosynthesis, electrons are taken from hydrogen atoms, an energy-poor molecule, and incorporated into sugar, an energy-rich molecule. The energy boost is supplied by light energy, although indirectly.

Mechanism of Photosynthesis

Photosynthesis takes place in two stages. In the first stage, called the **light-dependent (light) reactions,** light energy is used to convert ADP and **P** to ATP. In addition, in the predominant form of the light-dependent reactions, the electron carrier NADP is reduced to NADPH. The coenzyme NADPH, like NADH, is an energy-rich carrier of electrons. In the second stage, the **light-independent (dark) reactions,** these electrons are used along with energy from ATP to reduce CO_2 to sugar.

The Light-Dependent Reactions: Photophosphorylation

Photophosphorylation is one of the three ways of ATP production. It occurs only in photosynthetic cells. In this mechanism, light energy is absorbed by chlorophyll molecules in the photosynthetic cell, exciting some of the molecules' electrons. The chlorophyll principally used by green plants, algae, and cyanobacteria is *chlorophyll a*. It is located in the membranous thylakoids of chloroplasts in algae and green plants and in the thylakoids found in the photosynthetic structures of cyanobacteria. Other bacteria use *bacteriochlorophylls*. The excited electrons jump from the chlorophyll to the first of a series of carrier molecules, an electron transport chain similar to that used in respiration. As electrons are passed along the series of carriers, protons are pumped across the membrane, and ADP is converted to ATP by chemiosmosis.

The light driven reactions of photosynthesis are referred to as electron transport chain. When PS II absorbs photons of light, it is excited and the electrons are transported through electron transport chain of plastoquinone, cytochrome b6, cytochrome f and plastocyanin. The electrons released from PS II phosphorylate ADP to ATP. This process of ATP formation from ADP in the presence of light in chloroplast is called photophosphorylation.

Now, the PS II is in oxidised state. It creates a potential to split water molecules to protons, electrons and oxygen. This light dependent splitting of water molecules is called photolysis of water. Manganese, calcium and chloride ions play prominent roles in the photolysis of water. The electrons thus released are used in the reduction of PS II. Similar to PS II, PS I is excited by absorbing photons of light and gets oxidised. This oxidised state of the PS I draws electrons from PS II and gets reduced. The electrons released to PS I are transported through electron transport chain of ferredoxin reducing substrate, ferredoxin and ferredoxin NADP reductase to reduce $NADP^+$ to $NADPH_2$

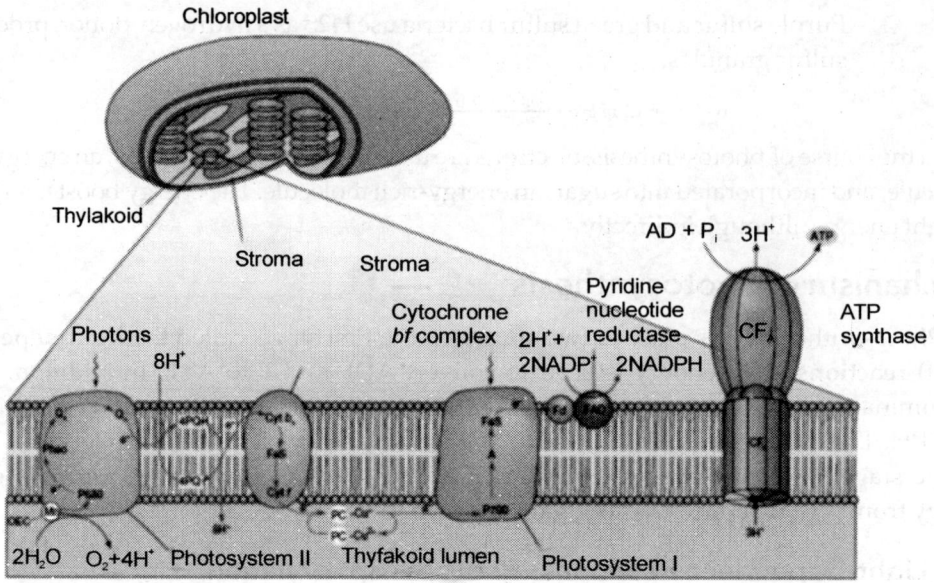

Figure 133 Photophosphorylation

Cyclic and noncyclic photophosphorylation

In chloroplasts, phosphorylation occurs in two ways – noncyclic photophosphorylation and cyclic photophosphorylation.

Noncyclic photophosphorylation

The process in which light energy is used to make ATP when electrons are moved from water to $NADP^+$ during photosynthesis; both photosystem I and photosystem II are involved.

When the molecules in the PS I are excited the electrons are released. So, an electron deficiency or a hole is made in the PS I. This electron is now transferred to ferredoxin to reduce NADP+. When the molecules in the PS II get excited, electrons are released. They are transferred to fill the hole in PS I through plastoquinone, cytochrome b6, cytochrome f and plastocyanin. When the electron is transported between plastoquinone and cytochrome f, ADP is phosphorylated to ATP.

The 'hole' in the PS I has been filled by the electron from PS II. Then the electrons are transferred from PS I to $NADP^+$ for reduction. Therefore, this electron transport is called noncyclic electron transport and the accompanying phosphorylation as noncyclic photophosphorylation. The noncyclic electron transport takes place in the form of 'Z'. Hence, it is also called Z-scheme.

Cyclic photophosphorylation

Under the conditions of (i) PS I only remains active (ii) photolysis of water does not take place (iii) requirement of ATP is more and (iv) nonavailability of NADP+ the cyclic photophosphorylation takes place. When the molecule in the PS I is excited, the electrons are released. The electrons are captured by ferredoxin through ferredoxin reducing substrate (FRS). Due to non-availability of NADP+, electrons from ferredoxin fall back to the molecules of PS I through the electron carriers - cytochrome b6, cytochrome f and plastocyanin. These electron carriers facilitate the down hill transport of electrons from FRS to PS I. During this transport of electrons, two phosphorylations take place - one between ferredoxin and cytochrome b6 and the other between cytochrome b6 and cytochrome f. Thus, two ATP molecules are produced in this cycle.

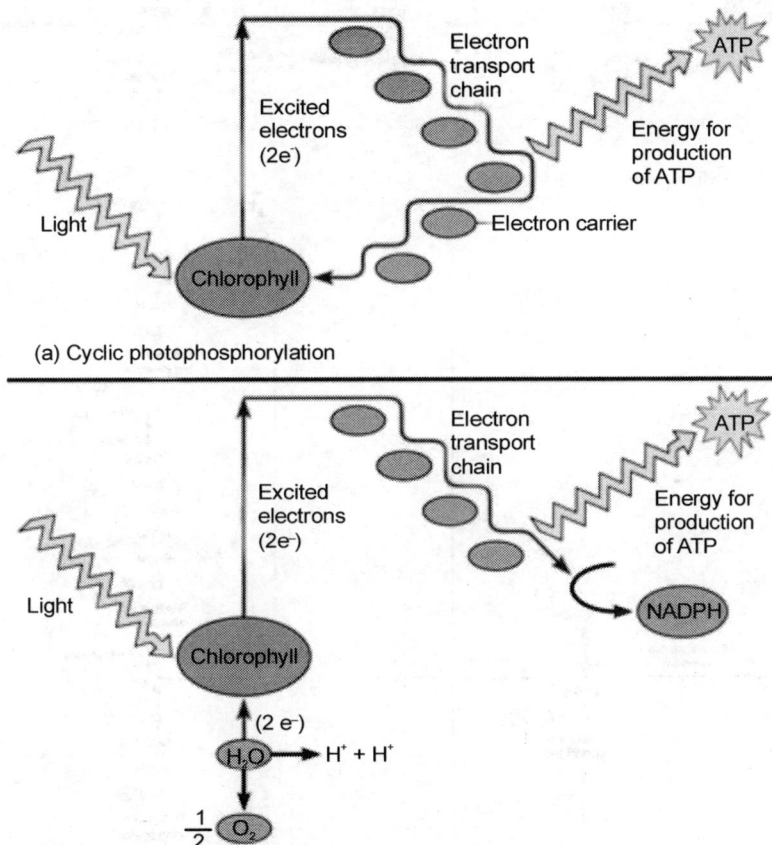

(a) Cyclic photophosphorylation

(b) Noncyclic photophosphorylation

Figure 134 Cyclic and noncyclic photophosphorylation

In **cyclic photophosphorylation,** the electrons eventually return to chlorophyll. In **noncyclic photophosphorylation,** which is used in oxygenic organisms, the electrons released from chlorophyll do not return to chlorophyll but become incorporated into NADPH . The electrons lost from chlorophyll are replaced by electrons from H_2O.

The Light-Independent Reactions (Dark reactions):
The Calvin-Benson Cycle

The reactions that catalyze the reduction of CO2 to carbohydrates with the help of the ATP and NADPH2 generated by the light reactions are called the dark reactions. The enzymatic reduction of CO2 by these reactions is also known as carbon fixation. These reactions that result in CO2 fixation take place in a cyclic way and were discovered by Melvin Calvin. Hence, the cycle is called Calvin cycle. Fixation of carbondioxide in plants during photosynthesis occurs in three stages –fixation, reduction and regeneration of RuBP.

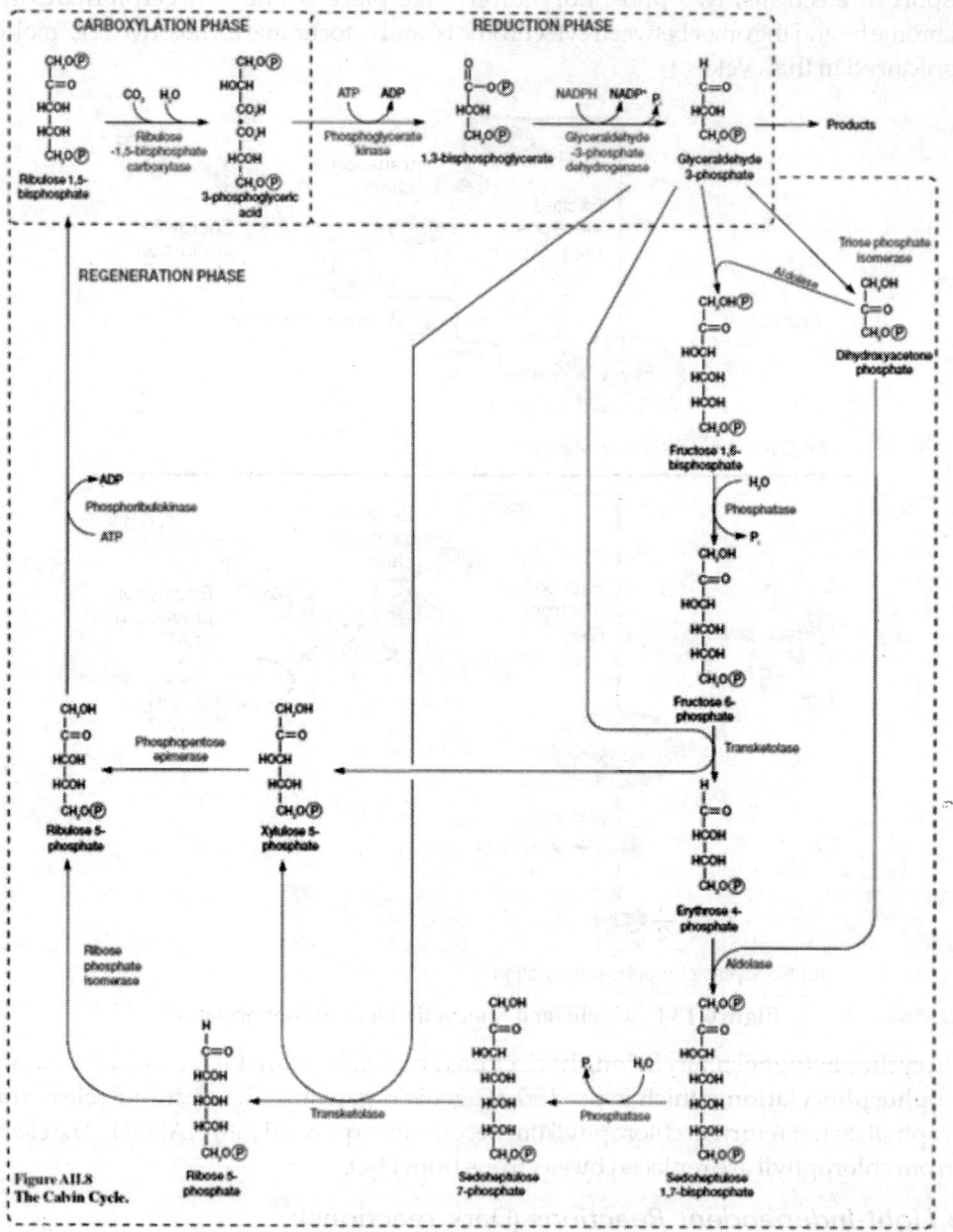

Figure AII.8
The Calvin Cycle.

Figure 135   Calvin cycle

The Carboxylation Phase

Carbon dioxide fixation is accomplished by the enzyme **ribulose 1,5-bisphosphate carboxylase** or ribulosebisphosphate carboxylase/ oxygenase (rubisco), which catalyzes the addition of CO_2 to ribulose 1,5-bisphosphate (RuBP), forming two molecules of 3-phosphoglycerate (PGA).

The Reduction Phase

After PGA is formed by carboxylation, it is reduced to glyceraldehydes 3-phosphate. The reduction, carried out by two enzymes, is essentially a reversal of a portion of the glycolytic pathway, although the glyceraldehyde 3-phosphate dehydrogenase differs from the glycolytic enzyme in using NADP_ rather than NAD.

The Regeneration Phase

The third phase of the Calvin cycle regenerates RuBP and produces carbohydrates such as glyceraldehyde 3-phosphate, fructose, and glucose. This portion of the cycle is similar to the pentose phosphate pathway and involves the transketolase and transaldolase reactions. The cycle is completed when phosphoribulokinase reforms RuBP. To synthesize fructose 6-phosphate or glucose 6-phosphate from CO_2, the cycle must operate six times to yield the desired hexose and reform the six RuBP molecules.

The glyceraldehyde 3-phosphate molecules are converted to RuBP through a series of reactions, which generate 4C, 6C and 7C phosphorylated compounds as intermediates. For better and easy understanding of these reactions, a simplified scheme of Calvin cycle considering three CO2 molecules fixation reactions is shown below.

The reactions of regeneration of RuBP are as follows.

1. Some of the Glyceraldehyde 3-phosphate molecules are converted to dihydroxy acetone phosphates.

2. Glyceraldehyde 3-phosphate combines with dihydroxy acetone phosphate to form fructose1,6-bisphosphate.

3. Fructose 1,6-bisphosphate undergoes dephosphorylation to form fructose 6-phosphate.

4. Fructose 6-phosphate combines with glyceraldehyde 3-phosphate obtained from the fixation of second molecule of CO_2 to form Ribose 5-phosphate (R5P) and Erythrose 4-phosphate (Er4P).

5. Erythrose 4-phosphate combines with DHAP obtained from the second CO_2 fixation, to form sedoheptulose 1,7-bisphosphate.

6. Sedoheptulose 1,7-bisphosphate undergoes dephosphorylation to form sedoheptulose 7-phosphate.

7. Sedoheptulose 7-phosphate combines with glyceraldehyde 3-phosphate obtained by the third CO_2 fixation, to form two molecules of 5C compounds – ribose 5-phosphate and xylulose 5-phosphate (Xy5P).

8. Ribose 5-phosphate and xylulose 5-phosphate molecules are transformed to ribulose 5-phosphate (Ru5P).

9. Ru5P molecules are then phosphorylated by ATP to form RuBP molecules, which again enter into the cycle of CO_2 fixation.

In the above illustration, three CO_2 molecules are fixed and the net gain is a 3C called DHAP. These triose phosphate molecules combine to form hexose phosphates, which are used to form sucrose. For every carbon fixation 3ATP and 2 $NADPH_2$ are consumed.

The Light Reaction in Green and Purple Bacteria

Green and purple photosynthetic bacteria differ from cyanobacteria and eucaryotic photosynthesizers in several fundamental ways. In particular, green and purple bacteria do not use water as an electron source or produce O_2 photosynthetically. It is a **anoxygenic process**. In contrast, cyanobacteria and eukaryotic photosynthesizers are almost always **oxygenic**. NADPH is not directly produced in the photosynthetic light reaction of purple bacteria. Green bacteria can reduce NAD directly during the light reaction. To synthesize NADH and NADPH, green and purple bacteria must use electron donors like hydrogen, hydrogen sulfide, elemental sulfur, and organic compounds that have more negative reduction potentials than water and are therefore easier to oxidize (better electron donors). Finally, green and purple bacteria possess slightly different photosynthetic pigments, **bacteriochlorophylls**, many with absorption maxima at longer wavelengths. Bacteriochlorophylls *a* and *b* have maxima in either at 775 and 790 nm, respectively.

There are four groups of green and purple photosynthetic bacteria, each containing several genera: green sulfur bacteria (*Chlorobium*), green nonsulfur bacteria (*Chloroflexus*), purple sulfur bacteria (*Chromatium*), and purple nonsulfur bacteria (*Rhodospirillum, Rhodopseudomonas*). Many differences found in green and purple bacteria are due to their lack of photosystem II. They cannot use water as an electron donor in noncyclic electron transport. Without photosystem II they cannot produce O_2 from H_2O photosynthetically and are restricted to cyclic photophosphorylation. Indeed, almost all purple and green sulfur bacteria are strict anaerobes. When the special reaction-center chlorophyll P870 is excited, it donates an electron to bacteriopheophytin. Electrons then flow to quinones and through an electron transport chain back to P870 while driving ATP synthesis. Note that although both green and purple bacteria lack two photosystems, the purple bacteria have a photosynthetic apparatus similar to photosystem II, whereas the green sulfur bacteria have a system similar to photosystem I.

Green and purple bacteria face a further problem because they also require NADH or NADPH for CO_2 incorporation. They may synthesize NADH in at least three ways. If they are growing in the presence of hydrogen gas, which has a reduction potential more negative than that of NAD, the hydrogen can be used di- rectly to produce NADH. Like chemolithotrophs, many photosynthetic purple bacteria use proton motive force to reverse the flow of electrons in an electron transport chain and move them from inorganic or organic donors to NAD. Green sulfur bacteria such as *Chlorobium* appear to carry out a simple form of noncyclic photosynthetic electron flow to reduce NAD.

In purple non sulfur bacteria, when light hitted photo system P870, excited and transfer energy to bacteriopheophytin (Bph) followed to Q, cytochrome b, c, Fes, Cytochrome C_2. This flow of electrons leads to synthesis of ATP (Figure 137). In green sulfur bacteria, photo system P840 excited when light hitted and transfered electron to bacteriopheophytin 663, (Bph) Fes, feredoxin (Fd), cytochrome b, and Cytochrome 232. This flow of electrons leads to synthesis of ATP (Figure 138).

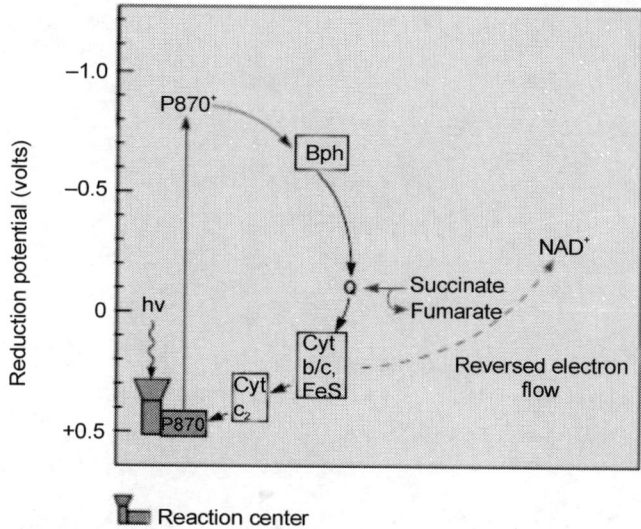

Figure 136 Purple non sulfur bacterial photosynthesis

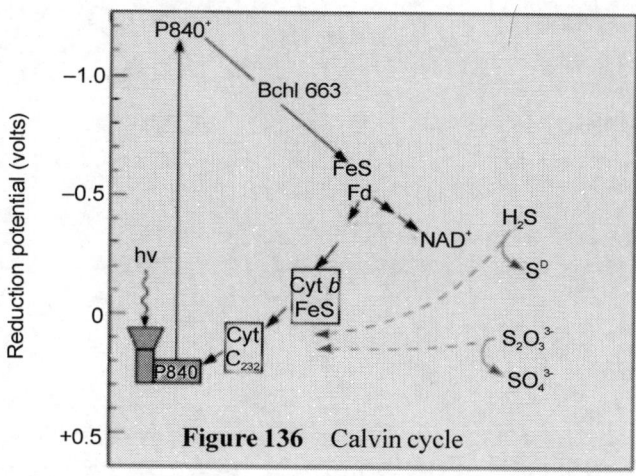

Figure 136 Calvin cycle

Figure 137 Green sulfur bacterial photosynthesis

L. ENZYMES

Refer Chapter 17 Pages - 654-658.

9

FOOD MICROBIOLOGY

Food is defined as a material, usually of plant or animal origin, which contains essential nutrients such as carbohydrates, fats, proteins, vitamins or minerals. It is ingested and assimilated by an organism to produce energy, stimulate growth and maintains life.

Food microbiology deals with the study of the microorganisms that inhabit, create or contaminate food. It also deals with the microorganisms involved in the spoilage, contamination and preservation of food.

A. SOURCES OF FOOD

Almost all foods are of plant or animal origin. Cereal grain is a stable food that provides more food energy. Maize, wheat and rice accounts for 87% of all grain production worldwide. Fungi and ambient bacteria are used in the preparation of fermented foods like leavened bread, alcoholic drinks, cheese, pickles and yogurt.

Seeds of plants are a good source of food for animals. It contains the nutrients necessary for the growth, including many healthful fats such as Omega fats. Edible seeds include cereals (maize, wheat, rice etc), legumes (beans, peas, lentils, etc) and nuts. Fruits are the ripened ovaries of plants that are attractive as a food source to animals. Some fruits, such as tomatoes, pumpkins are eaten as vegetables. Vegetables are a second type of plant matter that is commonly eaten as food. These include root vegetables (potatoes and carrots), leaf vegetables (spinach and lettuce), stem vegetables (bamboo shoots and asparagus) and other vegetables such as cabbage or cauliflower.

Animals are used as food either directly or indirectly by the products they produce. Meat is an example of a direct product taken from an animal. Food products produced by animals milk produced by mammary glands. Dairy products are cheese, butter etc. In addition, birds and other animals lay eggs, which are often eaten and bees produce honey.

B. CLASSIFICATION OF FOOD

Food materials are classified differently based on nutrition, dietary substances, food function and also based on food perishability.

1. Nutritional Classification

Foods are complex substances composed of chemical constituents called nutrients. According to the presence of such nutrient, foods has been classified as proteins, carbohydrate, fats, vitamins, minerals, water and roughage.

2. Dietary Sources Classifiction

Food can be obtained from animal as well as plant kingdom; that is organic as well as inorganic sources. Organic sources are meat, fish, egg, milk and milk products. Inorganic sources are cereals, pulses, vegetables etc.

3. Functional Classifiction

Food can also be classified according to the function it performs: They are as follows

- ↟ Body-building foods- Protein, Minerals
- ↟ Energy giving foods- Carbohydrates, Fats
- ↟ Protective foods- Vitamins, Minerals.
- ↟ Regulatory foods- Water, Roughage.

4. Classification of Food on the Basis of Perishability

The three categories of food are:

i.   **Perishable foods:** these food materials are cannot be stored for more than one or two days at room temperature. Milk is a good example of perishable food.

ii. **Semi-perishable foods:** It can be kept for a couple of weeks or even a month or two. They have a longer shelf life than perishable foods. Potato, arbi, onions, ginger, biscuits are some examples of semi-perishable food.

iii. **Non-perishable foods:** In the real sense, foods in this category are not really non-perishables, but they can be stored for much longer time as compared to perishables and semi-perishables. They can be stored for several months and the examples of such foods are cereals, pulses, dry fruits, spices, oil etc.

C. MICROORGANISMS IMPORTANT IN FOOD

Good numbers of microbes are involved in spoilage and ripening of food materials.

- ↟ *Mucor* are involved in the spoilage of many foods.
- ↟ *Mucor racemosus, Mucor rouxii* used in the saccharification of Starch.
- ↟ *Mucor* helps in the ripening of cheese.
- ↟ *Rhizopus stolonifer* is a **bread mold**, involved in the spoilage of berries, fruits, vegetables and bread.
- ↟ *Absidia* causes spoilage in fruits and vegetables.
- ↟ *Thamidium elegans* is found in meat in chilling storage, causing whiskers of meat.
- ↟ *Aspergillus repens* involved in food spoilage.
- ↟ *Aspergillus niger* produces black mold and also involved in ripening process.
- ↟ *Penicillium expansum* a blue green spored mold causes soft rot on fruits.
- ↟ *Penicillium digitatum*, olive or yellowish green mold causes soft rot of citrus fruit.
- ↟ *Penicillium italicum* called the blue contact mold causes rotting of citrus fruit.

- *Penicillium camemberti* produces ripening of Camembert cheese.
- *Penicillium roqueforti* aiding in the ripening of blue cheese Roquefort.
- *Trichothesium* is a pink mold causes rot in apple and peaches.
- *Geotrichum candidum* often called dairy mold gives white to cream coloured growth in dairy products.
- *Neurospora sitophila* termed the red bread mold grows on bread and also on bagasses.
- *Sporotrichum carnis* grows on chilled meat and causes white spot.
- *Botritis cinerea* causes diseases of grapes.
- *Cephalosporium acremonium* is a common species of food.
- *Cladosporium herbarum* causes black spot on different food.
- *Fusarium* often grows on food.
- *Endomyces* causes rot on fruits.
- *Monascus purpure* causes spoilage on dairy products.
- *Sclerotina* causes rots on vegetables and fruits.
- Blue mould rot in tomato caused by *Penicillium* sp. (also by Fusarium sp.)
- *Botrytis cinerea* causes Storage rot in grapes and strawberry
- Black mummy rot of grapes caused by *Guignardia bidwellii*
- Watery soft rot in apple caused by *Sclerotinia sclerotiorum*
- Blue mould on oranges caused by *Penicillium digitatum*.

Yeast

- *Schizosaccharomyces* found in tropical fruits and vegetables.
- *Saccharomyces* employed in leavening of bread, top yeast for ale and wine production.
- *Saccharomyces fragilis* and *Saccharomyces lactis* ferment lactose and important in dairy industry.
- *Zygosaccharomyces* involved in spoilage of honey, syrup and molassess.
- *Pichia* produces pellicle on beer and wine.
- *Debaryomyces* form pellicles on meat, brine and cheese.
- *Torulopsis* can spoil sweetened condensed milk and fruit juice.
- Lipolytic *Candida lipolytica* can spoil butter.
- *Rhodotorula* causes discolouration of food

Bacteria

- *Acetobacter* causes definite spoilage problem in alcoholic beverages.
- *Aeromonas* is pathogenic to fish and human.
- *Alcaligens viscolyticus* causes ropiness in milk.
- *Alcaligens metalcaligenes* gives slimy growth on cottage cheese.

- *Alteromonas* are potentially important in seafood.
- *Bacillus coagulans* employed for the manufacture of lactic acid.
- *Brevibacterium* causes orange red colouration in cheese.
- *Clostridium* causes anaerobic food spoilage.
- *Flavobacterium* involved in spoilage of shellfish, poultry, eggs, butter and milk.
- *Glucanobacter oxydans* causes ropiness of beer.
- *Halobacterium* causes discolouration of salted fish.
- *Lactobacillus* ferment sugar and converted to lactic acid. It is one of the importat bacterium in dairy industry.
- *Leuconostoc* involved in the production of flavoured products in dairy industry.
- *Listeria monocytogens* causes food borne diseases.
- *Pediococcus cerevisiae* has been used as a starter culture in fermented sausages.
- *Photobacterium* causes phosphorescence on meat and fish.
- *Propionibacteium* ferment lactic acid and convert to Propionic acid. It also causes colour defects in cheese.
- *Proteus* cause spoilage of meats, seafood and egg.
- *Pseudomonas* causes spoilage of food even at refrigerator temperature.
- *Salmonella* grows in food and cause food borne infections.
- *Serratia* produces red colouration on the surface of food.
- *Staphylococcus* produce enterotoxin on food.
- *Streptococcus faecalis* act as an indicator microorganism in food industry.
- *Streptomyces* cause undesirable flavours and appearance on food.
- *Vibrio* causes spoilage of fish and produce cholera in human.

Sources of microbial contamination in food

Sources of microbial contamination in food are as follows,

- Microflora present in soil and water
- Microflora present in air
- Microflora present on plant and plant products
- Microflora present on food utensils and equipments
- Microflora present in animal feeds
- Microflora present on animal hides
- Microflora present in Intestinal tracts of humans and animals
- Food Handlers,
- Packaging materials.

D. FACTORS AFFECTING KINDS AND NUMBER OF MICROBES IN FOOD

Various factors are related to the microbial growth in food. They are broadly categorized as intrinsic factors and extrinsic factors.

Intrinsic Factors

- ⋏ pH
- ⋏ Moisture Content
- ⋏ Water activity
- ⋏ Oxidation-reduction potential
- ⋏ Physical structure of the food
- ⋏ Available nutrients
- ⋏ Presence of antimicrobial agents

Extrinsic factors

- ⋏ Temperature
- ⋏ Relative humidity
- ⋏ Carbon dioxide or oxygen
- ⋏ Types and numbers of Microorganisms in the food

E. SPOILAGE OF DIFFERENT FOOD ITEMS

Food spoilage is the deterioration in the colour, flavour, odour or consistency of a food product. Food can deteriorate as a result of microbial growth, the action of enzymes or by the action of environmental factors.

Causes of the Spoilage of Food

Microorganisms, enzymes, atmospheric oxygen, insects, low temperature are the major causes of food spoilage.

Microbial Spoilage

There are three types of microorganisms that cause food spoilage is yeasts, moulds and bacteria.

Yeasts growth causes fermentation of food. There are two types of yeasts *true* yeast and *false* yeast. *True yeast* metabolizes sugar producing alcohol and carbon dioxide. This is known as fermentation. *False yeast* grows as a dry film on a food surface such as on pickle brine.

Moulds grow in filaments forming a tough mass which is visible as 'mould growth'. Mould can cause illness, especially if the person is allergic to molds.

Bacteria may grow under a wide variety of conditions. There are many types of bacteria that cause spoilage. They can be divided into: *spore-forming* and *nonspore-forming*. Eating spoiled food can cause food poisoning.

Enzymes are proteins found in all plants and animals. If uncooked foods are not used while fresh, enzymes cause undesirable changes in colour, texture and flavour. Enzymes are destroyed easily by heat processing.

Atmospheric oxygen can react with some food components which may cause rancidity or colour changes.

Infestations (invasions) by insects and rodents accounts for huge losses in food stocks.

Low temperature injury - the internal structures of the food are damaged by very low temperature.

Types of Food Decay

Most foods serve as good growth medium for many different microorganisms. Microorganisms will cause changes in appearance, flavour, odour and other qualities of foods. There are three types of food decay. They are Putrefaction, Fermentation and Rancidity

Putrefaction

Putrification is a set of anaerobic reactions with amino acids. It results in a formation of mixture of amine (e.g. cadaverine, putrescine, istamine), organic acids and strong-smelling sulfur compounds like mercaptans and hydrogen sulphide.

Protein foods + proteolytic microorganisms → amino acids + amines+ ammonia+ H_2S

Fermentation

Under anaerobic conditions, many species carry out fermentative metabolism. It results in various fermentation products, which are most harmful for the food taste.

Carbohydrate foods +Saccharolytic microorganisms → organic acids+alcohol+gases.

Rancidity

Rancidification is the decomposition of fats, oils and other lipids by hydrolysis or oxidation of food. This leads to the formation of free Fatty acids. These free fatty acids can then undergo further auto-oxidation. Oxidation primarily occurs with unsaturated fats by a free radical-mediated process and generate highly reactive molecules in foods and oils, which are responsible for producing unpleasant and noxious odours and flavours.

Fatty foods + lipolytic microorganisms → fatty acids + glycerol

Microbial Spoilage of Foods

Microorganisms cause spoilage not only by degradation of foods, also by synthesis of various products like pigments and polysaccharides leading to discolourations and formation of slimes.

Table 9.1   Types of food spoilage with some examples of casual organisms

S. No	Food	Types of Spoilage	Microorganisms involved
1	Bread	Moldy Ropy	*Rhizopus nigricans, Penicillium* *Bacillus subtilis*
2	Pickles,	Sauerkraut Film / Pink	*Yeasts- Rhodotorula*
3	Fresh Meats	Putrefaction	*Alcaligenes, Clostridium,* *Proteus vulgaris, Pseudomonas* *fluorescens*
4	Cured Meats	Moldy Souring	*Rhizopus, Aspergillus, Penicillium* *Pseudomonas, Micrococcus.*
		Greening, slime	*Lactobacillus, Leuconostoc*
5	Fish	Discolourations	*Alcaligenes, Pseudomonas,* *Flavobacterium*
6	Poultry	Slime, Odour	*Alcaligenes, Pseudomonas*
7	Eggs	Colourless GreenRots Off' flavor	*Alcaligenes, Proteus, Pseudomonas,* *P. fluorescens 'Acetobacter,* *Lactobacillus*
8	Juices		*Leuconostoc*
9	Fresh fruits	Soft rots	*Rhizopus, Erwinia*
10	Vegetables	Gray & black mold	*Botrytis, Aspergillus niger*
11	Milk/ Cream	Ropiness	*Alcaligenes viscolactis,* *Micrococcus, Enterobacter,* *Lactobacillus, Streptococcus, Bacillus*
		Decomposition of fats	*Proteus, Pseudomonas, Micrococcus,* *Bacillus, Clostridium* *Yeasts, Molds, Alcaligenes*
		Alkali formers	*Alcaligenes viscolactis, P.fluorescens*
		Flavour changes	*Lactobacillus, Streptococcus,* *Leuconostoc sp.*

Spoilage of Bread

- ⅄ Bread is one of the staple food.
- ⅄ It is included as a perishable food.
- ⅄ Freshness of bread is maintained only for few hours after baking.
- ⅄ Microbial attack along with physical and chemical changes of bread leads to spoilage.

- ⅄ Mold is the important spoilage causing organisms in bread.
- ⅄ Bread becomes contaminated after baking from the mold spores from the surrounding environment during slicing, cooling, packing and storage.
- ⅄ *Rhizopus* and *Aspergillus* causes spoilage of bread.
- ⅄ Ropiness is bacterial spoilage of bread that initially occurs as an unpleasant fruity odour, followed by enzymatic degradation of the crumb that becomes soft and sticky. It is caused by *Bacillus* sp.
- ⅄ *Rhizopus stolonifer* is a bread mould.

Spoilage of Fish

- ⅄ Fish is one of the highly perishable food.
- ⅄ Fish and fish products are less stable because of their high moisture content, rich nutrients and microorganisms.
- ⅄ Spoilage of fish is observed as change in colour, odour, texture, colour of eyes, colour of gills and softness of the muscle.
- ⅄ The main components of the fish are water, protein and fat.
- ⅄ Spoilage is caused by the action of enzymes, bacteria and the factors like high moisture content, high fat content, high protein content, weak muscle tissue, ambient temperature unhygienic handling also contributes the spoilage.
- ⅄ The process of spoilage involves rigor mortis, autolysis, bacterial invasion and putrefaction.
- ⅄ Rigor mortis means change in stiffness of muscle due to breakage of circulatory system and its related chemical signal leak.
- ⅄ Rigor mortis leads to cell **break** down due to enzyme action. It is called as autolysis.
- ⅄ Microbial invasion taken place after the death of the fish. Microbial attach damages protein and produce putrified odour.
- ⅄ Oxidative rancidity also caused odour.
- ⅄ Fish spoilage is reduced or prevented by drying, salting, chilling, canning and freezing.
- ⅄ *Pseudomonas, Moraxella, Acinetobacter, Shewanella, Flavobacterium, Aeroemonadaceae,* and *Vibrionaceae,* and Gram-positive bacteria such as *Bacillus, Micrococcus, Clostridium, Lactobacillus* and *Corynebacterium* are the major bacteria found in fish and cause spoilage.

Spoilage of Egg

- ⅄ It is a most important perishable food.
- ⅄ It is rich in protein, as well as various other nutrients.

Chicken eggs are the most commonly eaten eggs. They supply all essential amino acids for humans and provide several vitamins and minerals, including retinol (vitamin A), riboflavin (vitamin B2), folic acid (vitamin B9), vitamin B6, vitamin B12, choline, iron, calcium, phosphorus and potassium. Eggs are also a single-food source of protein.

The egg is one of the few foods to naturally contain vitamin D.

Cooked eggs are easier to digest, as well as having a lower risk of salmonellosis.

Spoilage causing microbes induce colour and odour change. Colourless rot is caused by *Alkaligens* and green rot is by *Pseudomonas*.

Spoilage of Meat

Meats are the perishable of all important foods because of abundance of nutrients and moisture content.

The **spoilage of meat** occurs, if the meat is untreated in a matter of hours.

Spoiled meat becoming unappetizing, poisonous or infectious.

Spoilage is caused by the practically unavoidable infection and subsequent decomposition of meat by bacteria and fungi, which are borne by the animal itself, by the people handling the meat and by their implements.

Fresh meats contain a large group of potential spoilage bacteria. *Acinetobacter* and *Moraxella* can grow rapidly and produce undesirable odours due to aminoacid metabolism.

Lactobacillus curvatus and *Lactobacillus* sake produce lactic acid and volatile fatty acids which impart a cheesy flavour in meat.

Heterofermentative *Leuconostoc carnosum* and *Leuconostoc gelidum* produce CO_2, and small quantity of lactic acid, causing accumulation of gas and liquid in the package.

Facultative anaerobic *Enterobacter, Serratia, Proteus* and *Hafnia* species metabolize amino acids while growing in meat to produce amines, ammonia, methylsulfides, mercaptans and cause putrefaction. Some strains also produce H_2S in small amounts to cause greening of the meat.

To reduce spoilage of' Fresh meats, storage at low temperatures (~ 0 to 1°C), modified atmosphere packaging and vacuum packaging are extensively used.

Spoilage of canned food

Canning is a method of preserving food in which the food contents are processed and sealed in an airtight container.

Canned goods are classified as (a) **low acid,** (b) **acid** or (c) **high acid** products.

A. Low Acid canned foods: pH >4.6 e.g. meat, milk, many vegetables etc. These kind of foods are heated to destroy most heat-resistant spores of pathogenic bacteria, *Clostridium botulinum*

B. Acid foods: pH 4.0-4-6 e.g. tomatoes, pears, etc.

C. High acid foods (pH <4.0) e.g. fruits, a few vegetables and fermented products like sauerkraut. For these food items heat treatment is given to kill all vegetative cells and some spores.

Canned food spoilage is due to both nonmicrobial (chemical and enzymatic reactions) and microbial reasons.

Production of hydrogen (hydrogen swell), CO_2, browning, corrosion of cans due to chemical reactions and liquification, gelation, discolouration of products due to enzymatic reactions are some examples of nonmicrobial spoilage.

Microbial spoilage is due to three main reasons:

1. Inadequate cooling after heating.
2. Inadequate heating.
3. Leakage (microscopic) in the cans.

Signs of canned foods spoilage is detected and described using any one of the following technical terms.

Soft Swell   A can that is bulged on both ends, but not so tightly that the ends can't be pushed in somewhat with a thumb press.

Hard Swell   A can that is so tightly bulged on both ends that the ends can't be pressed in. A can with a hard swell will generally "buckle" before it bursts.

Flipper   A can whose end normally looks flat, but "flips out" when struck sharply on one end.

Springer   A can with one end bulged out. With sufficient pressure, this end will flip in, but the other end will flip out.

Leaker   A can with a crack or hole in the container that has caused leakage.

Spoilage of Milk and Milk Products

Raw milk contains many types of microorganisms coming from different sources.

The average composition of cow's milk is 3.2% protein, 4.8% carbohydrates, 3.9% lipids, and 0.9% minerals. Casein and lactalbumin are a good N-source.

Milk fat can be hydrolyzed by microbial lipases with the release of small molecular volatile fatty acids (butyric,capric, and caproic acids).

Microbial spoilage of raw milk can potentially occur from the metabolism of lactose, proteinaceous compound, fatty acids (unsaturated) and the hydrolysis of triglycerides.

The spoilage will be predominantly caused by the Gram-negative psychrotrophic rods, such as *Pseudomonas, Alcaligenes, Flavobacterium sp.*, and some coliforms.

Pseudomonas will metabolize proteinaceous compounds to change the normal flavour of milk to bitter, fruity or unclean.

The growth of lactose-positive coliforms will produce lactic, acetic,and formic acids, CO_2, and H_2 leading to curdling and souring of milk.

Some *Alcaligenes* sp. and coliforms can also cause ropiness (sliminess) due to production of viscous polysaccharides.

Butter contains 80% milk fat and can be salted or unsalted.

Growth of bacteria (*Pseudomonas sp.*), yeasts (*Candida sp.*) and molds (*Geotrichum*) on the surface have been implicated in flavour defects (putrid, rancid, or fishy) and surface discolouration.

In unsalted butter, coliforms, *Enterococcus* and *Pseudomonas* can grow favourably in water-phase and produce flavour defects.

Spoilage of Fruits and Vegetables

Vegetables

Vegetables are considered as a moderately perishable food.

The main sources of microorganisms in vegetables are soil, water, air and other environmental sources.

Fresh vegetables are fairly rich in carbohydrates, low in proteins.

Microorganisms grow more rapidly in damaged or cut vegetables.

The presence of air, high humidity and higher temperature during storage increases the chances of spoilage.

The common spoilage defects are caused by molds belonging to genera *Penicillium*, *Phytophthora*, *Alternaria*, *Botrytis* and *Aspergillus*. Among the bacterial genera, species from *Pseudomonas*, *Erwinia*, *Bacillus* and *Clostridium* are important.

Microbial vegetable spoilage is generally described by the common term rot along with the changes in the appearance, such as black rot, gray rot, pink rot, soft rot and stem-end rot.

Refrigeration, vacuum or modified atmosphere packaging, freezing, drying, heat treatment, and chemical preservatives are used to reduce microbial spoilage of vegetables.

Fruits

Fresh fruits have high carbohydrate content (10% or more), very low protein (=10%) but have pH 4.5 or below.

Microbial spoilage of fruits and fruit products is confined to molds, yeasts, and acidophilic bacteria like lactic acid bacteria, *Acetobacter* and *Gluconobacter*.

Fresh fruits are susceptible to rot by different types of molds like *Penicillium, Aspergillus, Alternaria, Botrytis, Rhizopus* and others.

According to the changes in appearance, the mold spoilages are designated as black rot, gray rot, soft rot, brown rot and others.

Yeasts, Saccharomyces, Candida, Torulopsis, and *Hansenula* cause fermentation of some fruits such as apples, strawberries, citrus fruits and dates.

To reduce spoilage, fruits and fruit products are preserved by refrigeration, freezing, drying, and reducing aW, vacuum packaging and heat treatment.

Common spoilage defects of fruits and vegetables

- ⅄ Bacterial soft rot - *Erwinia carotovora*
- ⅄ Gray mold rot - *Botrytis cinerea*
- ⅄ Rhizopus soft rot - *Rhizopus nigricans*
- ⅄ Blue mold rot - *Penicillium sp*
- ⅄ Alternaria rot - *Alternaria sp*
- ⅄ Pink mold rot - *Trichothecium roseum*
- ⅄ Green mold rots - *Cladosporium, Trichoderma*
- ⅄ Brown rot - *Sclerotinia sclerotiorum*
- ⅄ Downy mildew - *Phytophthora, Bremia*
- ⅄ Sliminess or souring - *Saprophytic bacteria*

F. FOOD PRESERVATION

Food preservation is the processes involved in protecting food against microbes and other spoilage agents to permit its future consumption. The preserved food should retain a palatable appearance, flavour and texture, as well as its original nutritional value.

The following four points are the main reasons of food presevation:

- ⅄ To protect food against microbes and other spoilage agents
- ⅄ To ensure that food is safe for future consumption
- ⅄ To prolong food storage time
- ⅄ To allow many foods to be available year-round, in great quantity and the best-quality

Principles of Preservations

Micro-organisms, enzyme, chemical reaction of food components are the main causes of food spoilage.

Principles of preservations are as follows

- ⅄ Killing of micro-organisms
- ⅄ Inhibition of microbial growth

⋏ Removing micro-organisms

⋏ Destroying enzyme

⋏ Retardation of chemical changes

Conditions for the growth of micro-organisms

⋏ Suitable Temperature

⋏ pH value

⋏ Moisture (water activity of the food)

⋏ Nutrients

⋏ Time

Methods of food preservation

The following methods to be followed for the better preservation of food materials.

⋏ Heat treatment

⋏ Irradiation

⋏ Smoking

⋏ Drying and dehydration

⋏ Refrigeration

⋏ Freezing

⋏ Canning

⋏ Sugaring and Salting

⋏ Meat curing

⋏ Pickling in vinegar

⋏ Use of food additives

⋏ Filtration

Heat Treatment

Food is heated up or cooked. Heat kills micro-organisms and their spores, alters protein structure, destroys enzyme activity of micro-organisms in food .

Methods of Heating

Blanching

a. Mild heat treatment, usually applied to fruits and vegetables to **denature enzymes.**

b. Often used before freezing of fruits and vegetables.

Pasteurization

The process is named after the French chemist Louis Pasteur, who devised it in 1865 to inhibit fermentation of wine and milk. It destroys pathogenic microorganisms and extends the shelf life of a food.

Levels of pasteurization used to thermally processed milk are

1. **Low Temperature Long Time (LTLT)**: 63°C (145°F) for 30 minutes

2. **High Temperature Short Time (HTST)**: 72°C (161°F) for 15 seconds

3. **Ultra High Temperature(UHT)**: 138°C (280°F) (or higher) for 2 seconds

Milk is pasteurized by heating at a temperature of 63°C (145°F) for 30 minutes, rapidly cooling it and then storing it at a temperature below 10° C (50°F). Beer and wine are pasteurized by being heated at about 60°C (about 140°F) for about 20 minutes; a newer method involves heating at 70°C (158°F) for about 30 seconds and filling the container under sterile conditions.

Irradiation

Ionizing radiation or irradiation is used as a method to destroy enzymes and micro-organisms in food. It kills mould and delay ripening of fruits and vegetables. It inhibits sprouting in bulbs and tubers and remove insects from grain, cereal products, fresh and dried fruits and vegetables. It also destroys bacteria in fresh meats. Irradiated foods are **not** radioactive. Type of radiation used for food preservation is **ionizing radiation**. It includes **gamma rays, x-rays and electrons.**

Smoking

The smoke is obtained by burning hickory or a similar wood under low breeze/wind at about 93°C to 104°C. Preservative action is provided by such bactericidal chemicals in the smoke as formaldehyde(HCHO) and creosote(antiseptic obtained from wood tar) and by the dehydration that occurs in the smoke house.

Drying and Dehydration

Drying removes the moisture from the food so that bacteria, yeasts and moulds cannot grow and spoil the food. It also slows down the action of enzymes. Dried food items can be kept almost indefinitely, as long as they are not rehydrated. The process of **drying foods** removes roughly 80 to 90 percent of the water content of fruits and vegetables.

Dried Fruits

Dried fruits are unique, tasty and nutritious. It might be argued that dried fruits are even tastier than fresh fruits. They have been called nature's candy. Dried fruit tastes sweeter because the water has been removed thus concentrating the fruit's flavour. Dried fruit can be eaten as a snack or added to cereals, muffins or ice cream.

Refrigeration

Refrigeration is a process of lowering the temperature and maintaining it in a given space for the purpose of chilling foods, preserving certain substances, or providing an atmosphere conducive to bodily comfort. Storing perishable foods, furs, pharmaceuticals, or other items under refrigeration is commonly known as cold storage. Chilling slows down microbial activities and chemical changes resulting in spoilage. In chilling, food is kept at 0°C - 4°C.

Freezing

Freezing turns water in food to ice. Water is unavailable for reactions to occur, and for micro-organisms to grow. Freezing preserves food by preventing microorganisms from multiplying. In freezing, food temperature is reduced to about -17°C. Eg. Frozen meat, peas, vegetables, ice-cream

Canning (Appertization)

Canning is a process of preserving food by heating and sealing it in airtight containers. The process was invented (1809) by Nicolas Appert, a French confectioner. Heating destroys enzymes and micro-organisms. The sealing of cans ensures no micro-organism and oxygen can get in. Airtight containers make sure that no oxygen in the containers for bacteria to live and chemical changes. Vacuum seal prevents air from getting back into the product carrying with it naturally occurring microorganisms to recontaminate the food. Examples: All kinds of tinned foods such as soup, meat and beans.

Sugaring and Salting

Food is treated with salt, strong salt solution or strong sugar solution. After adding salt or sugar, the water potenial outside the micro-organisms is higher than that inside the micro-organisms. As a result water essential for enzyme action and microbial growth is removed by osmosis, the microbies can't continue to live.

Meat Curing

Meat is treated with salt or strong salt solution containing $NaNO_3$, KNO_3 and which may contain sugar and spices. **Salt and sugar** both cure meat by osmosis. In addition to drawing the water from the food, they dehydrate and kill the bacteria that make food spoil. **Nitrite or nitrate** are used for retarding rancidity, curing-pink colour and killing bacteria.

Wet curing and **dry** curing are two methods of curing.

Pickling in Vinegar

Food is kept in vinegar since micro-organisms can not grow well in low pH value solutions. This is called pickling.

Use of Food Additives

Food additives are natural and synthetic compounds added to food to supply nutrients, to enhance colour, flavour or texture and to prevent or delay spoilage. Some additives can inactivates or kill micro-organisms and retard chemical spoilage.

Table 9.2 Common food preservatives and their uses

Preservative	Effective Concentration	Uses
Propionic acid and propionates	0.32%	Antifungal agent in breads, cake, Swiss Cheeses
Sorbic acid and sorbates	0.2%	Antifungal agent in cheeses, jellies, syrups, cakes
Benzoic acid and benzoates	0.1%	Antifungal agent in margarine, cider, relishes, soft drinks
Sodium diacetate	0.32%	Antifungal agent in breads
Lactic acid	unknown	Antimicrobial agent in cheeses, buttermilk, yogurt and pickled foods
Sulfur dioxide, sulfites	200-300 ppm	Antimicrobial agent in dried fruits, grapes, molasses
Sodium nitrite	200 ppm	Antibacterial agent in cured meats, fish
Sodium chloride	unknown	Prevents microbial spoilage of meats, fish, etc.
Sugar	unknown	Prevents microbial spoilage of preserves, jams, syrups, jellies, etc.
Wood smoke	unknown	Prevents microbial spoilage of meats, fish etc.

Filtration

Filtration can remove microorganisms from the liquid. Drinks such as beer can be sterilized by filtration to avoid quality changes which may impose from treatments such as heat sterilization. The fluid is filter through millipore membrane which has pore diameter (e.g. 2 um) smaller than that of bacteria. Microorganisms can be effectively eliminated with this preservation procedure. Filtration is used for clearing fluids such as beer.

G.   FOOD POISONING

Food poisoning is a general term usually referring to a gastrointestinal disease caused by the ingestion of food contaminated by pathogens or their toxins. There are of two types of food poisoning. They are **food intoxications** and **food borne infection**. **Food intoxication** is a type of food poisoning caused by microbial toxins produced in a food prior to consumption. The presence of living bacteria is not required. **Food-borne infection** is a gastrointestinal illness caused by ingestion of microorganisms, followed by their growth within the host. Symptoms arise from tissue invasion and/or toxin production.

Food intoxication results from consumption of toxins (or poisons) produced in food by bacterial growth. Toxins, (not bacteria) cause illness. Toxins may not alter the appearance, odour or flavour of food. Common kinds of bacteria that produce toxins include *Staphylococcus aureus* food poisoning and *Clostridium botulinum* food poisoning.

The main factors responsible for the food borne illness are

a. Improper holding temperature during processing.

b. Inadequate cooling during storage.

c. Contaminated equipments and utensils.

d. Food from unsafe source.

e. Poor personal hygiene.

f. Adding contaminated ingredients to cooked foods.

Staphylococcal Intoxication

It is one of the most commonly occurring food poisoning.

It is caused by the enterotoxin produced by *Staphylococcus aureus*. The toxin is not detectable by taste or smell.

Staphylococcus is a gram positive, facultative anaerobic cocci shaped bacterium.

Enterotoxin producing strains produce golden or yellow colour pigment on solid medium and are coagulase positive.

The toxins of *Staphylococcus* are called as enterotoxin because it cause gastroenteritis.

Food become contaminated with *Staphylococcus* are from skin, nose and throat of most people, Infected wounds, pimples, boils and acne.

Staphylococcus is transmitted by people-to-food through improper handling, multiply rapidly at room temperature to produce a toxin that causes illness.

Symptoms of this poisoning include abdominal cramps, vomiting, severe diarrhoea. These usually appear within one to eight hours after eating staph-infected food and last one or two days.

Foods commonly involved in Staphylococcal intoxication include protein foods such as processed meats, chicken, sandwich fillings, cream fillings, potato and meat salads, custards, milk products and creamed potatoes.

The best treatments for these patients are rest, consuming plenty of fluids, and medicines to calm their stomachs. Antibiotics are not useful in treating this illness.

Washing hands and under fingernails vigorously with soap and water before handling and preparing food. Not prepare food, if a person has a nose or eye infection. Keeping kitchens and food serving areas clean and sanitized. Keeping hot foods hot (over 140°F) and cold foods cold (40°F or under) can prevent staphylococcal food poisoning.

Botulism

Botulism is a clostridial food poisoning.

Botulism is a serious illness caused by botulinum toxin (poison) produced by *Clostridium botulinum*.

This toxin affects the nerves and cause paralysis and respiratory failure.

Home packed food with low acid content, such as asparagus, green beans, beets and corn is major source for botulism.

C. botulinum is an anaerobic bacterium, which means it can survive and grow with little or no oxygen.

Symptoms of foodborne botulism are double vision and drooping eyelids, Slurred speech, dry mouth and difficulty in swallowing and Weak muscles.

Symptoms of foodborne botulism usually begin within 18 to 36 hours after eating contaminated food.

Some ways to prevent foodborne botulism are follow strict hygienic steps when home canning. Refrigerate oils with garlic or herbs. Keep baked potatoes wrapped in aluminum foil hot until served, or refrigerate them.

H.   FOOD BORNE INFECTIONS

Listeriosis

Listeriosis is one of the food borne infection. It is caused by *Listeria monocytogenes*.

It is a gram positive, a facultative anaerobic, motile, intracellular bacterium.

It is one of the most virulent foodborne pathogen.

L. monocytogenes is frequently carried by humans and animals. The organism can grow in the pH range of 4.1 to 9.6. It is a salt tolerant and relatively resistant to drying, but easily destroyed by heat.

Listeriosis primarily affects newborn infants, pregnant women, the elderly and those with compromised immune systems.

Symptoms of listeriosis are mild illness with fever, headaches, nausea, vomiting and diarrhoea.

Recent cases have involved raw milk, soft cheeses made with raw milk, and raw or refrigerated ready-to-eat meat, poultry or fish products.

Preventive measures for listeriosis include maintaining good sanitation, turning over refrigerated ready-to-eat foods quickly, pasteurizing milk, avoiding post-pasteurization contamination and cooking foods thoroughly.

Successful treatment with parenteral penicillin or ampicillin exist. Trimethoprim-sulfamethoxazole has been shown effective in patients allergic to penicillin.

Campylobacteriosis

Campylobacteriosis is an infectious disease caused by *Campylobacter jejuni, C. fetus,* and *C. coli.*

Humans can get infection from handling raw poultry, eating undercooked poultry food, drinking nonchlorinated water, raw milk, handling infected animal, human faeces.

Diarrhoea (often bloody), abdominal cramping and pain, nausea and vomiting, fever, tiredness are the symptoms of campylobacteriosis.

Campylobacteriosis usualy lasts for 2 to 5 days, but in some cases as long as 10 days.

Ciprofloxacin or azithromycin, erythromycin helps treatment of campylobacteriosis.

Campylobacteriosis is prevented by washing hands before preparing food, immediately after handling raw poultry or other meat, cook poultry products properly, drink pasteurized milk and chlorinated or boiled water, Wash hands after handling pet faeces.

E. coli Infection

Certain types of *Escherichia coli* can cause foodborne illness. Harmless strains of *E. coli* can be found widely in nature, including the intestinal tracts of humans and warm-blooded animals.

Pathogenic strains of *Escherichia coli* can cause enteric infection. They are ETEC, EPEC, EHEC, EIEC, EAEC.

Dangerous type of *E. coli* that cause food poisoning are enterohemorrhagic *E. coli* or EHEC.

EHEC often causes bloody diarrhoea and can lead to kidney failure in children or people with weakened immune systems.

E. coli and its toxins have been found in undercooked or raw hamburgers, salami, alfalfa sprouts, lettuce, unpasteurized milk, apple juice, apple cider and contaminated well water.

E. coli toxin can damage the lining of the intestine and cause symptoms including nausea, severe abdominal cramps, watery or bloody diarrhoea, tiredness, vomiting.

Symptoms usually begin from 2 to 5 days after eating contaminated food and may last for 8 days.

Most people recover from *E. coli* infection within 5 to 10 days without treatment. Antibiotics are usually not helpful and health care experts recommend not taking antidiarrhoeal medicines.

Some ways to prevent *E. coli* infection are eat only thoroughly cooked beef and beef products. Cook ground beef patties to an internal temperature of 160°F. Avoid unpasteurized juices. Drink only pasteurized milk. Wash fresh fruits and vegetables thoroughly before eating raw or cooking.

Salmonellosis

Salmonellosis is an infection caused by *Salmonella typhimurium* and *S. enteritidis.*

Salmonella bacteria can be found in food products such as raw poultry, eggs, beef and sometimes on unwashed fruit.

Salmonella infection frequently occurs after handling pets, particularly reptiles like snakes, turtles, and lizards.

Symptoms of salmonellosis are diarrhoea, fever, abdominal cramps, Headache.

Symptoms begin from 12 hours to 3 days after being infected.

Most cases of salmonellosis clear up within 5 to 7 days and don't require treatment. People with severe diarrhoea may need intravenous fluids and treat it with antibiotics such as ampicillin.

Ways to prevent foodborne salmonellosis are Drink only pasteurized milk. Cook poultry and eggs thoroughly. Don't eat foods containing raw eggs.

Shigellosis

Shigellosis is also called bacillary dysentery. It is an infectious disease caused by *Shigella dysenteriae, S. flexneri, S. boydii,* and *S. sonnei.*

Symptoms of shigellosis are fever, tiredness, watery or bloody diarrhoea, nausea vomiting and abdominal pain

Antibiotic such as ampicillin or ciprofloxacin are used for the treatment.

Ways to prevent shigellosis are wash hands with soap and water before preparing foods and beverages. Wash hands after using the bathroom or changing infant diapers.

Avoid swallowing swimming pool water.

Bacillus cereus

It is a Gram-positive, facultative aerobic spore forming rod shaped bacteria.

A wide variety of foods including meats, milk, vegetables and fish have been associated with the diarrhoeal type food poisoning.

Bacillus produce two types of infections. They are diarrhoeal type and vomiting type.

The vomiting-type outbreaks have generally been associated with rice products. It is due to emetic type low molecular weight heat stable peptide toxin.

The diarrhoeal type of illness is caused by a high molecular weight protein toxin that are heat labile.

The symptoms of *B. cereus* diarrhoeal type food poisoning are the onset of watery diarrhoea, abdominal cramps, and pain after 6-15 hours of consumption of contaminated food. Nausea may accompany diarrhoea.

The emetic type of food poisoning is characterized by nausea and vomiting within 0.5 to 6 h after consumption of contaminated foods. Occasionally, abdominal cramps and/or diarrhoea may also occur.

It is treated with Ampicillin.

Yersinia enterocolitica Gastroenteritis

Y. enterocolitica is a psychrotrophic small rod-shaped, non spore forming, motile, facultative anaerobic gram-negative bacterium. Strains of *Y.enterocolitica* can be found in meats (pork, beef, lamb, etc.), oysters, fish, and raw milk. The strains can grow between 0 and 44°C temperature, 5% NaCl.

Yersiniosis is the disease caused by *Y.enterocolitica*. Yersinia infections mimic appendicitis, but the bacteria may also cause infections of other sites such as wounds, joints and the urinary tract.

It is frequently characterized by symptoms such as gastroenteritis with diarrhoea and/or vomiting; however, fever and abdominal pain are the hallmark symptoms.

It is treated with ciprofloxazin, Azithromycin etc.

I.   PREVENTION OF FOODBORNE DISEASES

These are some basic ways to prevent food borne diseases.

- Wash hands carefully before preparing food.
- Wash hands, utensils, and kitchen surfaces with hot soapy water after they touch raw meat or poultry.
- Cook beef and beef products thoroughly, especially hamburger.
- Cook poultry and eggs thoroughly.
- Eat cooked food promptly and refrigerate leftovers within two hours after cooking.
- Wash fruits and vegetables thoroughly, especially those that will be eaten raw.
- Drink only pasteurized milk and juices and treated surface water.
- Wash hands carefully after using the bathroom, changing infant diapers or cleaning up animal faeces.

J.   MICROBIAL TESTING OF FOODS (Refer Lab Manual of same author for details)

The examination of foods for the presence of microorganisms is one of the important areas of food microbiology. Number of bacteria in foods is an indicator of their handling, processing and ultimately quality. Methods like SPC, MPN, MBRT and DMC are generally employed in bacterialogy laboratory to detect most of the food borne infections. Molecular method also available to detect food borne pathogens.

Standard Plate Count (SPC)

SPC is useful in determining the viable number of microorganisms in foods. The procedure consists of diluting the food/water in a series of sterile dilution blanks. From blanks, measured amounts of diluted sample are transferred by spread plating/pour plating over the appropriate agar plates and incubated for 24-48 h. Thereafter plates are examined and number of colonies are counted to calculate the number of viable cells or colony forming units per unit volume of material under test. Nutrient agar (or) plate count agar medium is generally used.

Most Probable Number (MPN) (Refer Lab Manual of Same Author for Details)

The method was introduced by McCrady in 1915. This is a qualitative method for determination of the presence of coliforms in potable water. The selective medium used are Lauryl tryptose broth, brilliant green 2% bile broth, EMB agar. These medium contains a bile salt. It is inhibitory for growth of non-intestinal lactose fermenting bacteria. Since the method is statistical in nature, MPN results are generally higher than SPC. The test can be performed in two formats- three tube and five tube method.

The advantages of the method are as follows:

1. It is a simple technique
2. Good agreement between results from different laboratories;
3. Identification of specific microbial group is possible by the use of appropriate selective and differential medium.

Dye reduction Method (MBRT)

Dye reduction test involve the use of redox dyes like methylene blue to determine the quality of milk. Methylene blue is reduced and loses its colour in the presence of actively growing bacteria. The time taken for the reduction of methylene blue is inversely proportional to the number of viable bacteria. The shorter the methylene blue reduction time poorer is the quality of the milk. The advantages of the method are

1. It is a simple and rapid method for grading bulk supplies of raw milk;
2. It is an inexpensive method;
3. No skilled personnel are needed to perform the test. Disadvantages of the test are it does not indicate the type and source of organism and variations in the degree of bacterial metabolism.

Direct Microscopic count (DMC)

Another quick and precise method for determination of microorganisms in liquid foods is the use of haemocytometer or Neubauer counting chamber. The method is quick and requires expertisation. In this method, appropriately diluted samples are transferred on the haemocytometer using a Pasteur pipette and counted directly under the microscope. The haemocytometer is itched with a grid system into number of squares (25 large squares). The number of microorganisms present in the large square are counted and multiplied with the factor for large square and dilution factor to determine the total number of microorganisms in the sample.

10

SOIL MICROBIOLOGY

Soil Microbiology deals with the study of the soil microorganisms and their activities in the soil.

A. SOIL

Soil is a thin layer of material on the Earth's surface in which plants grow. It is made up of weathered rock and decayed plant and animal matter.

B. SOIL FORMATION

Soil formation required long period of time. It can take 1000 years or more. It needs cooperation and interaction of many factors like air, water, plant life, animal life, rocks and chemicals. Soil is formed from the weathering of rocks and minerals. The surface rocks break down into smaller pieces through a process of weathering. Over a period of time weathering and mixture of organic matter creates a thin layer of soil. Further interaction of plants, animals and decaying process makes the soil thick and rich.

Weathering

Weathering is the process of the breaking down rocks. There are two different types of weathering. Physical weathering and chemical weathering. Physical weathering breaks the rock. Nature of the rock is hard. Chemical weathering also breaks the rocks, but it may change the rock as soft material.

Soils are a mixture of different things. It is made up of rocks, minerals, dead and decaying plants and animals. Soil can be very different from one location to another. It consists of organic and inorganic materials, water and air. The inorganic materials are the rocks that have been broken down into smaller pieces. It may appear as pebbles, gravel, or as small as particles of sand or clay. The organic material is decaying living matter. This could be plants or animals that have died and decay until they become part of the soil. The amount of water in the soil is closely linked with the climate and other characteristics of the region.

C. SOIL PROFILE

Soil Profile means to the layers of soil. The soil profile is an important tool in nutrient management. The soil profile is a vertical section of the soil that depicts all of its horizons. The soil profile extends from the soil surface to the parent rock material.

The regolith includes all of the weathered material within the profile. The regolith has two components: the solum and the saprolite. The solum includes the upper horizons with the most weathered portion of the profile. The saprolite is the least weathered portion that lies directly above the solid, consolidated bedrock but beneath the regolith.

Master Horizons of Soil

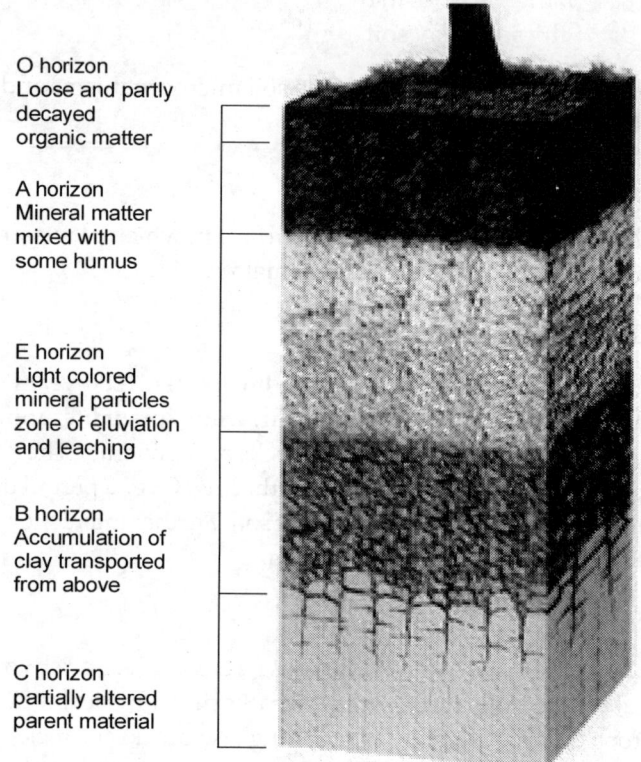

O horizon
Loose and partly
decayed
organic matter

A horizon
Mineral matter
mixed with
some humus

E horizon
Light colored
mineral particles
zone of eluviation
and leaching

B horizon
Accumulation of
clay transported
from above

C horizon
partially altered
parent material

Figure 138   Soil horizons

There are 5 master horizons in the soil profile. Not all soil profiles contain all 5 horizons. Soil profiles differ from one location to another. The horizons are represented by the letters: O, A, E, B and C.

O   The O horizon is a surface horizon that is comprised of organic material at various stages of decomposition. It is most prominent in forested areas.

A   The A horizon is a surface horizon that largely consists of minerals (sand, silt, and clay) and with appreciable amounts of organic matter. This horizon is predominant in grasslands and agricultural lands.

E   The E horizon is a subsurface horizon that has been heavily leached. Leaching is the process in which soluble nutrients are lost from the soil due to precipitation or irrigation. The horizon is typically light in colour. It is generally found beneath the O horizon.

B   The B horizon is a subsurface horizon that has accumulated from the layer(s) above. It is a site of deposition of certain minerals that have leached from the layer(s) above.

C   The C horizon is a subsurface horizon. It is the least weathered horizon. Also known as the saprolite, it is unconsolidated, loose parent material.

D.   SOIL TYPES

Depending on the size of the particles in the soil, it can be classified into these following types. They are Sandy soil, Silty soil, Clay soil, Loamy soil, Peaty soil, Chalky soil.

Sandy Soil   This soil type has the biggest particles. It creates better aeration and drainage. Sandy soil is formed by the disintegration and weathering of rocks such as limestone, granite, quartz and shale. Sandy soil is easier to cultivate if it is rich in organic material.

Silty Soil   Silty soil is considered to be one of the most fertile of soils. It can occur in nature as soil or as suspended sediment in water column. It is composed of minerals like quartz and fine organic particles.

Clay Soil   Clay is a kind of material that occurs naturally and consists of very fine grain material with very less air spaces. Due to this it is difficult to work with this soil.

Loamy Soil   This soil consists of sand, silt and clay. It is considered to be the perfect soil for gardening. The texture is gritty and retains water very easily.

Peaty Soil   This kind of soil is basically formed by the accumulation of dead and decayed organic matter. It is generally found in marshy areas. This soil is prone to water logging.

Chalky Soil   Chalky soil is very alkaline in nature and consists of a large number of stones.

E.   SOIL STRUCTURE

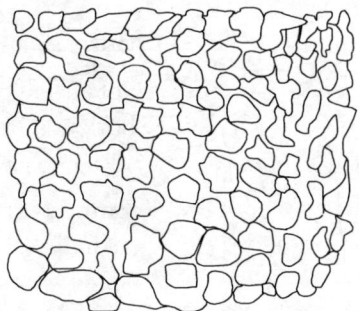

Figure 139   Granular soil

Soil structure is the arrangement of soil particles into groupings. These groupings are called peds or aggregates. It is found within certain soil horizons. Granular soil particles are characteristic of the surface horizon. Soil aggregation is an important indicator of the workability of the soil. Soils that are well aggregated are said to have "good soil tilth."

Types of Soil Structures in Soils

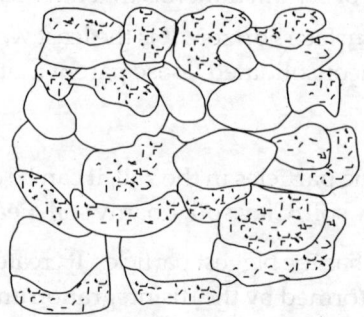

Figure 140 Blocky soil

Granular   it resembles coolie crumbs and is usually less than 0.5cm in diameter. Commonly found in surface horizons. (Figure 139)

Figure 141 Prismatic soil

Blocky   irregular blocks that are 1.5-2cm in diameter. (Figure 140)

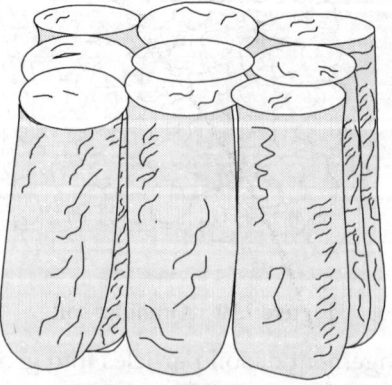

Figure 142 Columnar

Prismatic   it is a vertical column of soil. It is usually found in lower horizons. (Figure 141)

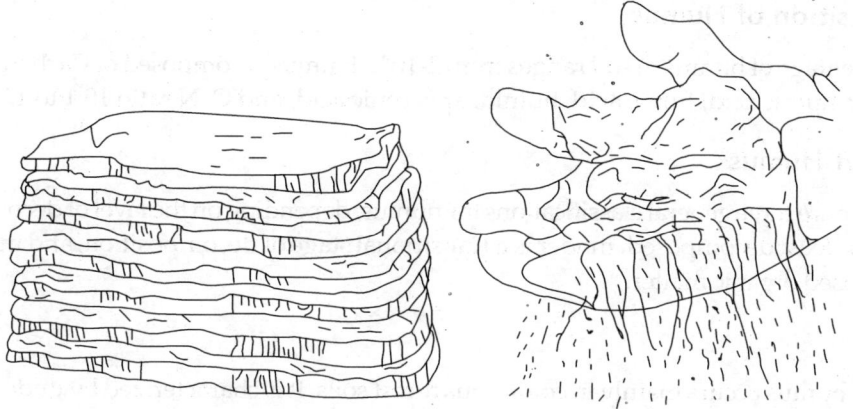

Figure 143   Platy soil   **Figure 144**   Single grained soil

Columnar	it is a vertical column of soil. It have a salt cap at the top. It is found in arid climates. (Figure 142)
Platy	it is of thin flat plates that are arranged horizontally. It is found in compacked soil. (Figure 143)
Single grained	soil is broken into individual particles that do not stick together. Commonly found in sandy soil. (Figure 144)

F. COMPONENTS OF SOIL

Soil is made up of five major components. They are organic matter, mineral matter, soil-air, soil water and soil microorganisms/living organisms. Dominant minerals found in the soil are silicon, aluminum and iron. Carbon, calcium potassium, manganese, sodium, sulphur, phosphorus etc. are also available in trace amount.

Organic matter is derived from organic residues of plants and animals. Organic matter serves as a food for microorganisms. It is the potential source of N, P and S.

Soil water comes from rain, snow, dew or irrigation. Soil water serves as a solvent and carrier of nutrients for the plant growth.

Soil air is a part of the soil volume. Pore spaces of soil are filled partly with soil water and partly with soil air. Soil aeration plays an important role in plant growth, microbial population, and microbial activities in the soil.

Varieties of Soil microorganisms are found in the soil. Bacteria are the most abundant and predominant organisms in soil. About 92% of soil microflora are bacteria.

G. SOIL HUMUS

Decomposed organic matter of the soil are called humus. Humus is the organic residue in the soil. It results from the decomposition of plant and animal residues in soil. It is the soft brown/dark coloured amorphous substance. It is composed of residual organic matter along with dead microorganisms. The process of forming humus are called humification.

Composition of Humus

Percentage of humus in soil ranges from 2-10%. Humus is composed of Carbon, Nitrogen acids like humic acid, fulvic acid, humin, apocrenic acid, and C: N ratio 10:1 to 12:1.

Types of Humus

There are three general classifications for humus, depending on the level of decomposition. Mor is the least decomposed, moder is a transitional stage of decomposition, and mull is fully decomposed organic matter.

Mor

Mor humus occurs mainly in coniferous forest soils. It is characterized by undecomposed or partially decomposed matter. It is a acidic humus. It has a carbon to nitrogen ratio of more than 20.

Moder

Moder humus is a transitional stage between mor and mull. It is found in hardwood forests and mountain grasslands. It is characterized by partial decomposition and shallow incorporation with the mineral soil. Moder humus has a carbon to nitrogen ratio of 15-25 and decomposition is caused by bacteria and invertebrates.

Mull

Mull humus consists of highly decomposed matter. Original litter is not recognizable in this humus. This humus is deeply mixed with the mineral soil. Mull humus is usually found under grass vegetation. It has a neutral pH and a carbon to nitrogen ratio of near 10. This is also the type of humus used for agriculture.

Functions of Humus

- It improves physical condition of the soil.
- It improves water holding capacity of soil.
- It serves as a store house for essential plant nutrients.
- It plays important role in determining fertility level of soil.
- It tends to make soils more granular with better aggregation of soil particles
- It prevents leaching losses of water soluble plant nutrients
- It improves biological activity in soil and encourage better development of plant-root system in soil
- It acts as buffering agent i.e. prevent sudden change in soil PH/soil reaction
- It serves as source of energy and food for the development of soil organisms
- It supplies both basic and acidic nutrients for the growth and development of higher plants
- It also improves aeration and drainage by making the soil more porous

H. SOIL MICROORGANISMS

Soil is an excellent culture media for the growth of various microorganisms. Aerobic bacteria are the predominant flora of soil (70%). It is followed by anaerobic bacteria and Actinomycetes (13% each) and fungi (3%). Algae, Protozoa and viruses comprises of 0.2-0.8%.

Bacteria

Morphologically, soil bacteria are divided into three groups. They are cocci, rod and spirllum. Bacilli are most numerous followed by cocci and spirllum in soil.

On the basis of ecological characteristics, bacteria are classified as Autochthonous (Indigenous species) and the Zymogenous (fermentative) flora. Autochthonous bacterial population is uniform and constant in soil. Zymogenous bacterial population in soil is low and requires an external source of energy, Eg. *Pseudomonas & Bacillus.*

On the basis of physiological activity or mode of nutrition, bacteria are classified as Autotrophs and Heterotrophs. All autotrophic bacteria utilize CO_2 as carbon source and derive energy either from sunlight or inorganic matter. Eg. *Chromatrum, Chlorobium, Rhadopseudomonas are photoautotrophs. Nitrobacter, Nitrosomonas, Thiaobacillus are chemoautotrophs.*

Bacteria present in soil are taxonomically included in the orders like Pseudomonadales, Eubacteriales and Actinomycetales. The most common soil bacteria belong to the genera are *Pseudomonas, Arthrobacter, Clostridium, Achromobacter, Sarcina, Enterobacter, Bacillus, Micrococcus, Chondrococcus, Archangium, Polyangium, Cyptophaga* etc.

Role of Bacteria in Soil

Bacteria brings biochemical transformations in the soil. They are involved in the following activities.

Cellulose decomposition	*Angiococcus, Cytophaga, Polyangium, Sporocytophyga, Bacillus, Achromobacter, Cellulomonas, Clostridium Methanosarcina, Methanococcus.*
Ammonification	*Bacillus, Pseudomonas.*
Nitrification	*Nitrosomonas, Nilrobacter Nitrosococcus.*
Denitrification	*Achromobacter, Pseudomonas, Bacillus, Micrococcus.*
Nitrogen fixation	*Rhizobium, Bradyrrhizobium, Azotobacter, Beijerinckia, Clostridium.*
Hemicellulose degradation	*Bacillus, Vibrio, Pseudomonas, Erwinia.*
Lignin degradation	*Pseudomonas, Micrococcus, Flavobacterium, Xanthomonas, Streptomyces.*
Pectin degradation	*Erwinia*

Fungi

Soil fungi possess filamentous mycelium composed of individual hyphae. The fungal hyphae may be aseptate /coenocytic (Mastigomycotina and Zygomycotina) or septate (Ascomycotina, Basidiomycotina and Deuteromycotina). Fungi are dominant in acid soils. Optimum pH range for fungi lies between 4.5 to 6.5. Soil fungi are aerobic and heterotrophic in nature.

Role of Fungi in Soil

Fungi plays significant role in plant nutrition.

They plays important role in the degradation of cellulose, hemi cellulose, starch, pectinin the organic matter added to the soil.

Lignin is resistant to bacterial decomposition but decomposed by fungi.

They also serve as food for bacteria.

Some fungi are predaceous in nature and attack on protozoa &nematodes in soil.

They also plays important role in soil aggregation and in the formation of humus.

Some soil fungi are parasitic and cause number of plant diseases such as wilts, root rots, damping-off and seedling blights. Eg. *Pythium, Phyiophlhora, Fusarium, Verticillium* etc.

Some soil fungi forms mycorrhizal association with the roots of higher plants and helps in mobilization of soil phosphorus and nitrogen. Eg. Glomus, Gigaspora, Aculospora, (Endomycorrhiza) and Amanita, Boletus, Entoloma, Lactarius (Ectomycorrhiza).

Algae

Algae are present in most of the soils where moisture and sunlight are available. They are photoautotrophic, aerobic organisms and obtain CO_2 from atmosphere and energy from sunlight.

Algae synthesize their own food.

They are unicellular, filamentous or colonial.

Soil algae are divided in to four main classes or phyla as follows:

1. Cyanophyta (Blue-green algae)
2. Chlorophyta (Grass-green algae)
3. Xanthophyta (Yellow-green algae)
4. Bacillariophyta (diatoms or golden-brown algae)

Blue-green algae and grass-green algae are more abundant in soil. Green-algae prefer acid soils while blue green algae are commonly found in neutral and alkaline soils. The most common genera of green algae found in soil are *Chlorella, Chlamydomonas, Chlorococcum, Protosiphon. Diatoms present in the soil are Navicula, Pinnularia. Synedra, Frangilaria.*

The dominant genera of BGA in soil are *Chrococcus, Phormidium, Anabaena, Aphanocapra, Oscillatoria* etc.

Some BGA posses specialized cells known as "Heterocyst". It is the sites of nitrogen fixation. BGA fixes nitrogen (non-symbiotically) in water logged paddy fields. There are certain BGA which possess the character of symbiotic nitrogen fixation in association with other organisms like fungi, mosses, liverworts and aquatic ferns Azolla, Eg. *Anabaena-Azolla* association fix nitrogen symbiotically in rice fields.

Role of Algae or BGA in Soil

Algae plays important role in the maintenance of soil fertility.

They add organic matter to soil.

Most of soil algae act as cementing agent in binding soil particles and thereby reduce/ prevent soil erosion.

Mucilage secreted by the BGA is hygroscopic in nature and thus helps in increasing water retention capacity of soil for longer time/period.

Soil algae through the process of photosynthesis liberate large quantity of oxygen in the soil environment and thus facilitate the aeration in submerged soils or oxygenate the soil environment.

They help in checking the loss of nitrates through leaching and drainage especially in un-cropped soils.

They help in weathering of rocks and building up of soil structure.

Protozoa

These are unicellular, eukaryotic, colourless organisms. They are larger than bacteria. They are abundant in surface soil. They can withstand adverse soil conditions due to the presence of "cyst". Some of the members are reproducing sexually by fusion of cells; rest of them reproduces asexually by binary fission. Most of the soil protozoa are motile by flagella or cilia or pseudopodia as locomotors organs. Amoeba belongs to rhizopoda motile by pseudopodia. Leishmania, Trypanosoma are motile by flagella. Members of ciliophora are move by using cilia. Eg. *Balantidium coli.*

Role of Protozoa

Protozoans derive their nutrition by feeding or ingesting soil bacteria.

They play important role in maintaining bacterial equilibrium in the soil.

Some protozoa have been recently used as biological control agents against phytopathogens.

Several soil protozoa cause diseases in human beings which are carried through water and other vectors, eg. Amoebic dysentery caused by *Entomobea histolytica.*

Factors affecting microbial flora of the Rhizosphere / Rhizosphere Effect

The rhizosphere is the zones of soil surrounding a plant root. The most important factors which affect / influence the microbial flora of the rhizosphere or rhizosphere effect are soil type and its moisture, soil amendments, soil pH, proximity of root with soil, plant species age of plant and root exudates.

I.   ROOT EXUDATES

It is a substance secreted by the root. It enhances the rhizosphere effect and activity of soil microorganisms. The following are the examples of root exudates.

Table 10.1   Root exudates and its composition

S. No	Root Exudates	Chemical Substances
1	Amino Acids	All naturally occurring amino acids.
2	Organic acids	Acetic, butyric, citric, fumaric, lactic, malic, propionic, succinic etc.
3	Carbohydrates / sugars	Arabinose, fructose, galactose, glucose, maltose, mannose, oligosaccharides, raffinose, ribose, sucrose, xylose etc.
4	Nucleic acid derivatives	Adenine, cystidine, guanine, uridine
5	Growth factors (phytohormones)	Biotin, choline, inositol, pyridoxine etc
6	Vitamins	Thiamine, nicotinic acid, biotin etc
7	Enzymes	Amylase, invertase, protease, phosphatase etc.
8	Other compounds	Auxins, glutamine, glycosides, hydrocyanic acid peptides, Uv-absorbing compounds, nematode attracting factors, spore germination stimulators, spore inhibitors etc.

J.   BIOGEO CHEMICAL CYCLING

Soil microorganisms are the most important agents in the cycling / transformation of various elements (C, N, P, K, S, Iron etc.) in the biosphere; where the essential elements undergo cyclic alterations between the inorganic state as free elements in nature and the combined state in living organisms. Life on earth is dependent on the cycling of nutrient elements from their elemental states to inorganic compounds to organic compounds and back into their elemental states. The microbes through the process of biochemical reactions breakdown complex organic compounds into simple inorganic compounds and finally into their constituent elements. This process is known as "Mineralization". Mineralization of organic carbon, nitrogen, phosphorus, sulphur and iron by soil microorganisms makes these elements available for reuse by plants.

The Carbon Cycle

The cyclic movements of carbon between the living organisms and the environment are referred as carbon cycle. It is broadly categorized in to three stages they are CO_2 fixation, CO cycling and CO_2 regeneration.

Major Carbon Sources

Carbon is available as organic form and also inorganic form. It is available in both living (Biotic) and nonliving (Abiotic) things.

Carbon in Living Things

Carbon is an important element of life. Carbohydrates, proteins, lipid, aminoacids are made up of carbon. Animals and plants need carbon for their survival. Energy needed for our survival is provided by the carbon containing carbohydrate.

Carbon in Non Living Things

Most actively cycling reservoir of carbon is carbon dioxide (CO_2). CO_2 dissolved in water and forming HCO_3; carbonate rocks (limestone, dolomite) deposits of coal, petroleum and natural gas derived from once-living things; dead organic matter. E.g., humus in the soil. Lithosphere contain carbon as diamond, graphite, mollusk shell and humus.

Nature of Carbon Recycling

Higher volume of noncycled carbon is available in fossil fuel (20,000,000 billion tons). Carbonaceous rock such as limestone is slowly dissolved by biologically produced acid with the release of CO_2 and HCO_3. Natural recycling is taken place by biological degradation of oil and other hydrocarbons. Most actively cycling reservoir of carbon is carbon dioxide (CO_2).

Rate of C cycle

The natural rate of C cycling in ocean and on land is close to a steady state. But human activities recently introduced changes in the C cycle. Rate of atmospheric CO_2 is affected by industrial CO_2 release. Burning of fossil fuel and other hydrocarbon accumulates larger volume of CO_2 and CO. Normal level of CO_2 in the environment is 0.032%. This level is now increasing due to imbalance between CO_2 Fixation and CO_2 regeneration. This may leads to green house effect.

CO_2 Fixation

Carbon enters the biotic world through the action of autotrophs are called CO_2 fixation.

Photosynthesis

Photoautotrophs like plants and algae uses light energy to convert carbon dioxide to organic matter through pentose phosphate or calvin cycle.

$$CO_2 + water \rightarrow glucose + oxygen$$

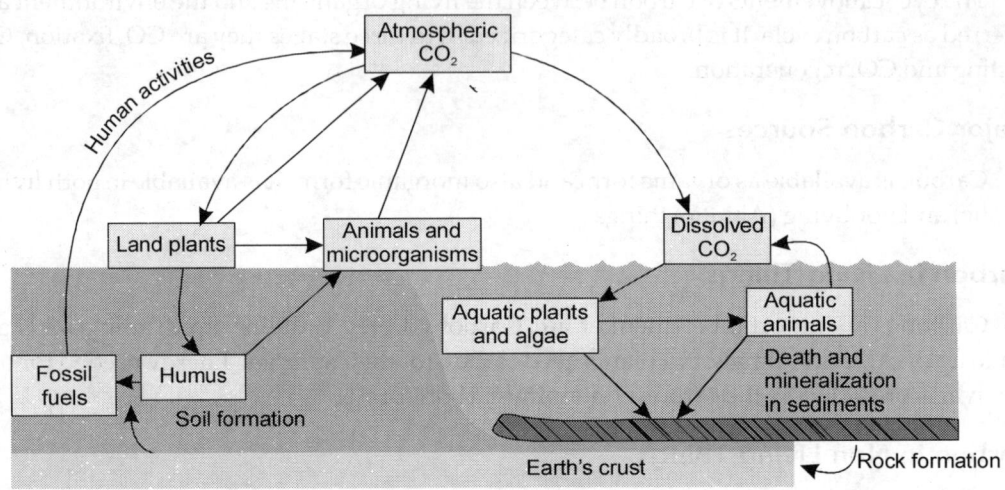

Figure 145   Carbon cycle

Chemoautotrophs converts CO_2 into organic matter through pyruvate carboxylase system.

Methanogenesis

Methanogenic archaebacteria plays an important role in the anaerobic reduction of CO_2. Methanogens converts dissolved CO_2 (HCO_3^-) using hydrogen into methane.

$$HCO_3^- + H^+ + 4H_2 \rightarrow CH_4 + 4H_2O$$

Methanosarcina barkeri is capable of converting methanol, acetate and methylamines to methane.

Acetogenesis

A group of facultative chemoautrophic anaerobes *Clostridium* and *Acetobacterium* capable of reducing CO_2 with H_2 to acetate.

$$2CO_2 + 4H_2 \rightarrow CH_3COOH + 2H_2O$$

Carbon Monoxide Cycling

Microbes also involved in the cycling of CO in both direct and indirect methods. Carboxybacteria like *Pseudomonas* are capable of converting CO to CO_2 and CO_2 is fixed by chemoautotrophs.

$$CO + H_2O \rightarrow CO_2 + H_2$$

Methanosarcina barkeri reduce CO by H_2 to methane.

$$CO + 3H_2 \rightarrow CH_4 + H_2O$$

Clostridium thermoaceticum converts CO to acetate

$$2CO + 2H_2 \rightarrow CH_3COOH$$

CO_2 Regeneration

Carbon is regenerated to the atmosphere and water by respiration (as CO_2) of living things, burning of coal and petroleum products, decaying of organic matter, Natural and accidental fire.

Microbial Degradation of Polymers

Micoroorganisms converts the polymers like starch, chitin, lignin, cellulose, lignocelluloses into a usable form and also releases CO_2. The uptake and return of CO_2 are not in balance.

Nitrogen Cycle

Nitrogen cycle is the sequence of biochemical changes from free atmospheric N_2 to complex organic compounds and further to simple inorganic compounds (ammonia, nitrate) and eventual release of molecular nitrogen (N_2) back to the atmosphere is called "nitrogen cycle".

Forms of Nitrogen

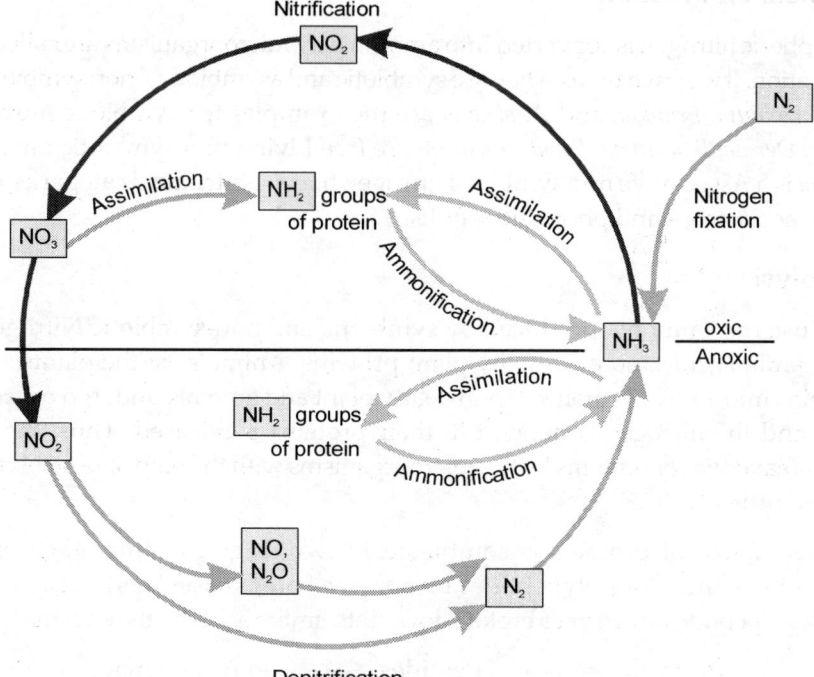

Figure 146   Nitrogen cycle

In nature, nitrogen exists in three different forms. They are gaseous / gas (78 to 80 % in atmosphere) form, organic (proteins and amino acids, chitins, nucleic acids and amino sugars) form and inorganic (ammonia and nitrates) form. Urea, Ammonia, Ammonium, Nitrate, Nitrite, Atmospheric Dinitrogen are also considered as a some forms of nitrogen.

Roles of Nitrogen

Plants and bacteria use nitrogen in the form of NH_4^+ or NO_3^-

It serves as an electron acceptor in anaerobic environment

Nitrogen is often the most limiting nutrient in soil and water.

Nitrogen is a key element for Amino acids, Nucleic acids (purine, pyrimidine), Cell wall components of bacteria (NAM).

Several biochemical steps involved in the nitrogen cycle are:

1. Biological nitrogen fixation
2. Proteolysis
3. Ammonification
4. Nitrification
5. Nitrate reduction and
6. Denitrification

1. Biological N$_2$ Fixation

Atmospheric nitrogen is converted into ammonia by microorganisms are called biological nitrogen fixation. It is of two types. They are symbiotic and asymbiotic / non symbiotic nitrogen fixation. *Rhizobium, Frankia,* and *Anabaena* are the examples for Symbiotic nitrogen fixers. *Azotobacter, Derxia, Bejerinkia, Rhodospirillum* are free Living non symbiotic nitrogen fixers. *Azospirillum* is a Associative non symbiotic nitrogen fixers. Nutritional categories of N$_2$ fixing bacteria are heterotrops and photoautotrophs.

2. Proteolysis

Plants use the ammonia produced by symbiotic and non-symbiotic Nitrogen fixers to make their amino acids and eventually plant proteins. Animals eat the plants and convert plant proteins into animal proteins. Upon death, plant and animals undergo microbial decay in the soil and the nitrogen contained in their proteins is released. Thus, the process of enzymatic breakdown of proteins by the microorganisms with the help of proteolytic enzymes is known as "proteolysis".

The breakdown of proteins is completed in two stages. In first stage proteins are converted into peptides or polypeptides by enzyme "proteinases" and in the second stage polypeptides / peptides are further broken down into amino acids by the enzyme "peptidases".

$$\text{Proteins} \xrightarrow[\text{Proteinases}]{} \text{Peptides} \xrightarrow[\text{Peptidases}]{} \text{Amino Acids}$$

The most active microorganisms responsible for elaborating the proteolytic enzymes (Proteinases and Peptidases) are *Pseudomonas, Bacillus, Proteus, Clostridium histolyticum, Micrococcus, Alternaria, Penicillium* etc.

3. Ammonification (Ammo acid degradation)

Amino acids released during proteolysis undergo deamination in which nitrogen containing amino (–NH$_2$) group is removed. Thus, process of deamination which leads to the production of ammonia is termed as "ammonification". The process of ammonification is mediated by several soil microorganisms. The processes of ammonification are commonly brought about by *Clostridium* sp, *Micrococcus* sp, *Proteus* sp. etc. and it is represented as follows.

$$CH_3CHNH_2COOH + \frac{1}{2}O_2 \longrightarrow CH_3COCOOH + NH_3$$
$$\underset{\text{Alanine}}{} \qquad\qquad\qquad \underset{\text{Pyruvic acid}}{} \quad \underset{\text{ammonia}}{}$$

4. Nitrification

Ammonical nitrogen / ammonia released during ammonification are oxidized to nitrates and the process is called "nitrification". Nitrification is a two stage process and each stage is performed by a different group of bacteria as follows.

Stage I: Oxidation of ammonia to nitrite is brought about by ammonia oxidizing bacteria viz. *Nitrosomonas europaea, Nitrosococcus nitrosus, Nitrosospira briensis, Nitrosovibrio* and *Nitrocystis*. The process of converting Ammonia to nitrite is known as nitrification. The reaction is presented as follows.

$$\underset{\text{Ammonia}}{2NH_3} + \frac{1}{2}O_2 \longrightarrow \underset{\text{Nitrite}}{NO_2} + 2H + H_2O$$

Stage II: In the second step, nitrite is oxidized to nitrate by nitrite-oxidizing bacteria such as *Nitrobacter winogradsky, Nitrospira gracilis, Nirosococcus mobilis* etc, and several fungi (eg. *Penicillium, Aspergillus*) and actinomycetes (eg. *Streptomyces, Nocardia*).

$$\underset{\text{Nitrite ions}}{NO_2 (-)} + \frac{1}{2}O_2 \longrightarrow \underset{\text{Nitrate ions}}{NO_3}$$

5. Nitrate Reduction

Several heterotrophic bacteria (*E. coli, Azospirillum*) are capable of converting nitrates to nitrites and nitrites to ammonia. Thus, the process of nitrification is reversed completely which is known as nitrate reduction. Nitrate reduction normally occurs under anaerobic soil conditions (water logged soils) and the overall process is as follows:

$$\underset{\text{Nitrate}}{HNO_3} + 4H_2 \xrightarrow[\text{Reductase}]{\text{Nitrate}} \underset{\text{ammonium}}{NH_4} + 3H_2O$$

Nitrate reduction leading to production of ammonia is called "dissimilatory nitrate reduction" as some of the microorganisms assimilate ammonium for synthesis of proteins and amino acid.

6. Denitrification

This is the reverse process of nitrification. During denitrification nitrates are reduced to nitrites and then to nitrogen gas and ammonia. Thus, reduction of nitrates to gaseous nitrogen by microorganisms in a series of biochemical reactions is called "denitrification". The process is wasteful as available nitrogen in soil is lost to atmosphere. The overall process of denitrification is as follows:

Nitrate ——→ Nitrite ——→ Nitric Oxide ——→ Nitrous Oxide ——→ Nitrogen gas

This process also called dissimilatory nitrate reduction as nitrate nitrogen is completely lost into atmospheric air. The most important denitrifying bacteria are *Thiobacillus denitrificans, Micrococcus denitrificans*, and species of *Pseudomonas, Bacillus, Achromobacter, Serrtatia paracoccus* etc.

Atmospheric nitrogen (N_2) is converted into ammonia and then to amino acids, proteins, nucleic acids, amino sugars etc. The proteins are then degraded to simpler organic compounds viz. peptones and peptides into amino acids which are further degraded to inorganic nitrogen compounds like ammonia, nitrites and nitrates. The nitrate form of nitrogen is mostly used by plants or may be lost through leaching or reduced to gaseous nitrogen and subsequently goes into the atmosphere, thus completing the nitrogen cycle.

Sulphur Cycle/Sulphur Transformations

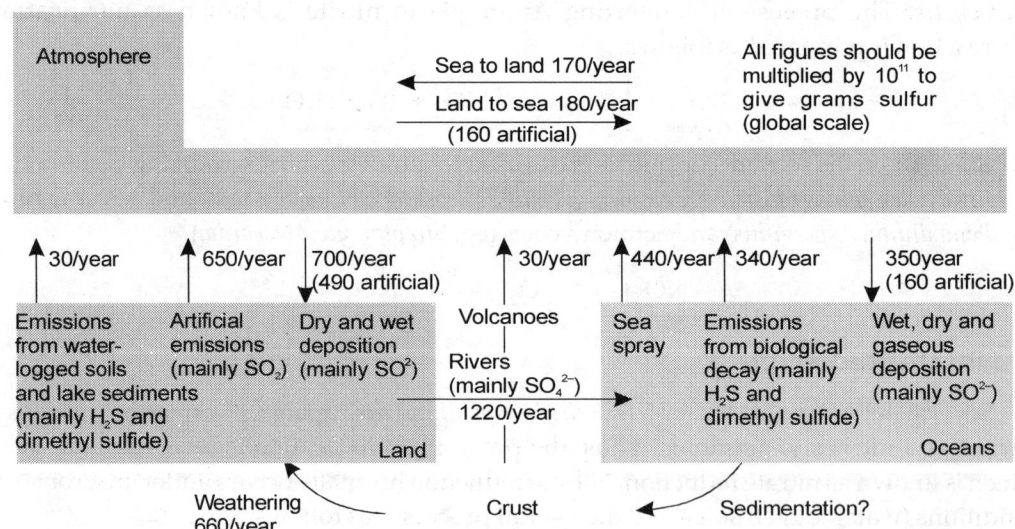

Figure 147 Sulfur cycle

Sulphur is the most aboundant and widely distributed element in the nature. It is found both in free as well as combined states. Sulphur is an essential element for all living systems. In the soil, sulphur is in the organic form as sulphur containing amino acids. Organic sulphur is metabolized by soil microorganisms to make it available in an inorganic form (sulphur, sulphates, sulphite, thiosulphale, etc) for plant nutrition. Of the total sulphur present is soil only 10-15% is in the inorganic form (sulphate) and about 75-90 % is in organic form. Cycling of sulphur between organic and elemental states and between oxidized and reduced states is brought about by various microorganisms, specially bacteria.

Various transformations of the sulphur in soil results mainly due to microbial activity. The major types of transformations involved in the cycling of sulphur are, 1. Mineralization, 2. Immobilization, 3. Oxidation and 4. Reduction

1. Mineralization is the breakdown of large organic sulphur compounds to smaller units and their conversion into inorganic compounds (sulphates) by the microorganisms.

2. Immobilization is the microbial conversion of inorganic sulphur compounds to organic sulphur compounds.

3. Oxidation of elemental sulphur and inorganic sulphur compounds (such as H_2S, sulphite and thiosulphate) to sulphate (SO_4) is brought about by chemoautotrophic and photosynthetic bacteria. When plant and animal proteins are degraded, the sulphur is released from the amino acids and accumulates in the soil which is then oxidized to sulphates in the presence of oxygen and under anaerobic condition. Organic sulphur is decomposed to produce hydrogen sulphide (H_2S). H_2S can also accumulate during the reduction of sulphates under anaerobic conditions which can be further oxidized to sulphates under aerobic conditions,

$$2S + 3O_2 + 2H_2O \longrightarrow 2H_2SO_4 \xrightarrow{\text{Ionization}} 2H(+) + SO_4 \, (\text{Aerobic})$$

$$H_2 + S + 2CO_2 + H_2O \xrightarrow{\text{Light}} H_2SO_4 + 2(CH_2O) \; (\text{Anaerobic})$$

    T. ferrooxidans and *T. thiooxidans* are the main organisms involved in the oxidation of elemental sulphur to sulphates. Heterotrophic bacteria (*Bacillus, Pseudomonas,* and *Arthrobacter*) and fungi (*Aspergillus, Penicillium*) and some actinomycetes are also reported to oxidize sulphur compounds. Green and purple bacteria (Photolithotrophs) of genera *Chlorbium, Chromatium, Rhodopseudomonas* are also reported to oxidize sulphur in aquatic environment.

4. Reduction of Sulphate: Sulphate in the soil is assimilated by plants and microorganisms and incorporated into proteins. This is known as "assimilatory sulphate reduction". Sulphate can be reduced to hydrogen sulphide (H_2S) by sulphate reducing bacteria (eg. *Desulfovibrio* and *Desulfatomaculum*).

Dissimilatory sulphate-reduction is favoured by the alkaline and anaerobic conditions of soil and sulphates are reduced to hydrogen sulphide. For example, calcium sulphate is attacked under anaerobic condition by the members of the genus *Desulfovibrio* and *Desulfatomaculum* to release H_2S.

$$CaSO_4 + 4H_2 \longrightarrow Ca(OH)_2 + H_2S + H_2O$$

Hydrogen sulphide produced by the reduction of sulphate and sulphur containing amino acids. Decomposition is further oxidized by some species of green and purple phototrophic bacteria (eg. *Chlorobium, Chromatium*) to release elemental sulphur.

$$CO_2 + 2H_2 + H_2S \xrightarrow[\text{Enzyme}]{\text{Light}} \underset{\text{Carbohydrate}}{(CH_2O)} + H_2O + \underset{\text{Sulphur}}{2S}$$

The predominant sulphate-reducing bacterial genera in soil are *Desulfovibrio*, *Desulfatomaculum* and *Desulfomonas*.

Phosphorus Cycle or Transformation

Phosphorus is only second to nitrogen as a mineral nutrient required for plants, animals and microorganisms. It is a major constituent of nucleic acids. It is essential for energy generation. This element is added to the soil in the form of chemical fertilizers, or in the form of organic phosphates present in plant and animal residues. Both inorganic and organic phosphates exist in soil and occupy a critical position both in plant growth and in the biology of soil.

Microorganisms are known to bring a number of transformations of phosphorus, these include:

i.   Altering the solubility of inorganic compounds of phosphorus,

ii. Mineralization of organic phosphate compounds into inorganic phosphates,

iii. Conversion of inorganic, available anion into cell components

iv. Oxidation or reduction of inorganic phosphorus compounds.

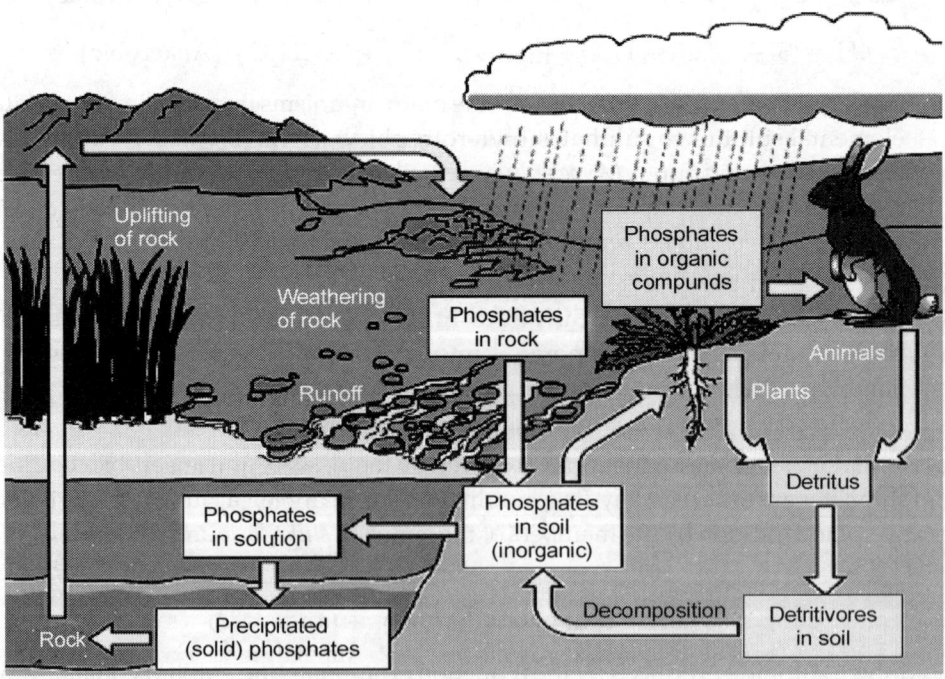

Figure 148   Phosphorus cycle

Insoluble inorganic compounds of phosphorus are unavailable to plants, but many microorganisms can bring the phosphate into solution. Soil phosphates are rendered available either by plant roots or by soil microorganisms through secretion of organic acids (eg. lactic, acetic, formic, fumaric, succinic acids etc). Phosphate solubilizing microorganisms (eg. species of *Pseudomonas, Bacillus, Micrococcus, Mycobacterium, Flavobacterium, Penicillium, Aspergillus, Fusarium* etc.) plays an important role in correcting phosphorus deficiency of crop plants. They may also release soluble inorganic phosphate (H_2PO_4) into soil through decomposition of phosphate-rich organic compounds.

Solubilization of phosphate by plant roots and soil microorganisms is substantially influenced by various soil factors such as pH, moisture and aeration.

In neutral or alkaline soils, solubilization of phosphate is more as compared to acidic soils. Many phosphates solubilizing microorganisms are found in close proximity of root surfaces and may appreciably enhance phosphate assimilation by higher plants. Fungi, bacteria and actinomycetes make available the organically bound phosphorus in soil and organic matter and the process is known as mineralization. The enzyme involved in mineralization (cleavage) of phosphate from organic phosphorus compound is collectively called as "Phosphatases".

The commercially used species of phosphate solubilizing bacteria and fungi are: *Bacillus polymyxa, Bacillus megatherium, Pseudomonas strita, Aspergillus, Penicllium avamori* and *Mycorrhiza*.

Iron Cycle or Transformation

Iron exists in nature either as ferrous (Fe^{++}) or ferric (Fe^{+++}) ions. Ferrous iron is oxidized spontaneously to ferric state, forming highly insoluble ferric hydroxide. Plants as well as microorganisms require traces of iron. Iron is always abundant in terrestrial habitats.

Soil microorganisms play an important role in the transformations of iron.

Many heterotrophic species attack on insoluble organic iron salts and convert into inorganic salts. Number of bacteria and fungi produce acids such as carbonic, nitric, sulphuric and organic acids which brings iron into solution. Under anaerobic conditions, the sulfides formed from sulphate and organic sulphur compounds remove the iron from solution as ferrous sulfide. Microbes liberate organic acids and other carbonaceous products of metabolism which results in the formation of soluble organic iron complex. Some bacteria are capable of reducing ferric iron to ferrous which lowers the oxidation-reduction potential of the environment (eg. *Bacillus, Clostridium, Klebsiella* etc). However, some chemoautotrophic iron and sulphur bacteria such as *Thiobacillus ferroxidans* and *Ferrobacillus ferrooxidans* can oxidize ferrous iron to ferric hydroxide which accumulates around the cells.

Most of the aerobic microorganisms live in an environment where iron exists in the oxidized, insoluble ferric hydroxide form. They produce iron-binding compounds in order to take up ferric iron. The iron-binding or chelating compounds/ligands produced by microorganisms are called "Siderophores". Bacterial siderophores may act as virulence factors in pathogenic bacteria and thus, bacteria that secrete siderophores are more virulent than non- siderophores producers. Therefore, siderophore producing bacteria can be used as biocontrol agents eg. *Pseudomonas fluoresence* used to control *Pythium*, causing damping-off diseases in seedlings. Recently Vascular - Arbusecular - Mycorrhiza (VAM) has been reported to increase uptake of iron.

K.   INTERACTIONS AMONG SOIL MICROORGANISMS

Soil is the largest terrestrial ecosystem. Wide variety of relationships exists between different types of soil organisms. The associations existing between different soil microorganisms. Some of the interactions are mutually beneficial, or harmful or neutral. The various types of possible interactions occurring among the microorganisms in soil are

a.   Beneficial interactions

    i.   mutualism

    ii.   commensalism and

    iii.   proto-cooperation

b.   Harmful interactions

    i.   amensalism,

    ii.   antagonism,

    iii.   competition

    iv.   Parasitism and

    v.   predation

a. Beneficial Association/Interactions

i. Mutualism (Symbiosis)

It is a positive relationship between two partners in which both the partners are benefited. When the benefit is in the form of exchange of nutrients is termed as "syntrophism". Lichen is the best example for this kind of interactions. Fungus provides protection and algae provides the nutrient to fungi through the photosynthesis.

Microorganisms may also form mutualistic relationships with plants. In this Rhizobium-legume association, Rhizobium bacteria are benefited by protection from the environmental stresses while in turn plant is benefited by getting readily available nitrate nitrogen released by the bacterial partner.

The Anabaena-Azolla is an association between the water fern Azolla and the cyanobacterium Anabaena. This association is of great importance in paddy fields, where nitrogen is frequently a limiting nutrient.

The protozoan-termite relationship is a classic example of mutualism in which the flagellated protozoa live in the gut of termites and wood roaches. The protozoa engulf wood particles, digest the cellulose and metabolize it to acetate and other products. Termites oxidize the acetate released by their flagellates. Because the host is almost always incapable of synthesizing cellulases, it is dependent on the mutualistic protozoa for its existence.

Many marine invertebrates (sponges, jellyfish, sea anemones, corals, ciliates) harbor endosymbiotic, spherical algal cells called zooxanthellae within their tissue. Because the degree of host dependency on the mutualistic alga is somewhat variable. The hermatypic (reef-building) corals satisfy most of their energy requirements using their zooxanthellae.

Pigments produced by the coral protect the algae from the harmful effects of ultraviolet radiation. Coral reefs are among the most productive and successful of known ecosystems.

ii. Commensalisms

In this association one organism/partner in association is benefited by other partner without affecting it. Eg, many fungi can degrade cellulose to glucose, which is utilized by many bacteria.

An example is nitrification, the oxidation of ammonium ion to nitrite by microorganisms such as *Nitrosomonas,* and the subsequent oxidation of the nitrite to nitrate by *Nitrobacter* and similar bacteria. *Nitrobacter* benefits from its association with *Nitrosomonas* because it uses nitrite to obtain energy for growth.

Commensalistic associations also occur when one microbial group modifies the environment to make it more suited for another organism. For example, in the intestine the common, nonpathogenic strain of *Escherichia coli* lives in the human colon, but also grows quite well outside the host, and thus is a typical commensal. When oxygen is used up by the facultatively anaerobic *E. coli*, obligate anaerobes such as *Bacteroides* are able to grow in the colon. The anaerobes benefit from their association with the host and *E. coli,* but *E. coli* derives no obvious benefit from the anaerobes.

iii. Proto-cooperation (synergism)

It is mutually beneficial association between two species. In this type of association one organism favour its associate by removing toxic substances from the habitate and simultaneously obtain carbon products made by the another associate/partner. Nutritional proto-cooperation between bacteria and fungi has been reported for various vitamins, amino and purines in terrestrial ecosystem and are very useful in agriculture.

Synthesis of a product which neither populations can perform on their own: For example, completion of a pathway. This type of synergism is known as syntrophy.

Close spatial relationships between microorganisms: For example bacteria are often seen on surfaces of algae due to chemotaxis.

Metabolism of toxic end-products: *Pseudomonas* produces organic end-products compounds from orcinol which are utilized by secondary microbes for growth. These end-products would have otherwise inhibited growth of *Pseudomonas.*

Production of degradative enzymes: *Arthrobacter* and *Streptomyces* (soil flora) produces enzymes which collectively degrade diazinon, an organophosphate pesticide (useful in the degradation of xnobiotics or recalciterant compounds).

b. Harmful Associations/Interactions:

i. Antagonism

It is the relationship in which one species is inhibited or adversely affected by another species. *Bacillus* species from soil produces an antifungal agent which inhibits growth of several soil fungi. Several species of *Streptomyces* from soil produces antibacterial and antifungal antibiotics.

Fatty acids produced on the skin by skin microflora restrict growth of unwanted pathogens. O_2 production by algae precludes growth of anaerobes. High concentrations of ethanol (eg wine production) precludes most microbes other than the yeasts. However, Acetobacter converts ethanol to acetate if O_2 is present. Lactate and propionate production in cheese manufacturing and acetic acid in vinegar inhibit spoilage causing microbes.

ii. Ammensalism

In this interaction one partner suppress the growth of other partner by producing toxins like antibiotics and harmful gases like ethylene, HCN, Nitrite etc.

Amensalism is the negative interaction. This is a unidirectional process based on the release of a specific compound by one organism which has a negative effect on another organism. A classic example of amensalism is the production of antibiotics that can inhibit or kill a susceptible microorganism. The attine ant-fungal mutualistic relationship is promoted by antibiotic producing bacteria that are maintained in the fungal garden system. In this case a streptomycete produces an antibiotic that controls *Escovopsis,* a persistent parasitic fungus that can destroy the ant's fungal garden. This unique amensalistic process appears to have evolved 50 million years ago in South America.

iii. Competition

Competition can be defined as "the injurious effect of one organism on another because of the removal of some resource of the environment". Competition for free space has been, reported to suppress the fungal population by soil bacteria. Therefore, organisms with inherent ability to grow fast are better competitors.

iv. Parasitism

It is an association, in which one organism lives in or on the body of another. The parasite is dependent upon the host and lives in intimate physical contact and forms metabolic association with the host. Bacteriophages are strict intracellular parasites of Chytrid fungi.

Some bacterial viruses can establish a lysogenic relationship with their hosts, and the viruses, in their prophage state, can confer positive new attributes on the host bacteria, as occurs with toxin production by *Corynebacterium diphtheriae*. Parasitic fungi include *Rhizophydium sphaerocarpum* with the alga *Spyrogyra*. Also, *Rhizoctonia solani* is a parasite of *Mucor* and *Pythium*, which is important in biocontrol processes, the use of one microorganism to control another.

v. Predation

Predation is an association in which predator organism directly feed on and kills the pray organism. The nematophagous fungi are the best examples of predatory soil fungi. Species of Arthrobotrytis and Dactylella are known as nematode trapping fungi.

Bdellovibrio penetrates the cell wall and multiplies between the wall and the plasma membrane, a periplasmic mode of attack, followed by lysis of the prey and release of progeny.

L. BIOFERTILIZERS

Biofertilizers are microbial inoculants or carrier based preparations containing living or latent cells of efficient strains of nitrogen fixing, phosphate solublizing and cellulose decomposing microorganisms. It is used to improve soil fertility and plant growth.

The need for the use of Biofertilizers has arisen primarily due to two reasons. They are

➤ Increased use of chemical fertilizers has caused serious concern of soil texture, soil fertility and other environmental problems.

➤ Use of Biofertilizers is both economical as well as environment friendly.

The following are the biofertilizers used in different fields

Table 10.2 Biofertilizers

N2 fixing Biofertilizers

1.	Free-living	*Azotobacter, Beijerinkia, Clostridium, Klebsiella, Anabaena, Nostoc,*
2.	Symbiotic	*Rhizobium, Frankia, Anabaena azollae*
3.	Associative Symbiotic	*Azospirillum*

P Solubilizing Biofertilizers

1.	Bacteria	*Bacillus megaterium* var. *phosphaticum, Bacillus subtilis Bacillus circulans, Pseudomonas striata*
2.	Fungi	*Penicillium sp, Aspergillus awamori*

P Mobilizing Biofertilizers

1.	Arbuscular mycorrhiza	*Glomus sp.,Gigaspora sp.,Acaulospora sp., Scutellospora sp. & Sclerocystis sp.*
2.	Ectomycorrhizae	*Laccaria sp., Pisolithus sp., Boletus sp., Amanita sp.*
3.	Ericoid mycorrhizae	*Pezizella ericae*
4.	Orchid mycorrhizae	*Rhizoctonia solani*

Biofertilizers for Micro nutrients

1.	Silicate and Zinc solubilizers	*Bacillus* sp.

Plant Growth Promoting Rhizobacteria

1.	Pseudomonas	*Pseudomonas fluorescens*

Diazotrophs

Microorganisms which pass independent life and fix atmospheric nitrogen are known as free living diazotrophs.

Rhizobium

Rhizobium is a soil bacterium. It fixes the atmospheric nitrogen symbiotically within the root of legume. All Rhizobium are root nodulating bacteria. They are gram negative, motile, non spore forming medium sized rod shaped cells. It contains the enzyme complex called nitrogenase. Rhizobia are predominantly aerobic chemoorganotrophs and grow well in the presence of oxygen. Yeast and mannitol are the most generally suitable nutrient for growth. Mannitol is often used as the carbon source. Fast-growing Rhizobia produce an acid reaction in mannitol containing media while slow growers produce alkaline reactions. Congo red can sometimes assist the recognition of Rhizobia amongst other kinds of bacteria. Rhizobium is responsible for a significant amount of nitrogen fixation. It can fix up to 220 pounds of N_2 per agricultural acre per year. YEMA medium is used for the cultivation of Rhizobium. It produce colourless glistening colonies.

Azotobacter

Azotobacter is a genus of free-living diazotrophic bacteria. It is primarily found in neutral to alkaline soils. It has several metabolic capabilties, including atmospheric nitrogen fixation. *A. chroococcum* happens to be the dominant inhabitant in soils capable of fixing N_2. The bacterium produces aboundant slime which helps in soil aggregation. *Azotobacter* species are Gram-negative, aerobic, motile bacteria. They are typically polymorphic in nature. Waksman medium is used for Azotobacter cultivation. It is a N-free Mannitol Agar Medium. In this medium *Azotobacter* grow as white glistening colony. *Azotobacter* have been found to produce some antifungal substance which inhibits the growth of some soil fungi like Aspergillus, Fusarium, Curvularia, Alternaria, Helminthosporium, Fusarium etc.

Azospirillum

Azospirillum lipoferum and *A. brasilense* are primary inhabitants of soil. They perform the associative symbiotic relation with the graminaceous plants. *Azospirillum* are N_2 fixing organisms isolated from the root and above ground parts of a variety of crop plants. They are Gram negative, spirillum having abundant accumulation of polybetahydroxybutyrate in cytoplasm. Five species of *Azospirillum* have been described. They are *A.brasilense*, *A.lipoferum*, *A.amazonense*, *A.halopraeferens* and *A.irakense*. The organism proliferates under both anaerobic and aerobic conditions but it prefer micro-aerophilic condition for growth. Apart from nitrogen fixation, growth promoting substance production (IAA), disease resistance and drought tolerance are some of the additional benefits due to *Azospirillum* inoculation. *Azospirillum* grows best when using N-free semisolid malic acid medium.

Cyanobacteria

The cyanobacteria were formerly called "blue-green algae" because of their ecology and their resemblance to the algae. Cyanobacteria are prokaryotes. They lack nuclear membrane and membrane-bound organelles. Cyanobacteria are the oxygen evolving photosynthetic diazhotrophs. They fix elemental (gaseous) nitrogen. Nitrogenase is present in cyanobacteria, which is responsible for nitrogen fixation. In low N_2 environments cyanobacteria will produce heterocysts to fix nitrogen. The ability of cyanobacteria to fix elemental nitrogen has made it a very important agricultural asset. They are used as nitrogen fertilizer in the cultivation of rice and beans. Nostoc, Anabaena, are the example for BGA. Anabaena symbiotically associated with Azolla and fixes atmospheric nitrogen. It is used as biofertilizer for wetland rice and it is known to contribute 40-60 kg N/ha per rice crop. Azolla, commonly known as water velvet. It is a small delicate free floating fern. Azolla and Blue Green Algae, both have been reported to be effective in improving the organic content of soil, P availability in soil as well as the soil physical properties.

Phosphate Solubilizing Microorganisms (PSM)

Several soil bacteria and fungi, notably species of *Pseudomonas*, *Bacillus*, *Penicillium*, *Aspergillus* etc. secrete organic acids and lower the pH of soil to bring about dissolution of bound phosphates in soil. Phosphatase is an enzyme secreted by microorganisms which also solubilize phosphate rock and available to the plant.

Mycorrhizae

Mycorrhizae are fungus-root associations. It was first discovered by A. B. Frank in 1885. The term "mycorrhizae" means fungus and roots. Mycorrhizae contributes to plant functioning in natural environments, agriculture, and reclamation. The roots of about 95% of all kinds of vascular plants are normally involved in symbiotic associations with mycorrhizae. There are three types of Mycorrhizae. They are ectomycorrhizae, endomycorrhizae and ectendo mycorrhizae. Mycorrhizal fungus belongs to the genus *Glomus*, *Gigaspora*, *Acaulospora*, *Sclerocysts* and *Endogone* which possess vesicles for storage of nutrients and arbuscles for funneling these nutrients into the root system.

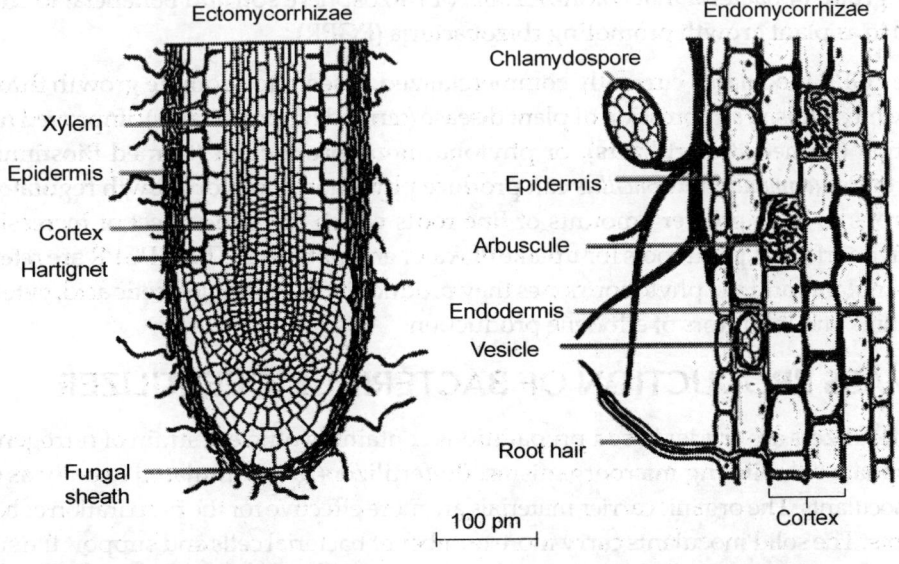

Figure 149   Ectomycorrhizae   **Figure 150**   Endomycorrhizae

Ectomycorrhizae form a sheath and the fungus grows between the plant cells, producing the "Hartig net." Bacteria also have relations with the mycorrhizal fungi. As the external hyphal network radiates out into the soil, a mycorrhizosphere is formed due to the flow of carbon from the plant into the mycorrhizal hyphal network and then into the surrounding soil. In addition, "mycorrhization helper bacteria" (MHB) can play a role in the development of the mycorrhizal relationships with the ectomycorrhizal fungus. One of the more important ectomycorrhizal fungi is *Pisolithus tinctorius.*

Endomycorrhizae penetrate cortical cells of most plants with two main types of structure, where they grow intracellularly and form coils, swellings, or minute branches. Endotrophic mycorrhizae are found in wheat, corn, beans, tomatoes, apples, oranges. A characteristic intracellular structure is the arbuscule, and thus endomycorrhizae often are called arbuscular mycorrhizal or AM fungi. Depending on the environment of the plant, mycorrhizae can increase a plant's competitiveness. In wet environments they increase the availability of nutrients, especially phosphorus.

Ectendomycorrhizae are formed by basidiomycetes. These have sheaths and intracellular coils.

Silicate Solubilizing Bacteria (SSB)

Microorganisms are capable of degrading silicates and aluminum silicates. During the metabolism of microbes several organic acids are produced and these have a dual role in silicate weathering. They supply H+ ions to the medium and promote hydrolysis and the organic acids like citric, oxalic, Keto and hydroxy carbolic acids form complexes with cations. They promote silicate removal and retention in the medium in a dissolved state.

Plant Growth Promoting Rhizobacteria (PGPR)

The group of bacteria that colonize roots or rhizosphere soil and beneficial to crops are referred to as plant growth promoting rhizobacteria (PGPR).

The PGPR inoculants currently commercialized. It seem to promote growth through at least one mechanism; suppression of plant disease (termed Bioprotectants), improved nutrient acquisition (termed Biofertilizers), or phytohormone production (termed Biostimulants). Species of *Pseudomonas* and *Bacillus* can produce phytohormones or growth regulators that cause crops to have greater amounts of fine roots which have the effect of increasing the absorptive surface of plant roots for uptake of water and nutrients. These PGPR are referred to as Biostimulants and the phytohormones they produce include indole-acetic acid, cytokinins, gibberellins and inhibitors of ethylene production.

M. MASS PRODUCTION OF BACTERIAL BIOFERTILIZER

Biofertilizers are carrier based preparations containing efficient strain of nitrogen fixing or phosphate solubilizing microorganisms. Biofertilizers are formulated usually as carrier based inoculants. The organic carrier materials are more effective for the preparation of bacterial inoculants. The solid inoculants carry more number of bacterial cells and support the survival of cells for longer periods of time.

The mass production of carrier based bacterial biofertilizers involves three stages. They are as follows,

- ⋏ Culturing of microorganisms.
- ⋏ Processing of carrier material.
- ⋏ Mixing the carrier and the broth culture and packing.

Culturing of Microorganisms

Rhizobium

Rhizobium is cultivated with the help of Yeast extract mannitol broth / agar (YEMA). Composition of Congo red yeast extract mannitol agar medium is as follows.

Mannitol	10g
K_2HPO_4	0.5g
$MgSO_4 7H_2O$	0.2g
NaCl	0.1g
Yeast extract	0.5g
Agar	20g
Distilled water	1000ml

Add 10 ml of Congo red stock solution (dissolve 250 mg of Congo red in 100ml water) to 1 liter after adjusting the PH to 6.8 and before adding agar. Rhizobium forms white, translucent, glistening, elevated and comparatively small colonies on this medium. Rhizobium colonies do not take up the colour of congo red dye added in the medium. Those colonies which readily take up the congo red stain are not rhizobia but presumably Agrobacterium.

Azospirillum

It is cultivated by making use of Dobereiner's malic acid broth with NH_4Cl (1g per liter)

Composition of the N-free semisolid malic acid medium

Malic acid	5.0g
Potassium hydroxide	4.0g
Dipotassium hydrogen orthophosphate	0.5g
Magnesium sulphate	0.2g
Sodium chloride	0.1g
Calcium chloride	0.2g
Fe-EDTA (1.64% w/v aqueous)	4.0ml
Trace element solution	2.0ml
BTB (0.5% alcoholic solution)	2.0ml
Agar	1.75g
Distilled water	1000ml

pH	6.8
Trace element solution	
Sodium molybdate	200mg
Manganous sulphate	235mg
Boric acid	280mg
Copper sulphate	8mg
Zinc sulphate	24mg
Distilled water	200ml

Azotobacter

Azotobacter is cultivated using the Waksman medium No.77. It is a N-free Mannitol Agar Medium for Azotobacter.

Mannitol	10g
$CaCO_3$	5g
K_2HPO_4	0.5g
$Mg SO_4.7H_2O$	0.2g
NaCl	0.2g
Ferric chloride	Trace
$MnSO_4.4H_2O$	Trace
N-free washed Agar	15g
pH	7.0
Distilled Water	1000ml

Phosphobacteria

Pikovskaya's Broth is used to cultivate phosphate solubilizing microorganisms.

Glucose	10g
$Ca_3(PO_4)_2$	5.0g
$(NH_4)2SO_4$	0.5g
KCl	0.2g
$MgSO_4. 7H_2O$	0.1g
$MnSO_4$	Trace
$FeSO_4$	Trace
Yeast Extract	0.5g
Distilled Water	1000ml

Under optimum conditions population level could be attained with in 4 to 5 days for *Rhizobium*; 5 to 7 days for *Azospirillum*; 2 to 3 days for phosphobacteria and 6-7 days for *Azotobacter*.

Inoculum Preparation

Prepare (appropriate) media in 250 ml, 500ml, 3litre and 5litre conical flasks and sterilize. The media in 250 ml flask is inoculated with efficient bacterial strain under aseptic condition and incubate at 37°C for 5-7days in rotary shaker (200 rpm). Observe the flask for growth and estimate the population. It serves as the starter culture. Inoculate 500ml flasks containing the media with starter culture and incubate appropriately. Similarly inoculate 3000ml and 5000ml flask and incubate properly to obtain larger volume of inoculums. Media is prepared in large quantities in fermentor, sterilized well, cooled and inoculated with the log phase culture grown in 5 litre flask. Usually 1 -2 % inoculum is sufficient. The cells are grown in fermentor by providing aeration and given continuous stirring. The broth is checked for the population of inoculated organism and contamination. The cells are harvested with the population load of 10^9cells ml-1. Grown culture is subjected for packaging.

Processing of Carrier Material

The use of ideal carrier material is necessary in the production of good quality biofertilizer. Peat soil, lignite, vermiculite, charcoal, press mud, farmyard manure and soil mixture can be used as carrier materials. The neutralized peat soil/lignite are found to be better carrier materials for biofertilizer production.

Ideal carrier material should be cheaper in cost, should be locally available, should have high organic matter content, should not have toxic chemicals, should have high Water holding capacity, should be easy to process, friability and vulnerability.

Preparation of Carrier Material

The carrier material (peat or lignite) is powdered to a fine powder so as to pass through 212 micron sieve. The pH of the carrier material is neutralized with the help of calcium carbonate. The neutralized carrier material is sterilized in an autoclave. Mix the carrier and the broth culture and pack properly.

Preparation of Inoculants Packet

The neutralized, sterilized carrier material is spread in a clean, dry, sterile metallic or plastic tray. The bacterial culture drawn from the fermentor is added to the sterilized carrier and mixed well by manual or by mechanical mixer. The culture suspension is to be added to a level of 40 - 50% water holding capacity depending upon the population.

The inoculant packet of 200 g quantities in polythene bags, sealed with electric sealer and allowed for curing for 2 -3 days at room temperature. Packed biofertilizers should be stored in cool and dry place.

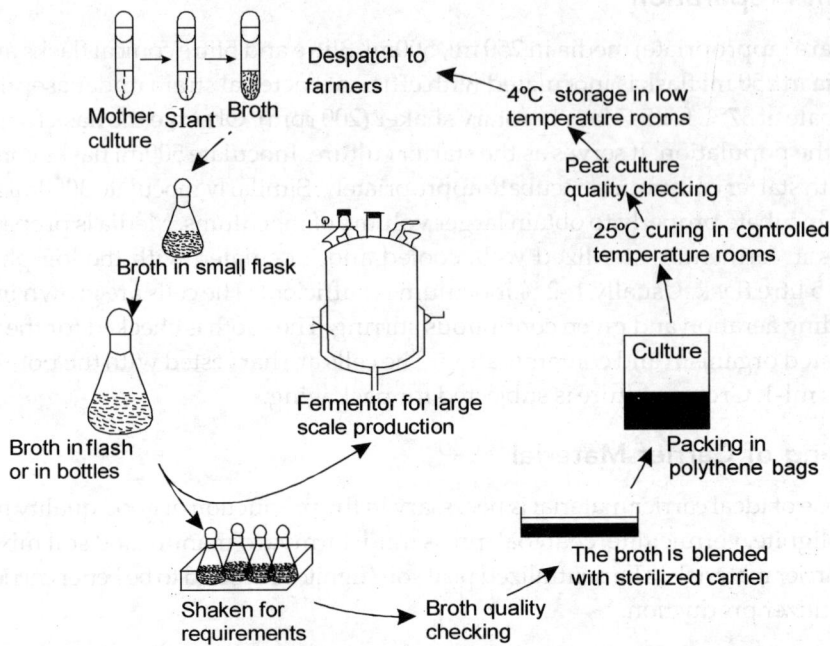

Figure 151 Biofertilizer production

N. MASS CULTIVATION OF CYANOBACTERIA BIOFERTILIZER

Anabaena azollae is a cyanobacteria lives in symbiotic association with the free floating water fern Azolla. *Anabaena azollae* can grow photoautotrophically and fixes atmospheric nitrogen. The inoculation of cyanobacteria in rice crop significantly influenced the growth of rice crop by secretion of ammonia in flood water. The use of neem cake coupled with the inoculation of Azolla greatly increased the nitrogen utilization efficiency in rice crop. Cyanobacteria add organic matter, secrete growth promoting substances like auxins and vitamins, mobilize insoluble phosphate and improve physical and chemical nature of the soil. Algalization has been shown to the saline alkali soils, help in the formation of soil aggregates, reduce soil compaction and narrow C:N ratio. These organisms enable the crop to utilize more of the applied nutrients leading to increased fertilizer utilizing efficiency of crop plant.

Mass Production

Cyanobacterial biofertilizers are cultivated by making use of the following methods. They are Cemented Tank method, Shallow metal troughs method and Polythene lined pit method and Field method (Refer chapter 17). The polythene lined method is most suitable for small and marginal farmers for the preparation of biofertilizer. In this method, small pits are prepared in field and lined with thick polythene sheets. For mass cultivation of Azolla, microplots (20m^2) are prepared in nurseries in which sufficient water (5-10cm) is added. For profuse growth of Azolla, 4-20 kg P$_2$O$_5$/ha is also amended. Optimum pH (8.0) and temperature

(14-30°C) should be maintained. Finally, microplots are inoculated with fresh Azolla (0.5 to 0.4kg/m₂). An insecticide (Furadon) is used to check the insect's attack. After 3 weeks, the mat azolla is ready for harvest and the same microplot is inoculated with fresh Azolla to repeat the cultivation. Azolla mat is harvested and dried to use as green manure.

Bio-fertlizer Application Methodology

There are three ways of using Bio-fertilizers. They are Seed treatment, Root dipping and Field application.

Seed Treatment

One packet of the inoculant is mixed with 200 ml of rice kanji to make slurry. The seeds required for an acre are mixed in the slurry so as to have a uniform coating of the inoculant over the seeds and then shade dried for 30 minutes. The shade dried seeds should be sown within 24 hours. One packet of the inoculant (200 g) is sufficient to treat 10 kg of seeds.

Seedling Root Dip

This method is used for transplanted crops. Two packets of the inoculant is mixed in 40 litres of water. The root portion of the seedlings required for an acre is dipped in the mixture for 5 to 10 minutes and then transplanted.

Main Field Application

Four packets of the inoculant is mixed with 20 kgs of dried and powdered farm yard manure and then broadcasted in one acre of main field just before transplanting.

Rhizobium is applied as seed inoculant. For transplantation crops, Azospirillum/ Azotobacter is inoculated through seed, seedling root dip and soil application methods. For direct sown crops, Azospirillum is applied through seed treatment and soil application. Phosphobacteria is inoculated through seed, seedling root dip and soil application methods.

Combined application of bacterial biofertilizers.

Phosphobacteria can be mixed with Azospirillum and Rhizobium. The inoculants should be mixed in equal quantities and applied by seed treatment or root tip application or field application.

O. ROLE OF BIOFERTILIZERS IN SOIL FERTILITY AND AGRICULTURE

Biofertilizers are known to play a number of vital roles in soil fertility, crop productivity. Some of the important functions or roles of Biofertilizers in agriculture are:

They supplement chemical fertilizers for meeting the integrated nutrient demand of the crops.

They can add 20-200 kg N/ha year under optimum soil conditions. They increases 15-25% of total crop yield.

They can at best minimize the use of chemical fertilizers.

Application of Biofertilizers results in increased mineral and water uptake, root development, vegetative growth and nitrogen fixation.

Some Biofertilizers (eg, *Rhizobium*, BGA, *Azotobacter sp*) stimulate production of growth promoting substance like vitamin-B complex, Indole acetic acid (IAA) and Gibberellic acids etc.

Phosphate mobilizing or phosphorus solubilizing Biofertilizers converts insoluble soil phosphate into soluble forms by secreting several organic acids and under optimum conditions.

Mycorrhiza enhance uptake of P, Zn, S and water, leading to uniform crop growth and increased yield. It also enhances resistance to root diseases and improve hardiness of transplant stock.

They liberate growth promoting substances and vitamins and help to maintain soil fertility.

They act as antagonists and suppress the incidence of soil borne plant pathogens and thus, help in the bio-control of diseases.

They are cheaper, pollution free and renewable energy sources

They improve physical properties of soil, soil tilth and soil health in general.

They improve soil fertility and soil productivity.

Blue green algae like Nostoc, Anabaena, and Scytonema are often employed in the reclamation of alkaline soils.

Bio-inoculants containing cellulolytic and lignolytic microorganisms enhance the degradation of organic matter in soil.

They plays important role in the recycling of plant nutrients.

P. SYMBIOTIC NITROGEN FIXATION

Introduction

The fixation of free nitrogen of air by microorganisms living symbiotically inside the plant is called symbiotic nitrogen fixation. The symbiotic nitrogen fixing bacteria belongs to the genus Rhizobium. It forms symbiotic relationship with the members of the family Leguminaceae, such as peas, beans, clovers etc. The actual site of bacterial symbiotic nitrogen fixation is the root nodules. Rhizobial genes responsible for nodule formation are termed as nod genes.

Process of Nodule Formation

A variety of microorganisms are present inthe rhizophere region. Host plant secretes *root exudates* in rhizosphere region of the soil. Root exudates contain growth stimulating substances like biotin, thiamine, amino acids etc. Root exudates attracts microorganisms present in the rhizosphere region. Rhizobium present in the rhizosphere region utilize aminoacids and other growth factors and produce Indole Acetic Acid. This IAA induce root hair growth, which leades to the formation of *curling of root hair*. It is looks like *Shepherd's Hook*. Root curling is also due to curling factor which includes cytokinin, polymixin B etc. *Rhizobium* aggregates at distinct sites on curled root hairs. The bacteria penetrate the relatively soft root hair tip with the help of cellulase / Pectinase like enzyme or invade damaged or broken root hairs and progress in as an *infection thread*. The infection thread grows towards root cortex and occupies throughout the central part of cortex. The Rhizobia are liberated either individually or in small groups enclosed by a membrane bound vacuole and enter the plant cell. The vacuoles are termed as *symbiosomes*. Rhizobium secretes nodulation factor, which initiates nodule formation. The first cells of developing nodule contains twice the number of chromosomes. These tetraploid cells gives rise to nitrogen fixing tissue. In the nodules the rhizobia are found to be enclosed in a membrane derived from hypertrophied cells of the host, where they multiply into 4 -6 in number and finally become enlarged and pleomorphic to form endosymbionts or bacteroids. Bacteroids appear as swollen, irregular, star shaped, club-shaped, branched or Y-shaped structures (Figure 152). The nodules grow rapidly, pushing its way to the surface of the root. The nodule contains a red pigment **leghaemoglobin.** The mature nodules remain connected with the root via vascular tissues through which exchange of fixed nitrogen of the bacteriods and the nutrient of host takes place.

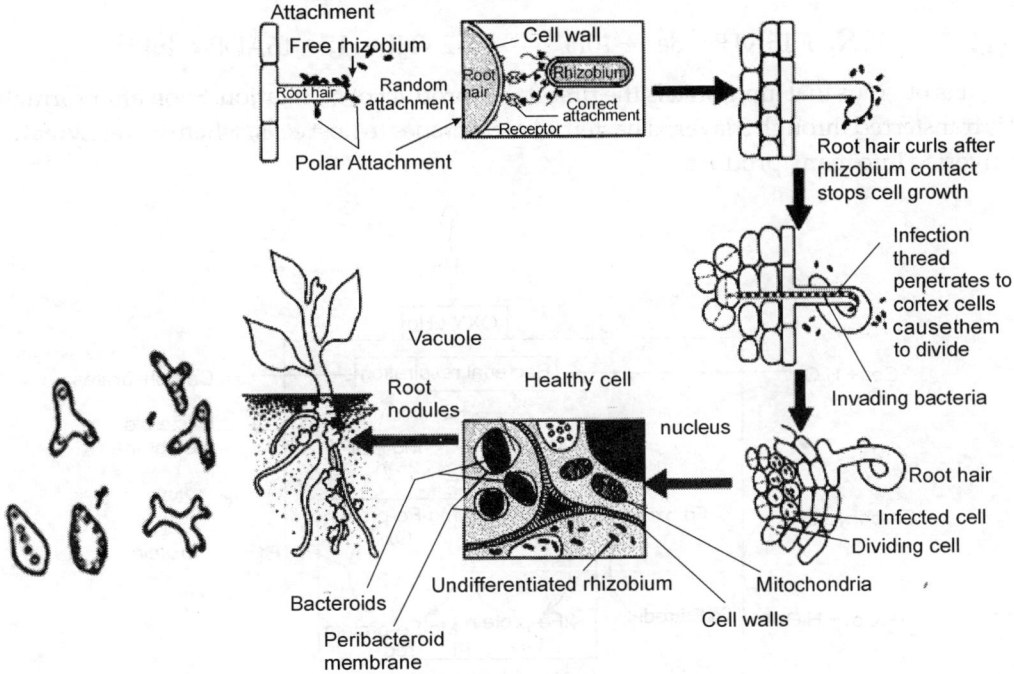

Figure 152   Process of nodulation

Nitrogen Fixation Process

Leghaemoglobin (LHb)

A characteristic feature of the healthy root nodules is the presence "leghaemoglobin. It has characteristics similar to haemoglobin found in animal. It is red in colour due to presence of iron. LHb is found only in healthy nodules. LHb is present outside the bacteroid membrane. It serves as an electron carrier, supplying oxygen to the bacteroids for the production of ATP and at the same time protecting oxygen sensitive nitrogenase system. LHb combines with O_2 to form oxyleghaemoglobin (OLHb) and makes it available at the surface of bacterial membrane.

Mechanism of Nitrogen Fixation

Nitrogenase (Mo-Fe-protein) and nitrogenase reductase (Fe-protein) are essential for nitrogenase activity. Fe-protein interacts with ATP and Mg++, and Mo-Fe-protein catalyses the reduction of N_2 to NH_3, H^+ to H_2 and acetylene to ethylene. The reduced ferredoxin or flavodoxin serves as a source of reductant for electron transfer during N_2 fixation. From reduced form of ferredoxin (Fd res.) electrons (e^-) flow to Fe-protein which reduce to Mo-Fe-protein (in nitrogenase enzyme complex) with subsequent release of inorganic phosphate, Pi. This enzyme complex gets energy from Mg ATP which in turn is produced after bacterial respiration. Finally, Mo-Fe-protein passes on the electron to reducible substrate i.e. N_2 (or other substrates). The equation of N_2 fixation in nodules of legumes may be written as :

$$N_2 + 16ATP + 8e^- + 10H^+ \longrightarrow 2NH_4^+ + H_2^+ + 16ADP + 16Pi$$

It is obvious that ammonia is the first stable product of N_2 fixation. Soon after formation, it is transferred through 3 layered bacteroid membranes to host cells, where it is enzymatically converted into many products.

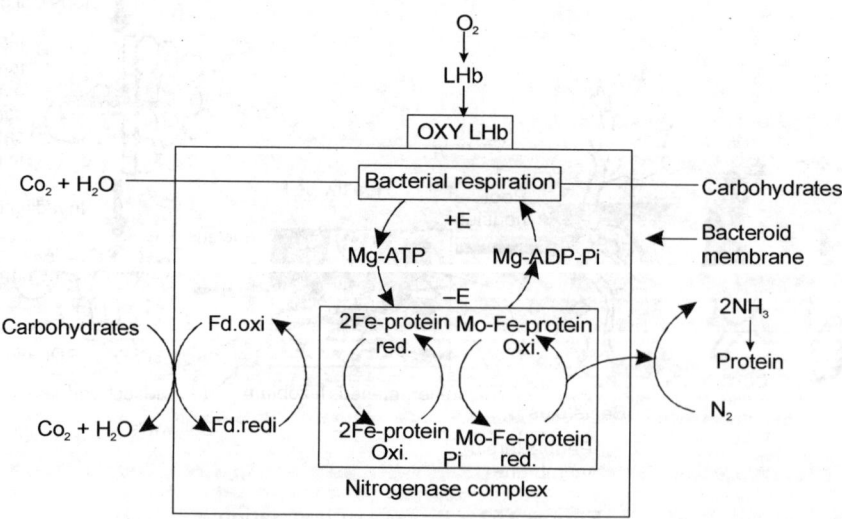

Figure 153 Mechanism of N_2 fixation

Q. BIOLOGICAL CONTROL

Biological control is the deliberate use of one species of organism to control the population of other species.

Microbial Biocontrol

The term 'Microbial control' was first used by E. A. Steinhaus (1949) to express the pest population management through disease causing micro-organisms. Microbial control includes all aspects of utilization of micro-organisms or their by-products in the control of pest species.

Definition

"Microbial control is a phase of biological control concerned with the employment of micro-organisms for the control and reduction of number of animals/plants in a particular area/given population".

The following agents acts as a microbial biocontrol agents

Table 10.3   Biocontrol agents

No.	Group of Microbe	Pathogens
1	Bacteria	*Bacillus papilliae* causing milky disease
		Bacillus thuringiensis
2	Viruses	Baculoviruses
		Granulosis viruses
		Nuclear Polyhedrosis Virus (NPV)
3	Fungi	Polymycetes
4	Protozoa	*Nosema locustae*
		Neogregarires
5	Rickettsiae	*Rickettsiella papilliae*
6	Nematodes	*Steinernema sp.*
		Heteromhbditus sp.

Advantages of Microbial Control

1. Host specificity
2. Ability of Multiply in their Target Hosts
3. No problem of Toxic Residues
4. No Evidence of Resistance
5. No Problem of Cross Resistance
6. Conventional methods for Application

7. Permanent Control of Pest
8. Ideally Suited for Integration with most other Plant Protection

Limitations or Disadvantages of Microbial Control

1. High Selectivity/host specificity
2. Requirement of additional control measures
3. Need the correct time of application
4. Delayed effect/Mortality
6. Storage problem
7. Difficulty of culturing in large quantities
8. Short Residual Effectiveness
9. Not legally protected and data needed for registration is highly expensive.

Bacterial Pathogens in Bio-controls of Crop Pest

Many bacteria are known for their pest control activity. Bacterial biocontrol agents are naturally present in the digestive tract of insects but seldom cause pathogenic infection because they lack the invasive power to penetrate through midgut wall but its toxin cause infection there by kill the pest.

Spore forming bacteria of the genus *Bacillus* known for its biocontrol activity. They are rod shaped, gram positive, motile bacteria. They form endospores in insect. They are a promising organisms for microbial control. There are three groups of *Bacillus* showing biocontrol activity. They are

i. Obligate pathogens: *Bacillus papillae* - milky disease causing organisms.
ii. Crystalliferous spore former: *Bacillus thuringiensis*
iii. Non crystalliferous facultative pathogen:*Bacillus cereus* and *B. sphaericus*.

1. Milky diseases *Bacillus papillae* and *B. lentimorbus*

Milky disease organisms are able to infect only beetles of the family Scarabacidae. About 62 species of insects have been found to pick up infection by injection. These bacteria multiply and sporculated readily in the haemolymph of the grubs. The turbidity due to accumulating spores lead to milkiness of their haemolymph and hence the "milky disease". This leads to the death of nymph.

2. *Bacillus thuringiensis* (BT)

It is a crystalline spore forming bacteria. It produce endospores and proteinaceous parasporal crystal in sporulation. The cells produce crystal which contains an endotoxin capable of paralysing the gut of larvae. The toxin is known as "Delta endotoxin". Generally susceptible insects are killed by toxic crystals using the mechanism called lethal septicemia. Recently genetic engineering has helped to incorporating genes coding for BT delta endotoxin into transgenic cowpea, cotton, tobacco and tomato plants so that internal production of toxic crystals by such plants can ward-off the attack of pests when visit for feeding.

BT is reported to be safe to honey bees and non-toxic to man. When ingested crystalline inclusion initially dissolve in the midgut releasing one or more proteins. These crystal proteins are activated by digestive gut enzymes with correct pH into toxic polypeptides. Smaller activated toxins bind to cell membrane lining the gut generating pores that disturb osmatic balance and lead to cellular swelling and lyses. Intoxicated insect larvae quickly stop feeding and eventually die. BT also produce other toxic metabolites such as beta exotoxin, alpha exotoxin and gamma exotoxin.

Viruses in Bio-controls of Crop Pest

Viruses are a non cellular sub microscopic infective entity. It multiples only as obligate intracellular parasite. Majority of viral pesticides infects the members of the order Lepidoptera, Hymenoptera, Diptera, Coleoptera, Neuroptera and Orthoptera.Baculoviruses, Nuclear polyhedrosis virus, Granulosis viruses, Entomopox viruses are example.

Baculoviruses (BV)

Baculoviruses are complex virus with single molecule of circular super coiled DNA. Virus protein particle contain 10-25 polypeptides of which 4-11 are associated with necleocapsides. Replication taken place in host cell nuclei, which results in death of larvae.

Nuclear Polyhedrosis Virus (NPVs)

NPV is the best known arthropod viruses. They show great promises for practical use in pest suppression. These viruses produce polyhedral inclusion bodies in host cell nuclei. It is a rod shaped virus with double stranded DNA. The NPV infect all insects' cells and cause death. It is highly host specific with no effect on beneficial fauna. It is safe to plant birds and higher animals and man. It enters through injection of plant material into insect gut through mouth and cuticle. Infected insect appear dull in colour and inceptive. Feeding rate of insect is reduced. In advanced stage integument fragile and rapture on slight disturbance emitting liquefied content (whitish fluid). Incubation period is 4-5 in 20 days. Earlier instars are more susceptible than 5th or 6th instar. Infected larvae hang invertedly from twigs.

Granulosis Viruses (GVs)

It has also show considerable promise as agents for insect pest suppression. They develop in either the nucleus or cytoplasm of host fat, tracheal matrix or epithelial cells. The virions are occluded singly in small inclusion bodies called capsules. The rod shaped virion contains DNA and are similar to NPV viruses. They usually oval occlusion bodies about 200×400 nm size. They enter through ingestion. The diseased larvae are less active, flaccid and fragile and period from infection to death is 6 to 20 days.

Cytoplasm Polyhedrosis Virus (CPV)

It is also a promising group for practical use. They develop only in the cytoplasm of host midgut epithelial cells. The spherical virions are occluded singly in polyhedral inclusion bodies and contain double stranded RNA. Infection by CPVs is not always lethal but shows larval growth reduction.

Entomopox Viruses (EPV)

It affects different insect orders. EPV replicate in the cytoplasm of host fat body cells and possible in heamocytes. Virions are ovoid cuboids and contain double stranded DNA. Two types of IBs (Infective Bodies) may be produced which kills the insects.

Fungi in Bio-controls of Crop Pest

Entomopathogenic fungi played an important role in the early development of Insect Pathology. Pathogenic fungal infections are referred to as mycoses. Insect mycoses are cause by fungi in the following classes:

i. Phycomycetes: Entomophthora
ii. Ascomycetes: Cordycepes, Nectria
iii. Basidomycetes: Septobasidium
iv. Fungi imperfecti : Cephelosporium, Metarahizium, Penicillium, Verticlillum.

The infective unit in most fungi is a spore-usually a conidium. Invasion is through the respiratory or alimentary tract. Conidia usually germinate on the cuticle and then penetrate. Enzymes and mechanical forces are involved. In most cases yeast like filaments of mycelium called hyphal bodies are produced which usually float free and apparently multiply in the haemocoel. Some strains produce sufficient toxins in this stage to cause death. Condidiophores are then produced which erupt through the cuticle and produce spores on the outside of the insect.

Protozoa in Bio-controls of Crop Pest

The protozoa subphyla Sporozoa and onidospora contains numerous entomophilic protozoans. The effects of protozoan infections are chronic. They may affect their hosts over a fairly long time period. The naturally occurring epizootics of protozoan disease in insect pest like European corn borer, some Lepidoptera, several species of flies, aquatic Diptera including mosquitoes and grasshoppers. *Mettasia grandis* is an important pathogen of the cotton bore weevil and showed considerable promise. *Mettaisa frogodermae* was studied and used in pest suppression programme of khapra beetle. Sporozoites are motile in the gut and soon penetrate the gut to the haemocoel and infect cells of susceptible tissues within 2 days. The diseased larvae die early within a week. *Nosema locustae* attacks grasshopper species.

Rickettsiae in Bio-controls of Crop Pest

Rickettsiae are obligate intracellular parasite. Rickettsiae develope and multiply on the cell cytoplasm where they fill the vacuolar areas. In ticks, they develop in cell nuclei. *Enterella* and *Rickettsiella* are act as a entamo pathogens. The *Enterella sp.* grows only intra-cellularly in the gut epithelium of the host and destroys it.Rickettsiella species primarily attack the fat body and cause generalized infection.

Nematodes in Bio-controls of Crop Pest

Nematodes are generally fusiform and vermiform shape with a terminal mouth situation on a rounded head and a tail tapered to pointed tip. They are invertebrates. Most of the nematodes that cause injury to their insect hosts are endo-parasitic, occurring in the haemocoel, gut lumen, malpighian tubules, ovaries or other organs. Infection may be either passive or active. Nematode activity, in the host may result in sub-lethal injury caused by nutritional depletion or organ disturbance. It may be expressed as retarded growth, reduced activity; lower fecundity, eventually sterility and even the production of inter sexes. On the other hand death generally from the mechanical destruction of host tissue.

11

ENVIRONMENTAL MICROBIOLOGY

A. INTRODUCTION

The term ecology is derived from two Greak words Oikos and logos. It means the law of the house hold. It is a science that explores the interrelationship between organisms and their biotic and abiotic environments. The word ecology was first defined by German biologist Ernst Haeckel (1866). He says that ecology is the total relationship of animals and its organic and inorganic environments. In modern days, ecology is also called as Environmental Biology. According to Odum (1969), ecology means the study of the interrelationships between organism and environment. The term microbial ecology came into frequent use only in the early 1960s. It means the science that explains the relationship between microorganisms and their biotic and abiotic environments.

B. SCOPE OF ECOLOGY

Microbial ecology emerged as an energetic and dynamic branch of science because microorganisms occupy a key position in the orderly flow of materials and energy through global ecosystem. Interference of human with the nature of microbial community leads to various problem. Accumulation of plastics, petrochemicals, fertilizers etc., disturbs microbial community. At the same time microbes play a crucial role in environmental problems. Microbes are actively involved in maintenance of natural resources, control of pollution, new source of food, restoring natural environment, indicates environment threat, human welfare, treatment of waste, relieve nitrogen fertilizer shortage, recover metals from ore and biocontrolling agent. These practical implications take microbial ecology a highly relevant and exciting subject for study.

C. RELATIONSHIP BETWEEN MICROORGANISMS IN DIFFERENT ENVIRONMENT

The ecosphere or biosphere means total area by which living organisms survive on earth and their abiotic surroundings. Abiotic components are nonliving chemical and physical factors such as temperature, light, water, and nutrients. Biotic components are living factors such as animals, plants and microorganisms. Biosphere can be divided into atmosphere, hydrosphere and lithosphere. Each ecosphere contains numerous habitats. A habitat is an ecological or environmental area that is inhabited by a particular species of animal, plant, or other type of organism. A place where a living thing lives is its habitat. It is a place where it can find food, shelter, protection and mates for reproduction. It is the natural environment in which an organism lives, or the physical environment that surrounds a species population. A habitat is made up of physical factors such as soil, moisture, range of temperature, and availability of light as well as biotic factors such as the availability of food and the presence of predators. A habitat is not necessarily a geographic area for a parasitic organism it is the body of its host, part of the host's body such as the digestive tract or a cell within the host's body. A microhabitat is the small-scale physical requirements of a particular organism or population. The monotypic habitat occurs in botanical and zoological contexts, and is a component of conservation biology. Some microorganisms are autochthonous or indigenous within a given habitat. Some microbes may be foreign and are called allochthonous. Allochthonous microbes are transient members of their habitat and do not occupy the functional niches of that ecosystem.

D. ATMOSPHERE

It is the gaseous envelope that surrounds the earth. It is retained by Earth's gravity. The atmosphere protects life on Earth by absorbing ultraviolet radiation, warming the surface through heat retention (green house effect) and reducing temperature extremes between day and night. The common name air is given to the atmospheric gases. It is used in breathing and photosynthesis. Air consists of 78.09% nitrogen, 20.95% oxygen, 0.93% argon, 0.032% carbon di oxide and small amounts of other gases. Air also contains a variable amount of water vapour, on average around 1% at sea level, and 0.4% over the entire atmosphere.

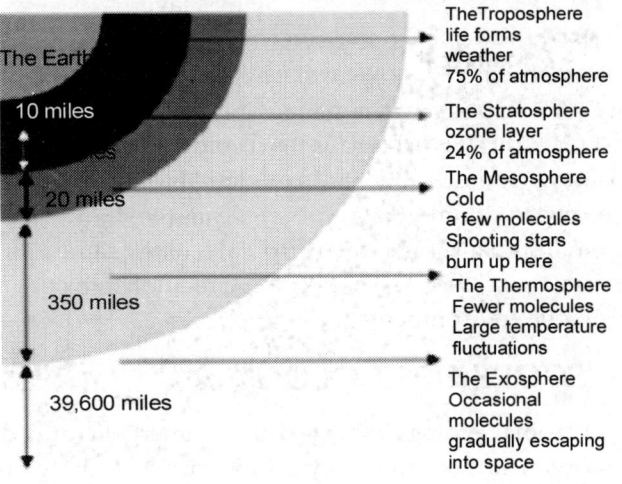

Figure 154 Layers of Atmosphere

Structure of the atmosphere

The atmosphere is divided into five layers on the basis of vertical distribution of temperature. They are Exosphere (700 to 10,000 km), Thermosphere (80 to 700 km), Mesosphere (50 to 80 km), Stratosphere (12 to 50 km) and Troposphere (0 to 12 km).

Troposphere

This layer is nearest to earth. It interferes with earth's hydrosphere and the lithosphere. It extends from Earth's surface to an average height of about 10-12 km. The temperature usually declines with increasing altitude in the troposphere. The lowest part of the troposphere (i.e. Earth's surface) is typically the warmest section of the troposphere. The troposphere contains roughly 80% of the mass of Earth's atmosphere. The troposphere is denser than all its overlying atmospheric layers. Nearly all atmospheric water vapour or moisture is found in the troposphere.

Stratosphere

The stratosphere is lying above the troposphere. It is separated from it by the tropopause. This layer extends from the top of the troposphere. This is 10-12 km above Earth's surface to 40-50 km. The atmospheric pressure at the top of the stratosphere is roughly 1/1000 at sea level. It contains the ozone layer. In stratosphere temperatures rise with increasing altitude. This rise in temperature is caused by the absorption of ultraviolet (UV) radiation by the ozone layer.

Mesosphere

The mesosphere is the third highest layer of Earth's atmosphere. It occupying the region above the stratosphere. It extends from the stratopause at an altitude of about 50 km to the mesopause at 80–85 km above sea level. It is characterize by decrease of temperature.

Thermosphere

The thermosphere is the second-highest layer of Earth's atmosphere. This atmospheric layer undergoes a gradual increase in temperature with height. The temperature of this layer can rise as high as 1500°C (2700°F). This layer is completely cloudless and free of water vapour.

Exosphere

The exosphere is the outermost layer of Earth's atmosphere. This layer is mainly composed of extremely low densities of hydrogen, helium and several heavier molecules including nitrogen, oxygen and carbon dioxide. The exosphere is located too far above Earth.

Other layers

The ozone layer is contained within the stratosphere. In this layer ozone concentrations are about 2 to 8 parts per million. It is mainly located in the lower portion of the stratosphere from about 15–35 km, though the thickness varies seasonally and geographically. About 90% of the ozone in Earth's atmosphere is contained in the stratosphere.

The ionosphere is a region of Earth's upper atmosphere, from about 60 km to 600 km altitude. It is distinguished because it is ionized by solar radiation. It plays an important part in atmospheric electricity and forms the inner edge of the magnetosphere.

Microorganisms in Air

Aeromicrobiology is the study of living microbes in the air. These microbes are referred to as bioaerosols. Air microbes have the opportunity to travel long distances with the help of wind and precipitation. These aerosols are ecologically significant because they can be associated with disease in humans, animals and plants. Typically microbes will be suspended in clouds. They are able to perform processes that alter the chemical composition of the cloud. It also induce precipitation.

Microbes in air available along with water droplets, dust particles and other matter. The source of the launching of airborne microbes is humans, animals and vegetation. Then they are transported and finally are deposited somewhere new. The atmosphere can have a variety of physical characteristics, and can be very extreme in terms of the relative humidity, temperature and radiation.

Many different microorganisms can be in aerosol form in the atmosphere, including viruses, bacteria, fungi, yeasts and protozoans. In order to survive in the atmosphere, it is important that these microbes adapt to some of the harsh climatic characteristics of the exterior world, including temperature, gasses and humidity. Many of the microbes that are capable of surviving harsh

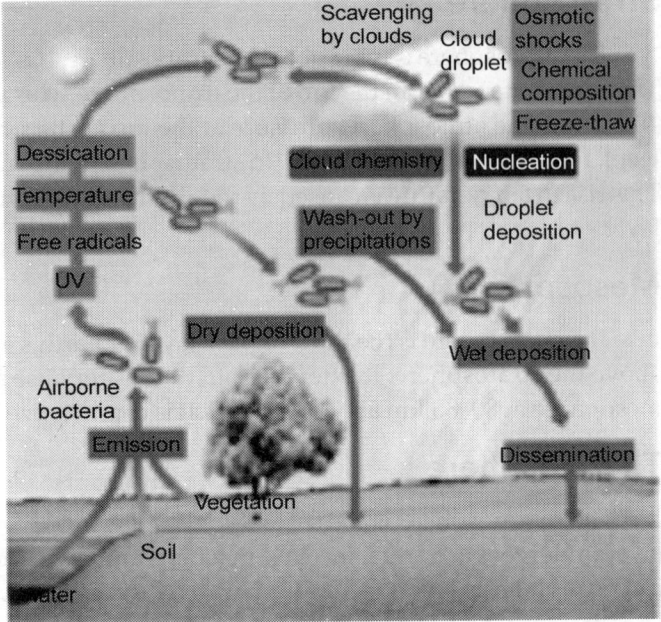

Figure 155 Cycle of microbes in air

condition conditions can readily form endospores, which can withstand extreme conditions. Several properties of spore contribute to their ability to withstand transport through the atmosphere. They are Low metabolic rates (do not require external nutrients), Spores are produced at high numbers (morethan 10^{12} spores per fruiting body per year), Spores contains extremely thick walls. Some spores are pigmented. (It protect against exposure of UV), Spores are in small size, Spores are relatively light, Spore contains gas vacuoles, they are in Variety of shapes passive liberation of spores in air current increase its survival. Pigmented microbes are easily protected from UV eg. Micrococcus. Many of these microorganisms can be associated with specific and commonly known diseases.

Sources of contamination

The major sources of contamination in the air are, automobile exhausts like incomplete combustion of fuel, agricultural sources like spraying of pesticides and insecticides, hospitals, industries like tannery industries, distilleries, nuclear power plants and chemical industries. All these industries provide smoke into the air. The smoke comprises of dust particles from the combustion of coal and oil. Automobile engines driven by petrol, produce many different substances in its exhaust. Some of these substances are serious air pollutants. Air pollutants are CO, CO_2, unburnt hydrocarbon, NO_2, SO_2 and lead from leaded petrol.

Air-borne diseases

The sources of microbes in air are soil, water, decaying bodies and diseased persons. There are several diseases caused by air-borne pathogens. They are found commonly in the air.

Detailed study of Airborne diseases – Refer Section Medical Microbiology and Virology

Bacterial

Bacterial flora that can resist environmental stresses is *Bacillus anthracis*. It is a gram positive rod shaped bacteria. It produce spore that resist environmental stresses. Spore makes *Bacillus anthracis* a highly resistant bacteria, allowing it can survive extreme temperatures, chemical contamination, and low nutrient environments. This bacteria is associated with Anthrax Examples are as follows.

Human Disease	Pathogens
Bacterial diseases	
Pulmonary tuberculosis	*Mycobacterial tuberculosis*
Pneumonia	*Klebsiella pneumoniae*
Pulmonary anthrax	*Bacillus anthracis*
Legionellosis	*Legionella spp.*
Whopping cough	*Bordetella pertussis*
Diphtheria	*Corynebacterium diphtheriae*
Fungal diseases	
Aspergillosis	*Aspergillus fumigatus*
Coccidioidomycosis	*Coccidioides immitis*
Viral disease	
Influenza	*Influenza virus*
Chicken pox	*Herpesvirus*
Common cold	*Picornavirus*

Fungal

Fungus that can resist environmental stresses is *Aspergillus fumigatus*. It is a major airborne fungal pathogen. This pathogen is capable of causing many human diseases when conidia are inhaled into the lungs.

Viral

An example of a viral airborne pathogen is the Avian Influenza Virus, which is a single stranded RNA viruses that can infect a broad range of animal species as well as humans and cause the Avian Influenza.

Aerosol

Aerosol can be defined as water droplets containing several types of microorganisms released into the air from various sources. Air currents may also bring the microorganisms from plant or animal surface into the air. Aerosols may form from a variety of sources, including splash from falling rain drops, spray from breaking waves, gas movements through water column. Some microbes from animals release droplets through coughing and sneezing.

Droplet

Sneezing, coughing or talking usually form droplets. Each consists of saliva and mucus. Droplets from diseased person contain several thousands of microbes.

Droplet nuclei

Droplets in a warm dry atmosphere tend to evaporate rapidly and become droplet nuclei. The residue of solid material, left over after drying up of a droplet is known as droplet nuclei.

Outdoor microflora

Microbes found outside the buildings is called outdoor microflora. The dominant microflora of outside air are fungi. The two common genera of fungi are *Cladosporium* and *Sporobolomyces*. Other genera found in air are *Aspergillus, Alternaria, Phytophthora* and *Erysiphe*. The outdoor air also contains basidispores, ascopores of yeast, fragments of mycelium and conidia of molds. Among the bacterial genera *Bacillus, Clostridium, Sarcina, Micrococcus, Corynebacterium* and *Achromobacter* are widely found in the outside air. The number and kind of microorganisms may vary from place to place.

Indoor Microflora

The air found inside the building is referred to as Indoor air. The commonest genera of fungi in indoor air are *Penicillium, Aspergillus*. The commonest genera of bacteria found in indoor air are *Staphylococci, Bacillus and Clostridium*.

Enumeration and assessment of microorganisms in air

There are several methods adapted to enumerate microorganisms in air. They require special devices and design. The most important methods are solid and liquid impingement devices, filteration, sedimentation, centrifugation and electrostatic precipitation.

Airborne bacterial cells, fungal cells and spores may be present in droplets as bioaerosols. They can be an important source of infection in medical facilities and can contaminate sensitive manufacturing operations. Microbiological monitoring of the air in facilities where pharmaceuticals and medical devices are produced is essential. In most countries international standards have been published for biocontamination control in clean rooms and other controlled environments. Airborne bacteria and fungi may be important in hospitals, in food factories, in office buildings and other working environments. For example, high levels of airborne fungal spores in bakeries may have a significant negative effect on product shelf life. Monitoring airborne microorganisms is therefore a key component of environmental monitoring in many sectors.

There are two principle means of monitoring the microbiological population of the air, passive monitoring and active sampling.

Passive monitoring

Passive monitoring is done by making use of settle plate technique. In this method, petri dish containing appropriate culture media are exposed to air for a given time and then incubated to allow visible colonies to develop and be counted. This method is effective in monitoring viable biological particles. They will not detect smaller particles or droplets suspended in the air. They cannot sample specific volumes of air. Settle plates are inexpensive and easy use, requiring no special equipment. They are also useful for directly monitoring airborne contamination of specific surfaces.

Active monitoring

Active monitoring requires the use of a microbiological air sampler to physically draw a known volume of air.

1. Impingers

Impingers use a liquid medium for particle collection. Sampled air is drawn by a suction pump through a narrow inlet tube into a small flask containing the collection medium. This accelerates the air towards the surface of the collection medium and the flow rate is determined by the diameter of the inlet tube. When the air hits the surface of the liquid, it changes direction abruptly and any suspended particles are impinged into the collection liquid. Once the sampling is complete the collection liquid can be cultured to enumerate viable micoorganisms. Since the sample volume can be calculated using the flow rate and sampling time, the result is quantitative.

| Anderson single stage viable impactor | Anderson two-stage viable impactor | Anderson six-stage viable impactor | Stage with Petridish |

Figure 156   Impactors

2. Impactors

Impactor samplers use a solid or adhesive medium, such as agar, for particle collection. It is a more commonly used in commercial applications. In a typical impactor sampler air is drawn into a sampling head by a pump or fan and accelerated, usually through a perforated plate (sieve samplers), or through a narrow slit (slit samplers). This produces laminar air flow onto the collection surface. (agar plate or contact plate filled with a suitable agar medium). The velocity of the air is

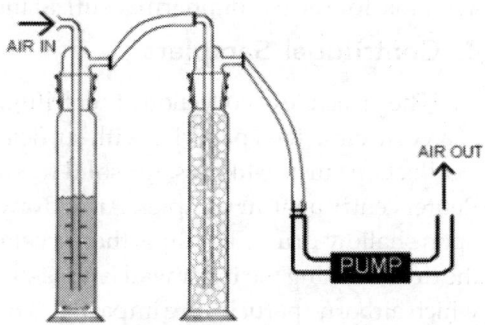

Figure 157   Impingers

determined by the diameter of the holes in sieve samplers and the width of the slit in slit samplers. When the air hits the collection surface it makes a tangential change of direction and any suspended particles are thrown out by inertia, impacting onto the collection surface. When the correct volume of air has been passed through the sampling head, the agar plate can be removed and incubated directly without further treatment. After incubation, counting the number of visible colonies gives a direct quantitative estimate of the number of colony forming units in the sampled air.

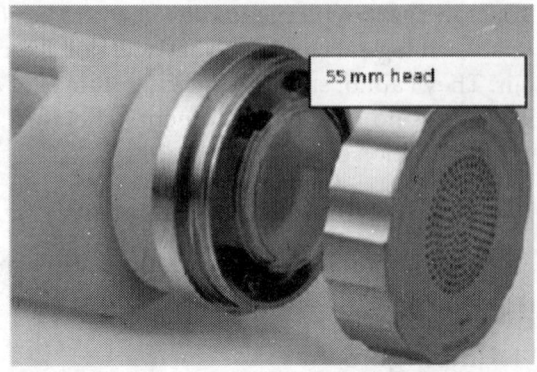

Figure 158 Sieve sampler

A wide variety of instruments have been developed using the impaction principle. One of the best known is the Andersen sampler, a multi-stage 'cascade' sieve sampler that uses perforated plates with progressively smaller holes at each stage, allowing particles to be separated according to size. Another well known instrument is the Casella slit sampler, in which the slit is positioned above a turntable on which is placed an agar plate. As air is drawn through the slit, the agar plate rotates, so that particles are deposited evenly over its surface.

3. Filteration

The most commonly used alternative method is filtration. In this method, air is drawn by a pump or vacuum line through a membrane filter. The filter medium may be polycarbonate or cellulose acetate. It can be incubated directly by transferring onto the surface of an agar medium. Isolated organisms were identified by culture or rapid methods. Filtration methods are accurate and reliable. Portable filtration samplers are also available for the use of pharmaceutical industry.

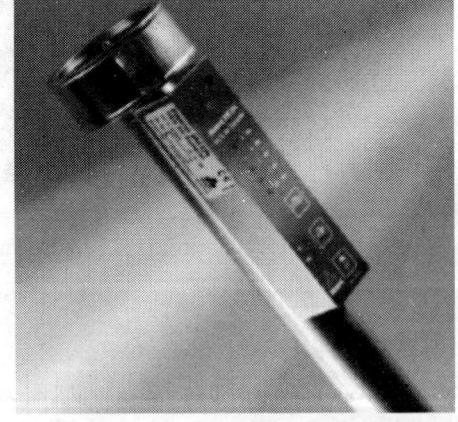

Figure 159 Centrifugal air sampler

4. Centrifugal Samplers

The principle of collection of centrifugal samplers is centrifugation. It involves the creation of a vortex in which particles with sufficient inertia leave the air stream to be impacted upon a collection surface like a semi-solid medium. The most frequent example of this type is the Reuter centrifugal air sampler. Air is drawn into the sampler by an impeller housed inside an open shallow drum. The air is then accelerated by centrifugal force toward the inner wall of the drum. Lining the inner wall is a plastic strip supporting a thin layer of agar medium, onto which airborne particles are impacted. These strips are subsequently removed and incubated. The motor for the impeller is battery operated and the whole sampler unit is small enough to be hand-held.

E. HYDROSPHERE

It is also called as aquatic ecosystem. Aquatic ecosystem is the most diverse ecosystem in the world. The first life originated in the water. Water is the most vital factor for the existence of all living organisms. Water covers about 71% of the earth of which more than 95% exists in oceans. Fresh water accounts for 2.5%, of this freshwater, 68.7% as ice and snow, 29.9 as ground water. Only 0.26% of the total freshwater is easily accessible. A very less amount of water is contained in the rivers (0.00015%) and lakes (0.01%), which comprise the most valuable fresh water resources.

Marine and freshwater are the components of hydrosphere. Hydrosphere is a more suitable habitat for microbial growth than atmosphere. Freshwater habitats are lakes, ponds, swamps, springs, streams and river. Marine habitats are oceans and the estuarine habitats. Aquatic habitats provide the food, water, shelter, and space essential for the survival of aquatic animals and plants. Aquatic biodiversity is the rich and harbours variety of plants and animals.

Fresh water

The study of freshwater habitats is known as limnology. Freshwater habitats are classified based on their physical and chemical properties. Those habits with standing water are called lenthic habitat Eg. Lakes and ponds. Those habitats with running water are called lotic habitat. Eg. River.

The uppermost layers of the hydrosphere are called *Neuston*. It is a surface microlayer. It is a interface between the hydrosphere and the atmosphere. It is characterized by high surface tension. This layer is a favourable habitat for *photoautotrophs*. Most primary producers and few secondary producers are common in this layer. Some mineral nutrients and metals became enriched in this layer.

Microbial numbers are 10 to 100 times higher than in the underlying the water column. Bubbles rising through the neuston

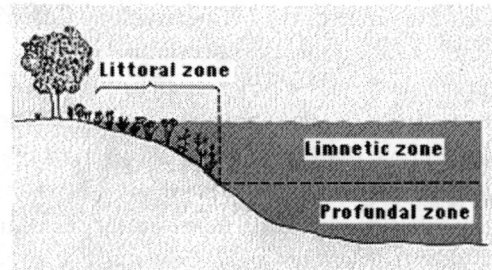

Figure 160   Zonation in lake

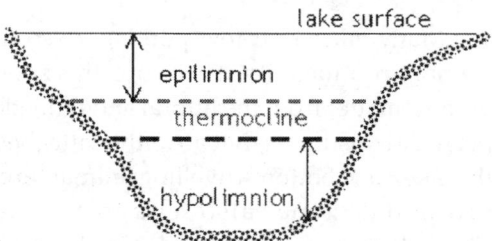

Figure 161   Thermal zonation in lake

layer and bursting play an important role in water to air transfer of bacteria and virus. *Pseudomonas, Caulobacter, Hypomicrobium, Acromobacter, Flavobacterium, Alcaligens, Brevibacterium, Micrococcus* are the autochthonous bacteria found in this area. Cyanobacteria (*Anabena, Microcystis*), Algae (*Chromulina, Nautococcus, Codosiga, Navicula*), Protozoa (*Vorticella, Clathrulina, Stylonychia*), Fungi (*Cladosporium*) and various yeast are found in top layer of water. Benthos include the organisms living at the bottom of the water mass. Nektonic animals are swimmers and are found in all aquatic system. *Plankton* also found in all aquatic ecosystem. Swamps and bogs are shallow aquatic environment dominated by plants. Swamps develop under various climatic conditions. Primary producers are high. Bogs develope only in cool, wet climates usually shallow rock pans. Dominat plant in this area are Sphaghum moss.

Lakes

Lake is one of the lenthic fresh water ecosystem. Lakes are divided into three zones based on the penetration of light. Deep lakes contain three distinct zones, each with its characteristic community of organisms. They are littoral zone, limnetic zone and profundal zone. (Figure 160)

Littoral zone

It is a zone close to shore. Here light reaches all the way to the bottom. The producers are plants rooted to the bottom and algae attached to the plants and to any other solid substrate. The consumers include tiny crustaceans, flatworms, insect larvae, snails, frogs, fish, and turtles.

Limnetic zone

This is the layer of open water where photosynthesis can occur. As one descends deeper in the limnetic zone, the amount of light decreases until a depth is reached where the rate of photosynthesis becomes equal to the rate of respiration. At this level, net primary production no longer occurs. The limnetic zone is shallower in turbid water than in clear and is a more prominent feature of lakes. Life in the limnetic zone is dominated by floating microorganisms - called plankton, actively swimming animals - called *nekton*. The producers in this ecosystem are planktonic algae. The primary consumers include zooplankton. The secondary (and higher) consumers are swimming insects and fish. These nekton usually move freely between the littoral and limnetic zones.

Profundal zone

Many lakes (but few ponds) are so deep. Light does not reaches here to support net primary productivity. Therefore, this zone depends for its calories on the drifting down of organic matter from the littoral and limnetic zones.The profundal zone is chiefly inhabited by primary consumers that are either attached to or crawl along the sediments at the bottom of the lake.Such bottom-dwelling animals are called the benthos.The sediments underlying the profundal zone also support a large population of bacteria and fungi. These decomposers break down the organic matter reaching them, releasing inorganic nutrients for recycling.

Lakes can be classified on the basis of their productivity and nutrient content. They are **oligotrophic lakes** and **eutrophic lakes**. Oligotrophic lake means those lake that contains low nutrient contents. They are deeper lakes. This type is characterized by lower primary productivity. They have larger hypoliminion and smaller epilmnion. High nutrient containing lakes are called eutrophic lakes. This type of lake having high primary productivity. This lake is shallower than other.

Thermal stratification of lakes (Figure 161)

The lake is stratified into epilimnion (Top layer), hypolimnion (Bottom region) and thermocline (Middle layer)based on the nature of temperature. The upper layer epilimnion has slightly higher temperature than bottom regions. The thermal stratification of lakes refers to a change in the temperature at different depths in the lake, and is due to the change in water's density with temperature. Cold water is denser than warm water and the epilimnion generally consists of water that is not as dense as the water in the hypolimnion. However, the temperature of maximum density for freshwater is 4°C.

Pond - Eco-System

Pond is a fresh water aquatic eco-system. It is a self-sufficient and self-regulating eco-system. It is an ideal ecosystem. Pond ecosystem is divided in to three, **based on the depth of water and types of living organisms. They are** (i) littoral; (ii) limnetic, and (iii) pro-fundal. The littoral zone is the shallow water containing rooted plants and this zone of the pond receives maximum light. The limnetic zone ranges from the shallow to the depth of effective light penetration and contains small crustaceans, rotifiers, algae, insects and their larvae. The pro-fundal zone is the deep water part where there is no effective light penetration and it is associated with organism like snails, mussels, crabs and worms.

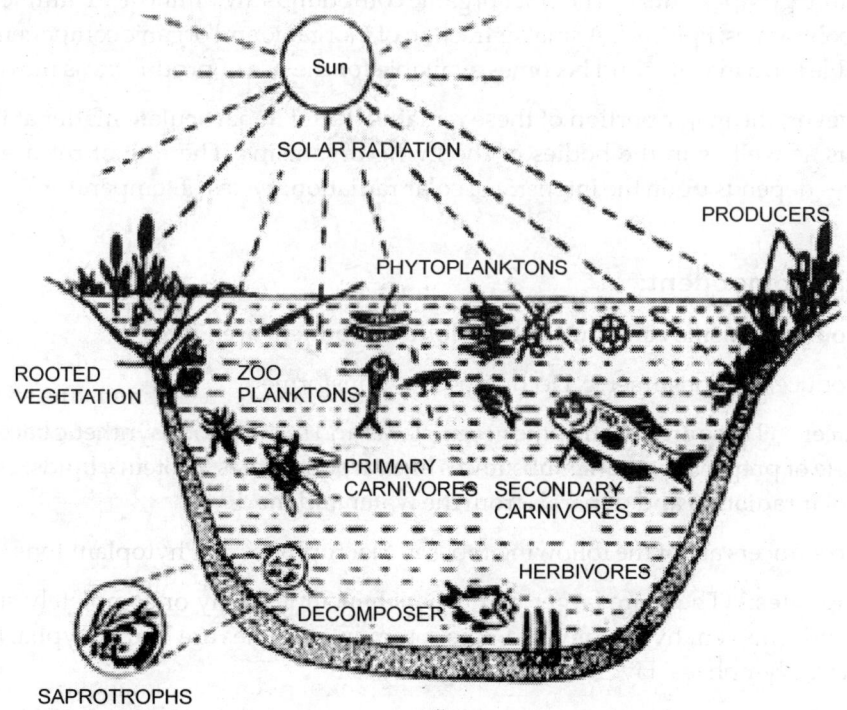

PONDS AS ECO-SYSTEM

Figure 162 Pond ecosystem

Thermal stratification of pond

The Pond is stratified into epilimnion (Top layer), hypolimnion (Bottom region) and thermocline (Middle layer)based on the nature of temperature. The upper layer epilimnion has slightly higher temperature than bottom regions. The thermal stratification of lakes refers to a change in the temperature at different depths in the lake, and is due to the change in water's density with temperature. Cold water is denser than warm water and the epilimnion generally consists of water that is not as dense as the water in the hypolimnion. However, the temperature of maximum density for freshwater is 4°C.

It has two main components: a. Abiotic component and b. Biotic component.

A. Abiotic Component

The abiotic component of pond consists of three sub-components:

i. Physical Components: The physical components influencing pond eco-system are heat, light and pH value of water.

ii. Inorganic Components: The basic inorganic compounds of a pond system are carbon dioxide, water, oxygen, nitrogen, phosphorous, calcium, etc.

iii. Organic Components: The chief organic compounds are amino acid, humic acid, fatty acid, carbohydrates, lipid, etc. A smaller fraction of inorganic and organic components remains in insoluble form in water and becomes available for the use of producers as nutrient.

However, the major portion of these remains stored in particulate matter at the bottom sediments as well as in the bodies of the living organisms. The rate of release of abiotic substances depends upon the intensity of solar radiation, cycles of temperature and climatic regimes,

B. Biotic Component:

The various organisms constituting the biotic component are:

a. Producer, b. Consumer, c. Decomposer or transformer.

a. Producer: These are autotrophic green plants and some photo-synthetic bacteria which are capable of preparing organic substances like carbohydrates, proteins, lipids, etc. with the help of solar radiation and minerals from the water and mud.

The producers are of the following types: i. Macrophytes. ii. Phytoplanktons.

i. Macrophytes: These are larger plants which include partly or completely submerged, floating and emergent hydro-phytes. Some common examples are Trapa, Typha, Eleocharis, Sagitattaria, Nymphaea, Hydrilla, Potamogeton, etc.

ii. Phytoplankton's: These are minute floating or suspended and non-rooted lower plants. These are distributed throughout the ponds as deep as light penetrates. These constitute the autotrophic component of pond and the life of heterotrophic component depends upon it. Some common examples are Volvox, Euglena, Algae, etc.

b. Consumers: In a pond eco-system, the primary consumers are the tadpole, larvae of frog, fish and other aquatic animals which consume green plants or algae as their food (herbivorous). These herbivorous aquatic animals become the food of secondary consumers. The examples of secondary consumers are frogs, fishes, snakes, crabs, etc. The secondary consumers become the food of tertiary consumers e.g. large fishes, turtles.

c. Decomposers and Transformers: There are a large number of heterotrophic bacteria, flagellates and fungi distributed throughout the pond specially more abundant in the mud. These micro-organisms attack the dead organism (plants and animals) and decompose the complex organic compounds into simple inorganic compounds and elements. These are also known as micro-consumers because during the process of decomposition, these absorb a fraction of organic compound. The rate of decomposition and transformation depends upon the physical factors like temperature. When the physical factors are favourable for the decomposers and transformers, the rate of decomposition and transformation from complex organic compounds to simpler inorganic compounds becomes faster.

River

It is a type of freshwater habitat. It is characterized by flowing water. They have zones of rapid water movement and pools with reduced currents. Zones of rapid movements tend to be shallower than pools. Pool zones allows deposition of slit due to decreased current velocity. Slit deposition is not reported in high current velocity rivers. Rivers have high degree of interfacing with the lithosphere along the river banks and there is a great deal of transfer of chemicals from lithosphere into river water through rainwater runoff and erosion of river bank.

The upper course of a river is usually characterized by swift flow, high degree of oxygenation and low temperature. Shading by forest generally keeps primary production in the upper course low and organic material input is derived mostly from the surrounding lithosphere. The middle course of a river is characterized by decreasing flow velocity, high temperature, less shading, resulting in significant intrinsic primary production. The lower course is subject to excessive silt deposition and tidal influences resulting in a die-off of stenohaline fresh water microorganisms and their replacement by salt tolerant estuarine organisms.

Most microbial and many microscopic organisms in rivers are attached to surface such as on submerged rocks. Dissolved nutrients are rapidly absorbed by these attached microorganisms and are liberated upon death and decay only to be absorbed again a small distance downstream. As a consequence, nutrients do not move with the speed of the current but exhibit a much slower movement. The cycling of nutrients in a river does not occur in place, but rather nutrient cycling involves some downstream transport before a cycle is completed. The path of a nutrient can be viewed as a spiral rather than cycle, a phenomenon known as nutrient spiraling.

Rivers often receive high amounts of effluents from industries and municipalities. Municipal sewage disposal into rivers introduces high concentrations of organic compounds. Oxygen concentration is generally depleted because of utilization of available oxygen during microbial decomposition of organic compounds. Industrial effluents, agricultural waste also introduced chemicals and heavy metals may adversely affect microbial activities and survival.

Microorganisms in river

Microbial populations in river by their nature always contains higher populations of alloththonous microorganisms. Members of the genera *Achromobacter, Flavobacterium , Brevibacterium, Micrococcus, Bacillus, Pseudomonas, Nocardia, Streptomyces, Micromonospora, Cytophaga, Spirillum* and *Vibrio* are reported in running water system. Stalked bacteria, such as *Caulobacter, Hyphomicrobium* are associated with submerged surfaces. Autotrophic bacteria are autochthonous members of micobiolta of running water play an important role in nutrient cycling. Photoautotrophs normally found in lake include the cyanobacteria and purple and green anaerobic bacteria. Chemolithotrophs play an important role in nitrogen sulfur and iron cycling within lakes. Cyanobacteria are found typically in high numbers near the surface, where light penetration is adequate to support their photoautotrophic metabolism . Microorganisms found in the sediment of fresh water lakes are usually different from those in the overlying waters. In shallow lakes and ponds, anaerobic photoautotrophic bacteria occur on the surface of the sediment, often conferring characteristic colour on these water bodies. Fungi are found on the debris that accumulates the sediment surface. These fungi produce cellulases and degrade cellulosic substrates. The principal ecological functions of microorganisms in fresh water environments. They decompose dead organic matter, liberating mineral nutrients for primary production, they assimilate and reintroduce into the food web dissolved organic matter, they perform mineral cycling activities, they contribute to primary production, they serve as a food source for grazers.

Marine habitats

The ocean occupies 71% of earth's surface. The huge water masses of the oceans have an important moderating effect on the climate of earth, being the ultimate reservoir and receptacle of the global water cycle. Evaporation and precipitation of water driven by the heat energy of

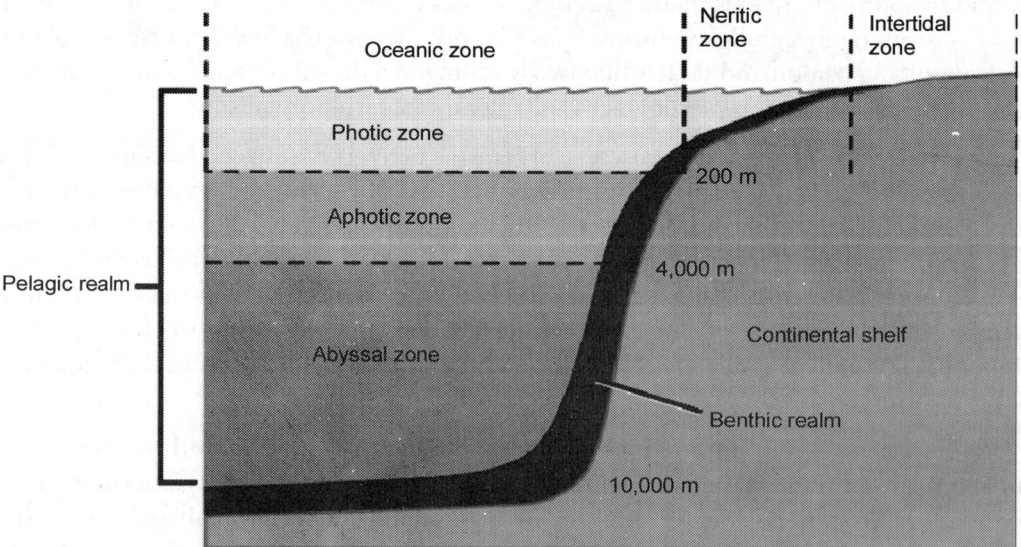

Figure 163   Zonation is marine ecosystem

the sun contribute heat from the equatorial to the polar zones. Environmental conditions in the marine ecosystem are remarkably uniform. This great uniformity is brought about by various mixing mechanisms including tidal movements, currents and thermohaline circulations. Tides are produced by the gravitational pull of the moon and the sun. ocean currents arise from the frictional drag of wind blowing across the surface of the water and the rotation of earth. The ocean contains almost every naturally occurring chemical element, but most are in extremely low concentrations. Major elements of ocean are sodium, chlorine, magnesium, sulfate, calcium and potassium. Minor components include carbon, bromine, strontium, boron, silica and fluorine. Salinity of marine habitat is in the range of 33-37%. The pH of seawater is 8.3-8.5. Temperatures below 100m of depth usually between 0 and 5°C.

Vertical and horizontal zones

The litteral zone or intertidal zone it is the interface between the marine ecosphere and the lithosphere. It occurs at the seashore. This zone is subjected to alternate periods of flooding and drying . The sublittoral zone extends from the low tide mark to the edge of continental shelf. This region also known as neritic or near shore region. The average depth of the neritic zone is less than 200m. The oceanic province extends seaward from the edge of the continental shelf. The term pelagic is used to designate open water or the high sea and include the portions of the neritic and the entirety of the oceanic provinces. The benthos or benthic region refers to the bottom zone.

The benthic region begins at the intertidal region and extends downward. The continental shelf is a gently sloping benthic region that extends away from the land mass. At the continental shelf edge, the slope greatly increases. The continental slope, also known as the bathyl region, drops down to the sea floor. The deep sea floor is known as the abyssal plain and usually lies at 4000m. The ocean floor is not flat but has deep ocean trenches and submarine ridges. The deep ocean trenches are known as the hadal region. Ocean trenches extended down to 11000m.

As with the freshwater environment, the upper most layer of the marine ecosphere is the surface tension layer, which interfaces the marine ecosphere with the atmosphere. The seawater air interface is the habitat for the pleuston, the marine equivalent of the neuston. It includes bacterial and algal inhabitants. *Pseudomonas* and pigmented bacteria are the major bacterial populations. Populations of primary producers like cyanobacteria, diatoms, Phaeophycomycota are sometimes found in pluston layer.

The marine ecosystem may also divided into vertical zones. The euphotic zone is the area of effective light penetration to the compensation level. Below the euphotic zoone can be divided into an epipelagic zone of 0 to 200m, which is typically euphotic and warm. A bathypelagic zone, extending from 200 to 6000m, which is normally disphotic and cold and hadal zone, below 6000m, which is cold and subjected to extreme pressure.

The level of light intensity in the ocean depends largely on turbidity. Recycling of mineral nutrients is extremely slow in the pelagic environment. Dead organisms from the epipelagic zone sink into the great depths of the bathypelagic and ultimately the benthic region. They carry with them essential nutrients, mainly nitrogen and phosphorus, that are liberated in the perpetual darkness of the deep ocean. From here they are returned to the surface water by upwelling. It takes several years for the average mineral nutrient molecule to be returned to the warm euphotic surface waters. Consequently, primary production in the euphotic zone of the pelagic environment is severely limited by the lack of mineral nutrients, where as the nutrient rich deep water lack light energy for photosynthetic primary production. Very low primary productivity was noted in these area. Planktonic algae is only a primary producer. This condition results in inefficient pelagic food chain. Upwelling phenomena often occur along the continental slope, caused by surface currents running rapidly away from the shore and being replaced by deeper, nutrient rich water.

Microbes in marine

The pelagic marine habitat is a unique environment for both macro and microorganisms. Primary production is done by algae and bacteria. Microbial numbers are relatively high in near shore, upwelling and estuarine waters but sink as low as 1-100/ ML in pelagic waters. Marine microbes should exhibit growth at salinities between 20 to 40 parts per thousand. True marine bacteria will not grow in the absence of sodium chloride. Marine bacteria require the ions in marine waters to maintain proper membrane function. For example sodium and chloride required for active transport. Some marine bacteria have multiple membranes surrounding their cells. Exposure to fresh water disturbs these membranes , causing loss of viability in these bacteria.

Estuaries

Estuary means tidal mouth of a great river, where the tide meets the current. Estuaries are among the world's most productive ecosystems. Estuaries are essential to the success of numerous fish and shellfish species of great commercial, ecological, and recreational importance. Estuaries are also the focal points for pollutants entering the sea from the land, and are now the most nutrient-enriched ecosystems on earth. The over abundance of nutrients, especially nitrogen, has caused profound changes in estuarine ecosystems. These include the proliferation of nuisance and noxious algal blooms, the reduction of oxygen availability, and the loss of essential fish and shellfish habitat. The severity of the effects of excess nutrients on estuarine ecosystem structure and function depends on the specific characteristics of each watershed and receiving estuary.

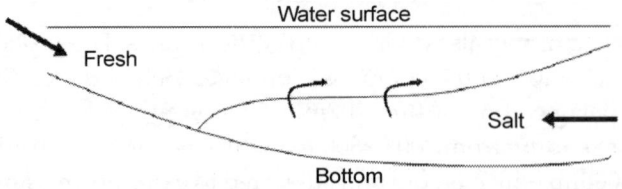

Figure 164   Esturine water flow

Freshwater runoff in the form of rivers and groundwater seepage interfaces with marine waters in estuaries, which are characteristically more productive than either the ocean or the fresh water input. They are areas of highly variable environmental parameters such as temperature, pH, salinity, organic loading and other factors. Estuaries are the ares of mixing of freshwater and sea water. It is a highly productive region because flow of freshwater bring organic matter at the river mouth estuaries are subject to tides and exhibit tidal tidal flushing. Photosynthesis in estuaries almost always exceeds respiratory activities. Larger portions of estuaries are occupied with semisubmerged plants. Saltmarsh estuaries may receive nutrients through upwelling of deeper water masses along the continental shelf, but the larger portion of nutrient input usually comes from the adjacent land in the form of runoff.

Within the estuaries, the plants and other primary producers (algae) convert energy into living biological materials. Detritus feeders, plant grazers, and zooplankton are the primary consumers, and the secondary consumers and tertiary consumers include estuarine birds, ducks, invertebrate predators, and fish. Excreta and detritus pass to the decomposer tropic level where microorganisms break down the material. At each stage in this trophic sequence matter and energy are consumed, and some of it is excrete as waste, or converted into body growth or heat after respiration.

Most of the bacterioplankton in typical estuary are closely related to surrounding freshwater or marine bacterial groups and belong to the phyla Proteobacteria, Bacteroidetes and Actinobacteria. Cyanobacteria play an important role as primary producers, study in a pelagic of a shallow estuary found that Oscillatoriales and chroococcoid colonies dominated the cyanoplankton biomass, whereas Synechococcus-like Cyanobacteria comprised 67.6–91.9% of the cyanobacterial biomass. Methanogenic Archaea are important for the mineralization of organic matter in anoxic estuarine environments.

Coral Reefs

Figure 165 Coral reef

Coral reefs are diverse underwater ecosystems. It is formed by the accumulation of calcium carbonate. A coral reef is a community of living organisms. Coral reefs are some of the most diverse ecosystems in the world. They are home to about 25% of all marine life. Coral reefs are built by colonies of tiny animals found in marine waters that contain few nutrients. Most coral reefs are built from stony corals. Corals consist of polyps. The polyps belong to a group of animals known as Cnidaria, which also includes sea anemones and jellyfish. Corals secrete hard carbonate exoskeletons which support and protect the coral polyps. Reefs grow best in warm, shallow, clear, sunny and agitated waters. Corals are also called as "rainforests of the sea". They occupy less than 0.1% of the world's ocean surface. They provide a home for fish, mollusks, worms, crustaceans, echinoderms, sponges, tunicates and other cnidarians. Reef-building corals live only in the photic zone (above 50 m), the depth to which sufficient sunlight penetrates the water, allowing photosynthesis to occur. Coral polyps do not photosynthesize, but have a symbiotic relationship with microscopic algae of the genus Symbiodinium, commonly referred to as zooxanthellae. These organisms live within the tissues of polyps and provide organic nutrients that nourish the polyp. The algae receive protection, CO_2 and mineral nutrients from the plankton capturing and heterotrophic metabolism of polyps. Symbiotic relationship between polyps and algae induce coral reefs grow much faster in clear water. Corals get up to 90% of their nutrients from their symbionts.

Coral rocks are made up of calcium carbonate secreted by living coral polyps. The coral organisms belong to the phylum Coelenterata. All reef building corals live as large colonies. Reef corals are typically shy, nocturnal feeders. During the day, polyps are withdrawn into skeletal cups and the corals appear more or less lifeless. But at night time the whole reef magically comes to life and coral polyps stretch out their tentacles, probing the waters for food. The reef looks like a field of flowers. Only in warmer waters, the coral polyps can extract calcium from the sea water and deposit it as calcium carbonate in their skeletons. Corals thrive only in crystal clear water. Small plants like Zooxanthella living in the coral tissue contribute to the yellow, brown and green colours of some reef forming corals. The brighter red and orange colours are created by pigment cells in the body wall.

Corals reproduce both sexually and asexually. An individual polyp uses both reproductive modes within its lifetime. Corals reproduce sexually by either internal or external fertilization. The corals spread by producing vast number of minute ciliated larvae called Planulae, through sexual reproduction. These larvae initially lead a free swimming life. Later they settle on rocks and start new colonies of polyps by repeated fission or budding. A coral reef is thus a result of the activity of millions of coral polyps over several thousand years. In India coral reefs occur in the Lakshadweep, Andaman and Nicobar Islands and in the south east-coast.

Economic importance

Some corals are highly prized for their decorative value. Precious corals like *Corallium nobile* are used in jewellery and ornaments. Corals are also important in building the coral reefs and islands, some of which are used as habitation by human and other animals. The organ pipe coral (Tubipora) is used in indigenous system of medicine in South India. Coral skeletons especially of species like porites are used in as building construction and for metalling their roads. Corals serve as raw materials for the preparation of lime mortar and cement because of their calcium carbonate content. Some older coral lime stones are rich in magnesium, hence they are of great value in making cement. Coral skeletons act as natural barriers against sea erosion and cyclonic storms. In several countries Fringing reef, Barrier reef and Atolls are helping in the tourism industry. They Protect shorelines from big waves by absorbing wave energy. Provide a safe place for fish to spawn (release eggs into the water). Provide habitats for a large variety of organisms. Provide food (**fish and** shellfish) for many people living along coastlines. They are a source of medication — some anti-cancer drugs and painkillers come from reefs. Help in the carbon cycle.

Microorganism Movement between Ecosystems

Microorganisms constantly are moving and being moved between ecosystems. This often happens naturally in many ways. Soil is transported around the Earth by windstorms and falls on land areas and waters far from its origins; rivers transport eroded materials, sewage plant effluents, and urban wastes to the ocean; insects and animals release urine, feces, and other wastes to environments as they migrate around the Earth. When plants and animals die after moving to a new environment, they decompose and their specially adapted and coevolved microorganisms (and their nucleic acids) are released. The fecal-oral route of disease transmission, often involving foods and waters, and the acquisition of diseases in hospitals (nosocomial infections) are important examples of pathogen movement between ecosystems. Each time a person coughs or sneezes, microorganisms also are being transported to new ecosystems.

Humans also both deliberately and unintentionally move microorganisms between different ecosystems. This occurs when microbes are added to environments to speed up microbially mediated degradation processes or when a plant-associated inoculum such as *Rhizobium,* is added to a soil to increase the formation of nitrogen-fixing nodules on legumes. One of the most important accidental modes of microbial movement is the use of modern transport vehicles such as automobiles, trains, ships, and airplanes. These often rapidly move microorganisms long distances.

Pathogens that are normally associated with an animal host are greatly affected by such movement because these microorganisms largely have lost their ability to compete effectively with microorganisms indigenous to other environments.

Upon moving to a new environment, the population of viable and culturable pathogens gradually decreases. However, more sensitive viability assessment procedures, particularly molecular techniques, indicate that **nonculturable microorganisms,** as observed with *Vibrio,* may play critical roles in disease occurrence.

Many studies have been directed toward learning why microorganisms which have coevolved with animals gradually die after being released to soils and waters. Among the possibilities are predation by protozoa, *Bdellovibrio*and other organisms, lack of space, lack of nutrients, and the presence of toxic substances. After many years of study, it appears that the major reason "foreign" microorganisms die out is that they can no longer compete effectively with indigenous microorganisms for the low amounts of nutrients present in the environment.

F.   MICROBIAL ADAPTATIONS IN EXTREME ENVIRONMENTS

Microbes grow in different ecosystem. They survive in a wide range of environmental conditions. They grow in varied pHs, temperatures, pressures, salinity, water availability and ionizing radiation. Stress factors have major effects on microbial populations and communities and can create an extreme environment. The microorganisms that survive in extreme environments are described as extremophiles. There are wide varieties of extremophilic organisms inhabiting different ecosystem. Thermophiles thrive at very high temperatures. Psychrophiles live in very cold environments. Alkaliphiles can live at high pH. Acidophile have an optimum pH at levels much lower than neutral. Halophiles live within extremely high salt concentration. and piezophiles who can live at pressures much higher than one atmosphere.

Halophiles

Many microbial genera have specific requirements for survival and functioning in extreme environments. A high sodium ion concentration is required to maintain membrane integrity in many halophilic bacteria, including members of the genus *Halobacterium*. *Halobacteria* require a sodium ion concentration of at least 1.5 M, and about 3 to 4 M for optimum growth. These organisms have the ability to grow at very high salt concentrations. In this case, the salt concentrations can be anywhere from 3% to 35%. Commonly, this group of extremophiles can be found in such environments as sea water, hypersaline lakes (the Dead Sea, the Great Salt Lake), and saline souls. Halophiles can also be divided into three different groups. They are as follows. Slight halophiles that grow at an optimum salinity 2% to 5%, moderate halophiles that grow at an optimum salinity of 5% to 20%, and finally extreme halophiles that grow at an optimum salinity of 20% to 30%. Also, some organisms are referred to as "halotolerant," meaning that the organism has the ability to grow in hypersaline environments. The salinity of water can start at 1M NaCl, but as times goes by the salinity can increase to over 5M NaCl. This causes natural fluctuations in the halophilic species that inhabit that particular body of water. For example, when water is around 1M to 3M NaCl, the environment tends to be filled with algae, protists and yeasts. However, when evaporation occurs, and the salinity increases 5M, those organisms die off because they cannot survive at such high salt concentrations. Organisms that can survive at these higher salt concentrations, such as red-orange halobacteria, drastically increase in numbers until the body is completely dried up or diluted back to a lower concentration. One of the biggest problems faced by halophiles in maintaining homeostasis is the balance of osmotic pressure. Since these organisms are in hypertonic solution, water diffuses out of the cells and into the surrounding environment. This even would cause non-halophilic organisms to plasmolyze or, if the organism does not have a cell wall, the organism would shrivel. Both of these reactions would be lethal to the organism. Usually, the organism would take up sodium ions to create equilibrium between the interior and the exterior cellular environments. However, since sodium ions at such high concentrations would be potentially lethal within a cell, most halophiles accumulate potassium ions while actively expelling sodium ions to create osmotic equilibrium. Halophiles also accumulate other non-disruptive solutes to maintain equilibrium. These can include amino acids, glycine, betaine, ecotine, and sucrose. In order to combat denaturation, aggregation, and precipitation of proteins at high salt concentrations, halophiles proteins often contain a high ration of acidic to basic amino acids, thus giving the surface of the proteins a negative charge. It is believed that this negative charge allows the proteins to be solvated in a high salt environment. Some halophiles make use of the protein bacteriorhodopsin. This compound and retinal is found in the membranes in lattice shaped areas, giving the membrane a purple colour. This protein act as a light dependant proton pump. This protein support phototropic growth. Halophiles also have novel gas vesicles to allow flotation of the organisms in liquid and into higher depths where more oxygen may be available, or where the salt concentration is at optimum range.

Thermophiles

Organisms survives in high temperature are called thermophiles. Eg. *Thermoplasma, Sulfolobus*. Its temperature optimum is around 70 to 80°C. Their cell wall contains lipoprotein and carbohydrate but lacks peptidoglycan. They grow lithotrophically on sulfur granules in hot acid springs and soils while oxidizing the sulfur to sulfuric acid. Oxygen is the normal electron acceptor, but ferric ion may be used. Sugars and amino acids such as glutamate also serve as carbon and energy sources. Thermoproteus is a long thin rod that can be bent or branched. Its cell wall is composed of glycoprotein. Thermoproteus is a strict anaerobe and grows at temperatures from 70 to 97°C and pH values between 2.5 and 6.5. It is found in hot springs and other hot aquatic habitats rich in sulfur. It can grow organotrophically and oxidize glucose, amino acids, alcohols, and organic acids with elemental sulfur as the electron acceptor. That is, Thermoproteus can carry out anaerobic respiration. It will also grow chemolithotrophically using H_2 and S0. Carbon monoxide or Co_2 can serve as the sole carbon source. Thermophiles are perhaps one of the most interesting varieties of the extremophilic organisms. These microorganisms are those that can thrive at temperatures over 50°C. Based on their optimal temperature, thermophiles can be subdivided into three groups: slight thermophiles with an optimal temperature between 50°C and 64°C and a maximum at 70°C, extreme thermophiles with an optimal temperature between 65°C and 85°C and finally hyperthermophiles with an optimal temperature above 85°C and a maximum above 90°C. It was previously believed that life could not thrive at temperatures above 113°C, however recent discoveries have found a microbe called strain 121 that is able to grow at 121°C and can survive at 130°C. In 2001, over sixty species of Bacteria and Archea have been isolated and grown between 80°C and 110°C. Of the thermophiles, there are a much higher number of anaerobes than aerobes. This is most likely due to the fact that oxygen is much less soluble at higher temperatures and therefore is not available for organisms to use in metabolic processes. Thermophiles can grow in both terrestrial and marine environments, including: solfataric fields, geothermal soils, volcanically heated surface waters, hot fumaroles, deep-sea vents, and even black smokers . These can also thrive in biotopes created by man, such as smoldering coal refuse and geothermal powerplants . Due to the hazards of living at such extreme temperatures, thermophiles have evolved a variety of mechanisms that allow them to survive at temperatures no other organisms can thrive at. These traits include unique membrane lipid composition, thermostable membrane proteins, and higher turnover rates for various protein enzymes. One of the most important attributes to the maintenance of homeostasis within the organism is that of the plasma membrane surrounding the organism. Archaean thermophiles, and also acidophiles, have membranes containing unique ether lipids. These tetraether lipids span the entire membrane forming a rigid monolayer that is impermeable to both ions and protons. Ether-type lipids are much stronger than the ester-type lipids found in non-thermophilic Bacteria and Eukarya. The lipid composition in the membranes of the thermophiles consists of more branched and saturated fatty acids than other organisms. Stronger lipid complex within the membrane helps the Archaean thermophiles to withstand higher temperatures. Thermophiles have developed distinct ways of heat stabilizing the proteins that are required for the maintenance of life. For one, the surface energy of the protein, along with the hydration of the non-polar groups that are exposed, are minimized. Also, hydrophobic regions are packed into a very dense core of the protein by charge-charge interactions between amino acids. There is also an increase in salt bridges and other networks, which help to stabilize the structures at higher temperatures. Finally, it has been shown that there is a distinct increase in the synthesis of chaperonin proteins after a heat shock. Chaperonins are proteins that unfold and help refold proteins that are not folded properly enough to perform their required function. Increasing the number of these during high temperatures, most likely allows the cells to have second chance at folding proteins that misfolded due to high heat.

Acidophiles

Organisms that inhabit the niche between pH0 and pH4 are termed acidophiles. These organisms often have the ability to grow at high temperatures as well. Organisms that grow such condition are called thermoacidophiles. Acidophiles can inhabit any niche within the bounds of low pH. The internal pH of the cell is maintained as close to neutral as possible, usually between pH5 and pH7, in order to avoid the denaturation of proteins and other molecules. *Picrophilus oshimae* has been recorded as having an internal pH of 4.6. Also, the cellular membranes have a very low protein permeability to keep stray protons from an acid out of the cytoplasm. In order to maintain the internal pH, acidophiles either actively excrete protons or use them in various metabolic reactions such as the reduction of oxygen in the membrane, before the acidic protons can cause internal cellular damage. Acidophiles also utilize non-energy processes to maintain internal pH. These include the maintenance of fixed negative charges on intracellular molecules and the upkeep of a proton diffusion potential. Protein enzymes must also be modified in order to keep from being denatured. Acidophilic enzymes have the charged amino acids replaced by neutral polar amino acids in their polypeptide chains. This reduces the electrostatic repulsion that occurs between charged groups at low pH, thus enhancing stability.

Alkaliphiles

Microbes that survives in higher pH are called alkaliphiles. These organisms thrive in environments with a pH between 10 and 12, with an optimum growth pH of about 9. Alkaliphiles also have the ability to live in neutral and even acidic environments. In order to survive at these levels, alkaliphiles have novel adaptations to cell wall structure. It has been shown that the cell wall of alkaliphiles contains a variety of acidic compounds, including: phosphoric acid, aspartic acid, galacturonic acid, glutamic acid, and gluconic acid. Having these negatively charged amino acids in the membrane allows the cells to better absorb sodium ions and hydronium ions (due to their positive charges),while at the same time repel the hydroxide ions which are in high concentrations at high pH levels. Having a membrane capable of this feat allows alkaliphiles to grow at pH levels higher than any other organism.

Barophiles

The bacteria found in deep-sea environments have different pressure requirements, depending on the depth from which they are recovered. These bacteria can be described as baro- or piezotolerant bacteria (growth from approximately 1 to 500 atm), moderately barophilic bacteria (growth optimum 5,000 meters, and still able to grow at 1 atm), and extreme barophilic bacteria, which require approximately 400 atm or higher for growth. Intriguing changes in basic physiological processes occur in microorganisms functioning under extreme acidic or alkaline conditions. Piezophiles are organisms that have the ability to grow at pressures higher than normal atmospheric pressure.

G. LITHOSPHERE

It is one of the main components of the ecosystem along with the atmosphere and hydrosphere. The lithosphere is made up of rocks. It contains all of the outer, thin shell of the planet, called the crust; and the uppermost part of the next-lower layer, the mantle. The thickness of the lithosphere varies; thickest below the continents and thinnest at mid-ocean ridges. Temperatures of the lithosphere is low near the earth's surface. Below the base of the lithosphere, rocks are hot enough that they actually deform by flowing, even though they remain solid due to the high confining pressure produced by the weight of the rocks above. That layer, on which the lithosphere rests, is known as the asthenosphere. The physical connection between the lithosphere and the asthenosphere generates a considerable amount of pushing and pulling on the lithosphere as the rocks below move around. In response, the lithosphere has broken into about a dozen large pieces, called lithospheric plates, or simply plates. The movement of the plates away from, towards and past each other, is known as plate tectonics. Under the ocean basins, the crust and mantle rocks of the lithosphere have more or less the same chemical composition. They are made up of minerals enriched in iron and magnesium, and deficient in silica.

Soil is a major part of lithosphere. The nature of soil and soil microbes are discussed in Chapter Soil Microbiology.

Biosphere found on the lithosphere are called Terrestrial Ecosystem. On the basis of the habitat conditions, the terrestrial eco-system can be divided into three sub-eco-systems.

These are, a. Grassland eco-system, b. Forest eco-system and c. Desert eco-system.

Grassland as an Ecosystem

One of the simplest and self-sufficient terrestrial eco-system is the grassland. It occupies approximately 19 per cent of the earth's surface. A grassland eco-system is composed of different components. They are as follows,

A. Abiotic Component: It consists of various nutrients present in soil or in aerial environment. Abiotic substances like carbon dioxide, water, nitrates, phosphates, sulphates, etc. supply the elements like C, H, O, N, S, P, etc. from air and soil. Some trace elements are also present in the soil.

B. Biotic Component: The various organisms constituting biotic components can be divided into the following headings:

a. Producers: The grasses and few herbs and shrubs are the autotrophs or producers of a grass-land eco-system. These prepare carbohydrate by the process of photo-synthesis in the presence of light, light trapping pigments (chlorophylls), carbon dioxide of the atmosphere and water from the soil. Some producers, species are Dicahanthiun, Cynodon, Desmodium, Digitaria, etc.

b. Consumers: There are mainly three types of consumers:

i. Primary Consumers: The primary consumers are herbivores mainly grazing animals like cows, buffalos, deer's, goats, sheep's, etc. In addition to the grazing animals some insects, termites and millipedes feed on the grasses.

ii. Secondary Consumers: These are the carnivores feeding on herbivores. Some common examples of secondary consumers are foxes, snakes, frogs, lizards, etc.

iii. Tertiary Consumers: These are the carnivore feeding on secondary consumers. Some common examples are snakes, hawks, etc.

c. Decomposers or Transformers: These are the microbes which decompose and transform the organic substances of dead organisms (plants or animals) into inorganic components. The inorganic components are subsequently absorbed by the producers for the preparation of food. The microbes are mainly fungi, some bacteria and actinomycetes.

Forest Ecosystem

A forest is a complete functioning ecosystem that supports in numerable plant and animal species as well as land, water and air subsystem. It is a heterogeneous complex of living and non-living elements which are interrelated. It may be small like a backyard or large like the planet earth which depends on the range of individual species or group of species, geology and other issues.

Different types of forest ecosystems and their characteristics are as follows:

A. Temperate Forests: Temperate forests are the regions which have seasonal variation in climate i.e., the climate changes a lot from summer to winter. The annual rain fall is about 750- 2000 mm and soil is rich. Such types of forests are found in western and central Europe, Eastern Asia and eastern North America.

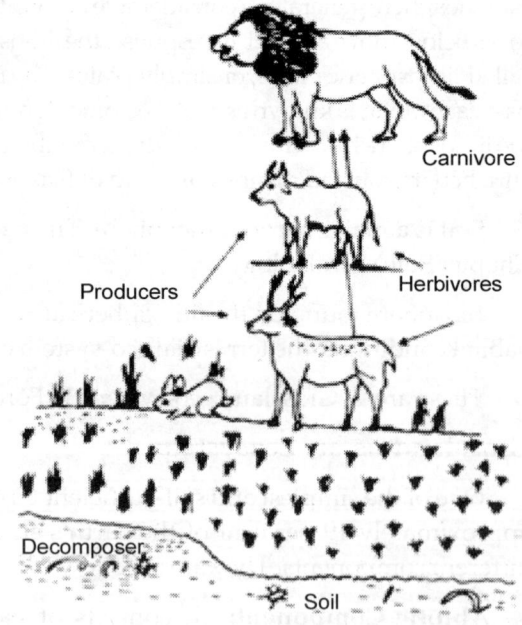

Figure 166 Food chain in terrestial ecosystem

These forests have deciduous trees (oaks, maples etc.) and coniferous trees (pines).

These forests contain abundant microorganisms, mammals (hares, deer, fares, coyotesetc). Birds (warblers, wood peckers, owls etc.) snakes, frogs, salamanders etc.

B. Tropical Rain forests: Tropical rain forests are special ecosystems which accommodate thousands of species of animals and plants. These contain densely packed tall trees. This prevents the growth of smaller plants. However, the areas where the sunlight can reach the surface become the place of growth of a number of interesting plants.

The annual rainfall in these regions is about 80 inches. The temperature remains almost same throughout the year. Such types of forests are found in Brazil of South America (Neotropic) and Central and West Africa. The area is always warm and muggy.

Structure of forest ecosystem

The abiotic components of this ecosystem include physical components (light, heat, etc.), inorganic components (carbon dioxide, water, oxygen, nitrogen, phosphorous, calcium etc.) and organic components (amino acids, humic acid, fatty ac-ids, carbohydrates etc.).

The various organisms constituting the biotic components are: (i) Producer, (ii) Consumer, (iii) Decom-poses or transformer.

i. Producers: The trees and other plants produce the basic food stuff (carbohydrate) and energy by the process of photosynthesis which are subsequently unassumed by other organisms within the food chains and food webs.

ii. Consumers: All animals including mammals, insects and birds are called consumers. The primary consumers eating only plants are termed as herbivores. Secondary consumers feed on herbivores, are termed as carnivores. Tertiary consumers feed on small carnivores, are also carnivores. Omnivores consume both plant and animals matters.

iii. Decomposers: The materials like leaves, needles, old branches, dead plants and dead animals are decomposed by worms, microbes, fungi, ants and other bugs. The decomposers break these items down in to their smallest primary elements to be used again i.e., the decomposers sustain the nutrient cycle of ecosystem.

Function of forest ecosystem

Mainly three important cycles are operating within forest ecosystem.

i. Energy cycle: The energy from the sun is converted in to biomass by the green plant which is subsequently consumed by other organisms. The biomass is converted in to other forms of energy by consumers and decomposers.

ii. Water cycle: Water cycle is operated with in forest eco-system. The water cycle collects, purifies and distributes world's water. The processes involved in water cycle are transpiration, evaporation, condensation, precipitation, infiltration etc.

iii. Nutrient cycle: Nutrient cycles operating in forest ecosystem regularly transform nutrients from the nonliving environment (air, soil, water, rock) to the living environment and then back again.

Desert Ecosystem

A desert is an area where evaporation exceeds precipitation. The annual precipitation in these regions is in between 25 mm and 50 mm, spread unevenly over the year. The desert gets heated during day time and temperature becomes high. The night can be quite cold since the lack of vegetation allows the heat from the ground to radiate away into atmosphere very quickly. The desert soil has very little organic matter but it is rich in minerals.

The desert plants have wax coated leaves, deep and widely spread shallow roots. These try to conserve water by having few or no leaves. The desert animals are usually small in size. They remain under cover during the day time and come out to feed at night. Many animals have thick external shell which reduces moisture loss due to evaporation.

The deserts differ from one another by their soil composition. Some deserts are made of very fine red sands and others consist of sand mixed with pebbles and rocks. The sands are mostly minerals and sometimes oils are found hidden deep within the rocks.

The different components of desert ecosystems are:

i. Abiotic Component: The abiotic component includes various nutrients present in the soil and arid environment. Interestingly, the abiotic component is having very little organic matter and water.

ii. Biotic component: The various organisms constituting the biotic components are:

i. Producers: The producers capable of producing food by photosynthesis are mainly shrubs or bushes, some grasses and a few trees. Most of the desert plants are succulents and others have seeds that remain dormant until rain awakens them. The desert plants include many species of cacti, desert rose, living rock, welwitchia etc.

ii. Consumers: The animals consuming the producers are insects, reptiles etc. There are also some rodents, birds, some mammalian vertebrates.

The desert insects include locust, a special type of destructive grasshopper, Yucca moth, darkling beethe etc. The desert reptiles may be snakes and lizards. The desert birds are sand grouse, gila wood pecker, road runner ostrich etc. The mammals residing in the desert are camels, horses, foxes, jackals etc.

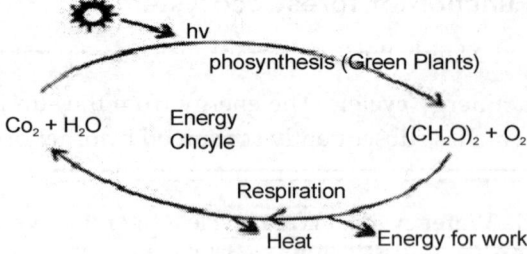

Figure 167   Water cycle

iii. Decomposes: The number of decomposes in the desert are very few because of poor vegetation leading to less organic matter. The usual decomposes are some bacteria and fungi which are thermophillic.

H.   MANGROVE

Mangroves are a diverse group of unrelated trees, palms, shrubs, vines and ferns that share a common ability to live in waterlogged saline soils subjected to regular flooding. They are highly specialized plants that have developed unusual adaptations to the unique environmental conditions in which they are found. There are around 80 species of mangroves found throughout the world. Most commonly they occur within tropical and subtropical sheltered coastal areas subjected to tidal influences.

Mangroves can be divided into two distinct groups: exclusive and non-exclusive. Exclusive mangroves are the largest group, comprising around 60 species. These mangroves are confined to intertidal areas and have not been found to exist within any other type of vegetation community. The remaining 20 plant species considered to be mangroves are referred to as non-exclusive. These plants are not restricted to the typical mangrove environment and are often found within drier, more terrestrial areas.

Mangroves are commonly found throughout the world between latitudes 32°N and 38°S. The upper and lower limits of this range are determined by temperature, while rainfall and the level of protection from wind and wave energy effect forest extent and diversity. Rainfall Areas, which have a great variety of mangrove species, are found along coasts that receive high rainfall, heavy run off and seepage into the intertidal zone from the hinterland. Such areas are commonly subject to extensive sedimentation, which provides a diverse range of substrate types and nutrient levels, which in turn are favourable for mangrove growth

The beneficial effects mangroves have on the marine ecology include:

Basis of a complex marine food chain.

Creation of critical habitat for fisheries and coastal bird populations.

Establishment of restrictive impounds that offer protection for maturing offspring.

Filtering and assimilating pollutants from upland run-off.

Stabilization of sediments and protection of shorelines from erosion.

Water and atmospheric quality improvements.

Contribute to the health of coral reefs.

I. UPWELLING

Upwelling is an oceanographic phenomena often occur along the continental slope. It is caused by surface currents running rapidly away from the shore and being replaced by deeper, nutrient rich water. The nutrient-rich upwelled water stimulates the growth and reproduction of primary producers such as phytoplankton. Due to the biomass of phytoplankton and presence of cool water in these regions, upwelling zones can be identified by cool sea surface temperatures (SST) and high concentrations of chlorophyll-a. The increased availability in upwelling regions results in high levels of primary productivity and thus fishery production. Approximately 25% of the total global marine fish catches come from five upwellings that occupy only 5% of the total ocean area. Upwellings that are driven by coastal currents or diverging open ocean have the greatest impact on nutrient-enriched waters and global fishery yields.

Upwelling in coastal systems also promotes increased productivity by conveying deep, nutrient-rich waters to the surface, where the nutrients can be assimilated by algae.

In the overall process of upwelling, winds blow across the sea surface at a particular direction, which causes a wind-water interaction. The major upwellings in the ocean are associated with the divergence of currents that bring deeper, colder, nutrient rich waters to the surface. There are at least five types of upwelling: coastal upwelling, large-scale wind-driven upwelling in the ocean interior, upwelling associated with eddies, topographically-associated upwelling, and broad-diffusive upwelling in the ocean interior.

Coastal upwelling is the best known type of upwelling, and the most closely related to human activities as it supports some of the most productive fisheries in the world. Deep waters are rich in nutrients, including nitrate, phosphate and silicic acid, themselves the result of decomposition of sinking organic matter from surface waters. When brought to the surface, these nutrients are utilized by phytoplankton, along with dissolved CO_2 (carbon dioxide) and light energy from the sun, to produce organic compounds, through the process of photosynthesis. Upwelling regions therefore result in very high levels of primary production (the amount of carbon fixed by phytoplankton) in comparison to other areas of the ocean. They account for about 50% of global marine productivity. High primary production propagates up the food chain because phytoplankton are at the base of the oceanic food chain.

J.   EUTROPHICATION

Eutrophication is one of the ecosystem disorder due to the accumulation of phosphates to an aquatic system. This leads to the formation of algal bloom. Algal bloom means excessive algal growth. Eutrophication arises from the oversupply of nutrients. It induces explosive growth of plants and algae. This consumes the oxygen in the body of water, thereby creating the state of hypoxia. The availability of phosphorus generally promotes excessive plant growth and decay, favouring simple algae and plankton over other more complicated plants, and causes a severe reduction in water quality. Phosphorus is a necessary nutrient for plants to live, and is the limiting factor for plant growth in many freshwater ecosystems. The source of this excess phosphate is detergents, industrial/domestic run-off, and fertilizers.

Factors control eutrophication

Nutrients

Human activities can accelerate the entry of nutrients into an ecosystem. Runoff from agriculture, pollution from septic systems and sewers, sewage sludge spreading, and other human-related activities increase the flow of both inorganic nutrients and organic substances into ecosystems. Elevated levels of atmospheric compounds of nitrogen can increase nitrogen availability. Phosphorus is often regarded as the main culprit in cases of eutrophication in lakes subjected to "point source" pollution from sewage pipes. These accumulated nutrients lead to excessive primary production.

Light and temperature

The intensity of light and prevailing temperature in the aquatic habitat affects algal growth. In the top layer optimum temperature and light intensity are exist, which induces excessive algal growth along with nutrients.

Effect of Eutrophication

Enhanced growth of aquatic vegetation and algal blooms disrupts normal functioning of the ecosystem, causing a variety of problems such as a lack of oxygen needed for fish and shellfish to survive. The water becomes cloudy, typically coloured a shade of green, yellow, brown, or red. Eutrophication also decreases the value of rivers, lakes and aesthetic enjoyment. Health problems can occur where eutrophic conditions interfere with drinking water treatment.

Terrestrial ecosystems are subject to similarly adverse impacts from eutrophication. Increased nitrates in soil are frequently undesirable for plants. Many terrestrial plant species are endangered as a result of soil eutrophication, such as the majority of orchid species in Europe.

Excessive agal growth leads to death of algae. They are decomposed and the nutrients contained in that organic matter are converted into inorganic form by microorganisms. This decomposition process consumes oxygen, which reduces the concentration of dissolved oxygen. The depleted oxygen levels in turn may lead to fish kills and a range of other effects reducing bio-diversity. Nutrients may become concentrated in an anoxic zone and may only be made available again during autumn turn-over or in conditions of turbulent flow.

Many ecological effects can arise from stimulating primary production, but there are three particularly troubling ecological impacts: decreased biodiversity, changes in species composition and dominance, and toxicity effects, Increased biomass of phytoplankton, Toxic or inedible phytoplankton species, Increases in blooms of gelatinous zooplankton, Increased biomass of benthic and epiphytic algae, Changes in macrophyte species composition and biomass, Decreases in water transparency (increased turbidity), Colour, smell, and water treatment problems, Dissolved oxygen depletion, Increased incidences of fish kills, Loss of desirable fish species, Reductions in harvestable fish and shellfish, Decreases in perceived aesthetic value of the water body.

Some algal blooms, otherwise called "nuisance algae" or "harmful algal blooms", are toxic to plants and animals. Toxic compounds enters in the food chain, resulting in animal mortality. Freshwater algal blooms can pose a threat to livestock. When the algae die or are eaten, neuro and hepatotoxins are released which can kill animals and may pose a threat to humans. An example of algal toxins working their way into humans is the case of shellfish poisoning. Biotoxins created during algal blooms are taken up by shellfish (mussels, oysters), leading to these human foods acquiring the toxicity and poisoning humans. Examples include paralytic, neurotoxic, and diarrhoetic shellfish poisoning. Other marine animals can be vectors for such toxins, as in the case of ciguatera, where it is typically a predator fish that accumulates the toxin and then poisons humans.

Control

Reduce flow of nutrients in to natural waters. Treat water with lime, which will remove phosphorus by precipitation. Nitrogen is removed by microbial denitrification. Excessive algal growth is cobtrolled by addinh copper sulphate in ponds. Use of predatory microorganisms for algar reduce algal bloom. Artificial destratification by means of physical mixing of lake can reduce the provbel of eutrophication.

K. FOOD CHAIN

Food chains were first introduced by the African-Arab scientist and philosopher Al-Jahiz in the 9th century and later popularized in a book published in 1927 by Charles Elton, which also introduced the food web concept.

In ecosystem, the biotic factors are linked together by food. For example, the producers form the food for the herbivores. The herbivores form the food for carnivores. "A food chain can be defined as a group of organisms in which there is a transfer of food energy, through a series of repeated eating and being eaten".

Producers → Herbivores → Carnivores.

The various steps in a fo od chain are called trophic levels.

Grass Land : Plants → Mouse → Snake → Hawk.

Forest : Plants → Goat → Lion.

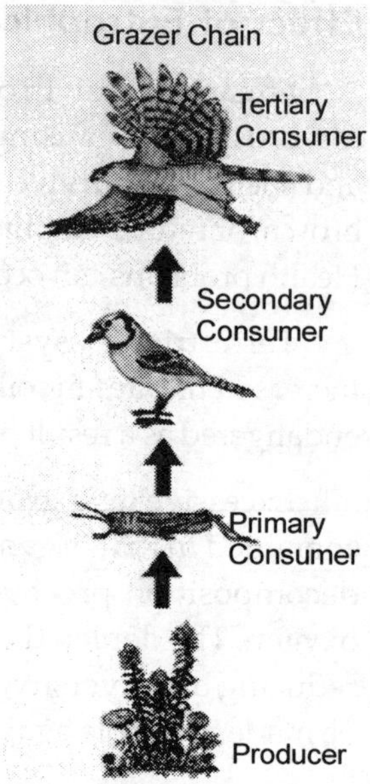

Grazer Chain

Tertiary Consumer

Secondary Consumer

Primary Consumer

Producer

Figure 168 Gazer food chain

Food Chain in Terrestrial Ecosystem

The sequence of food chains in the terrestrial ecosystem may be represented as follows. Grassland → Rodents → Snakes → Hawks.

Energy flows through food chain. The transfer of energy from one trophic level to another is called energy flow.

Types of Food Chains

The food chains are of two types, namely 1. Grazing food chain. 2. Detritus food chain.

Grazing Food Chain

The grazing food chain starts from green plants and ends in carnivores. This type of food chain depends on the autotrophs, which capture the energy from solar radiation.

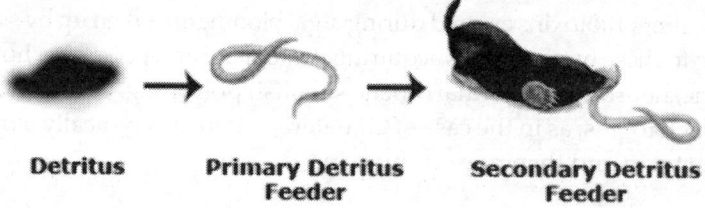

Detritus   **Primary Detritus Feeder**   **Secondary Detritus Feeder**

Figure 169   Detritus food chain

Detritus Food Chain

The organisms which feed exclusively on the dead bodies of animals and plants are called Detrivores. It includes bacteria, fungi, protozoan etc. These organisms ingest and digest the dead organic materials and convert it into CO_2 and wate

Food Web

In an ecosystem, the various food chains are inter-connected with each other to form a net work called Food Web. Simple food chain is very rare. So, each organism may obtain food from more than one trophic level. In grassland ecosystem, grasses eaten by grasshopper, rabbit and mouse. Grasshopper is eaten by garden lizard which is eaten by hawk. In addition hawk also directly eats grasshopper and mouse. Thus five lines are interconnected to form a food web. Food webs are very important in maintaining the stability of an ecosystem.

L. WASTE

Waste is an unwanted or useless material. It is also called as rubbish, trash, refuse, garbage, junk and litter. It is defined as materials that are not prime products for which the initial user has no further use. Wastes may be generated during the extraction of raw materials, the processing of raw materials into intermediate and final products, the consumption of final products, and other human activities.

Sources of Waste

Wastes are generated during different man made activities. Depends on the area from which it is generated the following wastes are described. Residential waste, industrial waste, Institutional waste, Construction and demolition waste, municipal services waste, manufacturing waste, Agriculture waste, Commercial waste.

Types of Waste

There are three types of waste. They are solid waste, semisolid waste and liquid waste.

Solid Wastes

Food wastes, paper, cardboard, plastics, textiles, leather, yard wastes, wood, glass, metals, ashes, special wastes, housekeeping wastes, packaging, construction and demolition materials, hazardous wastes, steel, concrete, dirt, Street sweepings; landscape and tree trimmings; general wastes from parks, beaches, and other recreational areas; sludge, agricultural wastes, hazardous wastes (e.g., pesticides).

Liquid Wastes

Liquid waste includes human waste, sullage, industrial waste and runoff. Human waste and sullage can arise from public institutions such as schools, as well as from individual households. Runoff is simply flood water that arises from rain or the release of collected water from a pond or dam. Sullage means Waste from household sinks, showers, and baths, but not toilets.

Waters often contain high levels of organic matter from industrial agricultural and human wastes. It is necessary to remove organic matter by the process of wastewater treatment.

M. SEWAGE/WASTEWATER/LIQUID WASTE TREATMENT

Every day Human activities, agricultural and industrial operations generate a tremendous volume of sewage and wastewater. This waste water is released into a river/

lakes/oceans. It also percolates to the ground water. It require treatment before discharge into waterways, because peoples uses these water for drinking, house hold uses, industrial uses and irrigation. Wastewater contains excessive amounts of nitrogen, phosphorus, and metal compounds, as well as organic pollutants. Natural water has an inherent self purification capacity. Organic nutrients are utilized and mineralized by heterotrophic aquatic microorganisms. Dense human population, community living patterns, large scale agricultural and industrial activities typically produce larger volume of contaminated liquid waste. These wastes are not naturally purified. Wastewater also contains chemical wastes that are not biodegradable, as well as pathogenic microorganisms that can cause infectious disease. Treatment of waste water is achieved in three stages. These stages are refered as primary, secondary and tertiary treatment.

Primary Treatment

This is the first step in sewage treatment process. This involves screening, trapping, skimming and sedimentation. Screening removes floating solid especially scum through a screen. Trapper also removes suspended solid materials by trapping. Magnets also used for trapping of metals. Skimming is a process by which floated solids are removed. Suspended solids are removed with the help of settling tank. The processes of removing suspended solids in settling tank are called sedimentation. Sedimentation is also carried out by adding coagulants like alum, ferric chloride, chlorinated copper. This type of chemical treated sedimentation is called chemical precipitation. The solids that sediment is strained off, and the sludge is collected to be burned or buried in landfills. Alternatively, it can be treated in an anaerobic digestion process. The liquid portion of the sewage containing dissolved organic matter is subjected to further treatment. Primary treatment removes 20-30% BOD from the sewage.

Secondary Treatment

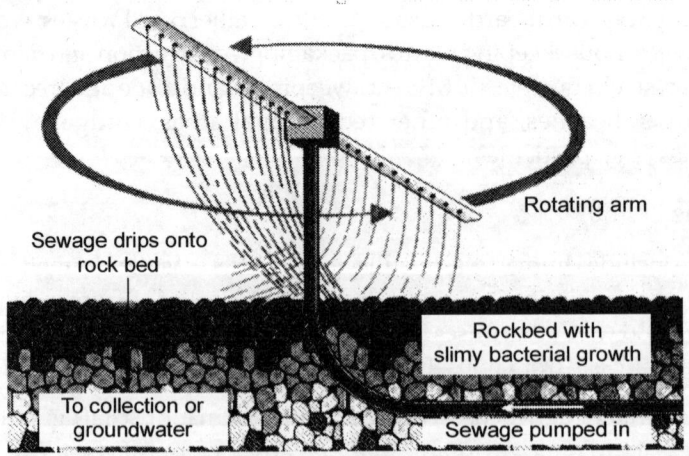

Figure 170 Trickling filter

This is also called as biological treatment. It is used for the removal of dissolved and fine or colloidal organic matter. This treatment step relies on the microbial activity. The treatment may be aerobic, anaerobic or pond process.

Aerobic Treatment

A simple relatively inexpensive film flow type aerobic treatment system is trickling filter. In this process, the liquid waste is sprayed over a bed of crushed rocks, tree bark, or other filtering material. Colonies of bacteria, fungi, and protozoa grow in the bed as slime matrix and act as secondary filters to remove organic materials. The microorganisms metabolize organic compounds and convert them to carbon dioxide, sulfate, phosphates, nitrates, and other ions. The material that comes through the filter has been 99% cleansed of microorganisms. *Zooglea ramigera* plays a major role in this process. The sewage maybe passed through three or more trickling filters or recirculated several times through same filter. Other aerobic treatment filters are Activated Biofilters, Submerged Filters, Biological Fluidized Beds and Rotating Biological Contractors.

Activated Sludge

Activated Sludge is a widely used aerobic method of sewage treatment. After primary settling, the waste stream is brought to an aeration tank. Air is put in and/or there is mechanical stirring which provides aeration of the waste. Sludge from a previous run is usually reintroduced to the tanks to provide microorganisms. This is why it is called activated sludge. During the period in the aeration

Figure 171   Activated sludge

tank, large developments of heterotrophic organisms occur. In the activated sludge tank the bacteria occur in free suspension and as aggregates or flocs. Extensive microbial metabolism of organic compunds in the sewage results in the production of new microbial biomass. Most of this biomass becomes associated with flocs that can be removed from suspension by settling. A portion of the settled sewage sludge is recycled and the remainder must be treated by composting or anaerobic digestion. Combined with primary settling, activated sludge reduces the BOD by 85% to 90%. It also drastically reduces the number of intestinal pathogens.

Anaerobic Digestors

Anaerobic digestors are large fermentation tanks which are continuously operated under anaerobic conditions.

Anaerobic Sludge Digestor

Anaerobic decomposition could be used for direct treatment of sewage, but it is

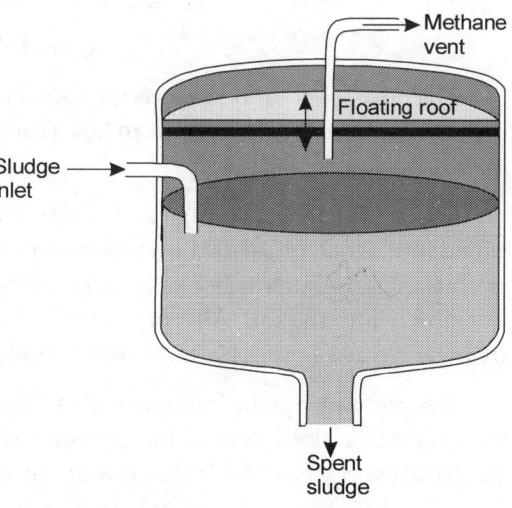

Figure 172   Anaerobic digester

economically favourable to treat the waste aerobically. Large-scale anaerobic digestors are usually used for processing of the sludge produced by primary and secondary treatments. It is also used for the treatment of industrial effluents which have very high BOD levels. The mechanisms for mechanical mixing, heating, gas collection, sludge addition and removal of stabilized sludge are incorporated into the design of large-scale anaerobic digestors. Anaerobic digestion uses a large variety of nonmethanogenic, obligately or facultatively anaerobic bacteria. In the first part of the process, complex organic materials are broken down and in the next step, methane is generated. The final products of anaerobic digestion are approximately 70% methane and 30% carbon dioxide, microbial biomass and a nonbiodegradable residue.

Septic Tank

In a septic tank, household sewage is digested by anaerobic bacteria, and solids settle to the bottom of the tank. Septic tanks are used with water carriage sanitation systems. The human waste is washed into the tank, where it is stored and partially treated. A septic tank is a watertight chamber, usually made of concrete, and is mostly under the surface of the ground. They have inlet and outlet pipes. Fibreglass, PVC or plastic tanks can also be used. The retention time of the wastewater in septic tanks should be a minimum of 19 hours but can be a great deal longer.

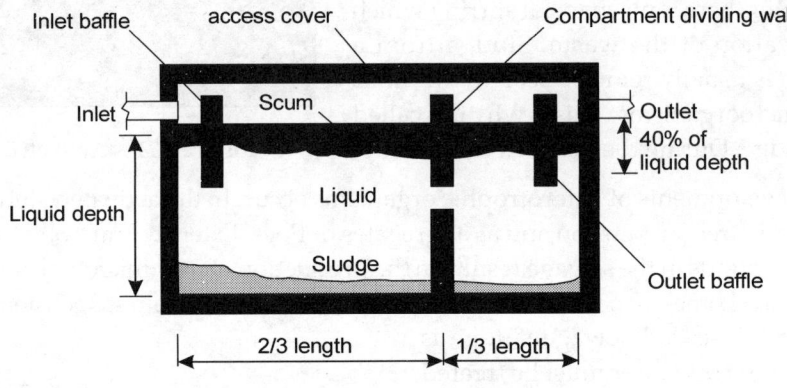

Figure 173   Septic tank

The purpose of septic tanks is for the solids to settle out of the wastewater and for anaerobic decomposition of organic solids to take place. The process of removing sludge from the septic tanks is called desludging.

Septic tanks are a storage and treatment unit. The effluent from septic tanks is usually piped into a soak pit, also known as a seepage pit. A seepage pit is lined with open-jointed or porous material such as bricks or stone without mortar, which allows the wastewater to seep out slowly into the soil. Alternatively the wastewater may be spread across a drainage field using an array of pipes buried below the surface.

A septic tank has the following advantages: can be built and repaired with locally-available materials; has a long service life; presents no problem of flies and odour. It does not require electrical energy because it uses gravity flow.

Oxidation Ponds

Oxidation Ponds are also known as stabilization ponds or lagoons. They are used for simple secondary treatment of sewage effluents. Within an oxidation pond heterotrophic bacteria degrade organic matter in the sewage which results in the production of cellular material and minerals. This supports the growth of algae in the oxidation pond. Growth of algal populations allows furthur decomposition of the organic matter by producing oxygen. The production of this oxygen replenishes the oxygen used by the heterotrophic bacteria. Typically oxidation ponds need to be less than 10 feet deep in order to support the algal growth. Oxidation ponds also tend to fill, due to the settling of the bacterial and algal cells formed during the decomposition of the sewage. Overall, oxidation ponds tend to be inefficient and require large holding capacities and long retention times. The degradation is relatively slow and the effluents containing the oxidized products need to be periodically removed from the ponds.

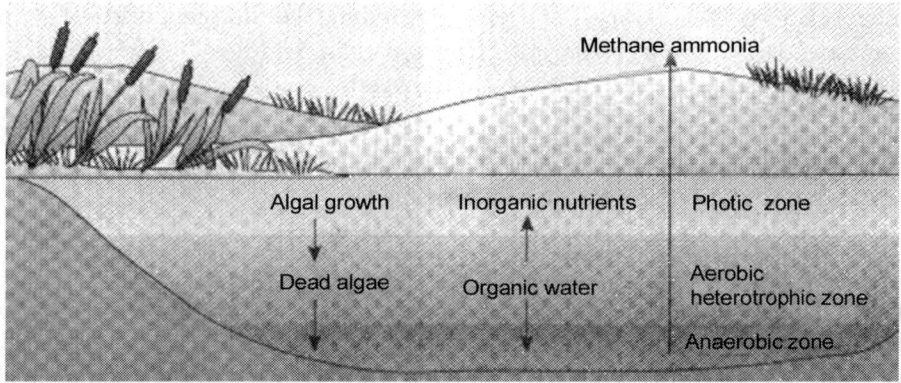

Figure 174   Oxidation pond

Tertiary Treatment

This treatment broadly involves the removal of suspended and dissolved solids, nitrogen, phosphorus and pathogenic microorganisms. It involved four major processes. They are solid removal, biological removal of nitrogen, biological Phosphorus removal and disinfection. Granular sand filteration and micro screening are used to remove suspended solids. Dissolved solids are removed by adsorption and ion exchange process. Adsorption to activated charcoal also removes many organic compounds such as polychlorinated biphenyls (PCBs), a chemical pollutant. Biological nitrogen removal is carried out by assimilization of nitrogen and denitrification. Phosphorus is removed by anaerobic phosphate striper system. It reduced the formation of eutrophication. Microorganisms are removed by disinfection.

Disinfection

Disinfection is the final step in the sewage treatment process. It is designed to kill enteropathogenic bacteria and viruses. Disinfection is commonly done by chlorination with chlorine gas or hypochlorite. Chlorine gas reacts with water to yield hypochlorous and hydrochloric acids which are the actual disinfectants.

N. SEMISOLID WASTE TREATMENT

Sewage sludge is a semi liquid mass that is produced during the waste water treatement process. Semisolid liquid sludge contains upto12% solid material. The various methods of sludge treatment and disposal as follows.

I. Preliminary Operations

It is performed to provide a constant and homogenous sludge for further processing.

Sludge Grinding

It is carried out to cut a large mass of sludge into small pieces.

Sludge Degriting

Removal of degrit is necessary for the effective treatment of sludge. Cyclone degritters are usually employed to separate grit particles from an organic sludge.

Sludge Blending

Sludge produced in primary and secondary and tertiary treatment of liquid waste / sewage treatment are mixed to produce uniform suspension.

II. Sludge Thickening or Concentration

Concentration of sludge may vary. It is necessary to concentrate the sludge by removing some amount of liquid. Gravity thickening is a most commonly used method for the thickening of the primary sludge. It canbe carriedout in a conventional sedimentation tank. Floatation thickening, centrifugal thickening, rotary drum thickening are the methods used for the thickening of sludge.

III. Sludge Stabilization

It is done to reduce the load of disease causing microorganisms. It is also performed to reduce the rate of putrefization and eliminate offensive odour. Lime stabilization, heat treatment, anaerobic treatment, aerobic treatment and composting are the physical, chemical and biological methods of stabilization.

IV. Conditioning of Sludge

Alum, lime, ferric chloride and organic polymers are used for the conditioning of sludge. Solids of sludge are coagulated with the release of absorbed water.

Heat treatment can stabilize the sludge. It is performed for a short period under pressure. It results in the coagulation of solids.

V. Disinfection

It results in the reuse of sludge on the land. In this process pathogens are completely removed. Pasteurization, heat drying, irradiation, high pH treatment, addition of chloride are some methods of disinfection.

VI. Dewatering

It involves reducing the moisture content of the sludge. It can be done by vacuum filteration, centrifugation, sludge drying beds.

VII. Heat Drying

Heat drying of sludge substantially reduce the water content. Mechanical heat drying can be carriedout by flash driers, spray dryers and rotary dryers.

VIII. Thermal Reduction of Sludge

It involves the total or partial conversion of organic solids to oxidized end products. It is associated with destruction of pathogenic microorganisms, detoxification of toxic compounds and reducing the volume of sludge. It is carried out by incineration or wet air oxidation.

IX. Disposal of Sludge

The spreading of sludge on or just below the soil surface is called land application of sludge. It may be used in agricultural or forest land. It can replace the use of expensive fertilizers. Sunlight and soil microbial action reduce the burden of pathogenic microorganism where as aerobic organic degradation creates unpleasent odour. There are two types land filling and lagooning.

Land Filling

It is a method of final disposal of sludge. The sanitary land fill is most suitable method. It is a low cost anaerobic method. In this method, the sludge is deposited in low lying and low value

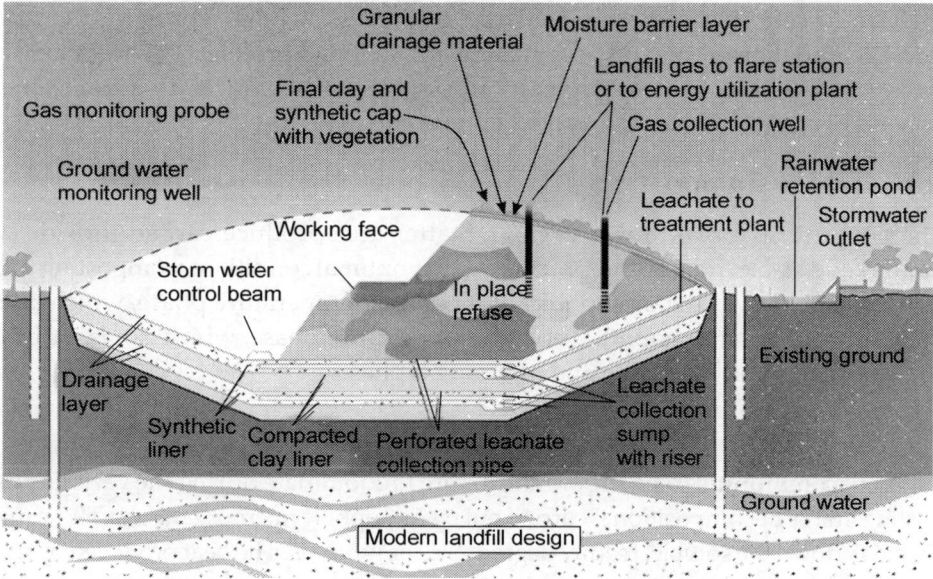

Figure 175 Land filling

sites. The deposition is done almost daily and the deposites are covered with the layer of soil. Two impermeable layer is constructed below the landfill to prevent the leakage of leachete to the surrounding land. The accumulated leachetecanbe taken out and treated by appropriate methods.

Lagooning

It is a shallow lake usually located near a river or sea. It is a convenient method of sludge disposal. In this method sludge is stabilized by aerobic or anaerobic decomposition. The stabilized solids settle to the bottom of the lagoon and accumulated.

O. SOLID WASTE TREATMENT

Solid waste can be treated with incineration, landfilling, composting, silage, pyrolysis, anaerobic digestion, gasification process.

Incineration

Incineration is a waste treatment process. It involves the combustion of organic substances contained in waste materials. It is one of the "thermal treatment" system. Incineration of waste materials converts the waste into ash, flue gas and heat. Waste volume after incineration is reduced to less than 5%. During incineration hydrocarbon vapour can be oxidised to carbon dioxide and Water. Incinerators are used in incineration process. Incinerators are designed to efficiently and safely burn waste at specified rates and temperatures.

Composting

Composting is the aerobic and thermophilic process of microbial decomposition of organic matter. It is a cost effective environmental friendly process. In other words, it is the process of biological degradation of solid organic material to stable end product. This technique is used effectively for the decomposition of domestic waste, agricultural and food wastes. The essential components of composting are substrate, aeration and moisture.

Microbiology of Composting

Microorganisms break down organic matter and produce carbon dioxide, water, heat,humus and stable organic end product. Under optimal conditions, composting proceeds through three phases: 1) the mesophilic, or moderate-temperature phase, which lasts for a couple of days, 2) the thermophilic, or high-temperature phase, which can last from a few days to several months, 3) cooling and maturation phase.

Mesophilic Stage

Different communities of microorganisms are predominate during the various phases of composting. Initial decomposition is carried out by mesophilic microorganisms. These microbes rapidly break down the soluble, readily degradable compounds. The heat produced by microbes causes the compost temperature to rapidly rise.

Thermophilic Stage

As the temperature rises above 40°C, the mesophilic microorganisms become less competitive and are replaced by thermophilic bacteria. At 55°C and above, many microorganisms especially pathogens are destroyed.

Cooling Stage

During the thermophilic phase, high temperatures accelerate the breakdown of proteins, fats, and complex carboydrates like cellulose and hemicellulose, the major structural molecules in plants. As the supply of these high-energy compounds becomes exhausted, the compost temperature gradually decreases and mesophilic microorganisms once again take over for the final phase of "curing" or maturation of the remaining organic matter.

Microorganisms in compost

Bacteria

Alcaligensfaecalis, Bacilllus brevis, B. circulans, B. coagulans, B. licheniformis, B. megaterium, B. sphaericus, B. stereothermophilus, B. subtilis, Clostridium thermocellum, Escherichia coli, Flavobacterium, Serratia sp.

Fungi

Aspergillusfumigates, Humicola grisea, Myriococcum thermophilum, Papulaspora thermophila, Sporotrichum thermophile.

Actinomycetes

Actinobifida chromogene, Microbispora sp., Nocardia, Pseudocardia, Thermomonospora, Streptomyces etc.,

Types of Compost

A mass of rotted organic matter made from waste is called compost. The compost made from farm waste like sugarcane trash, paddy straw, weeds and other plants and other waste is called farm compost. Composting is essentially a microbiological decomposition of organic residues collected from rural area, called rural compost or urban area, called urban compost.

Methods of Composting

There are three methods of composting. They are Coimbatore method, Indore method and Bangalore method. In Coimbatore method, composting is done in pits of different sizes depending on the waste material available. A layer of waste materials is first laid in the pit. It is moistened with a suspension of 5-10 kg cow dung in 2.5 to 5.0L of water and 0.5 to 1.0 kg fine bone meal sprinkled over it uniformly. Similar layers are laid one over the other till the material rises 0.75 m above the ground level. It is finally plastered with wet mud and left undisturbed for 8 to 10 weeks. Plaster is then removed, moistened with water, given a turning and made into a rectangular heap under a shade. It is left undisturbed till its use.

In the Indore method of composting, organic wastes are spread in the cattle shed to serve as bedding. Urine soaked material along with dung is removed every day and formed into a layer of about 15 cm thick at suitable sites. Urine soaked earth, scraped from cattle sheds is mixed with water and sprinkled over the layer of wastes twice or thrice a day. Layering process continued for about a fortnight. A thin layer of well decomposed compost is sprinkled over top and the heap given a turning and reformed. Old compost acts as inoculum for decomposing the material. The heap is left undisturbed for about a month. Then it is thoroughly moistened and given a turning. The compost is ready for application in another month.

In the Bangalore method of composting, dry waste material of 25 cm thick is spread in a pit and a thick suspension of cow dung in water is sprinkled over for moistening. A thin layer of dry waste is laid over the moistened layer. The pit is filled alternately with dry layers of material and cow dung suspension till it rises 0.5 m above ground level. It is left exposed without covering for 15 days. It is given a turning, plastered with wet mud and left undisturbed for about 5 months or till required.

Uses of Compost

Compost contains macro and micronutrients. Compost releases nutrients slowly as and when required. Compost buffers the soil, neutralizing both acid & alkaline soils, bringing pH levels to the optimum range for nutrient availability to plants. Compost helps sandy soil to retain water and nutrients. Compost loosens tightly bound particles in clay or silt soil so roots can spread, water drain & air penetrate. Compost alters soil structure. Compost can hold nutrients tight enough to prevent them from washing out, but loosely enough so plants can take them up as needed. Compost bacteria break down complex organic matter into simpler usable material plant available nutrients. Compost enriched soil have lots of beneficial insects, worms and other organisms that burrow through soil keeping it well aerated. Compost may suppress diseases and harmful pests. Compost encourages healthy root systems. Compost can reduce or eliminate use of synthetic fertilizers. Compost can reduce chemical pesticides.

Vermicompost

It refers to the process of compost formation by earth worm. Vermicomposting uses worms to transform organic waste into high-quality compost. The term vermicomposting means the use of earthworms for composting organic residues. Earthworms can consume practically all kinds of organic matter and they can eat their own body weight per day. The excreta (castings) of the worms are rich in nitrate, available forms of P, K, Ca and Mg. The passage of soil through earthworms promotes the growth of bacteria and actinomycetes. Actinomycetes thrive in the presence of worms and their content in worm casts is more than six times that in the original soil.

Silage

Silage is a fermented, high-moisture containing fodder. It can be fed to ruminants or used as a biofuel feedstock for anaerobic digesters. It is fermented and stored in a process called ensiling or silaging. It is usually made from grass crops, including corn, sorghum or other cereals, using the entire green plant (not just the grain). Silage is made either by placing cut green vegetation in a silo, by piling it in a large heap covered with plastic sheet, or by wrapping large bales in plastic film.

During the ensiling process, some bacteria are able to break down cellulose and hemicellulose to various simple sugars. Other bacteria break down simple sugars to smaller end products (acetic, lactic and butyric acids). The most desirable end products are acetic and lactic acid.

The following six phases describe what occurs during ensiling.

Phase I

Aerobic organisms predominate on the freshly harvested foliage surface. During the initial ensiling process, aerobic bacteria, respire and consume soluble carbohydrates. This phase reduces the oxygen to create the desired anaerobic conditions. Another important chemical change that occurs during this early phase is the breakdown of plant proteins. Proteins are first reduced to amino acids and then to ammonia and amines. Phase I ends once the oxygen has been eliminated from the silage mass.

Phase II

This is an anaerobic fermentation where the growth of acetic acid-producing bacteria occurs. These bacteria ferment soluble carbohydrates and produce acetic acid as an end product. Phase II lasts no longer than 24 to 72 hours.

Phase III

The lower pH enhances the growth and development of another anaerobic group of bacteria, those producing lactic acid.

Phase IV

Lactic-acid bacteria begin to increase, ferment soluble carbohydrates and produce lactic acid. Lactic acid is the most desirable for the fermentation and for efficient preservator. Phase IV is the longest phase in the ensiling process.

Phase V

The final pH of the ensiled forage depends largely on the type of forage being ensiled and the condition at the time of ensiling. Final pH of this phase is around 4.5.

Phase VI

This phase refers to the silage as it is being fed out from the storage structure. Phase VI occurs on any surface of the silage that is exposed to oxygen while in storage and in the feedbunk.

Table 11.1   Six phases of silage fermentation and storage.

	Phase I	Phase II	Phase III	Phase IV	Phase V	Phase VI
Age of Silage	0-2 days	2-3 days	3-4 days	4-21 days	21 days-	
Activity	Cell respiration; production of CO_2, heat and water	Production of acetic acid and lactic acid ethanol	Lactic acid formation	Lactic acid formation	Material storage	Aerobic decomposition on re-exposure to oxygen
Temperature Change*	69-90 F	90-84 F	84 F	84 F	84 F	84 F
pH Change	6.5-6.0	6.0-5.0	5.0-4.0	4.0	4.0	4.0-7.0
Produced microse involved		Acetic acid and lactic acid bacteria	Lactic acid bacteria	Lactic acid bacteria		Mold and yeast activity

P.   BIOFUEL

A biofuel is a type of fuel whose energy is derived from biological carbon fixation. Biofuels include fuels derived from biomass conversion, as well as solid biomass, liquid fuels and various biogases. Biofuels are gaining increased public and scientific attention, driven by factors such as oil price hikes, the need for increased energy security, concern over greenhouse gas emissions from fossil fuels, and support from government subsidies.

Bioethanol is an alcohol made by fermentation of carbohydrates. Cellulosic biomass are being developed as a feedstock for ethanol production. Ethanol can be used as a fuel for vehicles in its pure form. It is usually used along with gasoline. Gasoline is a petrol. Gasohol is a fuel mixture of 10% anhydrous ethanol and 90% gasoline.

Ethanol can be used in petrol engines as a replacement for gasoline; it can be mixed with gasoline to any percentage. An advantage of ethanol (CH_3CH_2OH) is that it has a higher octane rating than ethanol-free gasoline available at roadside gas stations which allows an increase of an engine's compression ratio for increased thermal efficiency.

Methanol is currently produced from natural gas, a non-renewable fossil fuel. It can also be produced from biomass as biomethanol. The methanol economy is an alternative to the hydrogen economy, compared to today's hydrogen production from natural gas.

Methanol is also called wood alcohol as it was produced by wood distillation in the past. Methanol is a clear, colourless, volatile liquid most commonly used as a solvent. It is also used in the chemical industry in the production of various compounds. Methanol is considered as the potential fuel for automobiles in the future. Natural gas is the main source for the production of methanol. It can also be produced from any gas that can be decomposed into hydrogen and carbon dioxide/carbon monoxide.

$$C + H_2O \longrightarrow CO + H_2$$
$$C + 2H_2O \longrightarrow CO_2 + 2H_2$$
$$CH_4 + H_2O \longrightarrow CO + 3H_2$$
$$CH_4 + 2H_2O \longrightarrow CO_2 + 4H_2$$

Natural gas is passed over a catalyst at high temperature and high pressure then treated with steam.

$$CO_2 + 3H_2 \longrightarrow CH_3OH + H_2O$$
$$CO + 2H_2 \longrightarrow CH_3OH$$

Methanol can be produced from direct oxidation of the hydrocarbons.

$$2CH_4 + O_2 \longrightarrow 2CH_3OH$$

Biodiesel is the most common biofuel. It is produced from oils or fats using transesterification and is a liquid similar in composition to fossil/mineral diesel. Chemically, it consists mostly of fatty acid methylesters. Feedstocks for biodiesel include animal fats, vegetable oils, soy, rapeseed, jatropha, mahua, mustard, flax, sunflower, palm oil, hemp, field pennycress, pongamiapinnata and algae. Pure biodiesel is the lowest emission diesel fuel. Although liquefied petroleum gas and hydrogen have cleaner combustion, they are used to fuel much less efficient petrol engines and are not as widely available. Biodiesel can be used in any diesel engine when mixed with mineral diesel. Biodiesel is an effective solvent and cleans residues deposited by mineral diesel, engine filters may need to be replaced more often. Biodiesel is also safe to handle and transport because it is as biodegradable as sugar, 10 times less toxic than table salt, and has a high flash point of about 300°F compared to petroleum diesel fuel, which has a flash point of 125°F.

Green diesel, also known as renewable diesel, is a form of diesel fuel which is derived from renewable feedstock rather than the fossil feedstock used in most diesel fuels. Green diesel feedstock can be sourced from a variety of oils including canola, algae, jatropha and salicornia in addition to tallow. Green diesel uses traditional fractional distillation to process the oils, not to be confused with biodiesel which is chemically quite different and processed using transesterification.

Oils and fats can be hydrogenated to give a diesel substitute. The resulting product is a straight chain hydrocarbon with a high cetane number, low in aromatics and sulfur and does not contain oxygen. Hydrogenated oils can be blended with diesel in all proportions. Hydrogenated oils have several advantages over biodiesel, including good performance at low temperatures, no storage stability problems and no susceptibility to microbial attack.

Many second generation biofuels are under development such as Cellulosic ethanol, Algae fuel, biohydrogen, biomethanol, DMF, BioDME, Fischer-Tropsch diesel, biohydrogen diesel, mixed alcohols and wood diesel.

Hydrogen - A New Fuel

Hydrogen is the simplest molecule present in the universe. Production and use of hydrogen represent a potential alternative source of fuel. It can be easily collected, stored (as gas, liquid or hydrides of metals) and transported (by trucks, ships or trains). Hydrogen can be piped or transmitted over wires for a distance of over 10 km. After the use, hydrogen does not pollute the environment.

For the production of hydrogen, water serves as a source of raw material. The bond between hydrogen and oxygen in water can be broken by providing necessary energy by heat, electricity or light photons as below:

$$H_2O \longrightarrow O_2 + H_2$$

Catalysts

Based on the types of energy used, the following categories of splitting water have been made:

i.　Electrolysis: Electrical splitting

ii.　Thermolysis : Splitting of water by heat,

iii.　Thermochemical lysis : Splitting of water by both heat and chemical catalysts.

iv.　Photolysis : Splitting of water by light.

The first three approaches have been already in practice, whereas photolysis of water has much future prospect as far as seeking of alternative source of energy is concerned.

Solar energy available on earth surface constitutes an abundant and free energy source. It can be converted directly into heat, mechanical energy, electricity or fuel. Nowadays several processes have been envisaged for the conversion of optical energy to chemical energy. These processes arc given below:

i.　Photo-chemical Process : Hydrogen is produced by using photocatalysts such as compound salts and photosynthetic dyes.

ii.　Photo-electro-chemical (PEC) Process : In this process semiconductor photocatalysts (e.g. Sr. Ti O3) are used for the production of hydrogen. Based on semi-conductor system PEC cells have been designed. PEC cells offer the unique advantages for converting optical energy to chemical energy. It is the most efficient chemical system designed so far.

iii.　PhotobiologicalProcess : This process involves the splitting of water by using natural or synthetic chlorophyll, algae and bacteria.

Photobiological Process of H_2 Production

Biophotolysis of water refers to break down of water and production of oxygen and hydrogen by biological process. Hydrogen production is brought about by the following means:

Hydrogenase and H_2 Production

In the early 1960s, production of hydrogen was demonstrated by using chloroplasts isolated from spinach (Spinaciaoleracea) in the presence of artificial electron donors and bacterial extracts containing hydrogenase. Electron donors (organic compounds) transfer electrons to photosystem I of the chloroplast from where electrons are received by electron carriers (e.g. ferredoxin). Hydrogenase accepted electrons from the electron carrier as shown below. The organic compounds which acted as electron donor also served as a source of hydrogen (Sasson, 1984).

Electron donor $\xrightarrow{e}$ Photo system I $\xrightarrow{e}$ Electron carrier H^+ $\xrightarrow{e}$ Hydrogenase $\xrightarrow{e}$ H_2

In the visible light, hydrogenase separates high energy electrons from ferredoxin and facilitates their transfer to H+; ultimately H_2 is evolved. Those plants which produce carbohydrates lack hydrogenase.

Table 11. 2   Algal Species Containing Hydrogenase

Groups	Species
Blue-green algae (Cyanobacteria)	*Anabeanaazollae*, *A. cylinderica*, *Anacystiselongata*, *Nostocmuscorum*, *Spirulinaplatensis, Synechococcuselongatus*
Green algae	*Chlamydomonasmoewusii, Chlorella fusca, C. homosphaera, C. kessleri, C. sorokiniana, Ulvalactuca*
Brown algae	*Ascophyllumnodosum*
Red algae	*Ceramiumrubrum*, *Chondruscrispus*, *CorallinaofficinalisPorphyrasp*, *Porphyridiumcruentum*

There are many bacteria which contain hydrogenase as they possess nitrogenase for nitrogen fixation.

Halobacteria

Halobacteria are rod-shaped and physiologically a unique bacteria, as they are highly halophilic (salt loving). They differ from other bacteria in respect of cell-wall and energy producing mechanisms. They require the high concentrations of salt (e.g. 3-4 M sodium chloride) and low amount of oxygen. They have the capacity to produce an enzyme hydrogenase.

Bacteriorhodopsin acts as a light driven proton pump and the changes occur in about 10 milli second. During photo-reaction cycle bR molecules take up the protons on the inner surface of membrane and release them on outer surface. Proton concentration gradient develops on cell membrane and electric potential is generated in the same way on inner and outer cell membrane. Consequently light energy is converted into electrochemical gradient (Stoeckenius, 1976).Each bR molecule pumps about 200 H+ per second under light condition.

Biogas

Biogas is produced by the biological breakdown of organic matter in the absence of oxygen. It is a type of biofuel. It is an odourless gas. It doesnot release smoke while burning. It burns with blue flame. Methane gas is also called as marsh gas and gobargas. Biogas is produced by anaerobic digestion of biodegradable materials. This type of biogas comprises primarily methane and carbon dioxide.

Process

Many microorganisms are involved in the process of anaerobic digestion, including acetic acid-forming bacteria (acetogens) and methane-forming archaea (methanogens). These organisms feed upon the initial feedstock, which undergoes a number of different processes, converting it to intermediate molecules, including sugars, hydrogen, and acetic acid, before finally being converted to biogas.

Process Stages

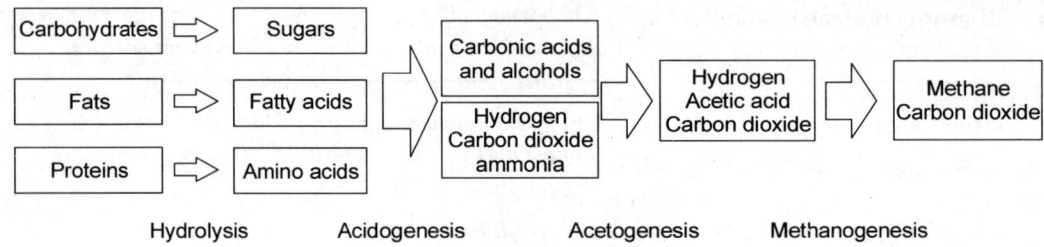

Figure 176 Process of Biogas

There are four key biological and chemical stages of anaerobic digestion:

- Hydrolysis
- Acidogenesis
- Acetogenesis
- Methanogenesis

Complex organic molecules are broken down into simple sugars, amino acids, and fatty acids by *hydrolysis*.

Acidogenesis results in breakdown of the Volatile Fatty acids (VAF). VFAs are converted into ammonia, carbon dioxide, and hydrogen sulfide, as well as other byproducts.

The third stage of anaerobic digestion is *acetogenesis*. Here, simple molecules created through the acidogenesis phase are further digested by acetogens to produce acetic acid, as well as carbon dioxide and hydrogen.

The terminal stage of anaerobic digestion is the biological process of *methanogenesis*. Here, methanogens use the intermediate products of the preceding stages and convert them into methane, carbon dioxide, and water. These components make up the majority of the biogas emitted from the system. Methanogenesis is sensitive to both high and low pHs and occurs between pH 6.5 and pH 8.

A simplified generic chemical equation for the overall processes outlined above is as follows:

$$C_6H_{12}O_6 \longrightarrow 3CO_2 + 3CH_4$$

Anaerobic digestion can be performed as a batch process or a continuous process. In a batch system biomass is added to the reactor at the start of the process.

Composition of Biogas

Biogas is a renewable fuel. Typical composition of biogas are 50-75% Methane, 25-50% Carbon dioxide, 0-10% nitrogen, 0-1% Hydrogen, 0-3% Hydrogen sulphide and 0-2% Oxygen.

Production of Biogas

There are two types of bio gas plants that are used in India. These plants mainly use cattle dung called "gobar" and are hence called gobar gas plant. Generally a slurry is made from cattle dung and water, which forms the starting material for these plants. The two types of bio gas plants are: They are

1. Floating gas-holder type and 2. Fixed dome type.

Floating Gas-Holder Type

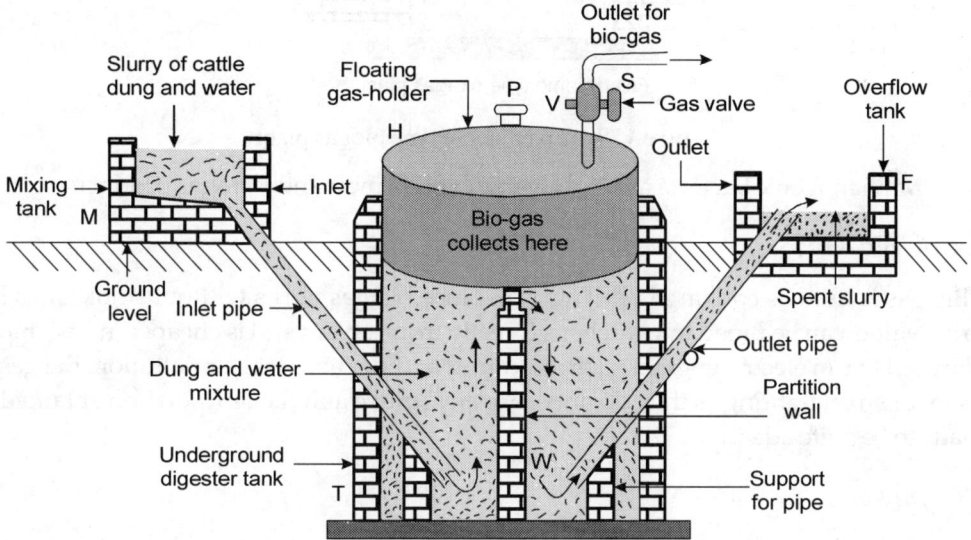

Figure 177   Floating gas-holder

A well is made out of concrete. This is called the digester tank T. It is divided into two parts. One side has the inlet, from where slurry is fed to the tank. The tank has a cylindrical dome H made of stainless steel that floats on the slurry and collects the gas generated. Hence the name given to this type of plant is floating gas holder type of bio gas plant. The slurry is made to ferment for about 50 days. As more gas is made by the bacterial fermentation, the pressure inside H increases. The gas can be taken out through outlet pipe V. The decomposed matter expands and overflows into the next chamber in tank T. This is then removed by the outlet pipe to the overflow tank and is used as manure for cultivation purposes.

Fixed Dome Type

A well and a dome are made out of concrete. This is called the digester tank T. The dome is fixed and hence the name given to this type of plant is fixed dome type of bio gas plant. The function of the plant is similar to the floating holder type bio gas plant. The used slurry expands and overflows into the overflow tank F.

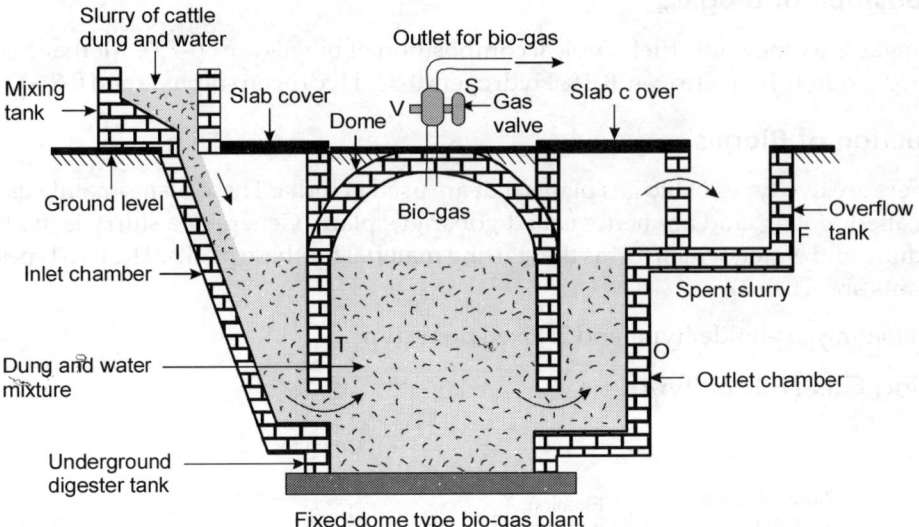

Figure 178　Fixed dome type biogas plant

Methanogenic bacteria are mainly involved in methane / biogas production.

Uses of Bio Gas

Bio gas is used as cooking fuel. This is because bio gas burns without smoke, has high calorific value, can be piped into kitchens directly from a plant and is cheaper in cost. Bio gas can be used to run electric engines such as pumps, as they cause less air pollution. Bio gas can be used for street lighting as they do not cause any smoke and the illumination obtained can be made to be quite adequate.

Pyrolysis

Pyrolysis is a thermochemical decomposition of organic material at elevated temperatures without the participation of oxygen. It involves the simultaneous change of chemical composition and physical phase. Pyrolysis typically occurs under pressure and at operating temperatures above 430°C.

Organic materials are transformed into gases, small quantities of liquid, and a solid residue containing carbon and ash. Several types of pyrolysis units are available, including the rotary kiln, rotary hearth furnace, and fluidized bed furnace. Pyrolysis is not effective in either destroying or physically separating inorganics from the contaminated medium. Pyrolysis treats and destroys semi-volatile organic compounds (SVOCs), fuels, and pesticides in soil. The process is applicable for the treatment of organics from refinery wastes, coal tar wastes, creosote-contaminated soils, hydrocarbons and volatile organic compounds (VOCs).

Gasification

Gasification is a process of thermal degradation of organic carbonaceous materials into carbon monoxide, hydrogen and carbon dioxide. This is achieved by reacting the material at high temperatures (>700 °C) with a controlled amount of oxygen. Gasification results in the formation of gas mixture. Gas formed in gasification are called syngas. Syngas means synthetic gas or producer gas. It is one of the renewable energy. It is a very good fuel.

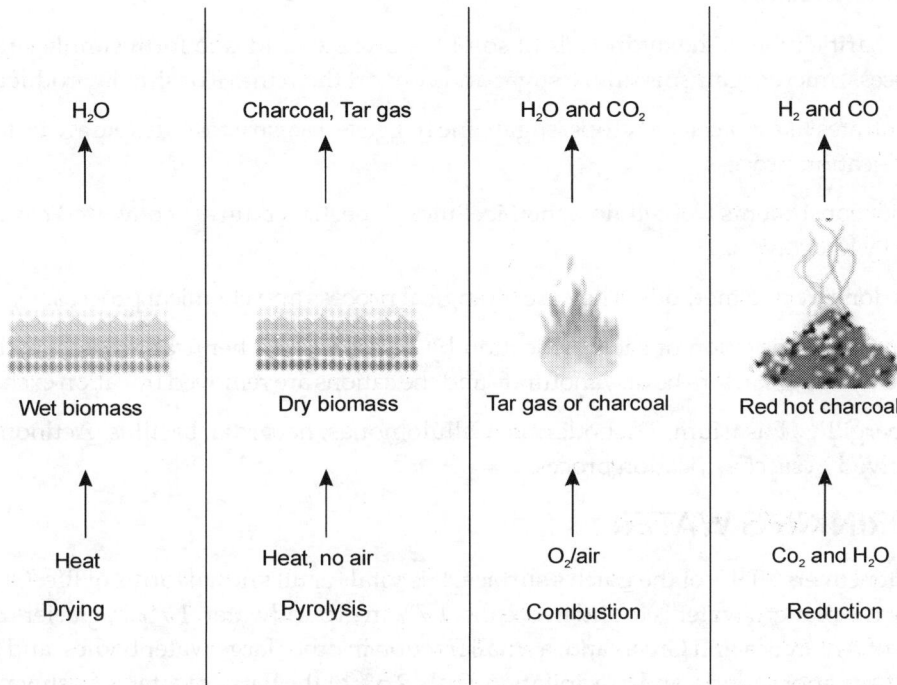

Figure 179   Gasification

The advantage of gasification is that using the syngas is potentially more efficient than direct combustion of the original fuel. Syngas may be burned directly in gas engines, used to produce methanol and hydrogen. Gasification of fossil fuels is currently widely used on industrial scales to generate electricity.

Gasifiers are involved in the gasification process. Several types of gasifiers are currently available for commercial use: counter-current fixed bed, co-current fixed bed, fluidized bed, entrained flow, plasma, and free radical.

In a gasifier, the carbonaceous material undergoes several different processes. They are drying, pyrolysis, oxidation and reduction.

Drying is carried out by heat, water evaporates from the wood.

Pyrolisis is done above 270°C. The wood structure breaks apart chemically. Long molecules are made smaller. Charcoal/char and tar-oil gases are created.

Combustion (oxidation) is also referred to as "flaming pyrolysis". It is carred in a gasifier. Part of the carbon (char) is oxidized (burned) to form carbon dioxide (CO_2) and Hydrogen (H). Hydrogen is oxidized to form water (H_2O).

Reduction -In the reduction area several key conversions take place, and these require significant HEATCarbon (char) reacts with CO_2 and converts it to carbon monoxide (CO). carbon also binds with H to create methane, and some CO reacts with H to form methane+ water.

Saccharification

Saccharification is the hydrolysis of soluble polysaccharides to form simple sugars. In this process, microorganisms utilize sugar and convert them into a desirable product.

Substrates like residues of crops, sugarcane baggase, forestry residues etc are utilized for saccharification process.

Major constituents like cellulose, hemicelluloses, pectin, chitin are converted to xylose or glucose by hydrolysis.

It is done by two methods. They are biological process and chemical process.

After the completion of saccharification by biological or chemical process, metals are removed by using calcium hexacyanofurate and the cations are removed by cation exchangers.

Aspergillus, Fusarium, Trichoderma, Cellulomonas, nocardia, bacillus, Actinomycetes are involved in saccharification process.

Q. DRINKING WATER

Water covers 70.9% of the Earth's surface. It is vital for all known forms of life. On Earth, 96.5% of the planet's water is found in oceans, 1.7% in groundwater, 1.7% in glaciers and the ice caps of Antarctica and Greenland, a small fraction in other large water bodies, and 0.001% in the air as vapor, clouds, and precipitation. Only 2.5% of the Earth's water is freshwater, and 98.8% of that water is in ice and groundwater. Less than 0.3% of all freshwater is in rivers, lakes, and the atmosphere.

Types of Natural Water

Natural water is commonly grouped into four wel marked classes. They are atmospheric water, surface water, stored water and ground water. Water present in the atmosphere either as a solid (snow, hail), liquid (rain) or gas (fog, mist) and called atmospheric water. Surface water includes fresh water on the earth's surface - lakes, ponds, standing water, rivers, streams and brooks. The stagnant land water found in dam, pond are called as stored water. During storage, in general, the number of microorganisms get reduced. Groundwater is water located beneath the earth's surface in soilpore spaces and in the fractures of rock formations. A unit of rock or an unconsolidated deposit is called an aquifer when it can yield a usable quantity of water.

Water Pollution

Water pollution is a major global problem which requires ongoing evaluation and revision of water resource policy at all levels. Water is the leading cause of disease and deaths. It accounts for the deaths of more than 14,000 people daily. An estimated 700 million Indians have no access to a proper toilet, and 1,000 Indian children die of diarrhoeal sickness every day.

1. Soil, vegetation, animal and human wastes, decomposing animal and vegetable matter are the sources of microorganisms in water.

2. Intestinal disease organisms spread by polluted water. These include organisms causing typhoid fever (*Salmonella enterica* ser. Typhi - aka "*Salmonella typhi*"), diarrhoea (*E. coli*, various serotypes of *Salmonella*), bacillary dysentery (*Shigella*), cholera (*Vibrio cholerae*), protozoan infections (amebic dysentery, giardiasis, cryptosporidiosis), viruses (poliomyelitis, hepatitis).

3. Eutrophication-Growth of algae and bacteria due to elevated levels of organic matter and chemicals such as nitrates and phosphates.

There are large number of water pollutants which may be dissolved, suspended or colloidal state. The pollutants may be categorized as organic pollutants, inorganic pollutants, microbiological pollutants and radioactive pollutants.

Water borne Pathogens

Water is an essential part of life. Water is also considered as an important source of microbial infection. Infected human carries pathogens and they can discharge pathogens in waste water. Some times waste water can mix with drinking water. Several species Bacterial and protozoan pathogens can survive in water and infect humans. Some of the GI infection may leads to death. The following table provides selected list of water borne pathogens along with the disease caused by them.

Table 11.3   Water borne pathogens and associated diseases

Microbial Agent	Disease
Bacteria	
Aeromonas hydrophila	Gastroenteritis
Campylobacter	Diarrhoea
Helicobacter pylori	gastritis, peptic ulcers
Legionella pneumophila	Legionellosis
Leptospira	Jaundice
Pseudomonas aeruginosa	Swimmers ear
Salmonella enteriditis	Gastroenterititis
Vibrio cholerae	Cholera
Vibrio parahaemolyticus	Diarrhoea
Yersinia enterocolitica	Gastroenteritis
Escherichia coli	Gastroenteritis
Salmonella typhi	Typhoid
Shigella sp.	Dysentery
Plesiomonas	Diarrhoea
Protozoa	
Acanthamoeba	Amoebic encephalitis
Cryptosporidium	acute enterocolitis
Cyclospora cayetanensis	Diarrhoea

Microbial Agent	Disease
Naegleria fowleri	Primary Amebic Meningoencephalitis
Entamoeba histolytica	Amoebiosis
Helminthes	
Ascaris lumbricoides	Ascariasis
Schistosoma sp.	Schistosomiasis
Fasciola hepatica	Fascioliasis
Taenia saginata	Taeniasis
Viruses	
Adenovirus	Respiratory diseases
Enterovirus	Gastroenteritis
Rotavirus	Gastroenteritis
Hepatitis A Virus	Infectious hepatitis

Details refer Chapter 12 and 14.

Measuring Water Quality/ Water Pollution

An important aspect of water microbiology, particularly for drinking water, is the testing of the water to ensure that it is safe to drink. Water quality testing can be done in several ways. One popular test measures the *turbidity* of the water. Turbidity gives an indication of the amount of suspended material in the water. Typically, if material such as soil is present in the water then microorganisms will also be present. The presence of particles even as small as bacteria and viruses can decrease the clarity of the water. Turbidity is a quick way of indicating if water quality is deteriorating, and so if action should be taken to correct the water problem.

The measurement of water quality is with the reference of organic matter especially carbon and microorganisms. To measure carbon removal in these wastewater treatment processes, several approaches can be used. Carbon removal can be measured (1) as total organic carbon (TOC), (2) as chemically oxidizable carbon by the chemical oxygen demand (COD) test, or (3) as biologically usable carbon by the biochemical oxygen demand (BOD) test. The TOC includes all carbon, whether or not it is usable by microorganisms. This is carried out by oxidizing the organic matter in a sample at high temperature in an oxygen stream and measuring the resultant CO_2 by infrared or potentiometric techniques. The COD gives a similar measurement, except that lignin often will not react with the oxidizing chemical, such as permanganate, that is used in this procedure. The BOD test, in comparison, measures only the portion of the total carbon that can be oxidized by microorganisms in a 5-day period under standard conditions. The biochemical oxygen demand is an indirect measure of organic matter in aquatic environments. It is the amount of dissolved O_2 needed for microbial oxidation of biodegradable organic matter.

The TOC, COD, and BOD provide different but complementary information on the carbon in a water sample. It is critical to note that these measurements, concerned with carbon and carbon removal, do not directly address concerns for removal of minerals such as nitrate, phosphate, and sulfate from waters. These minerals are having worldwide impacts on cyanobacterial and algal growth in lakes, rivers, and the oceans by contributing to the process of eutrophication. The removal of dissolved organic matter and possibly inorganic nutrients, plus inactivation and removal of pathogens, are important parts of wastewatertreatment.

Detection of Pathogenic Microorganisms

A wide range of viral, bacterial, and protozoan diseases result from the contamination of water with human fecal wastes.Assessment of microbial quality of water is a major concern of *sanitary microbiology*. Environmental microbiologists have generally used *indicator organisms*, as an index of possible water contamination by human pathogens. The following are among the suggested criteria for such an indicator:

1. The indicator bacterium should be suitable for the analysis of all types of water: tap, river, ground, impounded, recreational, estuary, sea, and waste.

2. The indicator bacterium should be present whenever enteric pathogens are present.

3. The indicator bacterium should survive longer than the hardiest enteric pathogen.

4. The indicator bacterium should not reproduce in the contaminated water and produce an inflated value.

5. The assay procedure for the indicator should have great specificity; in other words, other bacteria should not give positive results. In addition, the procedure should have high sensitivity and detect low levels of the indicator.

6. The testing method should be easy to perform.

7. The indicator should be harmless to humans.

8. The level of the indicator bacterium in contaminated water should have some direct relationship to the degree of fecal pollution.

Coliforms, including *Escherichia coli* is considered as indicator organisms.Enteric indicator bacteria are not detected in a water, the water is considered potable (fit to drink], or suitable for human consumption. *Coliforms* are defined as facultatively anaerobic, gram-negative, nonsporing, rod-shaped bacteria that ferment lactose with gas formation within 48 hours at 35°C. The coliform group includes *E. coli*, *Enterobacteraerogenes* and *Klebsiellapneumoniae*.

Presumptive test, confirmed test and completed tests were performed to confirm the potablity nature of drinking water. The presumptive test is carried out by means of Lauryl tryptose broth tubes inoculated with three different sample volumes (0.1ml, 1ml and 10ml) to give an estimate of the most probable number (MPN) of coliforms in the water. The confirmatory test uses Brillient green 2% bile broth and completed test uses EMB agar to detect indicator microorganisms. The complete process requires at least 4 days of incubations and transfers. Biochemical tests like indole, methyl red, Voges proskauer and Citrate tests are done to confirms the coliforms.

Faecal coliforms are coliforms derived from the intestine of warm-blooded animals, which can grow at the more restrictive temperature of 44.5°C. To test for coliforms and fecal coliforms, and more effectively recover stressed coliforms, a variety of simpler and more specific tests have been developed. These include the membrane filtration technique, the presence-absence (P-A) test.

The membrane filtration technique, has become a common and often preferred method of evaluating the microbiological characteristics of water. The water sample is passed through a membrane filter. The filter with its trapped bacteria is transferred to the surface of a solid medium or to an absorptive pad containing the desired liquid medium. Use of the proper medium allows the rapid detection of total coliforms, fecal coliforms, or fecal streptococci by the presence of their characteristic colonies. KF streptococcus agar is used to differentiate *Streptococcus faecalis*. Violet red Bile medium is used to differentiate faecal coliforms.

Molecular techniques are now used routinely to detect coliforms in waters and other environments, including foods.

Drinking Water Treatment

To purify water for drinking, a number of processes are conducted to reduce the microbial population and maintain that population at a safe level. Water purification is a critical link in controlling disease transmission in waters. Water purification can involve a variety of steps, depending on the type of impurities in the raw water source. Municipal water supplies are purified by a process that consists of at least three or four steps. Water is first routed to a sedimentation basin. Sand and other very large particles can settle in this basin. The partially clarified water is then mixed with alum and lime and moved to a settling basin where more material are precipitated. This procedure is called coagulation or flocculation and removes microorganisms, organic matter, toxic contaminants, and suspended fine particles. Water is

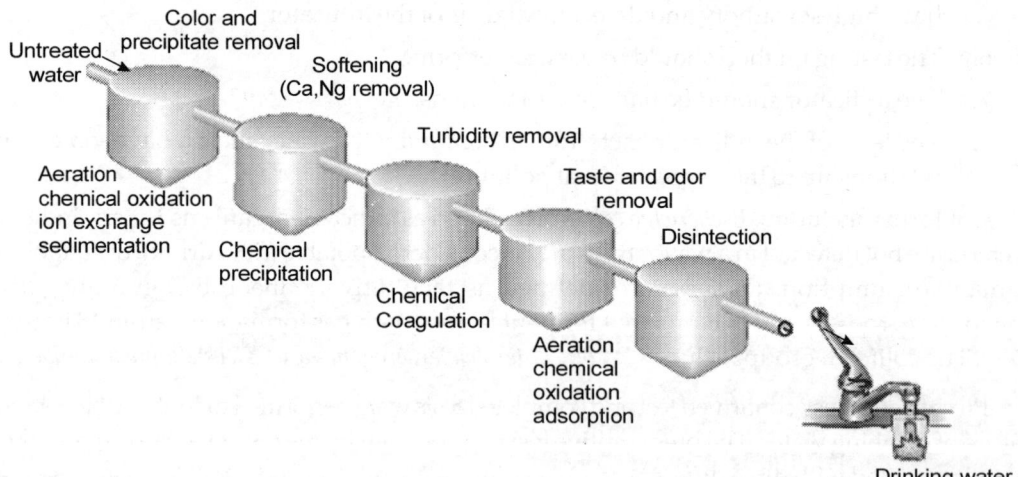

Figure 180 Drinking water treatment

further purified by passing it through a filtration unit. Rapid sand filters are used for physical trapping of fine particles and flocs. This filtration removes up to 99% of the bacteria. Slow sand filters are used to remove protozoans. This treatment involves the slow passage of water through a bed of sand in which a microbial layer covers the surface of each sand grain. Waterborne microorganisms are removed by adhesion to the gelatinous surface microbial layer. Viruses in drinking water also destroyed by Coagulation and filtration. After filtration the water is treated with a disinfectant. This step usually involves chlorination or ozonation.

Many communities then purify the water by chlorination. When added to water, chlorine maintains the low microbial count and ensures that the water remains safe for drinking purposes. Chlorine gas or hypochlorite (NaOCl) is used for chlorination purposes. The water is chlorinated until a slight residue of chlorine remains.

If the drinking water is taken from the river, steps used for the treatment for water treatment is illustrated in Figure 181.

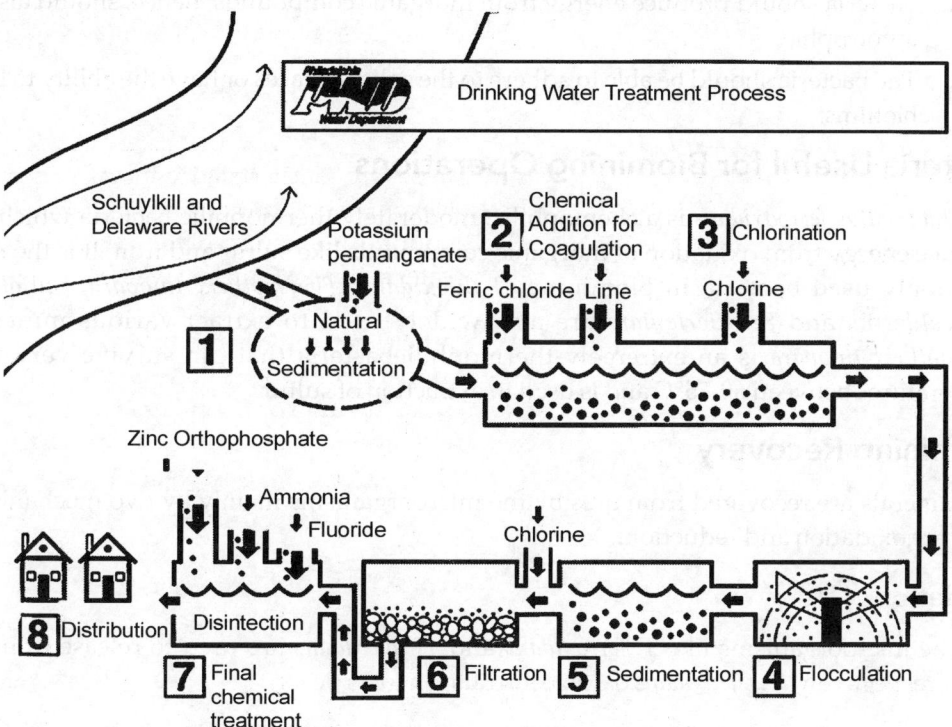

Figure 181 Drinking water treatment from river

R. BIOMINING AND BIOLEACHING

"Biomining describes the processing of metal-containing ores and concentrates using microbiological technology". Biomining has application as an alternative to more traditional physical-chemical methods of mineral processing. The modern era of bioleaching began with the discovery of the bacterium, *Thiobacillus ferrooxidans*.

Biomining is becoming popular because it is cheap, reliable, efficient, safe, and environmentally friendly. The efficiency of biomining can be increased either by finding suitable strains of microorganisms. It is also known as microbial leaching or bio-oxidation.

Microorganisms in Biomining

There are different types of bacteria present in nature that oxidize metal sulfides and solubilize minerals. Microbes helping in extraction of metals from the ores. Selection of suitable microorganisms is very important to ensure the success of biomining.

Characteristics of the bacteria used in biomining

1. Mineral extraction involves the production of high temperatures so the bacteria should be able to survive the heat, hence, they should be thermophilic.

2. Biomining involves the use of using strong acids and alkalis, hence, bacteria should be chemophilic.

3. Bacteria should produce energy from inorganic compounds, hence, should also be autotrophic.

4. The bacteria should be able to adhere to the solid surfaces or have the ability to form biofilms.

Bacteria Useful for Biomining Operations

Thiobacillus ferrooxidans is a chemophilic, moderately thermophilic bacteria which can produce energy from oxidation of inorganic compounds like sulfur and iron. It is the most commonly used bacteria in biomining. *T.thioxidans, Thermothrix thiopara, Sulfolobus acidocaldarius* and *S. brierleyiare* are also widely used to extract various minerals. *Thermothrix thiopara* is an extremely thermophilicbacteria that can survive very high temperatures between 60-75°C and is used in extraction of sulfur.

Biomining Recovery

Minerals are recovered from ores by the microorganisms mainly by two mechanisms. They are oxidation and reduction.

Oxidation

The microorganisms like *T. ferroxidans* and *T. thioxidans* are used to release iron and sulfur respectively. *T. ferroxidans* oxidize ferrous ion to ferric ion.

$$4Fe^{++} + O_2 + 4H^+ \longrightarrow Fe^{+++} + 2H_2O$$

The bacteria attach to the surface of the ore and oxidize by a direct and indirect method.

Direct Method

In this method the ore is oxidized by the microorganisms due to the direct contact with the compound.

$$2FeS_2 + 7O_2 + 2H_2O \longrightarrow 2FeSO_4 + 2H_2SO_4$$

Indirect Method

In this method the mineral is indirectly oxidized by an agent that is produced by direct oxidation. For example, the ferric ion produced by the direct method is a powerful oxidizing agent and can release sulfur from the metal sulphides. Thus production of ferric ion indirectly causes oxidation of metal sulfide resulting in the breaking of the crystal lattice of the heavy metal sulfide and separating the heavy metal and sulfur.

$$CuS + Fe^{+++} \longrightarrow Cu^+ + S + Fe^{++}$$

Reduction

Bacteria like *Desulfovibro desulfuricans* play an active role in reduction of sulfates which results in the formation of hydrogen sulphides.

$$4H_2 + H_2SO_4 \longrightarrow H_2S + 4H_2O$$

Types of Biomining

Biomining is commercially applied using three different engineered methods. They are heap bioleaching/biooxidation, stirred tank bioleaching/ minerals biooxidation and In situ bioleaching.

Heap Bioleaching

Biomining is commercially applied to heap leach copper sulfide ores and to "pre-treat" gold ores in which the gold is occluded in sulfide minerals. Heap bioleaching is widely practiced around the world for the extraction of copper from "secondary copper ores", which contain the mineral chalcocite (Cu_2S) and covellite (CuS). The ore is crushed to about 19mm or less and agglomerated in rotating drums with acidified water to condition the ore for the microorganisms and also to affix fine particles to the larger rock particles. The ore is conveyed to specially engineered pads where it is stacked. The pads are lined with high-density polyethylene(HDPE) and perforated plastic drain lines are placed on the pad to improve the drainage of copper-containing solution from the bottom of the ore heap. A coarse rock layer is placed above the drain lines and within this rock layer a network of perforated plastic air lines is arranged. Air is forced through the air lines and directed to the microorganisms in the heap by blowers external to the heap. The ore is stacked to a depth of 6-10 m most often with automated stackers. The ore is irrigated with acidic raffinate, the effluent from the solvent extraction facility where the copper is recovered from the solution and formed into cathodes. With acidic conditions and abundance of sulfide minerals and iron, naturally-occurring microorganisms develop within the ore heap (numbers exceed 10^6 per gram of ore), facilitating copper extraction. The maximum copper leached from heap bioleach operations is 80% - 90%.

Stirred Tank Biomining

This method is used for leaching from substrates with high mineral concentration. Copper and refractory gold ores are well suited for this type of method. Special types of stirred tank bioreactors lined with rubber or corrosion resistant steel and insulated with cooling pipes or cooling jackets are used for this purpose. Thiobacillus is the commonly used bacteria. Since it is aerobic, the bioreactor is provided with an abundant supply of oxygen throughout the process provided by aerators, pumps and blowers. This is a multi-step process consisting of large numbers of bioreactors connected to each other. The substrate moves from one reactor to another and in the final stage it is washed with water and treated with a variety of chemicals to recover the mineral.

In-situ Bioleaching

In this method the mineral is extracted directly from the mine instead of collecting the ore and transferring to an extracting facility away from the site of the mine. In-situ biomining is usually done to extract trace amounts of minerals present in the ores after a conventional extraction process is completed. The mine is blasted to reduce the ore size and to increase permeability and is then treated with water and acid solution with bacterial inoculum. Air supply is provided using pipes or shafts. Biooxidation takes place in-situ due to growing bacteria and results in the extraction of mineral from the ore.

Biomining of Copper

Copper was the first metal extracted by bioleaching. It is the metal most commonly extracted from oxide ores by this method. Copper is available in mines across the world in more than 350 types of ores. Copper is mainly present along with sulfur. Copper from low-grade ores like copper sulfide minerals is most commonly extracted by biooxidation (Heap).

Low grade copper ore is brought to the dump leaching site. The dump surface is wetted uniformly with water and sulfuric acid using sprayers to maintain acidity which helps the growth of acidophilic bacteria. Air is supplied to the dump through channels constructed for this purpose while building the dump. Biooxidation takes place over the course of time and copper is leached into the solution which is collected at the bottom of the heap. The leach solution rich in copper is treated chemically using electrowinning and solvent extraction techniques to extract pure copper.

Electrowinning

In this technique the leach solution containing copper leached from the dump is circulated through an electrowining cell and electricity is passed. Pure copper is obtained from the cell in the form of electro-won cathode. An Electrowining cell is basically a simple electro-voltaic cell with a lead or graphite anode and aluminum cathode with the leach solution being the electrolyte. When the electricity is passed through the cell, the copper ions present in the electrolyte are reduced to metallic copper and become deposited over the cathode.

Solvent Extraction

Copper solvent extraction systems consist of three loops. In the first loop the leach solution containing copper obtained from dump leaching is passed through the extraction chamber. Here the leach solution comes in contact with organic extractant which extracts copper.. Leach solution and organic extractant are passed through the leaching chamber for further leaching. The copper-rich organic extractant then enters the second loop and passes through stripping chamber. The stripping chamber consists of highly acidic electrolyte which strips copper from the organic extractant. The organic extractant is directed back to the extraction chamber in the first loop. The copper-rich acidic electrolyte enters the third loop and is subjected to electrowinning to extract pure copper. The spent electrolyte is directed back to the stripping chamber in the second loop.

Biomining of gold

Biooxidation of refractory gold ores to extract gold is carried out by a commercial procedure called BIOX developed by GENCOR S.A Ltd Johannesburg, South Africa.

Steps in BIOX

The BIOX process plant consists of two biooxidation stages. They are primary stage and secondary stage. Each stage uses three tanks. The slurry will be present for 2 days in the primary stage. In this stage growth of bacteria is enhanced and induce biooxidation. In the second stage the slurry will be present for a day.

1. Temperature of the plant is controlled to provide optimum temperature for the bacteria. Oxidation of sulfides generates high temperatures but the optimum temperature for bacteria is around 30°– 45° C.

2. Essential nutrients like nitrogen, phosphorus, and potassium are to be provided for better oxidation.

3. Oxygen should be provided throughout the process by means of agitators.

4. pH should be maintained between 1-2 throughout the process.

5. Carbon dioxide should be supplied in specified amounts to the slurry, serving as source of carbon to the bacteria.

Gold biomining is also carried out by heap/dump leaching of refractory gold ores using acidophilic and thermophilic oxidizing bacteria. The process is similar to that used for bioleaching of copper from low grade sulfide ores.

Biomining is also used to extract other minerals like nickel, zinc, molybdenum, uranium and cobalt.

Uranium Leaching

Uranium leaching is more important than copper, although less amount of uranium is obtained. For getting one tonne of uranium, a thousand tonne of uranium ore must be handled. In situ uranium leaching is gaining vast acceptance. However, uranium leaching from ore on a large scale is widely practiced in different countries including India.

Insoluble tetravalent uranium is oxidized with a hot H_2SO_4/Fe_3+ solution to make soluble hexavalent uranium sulfate at pH 1.5 – 3.5 and temperature 35°C.

$$UO_2 + Fe_2(SO_4)_3 \longrightarrow UO_2SO_4 + 2FeSO_4$$

Uranium leaching is indirect process. *T. ferroxidans* does not directly attack on uranium ore, but on the iron oxidant. The pyrite reaction is used for the initial production of Fe+ leach solution.

Microbially Enhanced Oil Recovery (MEOR)

Microbes are involved in the recovery of oil. Use of microbes improve the recovery process hence called microbially enhanced oil recovery (MEOR).

It was discovered in 1926 that microorganisms can be used in the petroleum industry to enhance oil recovery. Microbes can enhance the recovery of petroleum products directly or indirectly.

Direct Method

Oil recovery is enhanced directly by producing gases like nitrogen and carbon dioxide. Gases are produced by microorganisms. These gases cause the oil to swell due to the increase in pressure and free flow thus enhancing the recovery.

Indirect Method

In this method the bacteria produces chemicals like surfactants, solvents, and polymers. These chemicals increase the recovery of oil and various other petroleum products.

Steps in MEOR

Here are three steps in the MEOR.

1. Bacterial inoculum and nutrients are injected to the site (oil reservoir) below the surface of the earth.
2. Bacteria grows and multiplies utilizing nutrients and releases gases like hydrogen, carbon dioxide, and in some cases chemicals such as surfactants, solvents, and polymers.
3. These gases and chemicals mobilize oil in the reservoir increasing the oil flow, thus, enhancing its recovery.

MEOR has many advantages over conventional oil recovery.

S. BIOREMEDIATION

Bioremediation is the complete removal of pollutants by the use of micro-organism. Bioremediation can be also defined as any process that uses microorganisms or their enzymes to return the environment altered by contaminants to its original condition.

Bioremediation treatment technologies can be categorized as follows. They are phytoremediation, bioventing, bioleaching, landfarming, composting, biofilters, bioattenuation, bioaugmentation, rhizofiltration, and biostimulation.

Bioremediation can occur on its own (natural attenuation or intrinsic bioremediation) or can be spurred on via the addition of fertilizers to increase the bioavailability within the medium (biostimulation). Microorganisms used to perform the function of bioremediation are known as bioremediators.

Types of Bioremediation

Most important aspect of environmental microbiology is the effective management of hazardous and toxic pollution. It can be achieved in two ways. They are *In situ* and *ex situ* bioremediation. *In situ* bioremediation involves treating the contaminated material at the site, while *ex situ* involves the removal of the contaminated material to be treated elsewhere.

In situ Bioremediation

It involves a direct approach for the microbial degradation of xenobiotics at the site of pollution. It is successfully applied for clean up of oil spillages, beaches etc.,.There are two types of In situ bioremediation. They are intrinsic Bioremediation and engineered bioremediation. IntrinsicBioremediation is the natural bioremediation of contaminant by using indigenous microorganisms and the rate of degradation is very slow. Engineered bioremediation process used modified version of microorganisms and specific physiological processs for the bioremediation process.

In situ bioremediation is a long time consuming, cost effective process with minimal exposure to the public. Site of bioremediation is minimally disturbed and the degradation ability varies seasonally.

Ex situ Bioremediation

Ex situ bioremediation involves removal of waste materials and their collection at a specific place to facilitate microbial degradation. A specific microorganism in a controlled environment is used to clean or degrade the waste materials. Advantages are, It is a better controlled and more efficient process. Process can be improved and time required is short.

Disadvantages are, very costly and site of pollution is highly disturbed.

Types of Reactions in Bioremediation

Microbial degradation of pollutants primarily involved aerobic, anaerobic and sequential process.

Aerobic Process

It involves the utilization of oxygen from the oxidation of organic compounds. These compounds may servs as substrates for the supply of carbon and energy to the microorganisms. Two types of enzymes namely monoxygenases and dioxygenases are involved in aerobic biodegradation. Monoxygenases act on both aliphatic and aromatic hydrocarbons. Dioxygenases degrade only aliphatic hydrocarbons.

Anaerobic Process

It doesnot require oxygen. Hydrogenation and dehydrogenation of benzoates, phenols, dehalogenization of the polychlorinated biphenyl, chlorinated ethylenes; carboxylation and decarboxylation of toluene, cresol and benzoate are the examples of anaerobic reactions.

Sequential Bioremediation

In the degradation of xenobiotics both aerobic and anaerobic processes involved and are called sequential bioremediation. It is a effective way of reducing toxicity.

Phytoremediation

Heavy metals such as cadmium and lead are not readily absorbed or captured by microorganisms. The assimilation of metals such as mercury into the food chain may create very serious problem. Eg, Minamata disease. Phytoremediation is useful in these circumstances because natural plants or transgenic plants are able to bioaccumulate these toxins in their above-ground parts, which are then harvested for removal. The heavy metals in the harvested biomass may be further concentrated by incineration or even recycled for industrial use.

Bioremediation Mechanisms

The following mechanisms are involved in bioremediation process. They are biosorption, bioaccumulation, reduction, solubilization, precipitation and mehylation.

Criteria for Bioremediation / Biodegradation

For successful biodegradation of pesticide in soil, following aspects must be taken into consideration. i) Organisms must have necessary catabolic activity required for degradation of contaminant at fast rate to bring down the concentration of contaminant, ii) the target contaminant must be bioavailability, iii) soil conditions must be congenial for microbial / plant growth and enzymatic activity and iv) cost of bioremediation must be less than other technologies of removal of contaminants.

Strategies for Bioremediation

For the successful biodegradation / bioremediation of a given contaminant following strategies are needed.

a)   Passive/ intrinsic Bioremediation: It is the natural bioremediation of contaminant by indigenous microorganisms and the rate of degradation is very slow.

b)   Biostimulation:it is a process by which, nutrients are added to the environment to stimulate indigenous microorganisms in soil.

c)   Bioventing: Process/way of Biostimulation by which gases stimulants like oxygen and methane are added or forced into soil to stimulate microbial activity.

d)   Bioaugmentation: It is the introduction of microorganisms in the contaminated soil to facilitate biodegradation. It is also done by introduction of genetically modified organism into the contaminated site.

e)   Composting: Piles of contaminated soils are constructed and treated with aerobic thermophilic microorganisms to degrade contaminants. Periodic physical mixing and moistening of piles are done to promote microbial activity.

f)   Phytoremediation: it can be achieved directly by planting plants which hyperaccumulate heavy metals or indirectly by plants stimulating microorganisms in the rhizosphere.

g)   Bioremediation:Process of detoxification of toxic/unwanted chemicals / contaminants in the soil and other environment by using microorganisms.

h)   Mineralization: Complete conversion of an organic contaminant to its inorganic constituent by a group of microorganisms.

Bioremediation of Heavy Metals

Removal of heavy metal from the environment using biomethods are called as bioremediation of heavy metals. It is carried out by bacteria, algae, fungi, actinomycetes and higher plants. They accumulate high amount of heavy metals in their cells there by maximum quantity of heavy metals are removed. Removal of heavy metals using plants are called phytoremediation.

(i)   The species of algae like Chlorella, Anabaena inaequalis, Westiellopsis prolifica, Stigeocloniumtenue, Synechococcus sp. tolerate heavy metals. However, several species of Chlorella, Anabaena, marine algae have been used for the removal of heavy metals.

(ii) Fungi also are capable of accumulating heavy metals in their cells.Fungi produce several extracellular products which can complex or precipitate heavy metals. For example, many fungi and yeast release high affinity Fe-binding compounds that chelate iron. It is called siderophores. The Fe^{3+} chelates which are formed outside the cell wall are taken up into the cell. In Saccharomyces cerevisiae removal of metals is done by their precipitation as sulphidese.g. Cu^{2+} is precipitated as CuS.

Bioremediation of Xenobiotics

Removal of xenobiotics from the environment are called bioremediation of xenobiotics. Pesticides are the good examples for xenobiotics. Use of pesticides has benefited the modem society by improving the quantity and quality of the worlds' food production. Gradually, pesticide usage has become an integral part of modern agriculture system. Many of the artificially made complex compounds i.e. xenobiotics persist in environment and do not undergo biological transformation. Microorganisms play an important role in degradation of xenobiotics, and maintaining of steady state concentrations of chemicals in the environment. The complete degradation of a pesticide molecule to its inorganic components that can be eventually used in an oxidative cycle removes its potential toxicity from the environment.

T.   BIODEGRADATION

Environmental pollution is caused by toxic industrial byproducts, gases, and exhaust fumes from vehicles, chemicals, oil spillage, animal and human wastes. It is a major concern because of harmful effects like depletion of the ozone layer, green house effect, acid rain, global warming, contamination of water sources, health problems to humans, adverse effects on agriculture, and many more. Industrialization is the main cause of environmental pollution; however it is difficult to curtail it since it plays an important role in the economic development of countries. It is important, then, to take measures to minimize environmental pollution. Biodegradation of this pollution is more important and essential for better survival of human, animal and plant life. Pollutants may be organic or inorganic or toxic material or chemicals.

Decomposition of Cellulose

Cellulose is the most abundant carbohydrate present in plant residues/organic matter in nature. Rate of cellulose degradation is high when the cellulose is associated with pentosans (eg. xylans&mannans), but it is very slow when associated with lignin. Decomposition of cellulose occurs in two stages: (i) in the first stage the long chain of cellulase is broken down into cellobiose and then into glucose by the process of hydrolysis in the presence of enzymes cellulase and cellobiase, and (ii) in second stage glucose is oxidized and converted CO_2 and water.

$$1.\ \text{Cellulose} \xrightarrow[\text{Hydrolysis}]{\text{Cellulase}} \text{Cellobiose} \xrightarrow[\text{hydrolysis}]{\text{Cellobiase}} \text{Glucose}$$

$$2.\ \text{Glucose} \longrightarrow \text{Organic Acids} \longrightarrow CO_2 + H_2O$$

The intermediate products formed/released during enzymatic hydrolysis of cellulose (eg. cellobiose and glucose) are utilized by the cellulose-decomposing organisms or by other organisms as source of energy for biosynthetic processes. The cellulolytic microorganisms responsible for degradation of cellulose through the excretion of enzymes (cellulase&Cellobiase) are fungi, bacteria and actinomycetes.

Decomposition of Hemicelluloses

Hemicelluloses are water-soluble polysaccharides and consists of hexoses, pentoses, and uronic acids. They are the major plant constituents. The hydrolysis is brought about by number of hemicellulolytic enzymes known as "hemicellulases" excreted by the microorganisms. On hydrolysis hemicelluloses are converted into soluble monosaccharide/sugars (eg. xylose, arabinose, galactose and mannose) which are further convened to organic acids, alcohols, CO_2 and H_2O. Uronic acids are broken down to pentoses and CO_2. Various microorganisms including fungi, bacteria and actinomycetes both aerobic and anaerobic are involved in the decomposition of hemicelluloses.

Lignin Decomposition

Lignin is the third most abundant constituent of plant tissues. It accounts for about 10-30% of the dry matter of mature plant materials. Lignin content of young plants is low and gradually increases as the plant grows old. It is one of the most resistant organic substances for the microorganisms to degrade. Certain Basidiomycetous fungi are known to degrade lignin at slow rates. Complete oxidation of lignin result in the formation of aromatic compounds such as syringaldehydes, vanillin and ferulic acid. The final cleavages of these aromatic compounds yield organic acids, carbon dioxide, methane and water.

Protein Decomposition

Proteins are complex organic substances containing nitrogen, sulphur, and sometimes phosphorus in addition to carbon, hydrogen and oxygen. During the course of decomposition of organic matter, proteins are first hydrolyzed to polypeptides. These polypeptides are broken down to amino acids or ammonia and amides. The process of hydrolysis of proteins to amino acids is known as "aminization or ammonification", which is brought about by certain enzymes, collectively known as "proteases" or "proteolytic" enzymes. They are secreted by various microorganisms. Amino acids and amines are further decomposed and converted into ammonia. During the course of ammonification, various organic acids, alcohols, aldehydes are produced, which are further decomposed finally to produce carbon dioxide and water.

All types of microorganisms, bacteria, fungi, and actinomycetes are able to bring about decomposition of proteins. In acid soils, fungi are pre-dominant, while in neutral and alkaline soils bacteria are dominant decomposers of proteins.

Microbial Degradation of Xenobiotics

Xenobiotics are the artificial, foreign and synthetic chemicals. Pesticides, herbicides, solvents and other organic compounds are the examples for xenobiotics. Pesticides are of wide varieties of chemicals e.g. chlorophenoxyalkylcaboxylic acid, substituted ureas, nitrophenols, tri-azines, phenyl carbamates, organochlorines, organophosphates, etc.

Organophosphates like diazion, methyl par-athion and parathion are the most extensively used insecticides. Organomercurials like Semesan, Panodrench and Panogen have been practiced in agriculture since the birth of fungicides.The major fungicides used in agriculture are water soluble derivatives such as Ziram, Ferbam, Thiram, etc.Pentachlorophenol (PCP) is a broad spectrum biocide which has been used as fungicide, insecticide, herbicide, algicide, disinfectant and antifouling agent.

Members of the genus Pseudomonas are the most predominant microorganism that degrade xenobiotics. Mycobacterium, Alcaligens, Nocardia, Arthrobacter, Corynebacterium, Bacillus, Candida, Aspergillus sp., Xanthomonas, streptomyces, Fusarium etc.,

Co-metaboilis is the process associated with biodegradation of xenobiotics. Xenobiotics are not a source of carbon or energy.

Biodegradation of Pesticides

Pesticides and herbicides are used to prevent various plant diseases and used to improve crop yield. Common pesticides used in India are DDT, Aluminium Phosphide, Lindane, Methyl Bromide, Methyl Parathion, Sodium Cyanide, Methoxy Ethyl Merciru Chloride (MEMC) and Monocrotophos. Aliphatic acids, Phenylpyridazones, Quinolines, Sulfonylureas, Oxadiazolides, Amides, Benzoics, Carbamates, Thiocarbamates, Dithiocarbamates, Nitriles, Dintroanilins, Phenols, Phynoxy acids, Traizines, Cyclohexanedione, Imidazolines, N-phenylphthalamides are the examples for herbicides.

Most of the pesticides and herbicides are toxic and recalcitrants. Some of them are surfactants. Most commoly used herbicides and pesticides are aromatic halogenated compounds. Biodegradation of halogenated compounds are similar tononhalogenated aromatic compounds. Rate of biodegradation depends on the number of halogen atoms.

Dehalogenization of halogenate compound is an essential step in the detoxification of herbicides and pesticides. It is catalysed by dioxygenase. Most of the halogenated compounds are converted to Catechol or protocatechuate.

Azatobacter, Azospirillum, Bacillus and *E. coli* like organisms are actively involved in biodegradation along with Pseudomonas. Biodegradation through hydrolysis of p-o-aryl bonds by *Pseudomonas diminuta* and *Flavobacterium* are considered as the most significant organism in the detoxification of organophosphorus compounds. Biodegradation of pesticides are carried out by aerobic soil microbes.

Biodegradation of Hydrocarbon

It is a main pollutant from oil refinaries and oil spills. These are degraded by consortium of microorganisms. Eg. *Pseudomonas, Corynebacterium, Mycobacterium,* Arthrobacter, *Nocardia*.

The uptake of aliphatic hydrocarbon is a slow process due to their low solubility. Both aerobic and anaerobic process are involved in the degradation process.

Degradation of aromatic hydrocarbon is also performed by aerobic and anaerobic process. Pseudomonas maily involve in the degradation process. The following steps are taken place they are,

1. Removal of the side chain.

2. Opening of the benzene ring.

Most of the non halogenated aromatic compounds undergo serious of reactions to produce catechol and protocatechuate.

Biodegradation of toluene, L meandelate, benzene, phenol, anthranilate, naphthalene, phenanthrelene may leads to the formation of catechol.

Biodegradation of quinate, p hydroxyl L mandelate, benzene, phenol, vanillate leads to the formation of protocatechuate.

Catechol and protocatechuate undergo oxidative cleavage by ortho and meta cleavage pathway and converted to simpledegradable organic compounds.

Polychlorinated biphenyls (PCB) biodegradation

PCB have been implicated in cancer, damage to various organs and impaired reproductive function.

PCB is available in pesticides, in transformer, in paints and adhesives. Due to hydrophobic nature and high bioaccumulation potential, PCB is accumulate more in soil. They are resistant to biodegradability.

Aerobic and Anaerobic biodegradation is done by consortium of microorganisms like *Pseudomonas, Alkaligens, Corynebacterium, Acinetobacter.*

2, 4, 6 Trinitro toluene is detoxified by Pseudomonas and Clostridium. Nitrocellulose is degraded by hydrolysis process. Super bug (*Pseudomonas putida*) that carries Cam-OCT plasmid (Camphor and Octane), XLY (Xylene) and NAH (Naphthalene) plasmid are actively involved in biodegradation of xenobiotics.

Biodegradation of Pesticides in Soil

Pesticides reaching to the soil are acted upon by several physical, chemical, and biological forces. Microorganism's plays major role in the degradation of pesticides. Many soil microorganisms have the ability to act upon pesticides and convert them into simpler non-toxic compounds. This process of degradation of pesticides and conversion into non-toxic compounds by microorganisms is known as "biodegradation". Some chemicals that show complete resistance to biodegradation are called "recalcitrant".

The chemical reactions leading to biodegradation of pesticides fall into several broad categories, that are as follows.

a. **Detoxification**   Conversion of the pesticide molecule to a non-toxic compound are called detoxification.

b. **Degradation**   The breaking down/transformation of a complex substrate into simpler products leading finally to mineralization. Degradation is often considered to be synonymous with mineralization, e.g. Thirum (fungicide) is degraded by a strain of Pseudomonas and the degradation products are dimethlamine, proteins, sulpholipaids.

c. **Conjugation (complex formation or addition reaction)**   Conjugation or the formation of addition product is accomplished by those organisms catalyzing the reaction of addition of an amino acid, organic acid or methyl crown to the substrate, for e.g., in the microbial metabolism of sodium dimethlydithiocarbamate, the organism combines the fungicide with an amino acid molecule normally present in the cell and thereby inactivate the pesticides/chemical.

d. **Activation**   It is the conversion of non-toxic substrate into a toxic molecule, for eg. Herbicide, 4-butyric acid (2, 4-D B) and the insecticide Phorate are transformed and activated microbiologically in soil to give metabolites that are toxic to weeds and insects.

e. **Changing the spectrum of toxicity**   Some fungicides/pesticides are designed to control one particular group of organisms / pests, but they are metabolized to yield products inhibitory to entirely dissimilar groups of organisms, for e.g. the fungicide PCNB fungicide is converted in soil to chlorinated benzoic acids that kill plants.

Biodegradation of pesticides / herbicides is greatly influenced by the soil factors like moisture, temperature, PH and organic matter content, in addition to microbial population and pesticide solubility. Optimum temperature, moisture and organic matter in soil provide congenial environment for the break down or retention of any pesticide added in the soil. Most of the organic pesticides degrade within a short period (3-6 months) under tropical conditions. Metabolic activities of bacteria, fungi and actinomycetes have the significant role in the degradation of pesticides.

U.   BIODETERIORATION

Biodeterioration is the process of chemical or physical alteration of a manmade product of economic significance by microorganisms. It decreases the usefulness of that product. Various microorganisms are responsible for biodeterioration of valuable products e.g., paper, textiles, leather, paints, rubber, metal-pipes, wood, etc. It is also called as microbial deterioration.

Biodeterioration of paper

Undesirable changes observed in paper are called paper deterioration. If the deterioration occurs due to microorganisms are called paper biodeterioration. It is caused by vital activities of microorganisms. Indian pulp and paper industries produce 3 million tons of paper every year. Pulp-wood is used to manufacture paper. Paper Pulp is the raw material for paper manufacturing. Paper pulp is obtained from paper wood. They are treated physically or chemically for the purpose of separating and purifying cellulose fibrous in the form of fibrous pulp. This pulp is generally called "paper-pulp". Finished paper is the paper-sheet which is prepared by the refinement and fabrication of paper-pulp. It is also attacked by microorganisms.

Paper wood, paper pulp and finished paper are attached by different types of microorganisms. Those paper-pulps which are prepared by chemical treatments generally possess less nutrients for microorganisms and hence are less susceptible to microbial attack than the physically (mechanically) prepared paper-pulps. Microbial degradation of the paper-pulp may be encountered in the form of "paper-pulp slime". Paper-pulp slime is produced by the deposition of microorganisms and the subsequent enlargement of fibre, fines, and other debris from the water and compounds of the paper-making medium.

Basidiomycetous fungi are responsible for "white roots" and "brown rots" of pulp-wood. White rot is due to degradation of brownish lignin and leaving a white spongy cellulosic mases in the wood. Degradation of cellulosic materials and l leaving of lignin leads to the formation of brown rot. When the moist pulp-wood is stored, its surface is attacked and degraded by some ascomycetous and deuteromycetous fungi. This degradation is characteristically called "soft rots".

Bacteria, yeasts, moulds, algae, and protozoa have been isolated from pulp slimes. Bacteria, particularly capsulated bacilli such as *Enterobacter aerogenes* and *Bacillus* sp. represent the most important group of pulp slime producers. *Sphaerotilus natans*, the filamentous iron bacteria, can be found as part of the slime mass on those paper machines operating above pH 5.5. The bacterium *Alcaligenes viscosus* var. dissimilis has been obtained from pink pulp slime. Species of *Mucor, Penicillium, Trichoderma, Fusarium*, and yeasts (*Torula, Rhodotorula*) are the fungi that have been isolated from pulp slimes in various paper-making industries.

Finished paper, i.e., the paper-sheet which is prepared by the refinement and fabrication of paper-pulp is also attacked by microorganisms. Various fungi (*Penicillium sp., Aspergillus sp., Chaetomium*, etc.) and bacteria are the main attackers as cellulose. *Alternaria* colonize on paper and penetrate fibres and causes degradation. *Aspergillus* causes damage to paper texture. *Rhodotorulla* causes deterioration of waxy papers. Fungal mycelium readily penetrates and spread in the paper due to its mechanical destruction of fibres and cellulose. Bacteria are colonize only on the surface.

Biochemical mechanism of paper deterioration

Microbes produce cellulolytic enzyme which is responsible for biodeterioration of paper.

Cellulose → Oligosaccharides → Cellobiose + Celloterose → Glucose → Gluconic acid

Factors affect biodeterioration

1. Enviornmental (climatic Factors) factors like light, heat, humidity and moisture, dust and dirt, water. 2. Biological factors :- Microorganisms, insects and rodents. 3. Chemical factors 4. Human factors and 5. Disasters

Indicators of paper Deterioration

The different types of deterioration of the paper based materials are reflected in wear and tear, shrinkage, cracks, brittleness, warping, bioinfestation, discoloration, abrasion, hole, dust and dirt accumulation etc.

Control of Paper deterioration

The fumigating agents are used to control the growth of contaminated fungi deterioration of paper. Fumigating agents are ethyleme bromide, formaldehyde, phenol, sulphur di oxide

Prevention of biodeterioration of paper

1. Change the paper storage condition. 2. Moisture content of all materials should be close to 100% relative air humidity. 3. Fiber composition of paper should be high. 4. Humidity and temperature are maintained at low level. 5. Follow good housekeeping.

Biodeterioration of Leather

Leather can be defined as a durable and flexible material created by the tanning of animal rawhide and skin. It is an optimum medium for the growth of microorganisms because it contains proteins, lipids and other essential nutrients. Biodeterioration of leather can be seen as any undesirable change in the properties of leather caused by the vital activities of microorganisms.

Leather Production Processes

The leather manufacturing process is divided into three sub-processes. They are preparatory stages, tanning, crusting and surface coating

Leather Products

Leather shoes, leather garments and blazers, leather wallet, leather belts, leather handbags, phone cases, leather laptop bags and travel cases.

Microbes involved and Mechanism of biodeterioration

Implicated organisms in the biodeterioration of leather are *Aspergillus, Penicillium* and *Paecilomyces* sp. The formation of coloured imbleachable spots, perforation and loss in tensile strength are some of the spoilages resulting from the activities of these microorganisms. The spoilages result from the consumption of leather components such as proteins, collagen, minerals and lipids by the secretion of proteases, collagenase and lipases. Though microbial attack is very difficult to prevent because leather is an excellent organic substrate, clean storage conditions is one way of preventing it. Hides and leather can be damaged by bacteria, which are mainly responsible for the decomposition of untanned proteins (in raw hides and during soaking), and fungi, which thrive on tanned leathers containing carbohydrates, fats and proteins.

Prevention

A fungicide 2-(thiocyanomethylthio) benzothiazole is used as a biocidal agent.

Control measures of leather deterioration

Addition of fungicides and biocides. Antifungal substances like cresal and nitrophenol are used. Addition of sodium pentachlorophenol at the concentration of 0.1 to 0.2%. Addition of formaldehyde in the tanning process. Addition of bactericidal components in the soaking water.

Biodeterioration of metal

Microbial corrosion is also called bacterial corrosion, bio-corrosion, microbiologically influenced corrosion, or microbially induced corrosion (MIC). It is a corrosion caused or promoted by microorganisms, usually chemoautotrophs. It can apply to both metals and non-metallic materials. Microbial corrosion is a form of biodeterioration and is frequently referred to as biocorrosion. This degradative process primarily acts on metals, metalloids, minerals, and other rock-based materials. Bacteria, fungi, microalgae and naturally occurring organic/inorganic chemicals contribute to biocorrosion. Corrosion is the deterioration of a metal as a result of chemical reactions between it and the surrounding environment.

Microbes in Metal corrosion

Some sulfate-reducing bacteria produce hydrogen sulfide, which can cause sulfide stress cracking. *Acidothiobacillus thiooxidans* frequently damages sewer pipes. *Ferrobacillus ferrooxidans* directly oxidizes iron to iron oxides and iron hydroxides. In presence of oxygen, aerobic bacteria like *Acidithiobacillus thiooxidans, Thiobacillus thioparus* and *Thiobacillus concretivorus* are the common corrosion-causing factors resulting in biogenic sulfide corrosion. Anaerobic Bacteria, especially *Desulfovibrio* and *Desulfotomaculum* are common.

Mechanisms of action

In general, the process involves acidic degradation, electron movement, metal depolarization, polymerization and attachment of bifilm and mineral formation.

Biofilms

Biofilms are a primary tool used by bacteria to adhere to a metal surface and facilitate metabolism. Biofilms are an essential part of the degradation process. Bacteria create biofilms by producing extracellular polymeric substances such as lipids, polysaccharides, nucleic acids, and proteins. Fungal species can also reside within biofilms and cause enzymes and acids. This leads to metal corrosison.

Corrosion Prevention

Coatings: Paints and other organic coatings are used to protect metals from the degradative effect of environmental gases. Common organic coatings are Alykd and epoxy ester coatings.

Plating: Metallic coatings, or plating, can be applied to inhibit corrosion as well as provide aesthetic, decorative finishes.

Biodeterioration of Textiles

Materials produced from fibers and threads are classified as textile. These are fabrics, nonwoven materials, fur fabric, carpets and rugs, *etc.*. Textile fibers are the main raw material of the textile industry. Textile fibers are divided into natural and chemical ones. Natural fibers can be of plant (cotton, bast fibers), animal (wool, silk), or mineral (asbestos) origin. Chemical fibers are produced from modified natural or synthetic high-molecular substances.

Microbes in Textile biodeterioration

Textile materials are damageable by microorganisms, insects, rodents and other biodamaging agents. Annual losses due to microbiological damaging of fabrics reach hundreds of millions of dollars.

Fungi: *Aspergillus* sp. *Penicillium* sp., *Cladosporium* sp.; *Cheatomimum* sp.; *Alternaria* sp.; *Trichoderma* sp.; *Fusarium nivale; Myrothecium* sp.; *Memnoniella* sp.; *Stachybotrys* sp.; *Verticillum* sp.

Bacteria: *Cytophaga* sp.; *Cellulomonas* sp.; *Bacillus* sp.; *Clostridium* sp.; *Sporocytophaga* sp.; *Microbispora bispora; Pseudomonas* sp.; *Streptomyces* sp.

Mechanism

Fiber and fabric biodamaging by microorganisms is usually accompanied by the mass and mechanical strength loss of the material as a result of, for example, fiber degradation by microorganism metabolites: enzymes, organic acids, *etc.* The impact of microorganisms on textile materials that causes their degradation is performed in at least two main ways (direct and indirect). They are fungi and bacteria use textile materials as the nutrient source (assimilation) and textile materials are damaged by microorganism metabolism (degradation). Biodamages of textile materials induced by microorganisms and their metabolites are manifested in colouring, defects, bond breaks in fibrous materials, penetration deep inside, deterioration of mechanical properties, mass loss, change of chemical properties (cellulose degradation by microorganisms), liberation of volatile substances, and changes of other properties.

Microbial degradation of fabrics depends primarily on their chemical composition. The main component of plant fibres is **cellulose**. It is broken down by microorganisms through a process of enzymatic hydrolysis. Plant fibres also contain small quantities (up to 10%) hemicelluloses and lignin, which give the fibres rigidity, and pectins, which act as a kind of glue. Many microorganisms are capable of producing enzymes which decompose hemicelluloses and pectins. Lignin is the least rapidly decomposed component of plants.

Conditions favourable for biodegradation of fibres and fabrics

The rate of microbiological decomposition of fabrics is affected by environmental factors such as air relative humidity, temperature, light, and the properties of the fabrics, chiefly their chemical composition, fibre structure, density and thickness of weave, and the type of substances used in the finishing of the unwoven fabric.

Protection of fibres against microbial degradation

Control of environmental conditions during storage, transportation and use is an effective method for protecting fibres against biodeterioration. The temperature in storage rooms should be maintained at 18–20°C, and the air relative humidity should not exceed 60%.

Biocides are used to compat microorganisms. These are added at various stages of fibre production. Biocides make it possible to eliminate microorganisms effectively. Modern biocides are high effectiveness at low concentrations (of the order of ppm) against a wide spectrum of microorganisms

Biocides used in the textile industry

Inorganic compounds - metals such as silver, zinc, copper, metal oxides such as titanium dioxide, metal salts. They inhibits gram positive, gram negative bacteria, fungi, viruses.

Halogens and their compounds inorganic: chlorine, iodine, sodium and calcium hypochloride, sodium chlorate, chlorine dioxide; organic: chloroarylamides, halohydantoin, chloroisocyanuric acid inhibits gram positive, gram negative bacteria by denaturing of proteins, damage and dysfunction of cytoplasmic membrane and cell wall.

Biodeterioration of Wood

Wood is the most versatile and beautiful building material available to man. It has been with us since man first started to build his own shelter. Wood is a natural plant material. It is made up of a host of organic compounds. It is made from cellulose. Cellulose is a rich source of carbohydrate for both fungi and insects. They have evolved elaborate mechanisms for the digestion of cellulose. The cells also contain other compounds, such as starch, which is also a source of carbohydrate for these attacking agents

Microbes and Mechanism

It has been estimated that almost about 10% of all the paper wood cut is deteriorated by the microorganisms, particularly fungi. Basidiomycetous fungi are responsible for "white roots" and "brown rots" of pulp-wood. White rot is due to degradation of brownish lignin and leaving a white spongy cellulosic mases in the wood. Degradation of cellulosic materials and l leaving of lignin leads to the formation of brown rot. When the moist pulp-wood is stored, its surface is attacked and degraded by some ascomycetous and deuteromycetous fungi. This degradation is characteristically called "soft rots".

Bacteria, yeasts, moulds, algae, and protozoa have been isolated from wood. Major component of wood is cellulose. Mechanism of cellulose degradation involves a multistage decomposition of cellulose to glucose, brought about successively by the enzymes 1,4-endo- β-D-glucan cellobiohydrolase, endo-1-4-β-D-glucan glucanohydrolase and glucohydrolase of β-D-glucosides.

Preventive Measures

Three conditions must be available for the deterioration. They are oxygen, water and food. Keeping the wood totally immersed in water will cut off the oxygen supply, making it impossible for the attacking agents to survive and thus preserving the wood.

Timbers that are kept dry at all times will go a long way to prevent the attacks of many organisms, such as fungus.

Chemically treating the timber with a preservative containing a fungicide.

12

VIROLOGY

A. GENERAL PROPERTIES OF VIRUSES

Viruses are submicroscopic, unique group of infectious agents and are obligate intracellular parasites. They contain single type of nucleic acid. Viruses exist in two phases. They are intracellular and extracellular viruses. Size of the viruses ranges from 20 -14,000 nm in length. Complete virus particle is called as **virion.**

Definition

Luria and Darnell in 1968 defined viruses as entities whose genome is an element of nucleic acid either DNA or RNA and reproduces inside of living cells.

Novel properties of viruses

- Viruses are ultramicroscopic in size ranging from 10 nm to 450 nm. Smallest viruses are little larger than ribosomes. Largest viruses are like smallest bacteria.

- Viruses possess very compact structure.

- Viruses are acellular in nature. Because they do not independently fulfill the characteristic of life. They require host cell for their replication and protein synthesis.

- Viruses are crystalizable in nature.

- Basic structure of virus consists of protein capsid and nucleic acid.

- Capsid is made up of repeating unit called **capsomer**. It protects nucleic acid. Nucleic acid can be either DNA or RNA but not both.

- Nucleic acid types include single stranded RNA, double standard RNA, double standard DNA and single stranded DNA.

- Molecules on cell surface impart high specificity for host cell. Spike proteins of enveloped virus and capsid protein of non-enveloped virus are responsible for antigenicity and pathogenicity of viruses.

- Viruses lack machinery for synthesizing proteins. They are obligate intracellular parasites of bacteria, protozoa, fungi, algae, plants and animals.

- All the viruses are infectious in nature.

⮜   Viral genome is replicated within an appropriate host cell and directs the synthesis of viral components by host cellular system.

⮜   Progeny viruses are formed by assembly. Assembled viruses are disassembled during next progeny.

⮜   Viruses do not have any cell organelles like chloroplast, ribosomes, mitochondria etc.

⮜   A virus contains single genome only (either DNA or RNA).

Viruses differ from living cells by

1.   Their simple, acellular organization.
2.   Absence of both nucleic acid (RNA and DNA) in same virion.
3.   Their inability to reproduces independently..
4.   There is no cell division as like prokaryotes and eukaryotes.

B.   HISTORY OF VIROLOGY

The term virus is derived from Latin word. It means **"poison"**. The real history of scientific evaluation starts from 18th century. The prehistoric period starts from about 4000 BC itself. Ancient people were aware of virus infection and they also carried research into causes and prevention of diseases. A hieroglyph from Memphis. Memphis is the capital of ancient Egypt drawn in about 3700 BC depicts a temple priests, which shows clinical sign of paralytic poliomyelitis. Smallpox is endemic in China during 1000BC. Pharoh Ramresv was succumbed to smallpox in 1196 BC. He was mummified in Cairo museum. The pustule lesion is similar to recent patients.

Historical events of Virology

1976   Edward Jenner He is an English physician collects pustule fluid from Sarah Nelms a milkmaid. He suffered from cowpox and injected fluid into a teen age boy James Phipps. Boy develops fever and head ache and recovered fully. Then Jenner inoculated Phippes with live smallpox pustule. He did not developed small pox. This process is called vaccination. Thus he made the importance of vaccine.

1798	**Edward Jenner**	Small pox vaccine was developed
1885	**Louis Pasteur**	Developed rabies vaccine, described the term virus (Poison) and vaccination (to honor Jenner)
1886	**John Buist**	He described elementary bodies of small pox from skin lessions.
1886	**Adolf Mayer**	He demonstrated that gap of infected plant transmit disease to healthy plants.
1892	Dmitri Iwanowski **(1864-1920)**	He was a Russian botanist. He provided evidence for

virus causation of Mosaic Disease of Tobacco(TMV). He presented a paper to the St. Petersburg Academy of Science which showed that extracts from diseased tobacco plants could transmit disease to other plants. This was generally recognised as the beginning of Virology. Unfortunately, **Iwanowski** did not fully realized the significance of these results.

1898	**Martinus Beijerinck**	He proved that a virus particle causes the Tobacco mosaic disease and described Contagium Virus Fluidum or living infectious fluid. He extended Iwanowski's results on Tobacco mosaic virus and developed modern idea of the virus.
	Loeffler and Frosch	He discovered Foot and Mouth Disease virus (FMD). In 1898, **Fredrich Loeffler and Paul Frosch** showed that a similar agent was responsible for foot-and-mouth disease in cattle. These agents caused disease in animals as well as plants.
1900	**Walter Reed**	He demonstrated that yellow fever is caused by a virus. It was spreaded by mosquitoes.
1903	**Remlinger**	He discovered Rabies virus
1903	**A.Negri**	He observed inclusion bodies from rabies infected human brain cells called Negri bodies
1908	**Karl Landsteiner and Erwin Popper**	He proved that poliomyelitis is caused by Virus
1911	**Franas Peyton Rous**	He discovered causative agent for chicken cancer and Named Rous Sarcoma Virus. Awarded Nobel prize 1966
1915	**Frederic Twort**	He discovered viruses infecting bacteria.**Frederick Twort** (in 1915) and **Felix d'Herelle** (in 1917) were the first to recognize viruses which infect bacteria.
1917	**Felix D.Herelle**	He discovered the viruses of bacteria- and coined the term Bacteriophages (eaters of bacteria)
1932		Used mice as the host for virus
1933	**R.E.Shope**	He identified viruses causing cancer.
1935	**Wendel stanley**	He crystallized TMV and showed its infectious nature. Awarded Nobel prize in 1946
1938	**Max Theiler**	He developed live attenuated vaccine against yellow

fever. He was awarded Nobel prize in 1951. This discovery eventually enabled **Max Theiler** (1937) to propagate the virus in chick embryos and successfully produced an attenuated vaccine.

1939	Emory Ellis and Max Delbruck	They established concept of one-step virus growth cycle. They were awarded Nobel prize in 1969
1940	Helmut Ruska	He used electron microscope to take first viral pictures.
1941	George Hirst	He demonstrated that influenza virus agglutinates RBC.
1942	JJ .Bittner	He discovered mammalian RNA tumour virus (mouse mammary tumour virus).
1945	S.Luria and A.Hershy	He demonstrated bacteriophage mutation.
1947		Zika virus isolated from monkey
1949	John ender, T.Weller, F.Robbins	They demonstrated the growth of Polio virus using human tissue culture. They were the Nobel prize winners of the year 1954.
1950	A. Lwoff, L. Siminovuch and N.Kjeldgaard	Discovered lysogenic bacteriophage in *Bacillus megaterium.* Coined the term prophage. They were awarded the Nobel prize in 1965
1951	Theiler	He developed yellow fever vaccine
1952	Renato Dulbecco	He demonstrated that animal virus also form plaques in a similar way of phages. He was awarded Nobel prize in 1975.
1953	NP.Rowe	He discovered Adenovirus.
1955	FL Schaffer	Polio virus was crystallized
1957	Fraenkel-Contrat and Williams	They demonstrated that when mixtures of TMV protein and RNA were incubated together viral particles formed spontaneously. Discovered RNA as a genetic material of TMV,
1963	Baruch Blumberg	He discovered HBV. He was awarded Nobel prize in 1976.
1966	Edgar and Wood	He discovered pathways of Macromolecular assembly.
1967	Theoder Diener	He discovered viroids
1970	Howard Temin and David Baltimore	He discovered reverse transcriptase in Retro virus

1972	Paul berg	Created first rDNA in SV40 virus contain Lambda phage and galactose operon.
1976		Ebola virus discovered
1977	Frederick Sanger	Sequenced $\theta \times 174$ genome and demonstated it had 5375 nucleotides
1979		Small pox was officially declared as eradicated disease. Declared by WHO.
1981	Yorio Hinuma	He isolated T cell Leukaemia virus (HTLV)
1981	DG Kleid	FMD vaccine prepared by rDNA technology.
1982	Panicali and Paoletti	Vaccinia virus used as vaccine.
1983	Luc Montagnier and R.Gallow	Discovered HIV, causative agent for AIDS.
1986	Wang	HDV identified
		First HBV vaccine produced by genetic engineering method and approved for human use.
1989		HCV was identified.
1990		First human gene therapy was approved.
1994	Yuan Chang and P.Moore	Human Herpes Virus 8 was identified, responsible for Koposis sarcoma.
1995		Chicken pox vaccine was approver for US use
1997	S.Prusiner	Discovered Prions.
1999		Nucleotide sequence of large known virus genome *Paramecium bursaria chlorella* virus was completed
2002		Corona virus was identified as a causative agent of SARS.
2014		MERS-Middle East Respiratory Syndrome - a new viral disease is recognized

c.   **Classification of Virus-**Refer page 30- 33

d.   **Ultra structure of viruses-**Refer page 60- 64

e.   **Replication of viruses**

Reproduction or replication of viruses is also called as **Host-Virus interactions**. Viruses are **obligate intracellular parasite.** It reproduce only within a host cell. Viruses have the specific efficiency towards the cells. Certain groups of viruses may infect only specific groups of cells. The viruses make use of the metabolic machinery of the host cell to undertake replication.

Eg: Bacterial viruses only infect bacteria and Animal viruses only infect animal cells.

Depends upon tissue tropism, replication or reproduction of viruses may be divided into several stages. They are

➤   Adsorption of viruses,

➤   Penetration and uncoating,

➤   Replication of nucleic acid,

➤   Synthesis and assembly of viral capsids and

➤   Release of mature virions.

Adsorption of virion

It is a first step in multiplication process. It occurs through random collision on plasma membrane of host cell. Adherence is an important virulent mechanism of virus. Capsid protein in non-enveloped viruses and spike protein in enveloped viruses play a crucial role in attachment process. Eg: Haemagglutination spike of Influenza virus and Capsid protein in Parvovirus. Spike of non-enveloped virus also involved in attachment process. Eg. Adenovirus spike. Infection of the virus depends greatly on its ability to bind the cell. Viruses have special structure (some times called Ligands) binds to the receptors on the host cell surface. Receptor is a glycoprotein plays an important role in the tissue tropism and host specificity. Attachment is in most cases a reversible process - if penetration does not ensue, the virus can elute from the cell surface.

Examples

➤   Influenza A virus binds sialic acid receptor of respiratory track.

➤   Rabies virus adsorbs an Acetyl choline receptor of neurons.

➤   Vaccinia virus attach on epidermal growth factor receptor

➤   Rhino virus binds to intracellular adhesion molecules on the respiratory epithelial cells.

➤   HIV virus binds CD4 receptor

➤   HAV binds $\alpha 2$ macroglobin.

Penetration and uncoating

➤   Viruses penetrate the plasma membrane and enter a host cell shortly after adsorption. The mechanism of penetration and uncoating must vary with different viruses.

➤   Penetration of viruses involves 3 process. They are

Direct penetration, Fusion and Endocytosis.

1. Direct penetration (Translocation):   Naked virus (Eg. Poliovirus) undergoes conformational changes after adherence on plasma membrane and released only nucleic acid in to the cytoplasm.

2. Fusion: Enveloped virus fuses with plasma membrane with the help of fusion protein or F protein. It results in entry of capsid protein into the cytoplasm.

3. Endocytosis   Most of the enveloped viruses enter inside of host cells through receptor-mediated endocytosis and form coated vesicles. Immediately after entry uncoating taken place by making use of lysozyme or acidity formation within an endocytic vesicle.

Replication of nucleic acid

DNA viruses

Most of the DNA virus genome is replicated within a host cell nucleus with few exceptions. Eg. Small pox virus.

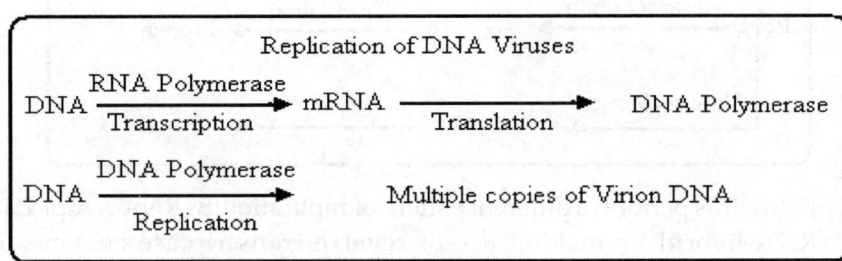

Upon entry, genome enters into a nucleus and perform transcription by using a enzyme RNA polymerase. In some viruses, two stages of transcription occur. They are early transcription and late transcription. During transcription mRNA will be formed, which is transferred to cytoplasm for translation process.

This process leads to the synthesis of protein responsible for DNA replication. DNA is replicated by using an enzyme DNA polymerase. This leads to the replication o genome.

Replication and transcription in RNA viruses

RNA viruses adopt four strategies for replication. The process of replication depends upon the nature of RNA.

1. Some RNA viruses use their RNA genome as a giant mRNA Eg. Poliovirus. These types of viruses are considered as + sense ss RNA viruses.

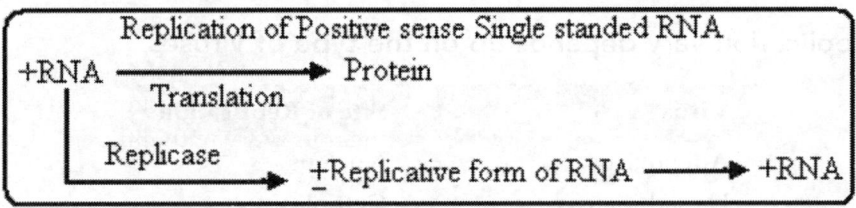

2. dsRNA viruses like Reoviruses carry a virus associated transcriptase and generates mRNA. RNA polymerase enzyme produces new dsRNA from +mRNA.

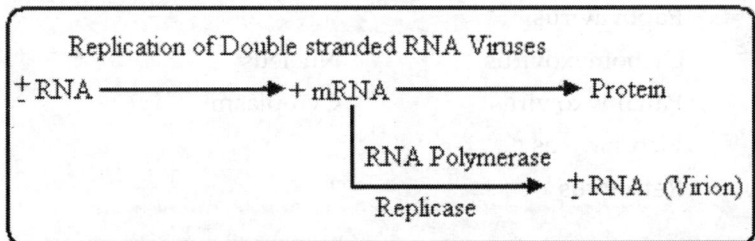

3. In case of negative sense ss RNA viruses, viral replicase converts the ss RNA into a double stranded RNA called the **replicative form**. This directs the synthesis of specific RNA genome.

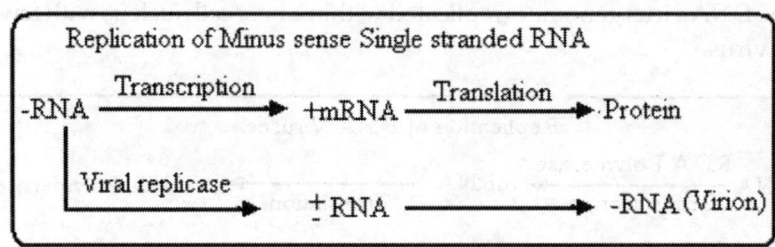

4. Retrovirus perform a different pattern of replication. Ss RNA is copied into DNA RNA hybrid by making use of reverse transcriptase enzyme. Then the Ribonuclease H degrades +RNA strand to leave – DNA. After the synthesis of – DNA, the reverse transcriptase copies this strand to produce a double stranded DNA. It is called proviral DNA, which can direct the synthesis of mRNA and multiple copies of new + RNA virion genome.

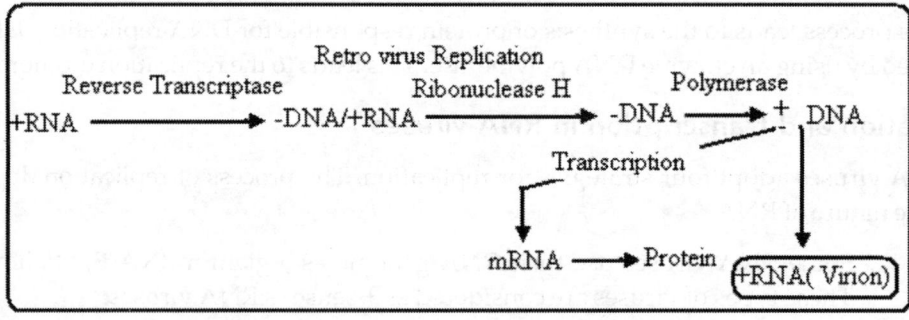

Site of replication vary depends up on the type of viruses

Virus	Site of Replication
Adenovirus	Nucleus
Hepadnovirus	Cytoplasm
Poxvirus	"
Parvovirus	"
Papovavirus	"
Orthomyxovirus	Nucleus
Panamyxo virus	Cytoplasm
Picornavirus	"
Retrovirus	"

Synthesis and assembly of capsid

Late genes of virus is responsible for the synthesis of structural proteins (capsid and spike protein). Viruses are assembled after the complete synthesis of structxural and functional protein. Assembly takes place in anyone of the adjacent places of plasma membrane, nuclear membrane, endoplasmic reticulum or golgi apparatus

Virus	Assembly Site
Adenovirus	Nuclear
Hepadnavirus	Cytoplasm
Poxvirus	Cytoplasm
Orthomyxovirus	Nucleus
Rhabdovirus	Cytoplasm

Virion Release

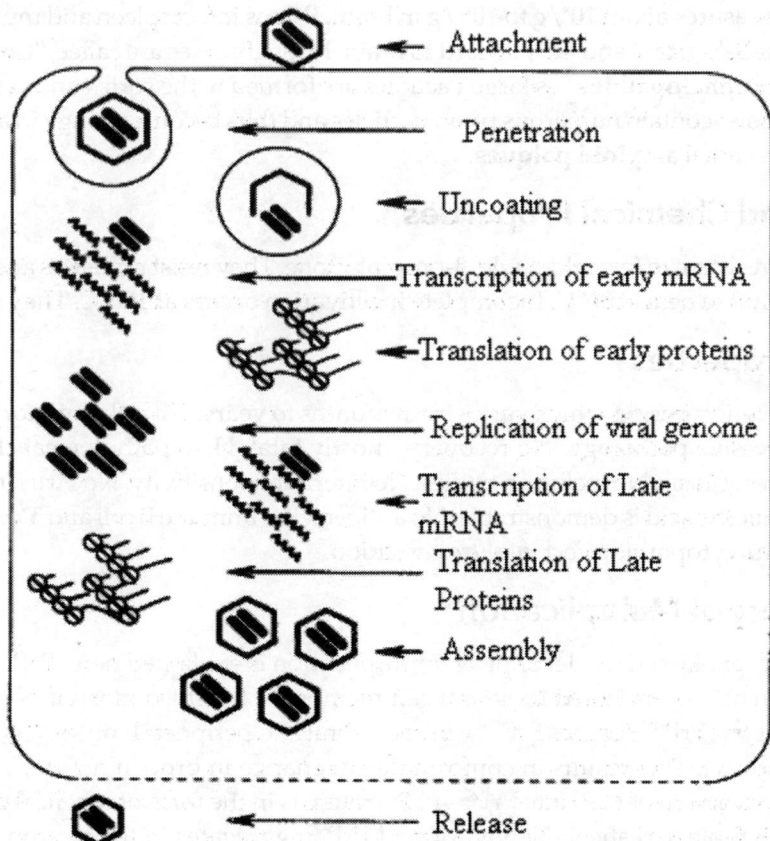

Figure 182 Virus multiplication

It differs between naked and enveloped viruses. There are three mechanisms available to release viruses from host cells. These are cell lysis or cytopathic effect, budding and cell degeneration. Most of the naked viruses are released by lysing the host cells and results in cytopathic effect. Enveloped viruses may receive envelope from golgi apparatus, endoplasmic reticulum or plasma membrane. If the virus receives envelop from plasma membrane. It releases by budding. Others may be released by cell lysis. Viruses such as Parvoviruses accumulate within the host cells and are released only after the death of the host cell, which follows the degeneration of host cell. Actin filament of host cell can aid in virion release.

F. PRIONS

Prions are proteinaceous infectious viral particle. They are unconventional viruses or slow viral agents. Prions were discovered by **Stanley B.Prusnier** in 1997.

Characteristic Features

Prion proteins are 30Kda in size. They are hydrophobic glycoproteins. Prion protein is known as **PrP^{sc}** .it is a modified form of cellular protein known as **PrP^c**. **PrP^c** is encoded by PrP gene. They are filterable (25 to 100nm average pore diameter). They contains 250 aminoacids. Protein titers measures about 10^8/g to 10^{12}/g in brain. Prions infect spleen and multiply in the reticuloendothelial system and later speard to brain. Prion diseases are called **"transmissible spongiform encephalopathies"** as large vacuoles are formed in the cortex and cerebellum of brain. The vacuoles contain numerous prion particles and thus become spongy in appearance. These areas are called **amyloid palques**.

Physical and Chemical Properties

Prions are resistant to formaldehyde, β-propiolactone. They resist proteases and nucleases. They are resistant to heat at 80°C. Incomplete inactivation occurs at 100°C. They resist UV.

Biologic Properties

Long incubation period which varies from months to years. No inflammatory response. Chronic progressive pathology. No recovery; mostly fatal. Histopathological changes like amyloid plaques, gliosis. No inclusion bodies. No interferon sensitivity. No virus interference. No infections nucleic acid is demonstrable. No antigenicity. Immune B cell and T cell functions are intact and no cytopathic effect *in- vitro* condition.

Prion Theory of Multiplication

Universally explained model of prion multiplication is explained here. PrP^C is a normal cellular protein and is anchored to neural cell membrane by gluco inositol phospholipids moiety. After entry, PrP^{SC} Reaches PrP^C by intracerebral and peripheral routes. It is aggregated and cleaves inositol. This results in conformational change in protein nature. This process also leads to conversion of PrP^C into PrP^{SC}. PrP^C protein is in the form of α helical coil. During conversion it is β-pleated sheet. Newly formed PrP^{SC} aggregates in the neuron and causes spongiform appearance of the nerve fibrils.

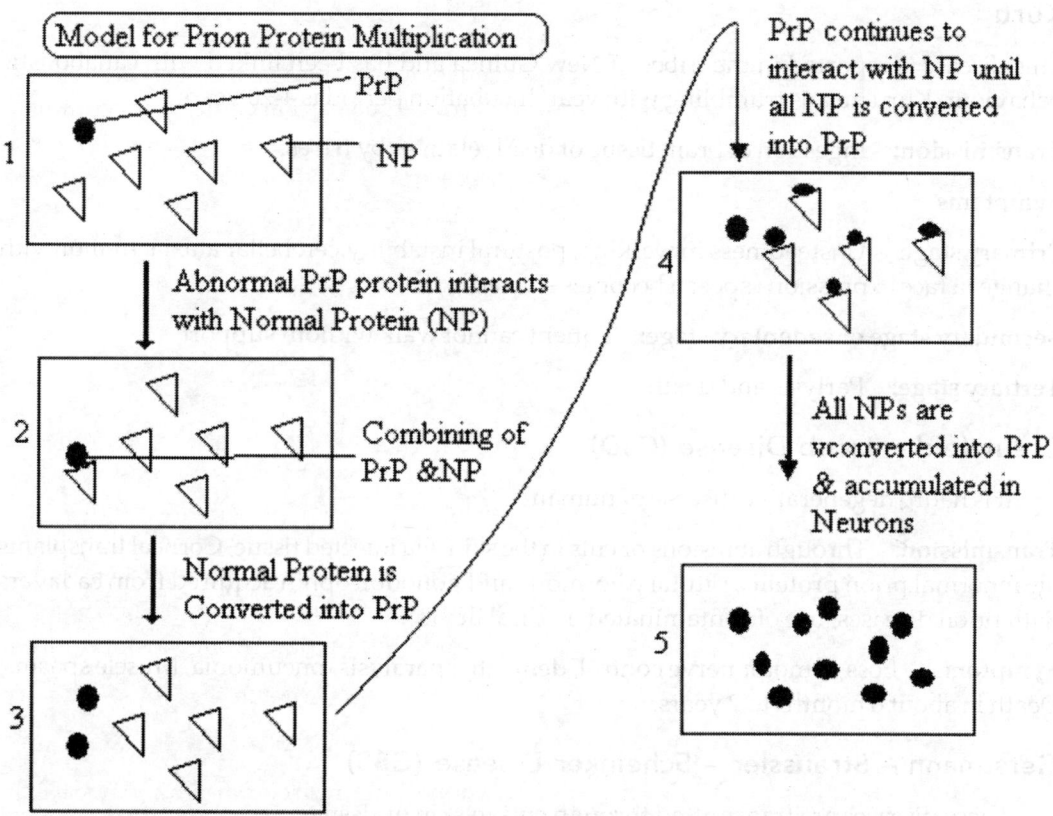

Figure 183 Prion multiplication

Prion Diseases

Scrapie

Scrapie is characterized by infected animals scrape themselves in the rock or tree. Host for scrapie are sheep and goats. Transmission is by Ingestion of contaminated food with infectious agent. Direct transmission from sheep to lamb. Symptoms are ataxia, tremor, tendency to rub constantly, paralysis and death.

Bovine Spongiform Encephalopathy (BSE)

The disease is also called **Mad Cow Disease**

Host: Cows

Transmission: Direct transmission from cow to calf. Human can be infected through consumption of beef of infected cow.

Symptoms: Depression, Unusual behaviour, Softening of brain tissue, Death in one years.

Kuru

This disease is reported in the tribes of New Guinea and has been linked with canabolistic behaviour. Kuru means trumbling with year. Incubation period is 4-20 years

Transmission: Ingestion of brain tissue of dead relatives by tribes.

Symptoms

Primary stage: Unsteadiness in walking, postural instability,cerebellar ataxia, tremor with change in face expression , speech becomes slurred.

Secondary stage or sedentary stage: Patient cannot walk without support

Tertiary stage: Parlysis and death

Creutzfeld – Jacob Disease (CJD)

It is neuro degenerative disease of human

Transmission: Through abrasions or cuts in the skin, via infected tissue, Corneal transplants by abnormal prion proteins, pituitary hormone and gonodotrophin acquired from cadavers with prion diseases, use of contaminated medical device.

Symptoms: Loss of motor nerve control, dementia, paralysis, pneumonia, Muscle spasms, Death in about 6 months to 2 years.

Gerstmann – Straussler – Scheinker Disease (GSS)

These diseases are transmitted through cuts in skin or tissues.

Transmission: Use of contaminated medical device, Ingestion of contaminated food.

Symptoms: Non inflammatory lesions with vacuoles, motor nerve problems.

Fatal Familial Insomnia (FFI)

It is a rare disease of prion, It is characterized by inflammatory lesions, amyloid plaque formation. Atrophy of thalamus occurs.

Lab Diagnosis: The only diagnostic method available is pathologic examination of brain revealing histological changes.No direct virus detection by electron microscope. No circulating antibodies. No serological testing or probe detection

G. VIROIDS

Characteristics of viroids

Viroids are novel subcellular entities made up of circular, ss RNA molecules. They lack protein component. They are abbreviated as Vd. Viroids were extensively studied by **Diener and Raymer. Diener** proposed the name "Viroids". More than 16 plant viral diseases are reported to be caused by Viroids. Viroids are obligate parasites of cell's transcriptional machinery. Viroids do not code for any protein. They replicate autonomously in susceptible cells. Viroids constitute about 246 to 375 nucleotides. They do not require any helper virus. They survive for 10 minutes at 90°C.

Taxonomy of Viroids

Family I	: Popsiviroidae
Genus I	: Popsiviroid -Potato Spindle Tuber Viroid
Genus II	: Hostuviroid -Hop Stunt Viroids
Genus III	: Cocadviroid -Coconut Cadang – Cadang Viroids
Genus IV	: Apscaviroid - Apple Scar Viroid.
Family II	: Avsunviroidae
Genus I	: Avsunviroid - Avocado Sunblotch Viroid
Genus II	: Pelamoviroid - Peach Latent Mosaic Viroid

The classification of Viroids is based on two factors

1. Presence of Conserved Central Region (CCR):

Members of Popsiviroidae contain CCR but Avsunviroid lack CCR.

2. Site of replication:

Popsiviroidae members replicate in nucleolus.

Avsunviroidae members replicate in chloroplast.

Viroids and their nucleotide numbers

Viroids	Abbreviation	No. of nucleotides
Potato Spindle Tuber Viroids	PSTVd	359
Chrysanthemum Stunt Viroid	CSVd	356
Hop Stunt Viroid	HSVd	297
Coconut Cadang –Cadaig Viroid	CCCVd	247
Hop Latent Viroid	HLVd	256

Structure of Viroids

The viroid has five domains. They are left terminal domain, pathogenic domain, Conserved Central domain, variable domain and right terminal domain.

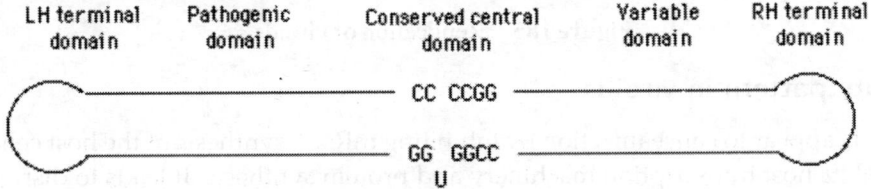

Figure 184   Structure of viroid

There are two **terminal domains** in the right and left ends of the Viroid molecule. They are implicated in Viroid replication and evolution.

The **pathogenic domain** has oligo (u) sequence. This domain contains structural elements that modulate symptom expression.

The central domain consists of 95 nucleotides. It has 7-nucleotide sequence that forms a stem loop structure. The **variable domain** is the most variable region in the molecule.

Replication of Viroids

Viroids replicate by asymmetrical model of replication.

1. Infecting +ss RNA acts as a template for synthesis of -ss RNA.
2. –ss RNA is synthesized completely by RNA dependent RNA polymerase.
3. -ss RNA is cleaned by unit length molecule.
4. The linear –ss RNA is circularized.
5. Circular ss RNA then serves as the template for synthesis of multimeric positive sense RNA by rolling circle mechanism.
6. +ss RNA is cleaved by ribonuclease and circularized by RNA ligase.

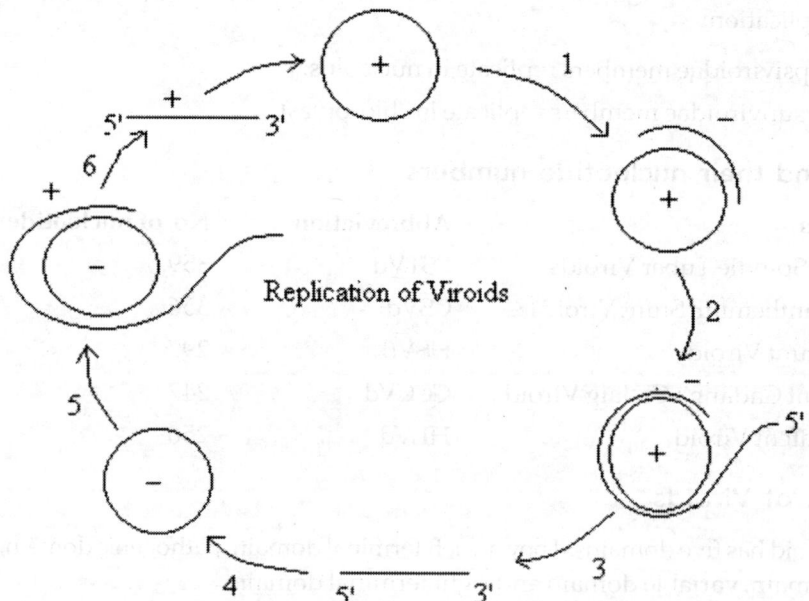

Figure 185   Replication of viroids

Infectivity pattern in viroids

Viroids appear to cause infection by inhibiting mRNA synthesis of the host cell. Viroid RNA inhibits host transcription machinery and protein synthesis. It leads to disruption of host cell metabolism and tissue degeneration.

Viroid transmission

Viroid transmission from cell to cell occurs through phloem channels and plasmodesmata. Viroids can be transmitted between plants through mechanical means, pollen or by vectors. Eg. PSTVd is reported to be transmitted by the aphid *Macrosiphum euphorbiae*.

Symptoms of viroid infection

Macroscopic symptoms

Stunting, Motting, Leaf distortion, Necrosis, Cytopathic effect, Viroid infection induces plasmalemmasome formation, Degenerative abnormalities occur in chloroplasts of viroid infected cell.

Origin of viroids

Precellular Theory Viroids originated early in precellular evolution when the primary genetic material consisted of RNA.

Viroids as primitive viruses Viroids originated from RNA and is capable of directing its own synthesis. They are primitive to viruses, as they do not have protein coat.

Viroids are escaped introns Viroids are believed to be introns. It is removed during post-transcriptional processing of mRNA in eukaryotes.

BACTERIOPHAGES OR BACTERIAL VIRUSES

H. T4 BACTERIOPHAGE

Phages that infect coliform group of bacteia are generally called as coliphages. The best-known member of large virulent bacteriophages is **T4**. It is included under the family **Myoviridae**.

Structure of T4 Phage

T 4 phage is a complex virus. It shows binal symmetry. It consist of three parts namely Head, Collar and Tail.

Head

Head is called as capsid protein. It is in the form of an Icosahedral shape. The structure is described as an elongated, bipyramidal, hexagonal prism. It is about 95nm long and 65nm wide. The head contains DNA, divalent cations (Mg^{2+}, Ca^{2+}), three internal proteins and ATP. The capsid is made up of 20,000 capsomers.

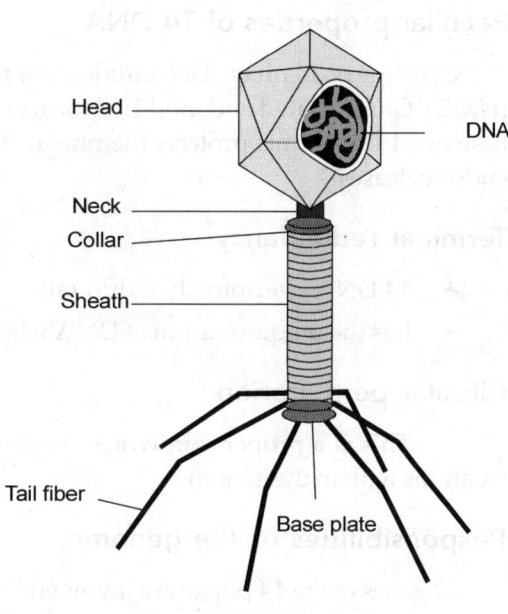

Head
DNA
Neck
Collar
Sheath
Tail fiber
Base plate

Figure 186 T4 phage

Neck

It is also known as head-tail connector or collar and it has been attached with whiskers.

Tail

The tail consists of an outer contractile tail sheath and an inner core or tube. The tail is about 80nm long and 18nm in diameter. The tail consists of 24 rings. Each protein ring contains 6 subunits. It is connected to the collar at upper end and base plate at the lower end.

Base plate

The base plate is hexagonal and has a tail pin at each corner.

Tail fibers

- A long thin tail fiber arises from each of the six corners of the base plate.
- Each tail fiber is about 130nm long and 2nm in diameter
- The tail fibers recognize specific receptor sites on the host cell wall during attachment.
- A single tail fiber is composed of two half fibers.
- The A half fiber (proximal) is attached to the base plate and BC half fibre (distal) has the attachment to host receptor.

DNA

Genome of T4 phage is Double stranded DNA and is 500µm long and has a molecular weight of 120×10^6 Daltons.

Peculiar properties of T4 DNA.

Cytosine of T4 phage DNA undergoes modification to form Hydroxyl Methyl Cytosine (HMC). Glucosylated and modified form of HMC is referred to as modified base or unusual base of T4 DNA. This protects the phage DNA from the action of host-encoded restriction endonucleases.

Terminal redundancy

- T4 DNA is terminally redundant
- It is the unique feature of DNA where same base pairs are found at both ends.

Circular permutation

This is a property in which terminally redundancy bases are present at different locations with in the genome.

Responsibilities of the genome

22 genes of the T4 phage are involved in DNA replication, 34 genes synthesize structural proteins and 19 genes are involved in assembly.

T4 Phage Multiplication cycle or life cycle of T4 Phage or Lytic life cycle of phage

T4 multiplication cycle can be divided in to two periods.

1. Eclipse period where nucleic acid replication and protein synthesis occurs and

2. Maturation period where morphogenesis (Assembly) of phage particles occurs.

T4 phage is also called as virulent phage. It undergoes lytic life cycle when it infecting coliforms (eg. *Escherichia coli*)

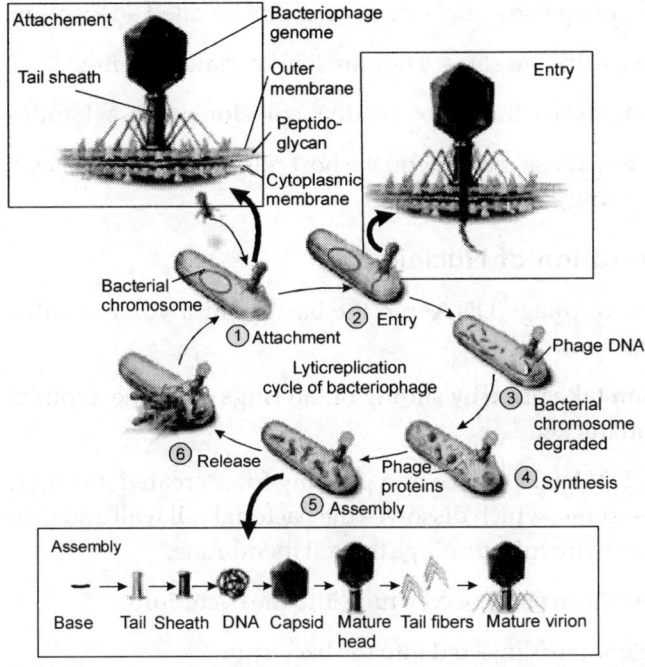

Figure 187 Life Cycle of T4 Phage

Lytic Life Cycle of T4 Phage

T4 phage is a virulent phage. It follows lytic life cycle during multiplication. On completion of the life cycle, they cause lysis or rupture of the bacterium. Hence the cycle is called lytic cycle.

Phage multiplication involves eclipse phas, maturation phase and Lysis or Raise period.

Eclipse period means time duration between virus infection and subsequent assembly of first virion.

Stages of Lytic Life Cycle

↣   Adsorption

↣   Entry or penetration of Nucleic acid

- ▲ Transcription and Translation of early genes.
- ▲ Breakdown of bacterial chromosome.
- ▲ Synthesis of structural proteins
- ▲ Replication of T_4 DNA
- ▲ Assembly or morphogenesis
- ▲ Release

1. Adsorption

Attachment of a phage on a surface of bacterium is called adsorption.

Attachment occurs in two steps. They are lending and pinning.

Phage attaches to the bacterium by random collision process (Landing).

Tail fibres selects maltose receptor on the host cell surface and spikes anchor firmly on the cell wall of host (Pinning)

2. Entry or Penetration of Nucleic acid

The penetration of phage DNA into the bacterium involves contraction, penetration, unplugging and injection.

Tail Contraction take place by sliding of tail rings over one another. So, the tail sheath becomes shorter and thicker.

Penetration of DNA takes place by a pushing force created during tail contraction. The phage produce Lysozyme, which dissolves the bacterial cell wall and drills a hoe in it. Due to this pushing force the core tube unplugs the cell membrane.

Unplugging results in entry of core tune into the bacterium.

Finally phage genome is injected into the bacterium.

The capsid remains outside.

Entry of phage genome into the host cell results in stoppage of bacterial cell activities.

3. Transcription and Translation of early genes

Early phage genes are transcribed into mRNA and translated into proteins.

Products of early genes are nuclease enzyme, Host DNA and RNA polymerase modifying enzyme.

4. Breakdown of Bacterial Chromosome

Super Helical DNA becomes relaxed due to helix destabilizing proteins of phage.

Nucleases digest the bacterial DNA into free nucleotides. The free nucleotides remains in the cytoplasm and used for phage replication.

5. Synthesis of Structural proteins

The phage utilizes host machinery for the synthesis of Capsid, Tail, Tail Fibres, Core and phage DNA replication.

Structure proteins are the products of late genes of phage.

This results in accumulation of viral components in bacterial cell Cytoplasm.

6. Replication of DNA

T^4 DNA replication begins at the 6th minute of infection.

The T_4 DNA has unusual Nucleotide 5-hyroxy methyl cytosine(HMC). It is synthesized by phage enzymes.

HMC is synthesized from the cytosine residue of host DNA.

Step involved in HMC synthesis are as follows.

Cytosine monophosphate (CMP)

$\downarrow$ T_4 hydroxy methylase

Di hydroxy methyl cytosine monophosphate (dhmcmp)

$\downarrow$ T_4 hydroxy methyl kinase

Di hydroxy methyl cytosine diphosphate (dhmcdp)

$\downarrow$ *E.coli* phosphase kinase

HMC triphosphate

Phage has produced two enzymes, namely dCTPase and deaminase to prevent incorporation of CTP.

DNA synthesis begins with the synthesis of RNA primer. Then DNA polymerase proceed the DNA synthesis by the addition of new nucleotides using RNA primer as template. At the end of this step, phage DNA molecules accumulate in the host cell.

7. Assembly or Morphogenesis

Build up of virus by fitting the viral proteins and DNA together is called viral assembly or morphogenesis.

Virus assembly occurs in two steps

- ⚑ Assembly of virus head, tail and tail fibres.
- ⚑ Assembly of virion.

Assembly of Heads

Head assembly takes place on the inner surface of the cell membrane.

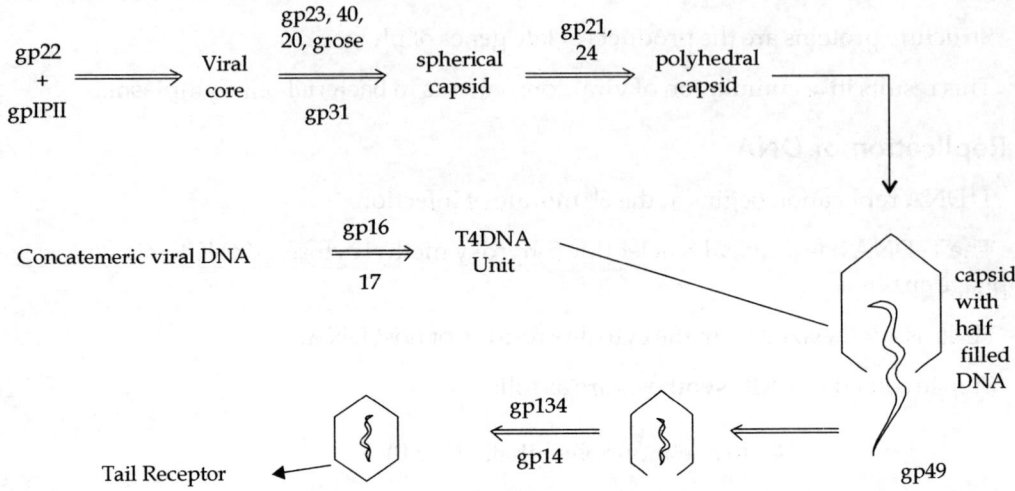

Figure 188 Assembly of head

Assembly of Tail

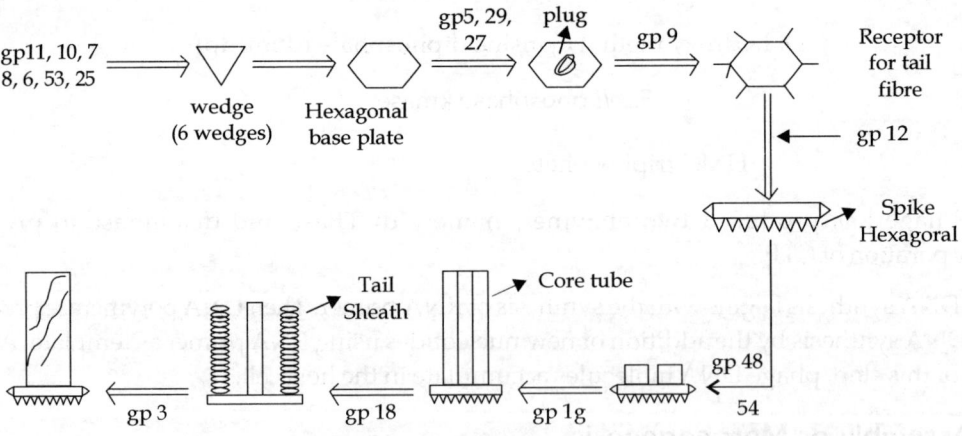

Figure 189 Assembly of tail

Tail

Six wedge shaped structure formes base plate and other structural protein like gp5, gp27, gp28, gp9, gp12, gp48, gp54 gp19, gp18, gp 3 joined one by one and finally forms a tail.

Assembly of Tail fibres

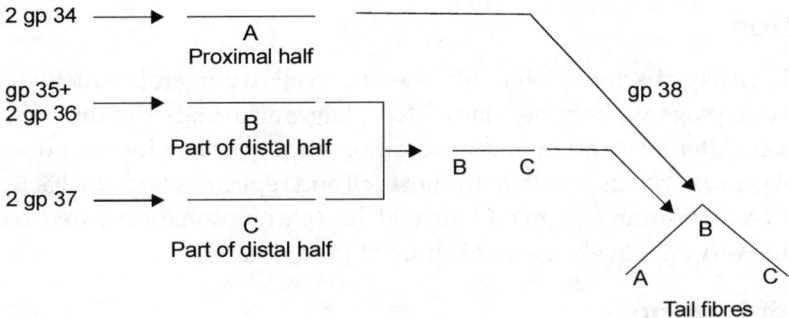

Figure 190 Assembly of tail fibres

Assembly of Virion

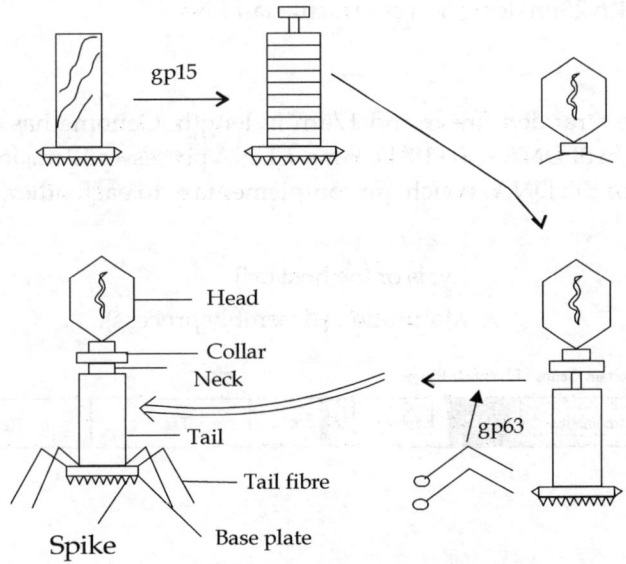

Figure 191 Assembly of virion

The virion assembly occur between 12 and 15 minutes after infection. The protein go 15 forms a neck. Head connects with the Tail. Finally 6 Tail fibres are attached to the base plate with the help of gp 64. Thus mature virion assembled inside the infected bacterium.

Release

This occurs 22 minutes after viral attachment to the host cell. Lysozyme synthesized by virus breaks cell wall, leading to cracks in the cell call. Lysis of host cell releases 150-300 progeny virion.

I. LAMBDA PHAGE (λ -TEMPERATE PHAGE)

Introduction

Normally virulent bacteriophages lyse their host cells during reproductive cycle. Lamda phages carry out lysogenic life cycle. Many DNA phages also establish a different relationship with their host. After adsorption and penetration viral genome does not destroy its host. Instead, viral genome remains with in the host cell and replicates with the bacterial genome. The phage DNA that is in an integrated form with host chromosome is referred to as **prophage**. Lambda phage very effectively uses K12 strain of *Escherichia coli*

Structure of λ phage

It belongs to Sophoviridae family. Head is icosahedral in shape and is about 55nm in diameter. Capsid contains 300 to 600 subunits of capsomers. Molecular weight of capid protein is 37, 500 D. Head is joined to tail by head – tail connector. Tail is long, flexible and is about 150x 8nm in size with 25nm long non contractile tail fiber.

λ DNA

DNA is double stranded linear and 17µm in length. Genome has 46, 500 base pairs. The molecular weight of DNA is 31x10⁶ Daltons. λ DNA possesses 12 nucleotides as extending units in ether side of the DNA, which are complementary to each other, and it is said to be cohesive ends.

S , R	Lysis of the host cell.
A	Maturation (Assembly process).

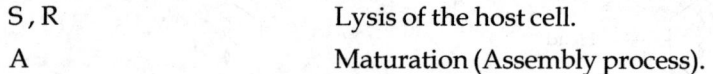

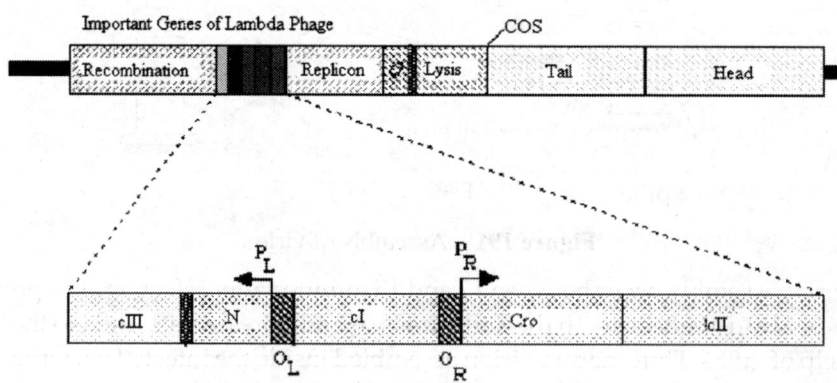

Figure 192 Geres of lambda phage

LIFE CYCLE OF λ PHAGE

λ phage is a temperate phage and is capable carrying out both lytic and lysogenic life cycle.

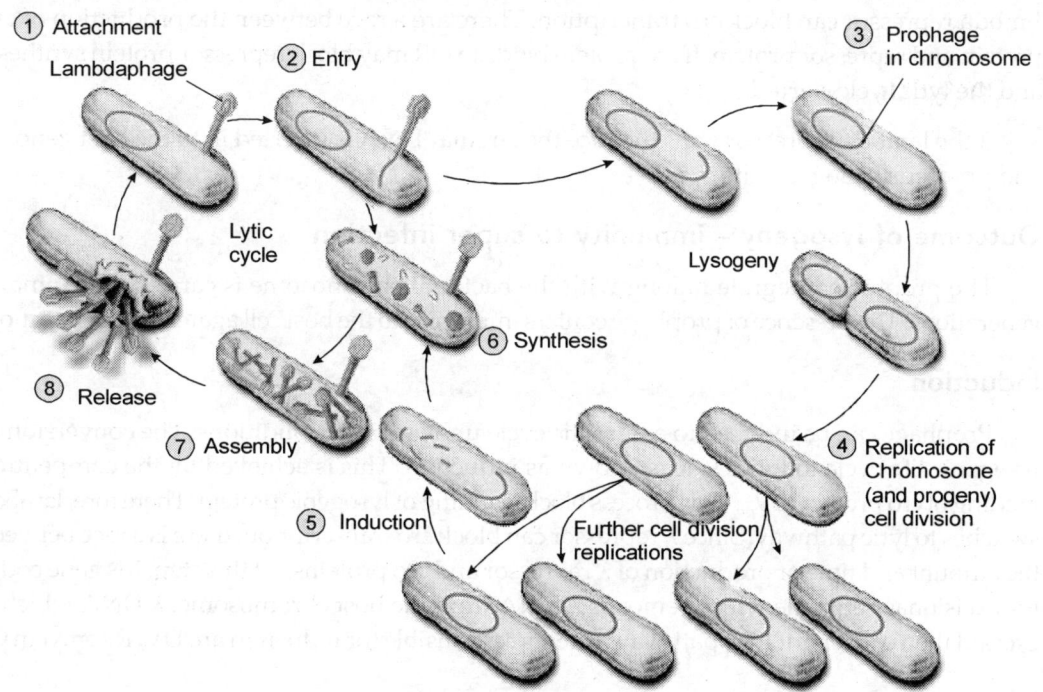

① Attachment

Lambdaphage

② Entry

③ Prophage in chromosome

Lytic cycle

Lysogeny

⑥ Synthesis

⑧ Release

⑦ Assembly

⑤ Induction

Further cell division/ replications

④ Replication of Chromosome (and progeny) cell division

Figure 193 Life cycle of phage

Refer T4 Phage Life Cycle-Refer 379-382

Lysogenic life cycle

After attachment, λ DNA is released into the cytoplasm of host cell. Suddenly λ DNA circularized and trascription has commenced, the cII and CIII protein accumulates.The cII protein binds with P_{RE} (RE-Repressor establishment) and stimulates RNA polymerase binding.The cIII protein protects cII from degradation by host enzyme.Lambda repressor cI rapidly synthesized and binds to O_R and O_L, thus turning off mRNA synthesis, and leads to stoping of cII and CIII protein synthesis. This initiates lysogenic cycle.CII protein stimulates the transcription of int gene and produced integrase enzyme. λ DNA has att site (attachment) that can base pair with a bacterial attachment site located between the gal operon and biotin operon.

Integrase enzyme aid integration and produced a prophage. Now the bacteria perform regular duties without any lysis of host cell but with new character. This is called as lysogenic life cycle. Lambda repressor is coded by cI gene. This repressor protein has 236 aminoacids and folds into a dumb bell shape.

Choice between lytic and lysogenic pathway

Cro protein accumulation leads to binding of this protein to O_R and O_L, turns off the transcription of the Repressor gene. It represses the P_{RM} (RM-Repressor maintanence) function.

Lmbda repressor can block cro trancription. There are a race between the production of cro protein and repressor protein. If cro protein binds to OR may block repressor protein synthesis and the lytic cycle started.

If the lambda repressor wins the race the circular DNA is inserted in to the host genome and proceeds Lysogenic life cycle.

Outcome of lysogeny – immunity to super infection

The prophage integrated along with the bacterial chromosome is carried on for many generations. The presence of prophage confers immunity to the host cell against super infection.

Induction

Prophage can be induced to enter lytic cycle under certain conditions. The conversion of lysogenic life cycle to lytic cycle is known as induction. This is achieved by the competiting binding of cro protein to P_{RE}. This process blocks binding of lysogenic protein. There fore, lambda switches to lytic pathway. Since λ repressor can block cro transcription, there is a race between the amount and time of production of λ repressor and cro proteins. At this step, Xis gene codes for excisionase enzyme which removes λ DNA from the host chromosome. λ DNA which is excised then carrier out lytic pathway. Agents responsible for induction are UV, Ritomycin C.

Lysogenic conversion

Temperate phages may induce phenotypic character of host cell by inserting phage gene into bacterial chromosome. The phenomenon in which prophage can confer new properties on the cells is called lysogenic conversion. This conversion involves alterations in bacterial surface characteristics and pathogenic properties of the host.

Expresion of lysogenic conversion

Synthesis of diphtheria toxin is induced when *Corynebacterium diphtheriae* is infected with β phage.*Clostridium botulinum* Synthesis its botulinum toxin because of the availability of new phage gene. Phage infected β-haemolytic *Streptococcus* produce Erythrogenic toxin . In *Staphylococcus aureus*, lipase activity is lost upon infection with phage L54 a owing to inactivation of the lipase structural gene by insertion of prophage. Pathogenic property of *Salmonella* is increased when it is infected by an epsilon phage (E_{15}). This phage modify the structure of its outer membrane lipoplysaccharide layer.

J.   M13 BACTERIOPHAGE

Introduction

M13 phage is a filamentous Coliphage, these bacteriophages are composed of helical capsids and single stranded DNA as a genome. It belongs to the family of Inoviridae. The name M13 derives from city of Munich where it was first isolated by Hotschneider in 1963. M13 designation may also be attributed to Messing who studied the use of M13 as cloning vector. M13 is known as male specific phages as its is capable of infecting only *E.coli* F⁺strains . F⁺ denotes the cell which contains F pilus also known as male cells or donor cells.

M13 phages are known as leaky phages as the phage particles are not released by lysing the host cell after infection and multiplication. Instead, M13 phages are extruded or leaked out through F pilus of the host cell. As a result M13 phages do not form clear plaques on the lawn of host cell (*E.coli* F⁺ strain).

Structure of M13

M13 phage is about 900nm long, 9nm in diameter. Capsid is made up of one major protein and four minor proteins. The major coat protein is the product of phage gene VIII (g8p). There are 2700 to 3000 copies of protein. The major protein is held together with four minor capsid proteins. Four minor capsid proteins are the product of the genes gIII, gVI, gVII and gIX. Minor capsid proteins are located at the ends of the filamentous M13 particle. gp8 proteins consist of approximately 50 aminoacid residues. It assumes the form of α helix and appears as a shot rod.

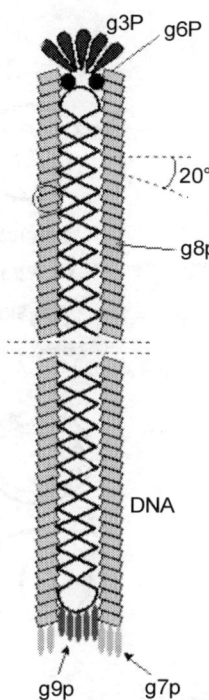

Figure 194   M13 phage

M13 has circular single strandeds DNA as a genome. It is 6,407 base pair long. The genome codes for a total of 10 genes and is named using Roman numerals I through X. Genome contains specific restriction sites for Bam H1, Hpa I, NspBII, BanII, AvaI, NaeI, BalI. It also contains unique origin of replication. Functions of various genes are presented as follows.

Replication

The filamentous phage infects only F⁺ cells of *Escherichia coli*.

Attachment and entry

- ➤ Adsorption of virus requires interaction between pilus and minor coat protein pIII, occurs at one end of the filamentous rod.
- ➤ As the rod shaped virus penetrated pIII interacts with host cell protein.
- ➤ This interaction mediates the removal of the major coat protein and allow the entry of viral ssDNA into the cytoplasm.

Replication of genome (Figure 195)

There are three stages of replication , which produce (-) strand DNA and (+) strand DNA. -stand serves as a template for (+) strand synthesis.

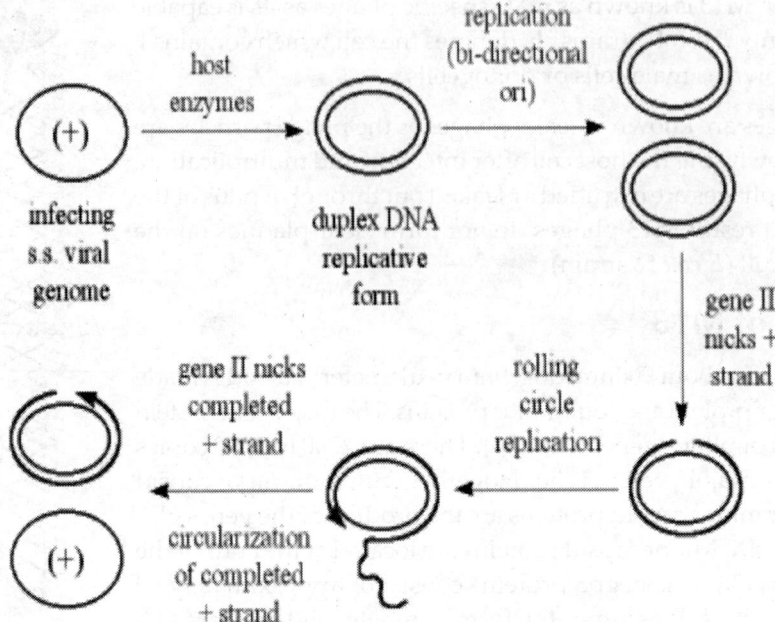

Figure 195 M13 phage genome replication

Stage I

Positive ssDNA is converted into DS circular form called Replicative form (RF) with the help of *E.coli* DNA ligase. Replicative form (RF) formed during stage I undergoes several steps to synthesize many RF molecules.

Stage II

Synthesis of RF

RF is initiated when gene II protein nicks the outer (+) strand at the origin. Using *E.coli* SSB (single strand binding) protein and DNA Pol III, the (+) strand is extended from 3'OH end. When the new(+) strand reaches the origin, it is cleaved again by geneII protein. The old (+) strand is freed and its 3' OH and 5'-P ends are joined by gene II protein. RF molecule replication is controlled by gene X which acts repressor.

Stage III

The new (+) strand is formed during stage II and this serves as ss molecule. Gene V protein binds to new (+) strand thereby preventing (-) strand synthesis.

Assembly and Release

Assembly of M13 phage occurs at the inner membrane of the host cell. The replicated DNA is moved to the cell membrane and g8p forms the capsid. Minor coat proteins are attacted at the respective ends and the complete M13 phage is extruded out form F pilus. M13 phages establish permanent infection without lysogeny and produce ~300 particles / infected cell.

Life cycle of filamentous phage

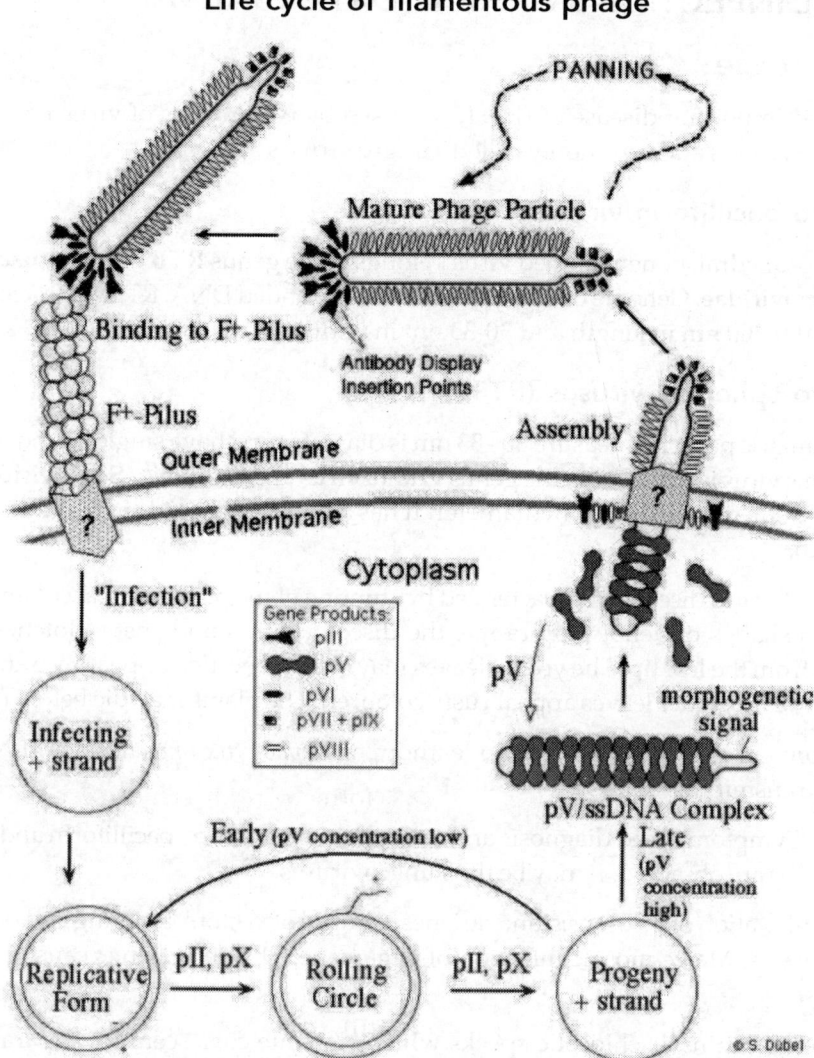

Figure 196   M13 phage life cycle

Application of M13 Phage

It acts as a cloning vector. Replicative form (RF) of M13 is the source for the vectors like M13MP1, M13 MP2, M13MP7, M13MP8. Insertion of Lac Z gene in M13RF leads to M13MP1. Site specific mutation of M13MP1 derived M13MP2. Other vectors are constructed from M13MP2 by using restriction enzymes and poly linkers. Larger volume of DNA segments are packed within M13 phage due to its filamentous nature. Host cells are not lysed while releasing mature phages because M13 is a leaky phage. M13 derived vectors also prepared by making use of other plasmids like pUC and are called phagmids eg.pEMBL8.

K.   VIRAL INFECTIONS OF RICE (*ORYZA SATIVA*)

Tungro Disease

This is the important disease of rice. It is caused by two groups of virus particles. It is causedby two viruses and commonly called Tungro viruses

Rice Tungro bacilliform viruses

It is an icosahedral, nonenveloped virus belongs to the genus **RTBV like viruses** and the family **caulimoviridae**. Genome of this virus is double stranded DNA. RTBV particles are rod-shaped and 100-300 nm in length and 30-35 nm in width. It contains DNA of 8.3 kb.

Rice tungro spherical viruses (RTSV)

It is a isometric particles measure 30 - 33 nm is diameter and have single stranded RNA as a genome. This virus is included in the genus **Waika virus** and the family **Sequiviridae**. RTSV particles are isometric and 30 nm in diameter. It has a polyadenylated single-stranded RNA of about 12 kb.

Symptoms: Tungro disease is characterized by stunting of the plant and leaf colours ranging from various shades of yellow to orange, the discolouration and rusty blotches spreads downwards from the leaf tip. The young leaves may show a mottled appearance and slightly twisted where as the older leaves appear rusty coloured. The plants may die before flowering.

Transmission: It is spreaded by green leafhopper *Nephotettix impicticeps* and *N.apicallis*. nymps also transmit this disease.

Diagnosis: Symptom based diagnosis and on electron microscopy bacilliform and spherical forms of viruses are observed. It may be the Tungroviruses.

Control: Cultivation of virus resistant varieties. Control of vectors. Rouging and destruction of infected plants. Maize mosaic infection of sugarcane. This infection is caused by Maize mosaic virus.

The disease begins as chlorotic specks which elongate coalesceni rows parallel to the midribs, giving mosaic blotting areas. The affected plants remain pale and stunted.

It can be transmitted by mechanical means and by Vector aphids - *Rhopalosiphum maidis*.

L.   SUGARCANE MOSAIC DISEASE

Poty virus -It is also called Sugarcane Mosaic virus

Chlorotic elongated and irregular stripes or streaks surrounded by normal green areas are the symptoms. It is spread by Aphids vectors in non persistant manner.

M. VIRAL INFECTIONS OF TOMATO

Tomato Mosaic Disease

The disease is caused by several viruses.

Causative agent: Tomato Acuba Mosaic Virus. (It is a strain of TMV), Cucumber mosaic virus, Nicotiana virus are the major causative agent.

Symptoms are Stunted growth. Young leaves are small, crinkled, deformed pale yellow and brittle. Mild to severe mosaic mottling with raised dark green areas. The necrosis of stem and fruit petioles are observed. Occasionally it may kills the plant.

TMV is transmitted through contact, soil debris. Sap transmission, Seed transmission, Aphid vector also responsible.

It is Control by Preventing contact of diseased plant with healthy plants, Rouging in seedbeds. Insecticides were also useful

Tomato leaf curl disease

Causative agent of this disease are *Gemini virus*. This virus has circular single stranded DNA as a genome. The main distinguishing feature of this virus is having two incomplete icosahedral structures.

Potato leaf roll virus. It is a positive sense single stranded RNA virus included in the family Luteoviridae genus Polerovirus . Virus is isometric icosahedral in shape.

Symptoms include the following

Vein clearing, stunting, marked reduction in leaf size with mild marginal curling of leaves, the internodes are shortened and plants are dwarfed. Complete yellowing and puckering of leaves, infected plants produce number of lateral branches, imparting a bushy appearance, induce complete or partial sterility.

Virus is Transmitted by White fly - *(B.tabaci)* and Through Grafting

Spraying with Ekatox and Roger at 10 day interval helps to reduce the incidence

N. BUNCHY TOP OF BANANA

Member of Luteovirus [18-20nm], Banana virus 1, Musa virus 1 are the causative agent of this disease. It is a positive sense single stranded RNA virus included in the family Luteoviridae. Virus is isometric icosahedral in shape. This disease is first originated in Australia. In India it is first reported in Kerala in 1940. it is a exported disease from Ceylon through suckers.

Symptoms are Leaves become short and narrow, bunched together and stand erect at the apex of the plant and produce typical symptom of the bunchy top disease, the margin of the leaves become wavy and later grow upward. Leaves show irregular nodular dark green to brown streaks on the underside of the lower portion of midrib, stunted.

The phloem cells contain many nuclei. Vascular tissues have abundance of chloroplast.

Aphid - *Pentalonia nigronervasa* is mainly responsible for transmitting this disease. Aphid is required to feed on the plant for few hours to 2 days for transmitting the disease. Incubation period is 30-40days. After this period the symptom starts. Sap also plays a vital role.

The disease can be excluded by strict quarantine regulations. Control insect vectors and aphids by spraying insecticides. Spray MCPA (2methyl, 4 chlorophenoxy acetic acid) to control aphids. Systemic eradication of infected plant must be carried out. Healthy suckers must be used to avoid infection

O.   COTTON LEAF CURL DISEASE

Gemini virus is a major causative agent. This virus has circular single stranded DNA as a genome. The main distinguishing feature of this virus is having two incomplete icosahedral structures.

Vein clearing of young leaves on a few twigs, which later may cover larger number of branches. Mosaic mottling marginal curling of lamina and puckering of leaves is prominent.

Grafting is the way of transmission.

P.   TOBACCO MOSAIC VIRUS

Introduction

Tobacco mosaic virus was the first plant virus to be discovered by Dimitri Iwanowsky and crystallized by Stanley in 1935.TMV causes mosaic disease of tobacco plants (*Nicotiana tabacum*). The disease is world wide in distribution. TMV also affects other plants, mostly of the family solanaceae.

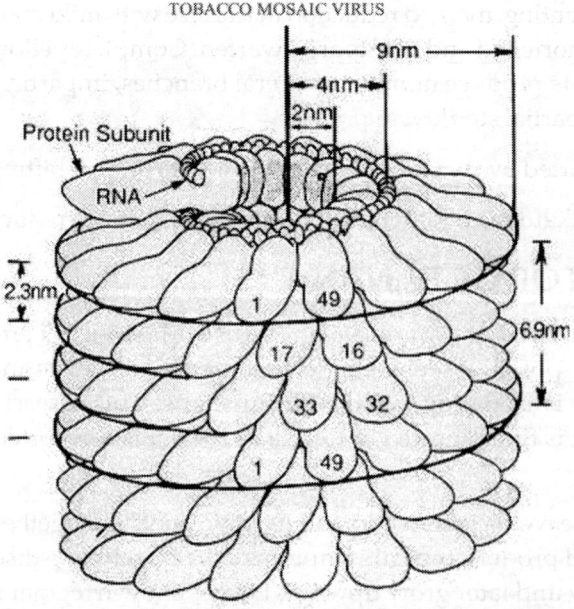

Figure 197   TMV

Morphology

TMV is the prototype of Tobamovirus group. It is a stiff, rod shaped virus with ss RNA as a genome. TMV particles are 300µm in length and 15µm in diameter. It has a centrally placed RNA molecules covered with a protein coat. Protein coat of this virus contains 2130 identical protein subunits each with a molecular weight of 18, 000 D. The protein subunits is composed of a single chain of 158 amino acid residues of known sequence. The aminoacid sequence starts with N-acetyl serine and ends with threonine. RNA of the virus has a molecular weight of 2.4kDa, which has 6400 nucleotides. RNA strand forms a helix with a radius of 40A⁰. The pitch of the helix is 23A⁰. Every three turns of the helix has 49 subunits, which are called as virus replicating units. RNA contains about 25.3%G, 29.8%A, 18.5%C and 26.3%U.

When TMV is treated with dilute alkali, it degrades into smaller oligomers with successive reduction of redimentation co efficient. TMV contains 5% nucleic acid, 95% protein and trace amount of metal ions.

Transmission

Virus is sap transmissible and enters the host through wounds. Virus is resistant to adverse environmental and climatic conditions and is infective at any period. Seed transmission of TMV has been reported in tomato. TMV can be transmitted by mechanical means, wind and water.

Replication of TMV

TMV is inoculated into the plant through natural openings, wounds etc., Soon after entry uncoating and activation takes place in cytoplasm. TMV does not produce any enzyme as the virus lacks an enzymatic system but it induce the host enzymatic system.. The viral RNA plays the role in transcription of mRNA.

TMV RNA transcription leads to the formation of mRNAs for Capsid proteins(CP) and Movement proteins(MP).TMV is replicated by making use of host enzymes like RNA polymerases, RNA replicase etc., and produced multiple copies of RNA genome.
TMV utilizes the amino acids, ribosomes and +RNA molecules of the host cell. Viral protein synthesis takes place in the cytoplasm and the multiplying rate is 1million / protoplast in 24hours. Movement of TMV from cell to cell takes place through Plasmodesmata and also with the help of MP proteins. Assembly of this virus takes place in cytoplasm.

Symptoms

The symptoms include various degrees of chlorosis, curling, mottling, dwarfing, distortion, and blistering of leaves, dwarfing of the entire plant, dwarfing, distortion and discoloration of flowers, and in some plants even development of necrotic areas on leaf.

The most common symptom on tobacco is the appearance of mottled dark-green and light-green areas on leaves. Stunting of young plants is common, and is accompanies by a light downward curling and distortion of leaves, that may become narrow and elongated rather than normal oval shape. Old leaves may not show symptoms, young ones develop typical symptoms.

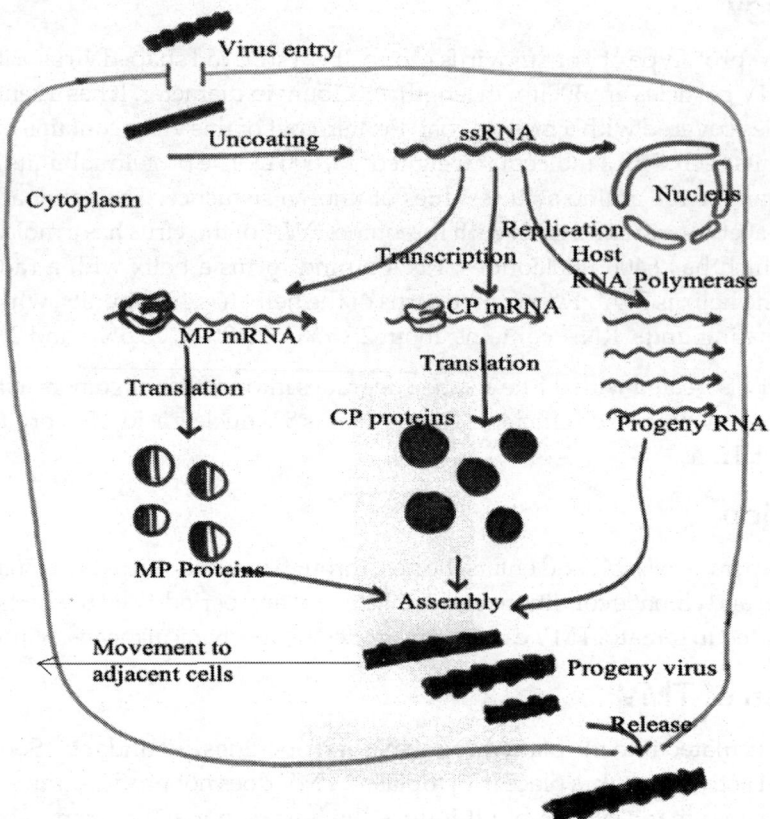

Figure 198 Replicate of TMV

Control

Growth of tobacco resistant varieties. Practicing field sanitations, Rouging of diseased plants and weeds, Destruction of diseases plants by chemical treatment. Workers in the field must wash hands with 3% trisodium phosphate or soap. Equipment and instruments used in plantations must be sterilized.

Q. CAULIFLOWER MOSAIC VIRUS (CAMV)

Introduction

Cauliflower Mosaic Virus (CaMV) is the prototype of caulimovirus group. CaMV is the first virus of higher plants with DNA as its genetic material. CaMV has widely been used in plant genetic engineering as a prominent vector. This virus was first reported in *Brassica campestris*.

Structure

CaMV is an icosahedral virus measuring about 50nm in diameter and have a molecular weight of 23x 10^6 Daltons. DS DNA is the genome of this virus. CaMV contains 35% nucleic acid, 65% protein and 0% lipid. The circular ds DNA molecule has a molecular weight of 4.9x 10^6 Daltons with a G + C content of 43%. The virus has a sedimentation coefficient of 208s upon $Cscl_2$ centrifugation. Virion has two major proteins and two minor proteins

Properties of virus in plant sap

Thermal inactivation Point is 70°C to 75°C for 10 minutes. Dilution end point is 1:2000, Longevity in vitro is 14 days. DNA is not sensitive to chloroform (1 to 5%)

Transmission and Host Range

CaMV is transmitted by mechanical methods, insects and through vegetative propagation. Aphids have been involved in persistent type of transmission of CaMV. Aphids retain the virus for 3 to 20 hours. CaMV has restricted host range. Only members of family Cruciferaceae are affected.

CaMV DNA and Genome Organization

Double stranded circular DNA with 8,025 bp. Sequence of CaMV contains eight closely packed reading frames. There are two small intergenic regions (IR) which do not code for any proteins. Gene II and VII are termed as non-essential genes. ds DNA is circular with 3 discontinuities at specific sites. The outer strand is U and the inner starnd is A. These discontinuities are termed as overhangs, which contain 8 to 20 nudeotides in which 3' and 5' ends overlap. The VIII gene is organized within Gene IV.

Replication of CaMV

After entry into the host cell, un coating and release of CaMV DNA occurs. Uncoated viral DNA is transported to the cell nucleus. There occurs removal of overhangs by host encoded DNases and closing of gaps by host encoded DNA ligase. Viral DNA becomes associated with the histone proteins of the host cell to form minichromosome configuration. Two RNA transcripts are generated as a result of transcription in the cytoplasm of the host cell. 35s RNA codes for structural proteins. 19S RNA is involved in replication of genome.

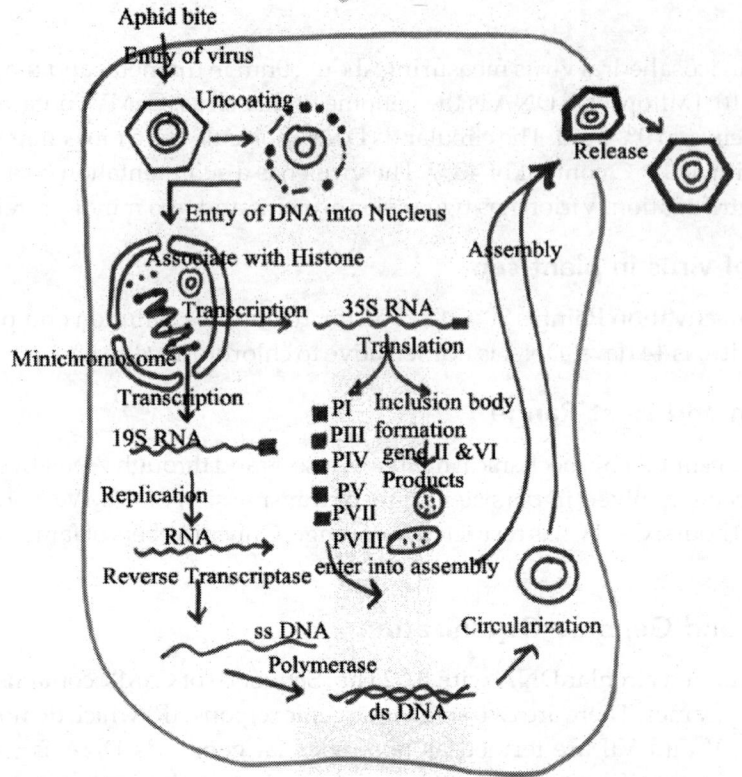

Figure 199 Replicate of CaMV

ds DNA virus is synthesized by the enzyme reverse transcriptase. Host RNA acts as a primer for the synthesis of viral DNA. Encapsidation occurs in the cytoplasm to form progeny virions. Inclusion body formation may occur by gene II and gene VI

Symptoms of CaMV infection

Vein clearing or chlorotic vein banding occurs *in Arabidopsis thaliana*, Brassica sp etc. The disease causes stunting of the entire plant.

Control

1. Control of insect vectors.
2. Elimination of diseased plants
3. Cultivation of virus-resistant varieties.

R. VIRUSOIDS AND SATELLITE RNAS

Virusoids of plants and Hepatitis Detta Virus (HDV) of animals are non-self replicating subviral agents. They are ss RNA molecules having about 220 to 1700 nucleotides. They can reproduce only in cells that have been infected by helper virus.

Virusoids require helper virus for replication and encapsidation. Helper virus is usually a RNA or DNA virus.

Virusoids replicate in the cytoplasm of the host cell using RNA dependent RNA polymerase. Virusoids do not interfere with the replication of helper virus but they modify the symptoms produced by helper virus infection. They can be spread by vegetative propagation with seeds or by insects.

Examples

Virusoids/Satellite RNA	Helper Virus
Brady Yellow Dwarf Virusoid	Luteoviurs
Tobacco Ringspot virusoid	Nepovirus.
Subterranean clover Hottle Virusoid	Sobemovirus.

Hepatitis delta virus (HDV)

It is the well-studied satellite virus of humans

It was first identified in 1970's in Australia as a nuclear antigen, the delta.(ä) antigen.

It is known to cause Hepatitis D infection.

Structure and Properties of HDV-Refer Page No. 441

ANIMAL VIRUSES

S. PICORNAVIRUSES

Introduction

Picornaviruses mean small RNA viruses (Pico means small rna indicates RNA). It is the most largest family of RNA viruses with important human and animal pathogenic viruses

Classification

The family *Picornaviridae* comprises five genera based on ICTV system of classification. *Enterovirus, Hepatovirus* and *Rhinovirus*, which infect humans: *Apthovirus* (foot-and-mouth disease virus), which infects cloven-hoofed animals and occasionally humans; and *Cardiovirus*, which infects rodents.

Enterovirus	Poliovirus	*-Poliomyeletis*
Rhinovirus	Human rhinovirus A	*-Common cold*
Hepatovirus	Hepatitis A virus	*-Hepatitis*
Cardiovirus	Encephalomyocarditis virus	*-rodent infection*
Aphthovirus	Foot-and-mouth disease virus	*-FMD*
Parechovirus	Human parechovirus	*-Meningititis*

Based on tissue tropism on human enteroviruses, found in human gut; the rhinoviruses, found in the upper respiratory tract and hepatovirus (Hepatitis A virus) in the intestine and liver.

Structure and Properties of Picorna Virus

The picornaviruses are small (22 to 30 nm) nonenveloped, single-stranded RNA viruses with icosahedral symmetry. The virus capsid is composed of four viral proteins VP1 to VP4, which form an icosahedral shell. The genome is a *single stranded RNA* (molecular weight, approximately 2×10^6 to 3×10^6). The RNA strand consists of approximately 7,500 nucleotides and is covalently bonded to a *noncapsid viral protein (VPg).* Picornaviruses

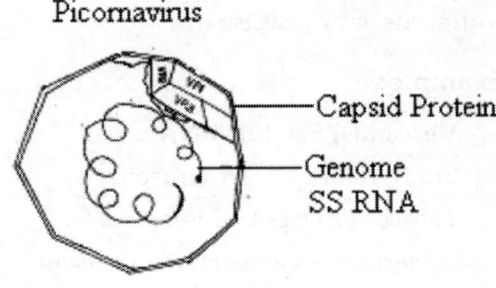

Figure 200

can survive for long periods in organic matter and are resistant to the low pH in the stomach (pH 3.0 to 5.0). Picornaviruses axre inactivated by pasteurization, boiling, Formalin, and chlorine. They are ether resistant due to absence of essential lipids. They replicate in the cytoplasm.

Multiplication of Picornavirus

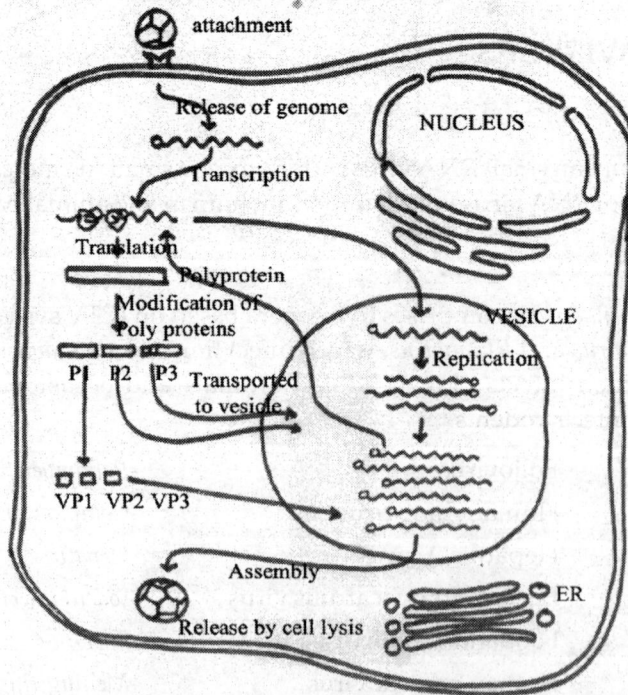

Figure 201-

The virus binds to a cellular receptor.   The mechanism of uncoating of RNA genome are unknown. Translation is initiated at an internal site of 741 nucleotides from the 5' end of the viral RNA and poly protein precursors are synthesized. Polyproteins are cleaved and produce individual proteins (P1,P2 and P3). P1 proteins contain viral structural proteins. P2 and P3 are responsible for proteases and RNA synthesis proteins. The proteins that involve in RNA synthesis are transported into the membranous vesicles. Positive sense RNA also transported into the vesicles. It is copied into minus sense RNA that is the template for the synthesis of positive sense RNA.Structural proteins are formed by partial cleavage of P1 precursor proteins. These proteins are transported to vesicles. Assembly takes place within the vesicles. Mature virions released after cell lysis.

Entero virus

Poliovirus, Coxsackieviruses, Echoviruses and Enteroviruses 68-72 are the examples for enteroviruses. Among these incidence of Poliovirus is more common and produce severe infection in human called **poliomyeletis**.

Poliomyeletis

Introduction: Polio is an ancient disease. In 1840 the German orthopedist Jocob Von Heine described the clinical features of Poliomyelitis and identified spinal cord as the problem area. Poliovirus has tropism for epithelial cells of the alimentary tract and cells of the central nervous system.

Causative Agent: All three Poliovirus serotypes (1 to 3) can give rise to paralytic poliomyelitis. Structure and Replication – Refer page 398 and 399

Symptoms: **Paralytic poliomyelitis** can occur without antecedent minor illnesses. A patient may suffer aseptic **meningitis with pains in the back and neck muscles** for several days.

Pathogenesis: Incubation period is about 7-141 days.   Humans are the only natural host of Poliovirus. They attach to a specific receptor on these cells, which in humans is encoded by a gene on chromosome 19. Primary replication of Poliovirus takes place in the oropharyngeal and intestinal mucosa (the alimentary phase). From here, the virus spreads to the tonsils and Peyer's patches of the ileum and to deep cervical and mesenteric nodes, where it multiplies abundantly (the lymphatic phase). Subsequently, the virus is carried by the bloodstream to various internal organs and regional lymph nodes (the viremic phase).

More concentrated damage results in flaccid paralysis of the muscles innervated by the affected motor nerves. Muscle involvement peaks a few days after the paralytic phase begins. Paralytic disease is called spinal poliomyelitis.

Diagnosis: Enteroviruses may be isolated from feces, pharyngeal swabs, saliva, and nasal aspirates and blood. Polio virus cultivation is performed with the help of tissue culture technique. The most specific of the conventional laboratory tests used to identify Picornavirus serotypes is the Neutralization test.

Control: The **Salk-type inactivate poliovirus vaccine (IPV)** consists of a mixture of three poliovirus serotypes grown in monkey kidney cell cultures and made noninfectious by Formalin treatment. It is given in two intramuscular injections. The **Sabin-type live attenuated Oral Poliovirus Vaccine (OPV)** is commercially available as trivalent antigen. The viruses are attenuated by multiple passages in monkey kidney or human diploid cell cultures, and the vaccine potency is stabilized with one Molar magnesium chloride or sucrose. In 1988, the World Health Assembly, the governing body of the World Health Organization, set the goal of global eradication of poliomyelitis by the year 2000 but the programme is extended for some more years.

Rhino viruses (Common Cold)

Tyrell and his collogues isolated Rhinoviruses in 1960 by inoculating specimens into monkey tissue culture. Rhinoviruses are named as JH, 2060,Salisbury virus or Mury virus before it gets its original name. Rhinoviruses cause **common cold** in human. Common cold probably the commonest infectious disease in man. Morethan 100 serotypes were identified.

Rhinoviruses are the members of Picorna group (rhino means nose). These viruses cause the lose of many million hours of mans work. Many people suffer from two to four colds every year. These viruses are infectious only for Human and Chimpanzees. Rhinovirus causes **common cold** in human. Common cold viruses are divided into two, according to tissue tropism. *M strains* grow and produce a cytopathic effect in monkey kidney cells. *H strains* are mostly isolated in human embryonic kidney cultures.

Structure and Replication – Refer page 398, 399

Symptoms: Incubation period is about 1-2 days. Usual symptoms in adults include irritation in upper respiratory tract, nasal discharge, headache, chill sensation, mild cold, malaise and cough. The nasal and nasopharyngeal mucosa becomes red and swollen. Secondary bacterial infection may produce otitis media, bronchitis etc.

Pathogenesis: The virus enters the human body via respiratory tract through droplet nuclei. It lodges on the respiratory epithelial cells and infect adjacent cells. Replication takes place inside of the cytoplasm and spreaded. After incubation period it produce toxicity, which cause stopping of ciliary action and cell death. Due to nonspecific immune response mucous secretion increases and inflammatory reaction also occurs. Infection stopped due to interferon release and antibody production.

Laboratory Diagnosis: Diagnostic method includes culture and serology

Culture: Specimen: Nasal washings. Human diploid fibroblast cells are used for cultivation and incubate at 33°C, recovery between 1-7 days. Identified by cytopathic effect, Differentiation by neutralizing sera.

Prevention and Control: Multiple serotypes and antigenic drift create major problem in Rhinoviral vaccine development. Formalin inactivated, parenterally administered vaccines induce antibody in serum but not in nasal secretions. There is no antiviral drugs are available to treat common cold. Avidone, Rhodanine, Disoxaril and their analogs blocks uncoating of virus. Exviroxine inhibits the polymerase.

Hepatitis A Virus and Foot and Mouth Disease virus are discussed separately in later part of this section.

T. ORTHOMYXO VIRUSES (INFLUENZA)

Introduction

The orthomyxovirus consists of three types of influenza viruses and are designated as A, B, and C. These viruses cause influenza, an acute respiratory disease with prominent systemic symptoms. The name myxo virus was proposed originally for a group of enveloped RNA viruses characterized by their ability to absorb into mucoprotien receptors on erythro-cytes, causing hemagglutination.

Classification and Antigenic Types

Influenza viruses are classified based on the presence of antigens the **nucleoprotein,** the **hemagglutinin,** and the **neuraminidase**. The nucleoprotein antigen is stable and is used to differentiate the three influenza virus types. The hemagglutinin and neuraminidase antigens are variable.

Structure and properties of Influenza virus

Influenza viruses are spherical shaped and 80-120nm in diameter. Antisense RNA genome occurs in 8 separate segments containing 10 genes. Nucleocapsid is enclosed in an envelope consisting of a lipid bilayer and two surface glycoproteins, a hemagglutinin and neuraminidase. Membrane protein is known as matrix protein or M protein. M2 protein projects through the envelope to form ion channel. It is responsible for gene transfer.

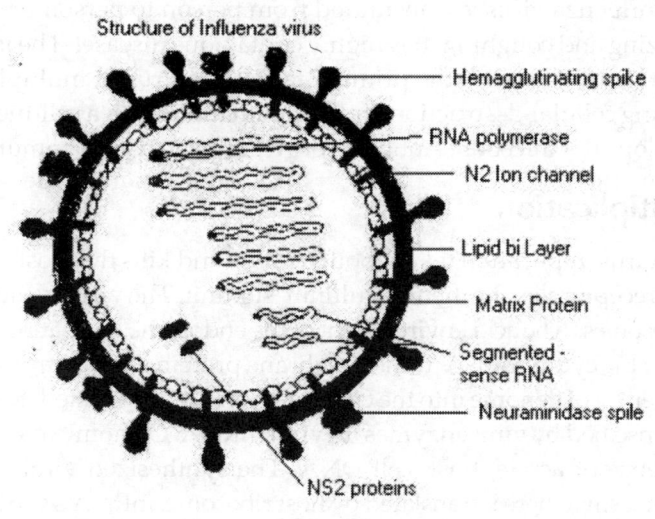

Figure 202

H gene is responsible for hemagglutinin spike and N gene responsible for Neuraminidase. Spikes measures about 10mm in length and molecular weight of 225000 Dalton. Type A usually responsible for the large outbreaks and is a constantly changing virus. New strains of type A virus develop regularly and results in a new epidemic every few years. Type B and C is fairly stable virus.

Type B causes smaller out breaks. Type C usually cause mild illness like common cold. Readily inactivated by non-polar solvents and by surface-active agents. Influenza C virus having 7 segments of RNA and only one sur-face protein. The virus inactivated by heating at 50°C for 30 minutes. It remains viable at 0-4° for about a week. Infectivity lost rapidly at 20°C. Preserved at –70° or by freeze - drying.

Gene Responsible for Viral Protein

10 genes from 8 segments of antisense RNA responsible for synthe-sis.

Symptoms

- Fever
- Cough with or without mucous
- Nasal discharge
- Muscle ache and stiffness
- Shortness of breath
- Sweating

- Stuffy and congested nose
- Sore throat
- Nosebleed
- Vomiting
- Joint Stiffness
- Loss of appetite

Pathogenesis

Influenza virus is transmitted from person to person primarily in droplets released by sneezing and coughing. It is highly contagious disease. The incubation period for influenza is 1-4 days. Alveoli is the primary target for virus. It multiplies in the respiratory mucosa, causing cellular destruction and inflammation. Both a cell-mediated response and antibody develop after infection. Antibody provides long-lasting immunity against the infecting strain.

Multiplication

Virus replication takes about 6 hours and kills the host cell. The virus attaches to sialic acid receptor via the hemaggultinin subunit. The virus is then engulfed by pinocytosis into endosomes. The acid environment of the endosome, uncoating the nucleocapsid and releasing it into the cytoplasm. A transmembrane protein derived from the matrix forms ion channel for entry of genome into the cytoplasm and is transported to the nucleus, where the genome is transcibed by viral enzymes to viral mRNA. Orthomyxo virus replication depends on the presence of active host cell DNA. The synthesized viral mRNA are transported to the cytoplasm, where it translated by host ribosome. mRNA's specifying viral membrane proteins (HA,NA,M) are translated by ribosome bound to endoplasmic reticulum and they undergoes glycosylation. The nucleocapsid is assembled in the nucleus. After the attachment of M1 protein to newly synthesized RNA, viral RNA synthesis is stopped and nucleocapsids are transported out. HA and NA proteins are transported to the cell surface and are incorporated into the plasma membrane. Virion nucleocapsids along with NS2 associate with regions of plasma membrane containing HA and NA proteins. After acquiring envelope and undergo maturation as they bud through the host cell membrane.

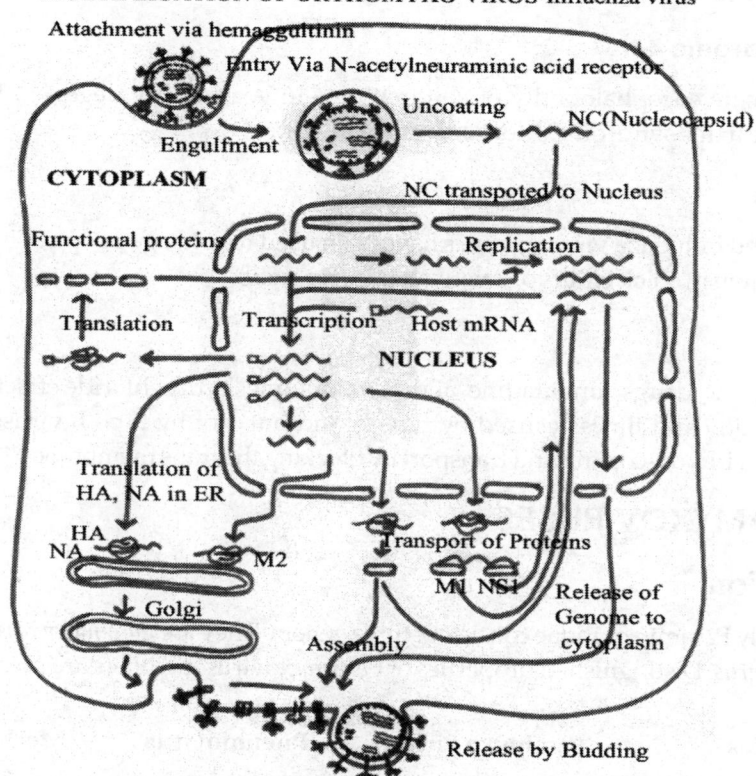

MULTIPLICATION OF ORTHOMYXO VIRUS-Influenza virus

Figure 203

Gene reassortment

Influenza virus genome is segmented, genetic reas-sorment can occur when a host cells is infected simultaneously with viruses of two different parent stains. **Antigenic shift** is the major antigenic change observed due to genetic reassortment. Smaller antigenic change is called **antigenic drift**.

Laboratory Diagnosis

Specimen: nasal wash and throat swab or sputum.

Methods Include

- ⅄ Isolation of virus
- ⅄ Cold agglutination
- ⅄ Influenza complement fixation
- ⅄ Immunoflourescent technique

Complications

Reye's syndrome

It is an acute encephalopathy of children of 2-16 years. Fatty degeneration of liver is associated with this syndrome. Mortality rate is 10-40%.

Prevention

Inactivated influenza virus vaccines have been used for old age people. The virus for the vaccine is grown in chick embryo, inactivated by formalin.

Treatment

The synthetic drugs Amantadine and Rimantadine hydrochloride effectively used to prevent infection and illness caused by type A and but not by type B viruses. The drugs interfere with virus un-coating and transport by blocking the trans membrane M2 ion channel.

U.   PARAMYXOVIRUSES

Introduction

The family Paramyxoviridae consists of three genera. They are *Paramyxovirus, Pneumovirus,* and *Morbillivirus.* Distinguished properties of Paramyxovirus members are presented in table.

Properties	Paramyxo virus	Pneumovirus	Morbilivirus
Hemagglutinin	Present	Absent	Present
Neuraminidase	Present	Absent	
Important human pathogen	Parainfluenza virus and Mumps virus	Respiratory syncytial virus	Absent Measles virus

Paramyxo virus

Structure of Parainfluenza virus & Mumps virus. Paramyxo viruses are enveloped particles with an average diameter of 120 to 300 nm. The complete virion consists of a nucleocapsid, an envelope and shows helical symmetry. It contains one molecule of single-stranded negative-sense RNA (molecular weight 5-8 x 10^6). Genome consists of the major nucleoprotein (NP), the phosphoprotein P and the L protein. The L protein is the RNA polymerase which is necessary for transcription of

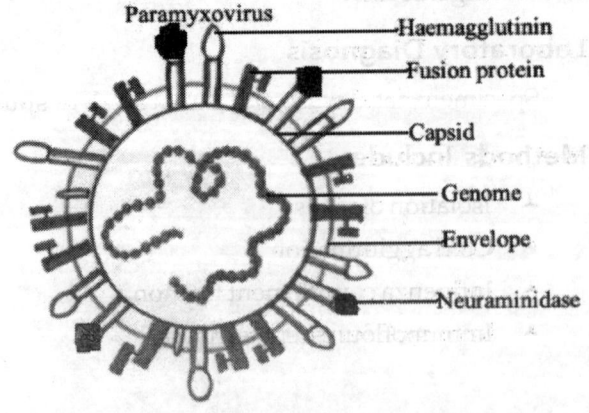

Figure 204   Paramyxo virus

viral RNA. The P protein facilitates RNA synthesis and the NP protein helps to maintain genome structure. The envelope is a double-layered membrane covered with spikes. It contains lipoproteins and glycoproteins. The nonglycosylated matrix protein (M) is attached to the inner side of the envelope. Spikes contain the hemagglutinin and neuraminidase (HN) and the cell fusion protein (F).

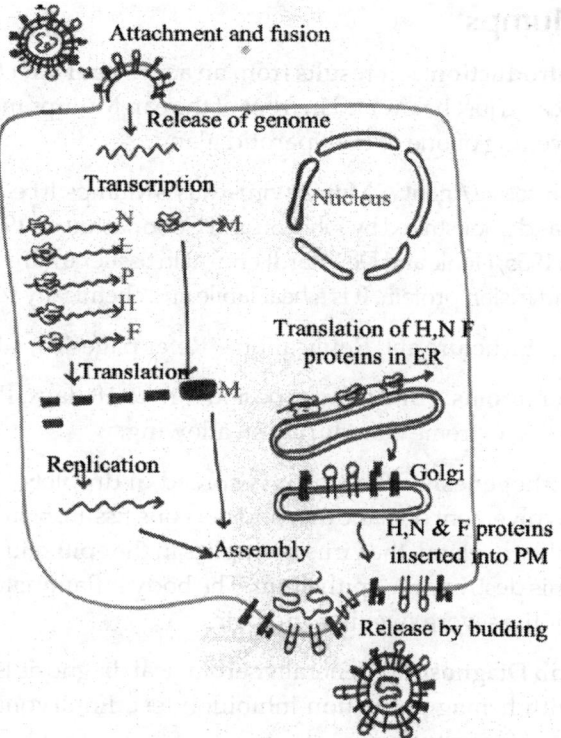

Figure 205 Replication of paramyxo virus

Symptoms

Common cold, bronchiolitis, and pneumonia.

Classification and Antigenic Types

Four human parainfluenza serotypes are now recognized: 1, 2, 3, and 4.

Multiplication

Parainfluenza viruses attach to the host cell by the hemagglutinin, which binds to the host cell neuraminic acid receptor, and then penetrate the cell by fusion with the cell membrane mediated by the F1 and F2 glycopeptides. The viral particles contain single-stranded negative-sense RNA, which cannot serve as a messenger. The virion transcriptase initiates transcription into 5-8 complementary messenger positive-sense RNA strands. They direct the viral protein synthesis and are copied into negative-sense RNA strands which are integrated in the new virions. After Assembly, progeny viruses are released by budding.

Pathogenesis

Transmission is by droplets or direct contact. The virus disseminates locally in the ciliated epithelial cells of the respiratory mucosa.

Diagnosis

Clinical symptoms are nonspecific. Laboratory diagnosis is made by detecting viral antigen, by isolating the virus, or by detecting a rise in antibody titer.

Control

No vaccine is available.

Mumps

Introduction: It results from an acute viral infection. Target of Mumps is parotid gland, is located just below and in front of the ear. Mumps means mumble. Mumps begins with painful swelling of one or both parotid gland.

Causative Agent: Mumps virus causes Mumps. It belongs to paramyxo viridae family. Viral etiology was demonstrated by Johnson and Goodpastuer in 1934. Hebel cultivated it in embroyonated eggs. In 1955, Henle and Deinhardt grew it in tissue culture. Virus posses hemagglutinin, neuraminidase and fusion protein. It is a heat labile and chemically sensitive virus.

Structure and Replication – Refer page 404, 405.

Symptoms: Incubation period – 16 to 18 days. Parotid swelling with pain is the first sign. Fever, extreme pain during swallowing.

Pathogenesis: The virus is spread in droplets. Primary infection consists of viremia and involvement of glandular and nervous tissue, resulting in inflammation and cell death. In the salivary gland, the virus multiplies in the epithelium of ducts that convey saliva to the mouth. This destroys the epithelium. The body inflammatory response to the infection is responsible for the severe swelling and pain.

Lab Diagnosis: Generally serological diagnosis is not necessary. This virus can be identified with hemagglutination inhibition test. Embryonated eggs and cell culture techniques are used for culturing.

Control: An effective vaccine is a variable and is often administered as part of the trivalent Measles, Mumps and Rubella (MMR) vaccine. It provides protection for at least 10 years.

Measles

Introduction: Measles is a highly contagious skin disease that is epidemic throughout the world. Measles virus, a member of the genus Morbili virus and the family Paramyxoviridae. Thomas Sydenham in 1690 gave the first clear and accurate description about Measles. Gold Berger and Anderson established the viral etiology of Measles in 1911 by transmitting the disease to Monkey through the inoculation of filtrates of blood and nasopharyngeal secretions.

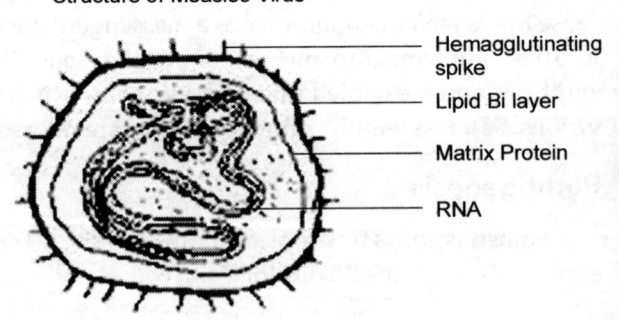

Structure of Measles Virus

- Hemagglutinating spike
- Lipid Bi layer
- Matrix Protein
- RNA

Figure 206

Structure: Measles is caused by a pleomorphic, medium sized (120-200 nm in dia) virus. It comes under Paramyxoviridae family. Its genome is RNA. It is an enveloped virus. It has two biologically active projections one 'H' is responsible for viral attachment to host cells and causes hemagglutinations. The outer 'F' is responsible for fusion of the viral outer membrane with the host cell. The M antigen also responsible for producing multi nucleated giant cells. The other name for multinucleated giant cells formed by measles is *war thin-finkeldey* cells. The virus is heat labile and readily inactivated by heat. UV rays, ether and formaldehyde.

Symptoms: Incubation period is about 10-12 days. It begins with fever, runny nose, cough and swollen weepy eyes. Within a few days, a fine red rash appears on the fore head and spread outward over the rest of the body.

Very rarely, Measles reactivation is observed after two to ten years and forms a disease called **Subacute Sclerosing Pan Encephalitis (SSPE)**, which is marked by slow progressive degeneration of the brain, resulting in death within two years.

Multiplication

Haemagglutinin and Neuraminidase spike proteins are responsible for viral attachment to host cell. F proteins mediate fusion of virus and host cell. This is followed by penetration of the nucleocapsid structures into the cytoplasm. The negative-sense RNA is transcribed with the help of host transcription factors. The order of gene transcription and translation are N, P, M, F, H and L. The virions RNA serves not only as a template for production of mRNA, but also for replication of intact RNA. After accumulation of genomic RNA and the different structural proteins in the cell cytoplasm, maturation takes place and the progeny virions are released by budding through cell membrane.

Pathogenesis: The respiratory route and conjunctiva acquire measles virus. It primarily replicates in the upper respiratory epithelium then spreads to lymphoid tissues and spreads throughout the body. Mucous membrane involvement is responsible for an important diagnostic sign **Koplicks spot** (small bluish white ulceration on the buccal mucosa). Damage to the respiratory mucous membrane leads to secondary bacterial infections. The skin rash of measles results from the cytopathic effect of measles virus replication in skin vascular endothelial cells and cellular immune response against the viral antigen in the skin.

Laboratory Diagnosis

Primary diagnosis is with the help of Kopliks spot formation.

Sample

Throat/*Nasopharyngeal swab*, Urine, Whole blood

Cytologic Diagnosis

Specimens should be fixed with formalin and stained with Hematoxylin and Eosin. Characteristic giant cells containing eosinophilic intranuclear and intracytoplasmic inclusions are observed for the first 2 or 3 days of the specimen.

Antigen Detection

It is with Immunofluorescence technique, immuno enzyme staining increase the sensitivity of the test.

Virus Isolation

Primary cultures of human embryonic kidney cell and monkey kidney cells are more sensitive for viral isolation.

Control

Children too young to be vaccinated. MMR vaccine is used.

Preschool children also vaccinated

V. ARBOVIRUS

Introduction

Arboviruses (Arthropod-borne viruses) are viruses of vertebrates biologically transmitted by insect vectors. Mosquitoes, ticks, flies and other insects transmit the virus. Arboviruses and rodent-borne viruses are placed among the Toga, Flavi, Bunya, Rhabdo, Arena and Filovirus Groups. There are more than 450 Arboviruses and Rodent-borne viruses of these about 100 are known pathogen for humans.

Characters and groups of arboviruses

Various features of arboviruses are summarized here

Characters of different arboviruses

Character	Togaviridae	Flaviviridae	Bunyaviridae	Reoviridae
Size	70nm	40-60nm	80-120nm	60-80nm
Shape	Spherical	Spherical	Spherical	Spherical
Availability of Envelope	Present	Present	Present	Absent
Genome	+SS RNA	+SS RNA	-SS RNA	DS RNA
Size of genome	9.7-11.8Kbp	10.7 Kbp	11-21Kbp	16-27Kbp
Number of Genome segments	Single	Single	Three	10-12 segments
Site of Replication	Cytoplasm	Cytoplasm	Cytoplasm	Cytoplasm
Site of Assembly	PlasmaMembrane	Endoplasmic Reticulam	PlasmaMembrane	PlasmaMembrane
Release	Budding	Cell lysis/ budding	Budding	Cell lysis
Examples	Alphavirus Chikunguya, EasternEquine Encephalitis, Venezuelan and Western Equine Encephalitis Virus, Sindbis Virus Rubivirus	Flavivirus Brazilian Encephalitis Dengue Japanese B Encephalitis Murray valley Encephalitis West Nile fever Yellow fever virus St. Louis Encephalitis virus	Bunya virus California encephalitis virus Guama virus Lacrosse virus Simbu virus Phlebovirus Sandfly fever virus Riftvalley fever virus Nairovirus	Orbivirus Coltivirus

Other arboviruses are Vesiculovirus, Arenavirus and Filovirus

Encephalitis

Encephalitis means inflammation of the brain. Many different viruses produce it. Alpha viruses, Flaviviruses, Bunya viruses may cause Encephalitis in human. This disease causing viruses are transmitted by Culex, Aedes and Iodes mosquitoes.

Characters of Causative agent: Refer Page No. 408

Pathogenesis

Pathogenesis of Encephalitis is not well studied. The virus enters in the human body through mosquito or tick bite. After entry the virus multiplies in non-neural tissue and is present in the blood 3 days before first sign of involvement of the CNS. Then the virus multiplies in the brain cells, destroyed the cell and encephalitis become apparent.

Symptoms

Incubation period is 4-21 days. Headache, chillness, fever, nausea, vomiting, generalized pain, within 24-48hours. Marked drowsiness also developed, Mental confusion, Tremors, Convolusions and coma. Mortality rate is 80%.

Lab Diagnosis

Recovery of virus

It is by the inoculation of serum with intracerebral inoculation of suckling mice. For some virus cell lines are developed.

Serology

Neutralizing hemagglutination-inhibiting antibodies are detected within few days. CF antibodies appear later.

Treatment

There is no specific treatment.

Dengue

Refer Medical Microbiology Details. Refer Page No. 553.

Rubella

Introduction: Rubella is a milder disease and often unrecognized that is difficult to diagnose. It is also called three days measles or German measles. It is a mild exanthematous fever characterized by transient macular rash and lymphadenopathy. The Rubella virus was isolated in tissue culture in 1962.

Morphology: The Rubella virus, a member of the **Togaviridae** family, causes Rubella. It is a small, pleomorphic, and spherical shaped, enveloped virus. Genome is positive sense ssRNA. It is included in the genus **Rubivirus (non arthropod borne virus).**

The virus is inactivated by ether, chloroform, formaldehyde, ß propiolactone. Destroyed at 56°C. Chemically the virion is composed of 75% protein, 18.8% lipid, 4% carbohydrate and 2.4% RNA. UV inactivates the virus within 40 seconds.

There are four types of antigens. Hemagglutinating antigens, CF antigens, Precipitating and Platelet aggregating antigens.

Symptoms: Malaise, headache, fever, mild conjunctivitis. Rash beginning on forehead and face, enlarged lymphnodes.

Incubation Period: 14 to 21 days.

Pathogenesis: The disease is transmitted via direct or droplet contact with respiratory secretions. Rubella virus multiplies in cells of the respiratory system; this is followed by viremic spread to target organs. Congenital infection is transmitted transplacentally Resultant fetal abnormalities are referred as the **congenital rubella syndrome**. It includes eye, brain damage, deafness, heart defects and low birth weight.

Laboratory Diagnosis: Hemagglutination inhibition test, EIA (Enzyme Immuno Assay), Latex agglutination, CFT and Neutralization test are used to diagnose Rubella.

Inoculation of tissue culture media with throat, blood or urine specimens are used for cultivation.

Prevention: Live attenuated Rubella vaccine is administrated to children's at 15 month of age. It was approved during the year 1969. The vaccine produces long lasting immunity. MMR also given.

Treatment: Gammaglobulin is used for treatment, but it is not a specific treatment.

W. RABIES VIRUS

Introduction

The family Rhabdoviridae consists of more than 100 single-stranded, negative-sense, nonsegmented viruses that infect a wide variety of hosts, including vertebrates, invertebrates, and plants. Common to all members of the family is a distinctive rod- or bullet-shaped morphology. Human pathogens of medical importance are found in the genera *Lyssavirus*. In Greek Lyssa means Rabies.

History

1885	- Human rabies vaccine
1903	- Diagnosis of negribodies
1940s	- Use of rabies vaccine for dogs
1954	- Addition of rabies immunoglobulin
1958	- Rabies virus grown in cultured cells.
1959	- Development of fluorescent antibody test.

Structure and properties of Rabies virus

Rabies virus is bullet shaped virus. Size of the virus is about 180 × 75 nm. Genome is negative sense single stranded RNA. Genome is nucleo capsid in nature. Matrix layer and outer envelope protects the genome. Matrix is made up of M protein. Outer envelope is made up of lipid bilayer as like plasma membrane. External envelope having spike, like projections. It is made up of glycoproteins. Spikes are responsible for pathogenic property of the virus. RNA dependent RNA polymerase is responsible for genome replication. L and P proteins control its activity. Rabies viruses of man and animals all over

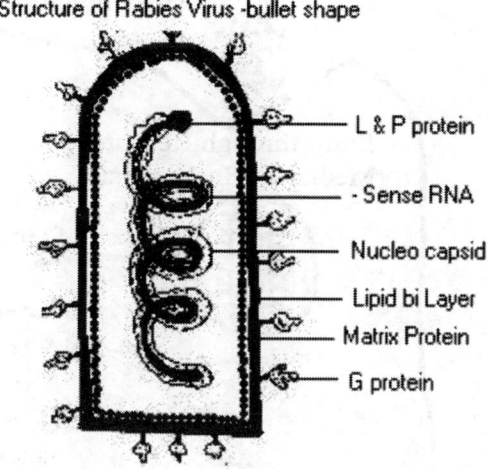

Structure of Rabies Virus -bullet shape

L & P protein
- Sense RNA
Nucleo capsid
Lipid bi Layer
Matrix Protein
G protein

Figure 207

the world appears to be of a single antigenic type. Antigens of Rabies viruses are,G protein, M protein,N protein &Hemaglutinin. Chemical compositions of the viruses are, 4% RNA, 67% protein,26% lipid & 3% carbohydrate

Multiplication

1. Virus binds to the cellular receptor and enters the cell via receptor mediated endocytosis.

2. The viral membrane fuses with the membrane of the vesicle releasing viral nucleocapsid.

3. This structure comprises negative sense RNA coated with nucleocapsid and a small number of L and P proteins, which catalyze RNA replication. Negative sense RNA is copied into 5 subgenomic mRNA by L and P proteins.

4. The N,P,M and L mRNAs are translated by free cytoplasmic mRNA

5. G mRNA is translated by ribosome's bound to the endoplasmic reticulum.

6. Newly synthesized P, N and L proteins involved in RNA replication. This process begins with positive sense RNA synthesis.

7. Positive sense RNA of the host serves as the template.

8. Some of the negative sense RNA enters to viral protein synthesis

9. G mRNA transcribes and synthesis glycoproteins.

10. G proteins travel to the plasma membrane.

11. Progeny nucleocapsid and M proteins are transported to the adjacent area of plasma membrane.

12. Assembly takes place and new viruses are released through budding process and infect new cells.

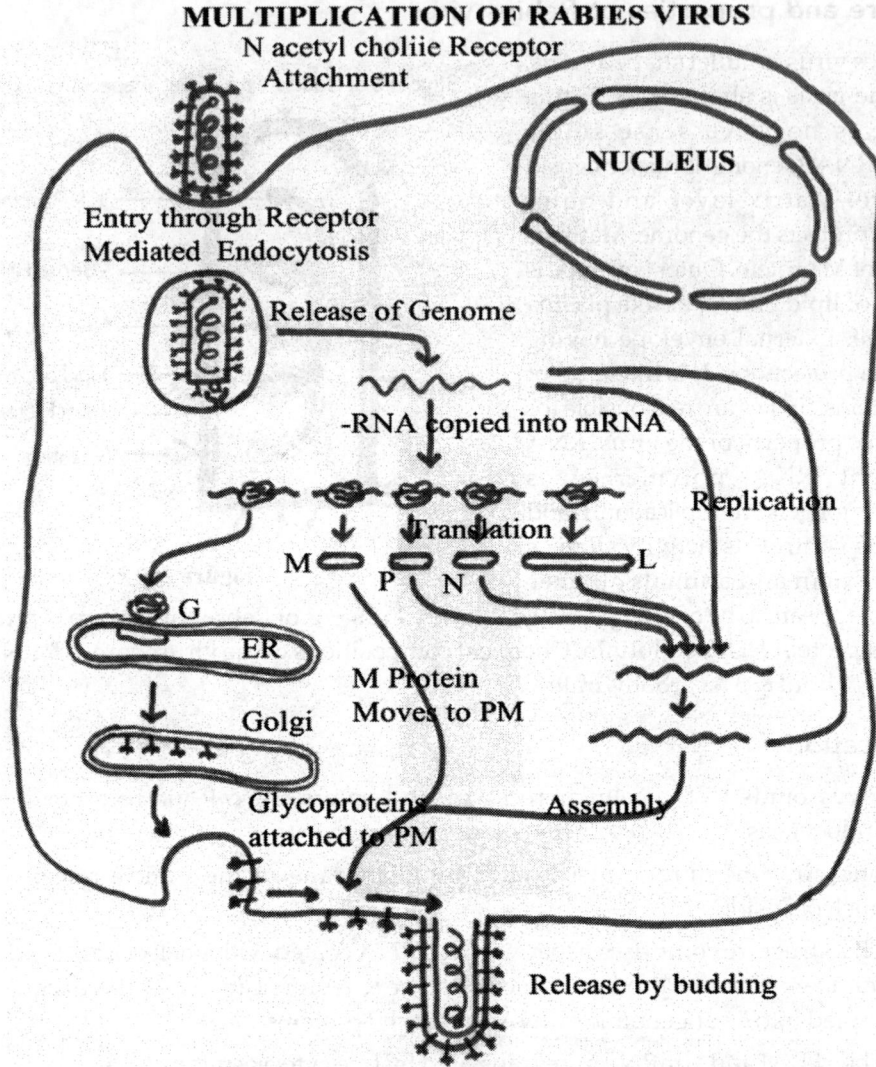

MULTIPLICATION OF RABIES VIRUS

Figure 208

Rabies

Rabies is acute, fulminant, fatal encephalitis. It is a madman disease that has instilled terror in human society. The reason are that, with a rare exceptions, all of the people who are bitten by a rabid animal, got rabies.

Rabies is an important zoonotic infection in which man is dead end of the infection. The word Rabies derived from the Latin word Rabidus, which means mad. It is an epidemic disease.

Causative agent: Rabies Virus Refer Page No. 411

Multification: Refer Page No. 411, 412

Symptoms: Incubation period is about 30-60 days. Clinical features in man. Hydrophobia, Lock jaw, Encephalitis, Hysteria, Acute polyneuritis, Polio myelitis. Clinical features in dog

Change in behavior of dog, Change in tone of bark, Change in feeding habit, Off feed, Eats abnormal objects, Fever, Vomiting, Excessive salivation,Paralysis of lower jaw, Restlessness, Convolescence, Paralysis leads to death

Pathogenesis: After deposition, rabies virus attaches to nicotinic acetylcholine receptors of tissue cells. Rabies virus may enter the peripheral nervous system directly and migrates to the brain, causing the symptoms of encephalitis. Characteristic inclusion bodies called **negri bodies**, form at the site of viral replication in the brain but the cells are not lysed. The immune response of the host probably plays an important role in pathogenesis since viral antigens are expressed on the surface of the infected cells.

Labdiagnosis

Specimen: Saliva / sputum, Skin biopsy, Hair follicle, Cerebrospinal fluid, Blood, Corneal swab, Urine, Brain

Laboratory tests: Negri body examination, Fluorescent antibody test, Complement fixation test, Enzyme Linked Immuno Sorbent Assay, Immuno peroxidase test and Hemagglutination test.

Negri body examination: Negri body is a Intracytoplasmic inclusion bodies, Cut small piece of the brain and place it on a glass slide, Stain the smear with Seller staining procedure. Examine under 100 X for the availability of negri bodies

Observation

Nerve cells	:	Blue cytoplasm and dark blue nucleus
Stroma	:	Pink
Erythrocytes	:	Copper colored
Negri bodies	:	Magenta to dark red with dark blue or black inner granules.

Freshly isolated strains of Rabies virus are called **street virus**.

Treatment

Management of wounds: Animal bite area is washed with soap initially. Apply 1- % cetrimonium bromide as antiseptic along with antirabies serum.

Post exposure immunization: Immunization Must be started at the earliest to ensure that the individual will be protected before the rabies virus invades central nervous system.

Two types of agents are employed to confer immunity to an individual who has been exposed to the Rabies virus: antirabies serum/ rabies immunoglobulin and anti rabies vaccine.

Antirabies vaccines available in India: Semples Sheep Brain Vaccine. Human Diploid Cell Vaccine (HDCV). Primary Chick Embryo Cell Vaccine (PCECV). Purified Vero Cell Rabies Vaccine (PVRV)

Site of vaccination: The ideal site for vaccination is the anterior abdominal wall, this area offers enough space to accommodate 10 injections at 10 different sites and cause least discomfort to the patient.

X. HERPES VIRUS

The Herpes virus family contains most important human pathogens. More than 50 herpes viruses are available in the animal world but only 8 are associated with human infections. The outstanding property of herpes viruses is their ability to establish **life long persistent infections** in their host and to undergo periodic reactivation.

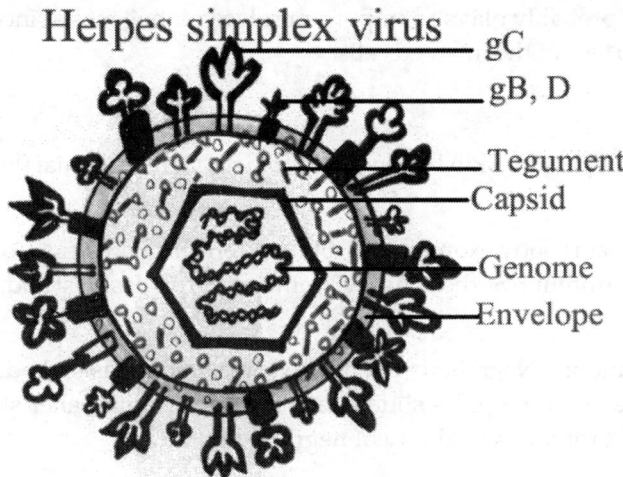

Figure 209

General properties of Herpes Viruses

Herpes viruses are large viruses. Virion is spherical shaped enveloped virus. Capsid is Icosahedral in shape. Linear double standard DNA, 124-235kbp. More than 35 proteins are available in virion. Virion Replicates is in the nucleus. Genome is large enough to code for at least 100 proteins.

Replication

Virion binds to the extracellular protein through gB and gC receptor. Another viral protein gD interacts with a second cellular receptor. This interaction mediates fusion of virus with host plasma membranes. The virus is uncoated, liberating tegument proteins and nucleocapsid into the cytoplasm. Viral nucleocapsid docks at the nuclear pore and release viral DNA into the nucleus, where the DNA circularizes. VP16 enhance transcription of viral genome and stimulate transcription of immediate early genes by host cell RNA polymerase II. Immediate early mRNAs are spliced and transported to the cytoplasm, where they are translated. Immediate early proteins (α proteins) are imported into the nucleus, where they activate the transcription of early genes. β protein genes are transported to the cytoplasm after transcription and are translated. β proteins are imported to the nucleus where they induce DNA replication and synthesis substrate for DNA synthesis. DNA replication produces long concatameric DNA molecules, the templates for late gene expression. Late mRNAs are transported to the cytoplasm and synthesis of gamma protein. These proteins are structural proteins and are needed for viral assembly. Some late proteins are inserted to ER and are transported to Golgi apparatus for glycosylation. Mature glycoproteins are transported to plasma membrane of the infected cell. Some gamma proteins are transported to the nucleus for assembly of nucleocapsid and DNA packaging. Newly replicated viral DNA is packaged into preformed capsids. These capsids, together with some tegument proteins bud from the inner nuclear membrane into the lumen of ER and acquire envelope. Enveloped virus then transported to the PM for release by Exocytosis.

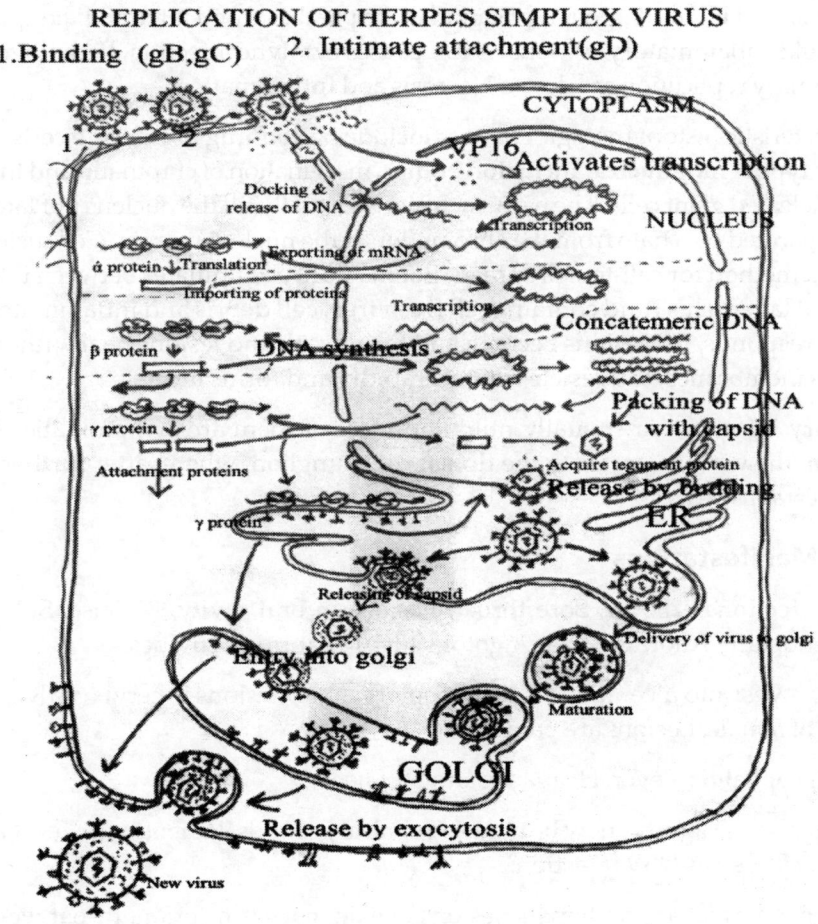

REPLICATION OF HERPES SIMPLEX VIRUS
1.Binding (gB,gC)   2. Intimate attachment(gD)

Figure 210

Latent infection occurs primarily in neurons found in sensory and autonomic ganglia. During this infection Latency Associated Transcript (LTT) promoter is synthesized and are involved in protein synthesis.

Simplex Virus

These viruses are extremely widespread in the human population. The virus is spread by direct contact with infected secretions. These are two distinct herpes simplex viruses, they are type 1 and type 2 (HSV 1 & HSV 2). Both types share many common antigens.

Herpes Simplex Virus type 1 and 2 infect epithelial cells and establishes latent infection in neurons. Type 1 is associated with respiratory tract. Type 2 primarily infects the genital tract and is mainly responsible for genital herpes. Both viruses can cause neurological complications.

Type 1 HSV is spread by contact with infected saliva, where as HSV-2 transmitted sexually.

Pathogenesis: HSV undergoes replication in the parabasal and intermediate epithelial cells, which invoke inflammatory response. HSV causes cytolytic infection. Pathological changes during primary replication are due to Necrosis and Inflammation.

Characteristic histopathologic changes include **ballooning of infected cells**, production of **cow dry type A intranuclear inclusion bodies**, margination of chromatin and the formation of multinucleated giant cells. The early inclusions virtually fill the nucleus but later condense and are separated by a halo from the chromation at the nuclear margin. Cell fusion provides an efficient method for cell-to-cell spread. Edema fluid accumulates between the epidermis and dermal layer. This fluid contains cell free virus, cell debris and inflammatory cells. As white cell responds, the lesions become and forms scab and lesions heal without scarring. In mucous membranes, the vesicle rupture rapidly and forms ulcers.

Primary infections are usually mild and most of them are symptomatic. After initial replication, the virus migrates to the dorsal root ganglion, where, after further replication latency is established.

Clinical Manifestations

Primary infections: Fever, Sore throat, Lesions in oral cavity, Malaise, Submandibular lymphadenopathy Anorexia, Pharyngotonsilitis, Keratoconjuctivities.

Genitalia: Vesiculo ulcerative lesion in penis of male, Lesions in cervix, vulva, vagina and perineum of female. Lesions are painful.

Brain: Encephalitis, Fever, Headache and Stiff neck.

Lab diagnosis: Diagnosis may be made on clinical grounds. Patients with fever and vesicles may be considered as HSV infection.

Specimens: Vesicular and hepatic lesions of skin, cornea or brain, throat washings, CSF and stool.

Isolation of virus: Virus isolation is one of the most reliable methods for confirmation of the clinical diagnosis. The specimens are transported through viral transport medium and inoculated into tissue culture. HSV has a wide host range, many cell culture system are susceptible. The appearance of typical cytopathic effects in cell culture in 2-3 days suggests the presence of HSV.

Serology: Several methods have been developed for rapid diagnosis. Antibodies of HSV are measured by Neutralization test, CF, ELISA, RIA and immuno fluorescence tests.

Treatment: Most of the drugs used for the treatment of HSV, inhibit the DNA synthesis. Vidarabine triphosphate inhibits DNA polymerase. Acyclovir targeting HSV infected cells and viral DNA polymerase. Topically applied Idoxuridine, Trifluridine, Vidarabine and Acyclovir have been used for Hepertic keratitis.

Varicella-zoster virus (Chicken pox)

VZV infection is a common childhood infection of the world. Varicella (chicken pox) is a mild, highly contagious disease, chiefly occurs in children, characterized by a generalized vesicular eruption of the skin and mucous membrane. Zoster (shingles) is a sporadic, disease of adults or immunocompromised individuals that is characterized by a rash limited in distribution to the skin.

Primary infection causes chickenpox and reactivation of the virus in the later life results in shingles.

Pathogenesis: The route of infection is the respiratory mucosa. The virus circulates in the blood and undergoes multiple cycles of replication and eventually localizes in the skin. Lesions of varicella infection are associated with cutaneous and mucosal endothelial cells. Swelling of epithelial cells, ballooning degeneration and the accumulation of tissue fluids results in vesicle formation. Eosinophilic inclusion bodies are found in the nuclei of infected cells. Multinucleated giant cells are common.

Zoster lesions are histopathologically similar to varicella. There is also an acute inflammation of the sensory nerves and ganglia. It is not clear what triggers reactivation of latent Varicella-Zoster virus infection in ganglia.

Symptoms: Incubation period is of 10-23 days. Malaise and fever are earliest symptoms. Earlier symptoms followed by rash characteristically begins on the scalp and trunk and spreads. Lesions appear on the Mouth, Rectum and Vagina. Other symptoms include Headache, Sore throat, and loss of appetite and Irritability. The most common complication of Zoster is **Post Herpatic Neuralgia (PHN).** Pain may be characterized by burning, itching or tingling sensations.

Lab Diagnosis: Demonstration of multinucleated giant cells and type A intranuclear inclusion bodies. CF, Neutralization test are used a serological test

Treatment: Vidarabine and Acyclovir are useful for treatment. Zoster Immuno Globulin (ZIG) also useful.

Cytomegalovirus (CMV)

This virus is ubiquitous in nature. They cause variety of human infections. CMV produces cell enlargement and was first described in 1904. In 1956, Smith first isolated human salivary gland virus and are later described as Cytomegalovirus by Weller.

CMV has the largest genetic content of the Human Herpes Viruses. Its DNA genome (240kbp) is significantly larger. It produces characteristic cytopathic effect and produces perinuclear cytoplasmic inclusions.

Virus is transmitted through person-to-person contact. Prevalence is related to socioeconomic condition. Infection rates increase in early childhood and peak at 1-2years of age.

Potential source of virus include saliva, urine, semen, breast milk, cervical and vaginal secretions.

Pathogenes: Incubation period is about 4-8 weeks. This virus causes infectious mononucleosis. Most CMV infections are sub clinical in nature and also establish life long latent infections. Salivary gland involvement is common and probably chronic. CMI is depressed with primary CMV infection. It may take several months for cellular responses to recover.

Symptoms: It causes infectious mononucleosis like syndrome. Malaise, Myalgia, Fever, Liver function Abnormalitis, Lymphocytosis are generalized symptoms.

Lab diagnosis

Specimens: Throat washings, Urine, Cervical swab and Blood

Techniques: Electron Microscopy, Immunohistochemical staining of tissues and Nucleic acid hybridization technique are used for diagnosis. Virus is isolated with the help of Human Fibroblast Cultures. Culture may be positive with in 1-2 weeks. Antibodies may be detected by Neutralization test, CF, RIA or immunofluorescence test.

Treatment and control: Ganciclovir and Acyclovir used to treat life-threatening infection. Vidarabine and Interferon's also used for treatment. Active and Passive immunization as well as antiviral agents are used for prevention.

Epstein-Barr Virus

Epstein Barr virus is a ubiquitous herpes virus, which causes infectious mononucleosis, and also induces nasopharyngeal carcinoma, Burkitts lymphoma and other lymph proliferate disorders.

In 1964, Epstein, Achong and Barr first discovered the virus. Henles developed serological methods for diagnose EBV.

EBV is a distinct human herpes virus and its genome consists of about 172kbp. Incubation period is 30-50 days. B Lymphocyte is one of the target cells for EB virus.

EBV requires close person-to-person contact for transmission. Saliva is the main source of infection. Viral replication occurs in epithelial cells of the pharynx and salivary glands. Following replication in epithelial cell, the virus infects B lymphoid cells, where it persists in a latent state. Infection of B-lymphocytes induces polyclonal proliferation and EBV antigen expression and a subset of these cells becomes permanently infected.

Symptoms: Burkitt's lymphoma – a tumour of the jaw, Nasopharyngeal carcinoma – cancer of epithelial cells, Infectious mononucleosis, Fever, Headache, Malaise, Fatigue, Sore throat, Pneumonitis, Hepatitis and Hematologic abnormalities.

Lab diagnosis: Saliva, Peripheral blood, Lymphoid tissue are used as Spermen.

Nucleic acid hybridization is the most sensitive means of detecting EB virus. Indirect Immunofluorescence and ELISA are also used for virus aborag detection.

Treatment and control: Acyclovir reduces EB virus shedding. There is no vaccine.

Y. ADENOVIRUS

Adenovirus was first isolated in 1953 from adenoids and tonsils. Adeno viruses are especially valuable systems for the molecular biology and biochemical studies of eukaryotic cell process. Adenovirus infections are common in young childrens. The name adenovirus is because it was first isolated from adenoids. This virus is associated with respiratory tract, conjunctiva and digestive tract. Adenoviruses are wide spread in nature and have been isolated from animal species ranging from frogs to humans.

The family adenoviride is subdivided into two genera. They are Mastadeno virus and Aviadeno virus

The human adenoviruses comprise 49 distinct serotypes that are grouped into 6 serogroups based on various immunologic, biologic and bio-chemical characteristics.

The name adenovirus was given during the year 1956. The adenoviruses were cultured and reported as unique viral agents in 1953.In 1963, Trentin demonstrated that Ad 12 caused tumours in rodents.

Structure

The adenovirus particle consists of an icosahedral protein shell surrounding a protein core that contains the linear, double-stranded DNA genome. The shell, which is 70 to 100 nm in diameter, is made up of 252 structural capsomeres. The 12 vertices of the icosahedron are occupied by units called pentons, each of which has a slender projection called a fiber. The 240 capsomeres that make up the 20 faces and the edges of the isocahedral shape are called hexons because they form hexagonal arrays.

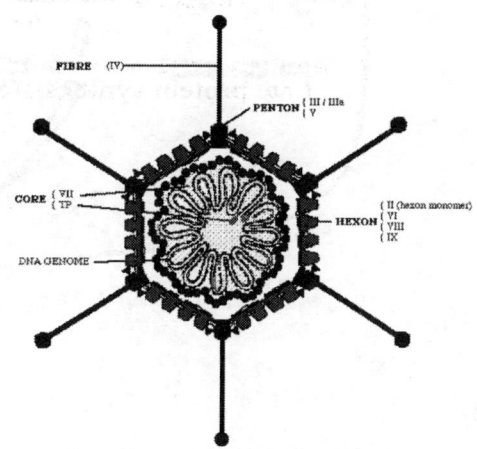

Figure 211   Adenovirus

The shell also contains some additional, minor polypeptide elements. The core particle is made up of two major proteins (polypeptide V and polypeptide VII) and a minor arginine-rich protein (μ). Genome of adenovirus has different types of genes, which are expressed in a sequenced way. The order of expressions are as follows.E1A, E1B, E2A, E2B, E3, E4, L1, L2, L3, L4 and L5. VARNA, IVA2.

Multiplication

Completion of adenovirus replication cycle requires 24-30 hours. Adenoviruses fiber attaches to a specific receptor on the cell surface (Epidermal growth factor). The virion is then taken into endocytic pits, where it enters the cytoplasm. Uncoating takes place and the viral genome associated with core protein VII is imported into the nucleus. Host cell RNA polymerase II transcribe the immediate early E1A gene. E1A mRNA is transported to the cytoplasm. E1A proteins are synthesized by the cellular translation machinery. These proteins are extensively modified by phosphorylation and are imported into the nucleus.

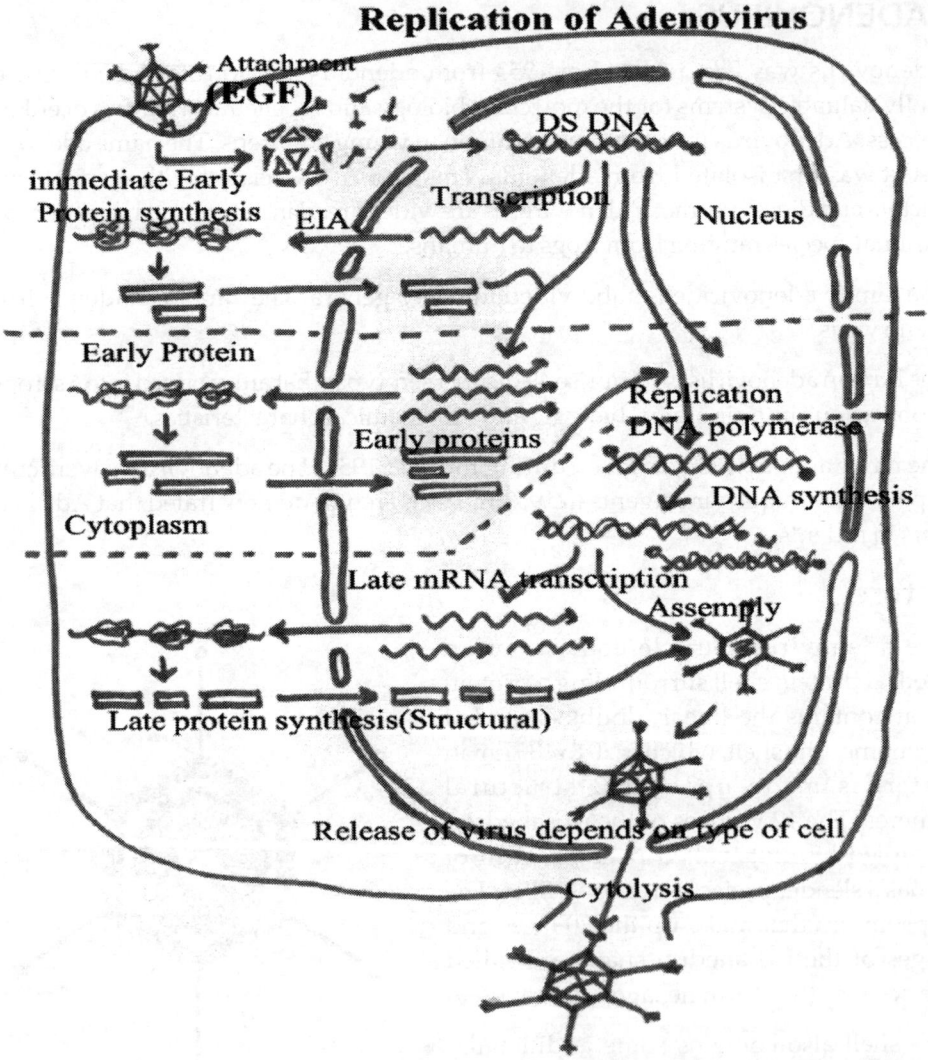

Replication of Adenovirus

Figure 212

The larger E1A proteins stimulates transcription of cellular early viral genes by cellular RNA polymerase. Early mRNAs are translated in to the early proteins. These early proteins include

DNA polymerase DNA binding proteins viral DNA replication initiation proteins. Early proteins cooperator in viral DNA synthesis. Replicated DNA serves as a template for further rounds of replication. DNA helps in transcription of late genes. Processed mRNAs are exported to cytoplasm.

Late mRNA efficient translation needs $_{VA}$RNA and $_{VA}$RNA I. Processed viral proteins are transported to the nucleus and helps in assembly. With in nucleus capsids are assembled and form non-infectious immature virion. By the action of protease mature virions are formed. Progeny virions are released usually upon destruction of the host cell.

Pathogenesis

Adenoviruses infect and replicate in epithelial cells of the respiratory tract, eye, gastro intestinal tract, urinary bladder, and liver. They usually do not spread beyond the regional lymph node.Some of the viruses persist as latent infection for years in adenoids and tonsils and are shed in the faeces. Infection may be productive, abortive, or latent. In productive infections, the viral genome is transcribed in the nucleus; mRNA is translated in the cytoplasm, and virions self-assembled in the nucleus. In latent infections, in transformed and tumor cells, viral DNA is integrated into the host genome. Virus-host DNA recombinants are also found in productive infections.

Adenovirus Host interaction may leads to the following types of infection in human.

- Productive infection – complete replication of infections virion & release of virus.
- Abortive infection – Synthesis of viral protein without production of infection virion.
- Semi permissive infection – Complete replication with low yield of infection virion.
- Malignant transformation – Associated with integration of viral DNA and differential viral & cellular gene expression.
- Viral latency – persistence of viral genome in the host cell.

Symptoms

Acute febrile pharyngitis, Pharyngo conjuctival fever, Pharyngitis, Fever, Cough, Pneumonia, Conjunctivitis, Gastroenteritis, Acute hemorrhagic cystitis and Urethritis.

Lab diagnosis

Virus is cultivated in primary humans embryonic kidney cells.

Complement fixation test and ELISA.

Control

There is no treatment. Whole-virus vaccines are not used because of the potential risk of oncogenesis. Other vaccines, including recombinant vaccines, are under development, but adenoviruses do not represent a serious health hazard.

Z.   HUMAN IMMUNO DEFICIENCY VIRUS(HIV)

Introduction

HIV is a human virus and the name HIV was given by International Committee on Virus Nomenclature in 1986. Montagnier and his colleagues first reported isolation of HIV in 1983 from the Pasteur Institute, Paris. HIV is classified as a retrovirus because it contains reverse transcriptase. It belongs to the genus Lentivirus . Two major antigenic types (HIV-l and HIV-2) have been identified and are readily distinguished by differences in antibody.

Structure

Virus is spherical in shape, about 90-120nm in size. HIV is an enveloped virus with cylindrical core inside. Envelope of this Virus contains 72 external spikes. Spike is made up of , gp 120 & gp41 proteins. The core contains two copies of ssRNA and several enzymes. Ten virus specific proteins have been discovered. The core of HIV 1 contains 4 nucleocapsid

proteins. The phosphorylated p25 polypeptide forms the chief component of the inner shelf of the nucleocapsid, where as the p17 contains 2 copies of single stranded RNA that is associated with the various preformed virus enzymes, including Reverse Transcriptase, Integrase, Ribonuclease and protease.

HIV is a thermolabile, being inactivated in 10 minutes at 50°C & in seconds at 100°C. At room temperature, in dried blood it may survive for upto 7 days. HIV is inactivated within 10 minutes by the treatment with 50% Ethanol, 3.5% Isopropanol, .5% Lysol, 0.5% Paraformaldehyde, 0.3% H_2O_2, 10% Household bleach. For treatment of contaminated medical instruments 2% solution of glutaraldehyde is useful.

Figure Refer Page No. 62.

AIDS

HIV causes very dangerous disease in human called AIDS. It is a condition where there is a deficiency in the Body's Natural Defense mechanisms or the Immune System. It is acquired because it is not a hereditary or due to long-term use of some medicines such as those for the treatment of cancer, because of certain behavioral pattern. Syndrome is a group of symptoms. When one gets AIDS there can be wide range of symptoms all due to the bodies diminished ability to fight disease. AIDS is one of the Sexually Transmitted Disease, worldwide distribution, and epidemic disease. It was first described in USA in 1981.

Transmission

HIV is believed to have originated in Central Africa. From here it is spread to the rest of the world. HIV is primarily transmitted by,

Homo and Hetero Sexual contact, Direct exposure of a person's bloodstream to body fluids, Mother to child through placenta, Intravenous drug abuse, Transfusion process, Invasive medical procedure, Drug abuse with needle sharing and New born can infected through Breast breeding.

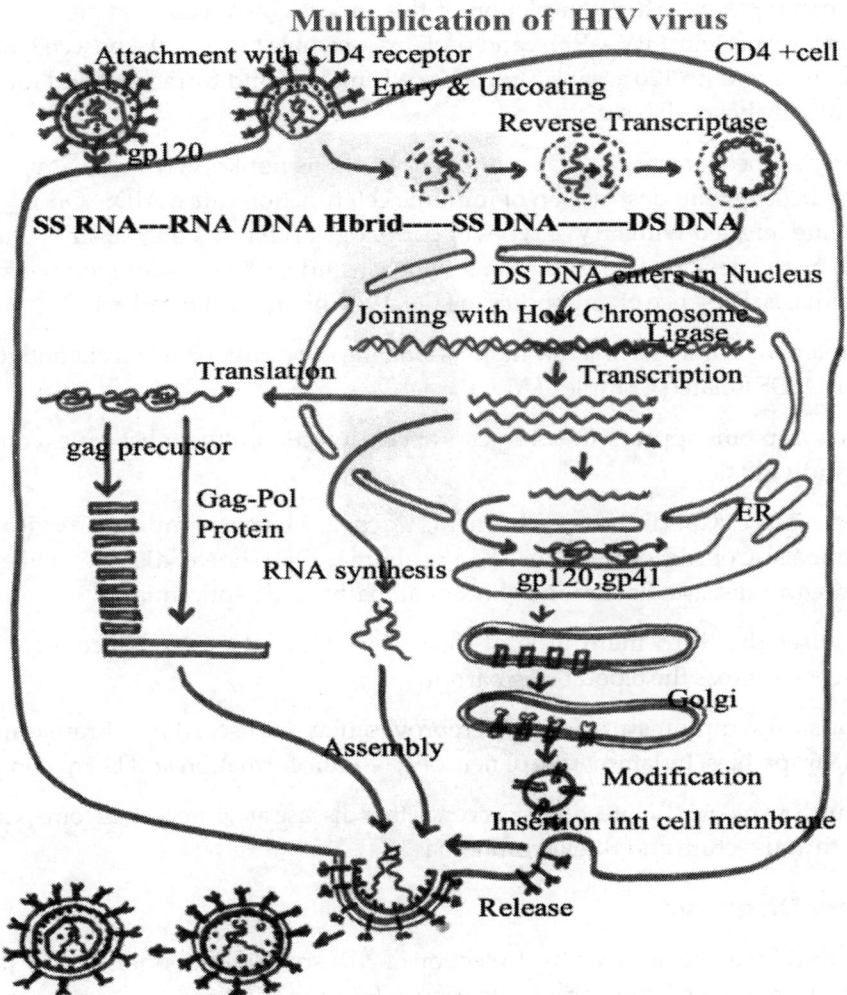

Figure 213

Pathogenesis

Once virus enters inside of the body the virus gp 120-envelope protein binds to the CD4 glycoprotein plasma membrane receptor on CD4+ cell. After the envelope has fused with the plasma membrane, the virus releases its core protein and 2 RNA strand in the cytoplasm. Inside the infected cell, the core protein remains associated with the RNA as it is copied into single stranded DNA by Reverse Transcriptase enzyme and the ss DNA is duplicated to from a dsDNA copy of the original RNA genome.

The viral dsDNA is then translocated to the nucleus and integrated into the host chromosomal DNA by the viral Integrase enzyme. This integrated viral DNA and chromosomal DNA is called **Provirus**. Then transcriptional factors stimulate transcription of proviral DNA into genomic ssRNA and after processing several mRNAs are formed. Then viral RNA is exported to cytoplasm. After completion of this process, host cell enzymes catalyses the synthesis of viral protein. HIV ssRNA and proteins assemble beneath the host cell membrane, into which gp41 and gp 120 are inserted. The cell enlarges and form bud. Bud forms a new virus. Eventually host cell lyses.

The precise mechanism of AIDS pathogenesis still is not known and many hypothesis exist. Many believe that, destruction of immune cell function cause AIDS. Once a human's CD+4 cells are infected with HIV, 4 types of pathological changes may ensure. First, a mild form of AIDS may develop with symptoms, which include, Fever, Lymphnode enlargement, Oral Candidiasis, Presence of antibodies to HIV, Weight loss and Head ache.

These symptoms occur in the first few months after infection, last for 1-3 weeks and recur. This is known as AIDS Related Complex (ARC).

Second symptoms appears after 2-8 years of HIV infection, although it varies considerable with each individual.

Symptoms are, Candidiasis of bronchi, trachea, Herpes simplex, Cervical cancer, Histoplasmosis, Coccodioidomycosis, Lymphoma, Diarrhoeal disease, Tuberculosis, Cytomegalovirus disease, Pneumonia, Encephalopathy and Septicemia.

HIV causes the third main type of disease involves the CNS, since virus infected macrophages can cross the blood brain barrier.

The classical symptoms are, Fever, Cerebrovascular disease, Ataxia, Brain tumor, Auto immune neuropathies, Inflammation of neurons, Nodule formation and Demyelination.

The fourth result of HIV infection is cancer. Other disease are, Kaposis sarcoma, Carcinoma of the mouth and rectum and B cell lymphoma.

Laboratory Diagnosis

HIV infection can be detected by, Detection of HIV specific antibodies (ELISA), Western blot, Polymerase chain reaction and Estimation of IgG and IgA level.

Prevention

Avoid sexual contact with HIV infected individuals. Don't share shaving materials. Avoid drug abuse. Follow Tamil culture. Screen blood before transfusion. Use condoms during sexual contact.

AA. BLUE TONGUE VIRUS

Blue tongue virus causes an infection in sheep. It change the color if the tongue (blue). This is the reason behind the naming of this virus. This virus is included under the family Reoviridae and the genus Orbivirus.

Structure

Size of the blue tongue virus 69nm in diameter. Virus is icosahedral in shape and has 32 capsomers. Capsid contain inner core and outer protein. Capsid contain two major proteins

(VP3 & VP7) and two minor proteins (VP2 & VP4). Genome is 10 segments of DS RNA. Outer protein coat certain two particle (VP2 & VP5)

Replication of Virus

Blue tongue virus adsorbs the host cell surface by VP2 and VP5 ligands. Upon entry uncoating taken place, which releases RNA genome in to the cytoplasm. RNA dependent RNA polymerase initiates the transcription of DS RNA and produce mRNA and functional protein is synthesized. Replication of DS DNA happens by making use of RNA polymerase (Replicase). Structural proteins are synthesized in the cytoplasm. Assembly of virus takes place by cell lysis.

Pathogenesis

Blue tongue viruses cause acute infection. Vector is responsible for transmission of this virus. Multiplication of virus take place in spleen, bone marrow, endothelial cell linings. Incubation period varies from 2 days to 2 weeks.

Symptoms

Fever, Inflammation and cyanosis of mucous membranes of mouth, nose and alimentary tract. Abortion and deformities of fetus may occur. Edema results from damage of blood vessel. Mortality rate is 5-30%.

Laboratory Diagnosis

Viruses isolated from RBC, Neutralization and fluorescent antibody tests are used to detect blue tongue virus. Molecular tests such as RNA finger printing and hybridization may be performed to detect blue tongue virus.

Prophylaxis

Insect vector control. Live attenuated vaccines are used for prevention. Animals may be protected by dipping or spraying anti viral compounds.

AB. RANIKHET DISEASE (NEWCASTLE DISEASE)

Newcastle disease (also called Ranikhet disease) is one of the major diseases limiting chicken production worldwide. The disease was first reported from Java in 1926 but outbreaks were soon reported from Newcastle in Britain, Ranikhet in India and Colombo in Sri Lanka. It soon spread, either naturally or by being transported between countries in refrigerated meat. It now has world-wide distribution.

Other names

Fowl pest, pseudofowl pest, new castle disease and avian pneumoencephalitis.

Causative agent

The disease, which affects birds, is caused by a **Paramyxovirus and the** Genus *Avulavirus.* The incubation period is between 2 and 18 days. It is a spherical shaped virus with 125-

250nm diameter. It is an enveloped virus with nucleo capsid. Envelop contain Haemagglutinin – Neuraminidase (HN) and fusion (F) protein. The virus has non segmented negative sense single stranded RNA genome. RNA is enclosed in the transcriptase complex. It contain nucleocapsid (NP), Large Polymerase (L) and phophoprotein.

Figure : Refer page No. 404

Replication

Refer Page No. 405

Pathogenisis

Ranikhet dion is a highly infectious disease of poultry. It affects chicken, turkeys, ducks, pigeons etc. Different strains of this virus able to infect birds with different clinical manifestation. Virus enters in to the respiratory tract of birds and affect mucosal cells. Virus spread from RT to other organs through blood. Poultry workers are high-risk group affected by this virus.

Symptoms

Diarrhoea with watery and green faeces; foul odor; discharge from the nose; coughing and sneezing;swelling of eye,swelling of the head; head and neck twisted to one side; drooping wings, **Decreased egg production** ,dragging legs; sleepiness; full, distended drop; convulsions and paralysis; death.

Ranikhet Disease is very dangerous and can kill chickens in large numbers. Once a chicken has been attacked by RD it cannot be treated.

Diagnosis

Clinical diagnosis are Sudden decrease in egg production, High morbidity and mortality and Characteristic signs and gross lesions

Lab diagnosis0

- ⋏ Isolation of virus is done by making use of chick embryo fibro blast technique.
- ⋏ Growth of this virus is identified by detecting cytopathic effect on cell culture.
- ⋏ Viruses are also detected by haemadsorption

Prevention and Control

- ⋏ No treatment.

Propylaxis

- ⋏ Development of virus free flocks.
- ⋏ Live viral vaccines prepared and administered intranacelly or intradermally.

AC. FOOT AND MOUTH DISEASE (FMD)

Foot and mouth disease mainly observed in cattles especially in sheeps. It is caused by **Aphthovirus** that belongs to the family **Picornaviridae**.

Structure

FMD is a non-enveloped, Icosahedral virus. Size of the virus is about 27nm. Positive sense single stranded RNA is a genome. Capsid protein contain four types of protein, that are VP1, VP2, VP3 and VP4. VP4 is associated with the genome. Sedimentation coefficient of this virus is 142-146S.

Figure Refer Page No. 398

Replication

Air, food and water allow the entry of virus into the animal. Virus attaches to host cell through receptor. Viral genome enters in to the host cell cytoplasm through unknown mechanism. After uncoating viral genome it self act as a mRNA and perform translation leads to the synthesis of poly protein. It is processed by viral protease and converted in to VP1, VP2, VP3, VP4. Replicase enzyme mediate the replication of genome, initially it produces – RNA. Which is happened special structure called vesicle. RNA acts as a template for synthesis of +RNA molecule & act a mRNA. Assemble taken place in the cytoplasm and the viral particles released by lysing the host cell (cytopathic effect).

Figure Refer Page No. 398, 399

Pathogenisis

FMD is an acute, highly infectious disease of cattle, goats, pigs, sheeps etc. Incubation period occurs upto 20days. Food is a main source for this virus. Virus mainly affect food and mouth mucosal and other epithelial cells.

Symptoms

Fever, Formation of vesicles (fluid-filled blisters) on mucous membranes of mouth & feet. Saliva contain viruses and also found in milk, blood, urine and faces leads to loss of productivity.

Lad diagnosis

Culturing of virus and Demonstration of antibodies by complement fixation test.

Control

Slaughter of infected animal, Disinfection of shed by sodium hydroxide solution. Submit vaccines are available to protect infection.

AD. RINDERPEST

Introduction

Rinderpest (RP) is a contagious viral disease of cattle, domestic buffalo, and some species of wildlife. It is characterized by fever, oral erosions, diarrhoea, lymphoid necrosis and high mortality. Rinderpest was established as an infectious disease in 1754.

Causative Agent

Rinderpest is caused by Rinderpest virus (RPV). It is a single-stranded RNA virus in the family Paramyxoviridae, genus *Morbillivirus*. It is immunologically related to canine distemper virus, human measles virus. There is only one serotype of rinderpest virus. Rinderpest virus is a relatively fragile virus.

Figure Refer Page No. 404

Physical and Chemical Properties of Virus

Sunlight is lethal, and the vaccine must therefore be kept in a brown bottle and protected from light; virus in a thin layer of blood is inactivated in 2 hours. The virus is very sensitive to heat, and both lyophilized and reconstituted virus should therefore be kept cold; lyophilized virus stored at -20° C is viable for years. Vaccine is more stable in a saline solution; reconstitution in a molar concentration of sulfate ions greatly increases resistance to heat. Rinderpest virus is rapidly inactivated at pH 2 and 12 (10 minutes); optimal for survival is a pH of 6.5-7. The virus is inactivated by glycerol and lipid solvents.

Symptoms 1

There is sudden onset of fever, followed by depression, loss of appetite, reduced milk production, nasal and eye discharges and laboured, rapid breathing. Irregular erosions appear in the mouth, lining of the nose and genital tract. Acute diarrhoea is a common feature of the disease. Most animals die 6–12 days after onset of clinical signs.

Epidemiology

Incubation period is 3-15 days. High morbidity rate, mortality rate is high with virulent strains but variable with mild strains

Hosts — Cattle, zebus, water buffaloes and many species of wild animals: African buffaloes, eland, kudu, wilde-beest, various antelopes, bushpigs, warthog, giraffes, etc. Sheep, goats are susceptible

Transmission — By direct or close indirect contacts

Laboratory diagnosis

Antigen detection

Agar gel immunodiffusion test ,Direct and indirect immunoperoxidase tests, Counter immunoelectrophoresis and Immunohistopathology

Virus isolation and identification

Virus isolation by cell culture technique , identification by Virus neutralization, Immunoperoxidase staining, Rinderpest-specific cDNA probes and Amplification by polymerase chain reaction (PCR)

Serological tests

ELISA and Virus neutralisation

Prevention and Control

No treatment

Sanitary prophylaxis

Isolation or slaughtering of sick and in-contact animals, Destruction of cadavers and Disinfection

Medical prophylaxis

Cell-culture attenuated virus vaccines are highly effective. The commonly used vaccine is an attenuated strain of rinderpest virus. In some countries a mixed rinderpest/contagious bovine pleuropneumonia vaccine is used. Immunity lasts at least 5 years and is probably life-long. Annual revaccination is recommended in order to obtain a high percentage of immunised animals in an area. Genetically engineered thermo stable recombinant vaccines are currently undergoing limited field trials. An experimental vaccinia-vectored vaccine containing the F and H genes of RPV has protected against challenge inoculation of virulent virus

Control strategy

Stamping out to remove the source of infection. Strict quarantine and movement controls to prevent spread. Decontamination to eliminate the virus. Tracing and surveillance to detect the extent of infection. Zoning to define infected and disease-free areas.

AE. SIMIAN VACUOLATING VIRUS (SV40)

Introduction

SV 40 viruses were discovered in 1960.It belongs to the family **Polyomaviridae**. Previously it is included in papovaviridae. The word papova is derived from three viruses namely Papilloma virus(PA), Polyomavirus(PO) and Simion Vacuolating Virus(VA). SV40 is a proto type of Polyomavirus. SV40 is known as **small tumor viruses**.

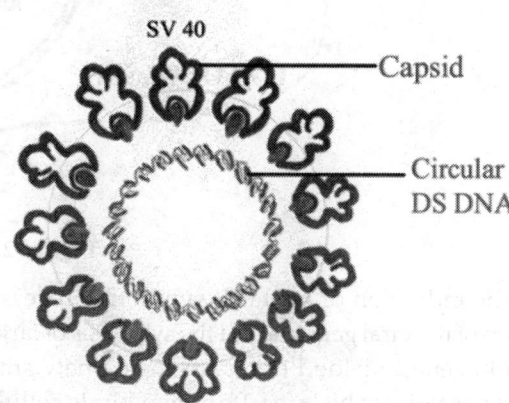

Figure 214

Structure

Polyomaviruses are icosahedral, 45-nm diameter particles, with three capsid proteins(VP1,VP2 and VP3) and is a non enveloped virus. They contain a 5-kbp circular, double-stranded DNA genome. The genome consists of two or three replicative genes (tumor antigens) encoded on one strand and three structural genes (capsid antigens) encoded on the other strand.

Small t and Large T is the functional gene, which synthesis functional protein (responsible for transformation and replication) and three structural genes, responsible for structural protein synthesis.

Multiplication

The virion first attaches to specific receptors on host cells with the help of VP1 protein. Virus penetrates the plasma membrane and is transported to the nucleus. Virus is uncoated in cytoplasm and the DNA is transported to the nucleus. SV40 uses cellular enzymes for its replication. Replication takes place by theta model.

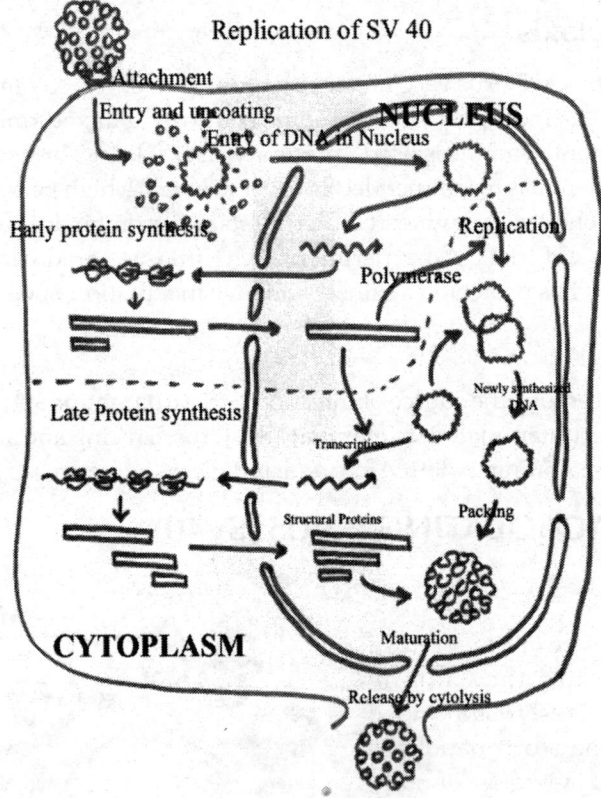

Figure 215

The induction of host cell synthetic processes depends on the expression of the early portion of the viral genome and the synthesis of large T antigen. Counter clock wise transcription of SV40 genes, yielded two proteins namely small t and large T protein(otherwise called Tumor protein). The large T antigen binds cellular tumor suppressor proteins p53 and Rb, and there by disrupts their normal cell cycle regulatory functions.

Clockwise transcription of late proteins leads to synthesis of structural proteins and replication, which leads to productive life cycle and cell lysis.

Outcome of host virus interactions

The most prominent phenotypic modifications associated with SV40-transformed cells include altered morpholosgy; altered growth patterns (increased growth rate, decreased requirement for serum growth factors, loss of contact inhibition, and enhanced ability to grow in semisolid medium [anchorage independence]); biochemical changes (increased metabolic rate, increased glycolysis, changes in properties of the cell membrane, synthesis of new antigens in the cell); and tumorigenicity (production of tumors when transformed cells are injected into appropriate test animals).

Pathogenesis

Human polyomaviruses (SV40) establish persistent infections in the kidneys; these infections may reactivate in immunosuppressed hosts and during some normal pregnancies. Progressive multifocal leukoencephalopathy is a rare disease of the central nervous system of some immunosuppressed patients.

Diagnosis

Clinical identification and the presence of antibodies are the best means of diagnosis.

Control

There are no known control measures.

SV40 is used in Molecular biology and Recombinant DNA technology as a model system or vector due to its simple gene organization.

AF. POX VIRUSES

Introduction

Pox virus is the largest virus. Poxviruses are assigned to genera on the basis of genetic and serologic relationships. The viruses are antigenically complex.

Classification

Poxviruses belongs to the family Poxviridae and the order chordopox virinae.

Structure of Pox Virus

Poxvirus is brick shaped virus. Orthopoxviruses are approximately 240 nm by 300 nm, with short surface tubules 10 nm wide. Parapoxviruses are narrower (160 nm) and have one long tubule that twisted around the virion.

Virions have a dumb bell-shaped core and two lateral bodies. The genome consists of one molecule of double-stranded DNA, from 130 kb (Parapox) to 260 kb (Fowlpox).

Figure Refer Page No. 62

Chemical composition

DNA - 3%, Protein-90%,Lipid- 5%.Virion contain more than 100 polypeptides. Many enzymes are available in core for un coating and replication.. Pox viruses are susceptible to UV and other irradiations, resistant to 50% glycerol and 1% phenol

Multiplication

Poxvirus replication takes place in cytoplasm. Virus entered into the cytoplasm through cellular pores. Upon the release of core into the hosts cytoplasm, the core synthesizes viral early mRNAs. Synthesis of early proteins induces a un coating reaction in which a nucleoprotein complex containing the genome is released from the core. Early protein mediates replication of viral genome. Newly synthesized viral DNA molecules can serve as template for additional cycles of genome replication. Some of the DNA serves as a template for transcription of viral intermediate gene. Activation of transcription of intermediate genes

Replication of Pox Virus

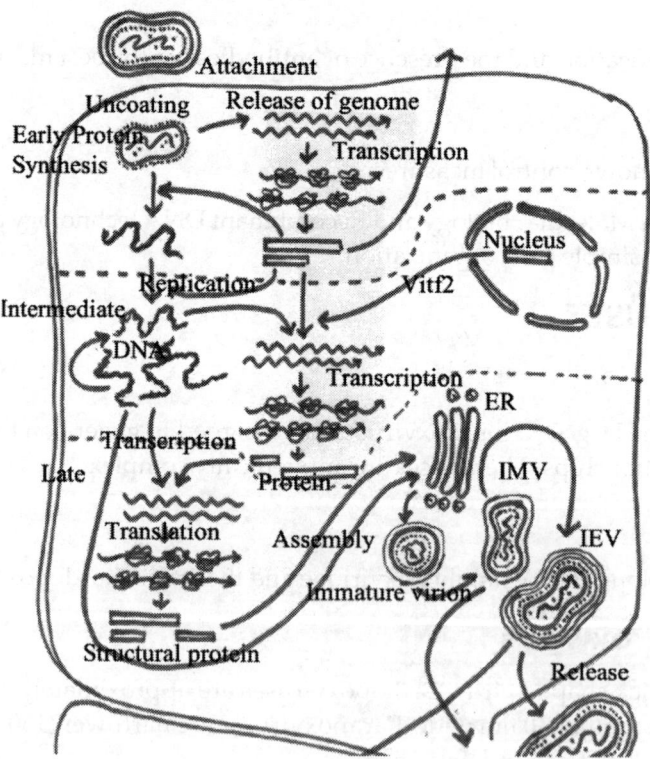

Figure 216

requires viral early protein. Vit f2 translocates from the nucleus, having polymerase activity, responsible for replication of genome. Translation of intermediate gene leads to transcription of late genes. Late genes synthesis structural and functional proteins such as early initiation protein, which must be incorporated into virion during assembly. Assembly of progeny virus particles begins, probably in association with internal membranes of the infected cells. Initial assembly leads to the formation of immature virion. Spherical shaped immature virion acquired double membranes from golgi network and converted to brick shaped intracellular mature virion (IMV). It may be released upon lysis. Some of the virion acquire another one layer from golgi and is converted into intracellular enveloped virus (IEV). IEV infects fresh cells. Several pox virus genes resemble mammalian genes.

Small pox

The last endemic case of smallpox occurred in 1977, total eradication was confirmed in 1980. Buist first microscopically demonstrated variola virusin 1887. Paschen in 1906 developed a staining technique to observe elementary bodies in smears from Smallpox lesions.

Causative agent structure and Replication Refer Page 431, 432

Symptoms: Poxvirus infections are characterized by the production of skin lesions. Lesion develops at the site of inoculation (usually the hand), and infection may be spread to other sites such as the face and/or genitals by scratching. The lesions of molluscum, usually multiple, are firm, pearly, flesh-colored nodules.

Pathogenesis: Infection is usually caused by invasion through broken skin and in most cases remains localized. Human monkeypox is acquired by contact or by airborne transmission to the respiratory mucosa. Initial viremia during the incubation period spreads infection to internal organs; a second viremia then spreads the virus to the skin.

Diagnosis: Electron microscopy of vesicle or scab material is an effective means of rapid diagnosis.

Immunofluorescence of infected cell cultures will differentiate morphologically similar poxviruses from different genera (e.g. *Orthopoxvirus* and *Yatapoxvirus*).

Poxviruses are easily isolated in tissue culture or chicken embryos. Cultivation allows identification by biological and serum neutralization tests.

Control: Control of this disease depends on health education and on breaking the link with the animal reservoir. Mass vaccination with Vaccinia virus, used for the vaccination for small pox.

AG. CANCER AND CARCINOGENIC VIRUSES

Cancer is a most serious problem in human. In Latin cancer means crab uncontrolled growth of existing cell is called as cancer.

Characters of cancer cells

Cancer cells arise from existing tissue or cells of the body. The growth of tumour is autonomous. Ana plastic in nature. Greater potentiality for growth & multiplication. Do not carry out normal cell function. Have large and irregular nuclei. Immortalization, Loss of contact inhibition, Reduced cellular adhesion. Loss of anchorage dependence, Lower serum requirement, Selective agglutination by lectins, Increase in negative charge in membranes. Increased sugar transport. Defective electrical communication. Disorganization of the cytoskeleton. Increased secretion of proteolytic enzymes. Increased rate of glycolysis. Chromosomal abnormality, Centriols are occur with long distance, Release excessive growth promoting factors.

Classification of cancer

On the basis of tissue from which it is generated cancer is classified as

- Carcinoma — Arising from epithelial cells
- Sarcoma — Connective tissue
- Fibroma — From fibrous tissue
- Glioma — Develop from the network of supporting connective tissues in the brain and CNS.
- Melanima — Pigmented tumour on Skin
- Lymphoma — From Lymph odes
- Teratoma — From morula stage of child development.

On the basis of severity it is classified as **benign tumour** and **malignant tumour**.

Causes of Tumour

Various agents may be responsible for cancer. They are chronic irritation, Atmospheric Pollution, Smoking, Radiation, Chemical and Viruses 20% of the cancer is due to viruses. Tumour viruses are classified as RNA tumour viruses and DNA tumour viruses.

DNA Tamour Viruses

Adenovirus, Polyoma virus, Simion virus 40, Herpes simplex virus, Papova virus, Pox viruses, Epstein Bar virus and Hepatitis B virus.

RNA Tumour Viruses

Human Immuno deficiency virus, Human T cell leukemia virus, Mouse memory Tumour virus, Avian leukemia virus and Human spuma virus.

Genes Responsible for cancer

Set of genes responsible for transformation of normal cell into cancer cell are called oncogenes. There are two types of oncogenes, that are cellular oncogens and viral oncogenes cellular oncogens are found in human. It is otherwise called as protoconcogenes. If it is activated by any physical, chemical or biological (virus) agents proto oncogens are converted into oncogenes

Proto Oncogenes

- ⋏ sis   -   Platelet derived growth factor.
- ⋏ fms   -   Receptor for colony stimulating factor.
- ⋏ erbB   -   Receptor for epidermal Growth factor.

Signal Transducers

- ⋏ Src   -   tyrosine kinase

Transcription Factor

- ⋏ Jun and for   -   transcripition factor
- ⋏ My ce   -   DNA binding protein

Genes that regulate programmed cell death

- ⋏ Bcl 2   -   suppresser of apoptosis

Tumour suppresser genes

- ⋏ Rb   -   suppresser of Retino blastoma
- ⋏ P53   -   Nuclear phosphoprotein inhibitor
- ⋏ P21   -   Lung cancer & colon cancer
- ⋏ DCC   -   Suppressor for colon carcinoma
- ⋏ WTI   -   Suppressor of wilms tumour.

Viral Oncogenes

20 genes were identified in viruses adeno virus

Adeno virus

- ⋏ E1A   –   Bind Rb gene and activate it. – cause res
- ⋏ E1B   –   Bind P53 gene

Human papilloma virus

- ⋏ E6   –   Bind P53 – genital tumour
- ⋏ E7   –   Bind RB – squamous cell carcinoma

Epstein bar virus - Nasopharyngeal caronoma

Homologous to bcl2 alter chromosome 14.

- ⋏ Butkits lymphoma
- ⋏ B cell lymphoma

⋏ HBV   -   Hepatocellular carcinoma

⋏ HTLV   -   T cell Leukemoa

⋏ Polyoma   -   Bind P53

Large T and Small T-SV 40 bind P53 and RB proto concogenes

Conversion of Proto Oncogens into Oncogens

Both viruses and mutagenic agents play an vital role in cancer causing process. In either process proto oncogenes are converted in to oncogenes. If the proto oncogenes are activated by viruses that genes are considered as viral oncogenes. Both gene conversion may leads to various cancer depends up on site and tissue involved.

Virus host interaction

Virus and host interaction may leads to following types of infection. Some type of interaction leads to cancer.

Productive infection – Complete replication of infections virion & release of virus.

Abortive infection – Synthesis of viral protein without production of infection virion.

Semi permissive infection – Complete replication with low yield of infection virion.

Malignant transformation – Associated with integration of viral DNA and differential viral & cellular gene expression.

Viral latency – persistence of viral genome in the host cell.

Host virus interaction may leads to the synthesis of Transformed proteins. Transformed proteins may induce cancerous cancerous growth. Mostly transformed

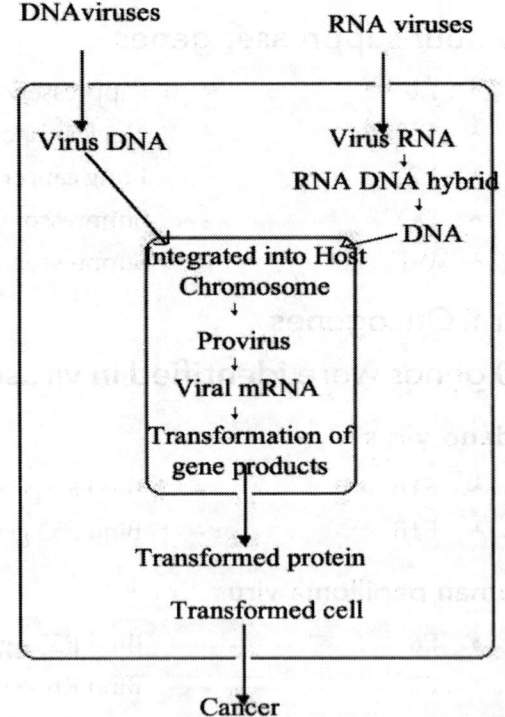

VIRUS HOST INTERACTION-CANCER

Figure 217

proteins are synthesized by cells infected with DNA viruses especially those virus multiplying with DNA nuclear virus integrated in to the human or animal chromosome host cell chromosome synthesis viral protein. This viral protein act on growth factor regulator gene or proto oncogenic part leads to cancer.

Metastasis

Clusters of cancer tissue dislodge from tumours and invade the blood or lymphatic vessel and are carried to other tissue where they proliferate. Now cancerous growth is called metastasis.

Diagnosis

Injecting reddish black drug (Hematopropyria derivative, HPD) in to a cancer available area. May identify or differentiable cancer cell from normal cell. Cancer cell retain the colour dye but normal cell loses the coloured dye.

Treatment

Magic bullet for B cell lymphoma is Ricin & Antibody against the viruses. Other methods are surgery and radiation.

AH. HEPATITIS VIRUSES

Introduction

Hepatitis is the term used for any condition where there is inflammation or necrosis of liver cells. Viruses, which damage the liver are called hepatitis viruses. Viral hepatitis has emerged as a major public health problem throughout the world. Various viral agents are responsible for Hepatitis. They are named as Hepatitis A virus (HAV), Hepatitis B virus (HBV),Hepatitis C virus(HCV), Hepatitis D virus(HDV) and Hepatitis E virus (HEV). **The only common feature of these Hepatitis viruses is their primary Hepatotrophism.** About 98% of Hepatitis is caused by Hepatitis A,B,C,D and E viruses. Other viruses are Cytomegaloviruses , Epstein barr viruses, Herpes simplex viruses and Yellow fever viruses.

HEPATITIS A VIRUS

Hepatitis type A causes subacute disease, occurring mainly in children and young adults. The term **infectious hepatitis** was coined in 1912 to describe the epidemic form of HAV infection.

Characters

Feinstone and his colleagues demonstrated HAV during the year 1973 from faecal matter through Immuno electron microscopy. It is a nonenveloped, icosahedral positive sense RNA virus. It is a member of Picornaviridae family. It is non cytopathic when grown in cell culture. It was classified into the Enterovirus during the year 1983. HAV genome comprises about 7500 nucleotides. The virus is relatively resistant to inactivation.

Pathogenesis

Clinical expression of HAV varies considerably. HAV enters body via ingestion of contaminated food, water and fecal oral route.

The virus spread from intestine to liver through blood stream. Incubation period for HAV is 3-5 weeks within a mean of 28 days.

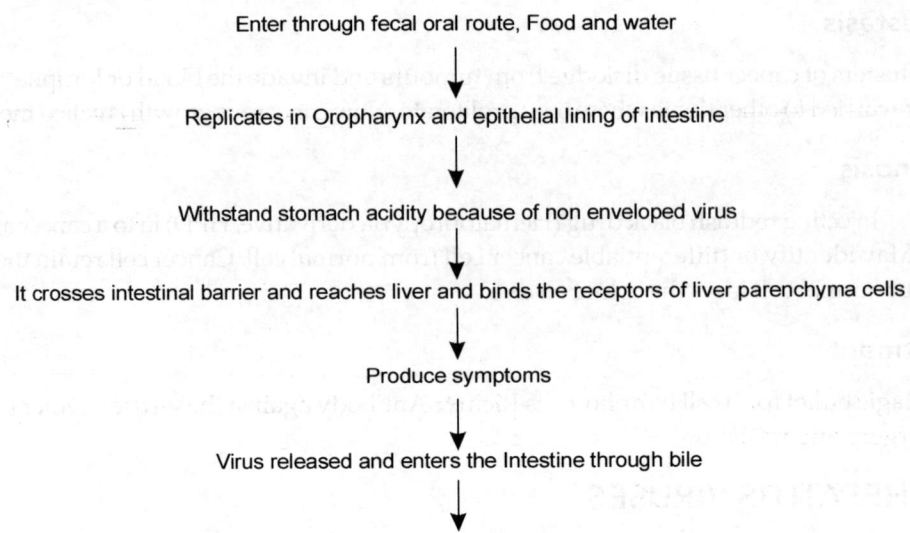

Enter through fecal oral route, Food and water

↓

Replicates in Oropharynx and epithelial lining of intestine

↓

Withstand stomach acidity because of non enveloped virus

↓

It crosses intestinal barrier and reaches liver and binds the receptors of liver parenchyma cells

↓

Produce symptoms

↓

Virus released and enters the Intestine through bile

↓

Release outer environment through feces 10 days before starting of symptoms

Symptoms

Fever, Malaise, Anorexia, Nausea, Vomiting, Liver tenderness, Jaundice (Yellow discoloration of sclera). Mortality is very low ranging from 0.1-1%. General symptoms subside with the onset of Jaundice.

Diagnosis

Serological techniques are available, that includes, Immuno electron microscopy, Complement fixation, Radio immuno assay, Enzyme immuno assay.

Control and Treatment

Improve sanitation. Prevent fecal contamination. Prevent direct contact with infected individuals. Available vaccines for control are, Inactivated vaccines (formaldehyde)- Intra muscular, Attenuated vaccine – Oral inoculation, Post exposure treatment include Human immunoglobulin and anti HAV.

Prevention

Drink only boiled or filtered water. Avoid eating fruits, salads or uncooked vegetables that have not been washed by you in potable water. Avoid eating of food or drinking beverages from street venders, especially if they are not covered. Take HAV vaccine.

HEPATITIS B VIRUS

HBV belongs to **Hepadna viridae**.It is the most important type among Hepatitis causing viruses. HBV causing Hepatitis is called **Serum Hepatitis.** HBV infection causes more than million-deaths per year worldwide.

In 1965, Blumberg reported a protein antigen in the serum of Australian patient. This antigen is called Australian antigen. The Australian antigen was shown to be associated with Serum Hepatitis. This was then considered as Hepatitis surface antigens (HbsAg).

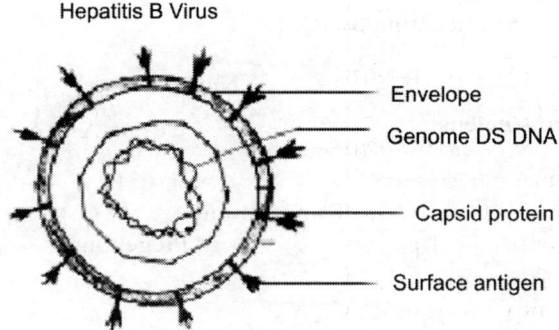

Hepatitis B Virus

Envelope

Genome DS DNA

Capsid protein

Surface antigen

Figure 218   Structure of HBV virus

HBV is a spherical virus. There are three envelope polypeptides are available and are designated as HbsAg, HbcAg & HbeAg. The nucleocapsid of the virion consist of the viral genome surrounded by the core antigen(27nm). The genome, which is approximately 3.2 KB in length, has an unusual structure and is composed by two linear strands of DNA held in circular configuration. Negative strand is complete but positive strand is incomplete. 3' end of genome is associated with a DNA polymerase molecule. There are 4 major Open reading frame (ORF) are available in the genome. They ORF-S (Codes for structural proteins of surface and core), ORF-P (Encodes polymerase that contains DNA polymerase & RNase H), ORF-X (Encoded transcriptional activator), ORF-C(Pre C protein). The virus is stable at 37°C for 60 minutes. Destroyed its antigen by 0.5% sodium hypochlorite for 3 minutes. HbsAg stable at pH2.4 for 6 hours. 2% glutaraldehyde destroy antigenicity within 3 minutes. HbsAg is resistant to UV. Virus did not grown in tissue culture medium.

Symptoms

Incubation period varies widely from 40days-6months, but is often about 2-3 months. Symptoms are Tiredness, Fever, Loss of appetite, Vomiting, Headache, Pain in abdomen with Diarrhoea, Jaundice, Itching sensation on skin, Inflammation in joints. Mortality rate is about 0.5-2% .

Complications are Pain in liver, Liver cancer, Hepato cellular carcinoma, Arthralagia, Polyarteries and Glomerulonephritis.

Pathogenesis

Pathogenesis involves 3 steps. They are 1. Entry of virus, 2. Multiplication and spread of virus and 3. Liver cell damage.

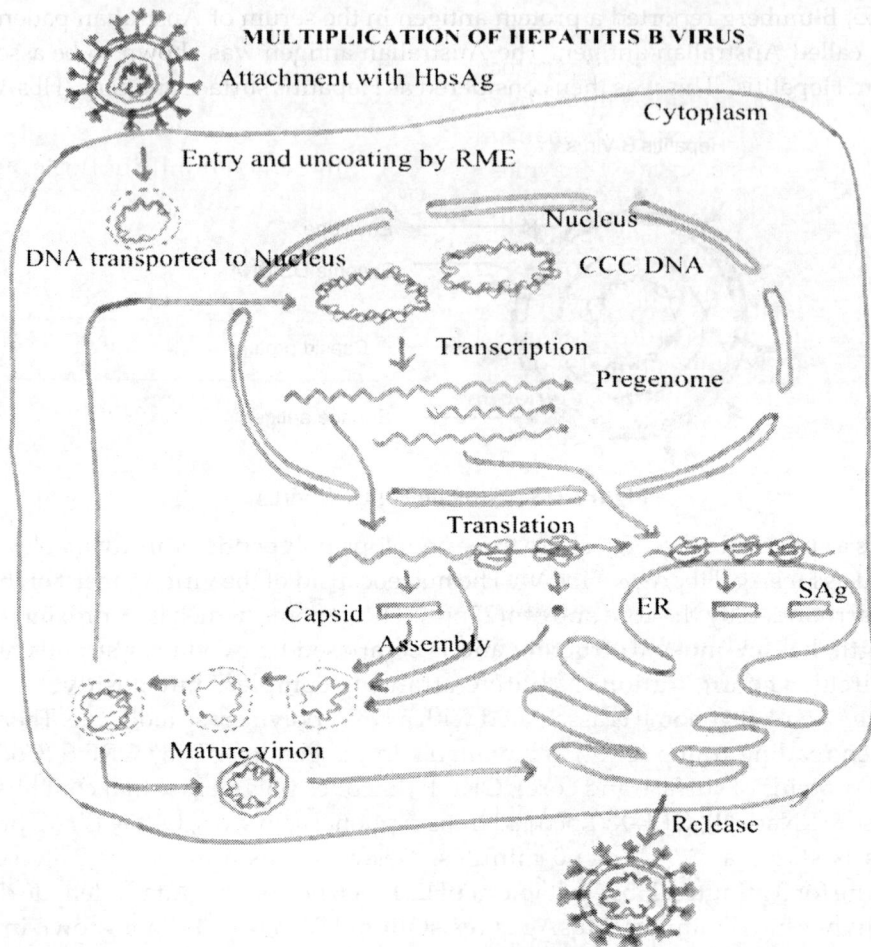

MULTIPLICATION OF HEPATITIS B VIRUS

Attachment with HbsAg

Cytoplasm

Entry and uncoating by RME

DNA transported to Nucleus

Nucleus

CCC DNA

Transcription

Pregenome

Translation

Capsid

ER

SAg

Assembly

Mature virion

Release

Figure 219   Replication of HBV virus

Entry

Viruses are transmitted through Blood transfusion, Sexual transmission, Neonates get infected from mothers and also through contaminated syringes and needles.

Multiplication and spread

The target of HBV is the hepatocyte (Liver cell). Virus attaches to the hepatocyte with the help of surface antigen (HbsAg). HBV enter inside of cytoplasm through receptor mediated endocytes (RME). DNA of the virion is transported into the nucleus. DNA transcription follows resulting in the formation of mRNA. Short mRNA transcripts undergoes translation in ER. Translation products are preS and S proteins. P protein translation occurs in the cytoplasm and it is associated with capsid protein formation. P and capsid proteins were assembled in cytoplasm and cover DNA. Replicated DNA undergoes nick formation resulting in dsDNA with staggered strand. Morphological changes occur on the surface of plasma membrane.

Complete virions released through exocytosis

Replication results in injury of hepatocytes and release of progeny virions into the blood stream. Cell injury is not only caused by cytopathic effect of virus and also caused by the activation of cytotoxic immune mechanisms. This will results in liver tissue degeneration and the release of liver associated enzymes into the blood stream.

This is followed by jaundice, the accumulation of bilirubin in the skin and other tissues with a resulting yellow appearance.

Laboratory Diagnosis

Blood tests are recommended for diagnosis of Hepatitis. Liver enzymes such as alanine aminotransferase and a aspartate aminotransferase are elevated in blood during the early stages of viral hepatitis. ELISA test is used for the detection of HbsAg . It is also used for the detection of Hepatitis antibodies.

Prevention

Prevented by active and passive immunization. Two types of vaccine currently available. They are, Recombinant HB vaccine-synthesized from yeast cells-safe and effective-provides 90% protection and Plasma derived vaccine. Vaccine injection was given in muscles of the upper and outer parts of the arm at birth, at 1st month and at 6th month. A booster dose is recommended at **5 years of age.**

Treatment

People with chronic and active i*nflammation of liver cells due to HBV are treated with Interferon. Interferon is prescribed in combination with ribavarin. Lamuvidine is recommended, if the patients develop cirrhosis in liver.

HEPATITIS C VIRUS

HCV belongs to the family **Flaviviridae** and the genus **Hepaca virus**. This virus is 30-60nm,spherical shaped, positive sense RNA virus. Genome has 10,000 nucleotides. HCV is the leading cause of **post transfusion hepatitis**. Frequency of HCV is greater than HBV. HCV is considered as the major risk factor because 80% of infections lead to chronicity. HCV rarely seems to cause fulminant hepatitis. All genomes contain a single large open reading frame, which is translated to yield poly proteins from which the viral protein were derived by post translational cleavage and other modification. Helicase, polyproteins and proteases were involved in RNA replication.

HEPATITIS DELTA VIRUS

HDV occurs only those who have HBV infection. In 1977, Rizzetto & colleagues in Italy identify a new viral antigen in the liver cell nuclei of patients infected with HBV. Later it was called HDV. HDV is coated with HbsAg, which is needed for release from the host hepatocyte and for entry in the next round of infection. HDV is a spherical, 36nm particle with an outer coat composed of HBV, surface antigen surrounding the circular ssRNA with 1.7-kb base pairs. The internal protein molecular weight 68,000. Closest relative of HDV is a satellite virus of plants.

Two types of infections were recognized. They are Co infection-HDV&HBV are transmitted together at the same time. Super infection-Delta infection occurs in a person already harboring HBV.

HDV has 1679 nucleotides. RNA replicates with the help of host RNA polymerase II. About 5% HbsAg carries worldwide are infected with HDV. Diagnosis is performed with immunofluorescence and ELISA. An IgM antibody appears 2-3 weeks after infection and is soon replaced by IgG antibody.

HEPATITIS E VIRUS

It is an single stranded linear RNA virus. It is included under the family **Calciviridae**. Incubation period ranges from 2-9 weeks with an average of 6weeks. Most cases occur in the young to middle aged adults (15-40year). HEV is a spherical non-enveloped virus, 30-32nm in diameter. It enters liver through intestine and blood. Identified by means of ELISA. There is no vaccine currently available for preventing HEV. Personal hygiene and sanitation are the only effective way of prevention.

Characters	HAV	HBV	HCV	HDV	HEV
Family	Picornaviridae	Hepadnaviridae	Flaviviridae	Defective virus	Calcivirus
Genus	Enterovirus	Hepadnavirus	Hepaca virus	Deltavirus	Hepevirus
Size	25–29nm	40–50nm	30–60nm	35nm	30–32nm
Capsid	SS RNA (+)	DS DNA	SS RNA (+)	SS RNA (-)	SS RNA (+)
Nucleic Acid	Icosahedral	Spherical	Spherical	Spherical	Icosahedral
Virion	Nonenveloped	Enveloped	Enveloped	Enveloped	Nonenveloped

AI. VIRAL VACCINES

Vaccine is defined as any microbial preparation that induce the immune response of animal/ human leads to the formation of memory cells. Vaccines may produce life long immune response. Viral vaccines are live or killed preparations of viruses or their components that are used in immunizations and offer immunity to the individual viral agents. **Vaccination** is a process by which vaccine is introduced into the human or animals. Vaccines protect the individual from the infectious diseases.

Origin of Vaccination

In1796, **Edward Jenner** extracted the contents of the pustules from cowpox-infected milkmaid **Sarah Nelmes,** and injected it into the arm of 8 year old James Philip. He developed immunity against smallpox virus when Jenner inoculated it Jenner practiced this to his patients and also French soldiers. Further work on immunization was carried out by Louis Pasteur (1822 – 1895) and produced chicken cholera vaccine , rabies vaccine etc. He gave the name vaccine to Honour Jenner. In Latin **Vacca** means **Cow**.

Requirements of Effective Vaccine are Safety, Induction of protective immune response, Biological stability, Ease of administration, Public acceptance and Low cost.

Immunization

A process of inducing immunity of an individual is called immunization. There are two types of immunization- Active immunization & Passive immunization

Active immunization

It refers to the stimulation of an immune response with the help of specific antigen.

Example:

Poliovirus vaccination leads to eradication of polio. It is of two types, natural active immunization and artificial active immunization. Natural active immunization refers to the development of resistance is due to natural infection. Artificial type may lead to the development of resistance by the action of vaccines.

Passive Immunization

It is defined as the development of resistance to an infection in a non-immune individual by the administration of sensitized antibodies or lymphocytes. It provides immediate production. If the immunity is transferred from the mother to child in a natural way is called natural passive immunization. If the antibodies and immune cells are transferred from immunized individual to non-immune individual is called artificial passive immunization.

Types of Vaccines

Live attenuated vaccines

Most live vaccines contain viruses that have been attenuated by laboratory manipulation. Attenuated viruses lose the pathogenic property but they have the property of multiplication with in a host cell. It produces life long immunity.

Examples: sabine polio vaccine, MMR vaccine.

Advantages

Produce humoral and cell-mediated immunity, Life long immunity, Generally requires only a single booster and Easy administration.

Disadvantages

Less stable and Virus may revert to virulent form

Inactivated or killed virus vaccines

Killed vaccine contains either whole virus particles, inactivated by chemical or physical agents. Killed vaccines are used along with adjuvant. Formalin, a propiolactone are the chemical agents used to inactive the viruses. UV radiation is a physical agent which may inactive the viruses. Adjuvants are the immuno stimulatory particle, helps in antigen presentation.

Example: Alum is one of the good adjuvant

Advantage: No mutation or reversion.

Disadvantages: Most booster dose is required.

High cost: Example: Salk polio vaccine,

Rabies vaccine and Hepatitis viral vaccine.

c. Subunit vaccine

Vaccines that employ components of a pathogenic virus are called sub unit vaccine.

Example: Gp 120 gene of HIV virus was cloned with a cell and produced gp120 protein. It is purified and used as a vaccine.

d. Recombinant antigen vaccine or recombinant vector vaccine

Gene of one virus is cloned with suitable cells and used as vaccines.

Example: Recombinant HBV. DNA sequence of HBsAg is cloned with yeast plasmid vector and transferred into a yeast cells. Recombinant yeast cells were selected with a suitable medium. Recombinant proteins were collected and used as vaccine.

Advantages

1. Stable
2. Stimulates both cellular and humoral immunity and
3. Small quantity is required.

Disadvantage

1. Cells may convert into virulent form.

e. Antiidiotype vaccine

Vaccine is made up of antibodies and stimulates specific antibody production against antigens. Antibody of one animal act as an antigen of another animal.

Example: targeted to gp 120 antibodies.

f. Synthetic peptide vaccine

Example: HBV

HBV surface antigen is made up of 48 amino acids. Synthetic peptides are synthesized based on the amino acid configuration of HBV surface antigens. Peptide molecules were encapsulated into microspheres and are directly introduced into the cells. This type vaccine results in high titer of antibody and T cell response against HBV surface antigen.

Advantages

Less toxic and Viral cultivation does not arised.

Disadvantages

Require Adjuvant and Less Immunogenic then whole virus vaccine.

g. DNA Vaccine

Instead of using viral components DNA of the directly used. DNA coding for viral antigen is coated with gold particle and injected into epidermis region by gene gun method. Antigen is expressed in human system and stimulates MHC-I and CD8$^+$ cytotoxic T cells. CD8$^+$ cell kill infected cell and releases viral antigen. Viral antigen expresses MHC-II, which will reach the formation of Antibody.

Example: It is a preferred method for immunization in infants against HBV and influenza.

h. Homologous Vaccine

Prepared from single pathogen and are used against the infection of same pathogen.

Example: Polio Vaccine

i. Heterologous Vaccine

Vaccine is prepared from are virus but used against from infection of a different virus.

Example: Cowpox virus is used to prevent Small Box.

AJ. ANTIVIRAL AGENTS

Antiviral agents include compounds that are used either as prophylactic agents or as inhibitors of virus multiplication. These antiviral agents have become the most important tool in the cure of viral diseases. There are two types of antiviral agents. They are Natural antiviral compounds – Eg., Interferons and Synthetic antiviral agents – Eg., Zidovudine.

Ideal antimicrobial agents must block viral production without causing lethal damage to uninfected cells.

Nature of ideal antiviral agents

Compound should block viral spread. It must block viral replication. It should be effective against the resistant forms, Molecular mechanism of the drug should be known. It should be safe, It should be inexpensive, Easy to formulate, Should be better than any competitive drug.

Stages in virus replication which are possible targets for chemotherapeutic agents

Attachment to host cell, Uncoating - (Amantadine), Synthesis of viral mRNA - (Interferon), Translation of mRNA - (Interferon), Replication of viral RNA or DNA - (Nucleoside anologues), Maturation of new virus proteins (Protease inhibitors) and Budding, release.

Synthetic Antiviral Agents

Antiviral drugs are Categorized according to their point of action in viral replication cycle.

Agents that block attachment uncoating or both

Amantadine and Rimantadine

These compounds inhibits fusion of viral envelop with endosome membrane. They prevent release of nucleocapsid in to the cytoplasm. They are used in treatment of influenza a infections. They act by binding to M2 protein.

Agents that inhibit DNA polymerase

Vidarabine, Idoxuridine

They are nucleoside analogues. They lack proper radicals for crosslinking to other nucleotides. When they act on a normal DNA chain, termination of DNA synthesis results.

Vidarabine: (Deoxyadenosine analoque)

It is a nucleoside purine analog, It is phosphorylated by kinases. Vidarabine phosphate act as a competitive inhibitor for DNA polymerase and act as a chain terminator. The tertiary structure has an extra hydroxyl group that prevents attachment of further residues. Thus when vidarabine acts on a nascent DNA chain, synthesis is inhibited. It is used to treat herpes simplex encephalitis. Available as a topical ophthalmic formulations. Ara –A, Vira –A are the trade names.

Idoxuridine(IDU)

It is a pyrimidine analogue. Phophorylated by cellular kinases. Idoxuridine triphosphate act as a competitive inhibitor of viral polymerase. It is used as a topical agent for herpes infection of cornea. Stoxil, Herpex are the trade names

Target Activated Nucleoside Analogues

Acyclovir

It resembles guanosine. But it is acyclic ie., sugar moiety is incomplete. It prevents cross-linking. It is irreversibly bound to DNA transcriptase and inactivate the enzyme leads to inhibition of viral mRNA synthesis. It can be administered orally and intravenously. It is used in treatment of herpes encephalitis.

Ganciclovir

It is highly reactive against cytomegalovirus (CMV) DNA polymerase. It is used in treatment of infections like retinits, encephalitis. It can be administered orally and intravenously.

Agents that inhibit reverse transcriptase

Zidovudine

It is also known as azidothymidine (AZT). It resembles thymidine except for substitution of hydroxyl group by nitrogen in 3' position of pentose ring. It competitively binds to reverse transcriptase and stop its action. It also results in chain termination because it lack 3' hydroxyl group essential for formation of phosphodiester linkage between two nucleotides. It is a potent drug for treating HIV infection. Side effects are Anaemia and neutropenia.

Protease inhibitors

Eg. Sanquinavir, Ritonavir, Indinavar, Nelfinavir

They are non-hydrolyzable synthetic peptides. They inhibit HIV protease, by altering aminoacid sequences. Approved drug for HIV. Commercial names are crixivan, norvir.

Broad spectrum nucleoside analogues

Ribavirin   It is a nucleoside analogue of guanosine. It inhibits enzymes needed for mRNA capping. It binds to RNA polymerase and blocks its activity. It is a broad spectrum drug. It is used to treat respiratory synecytial virus, Hepatitis C Virus, herpes simplex virus, Measles, Mumps and Lassa fever.

Forcarnet

It inhibits DNA chain elongation to pyrophosphate receptor on DNA polymerase. It inhibits reverse transcriptase by binding on to a specific affinity site. It is used to treat Herpes simplex infections in HIV- positive patients.

AK. INTERFERONS

Definition

Interferons are low molecular weight antiviral glycoproteins synthesized by virally sensitized calls and also in response to intracellular parasites.

Discovery

Interferons were discovered by Issaacs and Lindermann. The natural antiviral substance was named as interferon because it exhibits interference with viruses.

Nomenclature

Interferons are abbreviated as IFN.

Classification of Interferon

Five major classes of interferons are recognized based on the cells of origin. They are

- ⋏ IFN α – Synthesized by virus infected Leukocytes. It is made up of 20 different molecules.
- ⋏ IFN β - Synthesized by virus infected fibroblasts.
- ⋏ IFN γ – Produced by antigen stimulated T cells.
- ⋏ IFN ω and IFN $_T$ - Produced by placenta.

Characters of interferon

It is a soluble, non antigenic glycoprotein. Molecular weight ranges from 17-25KDa. Gene coding for IFN α and IFN β are found in chromosome 16 and the gene for gamma interferon is in 6th chromosome.

Biological effects if interferons

It is a multipotential regulatory factor. It is stimulate NK cell activity. Induce antibody production. It has antiviral property. Antitumour activity also observed in IFN. Induce phagocytic process. It also inhibits nonviral agents like Toxoplasma, Shigella like intra cellular parasites.

Immunological properties of Interferons

Interferons induce the action of B cells, Macrophages exhibits increased MHC expression, antitumour activity. T cells produce increased IL2 production and it also depresses viral replication.

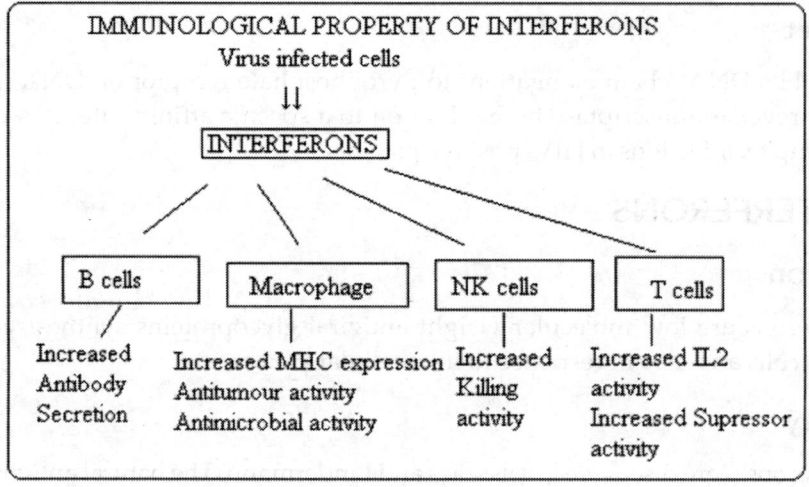

Figure 220

Inducers of interferons

Both DNA and RNA viruses can induce IFN synthesis. Pure DNA also induces interferons. Mitogenic agents, Endotoxin, Intra cellular parasites like. Listeria monocytogens. Chlamidiae, Rickettsias, Protozoa are also involved.

Interferons as therapeutic agents

Interferons can be used for the treatment of viral diseases and certain cancers. AIDS – IFN αn3. Genital warts – IFN αnB. Hepatitis B and C – IFN α2b. Multiple myelsma – IFN α2b. Papilloma virus infection – IFN αn3.

Side effects

Nausea, Fatigue, Fever, Chills, Lose of appetite and Muscular dysfunction.

To reduce the burden of IFN, patients should take plenty of water.

Interferons and Biotechnology

Artificially some of the interferons are synthesized by making use of biotechnological process. Alpha interferons are available in market as Intron A, Referon, wellferon. Human IFN gene is cloned into Chinese hamster overy cells and produced IFN β-1a and sold under the trade name Avonex. It is administrated intramuscularly at 30mcg concentrations. IFN β-1b is produced by cloning of INN- β gene in *E. coli* and marketed as Betaseron.

Mode of action of interferon

Path way I

Interferon binding on to cellular receptors leads to the production of an unusual intra cellular polymerase which links to ATP by 2′ to 5′ phospho diester linkages. This enzyme is also known as 2-5 linked Oligo adenylate synthetase. Upon activation this enzyme produces oligonucleotides from ATP. This olionucleotides activates the enzyme ribonuclease, which acts on the viral RNA or mRNA, and cleave it into free nucleotides.

Path way II

IFN also activates the enzyme Proteinkinase. Proteinkinases are the important components of growth and cell cycle. Active kinase phosphorylates erkanyotic initiation factor of protein synthesis (eIF2 α). Up on phosphorylation eIF2 α becomes inactive. This leads to inhibition of protein synthesis in virus and also in host.

13

IMMUNOLOGY

Survival requires both the ability to mount a destructive immune response against nonself and the inability to mount a destructive response against self.

-David Huston,*Biology of the Immune System*, JAMA 278

A. INTRODUCTION

Immunology is the study of all aspects of host defense against infection and of adverse consequences of immune responses. It is a relatively new science. Its origin is usually attributed to Edward Jenner. He discovered in 1796 that cowpox, or vaccinia virus induced protection against human smallpox, an often fatal disease. In human, the immune system begins to develope in the embryo. The immune system starts with haematopoietic (from Greek, "blood-making") stem cells. These stem cells differentiate into the major players in the immune system (granulocytes, monocytes, and lymphocytes). These stem cells also differentiate into cells in the blood that are not involved in immune function, such as erythrocytes (red blood cells) and megakaryocytes (for blood clotting). Stem cells continue to be produced and differentiate throughout the lifetime. By the time a baby is born, the immune system is a sophisticated collection of tissues that includes the blood, lymphatic system, thymus, spleen, skin, and mucosa.

B. IMMUNO HAEMATOLOGY

Immuno haematology refers to the study of blood groups, blood transfusion, blood group antigens, blood group Antigen and anti body reactions (Ag-Ab) and the blood group disorders.

Blood group systems

About 14 blood group system has been discovered in man. ABO blood group was first discovered in 1900 by Landsteiner.

ABO Blood group

It was discovered by Landsteiner in 1900. He found two types of antigens (Ags) on the RBC, they are Antigen A and Antigen B. Similarly there are two types of Antibodies (Abs) in the plasma called antibody a and antibody b. Based on the presence or absence of Ags and Abs in the blood of human, they are classified into 4 groups. They are A, B, AB and O. A blood group person contain antigen A and antibody b. B blood group person contain antigen B and antibody a. AB blood group person contain both Ag A and B and no antibody. O group contain no antigen but both antibodies a and b.

ISO antibodies

It is an ab produced by one individual that reacts with Ag of another individual of the same species. Anti a and b are iso antibodies.

ISO antigens

An antigen of an individual which is capable of eliciting an immune response in individuals of same species who are genetically different. Antigen A and antigen B.

Genetics of ABO blood group

The ABO blood group is inherited by a set of multiple alleles. The synthesis of antigen A is controlled by a dominant allele represented by A. Antigen B is controlled by B. The absence of antigens due to the presence of recessive alleles represented by O.

Transfusion of ABO blood group

The transfer of blood from one person to another is called blood transfusion. The person receiving blood is called recipient. The person donating blood is called donor. In blood transfusion the ABO blood grouping should be tested. When a person receives a wrong blood group agglutination of blood occurs in the body of the recipient and this leads to death.

A group person can receive blood from another A group person and O group person. He cannot receive blood from AB blood group person because A group person contains antibody b which reacts with the Ag B of AB blood group person.

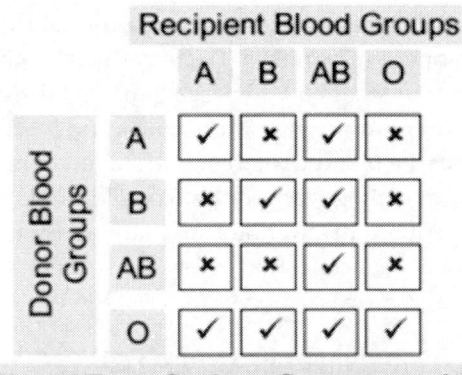

Blood Transfusion: Cross matching
✓: can be transfused
✗ : agglutination

Figure 221

B group person can receive blood from another B group and O group but not from AB group. AB group person can receive blood from all four group persons. O blood group person can receive blood from only O group person. AB blood group persons are called universal recipient and O blood group is called universal donor.

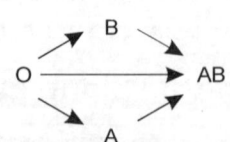

Figure 222

Distribution of ABO blood group

The frequency of ABO blood group varies in different countries. However O group is the very common group and AB is the rarest group.

H Antigen: RBC of all ABO group person possess a common Ag called H antigen. It is a precursor of antigen A and B.

Bombay blood group

Bhende *et al.*, in 1952 described rare blood group among the natives of Bombay. This is called Bombay phenotype or OH blood. Antigens A and B are completely absent and have anti a, anti b and anti h antibodies.

Rh blood group system

Landsteiner and Weiner discovered the special type of Ag in the RBC of Rhesus monkey and named as Rh factor. Based on the presence or absence of Rh factor, the human beings are classified into two groups. They are Rh positive (Rh⁺) and Rh negatve (Rh⁻)

Erythroblastosis foetalis

It is a haemolytic disease affecting babies of Rh⁻ mother and Rh⁺ father. It is a haemolytic disease. The Rh Ag of baby enters the blood of the mother. The blood of mother produces Rh antibody when this Ab enters the foetus, the foetus is affected. The Rh ab destroys the RBC of the foetus. The destruction of RBC leads to haemolytic anaemia, and jaundice. This caused death of the baby.

Prevention of Haemolytic disease

Selective marriages – When a lady happens to be Rh- she should marry only Rh- man and not Rh+ man. The destruction of Rh+ foetal cells in the material blood can be brought about by Rhogam Antibody treatment soon after birth of first baby.

Applications of Blood group

The study of blood groups bring enormous advantages in the medical field.

1. Blood transfusion.
2. Haemolytic disease.
3. Transplantation
4. Detection of Culprits
5. Disputed parentage

C. IMMUNITY

Immunity is concerned with resistance to infection. It is broadly classified into two types namely innate immunity and adaptive immunity or acquired immunity. Innate immunity is a non specific or natural or native immunity against all kinds of infection. Despite the multiple layers of the innate systems, some pathogens may evade the innate defenses.

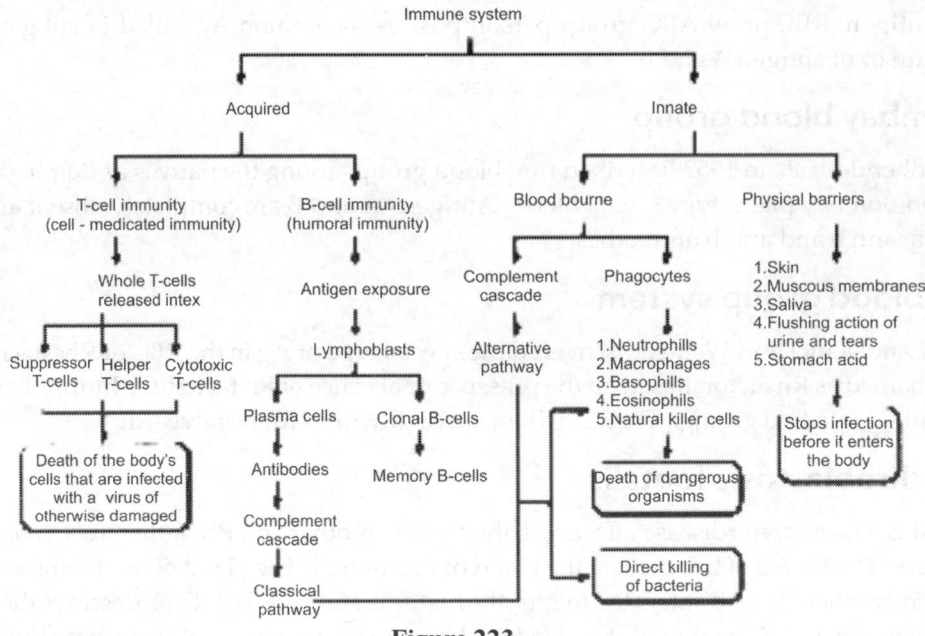

Figure 223

Innate immunity

All living organisms are naturally gifted with the resistance to certain infection from birth and this natural mechanisms are called as innate immunity / natural immunity / native immunity. This immunity act against all kinds of infection hence it is also called as non specific immunity.

The innate immune system includes

> Physical barriers / Factors / Mechanical factor
>
> Chemical Barriers / Factors
>
> Cellular Barriers / Factors
>
> Genetic Factors

Physical and Mechanical Factors

Skin Skin is one of the most important well known protective layer which contains outer stratum corneum, middle epidermis and inner dermis and subcutaneous connective tissue. Stratum corneum is impermeable to infectious micro organisms. Skin act as a very good mechanical barrier. It also contains sebaceous gland and sweat glant. Sebum of sebaceous gland prevent microbial growth. Sweat secretion contains Nacl which kill all microbes by creating osmotically stress environment.

Mucous Membrane Mucous membrane is present in respiratory tract, Genital tract, Urinary tract, Gastro intestinal system etc., which secretes mucous. Mucous traps microorganisms because of its sticky nature. Mucous also block adherence of microorganisms to the epithelial cells.

Cilia Cilia is present in the respiratory tract, which swept away the mucosal micro organisms by the constant movement of the cilia.

Coughing and Sneezing Coughing and sneezing help in driving out the foreign particles that enter digestive and respiratory tracts.

Peristalsis Peristalsis movement of intestine removes microorganisms in the intestine before they could invade and grow.

Washing action of Tear, Saliva and Urine Lacrymal gland of eye secretes tears. The flushing action of tear removes microorganism. Mouth is constantly bathed in saliva and the particles that enter the mouth are swallowed by the salivary secretion. The washing action of urine eliminates microbes from urethra.

Chemical Factors

Secretions of Skin Sebaceous gland secrets sebum which contain lipids and fatty acid. This creates acidic environment and kills microbes. Sweat gland secrets Nacl containing sweat that kills microorganisms by osmotic lysis.

Secretion of the digestive tract Oxyntic cells of stomach secrets hydrochloric acid. This creates acidic environment (pH^2) and lyse microbes present in stomach and enters the stomach.

Human Milk Human milk is rich in antimicrobial substances like lactoferritin, neuraminic acid. They fight against *E.coli*, *Streptococcus*, *Staphylococcus* etc.

Nasal Secretion and Saliva These secretions contain lysozyme, enzymes, IgA which kill bacteria and viruses.

Lysozyme Tears, nasal secretions, saliva, human milk and other body secretions contains lysozyme which have N acetyl muramidase activity, this kill most of the bacteria that contain peptidoglycon and proteins.

Interferon These are group of water soluble, non toxic glycoprotein synthesized naturally by the virally sensitized cells and prevent the growth of microbial cells. α Interferon, β interferon, γ interferon are produced only by activated T cells.

Complement Complement are group of 20 different protein found in the serum. Complement operates by two major mechanisms namely the classical and alternative pathways. The main function of complement is the opsonisaton of microorganisms and immune complexes there by promoting phagocytosis.

Properdin This is also a group of protein found in normal serum and involved in resistance to infection. Along with complement and Mg++, this causes lysis of gram negative bacteria and also inactivate certain viruses.

Secretions of Bacterial Flora

Skin *Propionibacterium acnes* present in skin which converts sebum (Lipid) into long chain aromatic fatty acid called oleic acid. It kill most of the pathogenic microbes present on the skin.

Intestine *E.coli* present in intestine secretes colicin and acid, which destroys pathogenic microbes present in the intestine.

Vagina *Lactobacilli* present in the vagina secretes microbicidal acid from glycogen, which prevent proliferation of pathogenic microorganisms.

Semen It contains bactericidal components, namely spermine and zinc.

Acute Phase Proteins This is a group of plasma proteins which increase very rapidly during infection. CRP protein binds C protein of Pneumcococci and act as a opsonins and activate complement which facilitates its killing by phagocytosis.

C. Cellular Factors

Phagocytosis It is a process of cell eating. Macrophages, neutrophils, Eosinophils, Basophils are actively involved in phagocytosis.

Process of Phagocytosis

The process of phagocytosis includes the following stages, Chemotaxis, attachment, Ingestion, Intracellular killing, digestion and release.

Chemotaxis It involves the movement by phagocytes to the site of infection or inflammation in response to the chemotactic factors produced by foreign particles or damaged tissue.

Attachment This is the 2nd stage of phagocytosis in which phagocytes attaches to the foreign particle. Attachment is promoted by opsonins (activated complement factor).

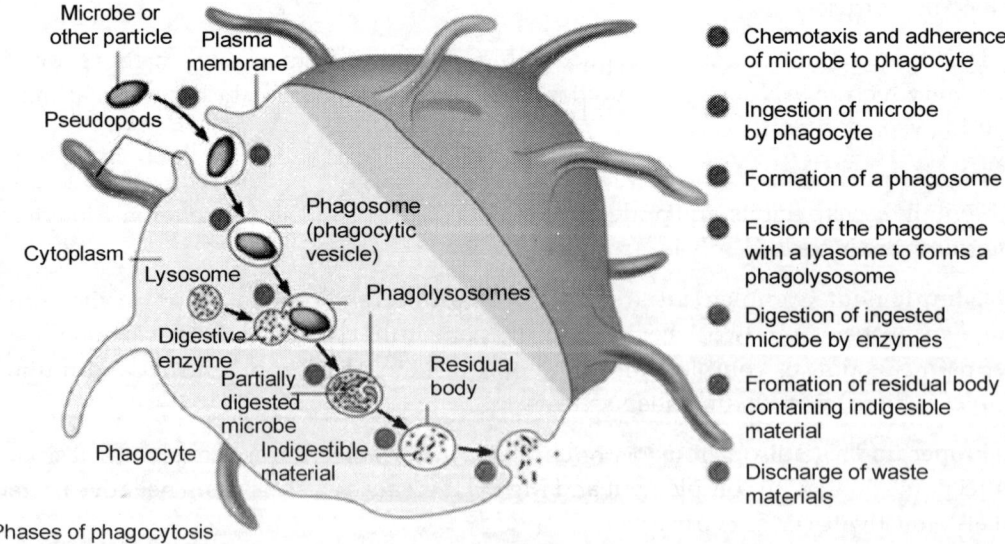

Figure 224

Ingestion Adherence induces membrane protrusions, called pseudopodia, to extend around the attached material. Fusion of the pseudopodia encloses the material with in a membrane bound structure called a phagosome. Phagosome soon fuses with the Lysosome of the phagocytic cell and results in the formation of phagolysosome.

Intracellular Killing Lysosome contains a variety of hydrolytic enzymes that digest the ingested material.

Digestion and Release Hydrolytic enzymes, H_2O_2, myeloperoxidase, lactoferitin digest the intracellular materials and combines with class I / II MHC molecules and released the digested material by exocytosis and the antigens are presented along with MHC, which will be recognized by T_4 cell and undergone CMI and HMI.

Natural Killer Cells

These are nonphagocytic lymphoid cells having large granules. They are also called large granular lymphocytes. These cells show natural cytotoxicity and they can kill a range of tumour cells and virally infected cells.

Genetic Factors

Natural immunity is also due to genetic factors, there by immunity differs at the level of species, races and individuals.

Species Immunity

Rats are resistant to diphtheria where as human and guinea pigs are susceptible to diphtheria. Measles virus attack only human beings.

Racial Immunity

Negroes are more susceptible to tuberculosis than whites, where as, for other infection negroes are highly resistant than white people.

Individual Immunity

Identical twins exhibits similar degree of resistance, where as heterogenous twins do not show such correlation.

Other Factors

Fever

A rise in body temperature following infection is a natural defence mechanisms. Rise in temperature destroys the infecting pathogens. It also stimulates the production of interferon.

Inflammation

Inflammation is an important non specific defense reaction to tissue injury. Acute inflammation is characterized by pain, heat, swelling and redness in injured area.

Acute inflammatory response begins when injured tissue cells releases chemical signals (inflammatory mediators) that activate the inner lining (endothelicum) or nearby capillaries.

Within capillaries selectins are displayed on the activated endothelial cells – first P selectin and then E selectin. These adhesions randomly attract and attach neutrophil to the endothelial cells, slow neutrophil down and cause them to roll along the endothelium. In this time, inflammatory mediators activate integrins (adhesion receptor of neutrophil). The neutrophil integrins then highly attach to endothelial adhesion molecules such as the intra cellular adhesion molecule 1 (ICAMI) and vascular cell adhesion molecule (VCAMI). This causes the neutrophil stick to the endothelium and stop rolling. The neutrophils now undergo dramatic shape changes squeeze through the endothelial wall (extravasation) into the interstitial tissue fluid, migrate to the site of injury and attack the pathogen. Other lymphocytes are also attracted to the infection site following neutrophils. The inflammatory mediators also raise the acidity in the surrounding extra cellular fluid.

This activates the extra cellular enzyme kallikrein, which splits bradykinin. Bradykinin binds to receptor on capillary wall, opening the Junction between cells and allowing fluid and leukocytes to leave capillary and enter the infected tissue. Similarly bradikinin binds to mast cells and activates the mast cell and releases histamine. It inturn makes the capillary wall wider so that more fluid leucocytes, kallikrein and bradykinin processor move out causing edema.

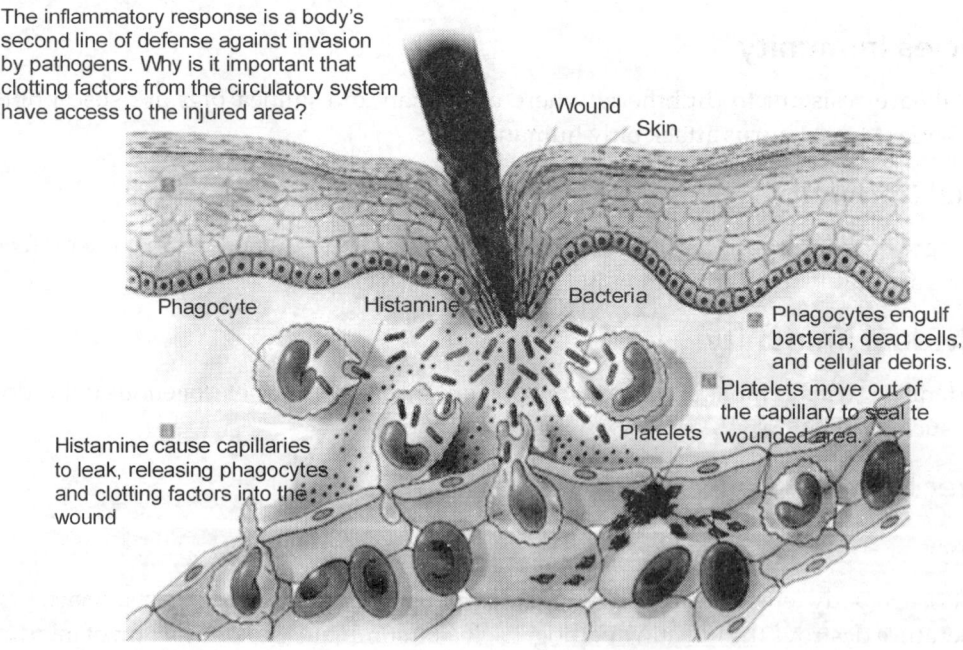

The inflammatory response is a body's second line of defense against invasion by pathogens. Why is it important that clotting factors from the circulatory system have access to the injured area?

Wound

Skin

Phagocyte   Histamine   Bacteria

Phagocytes engulf bacteria, dead cells, and cellular debris.

Platelets move out of the capillary to seal te wounded area.

Platelets

Histamine cause capillaries to leak, releasing phagocytes and clotting factors into the wound

Figure 225   Inflammation

Bradikinin then binds to nearby capillary cells and stimulates the production of prostaglandins to promote tissue swelling in the infected area. It (Prostaglandins) also bind to free nerve endings, making them fire and start a pain impulse. A arachidonic acid also produced by mast cell and take part in inflammation.

Events of inflammation are

Increase in blood flow and capillary dilation bring into the area more antimicrobial factors and leukocytes that destroy pathogens.

The rise in temperature stimulates the inflammatory response and inhibit microbial growth.

A fibrin clot often forms and may limit the spread of microbes.

Phagocytes collect in the inflamed area and phagocytose the pathogen. In addition, chemicals stimulate the bone marrow to release neutrophil and increase the rate of granulocyte production..

These process removes all damaged tissue and microbes from the site. This process leads to repair of damaged tissue and complete recovery.

Summary

1. Tissue damage caused release of vaso active and chemotactive factors that trigger a local increase in blood flow and capillary permeability.
2. Permeable capillaries allow an influx of fluid and cell.
3. Phagocytes migrate to site of inflammation.
4. Phagocytes and antibacterial exudates destroy bacteria.

Acquired Immunity

The resistance that acquired by an individual during his life is called acquired immunity. Antigenic stimuli induced the cell to produce antibodies or sensitized lymphocytes. This type of immunity is also called as specific immunity. Acquired immunity is of two types, they are active and passive. Both active and passive immunity may be natural or artificial.

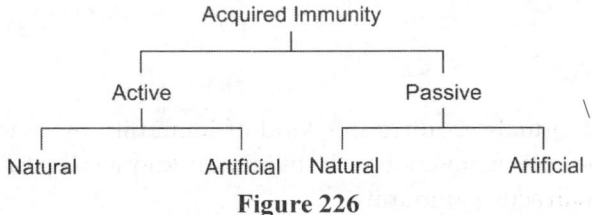

Figure 226

Active Immunity

Resistance developed by an individual in response to an antigenic stimulus. The antigen may gain entrance either by natural infection or through artificial vaccinations. This immunity results in the synthesis of antibodies (HI) or production of immunologically active cells (CMI).

Natural Active Immunity

Immunity is developed by the host in response to the antigen that enters by natural infection.

Eg. A person attacked with small pox virus develops natural active immunity.

At the end, person recovers from the infection. This will create permanent immunity, other examples are measles, mumps, chicken pox etc. Some of the microbial attack may provide immunity for a shorted duration. This is due to antigenic variation in microbial cells.

Artificial Active Immunity

Immunity is attained by the host in response to the antigen got by vaccination.

Vaccines are preparation of live (attenuated) or killed microbes or their products (Toxoids). Some vaccines administered orally, some by intradermal and intramuscular injection.

Examples

Attenuated live vaccines
    Anthrax vaccines
    BCG (Bacille Calmette – Guerin)
    Sabin Oral Polio Vaccine
    Measles Vaccines
Killed Vaccines
    Salk polio vaccines
    Influenza virus vaccines
    TAB (Typhoid AB)
    Pertussis vaccine
Toxoid
    Tetanus toxoid
    Diphtheria toxoid
    DPT

Passive Immunity

Non immune individuals acquires this kind of immunity by receiving antibodies or sensitized white blood cells from another individual is known as passive immunity. This immunity is inferior than active immunity.

Natural Passive Immunity

The immunity transferred from the mother to the child passively is known as natural passive immunity. Eg: In human, unborn baby receives IgG antibodies through placenta which protect unborn fetus from microbial attack. New born breast fed babies receives IgA antibodies orally in the form of colostrums (First milk produced by the mother after the birth of off spring). Ig A antibodies binds on the mucous membrane of offspring and protect the child from the invading microorganisms.

Artificial Passive Immunity

Transfer of immunity from an immunized donor to a non immune recipient by transferring antibodies or immunized lymphocytes is known as artificial passive immunity. Artificial passive immunity is brought about by using any one of the following sera.

i.   Hyper immune serum of animal or human

ii. Convalescent serum

iii. Pooled sera from different healthy individual.

Hyper Immune Serum

Anti Tetamus Serum (ATS) is prepared from animal and used for human treatment. It may create hypersensitivity. Hence it is discontinued.

Convalescent Serum

Serum collected from persons recovering from a particular disease contains high amount of antibodies for the specific antigen. This is used for immunization.

Polled sera

Serum collected from multiple individuals of the community infected with similar kind of infection and used for immunization.

Indian Immunization Schedule

Age	Vaccine
Birth	BCG
	OralPolio O dose
	Hepatitis B 1st dose
1 Month	Hepatitis B 2nd dose
6 weeks	Oral polio 1st
	DPT, Hib – 1st dose
10 weeks	DPT, Hib - 2nd dose
	Intramuscular Polio 1st
14 weeks	DPT, Hib 3rd dose
	Hepatitis B 3rd dose
	Oral polio 3rd dose
	Rota virus dose I
18 weeks	Intra muscular Polio 2nd
	Rotavirus dose2
9 Months	Measles vaccines
15-18 months	MMR
	DPT 1st Booster
	Hib 1st Booster
	Oral Polio 4th dose
	Varicella Vaccines
	Hepatitis A Vaccines
2 Years	Typhoid Vaccines
5 Years	DTP 2nd Booster dose
	Oral Polio Vaccine 5th dose
	Typhoid Vaccine
10 Years	TT Booster
	Typhoid Vaccine
15-16 Years	TT Booster
	Typhoid Vaccine

D.   ORGANS OF THE IMMUNE SYSTEM OR LYMPHOID ORGANS

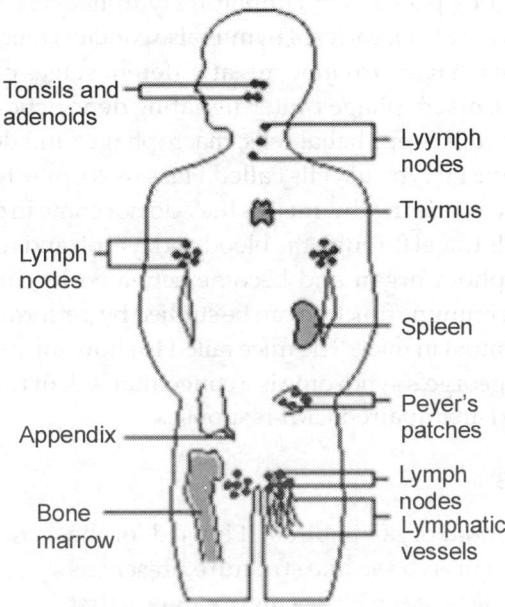

Tonsils and adenoids

Lyymph nodes

Thymus

Lymph nodes

Spleen

Peyer's patches

Appendix

Lymph nodes

Bone marrow

Lymphatic vessels

Figure 227   Lymphoid organs

Introduction

The organs concerned with immune reactions are called lymphoid organs. They contain lymphoid cells. The lymphoid organs are of two types. They are Central or primary lymphoid organs and Peripheral or Secondary lymphoid organs.

Primary Lymphoid Organs

Primary lymphoid organs are the major site of lymphopoiesis. Here lymphoid cells get proliferated become differentiated and mature into immuno competent cells without any antigenic stimulation

Thymus, bursa of fabricius in birds and bone marrow are the primary lymphoid organs.

T-lymphocyte

Capsule

Septum

Cortex

Medulla

Hassall's corpuscle

Epithellial reticular cell

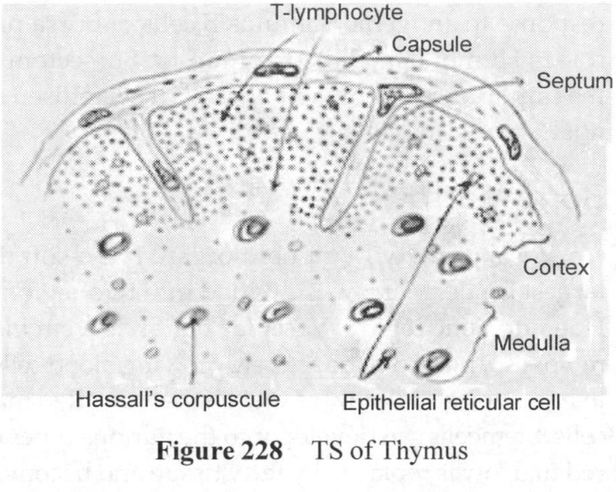

Figure 228   TS of Thymus

Thymus

It is a primary lymphoid organ. It is a flat organ situated above heart below the thyroid gland. The thymus is covered by a fibrous capsule.

Thymus is formed of two lobes. Each lobes contains lobules. Lobules are separated one another by trabaculae. Each lobule contains two parts. They are cortex and medulla. Cortex is a outer layer and is tightly packed with immature lymphocytes. Medulla contains mature lymphocytes. These are came from cortex. Thymus also contains reticular cells. Cortex contains 5 important types of cells that are thymocytes at different stages of maturation, network of stromal cells, nurse cells, macro phages, interdigitating dendrictic cells. Medulla consists of mature thymocytes, modullary epithelial cells, macrophages and dendritic cells. Along with this they also possess special type of cells called Hassals corpuscles. Lymphocytes are non functional when they are inside the thymus, as they do not come in contact with the antigens. These non functional cells travel through the blood and lymph and reaches thymus dependent area of secondary lymphoid organ and become activated by antigenic stimulation. The importance of thymus in immune function can be studied by performing neonatal thymectomy (surgical removel of thymus) in mice. The mice failed to show the production of Tcell and cell mediated immunity. Digeorge's syndrome is a congenital lack of the thymus and leads to the increase in infection and an impaired CMI response.

Bursa of Fabricius

It is a primary lymphoid organ in birds. It is used for the development of B cells. This is a sac like structure present as the outpushing in the cloaca of the bird. This organ was first described in 1621 by fabricius. Cross section of bursa showed the structure similar to thymus. It is formed of many lobes called follicles. Each follicle has outer cortex and inner medulla. The bursa contains three types of cell, namely lymphocytes, macrophages and plasma cells. The lymphocytes of bursa are called B cells. B cells of bursa migrate to the bursa dependent secondary lymphoid organs. There, the B cell produce plasma cells which later secrete antibodies in response to antigenic Stimulus B cells of bursa play a major role in Humoral mediated immunity. Bursectomy indicated the importance of bursa in birds. But sectonised chicken do not have plasma cells.

Figure 229   Bursa of Fabricius

Bone Marrow

It is a primary Lymphoid organ. It is a soft tissue within the cavities of bone. It is the largest tissue. Marrow is divided into two regions, namely vascular and adipose region & Haemopoietic region. Vascular region is a circulatory system that supplies nutrients and removes waste from the actively growing blood cells. Haemopoietic region is actively involved in haemopiesis is know as the red marrow. Red marrow contains totipotent cells called stem cells. Stemcells can develop into the various types of blood cell. In adult animals much of the red marrow is replaced by fatty tissue and become yellow marrow.

Functions

Haemopoietic stem cells arise in the bone marrow. Bone marrow function as both primary and secondary lymphoid organ.

Bone marrow is the primary source of B cell. During secondary immune response, large number of plasma cells are produced in the bone marrow.

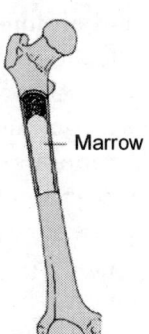

Bone marrow is a source of antibody synthesis. Some of the cells from marrow to thymus, where they develop into the T lymphocytes.

Secondary Lymphoid Organ

Figure 230 Bone marrow

Secondary lymphoid organs are a set of organs concerned with immune reactions. In this organ the lymphocytes are made functional. Here the cells are exposed to antigen and converted to sensitized cells / antibody producing cells. They are small and poorly developed at birth and they grow progressively with age. Lymphnodes, spleen, peyer's patches, Tonsils, (MALT) are the secondary lymphoid organs.

Lymph Node

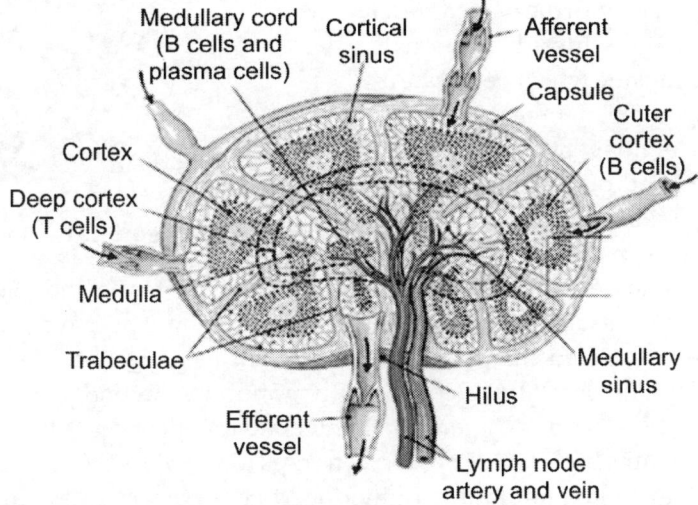

Figure 231 Lymph node

Lymph nodes are secondary lymphoid organs. They are complex, spherical or ovoid structures present along the lymphatic ducts.

One side of the lymph node has an indentation, the hilus, where the blood vessel enter and leave the node. Several afferent lymphatic ducts carry lymph to the Lymph node and one efferent. Lymphatic duct leaves the node. Numerous lymph nodes are present in the joints.

Lymph node is covered by a capsule. It penetrates in the node to form a septa called trabaculae. Lymph node is made up of three regions namely outer cortex, middle para cortex and an inner medulla. Cortex contains B cells, para cortex contains T cells and medulla contains both T and B cells. Large number of lymphoid cells are present in the cortex and grouped together to form nodules or primary follicles. Within these follicles, secondary follicles with germinal centers develop. The germinal centers are the sites of rapid multiplication of lymphoid cells. Cortex also contains dendrite macrophages. The medulla consists of lymphocytes arranged along strands of connective tissue fibres known as medullary cords. These are separated by medullary sinuses, which contains plasma cells. Cortical and medulary regions form B dependent area or bursa dependent area. Para cortical area forms the thymus dependent or T dependent area. Para cortex contains APC's with large quantities of MHC class I antigens on their surface. Lymph nodes are rich in phagocytic cells, they function as the centre of phagocytosis. Lymph Nodes are also responsible for the initiation and development of humoral and cell mediated immune response. Lymph node filter lymph.

Spleen

Spleen is a secondary lymphoid organ. It is a solid, encapsulated organ located in the upper part of the abdominal cavity behind the stomach and close to the diaphragm. It is deep red in colour. It has direct communication with the main arterial circulation. It filters blood. Spleen is surrounded by a capsule which penetrate as septa into the organ called trabaculae. There are two distinct regions, red pulp and an inner white pulb. Red pulp consist of large numbers of blood filled sinusoids in which phagocytes and plasma cells are found. Red

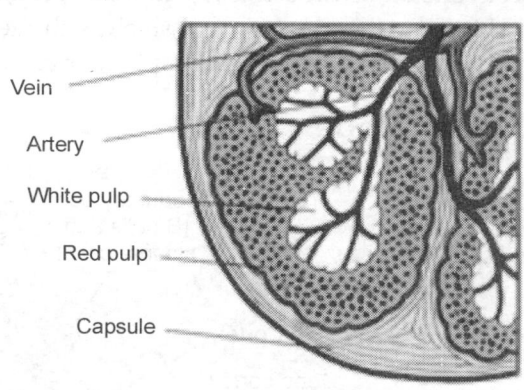

Figure 231 Spleen

pulp destroys old and dead RBC. Red pulp also a reserve site for haemopoiesis. White pulp consists of lymphoid tissue. The cellular arrangement of white pulp was first described by malphigi, hence it is called malphigian follicle. Central region of white pulp is called periarteriolar lymphatic sheet. PALS contains T lymphocytes. Secondary follicle is present in the white pulp which is formed by germinal center and mantle layer, which contains B cells. Marginal zone contains both T & B cells. Spleen traps blood borne antigen and initiate both humoral and cell mediated immunity. Individual do not possess spleen are susceptible to blood borne infection. Spleen function as Graveyard for dead blood cells. Reserve tank for RBC formation at times of necessacity. Act as a filter for traping foreign particle. It brings humoral and cell mediated immunity.

MALT

MALT is a mucosal associated lymphoid tissue. MALT is present in the mucosal layer of respiratory tract, alimentary canal, urino genital system. MALT is an uncapsulated tissue. In human peryer's patches, the tonsils and appendix are good example for MALT and it is associated with gut. It is also called as GALT (Gut Associated Lymphoid Tissue). Lymphoid tissue associated with respiratory tract are called bronchus associated lymphoid tissue (BALT) and lymphoid tissue associated with the epithelium of intestine are called GALT.

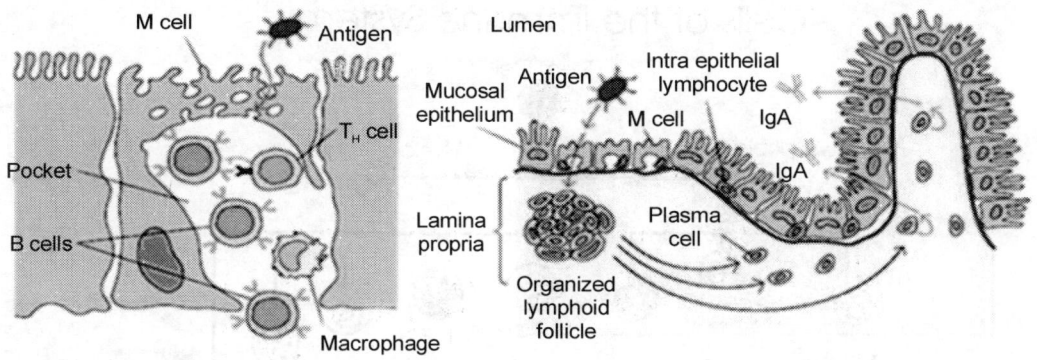

Figure 232   MALT

Peyers Patches

It is a secondary lymphoid tissue. Mucosal epithelium contains intra epithelial lymphocytes, many of which are T cells. Lamina propria is present under the mucosal epithelium. Lamina propria contains B cells, plasma cells, activated T_4 cells and macrophages. More than 15,000 lymphoid follicles are present in the lamina propria of healthy child. Lymphoid follicles or germinal enters. M cells have a deep invagination or pocket in the baso lateral plasma membrane which is filled with a cluster of B cells, T_4 and macrophages. Antigens in the intestine lumen are endocytosed into vesicles that are transported from the luminal membrane to the pocket membrane. The vesicles then fuse with the pocket membrane and the antigens activate B cell and then secrete IgA. This is used by the body to combat many types of infection.

Table 3.1   Differentiation / Comparison of Lymphoid organs

S.No	Primary Lymphoid Organ	Secondary Lymphoid Organ
1	Lymphoid stem cells proliferate differentiate and mature	The lymphoid cells become functional
2	Antigen cannot enter in	Antigen enter and stimulates the lymphoid cells
3	They atrophy with age	They increase in size with age
4	Contains either B cell or T cells	Contains both T cell and B cell
5	Thymus, bone marrow bursa of fabricius	Lymph nodes, spleen, MALT

E. CELLS OF IMMUNE SYSTEM

The immune system operates by producing cells. The cells of immune system are derived from the pluripotent stem cells in the bone marrow. They develop into two lines, Namely lymphoid lineage and myeloid lineage. In lymphoid lineage the stem cells develop into lymphoid progenitor. The lymphoid progenitor cells develops into T lymphocytes,

B lymphocytes and null cells. Natural killer cells and killer cells are the types of null cells. Lymphoid cells are mostly involved in specific immune response.

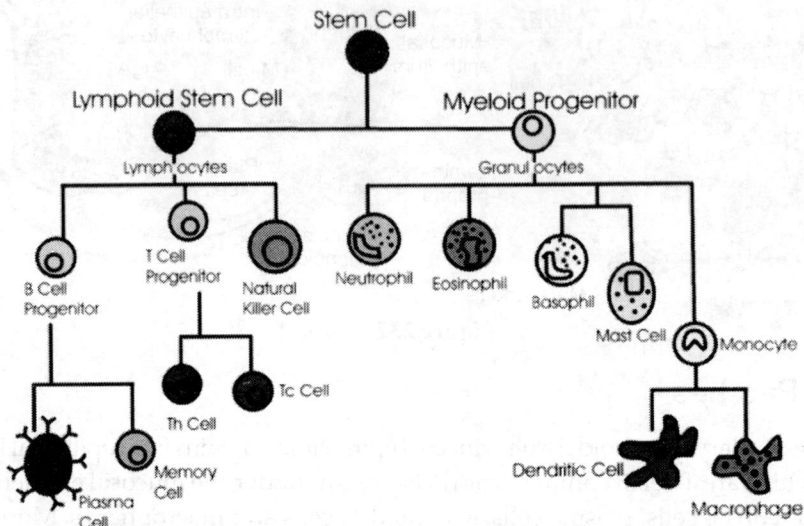

Cells of the Immune System

Figure 233

Myeloid cells

In myeloid lineage, the stem cells develops into myeloid progenitor cells. Myeloid progenitor cells are develop into erythrocyte, megakaryocyte, monocyte, granulocytes and mast cells. Myeloid cells are mostly involved in non specific immune response.

Monocytes

They are mono nuclear phagocytic cells. They are present in blood stream. They constitute approximately 4-10% of the nucleated cells in blood. They have horse shoe shaped nucleus. Cytoplasm contains azurophilic granules. Monocytes are converted to macrophages. They circulate in blood stream for about 8-10 hours

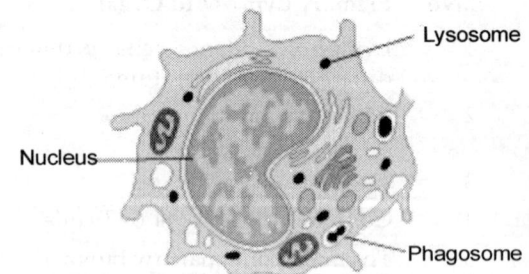

Figure 234 Monocyte

Macrophages

Macrophages are large, mononuclear phagocytic cells. They are derived from monocytes. Monocytes get enlarged and migrate into the tissue to become the tissue specific macrophages. Macrophage contains azurophilic granules that contains lysozyme, acid hydrolases, phosphatase, β glucouronidase, myeloperoxidase. Macrophages are actively participated in phagocytosis of various pathogens and tumour cells. Usually tissue macrophages are immobile but become active when stimulated by lymphokines. Macrophages are activated by gamma interferon. Various names of macrophages are as follows.

Histiocytes	– Soft tissue macrophages
Kupffer Cells	– Liver macrophages
Osteoclasts	– Bone macrophages
Microglial cells	– Brain / Nerve cell macrophages
Langerhans cells	– Epidermis macrophages
Glomerular Mesenglial cells	– Kidney macrophages
Pulmonary alveolar macrophages	– Lungs macrophages
Macrophages	– Spleen and lymphnodes

Macrophages are activated by IFN, and are effective in eliminating the engulfed pathogen and secrete inflammatory mediators. Macrophages are also called as Antigen presenting cells. They express high level of MHC class II molecules. Activated Macrophages also secrete Tumour Necrosis factor. Macrophages have surface receptor for C_3 component of complement (CD 35) and Fc component of antibody (CD16)

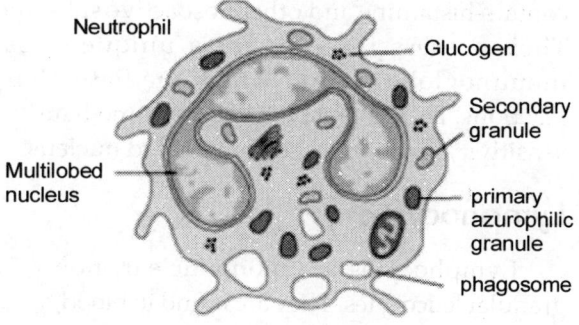

Figure 235

Granulocytes

The name granulocyte is due to granules present in the cytoplasm. Granules showed characteristic staining reaction. Staining characteristics differentiates these cells into Neutrophils, eosinophils and basophils.

Neutrophils

Neutrophils form the major part of the white blood cells. They contributed to about 40-70% of nucleated cells. They

Figure 236

have multi lobe nucleus with granulated cytoplasm. Granules do not take up acidic or basic stains and hence named neutrophil. They are also called as polymorpho nuclear leukocytes(PMN). They are important cell found predominantly at the site of inflammation. The Process by which the circulating neutrophils enter into the tissue space is called extravasation. Neutrophils are active phagocytic cells. They have two types of granules viz. primary granules and secondary granules. Primary granules like lysosomes, larger in size and contains peroxidase, lysozyme and hydrolytic enzymes. Secondary granules are smaller in size and contain enzymes like collagenase, lactoferrin and lysozyme. During phagocytosis, foreign materials are enclosed in a sac like structure called phagosomes. Then the primary and secondary granules are fuse together and eliminate foreign particle.Neutrophil also express a complement receptor (CD 35) and receptor for Fc portion of immunoglobulin(CD16)

Eosinophil

They have a bilobed nucleus. The granules are stained with acidic dyes such as eosin and hence its name. The granules are rich in hydrolytic enzymes. They are about 2-5% of the leucocytes in the healthy individual. Increased number of eosinophils are observed in allergic patients and are called Eosinophilia. Eosinophils are involved in the immunity against parasitic infections. They kill helminths by releasing chemical mediators stored in granules as they cannot phagocytose them since they are larger organisms.

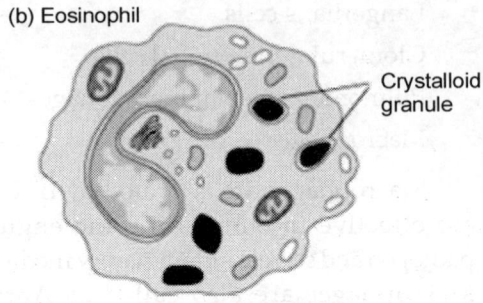

(b) Eosinophil

Crystalloid granule

Figure 237

Basophils

Granules of these cells are stained with basic dyes such as methylene blue, Hence the name basophils. They contribute about 0.2 to 0.4% to the total leucocyte population. Granules of basophil contain histamine and other vosoactive substances. They possess receptor for a unique class of immunoglobulin IgE, which mediate allergic reactions. They are responsible for immediate hyper sensitivity reactions. They have lobed nucleus.

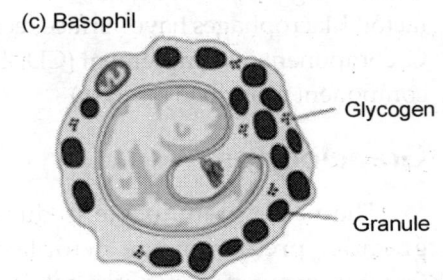

(c) Basophil

Glycogen

Granule

Figure 238

Lymphocytes

Lymphocytes are mononuclear, non granular leucocytes. They are found in blood, lymph and lymphoid tissue, such as spleen, lymph nodes, tonsils, peyer's patches etc. They are spherical or ovoid in shape.

About 20% leucocytes are lymphocytes. There are two types of lymphocytes. They are small and large lymphocyte.

Small lymphocyte have a large nucleus with a rim of cytoplasm. They do not contain

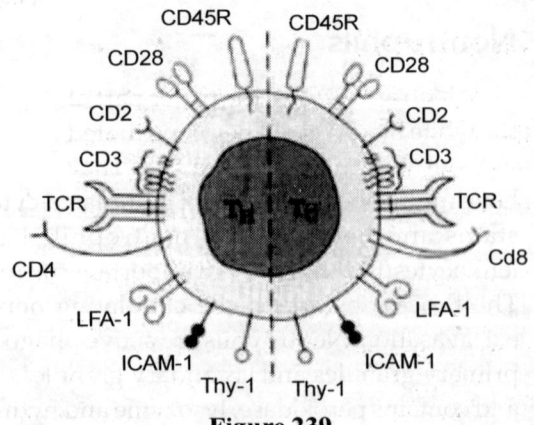

CD45R   CD45R
CD28   CD28
CD2   CD2
CD3   CD3
TCR   TCR
CD4   Cd8
LFA-1   LFA-1
ICAM-1   ICAM-1
Thy-1   Thy-1

Figure 239

endoplasmic reticulum. Small lymphocytes are further divided into two types based on their function that are T lymphocyte and B lymphocytes. Large lymphocytes have lower nuclear cytoplasm ratio, and are granulated. It is also called as large granulated lymphocytes (LGLs).

Lymphocytes normally posses specific receptors for antigens and thus mediate specific immunity.

T Lymphocytes

It is a mono nuclear, non granular leucocytes that matures in thymus. It is also called thymocyte, thymic dependent cells. T lymphocyte have a large nucleus with a rim of cytoplasm. The surface of T lymphocyte contains certain unique group of proteins called T cell surface markers. The following are the T cell surface markers.

Erythrocyte Receptor: It recognizes the sheep erythrocyte (SRBC), T Cell antigen receptor (TCR). It recognizes MHC class I & II antigens, The Ia protein receptor: It recognizes immune associated protein. Interleukin receptor: It contains IL1 & IL 2 receptor.

T lymphocytes are derived from haemopoietic stem cells of bone marrow. Thymic hormones are responsible for the formation of these cells. There are different types of T cells. They are the sub populations or subsets of T cells. They are the following.

T helper Cells (T$_H$ cells)

T suppressor cells (T$_S$ cells)

T cytotoxic cells (T$_C$ cells)

T delayed type hypersensitivity cells (T$_D$ cells)

T cells play two important functions – effector and regulatory. Th and Ts cells are regulator cells where as the Tc cells and T$_D$ cells are effector cells.

T helper cells

These cells help B cells and other T cells in immune response. They are regulator cells.

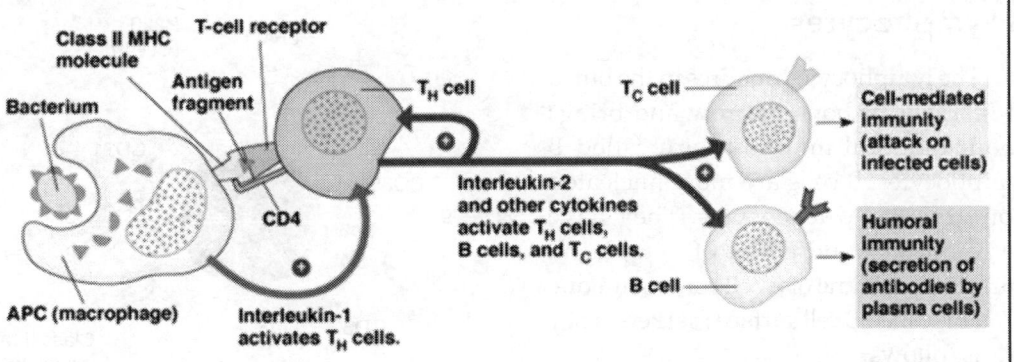

Figure 240

T Suppressor Cells

These cells suppress the activities of B cell and other T cells. They are regulatory cells. They inhibit ab production by B cells. They suppress the function of Th cell and Nk cells. These cells are responsible for immune tolerance by limiting the ability of the immune system to attack a persons own body tissue.

T cytotoxic cell

These cells kill micro organisms or microbes infected cells. These cells release cytotoxic substances directly into the attacked cells. These cells are lethal to tissue cells that have been invaded by viruses. T$_c$ cells also play an important role in destroying cancer cells.

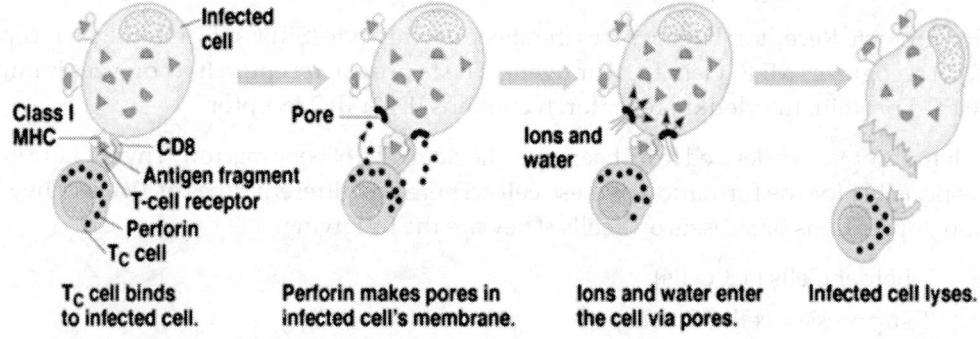

Class I
MHC
CD8
Antigen fragment
T-cell receptor
Perforin
T$_C$ cell

| Infected cell | Pore | Ions and water | |

T$_C$ cell binds to infected cell.　　Perforin makes pores in infected cell's membrane.　　Ions and water enter the cell via pores.　　Infected cell lyses.

Figure 241 Role of TC cell

T delayed type hypersensitivity cell

They are sub population of T lymphocytes. They bring macrophages to areas where delayed hypersensitivity occurs. T$_D$ cells secrete macrophage chemotoxin and macrophage migration inhibition factor. By secreting these factors the T$_D$ cells are directly involved in delayed hypersensitivity reaction.

B Lymphocytes

The lymphocytes matures in the bursa of fabricius or bone marrow and brings about humoral immunity are called B lymphocytes. B cells are mononucleated non granulated leucocytes. They have large nucleus and a rim of cytoplasm. Surface membrane of B cell contain unique proteins called B cell surface markers. They are as follows.

I a (immune associated) protein which binds to a Ia receptor of T cell.

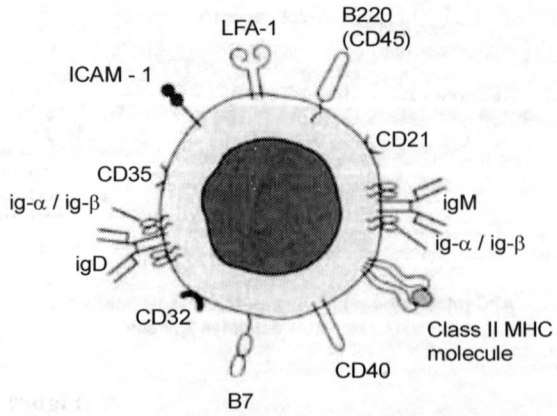

ICAM - 1
LFA-1
B220 (CD45)
CD21
CD35
ig-α / ig-β
igM
ig-α / ig-β
igD
CD32
Class II MHC molecule
CD40
B7

Figure 242 B lymphocytes

Fc receptor to bind with the Fc fragment of the immunoglobulin. CRI and CR 2 – receptors of complement system. Surface immunoglobulin (IgM & IgD)- specific receptor for antigen.

The B cells consist of two subsets. They are

T dependent cells – These cells require the help of T_H cells for the production of immunoglobulin.

T independent cells – These cells do not require the help of T cell for immunoglobulin production.

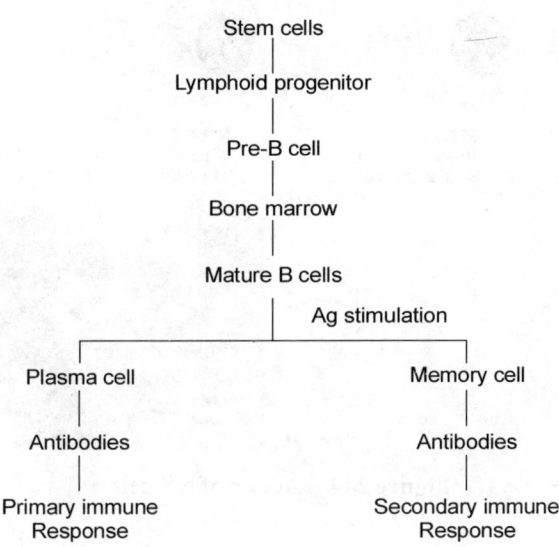

Figure 243   Fate of B cell

Plasma Cell

Cytokines are known to stimulate B cells to become Plasma cell. Plasma cells secrets different classes of antibodies such as IgG, IgA, IgM, IgD, IgE. Some plasma cells live for 2 to 3 days while others continue to produce antibodies for several weeks. Plasma cells are large lymphocytes with a large cytoplasm and small nucleus. They have basophilic cytoplasm and eccentric nucleus with heterochromatin in a characteristic Cart wheel arrangement. Plasma cells are end cells. The cytoplasm is completely filled with rough endoplasmic reticulum. Plasma cells are devoid of surface receptors but rich in immunoglobulin in the cytoplasm. The immunoglobulin is localized in the spaces of the endoplasmic reticulum, where it sometimes forms distinct aggregates termed Russel's nodes.

Large granular lymphocytes or null cells or third population cells

Apart form T cell and B cells, certain groups of lymphocytes do not express membrane bound molecules and receptors. They are larger than T cell and B cells. They are also called null cells. There are two types of null cells namely Natural killer cells and Killer cells.

Natural Killer Cells

They are a group of null cell. They have 2 or 3 large granules in the cytoplasm. Hence they are also called large granular lymphocytes. They kill the target cell without the aid of antibody or complement. So they are called antibody independent. NK cells are activated by virus infected cells through interferon. They recognize altered cell surface and brings about cytolysis and cytotoxicity.

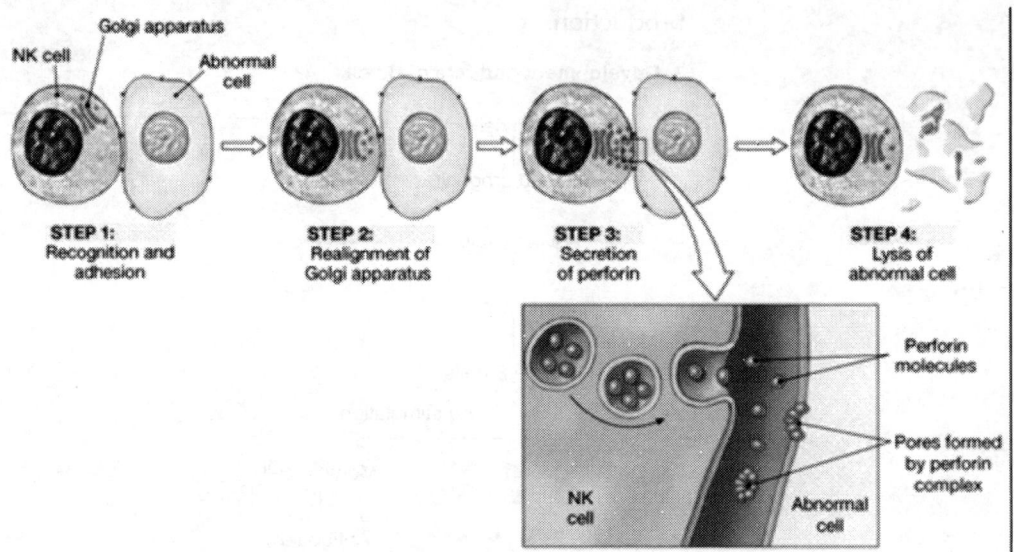

Figure 244   Action of NK cell

Killer Cells (K)

They are group of null cells. They are antibody dependent cells. These cells possess Fc receptors for binding IgG. These cells can bind with cells coated with IgG antibodies and can kill them. The reaction created by K cells are called ADCC reaction (Antibody dependent cell mediated cytotoxic reaction). By this cytotoxic property, K cells can kill a variety of cells such as tumour cells, bacteria, virus, fungi and parasites.

Mast Cells

Mast cells are sessile and are found in various tissues like skin, connective tissues of various organs and mucosal epithelial tissues. These cells available in a blood stream as undifferentiated cells. One they leave the blood and enter the tissues they become mast cells. These cells resemble the basophils in having large number of cytoplasmic granules that contain histamines and other pharmacologically active substance. Mast cells are similar to basophils in their appearance and function. There are two types of mast cell, that are mucosal mast cells and connective tissue mast cells. Serotonin and prostaglandins of mast cells brings about the inflammation.

Dendritic Cells

These cells are named because they resemble the dendrits of neuron on the cell surface extensions. They circulate in blood as immature cells and matured in tissue. Dendritic cells express high levels of both class I and Class II MHC molecule. They are potent Antigen Presenting cells. Dendritic cells process antigen and present it to T$_H$ cells. There are different types of dendritic cells, they are,

1. Langerhans cells – found in epidermins and mucous membrane.
2. Interdigitating dendritic cells – found in T cell areas of the secondary lymphoid organ.
3. Interstitial dendritic cells – found in heart, lungs, liver, kidney etc.
4. Veiled cells – circulating dendritic cells
5. Follicullar dendritic cells – found in lymph follicles.

They help in developing memory B cells with in a follicle.

Table 13.2   Difference between T cell and B cells

S.No.	T Cell	B Cell
1.	Matures in Thymus	Matures in Bursa of fabricius & Bone marrow
2.	Brings about cell mediated immunity	Brings about humoral mediated immunity
3.	Needs the cooperation of another T cells	Needs cooperation of T cells
4.	Produce Lymphokines	Produce antibodies
5.	Act on inter cellular hidden parasites	Act on intercellular parasites
6.	Brings delayed type hypersensitivity	Brings immediate type hypersensitivity
7.	On activation produce senitized T cell and memory cell	Produce plasma cell and memory cell
8.	Posses surface receptor	No surface receptor
9.	Posses thymus specific antigens	No antigens
10.	No fc fragment of Ig	Presence of Fc fragment for Ig
11.	LPS does not activates T Cells	LPS activates B cell

Lymphocytic Traffic

Lymphocytes that enter the blood from the primary lymphoid organ circulate through secondary lymphoid organ and circulate back to the blood via lumphatic system to be re circulated in the very same way. This continuous process of circulation and recirculation of lymphocytes is referred as lymphocytic traffic.

F.   HUMORAL AND CELL-MEDIATED IMMUNE RESPONSES

The humoral response (or antibody mediated response) involves B cells that recognize antigens or pathogens that are circulating in the lymph or blood ("humor" is a medieval term for body fluid). The response follows this chain of events:

Antigens bind to B cells.

 ⅄   Interleukins or helper T cells co-stimulate B cells. In most cases, both an antigen and a costimulator are required to activate a B cell and initiate B cell proliferation.

 ⅄   B cells proliferate and produce plasma cells. The plasma cells bear antibodies with the identical antigen specificity as the antigen receptors of the activated B cells. The antibodies are released and circulate through the body, binding to antigens.

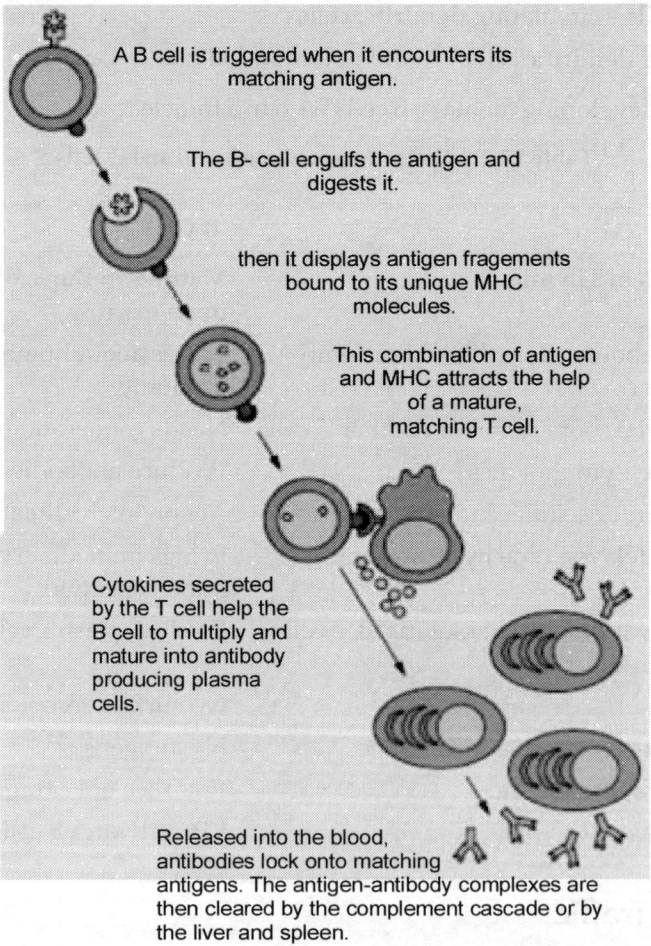

A B cell is triggered when it encounters its matching antigen.

The B- cell engulfs the antigen and digests it.

then it displays antigen fragments bound to its unique MHC molecules.

This combination of antigen and MHC attracts the help of a mature, matching T cell.

Cytokines secreted by the T cell help the B cell to multiply and mature into antibody producing plasma cells.

Released into the blood, antibodies lock onto matching antigens. The antigen-antibody complexes are then cleared by the complement cascade or by the liver and spleen.

Figure 245

B cells produce memory cells. Memory cells provide future immunity. The **cell mediated response** involves mostly T cells and responds to any cell that displays aberrant MHC markers, including cells invaded by pathogens, tumor cells, or transplanted cells. The following chain of events describes this immune response. Self cells or APCs displaying foreign antigens bind to T cells. Interleukins (secreted by APCs or helper T cells) costimulate activation of T cells. If MHC I and endogenous antigens are displayed on the plasma membrane, T cells proliferate, producing cytotoxic T cells. Cytotoxic T cells destroy cells displaying the antigens. If MHC II and exogenous antigens are displayed on the plasma membrane, T cells proliferate, producing helper T cells. Helper T cells release interleukins (and other cytokines), which stimulate B cells to produce antibodies that bind to the antigens and stimulate nonspecific agents (NK and macrophages) to destroy the antigens.

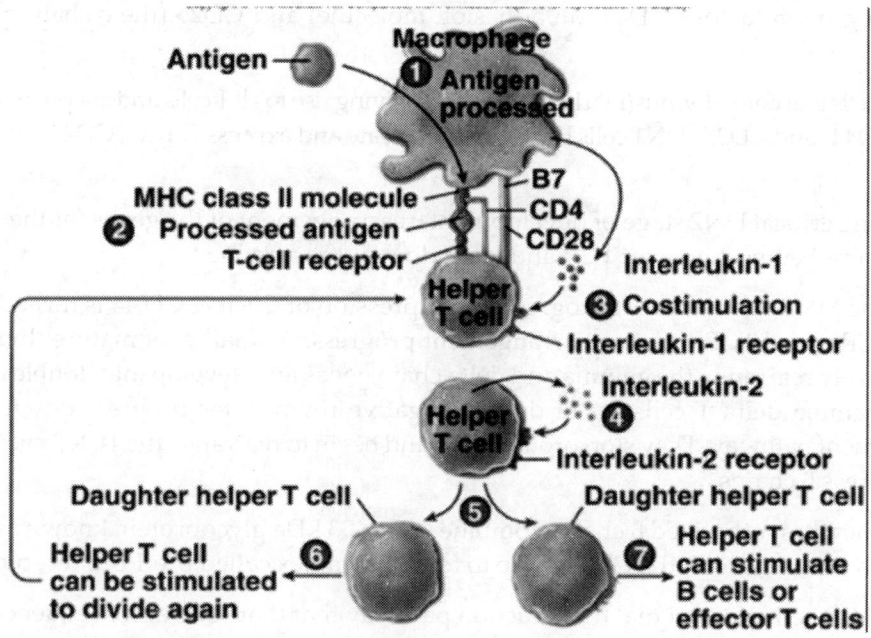

Figure 246

G. T – CELL MATURATION, ACTIVATION AND DIFFERENTIATION

The type of immunity that results from T cells coming into close contact with foreign cells or infected cells to destroy them; it can be transferred to a nonimmune individual by the transfer of cells. **Cellular (cell-mediated) immunity** is based on the action of specific kinds of T lymphocytes that directly attack cells infected with viruses or parasites, transplanted cells or organs, and cancer cells. T cells can lyse these cells or release chemicals (cytokines) that enhance specific immunity and nonspecific defenses such as phagocytosis and inflammation.

Progenitor T cells begins to migrate to the thymus from the early sites of haematopoiesis. T-cell maturation involves re arrangement of the germ line TCR genes and the expression of various membrane markers. In the thymus, developing T cells (Thymocytes) proliferate and differentiate along developmental pathways that generate functionally distinct sub populations of mature T-cells.

When precursor T cells arrive at the thymus, they do not express T-Cell receptor, the CD3 complex or the co receptors CD4 and CD 8. Recombination Activation Gene (R AG) 1 and RAG 2 are needed for re arrangement of TCR genes.

After arriving at the thymus, T cell precursors enter the outer cortex and slowly proliferate. During approximately 3 weeks of developments in the thymus, the differentiating T cell pass through a series of stages that are marked by characteristic changes in their cell surface phenotype. These cells are CD4 – and CD 8 –. It is also called as Double Negative (DN) cells. T cells are subdivided into four subsets (DN1 to DN4) characterized by the presence or absence of cell surface molecules. In addition to CD4 and CD 8 such as C-kit, (the receptor for stem cell growth factor), CD44 (an adhesion molecule) and CD25 (the c chain of the Il3 receptor)

The cells that enter thymus(DN1) are capable of giving rise to all T cells and are phenotypically C-kit⁺, CD44⁺ and CD25⁻. DN1 cells begins to proliferate and express C-Kit ⁺, CD44⁺ and CD25⁺ +(DN2)

During critical DN2 stage of development, rearrangement of the genes for the Tc1Rr, S and B chains begins, however TCR and locus does not re arrange.

As the DN 3 development progress, the expression of C Kit & CD 44 is turned off and TCRr, TCR α and TCR β chain rearrangement progress. A small % of mature thymocytes productively rearrange the gamma and delta chain genes and develop into double negative CD 3+ gamma delta T cells. Most double negative thymocytes progress down the α β development pathway. They stop proliferation and begin to rearrange the TCR β chain genes, then express β chains.

The newly synthesized β chains combine with a 33 kDa glycoprotein known as pre T α chain and associate with the CD 3 group to form a complex called pre T cell receptor.

Pre TCR activates a signal transduction pathway that that several consequences.

A cell has made a productive TCR β chain rearrangement and signals its further proliferation and maturation. It suppresses further rearrangement of TCR β chain genes resulting in allelic exclusion. It renders the cell permissive for rearrangement of the TCR β chain. It also induces developmental progression to the double positive state. After β chain rearrangement is completed, the DN3 cells quickly progress to DN4, the level of CD25 falls and both CD4 and CD8 co receptors are expressed. This double positive (DP) stage is a period of rapid proliferation.

This rapid proliferation generates clone of cells with single β chain re arrangement. Then DP cells stop proliferation and RAG2 protein level increases.

Each of cells then rearrange β chain gene, thereby generating a much more diverse population. This will leads to the development of immature single positive CD4⁺ thymocytes or single positive CD8⁺ thymocytes.

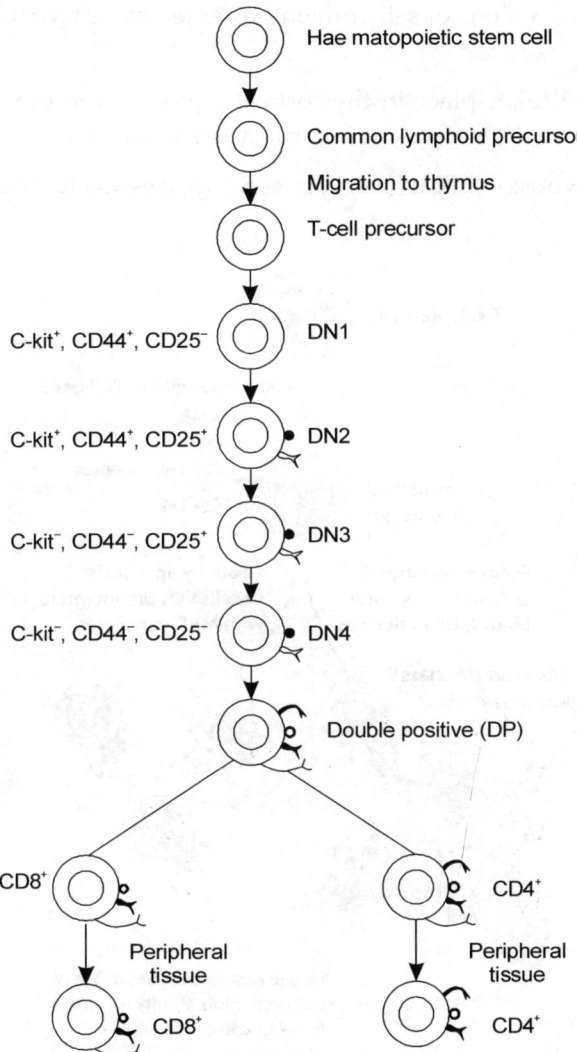

Figure 247   T cell maturation

Thymic selection of the T cells Repertoire

Random rearrangent within TCR germ line DNA combined with juctional diversity can generate an enormous T cell repertoire.

Thymocytes undergo two selection process in the thymus.

1. Positive selectin for thymocytes bearing receptors capable of binding self MHC molecules, which results in MHC restriction. Cells that fails positive selection are eliminated within the thymus by apoptosis.

2. Negative selection that eliminates thymocytes bearing high affinity receptors for self MHC molecules alone or self antigen presented by self MHC, which results in self tolerance.

Positive selection takes place in the cortical region of the thymus and involves the interaction of immature thymocytes with cortical epithelial cells.

Only those cells whose α β TCR heterodimer recognizes a self MHC molecule are selected for survival.

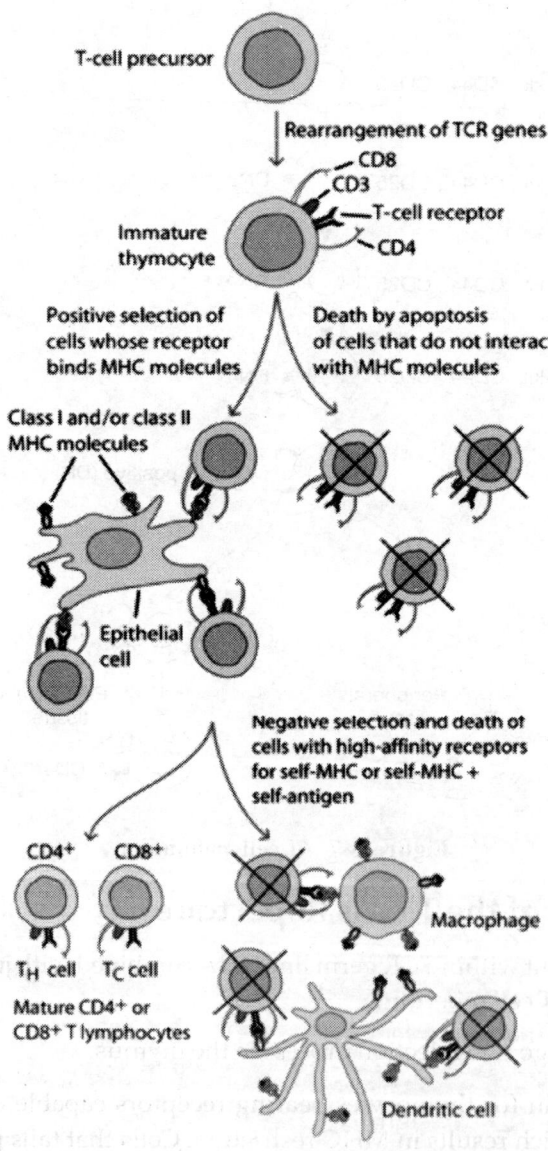

Figure 248 Thymic selection of T cell

Thymocytes with high affinity receptors were weeded out during negative selection via an interaction with thymic stromal cells. High affinity receptor recognizes self antigen.

Thymocytes with a range of affinities from low to high for self antigen presented by MHC molecule are selected by + selection process.

T-Cell Activation

Important event in the generation of cell mediated and humoral mediated immune response is the activation and clonal expansion of T cells.

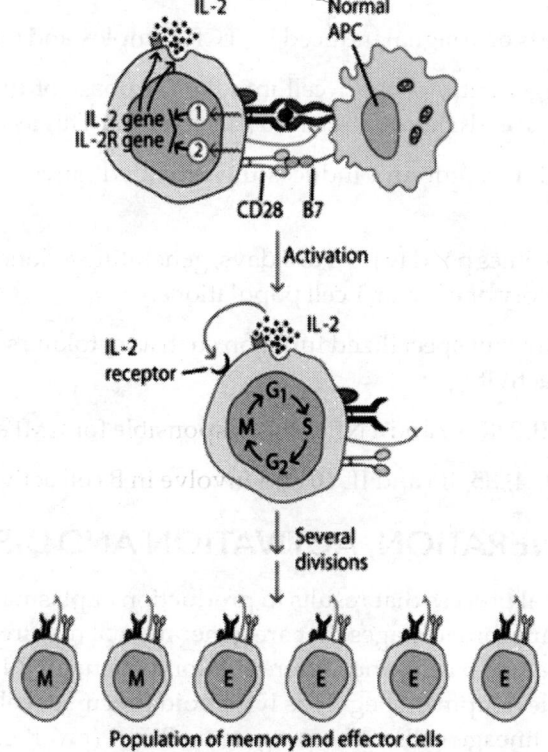

Figure 249   T cell maturation

T cell activation is initiated by interaction of the TCR-CD 3 complex(DP) with a processed antigenic peptide bound to either class I or class II MHC molecule on the surface of antigen presenting cells (APC).

Interaction of a T Cell with antigen initiates cascade of biochemical events that induces resting T cell to enter the cell cycle, proliferation and differentiating into memory cells or effector cells.

Activation of T cell by both signal 1 and co stimulatory signal 2 upregulates expression of IL 2 and the high affinity IL2 receptor, leading to the entry of the T cell into cell, cell cycle and several rounds of proliferation.

T.cell differentiation

CD 4 $^+$ and CD 8 $^+$ cells leave the thymus and enter circulation as resting cells in the Go stage of the cell cycle.

T cells that have not yet encountered antigen are called naive T cells.

If naive T cell recognizes an Ag-MHC complex on the target cell, it will be activated, initiating primary response.

The naive T cell enlarges into a blast cell and begins undergoing repeated rounds of cell division.

Activation depends on a signal induced by TCR complex and CD28-B7 interaction

These signals trigger entry of the T cell into the G1 phase of the cell cycle and induce transcription of IL2 gene and chain of the CD25 (Il2) receptor. This leads to production of IL 2.

IL 2 binds to IL 2 receptor and induces the activated naive T cell to proliferate and differentiated.

T cell divide 2 to 3 times per day for 4 to 5 days, generating a clone of progeny cells which differentiate into memory or effector T cell populations.

Effector T cells carryout specialized function such as cytokine secretion and B cell help and cytotoxic killing activity.

TH1 cells secrets IL2, IFNγ and TNFα. It is responsible for CMI and activates Tc cell.

TH$_2$ cells secrets IL4, Il5, Il6 and IL 10 and involve in B cell activation.

H. B CELL GENERATION, ACTIVATION AND DIFFERENTIATION

The developmental process that results in production of plasma cells and memory cells can be divided into three broad stages that are generation of mature, immuno competent B cells, activation of mature B cells and differentiation of activated B cells into plasma and memory cells. B cell development begins as lymphoid precursor cells differentiate into the earliest distinctive B lineage cells – the progenitor B cell (pro B cell), which expresses a transmembrane tyrosine phosphatase called CD45R. Pro B cells attached to the stromal cells.

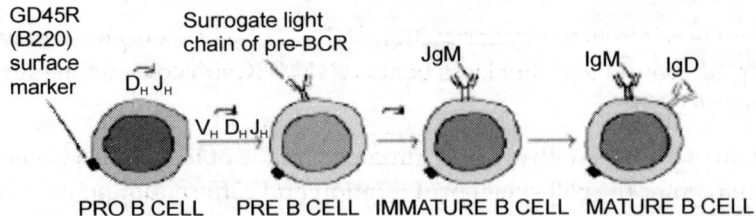

Figure 250 B cell differentiation

Stromal cells produces IL7. IL 7 supports conversion of pro B cell to Pre B cell. Continuation with the formation of pre B cell, heavy chain gene rearrangement, this results in the formation of immature B cells. Immature B cell express membrane IgM, IgD, IgB and forms B cell Receptor

(BCR), which is responsible for signaling the engagement of antigen. Full maturation is signaled by co expression of IgD & IgM on the membrane.

B cell Activation and proliferation

After export of B cells from the bone marrow, activation, proliferation and differentiation occurs in the periphery in response to antigen.

There are two types of B cell activation. They are thymus dependent (TD) and thymus independent (TI). TD requires direct contact with T_H cells. Antigen that can activate B cells in the absence of T_H cells are TI.

T-Independent and T-dependent antigens

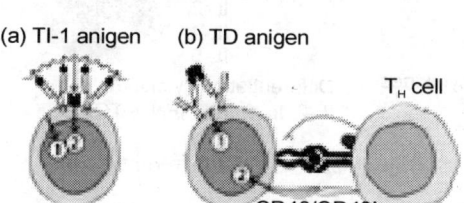

Figure 251   T dependent and Independent Antigens

Antigen Presentation

The transfer of processed Ag from the macrophage to B lymphocytes is called Ag presentation. Thymus dependent Antigen cannot activate B cell. They require cooperation of Th cells. Thymus independent Ag activate B cell directly. This Ag presentation activates the B cell. Th cell secretes a lymphokine called B cell stimulatory factor (BSF). The BSF activates B cells.

Triggering of the B Cell

Antigens (TD or TI) triggered B cell. This leads to change in metabolic activity of B cells.

Clonal proliferation

Triggered B cell, undergoes serious of division. This occurs in 5 days with 8 generation cell division. This results in a formation of clone of identical cells.

Production of Plasma and Memory Cells

The proliferating B cells finally produce two types of cells called plasma cells and memory cells.

Secretion of Immunoglobulins

Plasma cells are the immunoglobulin producing cells. They synthesize immunoglobulins and release them into the serum. As the antibody produced is specific to the Ag, it binds with it and is neutralized or destroyed.

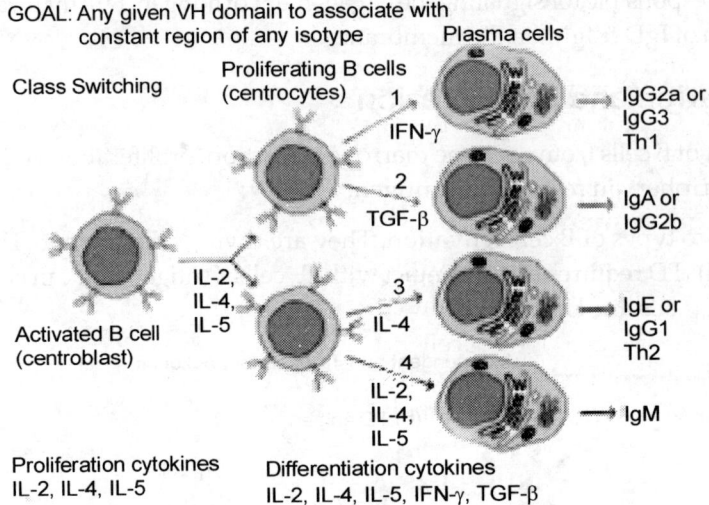

GOAL: Any given VH domain to associate with
constant region of any isotype

Class Switching

Proliferating B cells
(centrocytes)

Plasma cells

IFN-γ
1 → IgG2a or
IgG3
Th1

2
TGF-β → IgA or
IgG2b

Activated B cell
(centroblast)

IL-2,
IL-4,
IL-5

3
IL-4 → IgE or
IgG1
Th2

4
IL-2,
IL-4,
IL-5 → IgM

Proliferation cytokines
IL-2, IL-4, IL-5

Differentiation cytokines
IL-2, IL-4, IL-5, IFN-γ, TGF-β

Figure 252   B cell differentiation

I. THEORIES OF ANTIBODY PRODUCTION

Many theories have been proposed to explain the synthesis of antibody. They are the following.

1. Side chain theory
2. Direct template theory
3. Indirect template theory
4. Natural selection theory
5. Clonal selection theory
6. Two genes for one polypeptide chain theory
7. Class switch hypothesis.

1. Side chain theory

It is proposed by Ehrlich in 1897. The antibody forming cells have Ab molecules a side chain on their surface. When an Ag enters the host, it selects the appropriate side chain and binds with it. Then cells synthesize similar side chains repeatedly and are released into the circulation. They bind with the Ag specifically.

2. Direct template theory

It is proposed by Haurowitz in 1930. When an Ag enters the Ag producing cell, it acts as a template and directs the cell to assemble ab, specific to the Ag. This theory is not at all accepted.

3. Indirect template theory

When an Ag enters the Ab producing cell, modifies the DNA and instruct DNA to synthesis specific Ab.

4. Natural selection Theory

It was proposed by Jerney in 1955. According to this theory all the possible Abs are naturally present in the circulation. These are natural abs. When an Ag enters the circulation, it recognizes the specific natural Ab. By using this Ab as template, direct DNA of the Ab producing cell to produce more abs.

5. Clonal selection theory

It was proposed by Burnet in 1957. When an Ag enters the host, it select a lymphocyte having a specific receptor and binds with it. This binding stimulate lymphocyte, and proliferates. This produces a clone of cells. The activated lymphocyte produces two types of cells. They are plasma cell and memory cell. Plasma cells produce Abs. Memory cells become active and produce Abs when the Ag enters for the second time.

6. Two Genes one polypeptide hypothesis

It is proposed by Dreyer and Burnet in 1965. The synthesis of a polypeptide chain is controlled by two genes located in different regions of DNA. Of these two genes, one synthesizes variable region and the other gene synthesizes constant region.

7. Class switch Hypothesis

According to this, each B cell synthesizes at first, IgM Ab & then it switches over to IgG. This is called class switching. It is brought about by the cooperation to T cell. The T cell produces a factor called B cell differentiation factor.

J.   ANTIGEN

An **antigen** is a substance that stimulates the immune system. This substance can be a part of living organism (bacterium or parasite) but can also be a virus or a chemical toxin. When an antigen stimulates immune system, antibodies are produced that are specific to that antigen, and powerful white blood cells appear to destroy the invading antigen.

An **immunogen** is a specific type of antigen. An immunogen is a substance that is able to provoke an adaptive immune response if injected on its own. An immunogen is able to induce an immune response, whereas an antigen is able to combine with the products of an immune response once they are made.

Immunogenicity is the ability to induce a humoral and/or cell-mediated immune response. Antigenicity is the ability to combine specifically with the final products of the immune response (i.e. secreted antibodies and/or surface receptors on T-cells). Although all molecules that have the property of immunogenicity also have the property of antigenicity, the reverse is not true."

Allergen is a substance capable of causing an allergic reaction. The (detrimental) reaction may result after exposure via ingestion, inhalation, injection or contact with skin. Eg. some medicine, flower powder, seafood.

Super antigen is a class of antigens that cause non-specific activation of T-cells, resulting in polyclonal T cell activation and massive cytokine release.

Tolerogen is a substance that invokes a specific immune non-responsiveness due to its molecular form. If its molecular form is changed, a tolerogen can become an immunogen.

Immunoglobulin binding protein These proteins are capable of binding to antibodies at positions outside of the antigen-binding site. That is, whereas antigens are the "target" of antibodies, immunoglobulin-binding proteins "attack" antibodies. Protein A, protein G, and protein L are examples of proteins that strongly bind to various antibody isotypes.

Vaccine Antigens that induce a protective immune response against microbes and are used to prevent diseases.

Properties of immunogen

1. Immunogenicity- An ability of antigen which can stimulate the body to evoke a specific immune response.

2. Immune reactivity -An ability of antigen which can combine with corresponding Ab or sensitized lymphocyte.

Classification of Antigen

Based on Immunogenicity

Complete antigen is a substances with both immunogenicity and immune reactivity

Incomplete antigen (hapten) is asubstances only with immune reactivity ;

Hapten + carrier => complete antigen (immunogens); Hapten: Only possess immunoreactivity; Carrier: Make hapten to obtain the immunogenicity. Hapten-carrier conjugates are immunogenic molecules to which haptens have been covalently attached. The immunogenic molecule is called the carrier. Structurally these conjugates are characterized by having native antigenic determinants of the carrier as well as new determinants created by the hapten (haptenic determinants). The actual determinant created by the hapten consists of the hapten and a few of the adjacent residues, although the antibody produced to the determinant will also react with free hapten. In such conjugates the type of carrier determines whether the response will be T-independent or T-dependent.

Based on Chemical nature

Proteins Majority of immunogens are proteins (pure proteins or they may be glycoproteins or lipoproteins). Proteins are usually very good immunogens.

Polysaccharides Pure polysaccharides and lipopolysaccharides are good immunogens.

Nucleic Acids Nucleic acids are usually poorly immunogenic. However, they may become immunogenic when single stranded or when complexed with proteins.

Lipids In general lipids are non-immunogenic, although they may be haptens.

According to source of antigens

Xenoantigen; Alloantigen; Autoantigen; Heterophile antigen

Xeno antigen is an antigen that is found in more than one species. An antigen is something that is capable of inducing an immune response. The prefix "xeno-" means foreign or other. It comes from the Greek "xenos" meaning stranger, guest, or host. Pathogens: bacteria, virus, fungi, parasite.

Heterophile Ag (forssman Ag) is a common Ags shared by different species (between human and animal or microbes, between different species of microbe) (eg) M protein of *Streptococcus* bears common antigen determinant with basement membrane of kidney. This common Ag between bacteria and human being can causes poststreptococcal glomerulonephritis.

Alloantigen

Antigens of red blood cell - ABO system(blood typing) -very important in transfusion

Rh system-haemolytic disease of the newborn (HDNB)

HLA system(Human leukocyte antigen)

-relate to transplantation

-very important in immune regulation

Autoantigen

Release of sequestered Ag; Lens protein is released into blood to induce immune response to induce inflammation of lens. Change of molecular structure of auto-tissues. Denatured IgG becomes antigen to induce production of antibody (rheumatoid factor).

According to whether need the help of T cells when B cells produce Ab

T-independent Antigens

T-independent antigens are antigens which can directly stimulate the B cells to produce antibody without the requirement for T cell help In general, polysaccharides are T-independent antigens. The responses to these antigens differ from the responses to other antigens.

Properties of T-independent antigens

Polymeric structure

These antigens are characterized by the same antigenic determinant repeated many times.

Polyclonal activation of B cells

Many of these antigens can activate B cell clones specific for other antigens (polyclonal activation). T-independent antigens can be subdivided into Type 1 and Type 2 based on their ability to polyclonally activate B cells. Type 1 T-independent antigens are polyclonal activators while Type 2 are not.

Resistance to degradation

T-independent antigens are generally more resistant to degradation and thus they persist for longer periods of time and continue to stimulate the immune system.

T-dependent Antigens

T-dependent antigens are those that do not directly stimulate the production of antibody without the help of T cells. Proteins are T-dependent antigens. Structurally these antigens are characterized by a few copies of many different antigenic determinants.

Tumor specific Ag (TSA) - Only expressed on the tumor cells.

Tumor associated Ag (TAA) - Highly expressed on tumor cells but lowly expressed on normal cells.

Factors Influencing Immunogenicity

Contribution of the Immunogen

Foreignness

The immune system normally discriminates between self and non-self such that only foreign molecules are immunogenic.

Size

There is not absolute size above which a substance will be immunogenic. However, in general, the larger the molecule the more immunogenic it is likely to be.

Chemical Composition

In general, the more complex the substance is chemically the more immunogenic it will be. The antigenic determinants are created by the primary sequence of residues in the polymer and/or by the secondary, tertiary or quaternary structure of the molecule.

Physical form

In general particulate antigens are more immunogenic than soluble ones and denatured antigens more immunogenic than the native form.

Degradability

Antigens that are easily phagocytosed are generally more immunogenic. This is because for most antigens (T-dependant antigens, see below) the development of an immune response requires that the antigen be phagocytosed, processed and presented to helper T cells by an antigen presenting cell (APC).

Contribution of the Biological System

Genetic Factors

Some substances are immunogenic in one species but not in another. Similarly, some substances are immunogenic in one individual but not in others (*i.e.* responders and non-responders). The species or individuals may lack or have altered genes that code for the receptors for antigen on B cells and T cells or they may not have the appropriate genes needed for the APC to present antigen to the helper T cells.

Age

Age can also influence immunogenicity. Usually the very young and the very old have a diminished ability to mount and immune response in response to an immunogen.

Method of Administration

Dose

The dose of administration of an immunogen can influence its immunogenicity. There is a dose of antigen above or below which the immune response will not be optimal.

Route

Generally the subcutaneous route is better than the intravenous or intragastric routes. The route of antigen administration can also alter the nature of the response

Adjuvants

Substances that can enhance the immune response to an immunogen are called adjuvants. The use of adjuvants, however, is often hampered by undesirable side effects such as fever and inflammation.

Hapten

A substance that is non-immunogenic but which can react with the products of a specific immune response. Haptens are small molecules which could never induce an immune response when administered by themselves but which can when coupled to a carrier molecule. Free haptens, however, can react with products of the immune response after such products have been elicited. Haptens have the property of antigenicity but not immunogenicity.

Superantigens

When the immune system encounters a conventional T-dependent antigen, only a small fraction (1 in 10^4 -10^5) of the T cell population is able to recognize the antigen and become activated (monoclonal/oligoclonal response). However, there are some antigens which polyclonally activate a large fraction of the T cells (up to 25%). These antigens are called **super antigens**.

Examples of superantigens include: Staphylococcal enterotoxins (food poisoning), Staphylococcal toxic shock toxin (toxic shock syndrome), Staphylococcal exfoliating toxins (scalded skin syndrome) and Streptococcal pyogenic exotoxins (shock). Although the bacterial superantigens are the best studied. There are superantigens associated with viruses and other microorganisms as well. The diseases associated with exposure to superantigens are, in part, due to hyper activation of the immune system and subsequent release of biologically active cytokines by activated T cells.

Epitope or Antigenic Determinant

Portion of an antigen that combines with the products of a specific immune response. An epitope, also known as antigenic determinant, is the part of an antigen that is recognized by the immune system, specifically by antibodies, B cells, or T cells. The part of an antibody that recognizes the epitope is called a paratope. Although epitopes are usually thought to be derived from non-self proteins, sequences derived from the host that can be recognized are also classified as epitopes.

The epitopes of protein antigens are divided into two categories, conformational epitopes and linear epitopes, based on their structure and interaction with the paratope. A conformational epitope is composed of discontinuous sections of the antigen's amino acid sequence. These epitopes interact with the paratope based on the 3-D surface features and shape or tertiary structure of the antigen. Most epitopes are conformational. By contrast, linear epitopes interact with the paratope based on their primary structure. A linear epitope is formed by a continuous sequence of amino acids from the antigen.

Function

T cell epitopes

T cell epitopes are presented on the surface of an antigen-presenting cell, where they are bound to MHC molecules. T cell epitopes presented by MHC class I molecules are typically peptides between 8 and 11 amino acids in length, whereas MHC class II molecules present longer peptides, and non-classical MHC molecules also present non-peptidic epitopes such as glycolipids.

Cross-reactivity

Epitopes are sometimes cross-reactive. This property is exploited by the immune system in regulation by anti-idiotypic antibodies (originally proposed by Nobel laureate Niels Kaj

Jerne). If an antibody binds to an antigen's epitope, the paratope could become the epitope for another antibody that will then bind to it. If this second antibody is of IgM class, its binding can upregulate the immune response; if the second antibody is of IgG class, its binding can downregulate the immune response.

Epitope mapping

Epitopes can be mapped using protein microarrays, and with the ELISPOT or ELISA techniques. Intensive research is currently taking place to design reliable tools that will predict epitopes on proteins.

K. IMMUNOGLOBULINS

Immunoglobulins are immunologically active serum proteins formed in response to an antigen and react specifically with that antigen. Immunoglobulins are also called as Antibody and abbreviated as Ig. Immunoglobulins are immunological missles. It is produced by the body. They are found in the serum, body fluids and tissues. They are found only in vertebrates. They are synthesized by sensitized B cells. They are glycoprotein in nature.

Structure of Immunoglobulin

Rodyney porter in 1962 proposed basic structure of Immunoglobulins. It is a Y shaped molecule.

It is made up to two types of chains, they are light chain and heavy chain. Light chain is made up of 214 amino acid and each heavy chain is made up of 450 – 700 amino acids.

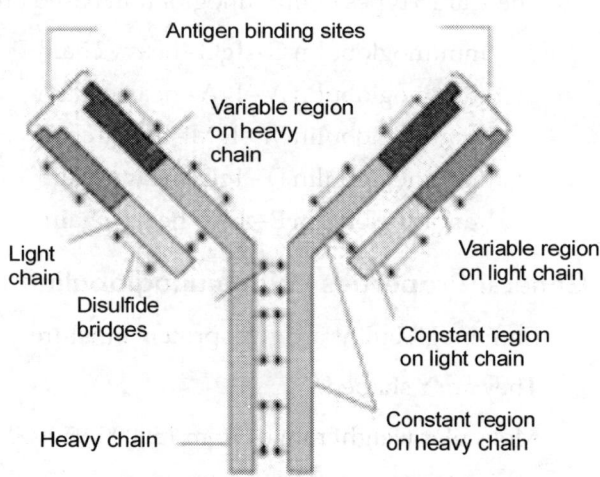

Light Chain

Two light chains are present in the immunoglobulin. There are two types of light chains, they kappa (K) and lambda (L). Types of light chain is based on the structure and amino acid sequence. Light chain is made up of 214 amino acids. Light chain has 2 intra chain disulphide bond.

Figure 253   Structure of Immunoglobulin

Heavy Chain

Each immunoglobulin contains two heavy chains. There are 5 heavy chains. Based on the nature of heavy chain, immunoglobulins are classified they are

1.   Gamma (γ) chain   - IgG
2.   Alpha (α) chain   - IgA

3. Mu (μ) chain   - IgM
4. Delta (δ) chain   - IgD
5. Epsilon (ε) chain   - IgE

The heavy chain has a central flexible region Hinge. Two heavy chains are linked by 1 to 13 inter chain disulphide bond. Heavy chain has 4 intra chain disulphide bond. Immumoglobin consist of two regions namely a variable region (V) and constant region (C). In constant region, the amino acid sequence remains constant but amino acid sequence of variable region shows variability. One end of the immunoglobulin chain is called amino terminal end or N terminal end and the other end is called carboxy terminal end or C terminal end. Based on the function, there are 2 regions recognized. They are Fab region or antigen binding fragment / region and F_c fragment or crystalizable fragment. Each light chain has two regions, namely variable light chain region or VL region and constant light chain region or CL region. Heavy chain has 4 regions. The first domain is called variable heavy chain region or VH region. It is located at the N terminal end. The other 3 domains are called constant heavy chain domains or CH domains(CH1, CH2, and CH3). CH3 is located at the C terminal end. The region which binds with Antigen are called paratope.

Classes of Immunoglobulin / Types of Immunoglobulin

There are 5 types of immunoglobulin based on their heavy chain. They are

Immunoglobulin G – IgG - heavy chain
Immunoglobulin A – IgA - heavy chain
Immunoglobulin M – IgM - heavy chain
Immunoglobulin D – IgD - heavy chain
Immunoglobulin E – IgE - heavy chain

General Properties of Immunoglobulin

Immunoglobulins are glycoprotein innature

They are Y shaped.

Molecular weight ranges from 150000-950000

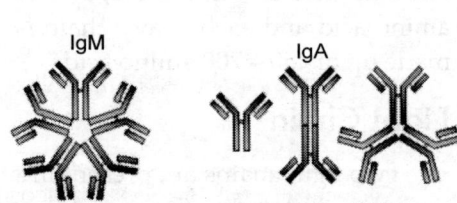

Figure 254   Types of Immunoglobulin

Half life varies from 2-23 days.

Carbohydrate content varies from 3-21%

They contains disulphide bond.

They are light and heavy chains.

They have defnite sedimentation coefficient.

They agglutinate antigens.

They precipitate with antigen.

They cross placenta

They involved in complement fixation.

IgG

It is a antibody. It is glycoprotein in nature. It is a monomeric form. It is Y shaped. This a predominant class of Ig and accounts for about 75-80% of the total Ig. Normal serum contains 8-16 mg/ml. Half life in serum is 23-25 days. They are found in Blood, Lymph and tissue. Sedimentation co efficient is 7 S. Carbohydrate content of IgG is 3%. Molecular weight of IgG is 150000. It fixes complement. It has single VH region and 3 CH region in heavy chain. It also as one VL and one CL region. It transfers through placenta. It enhanses phagocytic activity. It neutralizes toxin and viruses. It protects foetus. It activates the classical complement pathway.

IgA

It is an antibody. It is found in serum secretions etc. It is of two types. They are serum IgA and secretary IgA. It is abundance at about 10-15% among all Ig present in the human body. Two subclasses of IgA are IgA & IgA_2. Serum IgA is a monomer. Secrtary IgA is a dimmer. Molecular weight ranges from 160,000-500000.

Serum IgA

It is a monomer. It has a molecular weight of 160,000. The sedimentation coefficient is 7 S. It has a half life of 6-8 days. Serum has more amount of IgA, than IgA_2

Secretary IgA

It is a dimmer. Its molecular weight is about 5,000,00. Two pieces of IgA are linked by J chain.

Secreted by local epithelial cells.

Biological properties

Secretary IgA is termed as mucosal paint or antiseptic paint of the mucous membrane. It activates alternate complement pathway thereby neutralizes local toxins, promote phagocytosis and activate bacteriolytic activity. It also produces immunity against tape worms. IgA is present in the colo strum, which protects the baby from intestinal pathogens.

IgM

It is an antibody. It is the largest of the immunoglobulin. Hence it is often called as macroglobulin. It is a pentameric form. Individual immunoglobulin units are linked together by J chain or Joining chain. About 5 to 10% of total Ig are IgM. Half life of IgM is 5 days. The carbohydrate content is about 10 to 12%. Sedimentation Co efficient is 19S. There is no subclass in IgM. It is the earliest immunoglobulin synthesized by the fetus. Molecular weight is about 190,000-950,000. IgM shows 10 antigens binding sites.

Biological properties of IgM

It is the first antibody appeared in the immune response. It does not cross placenta. It enhances phagocytosis. It also shows properties such as opsonization, complement fixation, agglutination, cytolysis. It localizes in blood and protect animals from blood related infections.

IgD

It is one type of Antibody. It has two light chain a kappa or lambda type. Two heavy chains of IgD are of S type. It has a molecular weight of 180000. Sedimentation Co efficient is 7S. Carbohydrate content of IgD is 12%. Half life of IgD is 2-3 days. Subclasses of IgD are IgD_1 and IgD_2 It is associated with B cell surface.

Biological Properties

It is associated with the surface of B cells. It initiates immune response and has not been shown to have ab activity.

IgE

It is a monomeric antibody. Heavy chain is episilon type. It is found in traces in the serum and the average serum level is 0.002%. High IgE level is encountered during helminthic infections. Half life in serum is 2 days. There is no complement transfer. It does not crosses the placenta.

Molecular weight is about 190000. Sedimentation coefficient is 8S. Carbohydrate content of IgE ab is 12%. It is one of the skin sensitizing antibody. Originally is named as regain.

Biological Properties

IgE – Ag interaction on most cell result in degranulation of mast cell. It mediates type I hypersensitivity. It has effective role in helminthic infections.

General Functions of Immunoglobulin

The main function of immunoglobulins is protect body against the invading microbes. The protective role is carried out in two different ways namely, by direct attack on the invader and by activation of the complement system.

Agglutination–It is a process of clumping of particulate antigens. Eg RBC or bacteria. Haemagglutination / Bacterial agglutination.

Precipitation–Ig combines with soluble antigens and complex thus formed becomes insoluble and precipitates. Thus the toxin or antigen is made inactive.

Neutralization–The antibodies cover the toxic sites of the antigenic agents and neutralize them.

Lysis–Some antibody directly rupture the cell. Eg IgM molecule bring out haemolysis.

Opsonization–Some Abs coated the surface of bacteria and making the antigen more susceptible for phagocytosis. This process is termed as opsonization. These antibodies are known as opsonising antibodies.

Tissue fixation–The ability of Igs to attach to tissue cells is a well marked feature which is responsible for various hypersensitive reactions.

Selective Transport–IgG is selectively transported from mother to the foetus through the placenta which gives passive immunity to the new borne.

Activation of macrophages–Some Abs activates macrophages.

Activation of mast cells and Basophils–Some antibodies activate mast cells and basophils resulting in their degranulation.

Table 13.3 General Structure and Comparision of Immunoglobulin.

Character	IgG	IgA	IgM	IgD	IgE
Structure	Monomer	Dimer	Pentamer	Monomer	Monomer
% of Serum Antibodies	80%	10-15%	5-10%	0.20%	0.002%
Location of availability	Blood, Lymph, Intestine	Secretiones (Saliva, Tears); Blood, lymph	Blood, lymph, B cell surface	B Cell surface, Blood, lymph	Bound to mast cells, Basophils
Half life in serum	23 days	6 days	5 days	3 days	2 days
Complement Fixation	Yes	No	Yes	No	No
Placental Transfer	Yes	No	No	No	No
Chain structure	2 light chain and k; 2 heavy chain	2 light chain and k; 2 heavy chain	2 light chain and k; 2 heavy chain	2 light chain and k; 2 heavy chain	2 light chain and k; 2 heavy chain
Constant and Variable region	Single VH region and 3 CH region	Single VH region and 3 CH region	Single VH region and 3 CH region	Single VH region and 3 CH region	Single VH region and 4 CH region
Molecular Weight	150000	160000-500000	190000-950000	180000	190000
Sedimentation Co efficient	7S	7-15S	7-19S	7S	8S
Carbohydrate content	3%	10%	10-12%	12%	12%
Functions	Enhance Phagocytosis	Localized Protection	First Ab produced	Initiate immune Response in B Cells	Allergic reactions
	Protect Foetus	Protects new borns	Agglutinate antigens		Kills worms
	Neutralize toxins	Antiseptic paint	Kills Bacteria		
	Neutralize Viruses	Present in Colostrum			
		Immunity to tape worms			
Amino acid content in light chain	214 AA	214 AA	214 AA	214 AA	214 AA

Contd.

Character	IgC	IgA	IgM	IgD	IgE
Amino acid content in heavy chain	450-700 AA	450-700 AA	450-700 AA	450-700 AA	450-700 AA
General Structure	N terminal end; Carboxy terminal end; Hinge region; Fab region; Fc region; disulphide bond	N terminal end; Carboxy terminal end; Hinge region; Fab region; Fc region; Disulphied bond	N terminal end; Carboxy terminal end; Hinge Region; Fab region; Fc region; Disulphide bond	N terminal end; Carboxy terminal end; Hinge region; Fab region; Fc region; Disulphide bond	N terminal end; Carboxy terminal end; Hinge region; Fab region; Fc region; Disulphide bond

L. ANTIGEN – ANTIBODY REACTIONS

The interaction between antigen and antibody is called antigen and antibody reaction. It is abbreviated as ag – ab reaction. This reaction is the basis of humoral immunity.

Immune complex

$$Ag + Ab \rightarrow Ag - Ab \text{ complex.}$$

Specificity

The reaction between ag- ab is highly specific. Specificity refers to the discriminate ability of a particular antibody to combine with only one type of antigen. It is compared with lock and key system.

Binding sites of Ag and Ab

The part of the antigen which combines with the Ab are called epitope or antigenic determinant. The part of antibody which combines with the antigen are called paratope or antigen binding site. Most of the Abs have two binding sites. IgM having 5 – 10 binding sites.

Binding force of Ag and Ab

Binding between Ag and Ab in Ag-Ab reaction is due to three factors namely

Closeness between Ag and Ab.

Intermolecular force.

Affinity of antibody.

When the Ag and Ab are closely fit, the strength of binding is great. When they are apart, the binding strength is low. The intermolecular force which operate between Ag and Ab are non covalent forces and are of four types. They are

Electrostatic forces

Hydrogen bonding;

Hydrophobic bonding

Vander waals bonding.

Affinity refers to the strength of binding between particular molecule of Ab and a single antigenic determinant.

Avidity

It refers to the capacity of an antiserum containing various Abs to combine with the whole antigen that stimulated the production of Abs. It denoted that the overall capacity of abs to combine with multivalent Ag.

Avidity is the strength by the bond after the formation of Ag- Ab complexes.

Bonus effect

Some time two antigens are bridged by a single Ab. Such binding is week. Sometimes two Ags are bridged by one Ab, the binding will be strong. This phenomenon of giving extra strength to the Ag Ab complex are called bonus effect.

Cross reactivity

An antiserum raised against a given Ag may react with another closely related Ag. This reaction is called cross reaction

Detection of Ag – Ab reactions

In Vitro methods

Agglutination: Reaction between insoluble Ag and Ab are called agglutination. It produces a clump. Clumping is due to particulate antigen with its antibody.

Examples are

ABO blood grouping, Rh typing, Widal test, Weil felix reaction, Coomb's test, Brucella agglutination test

ABO blood grouping

RBC of a person agglutinate with specific antibodies. Based on blood grouping 4 blood group are recognized. A drop of blood sample is mixed with a drop of antiserum A and another drop of the blood sample is mixed with a drop of antiserum, B on a glass slide.

If the sample is clumbed with antiserum A, the sample belongs to group A; If the sample is clumped with antiserum B, the sample belongs to B; If the sample is clumped with both antiserum A and antiserum B, the blood sample belong to AB; If there is no agglutination the blood sample belongs to group O.

Rh typing

It is a type of agglutination reaction. It is performed by adding a drop of Anti D to a drop of blood on a slide. If clumping of blood occurs, the blood tested belong to Rh+. No clumping means Rh negative.

Widal test

It is used for the diagnosis of typhoid fever. Salmonella typhi, *Salmonella paratyphi A* and *S. paratyphi B* are the causative agents of typhoid.

Two types of antigens like O and H antigens are detected in the widal test. When antigens (known) are added to the patients serum, clumping occurs, which indicates the patient is suffering from typhoid fever.

Weil felix reaction

This test is used to detect typhus fever. Typhus fever is caused by the members of the genus Rickettsia.

Coomb's Test

It is devised by Coomb in 1945. It is used to detect anti Rh antibodies.

Brucella agglutination test

This test is used to detect brucellosis. This disease is caused by the bacterium Brucella.

Pregnancy test, cold agglutination, Haemagglutination test are some of the examples for passive agglutination

Precipitation Test

Precipitation is an immunological reaction between a soluble antigen and antibody resulting in the formation of an insoluble precipitate. The reaction is also called as precipitation reaction.

Simple double immuno diffusion

In this test, antigen and Ab are allowed to diffuse through agar to get Ag-Ab reaction. In this method, antigen and Ab are added to the separate well in agar plate and are allowed to diffuse each other and produce Ag Ab reaction. Hence it is called double immuno diffusion. When the Ag and Ab meet in optimal concentration precipitate is formed.

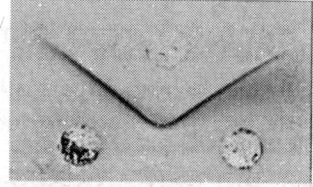

Figure 255 Simple immuno diffusion

Radial immune diffusion

In this test, Ab is incorporated in the agar and the antigen is added to the well. Antigen in the well diffuses radially and reacts with the Ab and produced precipitin ring where the Ag and Ab concentration is optimum.

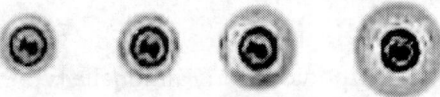

Figure 256   Radial Immuno diffusion

It is used to estimate the antigen quantitatively.

Immuno electrophoresis

It consists of two methods that are combined, namely electrophoresis and immune diffusion. Electrophoresis is defined as the migration of charged particles under the influence of electric field.

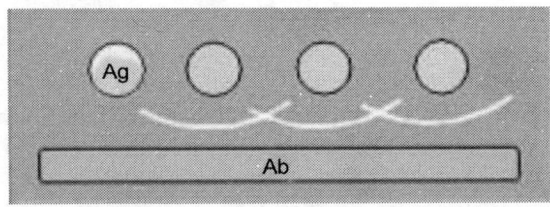

Figure 257   Immuno electrophoresis

In immune diffusion, separated compounds are allowed to diffuse through agar.

Antigens are separated using electrophoresis initially.

Add Abs to the Adjacent area of Ag separation.

Both Ag-Ab are allowed to separate.

Precipitation ring is formed where ever optimum concentrations of Ag and Ab meet.

It is used to test proteins present in the urine or body fluids.

Counter current immunoelectrophoresis

Antigen and Ab are allowed to migrate simultaneously on agar by applying electric field. Precipitin line is formed at the center.

Rocket immune electrophoresis

It is a combination of electrophoresis and immune diffusion. The precipitin curve has the appearance of a rocket. Hence this experiment is called rocket immuno electropheresis.

Antibody is incorporated in the agar. Antigens of increasing concentration are kept in wells along one edge of the agar. On electrophoresis the Ag move towards the anode forming a rocket shaped precipitin arc. The height of the rocket is proportional to Ag concentration.

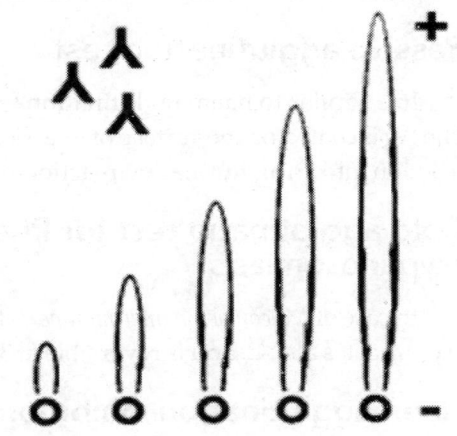

Figure 258   Rocket immune electrophoresis

Immuno fluorescence

When Abs are mixed with fluorescent dyes such as fluorescein or rhodamine, they emit radiation. This phenomenon of emitting radiation by antibodies labelled with fluorescent dye is called immuno fluorescence. It is observed by a fluorescent microscope.

It is used the locate and identify antigens in tissues. Certain pathogenic bacteria can be identified. Antibodies directed to tissue, Ag can be identified.

There are 3 methods of Immuno fluorescence. They are direct immuno fluorescence, indirect immuno fluorescence and sandwich immunofluorescence.

Direct method

In this method, the antibody labelled with fluorescent dye is applied directly on the tissue section. The labelled Ab binds with specific Ag. This can be observed under the fluorescent microscope.

Indirect method

Unlabelled antibodies directly applied on the tissue section. Unlabelled Anti antibodies binds with its specific Ag. Then Anti body labelled with fluorescent dye added to the tissue. Anti antibody specifically binds with already added / linked unlabelled ab.

Sandwich method

This is an immuno fluorescence method used to test the number of cells producing antibodies for a specific antigen.

Take Lymphocyte and fix with ethanol.

Treat fixed cells with polysaccharide Ag of Pneumococcus.

This Ag combine with those lymphocyte which have the capacity to produce Ab against pneumococcal Ag.

Now fluorescent ab is added.

Antigen is sandwiched between Ab.

Passive agglutination test

It is similar to haemagglutination test but the physical nature of the reaction is altered. The Ag is coated on the surface of a carrier particle and thereby helps to convert a precipitation reaction into an agglutination reaction making the reaction more sensitive.

Cold agglutinatin test for Pneumonia, Malaria and Trypanosomiasis

In case of *Mycoplasma pneumoniae* detection, patients serum agglutinates human O group erythrocytes at 4°C, and is reversible at 37°C.

Haemagglutination inhibition test

This test is used for the diagnosis of certain viruses like Influenza and Mumps virus and some parasitic diseases. If a person is suspected having Influenza, his serum is mixed with a known influenza virus and then RBC cells. If Ab is present, it binds to virus and prevents agglutination of RBC. If Ab is absent, haemagglutination occurs and the person is +ve for influenza.

Complement Fixation Test

Complement fixation test (CFT) was first introduced by Wasserman in 1909 for the diagnosis of Syphilis. Now this technique is also used for the diagnosis of viral infections. CFT is convenient and rapid to perform, demand for equipment and reagent is small and varieties of test antigens are readily available.

It is a simple test and has two antigens -antibody reactions, one of which is the indicator system.

Two test tubes are taken

Add antigen to both tubes.

Test serum is added to one tube.

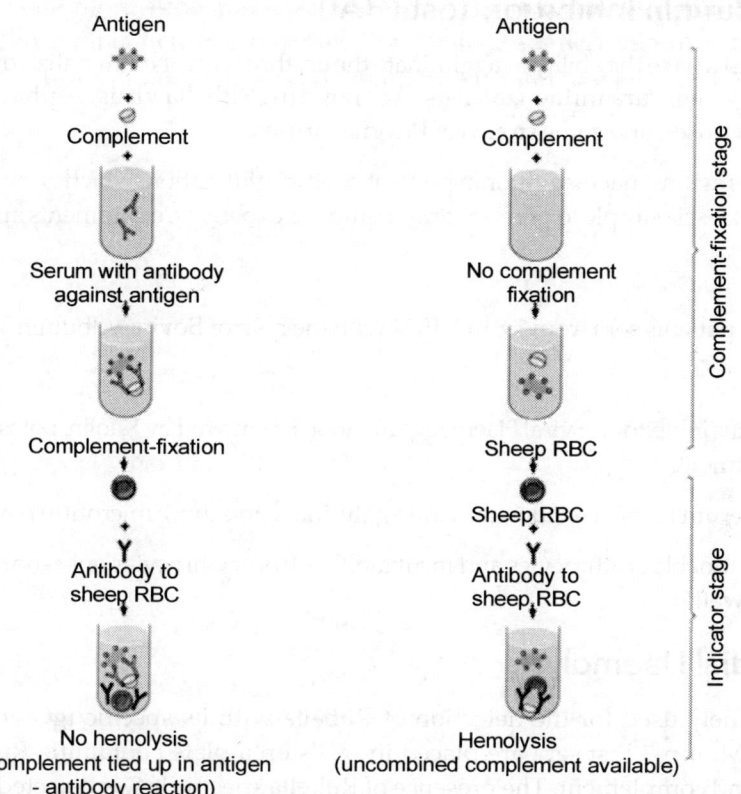

(a) Positive test. All available complement is fixed by the antigen-antibody reaction; no hemolysis occurs, so the test is positive for the presence of antibodies.

(b) Negative test. No antigen-antibody reaction occurs. The complement remains, and the red blood cells are lysed in the indicator stage; so the test is negative.

Figure 259   CF test

If antibody present Ag-Ab complex will form.

When complement is added, Ab is present, it fix complement and consume it.

Indicator cells and antierythrocytic antibodies are added.

If complement is present the indicator cell will lysis (negative result)

If the complement is consumed, no lysis of cells (positive result)

First reaction taken place between a known virus and a specific antibody in the presence of complement. Complement is fixed by antigen antibody complex.

Second antigen -antibody reaction consists of reacting sheep RBC with haemolysin. When this indicator system is added to the reactants, the sensitized RBCs will only lyse in the presence of free complement.

Haemagglutinin Inhibition Test (HAI)

Some viruses have the ability to agglutinate the erythrocytes of mammalian or avian species. Eg. Influenza virus, Para influenza virus, Adenovirus, Rubella virus, Alphaviruses, Bunya viruses, Flaviviruses, and some strains of Picorna viruses.

Antibodies against haemagglutinin prevent haemagglutination, which is the main principle of this test HAI test is simple to perform and requires inexpensive equipments and reagents.

Procedure

Dilute the patients sera from 1:8 to 1:1024 with the help of Bovine Albumin Vernol Buffer

(B AVB)

Non-specific inhibitors of viral Haemagglutination is removed by Kaolin, potassium periodate or by heat treatment.

Add the serum to the fixed dose of viral agglutinin containing microtitre plate.

Add agglutinable erythrocytes and incubate. Control erythrocytes only shows button at the bottom of the well.

Single Radial Haemolysis

It is routinely used for the detection of Rubella with its specific IgG and also for the diagnosis of Mumps. Test sera are placed in wells on a plate containing Rubella antigen-coated RBC and complement. The presence of Rubella specific IgG is detected by the lysis of rubella antigen -coated RBC. The zone of lysis around the well is depend on the level of specific antibody present.

Radio Immuno Assay

One of the most sensitive techniques for detecting antigen or antibody is Radio Immuno Assay (RIA). S.A.Berson and Rosalyn Yalow developed this technique in 1960. Microtitre wells are coated with a constant amount of antibody specific for antigen. A serum and specific radiolabelled antigen are then added. After incubation, the supernatant is removed and the amount of radioactivity bound to the antibody is determined. If the sample is infected, the amount of label bound will be less than in control with un infected serum

Gamma emitting isotope ^{125}I and beta emitting isotope ^{3}H are commonly used for radio immuno assay.

Enzyme Linked Immuno Sorbent Assay (ELISA)

Engval and Perlmann developed it in 1970. There are three important types of ELISA

Competitive ELISA

Sandwich ELISA

Indirect ELISA

Competitive ELISA

It is used for the detection of antigens. Antibody is first incubated with a sample-containing antigen. The antigen and antibody complex is added to the antigen coated microtitre well. If more antigen present in the sample, the less free antibody will be available to bind to the antigen coated well. Addition of an enzyme conjugated secondary antibody specific to the primary antibody can be used to determine the amount of primary antibody bound to the well. It is a quantitative test for the Antigen detection.

Sandwich ELISA

Antigen can be detected by this method. Antibody is immobilized on a microtitre plate. A sample-containing antigen is added and allowed to react with antibody. After washing enzyme linked second antibody is added and allowed to react with bound antigen. Substrate is added to measure colour reaction.

Indirect ELISA

Antibody can be detected with an indirect ELISA. Serum containing antibody can be detected by adding antigen to coated microtitre well and allowed to react with antigen. After washing the non bound primary antibody enzyme conjugated secondary antibody is added, which binds to the primary antibody. Free secondary antibodies are washed and a substrate for the enzyme is added. The amount of colour is directly proportional to the quantity of antibody present in the serum.

Flocculation In these tests Ag-Ab reaction results in visible floccules which do not sediment but remain dispersed in the medium. Diagnosis of syphilis is based on flocculation. The spirochaete of syphilis is a heterogenic Ag and shared by beef heart. This is prepared in cardiolipin and when patients serum having Ab is added, flocculation occurs. The test may be carried out on slides.

Opsonization The name opsonin is given by Wright (1903). It refers to a heat labile substance present in fresh sera, which facilitates phagocytosis. This factor was later identified as complement. Opsonization is a process by which the particulate Ag becomes more susceptible to phagocytosis after getting coated by opsonin Mechanism: The Ab combines with the surface Ag of bacteria. This Ag-Ab complex activates C3 component of complement system resulting in Ag-Ab-complement complex. Phagocytic cells have receptors for C3 and also for the Fc component of the Ab. As a result, the pahgocytic cells take up the Ag and cause it lysis.

Neutralization test Neutralization of a virus is defined as the loss of infectivity through reaction of the virus with specific antibody. Virus and the serum are mixed under appropriate condition and then inoculated into cell culture, eggs or animals. The presence of un-neutralized virus may be detected by reactions such as cytopathic effect, haemadsorption, interference and immunofluorescence tests.

In-vivo methods

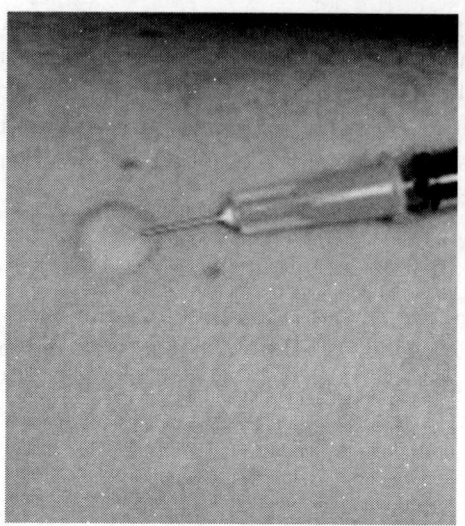

Figure 260   Skin test

Skin test

Mantoux Test -The Mantoux skin test is given by using a needle and syringe to inject 0.1 ml of 5 tuberculin units of liquid tuberculin between the layers of the skin (intradermally), usually on the forearm (Figure 260). A tuberculin unit is a standard strength of tuberculin. The tuberculin used in the Mantoux skin test is also known as **purified protein derivative**, or **PPD**. For this reason, the tuberculin skin test is sometimes called a **PPD skin test**. With the Mantoux skin test, the patient's arm is examined 48 to 72 hours after the tuberculin is injected. **Most people with TB infection have a positive reaction to the tuberculin.** The reaction is an area of induration (swelling that can be felt) around the site of the injection. The diameter of the indurated area is measured across the forearm; **erythema** (redness) around the indurated area is not measured, because the presence of erythema does not indicate that a person has TB infection.

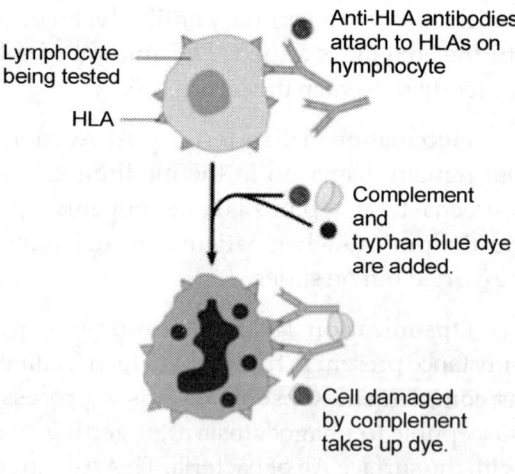

Figure 261   HLA typing

Human Leukocyte Antigen (HLA) typing

Human Leukocyte Antigen (HLA) typing is a testing process that is used to match patients and donors for cord blood or bone marrow transplants. HLA antigens are proteins found in most cells in our body. Our immune system uses these proteins or markers to recognize which cells belong in our body and which do not. If the immune system determines a cell as not belonging to the body, the cell is attacked. Thus, HLA typing is done to reduce the risk of the transplanted stem cell being attacked by the immune system of the recipient. A close match between the patient's HLA antigens and the donor's can reduce the risk of the patient's immune cells attacking the donor's cells or vice versa. HLA typing is usually done for all allogeneic transplants, using a blood sample.

The Advantages of HLA Typing

HLA typing improves the chances of **a** successful transplant. Matching HLA tissue traits before a transplant is mandatory, as it: Promotes Engraftment - Engraftment occurs when the donated cells that were transplanted begin to grow and make new cells. For the transplant to succeed, the donated cells need to engraft quickly, as the patient is at a high risk of infection.

M. TRANSPLANTATION

Transplantation refers to the implantation of a tissue from one individual to another. The implanted tissue is called graft or transplant.

TYPES OF GRAFT

It is classified into 4 types. They are Auto graft or autogenic graft.

Syn graft or syngenic graft.

Allo graft or allogenic graft.

Xeno graft or xenogeaic graft.

Autograft   It refers to a graft takes place within the same individual. Tissue of an individual is removed from one place and is implanted in another place of the same individual.

Syngraft   It is a graft that takes place between two genetically identical individual of the same species. It is also called as Isograft. If MHC antigens are identical, the graft survives and is not rejected.

Allograft   Transfer of tissues between two genetically distinct members of the same species. It is also called as homo graft.

Xeno graft   Transfer of tissue between two individual of two different species. It is also called as heterograft.

Graft Acceptance   When transplantation is made between genetically identical individuals the graft survives and lives as healthy as it is in the original places. When the graft tissue remains alive, it is said to be accepted and the process is called graft acceptance.

Graft rejection   When transplantation is made between genetically distinct individual, the graft tissue dies and decays. When the graft tissue dies, the graft is said to be rejected and the process is called graft rejection. It is of two types. They are host verses graft reaction and graft verses host reaction.

Host Versus Graft reaction (HVG)

The graft tissue antigens induce an immune response in the host. This type of immune response is called host versus graft reaction.

Allograft Rejection

Types of Allow graft rejection.

1. Acute rejection
2. Hyper active rejection
3. Insidous rejection.

Acute rejection: It is a quick graft rejection. Early acute rejection – It is due to the stimulation of T lymphocytes and CMI. Late acute rejection – Due to the stimulation of B lymphocytes and HMI.

Hyperacute Rejection: It is a very quick rejection. It is due to pre existing humoral Abs in the serum of the host as a result of presensitization with previous grafts.

Insidious Rejection: It is a secret rejection due to deposition of immune complex on tissues like glomerular membrane that can be demonstrated in kidney by immune fluorescence.

MECHANISM OF ALLOGRAFT REJECTION

Immunological contact

When tissue is implanted, the graft as can pass into local lymphnodes of the host. The graft Ags then make contact with the lymphocytes of the host.

Production of sensitized T cell and Cytotoxic Abs

When the contact is established between graft Ag and host Lymphocytes, sensitized T cells and Cytotoxic Abs are produced in the host. This bring about graft rejection.

First set Rejection

When graft is made between genetically different individuals, the graft gets blood supply from the host and it appears to be normal for the first 3 days. But on the 5th day, sensitized T cells, macrophages and a few plasma cells invade the graft. Inflammation starts in the graft. This leads to necrosis. It is similar to the primary immune response to an Ag.

Second set rejection

It is similar to secondary immune response. When a graft is implanted in an animal which has already rejected a graft, the 2nd graft is rejected in an accelerated fashion in about 6 days. The rejection of the second graft in an accelerated pattern is called 2nd set rejection. B cell is involved in this reaction.

Cell mediated Cytotoxic reaction

The 1st set rejection of allograft is brought about mainly by CMI response. During this reaction, APC bearing. Class I and Class II molecules. They carry the Ag of the graft to both CD4 helper T cell and CD8 Cytotoxic T cells to initiate the process.

On stimulation, T$_H$ cells in association with IL1, IL2, γIFN causes lysis of the graft.

Antibody mediated Cytotoxic reaction

The 2nd set rejection of graft is brought about mainly by HMI response. This is one of hyperacute rejection brought about by Abs.

Complement, K cells, Macrophages, mast cells, platelets, B cells brought out this rejection.

Graft Versus Host Reaction (GVH)

In some instance, the graft tissue elicits an immune response against the host Ags. This immune response is called graft versus host reaction. It occurs in following occasions.

Graft must remain inside the host and the host should not reject the graft.

The graft should have immune competent T cell.

The transplantation Ags of the host should be different from that of the graft.

Mechanism of GVH

The graft lymphocytes aggregate in the host lymphoid organs and are stimulated by the lymphocytes of the host. The stimulated lymphocytes of graft, produce lymphokines. The lymphokines activate the host T cell. Activated T cell activates B cell. The activated B cell react with the self Ag and cause damage.

Clinical symptons

Skin rash, Retarded growth, Diarrhea, Hepatomegaly, Splenomegaly, Anaemia.

Prevention of Graft Rejection

Should perform blood grouping and Rh typing before performing transplantation.

Test cytotoxic antibodies in the serum of the host before performing transplantation.

HLA typing should be done.

Use immune suppressive drugs like azothoprine, stenoids etc.

Suitable graft should be used.

N. HYPERSENSISTIVITY

Hypersensitivity is defined as the violent reaction of the immune system leading to severe symptoms and even death in a sensitized animal when it is re exposed to the same Ag for the second time.

It is also called as allergy.

Hypersensitivity is the changed reactivity of the immune system.

It is a beneficial protective system gone out of order.

It does not occur in all human beings. Only about 10% of human population suffers from hypersensitivity. The factors causing hypersensitivity are called allergen.

Factors Causing Hypersensitivity

Numerous factors or allergens are involved in hypersensitivity.

Drugs – penicillin, sulphamide, aspirine

Airborne particles – Pollen, house dust, mite, spares, animal feathers.

Food stuffs - shell fish, strawberries, brinjal.

Infectious microorganisms

Blood transfusion of mismatched blood.

Classification of Hypersensitivity

Based on time

Based on the time required for a sensitized host to develope clinical reactions upon re exposure to the antigen. They are immediate hypersensitivity and delayed Hypersensitivity.

Different Mechanisms of pathogenesis

Cosbs and Gell in 1963 classified hypersensitivity in to 4 types. They are

Type I -IgE mediated anaphylactic reactions.

Type II-Antibody mediated cytotoxic reactions.

Type III-Immune complex mediated reaction

Type IV-Cell mediated delayed type hypersensitivity

Type I Hypersensitivity

It is also known as immediate or anaphylactic hypersensitivity.

Anaphylaxis is defined as an allergic reaction to a foreign substance to which it has previously become sensitized resulting from the release of histamine, serotonin and vasoactive substances.

Factors causing Type I Reaction

Drugs, dust, mite, pollen, food allergy. Mechanism of Anaphylaxis. It is mediated by IgE antibody. When a person receives the allergens for the first time, the allergen get attached to the B cell. Allergen stimulate the B cell to produce IgE Abs. IgE abs have strongs affinity for its receptor mast cells and Basophils. IgE coats the surface of Basophils and mast cells. When the animal is exposed to the same antigen for the second time, the allergens cross link the IgE antibodies found on the mast cells. This cross linking of IgE triggers the mast cell and series of enzymatic reactions. As a result the mast cells release granules. The process of releasing of granules from mast cell is called degranulation. These substances that are responsible for various diseases. The mediators and their role in the anaphylactic reactions are as follows.

Molecule	Effects
Primary mediators Histamine	Vascular permeability, smooth muscle contraction.
Serotonin	Vascular permeability, smooth muscle contraction.
EcF-A(eosinophil chemotactic factor – Anaphylxis)	Eosinophil chemotaxis
NCF	Neutrophil chemotaxis
Proteases	Mucous secretion, connective tissue degradation
Secondary mediators Leukotriens	Vascular permeability, smooth muscle contraction.
Prostaglandins	Vasolidation, platelet activiation, smooth muscle contraction.
Bradykinin	Vascular permeabilitySmooth muscle Contraction.
Cytokines	Number of effects including activation of vascular endothelium, eosinophil recruitment and activation.

Complement activation also releases anaphylotoxins. These mediators bring about the anaphylactic reactions such as burning, itching sensations, vasodilatation, capillary permeability, oedama etc.

Symptoms

Shock, Very low blood pressure, Decreased respiratory vate, **Urticaria** (It is a skin eruption characterized by profound itching, red circular or irregularly shaped erutions on any part of the body), **Atopy** (It refers to naturally occurring familary hypersensitivity such as hay fever and asthma), **Hay Fever** (Rhinitis with inflammation of the nasal membranes sneezing, nasal congestion, nasal itching and rhinorrhoea), Diarrhoea, Vomiting and severe allergic due to penicillin injection may leads to death.

Test

Skin test–Allergens are introduced into the skin. Redness, itching and a raised wheal appear within 20 minutes. This indicates positive reaction.

Radio allergo sorbent test: IgE levels.

Differential leucocyte count.

ELISA.

Prevention and Treatment

Avoiding contact with Allergens.

Inhibition of Histamine receptors – Antishistamine

Blocking the release of Histamine – Theophyllin.

Stabilizing mast cell – isoprenaline , sodium chromogycates.

Steroids inhibit the release of amines from mast cells.

Type II Hypersensitivity

It is also known as cytotoxic hypersensitivity or antibody mediated cytotoxic reactions.

It affects a variety of organ and tissues.

Mechanisms – Three different mechanisms

Antibodies are attached to the antigens surface. Then the macrophages bind to the antigen (Ab coated). The macrophage engulf and destroy antibody coated cells.

The antibody coated antigen bind to the phagocytic cells through C_3b and activates complement – classical pathway and produce MAC (membrane attacking complex), which lyses the antibody coated antigens.

Antibody coated Antigens triggers Fc receptors of NK cells which lyses antigens by ADCC reaction.

Types of Cytotoxic hypersensitivity

There are two types of Cytotoxic hypersensitivity. They are isoimmune reactions and autoimmune reactions.

Isoimmune reactions

It is due to the involvement of iso antigens anitbodies. Iso antigens are the Ags of the same species differing in their antigenic properties.

Eg. Transfusion reactions.

Erythroblastosis foetalis.

Transplant rejection reaction.

Transfusion ABO system. Transfusing a patient with the incorrect ABO blood group may have fatal consequences. Donor red cells may be destroyed by an antibody (IgM) in the recipients plasma. This leads to disseminated coagulation.

Erythroblastosis foetalis It is a severe haemolytic disease of new born. It occurs in Rh-women carrying Rh+ baby. Upon birth, when placenta separates, mother is exposed to Rh+ blood of the infant and elicits of humoral response leading to memory B cells. In second pregnancy, with another Rh+ child, some of Rh+ antigens are recognized by the memory B cell and IgG is produced. This crosses the placenta and binds foetal RBC. This will lead to the destruction of RBCs. Mild to severe anaemia may result and can even cause brain damage. Protection is provided with Rhogam antibody. It is given within 72 hours of the first delivery.

Transplant Rejection Reaction: A transplant can produce Abs to leucocyte Ags or transplantation Ags or HLA Ags. These Abs are cytotoxic to graft tissues. Such damaged tissues are attacked by phagocytic and lymphoid cells.

Autoimmune haemolytic anaemia

It is the production of Abs to one's own RBCs that can lead to Ab mediated complement lysis that may lead to anaemia.

Type III Hypersensitivity

It is also called as immune complex mediated hypersensitivity.

In some instances, huge numbers of Ags enter the body. In response, the body produces higher concentration of Abs. These Ag and Abs combine together to form an insoluble precipitate called Ag-Ab complex or immune complex. These complexes get attached in and around minute blood vessel in the regions of glomerulus, skin etc. They cause tissue damage leading to hypersensistivity.

Eg. Arthus rection, serum sickness.

Mechanism

When huge volume of Ag enters the body, B cell produced large amount of IgG or IgM. These abs bind with Ag and form immune complex. These get acttached with minute capillaries. These complexes bind to the complement and produced Anaphylotoxin and chemotoxin.

Anaphylotoxin triggers mast cell and produce Amines. Chemotoxin attracts the polymorphs and promotes phagocytosis, which results in the release of hydrolytic enzymes. This will damage tissue.

Type IV cell mediated Delayed hypersensitivity

It is caused by the interaction between Ags and sensitized T cell.

This reaction leads to inflammatory reaction and causes tissue damage.

Antibodies are not involved in this hypersensitivity. The symptom on the skin appear only after 24-72 hours.

The T cell on contact with the Ag produce a soluble protein called lymphokine, which is responsible for this type hypersensitivity.

It is caused by infectious pathogens like bacteria, viruses, fungi, parasites and nical salts, neomycin ointment etc.

Common examples are tuberculin reaction and contact dermatitis.

Mechanism

When T cell contact with the antigen for the 2nd time, cells release soluble protein called lymphokines. This activate macrophages to kill intracellular bacteria leads to the formation of inflammatory cells like giant cell and epitheloid cells.

Bacteria + Ag → T cells → lymphokines → Bacterial infected cell → Bacterial killed.

Tuberculin Reaction (Mantoux Reaction)

It is type IV hyprsensitivity

It is due to the interaction of sensitized T cell and tubercle bacterium.

The reaction manifest on the skin very late only after 48-72 hrs.

When a small dose by PPD (Tuberculin) is injected intradermally in an individual already having tubercle bacilli, the reaction occurs.

Arthus Reaction

It is first observed by Arthus.

It is a type III hypersensitivity.

It is a local immune complex reaction produced on the skin.

Antigens which induce the Arthus reactions are horse serum, egg albumin etc.

It is characterized as erythema, induration, oedema, haemorrhage and necrosis.

This reaction occurs in ab excess.

It appears in 2-8 hours after injection and persist for about 12-24 hours.

In human beings, Arthus reaction is produced by the intra dermal injection of respective Ags in patient with aspergillosis.

Serum sickness

It is an immune complex disease caused by the enormous amount of foreign serum such as antidiphtheria antiserum, antitetanus Antiserum.

It is discovered by Von prequel and Shick in 1905.

It is a systemic form of type III hypersensitivity symptoms include fever, lymphadenopathy; splenomegali, endocarditis, vasculities, urticarial rashes, abdominal pain, nausea and vomiting.

The ab involved are IgG type.

The complex activate complement to produce Anaphylotoxin and chemotoxin.

Single dose is enough to cause infection.

Serum sickness appears 7 to 12 days after injection.

Tuberculin reaction is characterized by local erythema and induration.

This reaction is caused by T cell.

Contact dermatitis

It is the inflammation of the skin due to contact with a substance to which the person is allergic.

The substance causing this allergy include cosmetics, disinfectant, insecticides, rubber, leather etc.

Dinitrochloro benzene is a potent skin sensitizer. It cause dermatitis in 95% of people.

Reaction is characterized by oedemo of the epidermis.

It appears 12-15 hrs after contact.

Stimulatory Hypersensitivity

It is caused by the interaction of abs with cell surface antigen leading to stimulation of cells.

This phenomenon of stimulation occurs in Graves disease.

Throtoxocosis or Grave's disease

It is disease due to over activity of thyroid gland.

It is caused by an ab called long acting thyroid stimulator (LATS)

It is an IgG type.

It acts on the thyroid cell surface Ag, which is basically a receptor for thyroid stimulating hormone produced by pituatary gland.

This causes the release of thyroxin in higher dose from the thyroid gland and causes Graves disease.

Symptoms include fatigue, weight loss and rapid heart beat.

Myasthenia gravis

It is a neuromuscular disease.

It leads to fluctuating weakness.

It is caused by circulating Abs that block acetyl choline receptor at the post synaptic neuromuscular junction, inhibiting effect of acetlycholine.

Symptoms are muscle weakness of all organs. Treated with immunosuppression drugs.

O.   MAJOR HISTOCOMPATIBILITY COMPLEX (MHC)

It refers to a cluster of genes responsible for immune response, transplantation Antigen and proteins of the complement

It is abbreviated as MHC.

It was discovered by Goer in 1930.

MHC produces a set of proteins called histocompatibility molecules.

MHC molecules are located on the cell membrances of nucleated cells. These molecules are responsible for immune recognition, complement level, graft rejection etc.

MHC is present in all mammals. The MHC of mouse is called H-2 and man is called HLA.

HLA (Human Leucocyte Antigen)

The MHC of man is called HLA. It is a clusture of structural gene responsible for the production of Ags located on the nucleated cells and components of complements.

The HL complex of man is located in the short arm of chromosome No. 6. It has six loci and are named as A, B, C, D, S and + la.

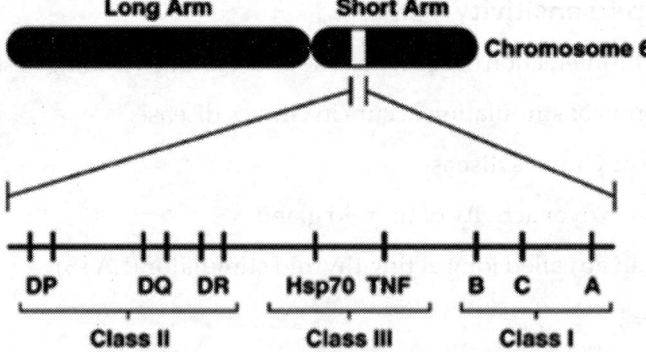

Figure 262   HLA of Human

The locus D is further subdivided into DP, DQ, DR locus S is further subdivided into C_2, C4 and Bf. Loci A,B,C are called class I genes. They also called as transplantation Antigen. They cause transplantation reactions. Loci DR, DQ, and DP belong to class II genes. They produce antigen called immune associated Antigen (Ia). These are associated with regulation of Immune response. They are present on the surfaces of B cell, macrophages, Monocytes, Antigen presenting cells and activated T cells. Locus S is called class III. They are responsible for complement components C_4, C3 and factor B. Locus Tia is called class IV genes. It is located adjacent to A. It is associated with antigens present on T cells by leukemia.

Histocompatible molecules (MHC molecules). The molecules produced by MHC genes are called Histocompatible antigens. They are located on the cell membrane of nucleated cells of body or in the blood serum. They are classified in to 4 types

Class 1 Molecules

Class 2 Molecules

Class 3 Molecules

Class 4 Molecules

Class 1 Moleucles These antigens are found on the nucleated cells especially lymphocytes and platelets. These antigens are called HLA. They are responsible for graft rejection. Hence these Ags are called transplantation Ags or histocompatible Ags.

Class 2 Molecules These Ags are present on the surface of B cell, macrophages, monocytes, APC & activated T cells. This molecule plays a vital role in Antigen presentation, T-B cell cooperation and macrophage B cell cooperation. These antigens are associated with the regulation of immune response. It is also called as Immune associated antigens.

Class 3 Moleclues These molecule include complements like C_2, C4 and factor B.

Class 4 Molecules These antigens are present on T cells of leukemia as well as on immature thymocytes which lose Tia during differentiation into mature T cell.

Function of MHC

Production of HLA: The MHC genes control the production of Ag located on the nucleated cells such as lymphocytes.

Production of Immune Associated Ags: The MHC genes control the production of immune associated Ag located on the surfaces of B cells etc.

Control of the level of complement components: MHC genes control the level of C_2, & C4 is classical pathway and factor B of alternative pathway.

HLA Complex help in T cell recognition: T Cell recognize foreign Ag only in the presence of Class I or Class II MHC molecules. MHC restriction is the recognition foreign ag located on the surface of cell.

Graft rejection: It is an immunological reaction. It is mediated by lymphocytes. The MHC is responsible for intense graft reaction.

Non Immunological Function: It control body weight in mice, egg production in chickens etc.

P.   CYTOKINES

Cytokines is the word derived from cyto-, cell; and -kinos, movement. They are small cell-signaling protein molecules that are secreted by numerous cells especially those involved in innate and adaptive immunity. Cytokines are Low molecular weight proteins (30 KDa), they Bind receptors, alter gene expression; Cytokines regulate immune responses; Cytokines can activate many cells. The different types of cytokines are

Lymphokines - produced by lymphocytes

Monokines - cytokines made by monocytes.

Chemokines - cytokines with chemotactic activities

Interleukines - cytokines made one leucocyte that act on other leucocyte. A cytokine can be pleiotropic, redundant and multifunctional.

Pleiotropic means different effect on different cells.

Redundant refers to the ability of a number of different cytokines carryout same functions.

Multifunctional means the same cytokine is able to regulate a number of different functions. There are three categories of cytokines

Cytokines that regulate innate immune response

Cytokines that regulate adaptive immune response

Cytokines that stimulate haematopoiesis.

Cytokines that regulate innate immunity

They are produced primarily by mononuclear phagocytes such as macrophages and dendritic cells. It is also be produced by T-lymphocytes, NK cells, endothelial cells and mucosal epithelial cells. They are produced primarily in response to pathogen-associated molecular patterns (PAMPs) such as LPS, peptidoglycan monomers, teichoic acids, and nucleic acids. Examples IL-1, IL-6, IL-8, and TNF-alpha, type I interferons.

Tumor necrosis factor-alpha (TNF-alpha) - Functions include acting on endothelial cells to stimulate inflammation and the coagulation pathway;

Interleukin-1 (IL-1) -it mediates acute inflammatory responses. It also works synergistically with TNF to enhance inflammation. Functions of IL-1 include stimulating the liver to produce acute phase proteins, catabolism of fat for energy conversion,

Chemokines - It enable the migration of leukocytes from the blood to the tissues at the site of inflammation.

Interleukin-12 (IL-12)-IL-12 is a primary mediator of early innate immune responses to intracellular microbes. It is also an inducer of cell-mediated immunity.

Type I Interferons - Interferons modulate the activity of virtually every component of the immune system. Type I interferons include 13 subtypes of interferon-alpha, interferon-beta, interferon omega, interferon-kappa, and interferon tau. The most powerful stimulus for type I interferons is the binding of viral DNA or RNA to toll-like receptors

Interleukin-6 (IL-6)-IL-6 functions to stimulate the liver to produce acute phase proteins; stimulates the proliferation of B-lymphocytes; and increases neutrophil production.

Interleukin-10 (IL-10)-IL-10 is an inhibitor of activated macrophages and dendritic cells and regulates innate immunity and cell-mediated immunity.

Interleukin 15 (IL-15)-IL-15 stimulates NK cell proliferation and proliferation of memory T8-lymphocytes.

Interleukin-18 (IL-18) - IL-18 stimulates the production of interferon-gamma by NK cells and T-lymphocytes and thus induces cell-mediated immunity. It is produced mainly by macrophages.

Cytokines that Regulate Adaptive Immune Responses (Humoral Immunity and Cell-Mediated Immunity)

Cytokines that regulate adaptive immunity are produced primarily by T-lymphocytes that have recognized an antigen specific for that cell. These cytokines function in the proliferation and differentiation of B- lymphocytes and T-lymphocytes after antigen recognition and in the activation of effector cells.

Interleukin-2 (IL-2) - IL-2 is a growth factor for NK cells and antigen-stimulated T-lymphocytes and B-lymphocytes.

Interleukin-4 (IL-4) - IL-4 is a major stimulus for production of IgE and the development of T_h2 cells for defense against helminths and arthropods. IL-4 is produced mainly by T_h2 cells and mast cells.

Interleukin-5 (DL-5) - IL-5 is a growth and activating factor for eosinophils as a defense against helminths and arthropods. It also stimulates the proliferation and differentiation of antigen-activated B-lymphocytes and the production of IgA. IL-5 is produced mainly by T_h2 cells.

Interferon-gamma (IFN-gamma) - Interferons modulate the activity of virtually every component of the immune system. IFN-gamma is the principal cytokine for activating macrophages. It also induces the production of MHC-I molecules, MHC-II molecules, and co-stimulatory molecules by APCs in order to promote cell-mediated immunity and activates and increases the antimicrobial and tumoricidal activity of monocytes, macrophages, neutrophils, and NK cells.

Transforming growth factor-beta (TGF-beta) - TGF-beta functions to inhibit the proliferation and effector function of T-lymphocytes; inhibit the proliferation of B-lymphocytes; and inhibits macrophage function. It also promotes tissue repair. TGF-beta is produced by T-lymphocytes, macrophages and other cells.

Lymphotoxin (LT) - LT plays a role in the recruitment and activation of neutrophils and in lymphoid organogenesis. Being chemically similar to TNF, LT is also a mediator of acute inflammatory responses. LT is made by T-lymphocytes.

Interleukin-13 (IL-13) - IL-13 increases the production of IgE by B-lymphocytes, inhibits macrophages, and increases mucus production. IL-13 is made primarily by T_h2 cells.

Cytokines that Stimulate Hematopoiesis

Produced by bone marrow stromal cells, these cytokines stimulate the growth and differentiation of immature leukocytes. Examples include:

Colony-stimulating factors (CSF) - Promote the production of colonies of the different leukocytes in the bone marrow and enhance their activity. Examples include granulocyte macrophage colony stimulating factor (GM-CSF), granulocyte colony stimulating factor (G-CSF), and macrophage colony stimulating factor (M-CSF). In addition to their role in promoting production of leukocyte colonies, the CSFs also appear to promote their function.

Stem cell factor - Stem cell factor makes stem cells in the bone marrow mor responsive to the various CSFs. It is made mainly by bone marrow stromal cells.

Interleukin-3 (IL-3) -IL-3 supports the growth of multilineage bone marrow stem cells. IL-3 is made primarily by T-lymphocytes.

Interleukin-7 (IL-7) - IL-7 plays a role in the survival and proliferation of immature B-lymphocyte and T-lymphocyte precursors. IL-7 is produced mainly my fibroblasts and bone marrow stromal cells.

Some viruses cause infected host cells to secrete molecules that bind and tie up cytokines, preventing them from binding to normal cytokine receptors on host cells. Poxviruses cause infected host cells to secrete molecules that bind interleukin-1 (IL-1) and interferon-gamma (IFN-gamma). Cytomegaloviruses (CMV) cause infected host cells to secrete molecules that bind chemokines.

Lymphokines are a subset of cytokines that are produced by a type of immune cell known as a lymphocyte. They are protein mediators typically produced by T cells to direct the immune system response by signalling between its cells. Lymphokines have many roles, including the attraction of other immune cells, including macrophages and other lymphocytes, to an infected site and their subsequent activation to prepare them to mount an immune response. Circulating lymphocytes can detect a very small concentration of lymphokine and then move up the concentration gradient towards where the immune response is required. Lymphokines aid B cells to produce antibodies.

Important lymphokines secreted by the T helper cell include: Interleukin 2, Interleukin 3, Interleukin 4, Interleukin 5, Interleukin 6, Granulocyte-macrophage colony-stimulating factor, Interferon-gamma

Chemokines are a family of small cytokines, or proteins secreted by cells. Their name is derived from their ability to induce directed chemotaxis in nearby responsive cells; they are chemotactic cytokines.

Interleukins are a group of cytokines (secreted proteins/signaling molecules) that were first seen to be expressed by white blood cells (leukocytes). The term interleukin derives from (inter-) "as a means of communication", and (-leukin) "deriving from the fact that many of these proteins are produced by leukocytes and act on leukocytes".. The majority of interleukins are synthesized by helper CD4+ T lymphocytes, as well as through monocytes, macrophages, and endothelial cells. They promote the development and differentiation of T, B, and haematopoietic cells.

Table 13.4   Summary of interleukin functions

Name	Source	Target cells	Function
		T helper cells	co-stimulation
		B cells	maturation & proliferation
IL-1	macrophages, B cells	NK cells	activation
		macrophages, endothelium, other	inflammation, small amounts induce acute phase reaction, large amounts induce fever
IL-2	Thl-cells	activated T cells and B cells, NK cells, macrophages, oligodendrocytes	Stimulates growth and differentiation of T cell response. Can be used in immunotherapy to treat cancer or suppressed for transplant patients.
IL-3	activated T helper cells, mast cells, NK cells, endothelium, eosinophils	hematopoietic stem cells	differentiation and proliferation of myeloid progenitor cells to e.g. erythrocytes, granulocytes
		mast cells	growth and histamine release
IL-4	Th2 cells, just activated naive CD4+ cell, memory CD4+ cells, mast cells, macrophages	activated B cells	proliferation and differentiation, IgGl and IgE synthesis. Important role in allergic response (IgE)
		T cells	proliferation
		endothelium	
IL-5	Th2 cells, mast cells, eosinophils	eosinophils	production
		B cells	differentiation, IgA production
IL-6	macrophages, Tn2 cells, B cells, astrocytes, endothelium	activated B cells	differentiation into plasma cells
		plasma cells	antibody secretion
		hematopoietic stem cells	differentiation
		T cells, others	
IL-7	Bone marrow stromal cells and thymus stromal cells	pre/pro-B cell, pre/pro-T cell, NK cells	induces acute phase reaction, hematopoiesis, differentiation, inflammation differentiation and proliferation of lymphoid progenitor cells, involved in B, T, and NK cell survival, development, and homeostasis, proinflammatory cytokines

Name	Source	Target cells	Function
IL-8 or CXCL8	macrophages, lymphocytes, epithelial cells, endothelial cells	neutrophils, basophils, lymphocytes	Neutrophil chemotaxis
IL-9	Th2 cells, specifically by CD4+ helper cells	T cells, B cells	Potentiates IgM, IgG, IgE, stimulates mast cells
IL-10	monocytes, Th2 cells, CD8+ T cells, mast cells,	macrophages	cytokine production
		B cells	activation
		mast cells	.
	macrophages, B cell subset	Th l cells	inhibits Th l cytokine production (IFN-v, TNF-Bl IL-2)
		Th2 cells	Stimulation
IL-11	Bone marrow stroma	bone marrow stroma	acute phase protein production, osteoclast formation
IL-12	dendritic cells, B cells, T cells, macrophages	activated [35] T cells,	differentiation into Cytotoxic T cells with IL-2,[35] IFN-Y, TNF-a, 1 IL-10
		NK cells	IFN-y, TNF-a
IL-13	activated Th2 cells, mast cells, NK cells	TH2-cells, B cells, macrophages	Stimulates growth and differentiation of B cells (IgE), inhibits THl-cells and the production of macrophage inflammatory cytokines (e.g. IL-1, IL-6), I IL-8.IL-10, IL-12
IL-14	T cells and certain malignant B cells	activated B cells	controls the growth and proliferation of B cells, inhibits lg secretion
IL-15	mononuclear phagocytes (and some other cells), especially macrophages following infection by virus(es)	T cells, activated B cells	Induces production of Natural killer cells
IL-16	lymphocytes, epithelial cells, eosinophils, CD8+ T cells	CD4+ T cells (Th-cells)	CD4+ chemoattractant
IL-17	T helper 17 cells (Th 17)	epithelium, endothelium, other	osteoclastogenesis, angiogenesis,] inflammatoryl cytokines

Name	Source	Target cells [33]	Fnction
IL-18	macrophages	Thl cells, NK cells	Induces production of IFNy, f NK cell activity
IL-19	-		-
IL-20	-		regulates proliferation and differentiation ofl keratinocytes
IL-21	activated T helper cells, NKT cells	All lymphocytes, dendritic cells	costimulates activation and proliferation of CD8+ Tl cells, augment NK cytotoxicity, augments CD40-driven B cell proliferation, differentiation and) isotype switching, promotes differentiation of Thl7j cells
IL-22	-		Activates STAT1 and STAT3 and increases) production of acute phase proteins such as serum amyloid A, Alpha 1-antichymotrypsin and haptoglobin in hepatoma cell lines
IL-23	-		Increases angiogenesis but reduces CD8 T-cell infiltration
IL-24	-		Plays important roles in tumor suppression, wound healing and psoriasis by influencing cell survival.
IL-25	-		Induces the production IL-4, IL-5 and IL-13, which stimulate eosinophil expansion
IL-26	-		Enhances secretion of IL-10 and IL-8 and cell surface expression of CD54 on epithelial cells
IL-27	-		Regulates the activity of B lymphocyte and T lymphocytes
IL-28	-		Plays a role in immune defense against viruses
IL-29	-		Plays a role in host defenses against microbes
IL-30	-		Forms one chain of IL-27
IL-31	-		May play a role in inflammation of the skin
IL-32	-		Induces monocytes and macrophages to secrete TNF-a, IL-8 and CXCL2
IL-33	-		Induces helper T cells to produce type 2 cytokine
IL-35	regulatory T cells		Suppression of T helper cell activation

Q. COMPLEMENT

The complement is a group of 20 serum proteins. Each protein within the system is assigned a number and they react in sequence once the system has been activated. Many of the proteins are pro-enzymes that require proteolytic cleavage in order to become active. The complement cascade forms part of the body's innate immune system and is involved in host defense against infection, the initiation of an inflammatory response and the destruction of certain bacteria and viruses.

The complement cascade can be activated in 3 different ways, all of which lead to the formation of a convertase that cleaves C3 to form C3a and C3b. C3b in turn activates C5 and the remainder of the cascade which leads to the formation of the Membrane Attack Complex.

It is a heat-labile serum component that was able to lyse bacteria. Complement can opsonize bacteria for enhanced phagocytosis.

Complement comprises over 20 different serum proteins that are produced by a variety of cells including, hepatocytes, macrophages and gut epithelial cells. Some complement proteins bind to immunoglobulins or to membrane components of cells. Others are proenzymes that, when activated, cleave one or more other complement proteins. Upon cleavage some of the complement proteins yield fragments that activate cells, increase vascular permeability or opsonize bacteria.

II. PATHWAYS OF COMPLEMENT ACTIVATION

Complement activation can be divided into four pathways the classical pathway, the lectin pathway, the alternative pathway and the membrane attack (or lytic) pathway. Both classical and alternative pathways lead to the activation of C5 convertase and result in the production of C5b which is essential for the activation of the membrane attack pathway.

Sequential **activation** of the protein components of the **complement cascade** upon cleavage by a *protease*, leads to each component's becoming, in its turn, a *protease*. Three pathways are involved in complement attack upon pathogens:

- ⅄ classical pathway
- ⅄ alternative pathway
- ⅄ mannose-binding lectin pathway (MBL -MAPS)

The **classical pathway** utilizes C1, which is activated by binding of an antibody to its antigen.

Inactive **C1** circulates as a serum molecular complex comprising 6 C1q molecules, 2 C1r molecules, and 2 C1s molecules. Constant regions in some immunoglobulins specifically bind C1q, activating C1r and C1s. The mu chains of IgM and some gamma chains of IgG contain specific binding sites, though IgM is far more effective than IgG.

Activated **C1s** is a *serine protease* that cleaves **C4** and **C2** into small inactive fragments (C4a, C2a) and larger active fragments, **C4b** and **C2b**. The active component C4b binds to the sugar moieties of surface glycoproteins and binds noncovalently to C2b, forming another *serine protease* C4b•C2b, which is called *C3 convertase* because it cleaves **C3**, releasing an active C3bopsonin fragment.

Macrophages and neutrophils possess receptors for C3b, so cells coated with **C3b** are targetted for phagocytosis (opsonization). The small **C3a** fragment is released into solution where it can bind to basophils and mast cells, triggering histamine release and, as an **anaphylatoxin**, potentially participating in anaphylaxis.

C3 amplifies the humoral response because of its abundance and its ability to auto-activate (as a*C3 convertase*). Breakdown of C3b generates an antigen-binding C3d fragment that enhances antigen uptake by dendritic cells and B cells .

Binding of **C3b** to **C5** induces an allosteric change that exposes C3b•C5 to cleavage by C4b•C2b, which is now acting as *C3/C5 convertase*. The alternative pathway possesses a distinct *C5 convertase*, so the two pathways converge through C5.

⅄ Cleavage of C5 by the C3/C5 convertase releases

⅄ **anaphylotoxic C5a**, which promotes chemotaxis of neutrophils

C5b, which complexes with one molecule of each of C6, C7, and C8.

The resultant **C5b•6•7•8 complex** assists polymerization of as many as **C9** molecules to form a cytolysis-promoting **pore (membrane attack complex)** through the plasma membrane of the target cell, which then suffers osmosis-induced cytolysis.

Another cytolytic mediator utilized by CTLs and NK cells is **perforin**, which is a 534 aa glycoprotein with sequence homology to the membrane attack component of complement C9. Like C9, **perforin** integrates into the target cell membrane, forming polyprotein pores up to 20nm in diameter comprising 12 – 18 perforin monomers, which breach membrane integrity and permit cytolytic cell death.

The **alternative pathway** is *not* activated by antigen-antibody binding, but instead relies upon **spontaneous** conversion of C3 to **C3b**, which is rapidly **inactivated** by its binding to **inhibitory proteins** and **sialic acid** on the cell's surface. Because bacteria and other foreign materials lack these inhibitory proteins and sialic acid, the C3b is not inactivated and it forms the C3b•Bb complex with **Factor B**. The C3b.Bb complex acts as a *C3 convertase*, forming C3b•Bb•C3b, which acts as a *C5 convertase* that can ititiate assembly of the membrane attack complex. C3b•Bb, acting as a *C3 convertase*, provides a positive feedback loop that amplifies production of C3.

Opsonization - To make bacteria or other cells more susceptible to phagocytosis (by neutrophils and monocytes/macrophages) by coating their cell membrane surfaces with various proteins, e.g., antibodies or elements of the complement system, e.g., activated C3 ("C3b").

Chemotactic agent - Any chemical, usually a microbial protein or an immune regulatory substance, which induces the directed migration (but not the rate of movement) of leukocytes along its concentration gradient to the area(s) of higher concentration; histamine, various lymphokines, neutrophil products, and several of the activated complement proteins (C3a, C5a) serve this function.

Cytolysis - The rupture and destruction of cells by the breakdown of their outer lipid cell membrane = plasmalemmae; various immune molecules have this ability including the membrane attack complex of the complement cascade (activated C5, C6, C7, C8, and C9).

Membrane Attack Complex - The series of protein molecules achieving the completion of the complement cascade using a common set of steps in which C5, C6, C7, C8, and C9 are activated in sequence, by either the classical or alternate pathways, to produce the molecular structure capable of cytolysis — one unit each of activated C5, C6, C7, and C8 ("C5b678") will bind from 10-16 molecules of C9 to assemble a hydrophobic pore (a very short tube) which crosses the cell membrane and lets ions, water, and other small molecules to pass in either direction, based upon concentration gradients; as a result, the attacked cell, which receives from hundreds to thousands of these pores from activation of the complement cascade, swells, ruptures, and dies.

R. IMMUNITY TO VIRAL INFECTIONS

Resistance to virus infections involves humoral immunity, interferon sensitization of host cells, and cell-mediated immunity.

1. Antibodies can neutralize viruses by combining with them and interfering with their adsorption and entrance into cells.

2. Antibodies enhance phagocytosis and destruction of viruses in much the same way as for bacteria.

3. Interferons are important in resistance when the target cell is reached immediately as in the case of colds and influenza. Interferon-stimulated cells shut down protein synthesis and destroy viral mRNA. Some interferons also stimulate the activity of T cells and natural killer cells, thus accelerating the immune response to a viral infection.

4. Cell-mediated immunity to viruses is a major resistance mechanism when enveloped viruses modify host cell membranes and bud off from the surface (e.g., herpesvirus, poxvirus, influenza, mumps, measles, rabies, and rubella viruses). Activated lymphocytes can recognize and destroy virus-infected cells because the target cell's plasma membrane has been altered.

    ⋏ CTLs destroy virus-infected cells through FasL, and the production of granzymes and perforin that form channels through the plasma membrane of infected cells causing apoptosis and cytolysis. The class I MHC proteins are involved in T-cell recognition of infected cells. Cells displaying both viral antigens and the proper class I MHC will be destroyed. CTLs are also involved in the destruction of cancer cells (**immune surveillance**).

    ⋏ Natural killer cells (NK cells) are nonB, nonT lymphocytes that are active without any prior antigen exposure. Interferon and antibodies, however, will stimulate them to greater activity. They also are capable of destroying virus-infected and cancer cells (*see figure 31.21*). They are components of innate immunity and possess no antigen receptors.

S. IMMUNITY TO BACTERIAL INFECTIONS

Humoral immunity appears to be more important than cellmediated immunity in the defense against most bacterial pathogens.

Antibodies and complement attack pathogens in many ways.

1. IgG, and the C3b and C4b components of complement are opsonins. That is, they aid in the phagocytosis of bacteria by macrophages and granulocytes through the process called opsonization.

2. IgM and IgG will agglutinate bacterial pathogens, thus limiting their spread and enhancing the efficiency of phagocytosis.

3. Antibodies can trigger complement attack on the wall of gram-negative bacteria by the classical pathway. Once this pathway is activated, it forms a C5b-9 membrane attack complex that forms channel or pore in the bacterial cell, leading to lyses.

4. Complement C3a, C5a, and C5b67 also attract neutrophils and macrophages to a site of infection.

5. Antibodies called antitoxins bind to bacterial exotoxins and block their action (neutralization).

6. Cell-mediated immune responses by activated macrophages and T-cells are also important, particularly in resisting intracellular bacterial pathogens. Activated T cells secrete several cytokines, which have a variety of effects.

    ➤ The macrophage activating factor stimulates macrophages to become "angry" macrophages and more effectively phagocytose and destroy pathogens. Interferon-_ is a major macrophage activating factor.

    ➤ The macrophage chemotactic factor and migration inhibition factor attract more macrophages and keep them in the area of infection after arrival.

    ➤ Interleukin-2 (IL-2) stimulates the proliferation of activated T cells to increase the population of cells involved in the cell-mediated immune response. It also increases the effectiveness of cytotoxic T cells and NK cells by promoting the synthesis of immune interferon by T cells.

T. IMMUNIZATIONS

Those substances which induce an immunity of an individual without cusing a disease are called vaccine. A process of inoculating vaccine ace called vaccination. The purpose of vaccination or immunization is to intentionally expose the human immune system to a foreign infectious agent so that it forms a memory of that agent. Passive immunization refers to injection of prepared antibodies into a person who has either already been infected or is at risk of acquiring an infection. In this case the infected person's immune system is not actively protecting the body, hence the name passive immunization. Examples of passive immunization include Rabies and Hepatitis A.

Types of Vaccines

Vaccines are of two general types: **In live attenuated** vaccines, the organism in the vaccine is alive but unable to infect a person with a normal immune system. Patients with impaired immunity-such as those with immune deficiencies, on chemotherapy for cancer, or with AIDS-and pregnant women must not be given live vaccines. Examples of live attenuated vaccines are measles, mumps, rubella and oral polio.

Inactivated or killed vaccines contain dead, but intact, organisms, so the immune system can still recognize them. Most vaccines are inactivated.

Vaccines are usually given at multiple intervals because the immune system needs several reminders to "boost" immunity.

Other types of vaccines are described in Virology Part

Specific vaccines are available for the following infections:

Diphtheria

Disease: Diphtheria is a bacterial illness acquired through inhalation of infected particles (*Corynebacterium diphtheria*). It causes a severe sore throat and possibly heart and nerve damage. The bacteria live in the airways of healthy or recovering humans. Toxin mediates this infection.

Vaccine: DPT

Interval: 2-, 4-, 6-, and 15 months and 5 years of age.

Pertussis

Disease: A bacterial illness acquired through inhalation of the infected particles (*Bordetella pertussis*). It causes severe, life-threatening coughing spells (whooping cough), and possibly seizures and brain damage. The bacteria usually live in the airways of adults with no or minimal cough. Toxin mediates this infection.

Vaccine: DPT

Interval: 2-, 4-, 6-, and 15 months and 5 years of age.

Tetanus

Disease: Tetanus is a bacterial infection acquired through dirty wound infection (*Clostridium tetani*). Tetanus causes severe and painful muscle contractions. The bacteria are abundant in the soil. Toxin mediates this infection.

Vaccine: DTP

Interval: 2-, 4-, 6-, and 15 months, 5 years of age. The vaccine must be repeated every 10 years.

Polio

Disease: Polio is a viral infection involving the mouth and throat, and later the blood and spinal cord. Approximately 10% of the infected people develop spinal cord infection, causing muscle paralysis, usually one-sided.

Vaccine: OPV (oral=live) and Injected (inactivated); inactivated vaccine is given to children with immunodeficiencies.

Interval: 2-, 4-, 6-, and 18 months and 5 years of age.

Measles

Disease: Measles is a viral infection acquired through breathing infected particles. It causes rash, croupy cough, pneumonia, diarrhea, and possibly brain infection and bleeding.

Vaccine: MMR

Interval: 15 months and 12 years.

Mumps

Disease: Mumps is a viral infection acquired through breathing infected particles. It causes painful swelling of the Parotid gland, testes, and pancreas gland.

Vaccine: MMR

Interval: 15 months and 12 years.

Rubella (German Measles)

Disease: Rubella is a viral infection acquired through inhalation of infected particles. It causes rash, fevers, and enlarged lymph nodes. If a pregnant woman becomes infected, the fetus could be severely and permanently damaged.

Vaccine: MMR

Interval: 15 months and 12 years.

Haemophilis influenza type b (Hib)

Disease: Haemophilis influenza type b is a bacterial infection acquired through inhalation of infected particles or through contact with infected objects. It causes life-threatening conditions such as meningitis (infection of the lining of the brain), throat swelling, and joint infection.

Vaccine: Hib. It will be given along with DPT

Interval: 2-, 4-, 6-, and 15 months.

Influenza Virus

Disease: Influenza is a viral infection of the upper- and lower respiratory tract. It can be fatal in people with heart, lung, and other chronic diseases.

Vaccine: Flu shot;

Interval: yearly

Pneumococcal pneumonia

Disease: Pneumococcal pneumonia is a bacterial illness causing pneumonia.

Vaccine: Pneumococcal vaccine; recommended for people with heart, lung, or other chronic illnesses. Pneumococcal conjugate vaccine against pneumococcal pneumonia – three primary doses at 6, 10, and 14 weeks, followed by a booster at 15-18 months.

Varicella-Zoster (Chickenpox)

Varicella-Zoster is a viral infection acquired through inhalation of infected particles. It causes painful blistering and later crusty rash and fevers. Rare complications include infections of the brain, joints, and kidneys and/or hemorrhaging. Vaccination is recommended for children with immunodeficiencies, but it is safe and frequently given to healthy children.

Hepatitis B

Hepatitis B is a viral infection of the liver. It is acquired through exposure to blood (such as in a transfusion), through sexual intercourse, and from a mother to her fetus. Vaccination is recommended in high-risk patients, especially the health care providers.

Recombinant HB vaccines are available in the market.

Cholera

Cholera is a bacterial infection of the small intestine. It causes severe watery diarrhea and dehydration that could lead to death. The vaccine is recommended for travelers to Africa, Middle East, and the Far East.

Plague

Plague is a bacterial infection carried by rodents. It causes fever, skin sores, enlarged lymph nodes, and if not treated, death. Humans are accidentally infected by fleas that feed off the infected rodents. The vaccine is recommended for people traveling to or working in areas where plague is prevalent.

Typhoid fever

Typhoid is a bacterial infection caused by Salmonella. It causes diarrhea, fevers, and if left untreated, death. The vaccine is recommended for travelers to Africa, South America, the Middle East, and the Far East **TAB vaccine is given at the age of 2 years**

Rabies

Rabies is a viral infection acquired through the bite of an infected mammal. It causes fevers, headaches, restlessness, seizures, coma, and death. Immunization is passive by injection of anti-Rabies antibodies.

Lyme Disease

Lyme Disease is a bacterial infection acquired through the bite of a tick that feeds on deer. It causes rash, fever, and, left untreated, possible neurological or heart damage. Vaccination is recommended to those who live in areas where deer population is large and in contact with the human population.

Rotavirus vaccine against infant diarrhoea due to rotavirus – two doses 4 weeks apart usually given at 6 weeks and 10 weeks

HPV vaccine against Human Pappova Virus infections known to cause cervical cancer – it is as yet given to girls above age of 10 years; three doses ; 0, 1 month and 6 months.

14

MEDICAL MICROBIOLOGY

A. NORMAL FLORA OF THE HEALTHY HUMAN HOST

Introduction

Those microorganisms present in our body without affecting normal functioning of the body are called **normal flora**. Micro organisms like bacteria and yeast are prevalent on and in the body and are considered as normal flora.

Skin

It is one of the most understandable regions in the body. It provides good barrier system and is considered as a first line defence system in the body. It has epidermis, dermis, subcutaneous connective tissue, sweat glands, sebaceous glands and hair follicles etc. The skin varies widely in structure and function, depending on its location on the body. Coagulase negative – *Staphylococcus, Staphylococcus aureus, Propionibacterium acnes, Bacillus, Streptococci, Malassezia furfur, Candida* and *Pityrosporum* are the normal flora of skin.

Eye

Lining of the eyelids and covering the eyeball is protected by a delicate membrane called conjuctiva. It is a continuation of the skin. Flow of tears, removes microorganisms from eye. Lysozyme, a microbicidal substance present in tears that inhibits the growth of microbes. At birth and throughout the human life, bacterial commensals are found on the conjuctiva of the eye. The predominant bacterium is *Staphylococcus epidermidis, Staphylococcus aureus, Corynebacterium, Streptococcus pneumoniae, Neisseria, Moraxella, Branhamella spp, Escherichia, Klebsiella, Haemophillus, Proteus, Enterobacter, Bacillus spp* and few anaerobic bacteria.

Respiratory Tract

The respiratory tract includes nose, tonsils, nasopharynx, throat, trachea, bronchi and lungs. The mucous membrane of the upper respiratory tract is moister than skin, nevertheless, they can create more problems to microorganisms. Normal flora of nose includes *Coagulase negative –Staphylococcus, Viridance Streptococcus, Staphylococcus aureus, Neisseria, Haemophillus, Streptococcus pneumoniae.*

Lower Respiratory Tract

Normally lower respiratory tract do not have any normal microbiota, because of the efficient mechanical removal by mucous and cilia. Most of the bacteria usually are engulfed and destroyed by phagocytic action of alveolar macrophages.

Oral Cavity

Normal microbiota of the oral cavity contains those organisms that are able to resist mechanical removal by adhering to surfaces like gums and teeth. Oral cavity serves as a ideal environment for microbes because it provide moisture, and food material. However continuous desquamotion of epithelial cell, flow of saliva and mechanical flushing action removes microbes from the oral cavity. *Staphylococcus, Viridance Streptococcus, Neisseria, Haemophillus, Streptococcus pneumoniae, Veillonella, Fusobacterium, Treponema, Porpyromonas, Prevotella, Branhamella , Diphtheroids, Candida* are considered as a normal flora of mouth.

Gastro Intestinal Tract

Stomach, small intestine and large intestine are the part of gastro intestinal tract.

Stomach

Stomach receives different transient organisms from the oral cavity. The stomach usually contains less than 10 viable bacteria per milliliter of gastric juice. This is due to the bactericidal effect of gastric hydrochloric acid and digestive enzymes. Organisms present in stomach include *Staphylococci, Streptococci, Lactobacillus, Peptostreptococcus* and yeast.

Small Intestine

The small intestine is divided into three areas: Duodenum, Jejunum, and illeum. *Enterococcus faecalis, Lactobacilli, Diptheriods* and *Candida albicans* are occasionally found in the jejunum.

Large Intestine (Colon)

Colon has the largest microbial population in the body. It has been estimated that the number of microorganisms in stool specimen is 10^{12} organisms per gram wet weight, which means that about 25% of feces is made up of microorganisms. Over 500 different bacterial species have been isolated from human feces. It has been estimated that an adult excretes 30 million bacterial cells daily through defecation. Some examples are *Lactobacillus, Bacteroides, Clostridium, Mycobacterium, Enterococci, Enterobacteriaceae members, Peptostreptococcus, Actinomyces, Streptococcus, Pseudomonas* and *Entamoeba coli.*

Genito Urinary Tract

The upper genitourinary tract (kidney, ureters and urinary bladder) is usually free of microbiota. The regions of the genitourinary tract that harbour microflora are the vagina and outer opening of the urethra in female and anterior opening in the males. Normal microbiota of the urethra are coagulase negative *Staphylococci, Diptheroids, Streptococci, Mycobacterium, Bacteriods, Fusobacterium, Peptostreptococcus spp.* etc. Normal microbiota of the vagina are *Lactobacillus, Diptheroids, Peptostreptococcus, Streptococcus, Clostridium, Candida* and *Gardinerella vaginalis*.

External Ear

Normal microbiota resemble to that of skin with coagulase negative *Staphylococci* and *Corynebacterium*. Less frequent founders are *Bacillus, Micrococcus* and *Neisseria* spp. Gram negative rods such as *Proteus, E.coli, Pseudomonas* are occasionally present.

B. HOST MICROBE INTERACTIONS

Human is the most important hosts to microbes. Human harbours *billions* of microbes on their body as commensals. The relationship between normal flora and host is called symbiosis. The interaction between pathogenic microbes and host is called parasitism. During this type of interactions microbes utilize nutrients from host and interferes its metabolism.

The outcome of host microbe interactions depends on three main factors or to induce an infectious disease; a pathogen must be able to perform the following functions. They are Invasiveness, Infectivity and Pathogenic potential

i. Invasiveness is the ability of the pathogen to spread adjacent tissues.

ii. Infectivity is the ability of an organism to establish an infection.

Pathogenic potential refers to the degree that the pathogens cause morbid symptoms. It mainly depends on toxigenicity. It is the pathogens ability to produce toxins, chemical substance that will damage the host and produce disease.

Transmisibility, attachment, colonization and entry of pathogen to the host depend on its degree of invasiveness. Growth and multiplication depends on its degree of infectivity. Symptom producing nature depends on its toxigenicity.

Transmissibility of the Pathogen

An essential feature in the development of an infectious disease is the initial transport of the pathogen to the host. The organism is transmitted to the host by means of Airborne transmission, Contact transmission, Vehicle transmission, Vector borne transmission, Carrier transmission, Waterborne transmission and Food borne transmission methods.

Attachment and Colonization of the Pathogens

After transmission pathogen must attach on host cells and tissues to establish infection. Colonization depends on the ability of pathogen to compete successfully with the host normal flora for essential nutrition. Adherence and colonization depends on the following virulent factors. They are *capsule, fimbriae, lectin, ligand, pili, slime, mucous gel, teichoic acid and cell wall antigens*.

Entry of the Pathogens

After attachment, a pathogen sometimes enters inside of the host cell or it may grow on the surface. This may be accomplished through the following lytic substance. Coagulase, collagenase, deoxyribonuclease, alkaline protease, exotoxins, hemolysins, hyaluranidase, hydrogen peroxidase, lecithinase, leucocidines, porins, protein A, and streptokinase. These factors alter the host tissue by penetrating the cells passively by Small breaks, lesions or ulcers in mucous membrane that permit initial entry and Wounds, abrasions or burns on the skin surface. It also through mosquito biting wound, tissue damage caused by other pathogens.

After entry the pathogen may penetrate to deeper tissues and continue disseminating throughout the body.

Growth and Multiplication of the Pathogens

Successful growth and reproduction of pathogen requires an appropriate environment within the host. During growth pathogen produces some metabolic end products, which are often toxic and produce a condition known as *septicemia*.

Toxigenicity

Two distinct categories of the disease can be recognised based on pathogens role in the disease causing process, which are infection and intoxications. An infectious disease results partly from the pathogen growth and reproduction that often produce tissue alteration. *Toxemia* refers to the condition caused by toxins. Microbial toxins are divided into two types; they are exotoxins and endotoxins. Exotoxins are soluble, heat labile proteins that are released by growing microbial cell to the surrounding medium. Exotoxins maybe divided into three main categories on the basis of mode of actions.

i.   Neuro toxins affect nerve cells

    Eg. Tetanus toxin

ii. Entero toxins affect intestinal mucosa.

    Eg. Cholera toxin

iii. Cyto toxins affect epithelial cells

    Eg. Pseudomonas exotoxin A

Mode of action of exotoxins are

i.   Inhibition of protein synthesis

ii. Inhibition of nerve synapse function

iii. Disruption of membrane transport

iv. Damage plasma membrane.

Endotoxins

More gram-negative bacteria have a lipopolysaccaride in the outer membrane of their cellwall. Under certain circumstances, it is toxic to hosts, this lipopolysaccarides are called endotoxins. Endotoxins initially activates hageman factor, which inturn activates upto four humoral systems. They are Coagulation, Complement, Fibrinolytic function and Kininogen sysem activation. Endotoxins also indirectly induce fever in the host by inducing macrophages, which release endogenous pyrogens that reset hypothalamic thermostat. Recent evidences indicates that one important **endogenous pyrogen** is the lymphokine, interleukn I, TNF and IL6.

Pathophysiological reactions of endotoxins are fever, changes in white blood cell count, disseminated intravascular coagulation, tumour necrosis, hypotension, shock etc.

Table 14.1 Difference between exotoxins and endotoxins.

S.No.	Exotoxins	Endotoxins
1	Synthesised by specific pathogens that often have plasmids.	Found on gram negative pathogens.
2	Heat labile proteins inactivated at 60°-80°C.	Heat stable.
3	Very small amount is required to creating toxic condition.	Toxic only at high doses.
4	Highly immunogenic and stimulates the production of neutralizing antibodie.	Weekly immunogenic.
5	Easily inactivated by formaldehyde, iodine and other chemicals and form toxoids.	Inactivated by using the temperature 250°C for 30 minutes.
6	Unable to produce a fever.	Able to produce a fever
7	Often given the name of the disease they produced.	There is no specific name.

C. STAPHYLOCOCCUS INFECTIONS

Staphylococci are Gram positive spherical bacteria. It was first observed by Von Reckling Heusen in 1871. Sir Alexander Ogston demonstrated the causative role of *Staphylococcus* in 1880. *Staphylococcus* is in the family staphylococcaceae as per edition II of bergeys manual systematic bacteriology. Most significant pathogens of this genus is *S.aureus*.

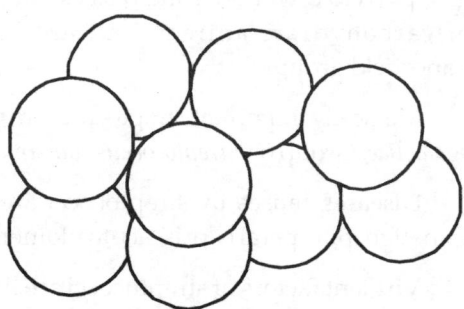

Figure 263 Grape like clusters

Staphylococci are non motile, non spore forming, facultative anaerobic bacteria. They are catalase positive and oxidase negative. Pathogenic bacteria produce coagulase enzyme. They utilize potassium tellurite and reduce it into telluramine. *S. aureus* produce golden colonies on Nutrient Agar and β haemolytic colony on Blood agar. It also produces black colour colonies on Baired Parker agar and Vogel Johnson medium.

It spreads from person to person by aerosols. Babies colonize Staphylococci from their surrounding. Carriers also responsible for transmission.

Staphylococcus attaches to the host epithelical cells and endothelical cells (Laminin and Fibronectin) with the help of protein. Adhesions promotes attachment to collagen.Invasions of *Staphylococci* produces toxins like Alpha toxin, Beta toxin, Delta toxin, Gamma toxin, Coagulase, Leucocidin, Staphylokinase, DNAase, Lipase, Protease, Fatty Acid Modifying Enzyme (FAME) at the site of invasions and causes the followimg infections Folliculitis Furuncle, Carbuncle, Cellulitis, Style (Infection in eye lid), Sinusitis,Pneumoniae, Impetigo,Bacteremia, Endocarditis, Osteomyelitis, Scalded Skin Syndrome (SSS), Toxic Shock Syndrome (TSS),Cystitis, Diarrhoea etc,.

Cloxacillin, Dicloxacillin is highly effective for skin infections. Erythromycin is used for the respiratory infections. Nafcillin or oxacillin is recommended for systemic infections. Vancomycin or Cephalosporins are suitable for allergic patients.

D. STREPTOCOCCUS INFECTION

Streptococci are Gram-positive cocci that grow in pairs and chains. Streptococci were first described by Billroth in 1874. Rosenbach (1884) isolated the cocci from human supportive lesions and named as *S.pyogenes*. Streptococcus are classified as α haemolytic Streptococci, β haemolytic Streptococci, γ haemolytic Streptococci based on haemolytic pattern on blood agar medium.

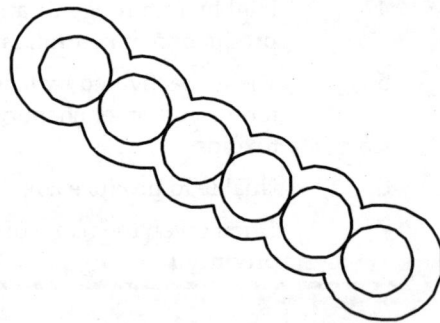

Rebecca Lancefield (1933) serologically classified β haemolytic streptococci into 19 sero groups (A to U without I and J) based on the nature of carbohydrate antigen and are known as Lancefield groups.

Figure 264   Chain of cocci

Medically Important Streptococci are group A- *Streptococcus pyogenes*, group B- *Streptococcus agalactiae,* Group C- *Streptococcus equisimilis,* group D- *Streptococcus anginosus.*

Diseases causes by streptococci are pharyngitis, scarlet fever, septicemia, erysipelas, impetigo, puerperal sepsis, acute glomerulo nephritis and endocarditis.

Virulent factors of streptococci are hyaluranic acid (capsule), peptidoglycon, M Protein, C_5a peptidase, lipoteichoic acid, Streptolysin O, streptolysin S, pyogenic exotoxin, esterases, streptokinase, phosphatase, proteinase,   DNAase, ATPase, nuclease, cardio hepatic toxin.

Streptococci enter into the human body by various direct and indirect mechanisms. Organisms adhere to epithelial cells via lipoteichoic acid and enhance the colonization. Organism may invade adjacent tissues or distant tissues via blood stream. During infection Streptococcus produce various types of virulent factors which may induce infections like Suppurative (Tonsillar Abscess, Otititis, Septicemia , Osteomyelipis) and non suppurative (acute Rheumatic fever , Glomerulonephritis) infections

Pharyngitis

Streptococcus attaches to the buccal epithelial cells through lipoteichoic acid. Virulent strains resist the activities of immune response and induce inflammatory response and produce Strep throat, Fever chills, Headache, Malaise, Nausea and Vomiting. Pharynx may be mildly erythematous or beefly red with greyish yellow exudates and may be bleeding.

Scarlet Fever

Scarlet means Strep throat with a red skin rash. It is closely related to Erythrogenic toxin of Streptococcus. It can accompany pharyngitis. A fine papular, red rash disseminates the whole body.

Impetigo

Superficial infection that usually begins as small vesicles, progressing to sweeping lesions with amper crust and slightly cloudy purulent exudates.

Erysipelas

Spreading infection of the Skin or Mucous membrane, are usually seen in face. Lesions are characterized by erythema, oedema and induration .

Cellulitis

It is an inflammation of the skin and underlying connective tissue in diabetics patients cellulitis may lead to gangrene.

Acute Rheumatic Fever

It is a non-suppurative inflammatory response. It is manifested by arthritis, carditis, chorea, erythema marginatum or subcutaneous nodules. Symptoms occur within 2-3 weeks. Pathogenesis is poorly understood. Antigenic cross reactivity between Streptococcal antigens and heart tissue, direct toxicity due to exotoxins, actual invasion of heart by Streptococcus are the reason for Rheumatic fever. Prolonged bed rest needed for recovery. Salicylates and Corticosteroids are used for treatment.

Gram staining of specimen, growth on blood agar, ASO test, Dick test are used for the diagnosis. Specimen is streaked on blood agar and incubated at 37°C. Colonies are less than 1mm, grey white or colourless, dry or shining and usually irregular. *Streptococcus pyogenes* produce Beta hemolytic colonies. Kanamycin blood agar is recommended for group B Streptococcus.

Erythromycin, Clindamycin, Cephalexin, Penicillin, Vancomycin, Streptomycin are used for the treatment.

E. PNEUMONIA

Pneumonia is an infection of one or both lungs which is usually caused by a bacteria, virus and fungus. It is an endogenous infection. Pneumoniae is caused by bacteria's like *Klebsiella pneumoniae, Mycoplasma pneumoniae, Streptococcus pneumoniae, Legionella pneumophilla.* But most common causative agent is *Streptococcus pneumoniae.*

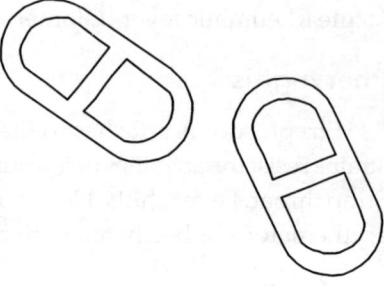

Symptoms of pneumonia are shortness of breath, Fever, chills, cough, chest pain, pain during inhalation, fluid around lungs, abscess, empyema, rust coloured sputum.

Figure 265   Diplococci

Breathing of droplet nuclei allows the entry of bacteria to upper respiratory tract. Pneumococci resist the defense mechanism and enters to the lungs. They usually settle in the air sac of the lung and rapidly grow. Growth induces immune response. It results in inflammatory response. This leads to blood vessel damage and in increase of vascular permeability, leakage of blood, vasodilation, extravagation, excess mucous secretion, accumulation of mucous, lung tissue damage, disturbance in gas exchange that leads to suffocation.

Laboratory diagnosis is done by gram staining, observation of rust coloured sputum and observation of alpha haemolytic colonies on blood agar indicates the presence of *Streptococcus pneumoniae.*

Cell wall inhibitors like Penicillin, Ampicillin are the drug of choice. Erythromycin, Chloramphenicol also used in some instances.

F. DIPHTHERIA

Diphtheria is an upper respiratory tract illness characterized by sore throat, low-grade fever, and an adherent membrane of the tonsil. Pseudomembrane formation is the characteristic feature of this infection. It is due to damage of blood vessel that results in plasma leakage. Plasma produce network of fibrin, which leads to the formation of pseudomembrane.

Corynebacteria are Gram-positive, aerobic, nonmotile, rod-shaped bacteria related to the Actinomycetes. They have the characteristic of forming irregular shaped,

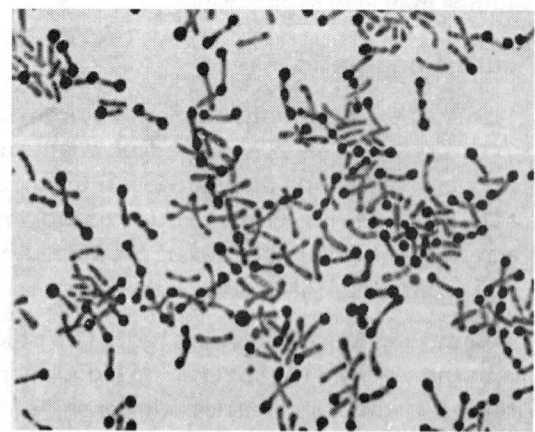

Figure 266   Club shaped bacilli

club-shaped or V-shaped arrangements in normal growth. They undergo snapping movements just after cell division, which brings them into characteristic arrangements resembling Chinese letters.

Three strains of *Corynebacterium diphtheriae* are recognized by Mc.Leod on the basis of growth and other characters, gravis, intermedius and mitis.

Transmission occurs by droplet spread through contact with a patient or carrier, or articles soiled with discharges from infected lesions.

Incubation period is usually two to five days.

The pathogenicity of *Corynebacterium diphtheriae* includes two distinct phenomena. They are Invasion of the bacteria in local tissues of the throat and toxin production. The diphtheria toxin causes the death of eukaryotic cells and tissues by inhibition of protein synthesis. This leads to plasma leakge and pseudomembrane formation.

Alginate and cotton throat swab are collected and inoculated on tellurite blood agar. Corynebacterium produces beta haemolytic black colonies on tellurite blood agar. Clinically it is confirmed by observing whitish membrane on tonsil.

It is treated using Antitoxin. Antibiotics commonly used are Oral erythromycin, IM penicillin. DPT vaccine will prevent the infection. It is given at 2nd, 4th, 6th, 18th months and 5th years and also at 15 years and every 10 years thereafter.

G. MENINGITIS

Meningitis is the infection of meninges of brain or spinal cord. It is caused by a variety of bacteria and viruses which includes *Haemophilus influenzae*, *Neisseria meningitidis*, *Escherichia coli*, *Staphylococcus aureus*, *Streptococcus pneumoniae* and *Streptococcus pyogenes*.

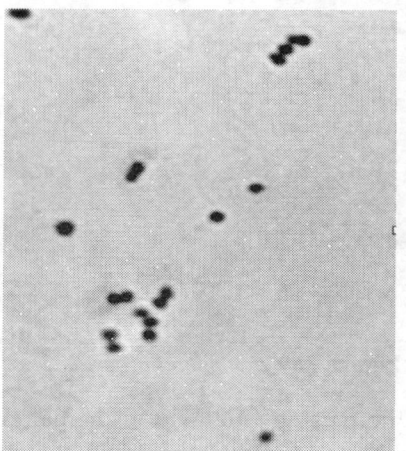
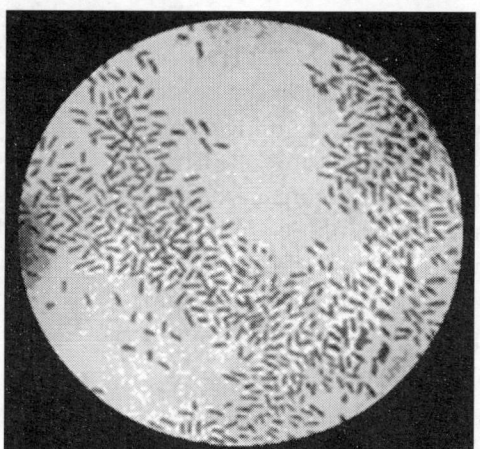

Figure 267 *Neisseria meningitidis*   **Figure 268** *Haemophillus influenzae*

Haemophilus influenzae is a small, nonmotile Gram-negative bacterium. *Haemophilus influenzae* prefers a complex medium and requires X factor (i.e., hemin) and V factor (NAD or

NADP). In the laboratory it is usually grown on chocolate agar. The bacterium grows best at 35-37°C and has an optimal pH of 7.6. *Haemophilus influenzae* is generally grown in the laboratory under aerobic conditions or under slight CO_2 tension (5% CO_2).

Neisseria meningitidis is also called Meningococcus. It is a Gram negative coccus, usually seen in pairs. *Neisseria meningitides* is usually cultivated in a peptone blood based medium in a moist chamber containing 5-10 % CO_2. New york city agar medium and Modified Thayer Martin agar act as a selective medium

Meningititis by *H. influenzae* and meningococci are usually begins in the upper respiratory tract. They attach to the nonciliated columnar epithelial cells of the nasopharynx and multiplied specifically using virulent factors. After multiplication the bacteria enters Central Nervous System via blood and lymphatics and causes meningititis.

Symptoms of meningititis are Irritability, Fever, head ache, vomiting, photophobia, chills, myalgia, weakness.

The recommended treatment for *H. influenzae* meningitis is Ampicillin and a third-generation Cephalosporin or Chloramphenicol. Amoxicillin plus clavulanic acid (Augmentin) is effective against ß-lactamase producing strains.

Current recommendations for vaccination of infants require parenteral administration of penta valent vaccines, which is used for the prevention of 5 disease. Diphtheria-Tetanus-Pertussis (DPT), *Haemophillus influenzae* b strain (Hib) conjugate, and Hepatitis B.

H.   WHOOPING COUGH

Whooping cough is also called as pertusis. It is a highly contagious disease of respiratory tract. Pertussis means "Intensive Cough". This disease is caused by *Bordetella pertussis*. It was first isolated in 1906 by Bordet and Gengou. It is a small ovoid coccobacilli, non motile and non sporing bacterium. It contains bipolar metachromatic granules, which is demonstrated by Toludine Blue staining. It is a fastidious bacterium and grows on rich media supplemented with blood.

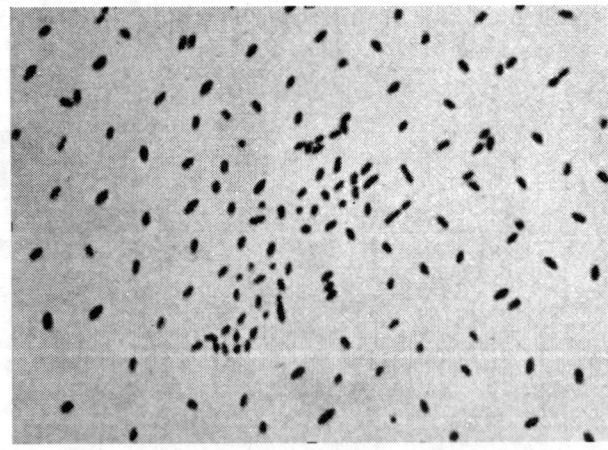

Figure 269   *Bordetella pertussis*

Bacteria enters upper respiratory tract and adhere using "filamentous hemagglutinin" (FHA) and cell-bound pertussis toxin (PTx). It produces a highly lethal toxin, which causes inflammation. It produces the Tracheal cytotoxin, which is toxic for ciliated respiratory epithelium and which will stop the ciliated cells from beating. It produces the Pertussis toxin. This activates adenylate cyclase enzyme and disrupt cellular function and leads to specific symptoms.

Pertussis begins as a mild upper respiratory infection. Symptoms resemble those of a common cold, runny nose, slight fever and a mild cough. Within two weeks, the cough becomes more severe and is characterized by rapid coughing followed by a high pitched whoop. The incubation period is usually five to 10 days, but may be as long as 21 days.

On the basis of symptoms produced by the organisms, the disease stage was divided into three stages. They are Catarrhal stage, Paroxymal stage and Convolesent stage

Cough plate method, Post Nasal Swab, Pernasal Swab were inoculated on levinthol medium and is incubated at 35-36°C for 48-72 hours. Colonies are small dome shaped, smooth opaque greyish white, refractile and glistening, resembling mercury drops.

Stages	Incubation period	Catarrhal	Paroxymal	Convolescent
Duration	7-10 days	1-2 weeks	2-4 weeks	3-4 weeks
Symptoms	None	Low grade fever, rhinorrhoea and progressively worsening cough and also mucous membrane inflammation	Rapidly continuous cough followed by the characteristic whoop.	Diminished paroxymal cough, pneumonia
Culturing	None	Possible	Possible	Rare

Treatment with Erythromycin, will eliminate viable *B pertussis* organisms from the respiratory tract within a few days. DPT vaccination is given in five doses at 2, 4, 6, 12-18 months and 4-6 years of age.

I.   TUBERCULOSIS

Tuberculosis is one of the lower respiratory tract infection. It is a progressive granulomatous disease of the lungs. It is caused by *Mycobacterium tuberculosis*. It is a Acid-Fast Bacilli. The causative agent was first described by Robert Koch during the year 1882.

Acid-Fastness of tubercle bacilli is due to Mycolic acid content of the cell wall. It is fairly large nonmotile rod

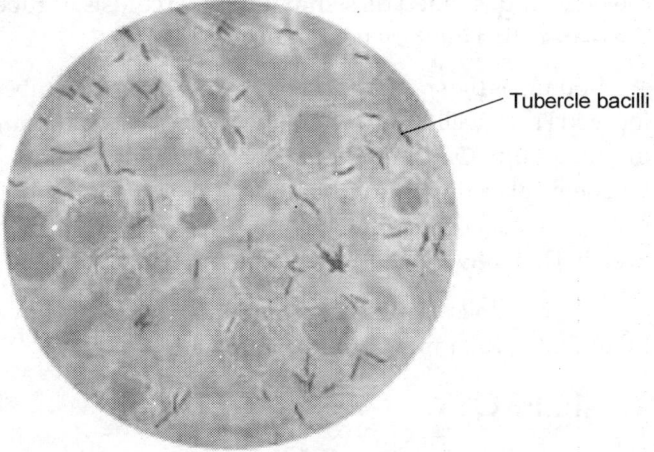

Tubercle bacilli

Figure 270   AFB-*Mycobacterium tuberclosis*

shaped, nonspore forming and non capsulated bacterium. In tissue it is thin straight rods, occurring singly, in pairs or in clumps. It is an obligate aerobic, slow growing bacterium. Generation time of this bacterium is 6-12 hours. It grows in the media containing egg, asparagine, potatoes and serum.

Prominent symptoms of tuberculosis are chronic productive cough, low-grade fever, night sweats, easy fatigability, and weight loss.

M tuberculosis infections occur by airborne transmission of droplet nuclei. After the inhalation, the bacilli are deposited in the alveolar spaces of the lungs. It resists intracellular destruction. It also multiplied within macrophages and kills it. In this stage lymphocytes begins to infiltrate in the lung field. This results in T cell activation and release of cytokines and other factors. This activates resting macrophages and kill infected macrophages. Activation of macrophages can results in bacterial killing. TH1 cells produce IL-2 & gamma interferon, which promote the inflammatory reactions and CMI. TH2 cells produce IL-4, IL-5 and IL-10 and promote antibody production.

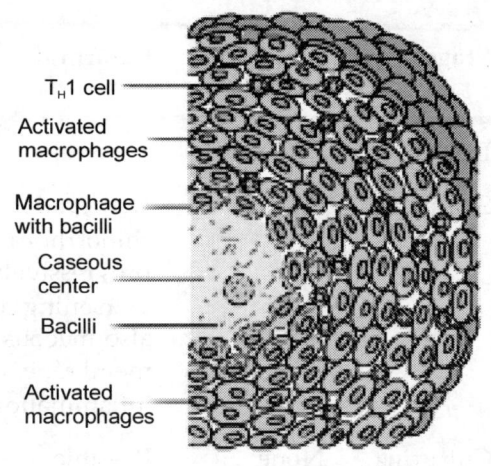

Accumulation of Mycobacteria stimulates an inflammatory focus. It is characterized by a mononuclear cell infiltrate surrounding a core of degenerating epithelioid and multinucleated giant (Langhans) cells. This lesion is called a **tubercle.**

Figure 271   Tubercle

Rupturing of tubercle leads to release of tubercle bacilli. This passes tubercle bacilli throughout the body and is called disseminated tuberculosis. If tubercle bacilli spreads within the lung fileld are called military tuberculosis.

Sputum is the specimen used for the diagnosis of tuberculosis. Sputum smears is examined for Acid Fast Bacilli for the preliminary diagnosis. Sputum treated with 40% KOH is inoculated in LJ medium. On LJ medium tubercle bacillus grown as cabbage colored rough colonies. Biochemical testing, cord formation and gene probe analysis confirms the identity. Mantoux test is also called tuberculin skin test used for the diagnosis of tuberculosis using Purified Protein Derivative (PPD).

Prevention is by BCG vaccination. Treatment is by using combined therapy Isoniazid, Ethambutol, Rifampin and Streptomycin.

J.   LEPROSY

Leprosy is a chronic granulomatous disease of man. It involves the skin, peripheral nerves and nasal mucosa. It is defined as a hypopigmented or reddish skin lesion with defnite loss of sensation. Leprosy is also called Hanson's disease because Hanson first observed Lepra bacilli in 1868. In 1970 it was found that *M.leprae* causes a systemic infection in the nine-banded armadillo.

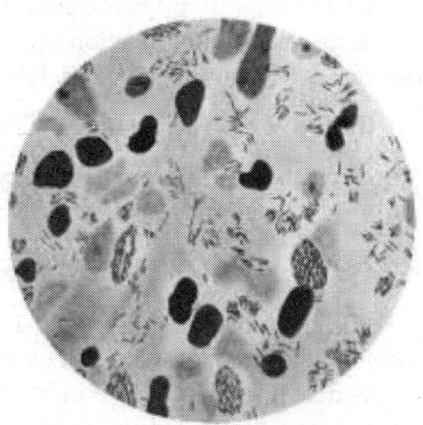

Figure 272 Lepra bacilli

M.leprae causes leprosy. It was the first bacilli isolated from human. It is one of the least understood bacterium. It is a straight or slightly curved rods showing considerable morphological variations. Organisms are found singly or in large masses termed globi. Large numbers of bacilli may be packed in the cells in an arrangement that suggests packets of cigars. It is one of the Acid-Fast Bacilli.

Organism enters to the human body through respiratory route or through skin. It mainly attacks nerve cells and grows very slowly in mononuclear macrophages especially the histiocytes of skin and Schwann cells of the nerves. Attack of immune cells against affected nerve cells produces nerve damage leading to deformity. On the basis of immunological findings, histopathology and clinical findings Ridley and Jopling have established a classification scheme consisting of five terms of leprosy. They are as follows (Table 14.2).

Table 14.2   Categories of leprosy

S.No	Type of disease	Immune response	Bacilli in skin	Bacilli in nasal mucosa
1	Tuberculoid (TT)	Good	None	None
2	Boderline Tuberculoid (BT)	Less good	None for few	None
3	Boderline (BB)	Partial	Few	None
4	Border line lepromatous (BL)	Poor	Moderate	Few
5	Lepromatous (LL)	Nil	Many with globi	Many with globi

Among these five forms, two forms are stable. They are Tuberculoid and Lepromatous.

Leprosy is divided into two groups, as per WHO report 1982, they are Paucibacillary (lesion contains few bacteria) which include all cases of tuberculoid types and some cases of borderline type (BT, TT). Multibacillary (they contains large numbers of bacilli) includes all cases of Lepromatous types and some cases of Borderline (BB, BL and LL).

Common symptoms include progressive nerve damage, chronic skin lesions, and ulcerative lesions of mucous membrane, deformed faces, loss of fingers and toes.

Acid fast staining is the staining used for the diagnosis. Lepromin test also used to diagnose the infection.

Bacteriological index is defined as the number of viable bacilli in a lesion. It is used to assess the severity of infection Morphological index is the percentage of uniformly stained cell versus total number of cells. It is used to assess the effect of antibiotic treatment.

Culture is done using inoculation of footpads of mouse, rat and Nine Banded Armadillo.

Dapsone (4, 4 diamino diphenyl sulphone) is an effective monotherapy. Rifampicin, dapsone, mimocycline and clarithromycin are also used for the treatment.

K.   BACILLARY DYSENTERY

It is also called as shigellosis, **Dysentery** is an intestinal inflammation in the colon. It can lead to severe diarrhoea with mucus or blood in the faeces. Dysentery is due to rod shaped bacterium and are called Bacillay dysentery.

It is caused by Gram negative, rod shaped, nonmotile, facultative anaerobic bacterium Shigella. The genus Shigella is differentiated into four species. They are *S. dysenteriae, S .flexneri, S. boydii* and *S sonnei.*

Symptoms are Vomiting, Fever, Mucous and Blood stool with Mild dehydration and abdominal pain (cramps and tenesmus), Weight loss, HUS, Reiter syndrome.

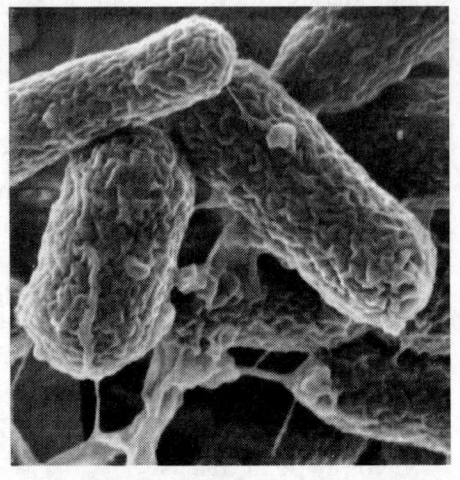

Figure 273   EM view of Shigella

Reiter syndrome is an arthritis that affect joint tissues, leading to inflammation of the joints. Hemolytic Uremic Syndrome (HUS) is due to Shiga toxin. This taxin primarily damage blood vessels, vascular tissue causes renal failure leads to neurological complication, and kidney failure.

Human is the primary reservoir of Shigella sp. Organism enters through food and water. After entry into the intestine, bacteria binds to host cells surface and engulfed by M cells. Ingested bacteria escape endocytic activity, multiply in cytoplasm. During Multiplication Shigella produce enterotoxin called Shigatoxin. Shigatoxin act as Enterotoxin (Provokes fluid loss); act as Neurotoxin (Causes paralysis); act as Cytotoxin (Induces apoptosis). This toxin activity leads to specific symptoms.

Patients presenting with watery diarrhoea and fever should be suspected of having Shigellosis. Stool sample is primarily enriched with GN broth and inoculated on media like MacConkey, Haektoen Enteric Agar, and Salmonella-Shigella (SS) Agar. Shigella produces NLF colony on Mac conkey agar and colourless colonies on SS agar and haektoein enteric agar.

Absorbable drugs such as Ampicillin, Trimethoprim, Sulfamethoxazole and Ciprofloxacin are used for the treatment. It is prevented by personal hygiene and environmental sanitation. Give exclusive breast-feeding to the children for at least one year. Avoid bottle-feeding. Immunize the child for Measles.

L.   CHOLERA

Cholera is a severe diarrhoeal disease caused by *Vibrio cholerae*. It belongs to the family Vibrionaceae. It is an acute illness.   *Vibrio cholerae* is a gram-negative comma shaped bacilli with single polar flagellum. It was first isolated in pure culture by Robert Koch in 1883. Generation time for *Vibrio cholerae* is less than 30 minutes.

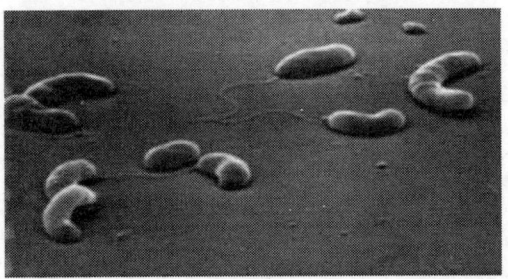

Figure 274   EM structure of *V cholerae*

Cholera is acquired by drinking **contaminated water and food. Organism enters the stomach, resists acidity and** reached small intestine. It attaches to the intestinal mucosa, colonize and produce cholera toxin. It is a Heat-labile enterotoxin. It causes excessive secretion of water and electrolytes by the intestinal epithelium. Incubation Period is 6-48 Hours.

Symptoms are characterized by profuse watery diarrhoea, vomiting, leg cramps and resulting in acidosis and hypervolumic shock. The watery diarrhoea is also called "rice-water stool" and contains enormous numbers of vibrios. Untreated cholera frequently results in death.

Presence of dartling motility in stool sample indicated the infection is due to *Vibrio*. Stool sample is inoculated on Tri Sodium Citrate Bilesalt Sucrose agar (TCBS). It produces Circular yellow colour colonies.

Treatment of cholera involves the rapid intravenous replacement of the lost fluid and ions. Most antibiotics and chemotherapeutic agents have no value in cholera therapy, although a few (e.g. tetracyclines) may shorten the duration of diarrhoea and reduce fluid loss.

M. TYPHOID FEVER

Typhoid fever is also called as Enteric fever. It is caused by *Salmonella typhi, Salmonella paratyphi A, Salmonella paratyphi B, Salmonella paratyphi C*. They are gram negative, straight rods, motile by peritrichous flagella and facultative anaerobes. Some strains produce hydrogen sulfide. Salmonella grow on ordinary media & also on selective cum differential media. On Mac Conkey agar it produce small, circular, translucent, NLF colonies. In Wilson Blair Bismuth Sulfite Medium, the colonies are jet black with metallic sheen in nature. On Salmonella-Shigella agar black centered colonies were observed.

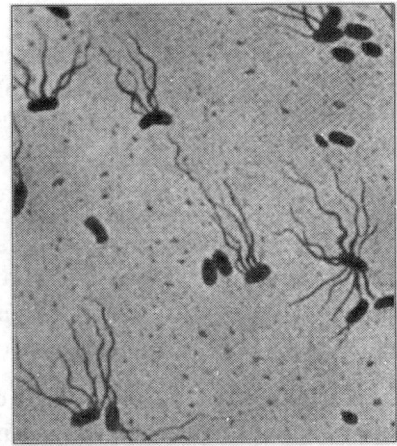

Fever, rash, coated tongue, head ache, pain in abdomen, vomiting, diarrhoea are the major symptoms. **Figure 275**   EM view of Salmonella

It is transmitted through faecal oral route and contaminated food and water. In the small intestine, the bacilli attack epithelial cells of the villi and pass through the submucosal coat, where they are phagocytosed by Neutrophils and Macrophages. The virulent bacilli resist intracellular killing and multiply within these cells and cause bacteremia. During this period some organisms undergo lysis, thereby liberating endotoxin in the circulation. The bacteremia and toxemia cause pyrexia and other symptoms. Some organisms localize in organs like gallbladder, liver, spleen etc,. Normally duration of infection varies from 7-14 days in Typhoid fever, 4-5 days for Paratyphoid fever. Typhoid fever is severe than Paratyphoid fever.

Widal test is done to diagnose typhoid fever, stool, urine, blood culture is also done.

Sanitary measures and vaccines are the two main ways to prevent Enteric Fever. Chloramphenicol, ampicillin, amoxicillin, Ciprofloxacin, norfloxacin are used for the treatment. TAB vaccine is used for prevention.

N. GONORRHOEA

Gonorrhoea is a venereal disease. Galen first coined the name gonorrhoea in 130 A.D. The word gonorrhoea means *flow of seed*. The disease is a sexual contact disease.

It is caused by *Neisseria gonorrhoeae*. It is also called Gonococcus. Neisser first described it in gonorrhoeal pus in 1879. It is a Gram negative coccus. It is an aerobic bacterium and uses glucose as the principal carbon and energy source. Some strains require 5-10 % carbon-di-oxide. A popular selective medium is the Thayer Martin Medium, the medium contains Vancomycin, Colistin, Nystatin and Trimethoprim lactate. Vancomycin inhibits Gram positive bacteria. Colistin inhibits Gram

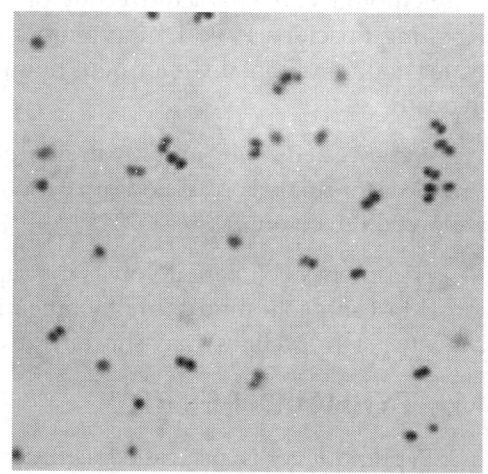

Figure 276   Gonococci

negative bacteria except *Proteus*. Nystatin inhibits yeast cells. Trimethoprime lactate inhibits *Proteus*. Colonies are small round, translucent, convex or slightly umbonate with finely granular surface and lobate margin. Gonococci only ferment glucose and not maltose.

It is only transmitted from person to person by direct contact. It can colonize on mucosal surfaces and columnar epithelial cells of cervix and urethra. Gonococci attaches, colonizes in the initial entry area and triggers inflammatory response. Activities of inflammatory response lead to lysis of macrophages and form purulent discharge.

Incubation period is about 2-8 days. The male patient presents with burning on urination and yellow purulent discharge (urethral discharge) that indicates urethritis. This leads to urinary tract infection, sterility, and arthritis.

In women, symptomatic infections usually begin with painful urination and vaginal discharge. It leads to Pelvic Inflammatory Disease (PID). Scar tissue formed as a result of infection in fallopian tubes may block normal passage of ova through the tubes causing sterility. Scaring of fallopian tube can also lead to a dangerous complication - Ectopic pregnancy in which the ovum is fertilized and develops in the fallopian tube and leads to life threatening internal hemorrhaging.

Opthalmia neonatorum is one of the eye infection occurs during childhood days. Gonococcal vulvovaginitis occurs in girls at 2-8 years of age.

Urethral discharge, Vaginal discharge are subjected to gram staining and culturing. Thayer Martin agar and Newyork city agar medium used for isolation. Gonococcus produce small grey, entire, sticky colonies formed on Chocolate agar.

Ceftriaxone plus Doxycycline are the drug used for the treatment. Erythromycin may be substituted for Doxycycline in pregnant womens.

O. SYPHILIS

Syphilis is an epidemic disease. Syphilis is caused by the spirochaete *Treponema pallidum*. *Treponema* is a Gram-negative, thin, motile, spiral shaped bacterium in the order *Spirochaetales*. The name Spirochaete is derived from the Greek word "coiled hair."

Syphilis results in the formation of lesions throughout the body. The bacteria usually enter the body during sexual intercourse, through the mucosal membranes of

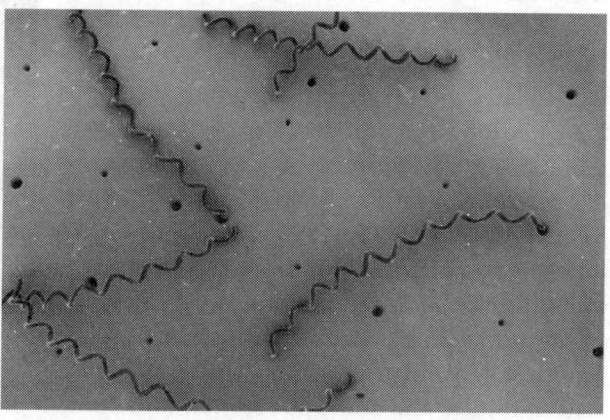

Figure 277　*Treponema pallidum*

the vagina or urethra, but may rarely be transmitted through scar wound or scratches. *T. pallidum* tends to invade the interstitial spaces of tissue at the site of infection. *T. pallidum* penetrates intact mucous membranes or broken skin, begins to divide slowly and disseminates and produce symptoms.

Syphilis has three distinct stages that are primary, secondary and tertiary syphilis with a latency stage between the second and third stages.

Primary Syphilis is observed about 10-60 days after exposure to Syphilis. The area of infection is marked by the appearance of a hard, slightly elevated, round, ulcerous development called a "chancre" (a destructive sore). The chancre represents the primary site of initial replication. The chancre usually appears on the penis, labia, cervix, anorectal region, or around the mouth.

One to six weeks after the healing of chancre, a pale red rash appears usually on the palms or soles of feet, but may occur over the entire body. These rashes are accompanied by fever, sore throat, headache, joint pains, poor appetite, weight loss, and hair loss and are called Secondary Syphilis.

Next stage is the Latency Stage. In this stage there are no obvious symptoms but *T. pallidum* is still present.

Tertiary Syphilis is a complicated stage gumma lesions are also present in this stage and lead to destruction of soft tissue or bone and finally death.

Diagnostic tests for Syphilis are dark field illumination, direct fluorescent antibody staining techniques, Serological tests, Treponemal and nontreponemal tests. FTA-AB test also important.

Penicillin is the drug of choice for untreated infections with *T. pallidium;* Penicillin G is recommended for congenital and late Syphilis. Tetracycline, Erythromycin, and Chloramphenicol can be used as alternative antibiotics for patients allergic to Penicillin.

P. TETANUS

Tetanus is infection of the nervous system. It is caused by *Clostridium tetani*. It is also called tetanus bacilli. Arthur Nicholaire discovered the Tetanus bacillus in 1884. The organism was isolated in pure form by Kitasato in 1889. It is quite long and thin organism. It is usually Gram positive and produce terminal spore, so it appears like drumstick. Motility is by Peritrichous flagella and is an obligate anaerobic bacterium. Optimum temperature is 37°C and pH 7.4. It is a moderately fastidious bacterium requires vitamins and amino acids for growth. It grows in Cooked Meat Medium

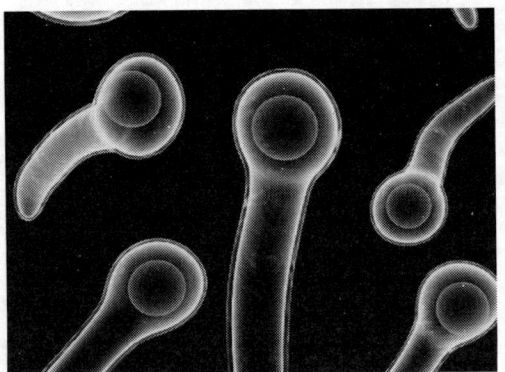

Figure 278 Terminal spore-*C.tetani*

with turbidity and some gas formation within 48 hours, there is no digestion of meat, and black colour is formed during prolonged incubation.

Spores of the bacteria *C. tetani* live in the soil. Infection begins when the spores enter the body through an injury or wound. The spores germinates, grows and produce 2 distinct toxins, they are Tetanolysin and Tetanospasmin. The toxin blocks nerve signals from the spinal cord to the muscles, causing severe muscle spasms. The spasms can be so powerful that they tear the muscles or cause fractures of the spine. Incubation period is 7 to 21 days.

Tetanus often begins with mild spasms in the jaw muscles (lockjaw). The spasms can also affect the chest, neck, back, and abdominal muscles. Sometimes the spasms affect muscles that help with breathing , which can lead to breathing problems. Prolonged muscular action causes sudden, powerful, and painful contractions of muscle groups. This is called tetany. These episodes can cause fractures and muscle tears. Other symptoms include drooling, excessive sweating, fever, hand or foot spasms, irritability, swallowing difficulty, uncontrolled urination or defecation.

Microscopic examination of pus (Gram staining) is useful to demonstrate the presence of bacilli. Cultivation is with the help of Robertson Cooked Meat Medium.

The nature of prophylaxis depends largely on the type of wound and the immune status of the patient. The available methods are surgical attention, antibiotics and immunization. Surgical prophylaxis aims at removal of foreign bodies, to avoid providing an anaerobic environment favourable for Tetanus bacillus. Treatment may include Antibiotics, including penicillin, clindamycin, erythromycin, or metronidazole. Injections of Tetanus toxoid are prophylactic. Human Tetanus Immunoglobulin (HTIG) in a dose of 250 IU intramuscularly. Routine immunization with Tetanus toxoid should begin at 1-3 months of age by DPT vaccine.

Q. CHANCROID

Chancroid is a sexually transmitted infection. It is characterized by painful sores on the genitalia. It is transmitted from one individual to another only through sexual contact. It is also known as soft chancre and ulcus molle. It is caused by the fastidious Gram-negative streptobacillus *Haemophilus ducreyi*. This disease is found primarily in developing countries.

H. ducreyi enters skin through microabrasions incurred during sexual intercourse. A local tissue reaction leads to development of erythematous papule, which progresses to pustule in 4–7 days. It then undergoes central necrosis to ulcerate.

These are only local symptoms and no systemic manifestations. Size of the ulcer Ranges from 3 to 50 mm. It is painful. It has sharply defined borders. Painful lymphadenopathy occurs in 30 to 60% of patients. Dysuria (pain with urination) and dyspareunia (pain with intercourse) are observed in females. About half of infected men have only a single ulcer. Women frequently have four or more ulcers, with fewer symptoms.

Chancroid is diagnosed by looking at the ulcer(s) and checking for swollen lymph nodes. There are no blood tests for chancroid. *H. ducreyi* can be identified by PCR based methods. Patient has one or more painful genital ulcers. The combination of a painful ulcer with tender adenopathy is suggestive of chancroid.

The infection is treated with antibiotics like azithromycin, ceftriaxone, ciprofloxacin, and erythromycin. Large lymph node swellings need to be drained, either with a needle or local surgery. Chancroid is a bacterial infection that is spread by sexual contact with an infected person. Avoiding all forms of sexual activity is the only absolute way to prevent a sexually transmitted disease.

R. VAGINITIS

Vaginitis is an inflammation of the vagina. It can result in discharge, itching and pain. It is associated with an irritation or infection of the vulva. Candidiasis, Bacterial vaginosis, Trichomoniasis vaginitis, Chlamydia vaginitis, Viral vaginitis and Non-infectious vaginitis are the types of vaginitis.

Causative agents of vaginitis are as follows; Candidiasis is caused by *Candida albicans* (a yeast). Bacterial vaginosis is caused by *Gardnerella vaginalis*.Trichomoniasis vaginitis is caused by *Trichomonas vaginalis*. Chlamydia vaginitis is caused by Chlamydia. Viral vaginitis is caused by Herpes simplex Virus. Non-infectious vaginitis is caused by improper hygiene.

Other less common vaginitis is caused by *Neisseria gonorrhea* and *Mycoplasma*.

A woman with vaginitis may have itching or burning and may notice a discharge. The following symptoms may indicate the presence of infection, which should be followed up with a professional health care practitioner for diagnosis and treatment: irritation and/or itching of the genital area, inflammation (irritation, redness, and swelling caused by the presence of extra immune cells) of the labia majora, labia minora, or perineal area, vaginal discharge, foul vaginal odour, pain/irritation with sexual intercourse

Diagnosis is made with microscopy (mostly by vaginal wet mount) and culture of the discharge after a careful history and physical examination have been completed. The color, consistency, acidity and other characteristics of the discharge may be predictive of the causative agent.

Candida vaginitis — usually causes a watery, white, cottage cheese-like vaginal discharges. The discharge is irritating to the vagina and the surrounding skin.

Bacterial vaginitis — *Gardnerella* usually causes a discharge with a fish-like odor. It is associated with itching and irritation, but not pain during intercourse.

Trichomonas vaginalis — can cause a profuse discharge with a fish-like odor, pain upon urination, painful intercourse, and inflammation of the external genitals.

Chlamydia is another sexually transmitted form of vaginitis. Unfortunately, most women with chlamydia infection do not have symptoms, which make diagnosis difficultIf left untreated, chlamydia can cause damage to a woman's reproductive organs, and can make it difficult for a woman to become pregnant.

Several sexually transmitted viruses cause vaginitis, including the herpes simplex virus and the humanpapilloma virus (HPV). The primary symptom of herpes is pain associated with lesions or "sores." These sores usually are visible on the vulva or the vagina.

Trichomoniasis is treated with metronidazole, or tinidazole. Bacterial vaginosis is by clindamycin. Yeast infections are by azole.

Good hygiene, drying completely after bathing, wearing fresh undergarments, and wiping from front to rear after defecation all help to prevent contamination of the vagina with harmful bacteria. Prevention of trichomoniasis revolves around avoiding other people's wet towels and hot tubs, and safe-sex procedures, such as condom use. Consumption of food like yogurt, sauerkraut and probiotic supplements may prevent candidiasis.

S. ANTHRAX

Anthrax is an acute disease caused by the bacterium *Bacillus anthracis*. Most forms of the disease are lethal and it affects both humans and animals. *Bacillus anthracis* is discovered by

Robert Koch. It is a rod shaped spore forming bacterium. It is one of the largest pathogenic bacterium. In tissue it is found singly, in pair, or in short chains. The entire chain being surrounded by a capsule. It is a catalase positive, nonmotile bacterium.

Anthrax can be transmitted to human by contact with infected animals. Anthrax does not spread from person to person. Anthrax can infect humans in three ways. The most common is infection through the skin, which causes an ugly sore that usually goes away without treatment. Humans and animals can ingest anthrax from carcasses of dead animals that have been contaminated with anthrax. Ingestion of anthrax can cause serious, sometimes fatal disease. The most deadly form is inhalation anthrax. If the spores of anthrax are inhaled, they migrate to lymph glands in the chest where they proliferate, spread, and produce toxins that often cause death. The incubation period may be relatively short, from one to five days.

Three forms of human anthrax disease are recognized based on their portal of entry. Cutaneous, the most common form (95%), causes a localized, inflammatory, black, necrotic lesion. Pulmonary, the highly fatal form, is characterized by sudden, massive chest oedema followed by cardiovascular shock.

Gastrointestinal, a rare but also fatal (causes death to 25%) type, results from ingestion of spores.

Cutaneous Anthrax

Cutaneous (on the skin) anthrax infection in humans presents as a boil-like skin lesion that eventually forms an ulcer with a black center (eschar). The black eschar often shows up as a large, painless necrotic ulcer at the site of infection. In general, cutaneous infections form within the site of spore penetration between 2 and 5 days after exposure. cutaneous anthrax infections normally do not cause pain.Cutaneous anthrax is typically caused when Bacillus anthracis spores enter through cuts on the skin. This form of Anthrax is found most commonly when humans handle infected animals and/or animal products . Symptoms include muscle aches and pain, headache, fever, nausea, and vomiting. The illness usually resolves in about six weeks, but death may occur if patients do not receive appropriate antibiotics.

Inhalation Anthrax

Respiratory infection in humans initially presents with cold or flu-like symptoms for several days, followed by severe (and often fatal) respiratory collapse.The first symptoms are subtle, gradual and flu-like (influenza). In a few days, however, the illness worsens and there may be severe respiratory distress. Shock, coma, and death follows. Inhalation anthrax does not cause a true pneumonia. In fact, the spores get picked in the lungs up by macrophages. Most of the spores are killed. Unfortunately, some survive and are transported to lymph nodes. In the lymph nodes, the spores that survive multiply, produce deadly toxins, and spread throughout the body. Severe hemorrhage and tissue death (necrosis) occurs in these lymph nodes in the chest. From there, the disease spreads to the adjacent lungs and the rest of the body. Inhalation anthrax is a very serious disease, and unfortunately, most affected individuals will die even if they get appropriate antibiotics. The antibiotics are effective in killing the bacteria, but they do not destroy the deadly toxins that have already been released by the anthrax bacteria.

Gastrointestinal Anthrax

Now rare, anthrax of the bowels (gastrointestinal anthrax) is the result of eating undercooked, contaminated meat. The symptoms of this form of anthrax include nausea, loss of appetite, bloody diarrhea and fever followed by abdominal pain. The bacteria invade through the bowel wall. Then the infection spreads throughout the body through the bloodstream (septicemia) with deadly toxicity. Gastrointestinal infection in humans is most often caused by consuming anthrax-infected meat and is characterized by serious gastrointestinal difficulty, vomiting of blood, severe diarrhea, acute inflammation of the intestinal tract, and loss of appetite. Lesions have been found in the intestines and in the mouth and throat. After the bacterium invades the bowel system, it spreads through the bloodstream throughout the body, while also continuing to make toxins. Gastrointestinal infections can be treated but usually result in fatality rates of 25% to 60%, depending upon how soon treatment commences.

Pathogenesis

B. anthracis possesses a capsule that is antiphagocytic and is essential for full virulence. The organism also produces three plasmid-coded exotoxins. They are oedema factor (a calmodulin- dependent adenylate cyclase), lethal toxin and protective antigen. Oedema factor causes elevation of intracellular cAMP and is responsible for the severe edema usually seen in B. anthracis infections; lethal toxin is responsible for tissue necrosis; protective antigen (so named because of its use in producing protective anthrax vaccines) mediates cell entry of edema factor and lethal toxin.

Culture

On ordinary media anthrax bacilli grown as medusa head appearance. Selctive media is PLET medium, (Polymyxin, Lysozyme, EDTA and thallocon acetate are added to heart infusion agar). Blood or pus smear is taken and performed gram and spore staining to demonstrate anthrax bacillus.

Prevention and treatment

A number of anthrax vaccines have been developed for preventive use in livestock and human. Infections with *B. anthracis* can be treated with β-lactam antibiotics such as penicillin and others which are active against Gram-positive bacteria. Penicillin-resistant B. anthracis can be treated with fluoroquinolones such as ciprofloxacin or tetracycline antibiotics such as doxycycline.

T.   SMALL POX - 433

U.   COMMON COLD - 400

V.   INFLUENZA - 401-404

W. MEASLES – 406- 408

X. MUMPS – 406

Y. POLIO – 399, 400

Z. RABIES – 412, 413

AA. HEPATITIS – 437-442

AB. AIDS – 422-424

AC. CHICKEN POX – 417

AD. DENGUE

Dengue is one of the most important human infections. It is caused by Dengue virus. It belongs to the family Flaviviridae. Dengue fever is an acute viral infection caused by at least four different strains of Dengue virus. Dengue is transmitted by the bite of an infected mosquito known as *Aedes aegypti*. This mosquito breeds in clean, stagnant water and has a flight range of 100 – 200 metres. The mosquito gets the Dengue virus after biting a human being infected with Dengue virus. Incubation period is 3 – 14 days. Dengue fever is a febrile illness that affects infants, young children and adults.

Figure 279   Dengue virus

Classic dengue begins suddenly with an influenza-like syndrome consisting of fever, malaise, cough and headache. Severe pains in muscles and joints (breakbone) occur. Enlarged lymph glands, rash and low WBC counts and platelet are common.

In its serious form Dengue causes two important manifestations. Dengue hemorrhagic fever and Dengue Shock Syndrome. Dengue haemorrhagic fever (DHF) is a **severe form of Dengue** fever. The patient suffering from Dengue haemorrhagic fever develops **bleeding from nose, gums or skin.** Sometimes, the patient may have **coffee coloured vomiting or black stools.** This indicates bleeding in gastro intestinal tracts which is serious. During Dengue Shock Syndrome (DSS), the patient suffering from dengue may manifest rapid & weak pulse, low blood pressure, cold clammy skin and restlessness.

The diagnosis can be confirmed by isolation of **virus in cell culture** (direct evidence) or by serological test that demonstrate the presence of **IgM antibody** or **four fold rise in antibody titer** during disease (indirect evidence). The tests for confirmation of Dengue should be done in reliable laboratories.

Aedes aegypti can be easily recognized by its peculiar **white spotted body** and legs. The female mosquito lays her eggs in water containers in and around the homes, and other dwellings. It is a **day biting mosquito.**

Control of Dengue is by controlling vectors. To prevent the mosquitoes from breeding, drain out the water from air coolers, tanks, barrels, drums, buckets etc.

There is no specific treatment for Dengue. Paracetamol is the drug of choice to bring down fever and relieve joint pain.

AE. YELLOW FEVER

Yellow fever is an acute viral hemorrhagic disease. Yellow fever is one of the most dangerous infectious diseases.Yellow fever is caused by the yellow fever virus. It is a 40 to 50 nm wide enveloped RNA virus. It belongs to the family Flaviviridae. The genome is the positive sense single-stranded RNA.

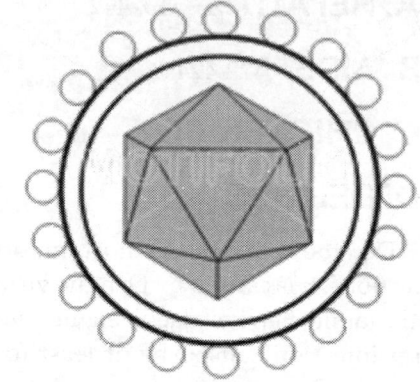

Symptoms of this disease include fever, chills, anorexia, nausea, muscle pain and headache, which generally subsides after several days. In some patients, a toxic phase follows, in which liver damage

Figure 280   Yellow fever virus

with jaundice (inspiring the name of the disease) can occur and lead to death. Because of the increased bleeding tendency (bleeding diathesis), yellow fever belongs to the group of hemorrhagic fevers.

Yellow fever begins after an incubation period of three to six days. The viruses infects monocytes, macrophages and dendritic cells. They attach to the cell surface via specific receptors and are taken up by an endosomal vesicle. Inside the endosome, the decreased pH induces the fusion of the endosomal membrane with the virus envelope. The capsid enters the cytosol, decays, and releases the genome. Receptor binding as well as membrane fusion are catalyzed by the protein E, which changes its conformation at low pH, causing a rearrangement of the 90 homodimers to 60 homotrimers.After entering the host cell, the viral genome is replicated in the rough endoplasmic reticulum (ER) and in the so-called vesicle packets. At first, an immature form of the virus particle is produced inside the ER, whose M-protein is not yet cleaved to its mature form and is therefore denoted as prM (precursor M) and forms a complex with protein E. The immature particles are processed in the Golgi apparatus by the host protein furin, which cleaves prM to M. This releases E from the complex which can now take its place in the mature, infectious virion.

The yellow fever virus is mainly transmitted through the bite of the yellow fever mosquito *Aedes aegypti*. Tiger mosquito (*Aedes albopictus*) can also serve as a vector for the virus. After transmission, the viruses replicate in the lymph nodes and infect dendritic cells and infect hepatocytes. It leads to eosinophilic degradation of these cells and to the release of cytokines. Necrotic masses known as Councilman bodies appear in the cytoplasm of hepatocytes. Fatality may occur when cytokine storm, shock and multiple organ failure follow.

Serologically, an enzyme linked immunosorbent assay can confirm yellow fever. Liver biopsy can verify inflammation and necrosis of hepatocytes and detect viral antigens.

Personal prevention of yellow fever includes vaccination as well as avoidance of mosquito bites in areas where yellow fever is endemic. Vaccination is the most important preventive measure against yellow fever. There is no cure for yellow fever. A symptomatic treatment includes rehydration and pain relief with drugs like paracetamol. Aspirin should not be given because of its anticoagulant effect.

AF. COLD SORE

It is also called as Herpes labialis, herpes simplex labialis or orolabial herpes. It is a type of herpes simplex viral infection. Symptoms occurring on the lip. The infection is caused by herpes simplex virus (HSV). It causes small blisters or sores on or around the mouth commonly known as cold sores or fever blisters. The sores typically heal within 2–3 weeks, but the herpes virus remains dormant in the facial nerves, following orofacial infection, periodically reactivating (in symptomatic people) to create sores in the same area of the mouth or face at the site of the original infection.

Cold sore has a rate of frequency that varies from rare episodes to 12 or more recurrences per year. People with the condition typically experience one to three attacks annually. The frequency and severity of outbreaks generally decreases over time.

The main symptom of oral infection is inflammation of the mucosa of the cheek and gums — known as acute herpetic gingivostomatitis — which occurs within 5–10 days of infection. Other symptoms may also develop, including headache, nausea, dizziness and painful ulcers — sometimes confused with canker sores — fever, and sore throat. Primary HSV infection in adolescents frequently manifests as severe pharyngitis with lesions developing on the cheek and gums. Some individuals develop difficulty in swallowing (dysphagia) and swollen lymph nodes (lymphadenopathy).

Cold sore outbreaks may be influenced by stress, menstruation, sunlight, sunburn, fever, dehydration, or local skin trauma.

HSV-1 can in rare cases be transmitted to newborn babies by family members or hospital staff who have cold sores; this can cause a severe disease called Neonatal herpes simplex.

People can transfer the virus from their cold sores to other areas of the body, such as the eye, skin, or fingers; this is called *autoinoculation*.

Docosanol, a saturated fatty alcohol, is a safe and effective topical application that has been approved by the United States Food and Drug Administration for herpes labialis in adults with properly functioning immune systems. The duration of symptoms can be reduced by a small amount if an antiviral, anaesthetic or non-treatment cream (such as zinc oxide or zinc sulfate) is applied promptly.Effective antiviral medications include acyclovir and penciclovir, which can speed healing by as much as 10%. Famciclovir or valaciclovir, taken in pill form, can be effective using a single day, high-dose application and is more cost effective and convenient than the traditional treatment of lower doses for 5–7 days.

Avoiding touching an active outbreak site, washing hands frequently while the outbreak is occurring, not sharing items that come in contact with the mouth, and not coming into contact with others (by avoiding kissing, oral sex, or contact sports) can reduce the likelihood of the infection being spread to others.

AG. SUPERFICIAL MYCOSIS

Pityriasis versicolor

It is also called *Tinea versicolor*. It is a superficial skin infection. It is a chronic and less serious infection of the skin.This infection involves stratum corneum characterized by discolouration or depigmentation of skin. It is caused by *Malassezia furfur*. It is a dimorphic fungus.It infects only skin. Chest, abdomen, upper limps and back are the susceptible area to *M.furfur*. It affects external appearance of an individual. Fungus interferes melanin production by the production of Dicarbolic acid. It inhibits the activity of *Tyrosinase*, which is responsible for the synthesis of melanin.Some individuals develop folliculitis.

Pus/skin scrapings are subjected to *Microscopy*. Direct microscopic observation of the specimen with 10-20% KOH showed Short unbranched hyphae and spherical cells. Lesions also fluorescence under woods lamp test. Modified SDA supplemented with olive oil with antibiotics is used for cultivation. Shiny or pasty white to cream coloured colonies are formed after 1-2 weeks.It is treated by the application of *Selinum sulfide*

Tinea Nigra

It is a superficial, chronic and asymptomatic infection of the Stratum Corneum. It is caused by *Exophiala werneckii*. Lesions usually consist of a solitary macuole with sharply defined margins. The brownish colour of the lesion is darkest at the advanced periphery, where most of the actively growing organisms are located. Many cases involves the palm, other parts of the skin may also be infected, including fingers and face.

Specimens are mounted by 20% KOH or Calcofluor white stain. Observation reveals brown pigmented septate hyphae and budding yeast cells. Culture with Sabuourds Dextrose Agar with or without antibiotics recover the organism. Colony is shiny, moist and often white to gray in colour initially. Within a few days, the colony darkens and becomes olive to black. Later, mycelium will develops and the colony appears dull and fuzzy. It is treated by salicylic acid or tincture of iodine.

Piedra

There are two types of Piedra. Black piedra and white piedra

Black Piedra

Black Piedra is a Nodular Infection of the Hair shaft. It is caused by *Piedraia hortae*.The organism is a plant parasite.Human infection caused by this parasite is called black piedra. Black piedra consist of slow growing brown to reddish black mycelia. It produces spindle shaped ascospores.Hard nodules found along with the infected hair shaft. Nodules have a hard carbonaceous consistency. Hair is subjected to diagnosis. SDA medium is used for cultivation. Colour of the colony on SDA is greenish black or black.Asci and ascospores are rarely seen in culture.

White Piedra

Trichosporan beigelii is a causative agent. It present as larger, softer, yellowish nodules on the hair. *Trichosporan beigelii* is sensitive to Cycloheximide. It is a dimorphic fungi. Yeast cultures are white and have a pasty consistency. As the culture ages, colonies develop deep radiative farrows and take on a yellowish colouration with the creamy texture.Microscopic examination reveals septate hyphae that fragments rapidly to form arthroconidia.The Arthroconidia rapidly roundup and many cells form Blastoconidia. It affects hairs of the scalp, mustache and beard.It is characterized by the development of cream coloured soft pasty growths along infected hair shaft.Piedra is endemic in tropical underdeveloped countries. Treatment is done by Selenium disulphide, Hyposulfite, Thiosulphite of salicylic acid. Miconazole nitrate is also used. Shaving or Cropping of infected hairs close to the scalp surface achieves effective therapy.

Dermatophytosis

Dermatophytosis is a fungal infection. It involves the superficial areas of the body. Causative agent invades the keratinized portion of the hair, skin and nails. Dermatophytes are transmitted by close contact and the organisms may spread rapidly within families and enclosed communities. It is also called Tinea or Ringworm infection. The lesions of the Tineas are often appear as pink circular lesions and gradually advance to form new borders. Scaling and peeling is common in affected areas.The lesions are rarely painful but can be very itchy. Infection of the nail is chronic and produces discolouration and thickening. Scalp infection leads to scaling and inflammation with hair lose which may sometimes associated with scarring.

Three different genera are recognized as dermatophytes. They are Epidermophyton, Microsporum andTrichophyton.

These fungi invade only dead corinified layers of the skin, hair and nails. Only a few organisms are necessary to induce an infection following trauma.Dermatophytes grow in filamentous form within the stratum corneum. The inflammatory response results in a specific symptoms.

Samples used for diagnosis are Hair plug, nail cut and skin scrape. The specimen is mounted in a small amount of 10% or 20% Potassium Hydroxide or calcofluor white. The KOH slides are gently heated and allowed to clear for 30 to 60 minutes before examining on a light or phase contrast microscope. Calcofluor white slides are examined on a fluorescent microscope.

Treatment is with topical ointments such as miconazole, tolnaftate or clotrimazole for 2 to 4 weeks. Griseofulvin and itraconazole are the only oral fungus agents currently approved.

AH. SUBCUTANEOUS MYCOSIS

Sporotrichosis

It is a chronic granulomatous infection of human skin. It is caused by *Sporothrix schenckii*. It is a dimorphic fungus. This disease is also called 'Rose Thorn Disease'. *Sporothrix* is implanted into the deep layers of the skin through trauma. Initially small hard painless nodule appear at the site of injury and it enlarges into a fluctuant mass that eventually breakdown and ulcerates.

Aspirated fluid, Pus, Biopsy tissue are examined directly by KOH mount. Calcofluor white stain also used. In infected tissue, the fungus is seen as cigar shaped yeast cells without mycelium. *Asteroid bodies* are seen in lesions, composed of central fungus cells with eosinophilic materials radiating extensions. It is an antigen and antibody complex along with complement.

It is treated with Oral solution of saturated Potassium Iodide. Surface lesions are treated with 2% Potassium Iodide in 0.2% iodine.Other antifungal agents involved in the treatment purpose are Griseofulvin, Amphotericin B. Flucystosine, Dihydroxystilbamide, Ketaconazole

Chromoblastomycoses

The fungus responsible for this disease produce melanin like pigments. Causative agents of this disease are *Phialophora veruucosa, Cladosporium carrionii, Fonsecae petrosoi, Fonsecae compacta, Rinocldiella aquaspersa*

Fungi are introduced in the skin by trauma. Primary lesions become verrucous and wart like extension along the draining lymphatics. Cauliflower like nodules are formed which cover the area. Small ulceration also formed within short duration, these lesions become raised and appears scally and dull red or grayish in colour. It may be due to melanin pigment produced by the causative agent. 10% KOH mount is performed to detect sclerotic body. Surgical exision with wide margin is the therapy. Fluconazole, Itaconazole is also recommended.

Mycetoma

Mycetoma is a chronic subcutaneous infection induced by traumatic inoculation with any of several saprophytic fungus. The disease is first observed in Madurai district of Tamilnadu. So this disease is also called as Madura foot and Maduramycosis. The clinical features defining Mycetoma are local swelling and interconnecting, often draining sinuses that contains granules, which are Microcolonies of the agent embedded in tissue material.

Based on etiological agent Mycetoma is classified into three. They are Actinomycetoma (caused by Actinomycetes). Eumycetoma (caused by a Fungus). Botryomycosis (caused by group of bacteria).

The fungal agents of Mycetoma are*Madurella grisea, Phialophora jeanselinei etc.*

Mycetoma develops after traumatic inoculation of any one of the agents of Mycetoma. Subcutaneous tissues of the foot, lowed extremities, hands and exposed areas are most often involved in Mycetoma formation.Pathology is characterized by supporation and abscess formation, granulomas and the formation of draining sinuses containing the granules.

Pus, Exudates or Biopsy material are the specimens. Specimen is inoculated into SDA with antibiotics.

Amphotericin B is recommended. Katoconazole, Nystatin, Flucytosine, KI and Miconazole also used.

AI. SYSTEMIC MYCOSIS

Coccidioidomycosis

Coccidioidomycosis is an infection caused by the dimorphic fungus called *Coccidioids immitis*. The Infection is acquired through inhalation of dust containing arthrospores of the fungus.

In the infected host, *Coccidioids immitis* exist as spherules, a spherical thick walled structure, that are filled with a few to several hundred endospores. As the spherule enlarges, the nuclei undergoes mitosis, the cytoplasm condenses around the nuclei and cellwall forms around each developing endospores. At maturation, the spherules ruptures to release its endospores. These endospores enlarge and to form mature spherule. It is observed in tissues and may appear in sputum of patients with Coccidioidal cavities in lungs.

Inhalation of arthroconidia leads to a primary infection. Respiratory tract is a major site for the infection. 60% of the infections are asymptomatic.

Symptoms are Fever, Chest pain, Cough, Weight loss, Extra pulmonary infection involves the meninges, skin or bone.

Clinical exudates should be examined directly in 10% or 20% KOH or Calcofluor white stains. Tissue specimens are stained with hematoxylin and eosin. Microscopic examination shows spherules and endospores.Clinical specimens are inoculated into the Inhibitory Mold Agar or Sabourauds Agar with antibiotics like Cycloheximide, Chloramphenicol and Gentamycin. It produces a white, grey or brownish colour, powdery to cottony texture colonies on mycological medium.. Microscopically colonies are observed as hyaline branching septate hyphae and as the culture ages, characteristic arthroconidia are produced.

The Coccidioidin skin test also used for diagnosis.

Patients with severe disease require treatment with Amphotericin B, which is administered intravenously

Histoplasmosis

Histoplasmosis is the most prevalent pulmonary mycosis of humans and animals. It is caused by the dimorphic fungi *Histoplasma capsulatum*. It is distributed throughout the world. This disease is also called Darlings disease.

Histoplasma capsulatum is a facultative intracellular parasite. In tissue, yeasts are seen within macrophages

Etiological agent *Histoplasma capsulatum* is entered into the lungs through inhalation.Yeast cells are engulfed by alveolar macrophages. Within macrophage, yeast cells are able to multiply and are disseminated to reticuloendothelial tissues such as the liver, spleen, bone marrow and lymph nodes through blood stream.

Smears of infected specimens are fixed with methanol and stained with Wright or Giemsa stain will reveal characteristically ellipsoidal yeast cells inside of macrophages. Purulent portion of the specimen should be selected for culturing.*H .capsulatum*is identified by characteristic macroconidia at 25- 30°C and observation of yeast cells at 37°C.Histoplasmin skin test is a valuable tool in diagnosis.

Amphotericin B is a drug of choice.

Blastomycosis

It is a chronic infection caused by *Blastomyces dermatitidis*. This disease is also called North American blastomycosis. Blastomycosis is first described by Gilchrist. Hence it is also known as Gilchrists or Chicago disease. Blastomycosis is primarily a pulmonary infection characterized by spread to the skin and other parts of the body.

On sabourauds glucose agar at 25°C, the organism grows as a mould, producing a colony of uniform hyaline, septate hyphae and conidia. Colony development requires at least 2 weeks. Many strains produce a cottony mycelium that becomes tan to brown with age.Microscopically the mycelial form produces abundant conidia from the aerial hyphae and lateral conidiophores. The conidia are spherical, ovoid or pyriform in shape and are 3-5m in diameter.

Blastomycosis are acquired by inhalation of exogenous, infectious particles.Initial site of infection is lungs.In the alveoli *Blastomyces dermatitidis* induces an inflammatory response characterized by the infilteration of both macrophages and neutrophiles and subsequent formation of granulomas.

Symptoms are fever, malaise, night sweats, and cough.

Amphotericin B is effective against blastomycosis

Candidiasis

It is caused by *Candida albicans.* It is capable of producing yeast cells with pseudohyphae and true hyphae. It is demonstrated by germ tube technique. On xmost media they produce raised, cream coloured opaque colonies within 24-48 hours. It produces ellipsoidal or spheical budding yeasts about 3-6m in size.

Yeast invasion of the vaginal mucosa leads to vulvovaginitis characterized by irritation, pruritus and vaginal discharge.

Numerous systemic manifestations of Candida may follow introduction of Candida in to the blood stream. They are as follows Esopharyngitis, intestinitis, infant diarrhoea , bronchopulmonary candidiasis, pylonephritis, cystitis, endocarditis , myocarditis , endoopthalmitis, meningitis, orthritis, osteomyeletis, peritonitis, macronodular skin lesions.

Cutaneous candidiasis treated with topical antimycotic substances like ketoconazole, nystatin, miconazole. For the treatment of systemic candidiasis amphotericin B , flucytosine are recommended.

Cryptococcosis

Cryptococcosis is usually associated with immuno suppression. Disease is world wide in distribution. It is caused by encapsulated yeast *Cryptococcus neoformans*. Visible colonies of *Cryptococcus neoformans* develop on routine laboratory media within 36-72 hours. They are white to cream coloured opaque and may be several millimeters in diameter. Colonies are typically mucoid in appearance and the amount of capsule produced can be judged by the degree of colony wetness.

The high prevalence of *Cryptococcus neoformans* in nature and relatively low frequency of disease suggests that many persons are probably exposed without any symptoms. Cryptococcosis is initiated in the lungs after inhalation of yeast cells of *Cryptococcus neoformans*. Based on symptoms there are two types of Cryptococcosis. They are Pulmonary Cryptococcosis and Disseminated Cryptococcosis

Cryptococcosis is treated with both Amphotericin B and Flucytosine.

AJ. AMOEBIOSIS

Amoebiosis is one of the intestinal disorder. It is caused by *Entamoeba histolytica*. Amoebas are unicellular organisms common in the environment. ***Entamoeba histolytica*** are distributed Worldwide. It is a endoparasite. It Feeds dissolved tissues, bacteria and RBCs. It causes fatal and serious disease. Infected individual discharge mucous and blood in their stool.

Trophozoite and Cyst are present in Enatmoeba. The Trophozoite is the actively metabolizing, mobile stage, and the cyst is dormant and environmentally resistant. They are actively motile using pseudopods.

The cyst is a spherical structure with a thin transparent wall. Fully mature cysts contain four nuclei with the characteristic amebic morphology.

Observed Symptoms are Fever, Amebic dysentery, other clinical symptoms include fulminant ulceration , Non dysentery gastroenteritis, Amoeboma formation, Amoebic colitis ,Hepatomegaly, Amoebic abscess.

Quadrinucleate cysts of *E histolytica* is ingested through fecal-oral transmission or food and water. Ingested excyst in the small intestine. Trophozoites are carried to the colon, where they mature and reproduce. Excystment occurs after ingestion and is followed by rapid cell division to produce four amoebas which undergo a second division. Each cyst thus yields eight tiny amoebas (Trophozoites). These trophozoits invad the tissue and cause symptoms.

Stool samples are subjected to saline and iodine wetmount, which clearly describes the cyst and trophozoites of Entamoeba.

Acute intestinal disease is treated with Metronidazole. Amoebic liver abscess is best treated with Metronidazole, Dihydroemetin, Chloroquine or Dehydroemetine.

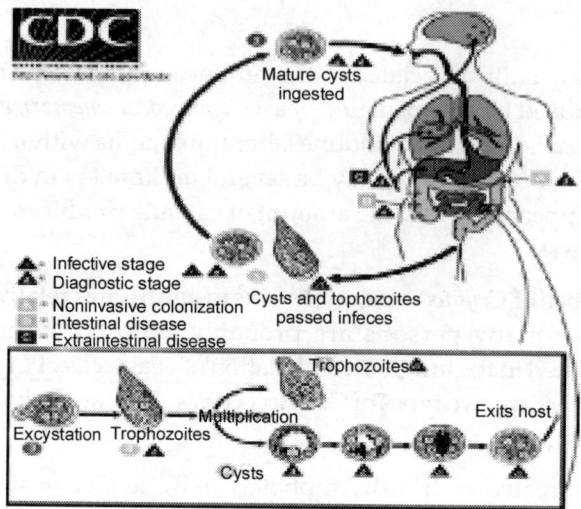

Figure 281   Life cycle of Entamoeba

AK. GIARDIASIS

It is one of the intestinal infection. It is caused by a protozoa called Giardia lamblia. The *Giardia* life cycle involves Trophozoite and the Cyst.

Trophozoite - It is looks like tennis racket. The dorsal surface is convex and the ventral surface is concave with a sucking disc, and has two nuclei that resemble eyes, structures called median bodies that resemble a mouth, and four pairs of flagella that look like hair; these combine to give the stained trophozoite the eerie appearance of a face.

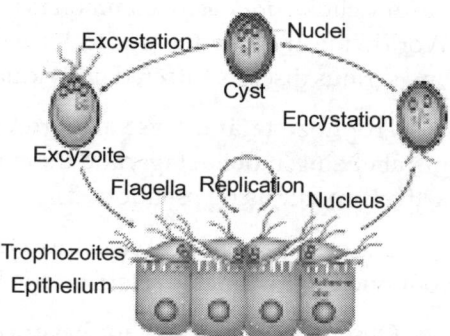

Figure 282   Life cycle of Giardia

The *Giardia* cyst is usually seen in the faeces. It is ovoid, 6 to 12 μm long, and can often be seen to contain two to four nuclei at one end and prominent diagonal fibrils. Flagella and sucking disc are seen inside of cytoplasm.

Giardia infection is acquired by ingesting cysts. The exposure of cysts to host stomach acidity and body temperature triggers excystation, trophozoite is formed. The trophozoites multiplies by binary fission. Trophozoites are again converted to cyst.

Diarrhoea or loose, foul-smelling stools, steatorrhoea (fatty diarrhoea), malaise, abdominal cramps, excessive flatulence, fatigue and weight loss are the symptoms.

Diarrhoeal specimens contain trophozoites/Cyst. It is detected using saline and iodine wetmount. The drug of choice for treating *Giardia* infections is Quinacrine hydrochloride and metronidazole.

AL. MALARIA

Malaria has been a major disease of human. It is caused by *Plasmodium falciparum, P vivax, P ovale* and *P malariae*. Malaria is spread to humans by the bite of female mosquitoes of the genus *Anopheles*.

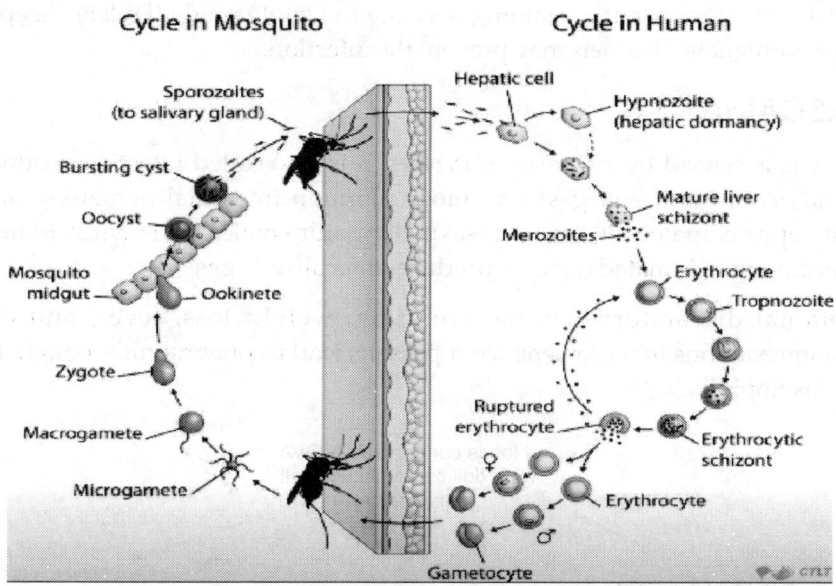

Figure 283 Life cycle of Malaria

Sporozoites are produced by sexual reproduction in the midgut of vector anopheles mosquitoes and migrate to the salivary gland. When mosquito bites, inject sporozoites into small blood vessels and sporozoites enters liver parenchymal cells. In the liver, the parasite develops into a spherical, multinucleate liverstage schizont. The mature schizonts rupture, releasing thousands of uninucleated merozoites into the bloodstream. Each merozoite can infect a Red Blood Cell. Within the red cell, the merozoite develops to form either an erythrocytic stage (blood-stage) schizont. The mature erythrocytic stage schizont contains 8 to 36 merozoites, which are released into the blood when the schizont ruptures. These merozoites proceed to infect another generation of erythrocytes.

The gametocyte, which is the sexual stage of the *Plasmodium*, is infectious for mosquitoes that ingest it while feeding. Within the mosquito, gametocytes develop into female and male gametes (macrogametes and microgametes), which undergo fertilization and then develop over 2 to 3 weeks into sporozoites.

The most characteristic symptom of Malaria is fever. Other common symptoms include chills, headache, myalgias, nausea, and vomiting. diarrhoea, abdominal pain. In all types of Malaria the periodic febrile response is caused by rupture of mature schizonts. In *P vivax* and *P ovale* malaria, a brood of schizonts matures every 48 hours, so the periodicity of fever is

Tertian ("tertian malaria"), whereas in *P malariae* disease, fever occurs every 72 hours ("quartan malaria"). The fever in Falciparum Malaria may occur every 48 hours, but is usually irregular.

Diagnosis of Malaria generally by looking Giemsa-stained thick and thin blood smears.

Drugs used for treatment are Primaquine, Chloroquine, Mefloquine, Quinine, Quinidine, Pyrimethamine-sulfadoxine, Doxycycline.

Use of an insect repellent containing N,N-diethyl *m*toluamide (DEET). Sleeping under insecticide-impregnated bednets may prevent the infection.

AM. ASCARIASIS

Ascariasis is caused by *Ascaris lumbricoides*. It is also called intestinal Round Worms. *Ascaris lumbricoides* is the largest and most common intestinal nematode of humans. Females are approximately 30 cm long; sexually mature males are smaller. Mated females produce fertile eggs. Unmated females produce unfertilized eggs.

Abdominal discomfort, nausea, vomiting, weight loss, fever, and diarrhoea. Allergic manifestations in hypersensitized persons lead to pneumonitis, cough, low-grade fever and eosinophilia.

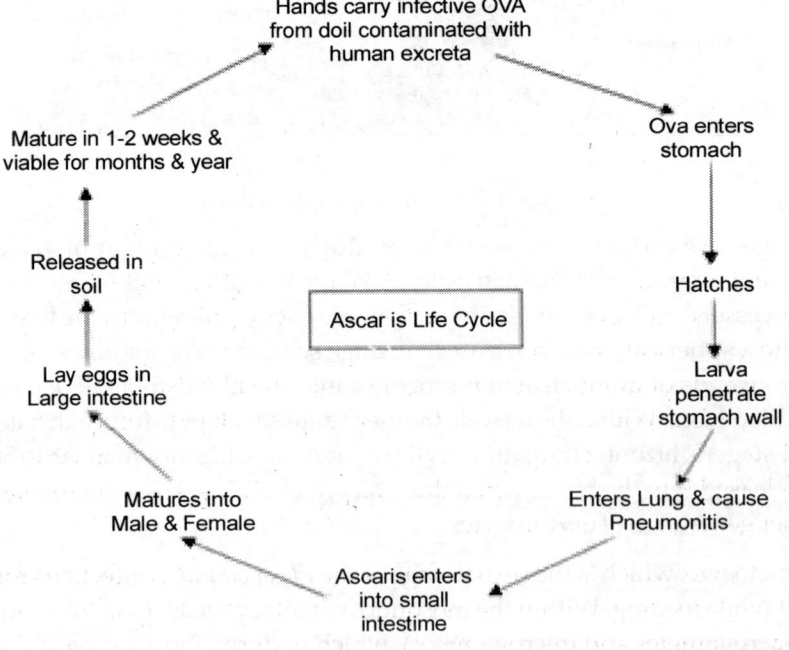

Figure 284   Life cycle of Ascaris

Infection begins when the embryonated eggs are ingested with food or drink. After ingestion by a human, the eggs pass to the duodenum where they hatch; the released larvae penetrate the intestinal mucosa, enter the lymphatics and portal system, and are carried to the liver, heart, and lungs. This migratory phase requires a few days. The larvae then break out of the capillaries into the alveoli, pass up the respiratory tree and are swalloved. They reach the intestine and continue their development. They become sexually mature adults. The adults live for about a year and are subsequently passed in the faeces. Females produce as many as 240,000 eggs per day.

Saline and iodine wet mount assesses the worm burden of an individual.

The most effective method to control Ascariasis is sanitary disposal of faeces. Mebendazole is effective against numerous intestinal nematode. Levamisole is also useful.

AN. FILARIOSIS

It is also called Elephantiasis. It is caused by *Wuchereria bancrofti*. It is a Filarial worm. The Filariae are thread-like parasitic Nematodes (roundworms) that are transmitted by arthropod vectors. The adult worms inhabit specific tissues where they mate and produce Microfilariae. Microfilariae is the characteristic tiny, thread-like larvae.

Adult *Wuchereria* is a elongated and slender in nature. Males are about half the size of females.

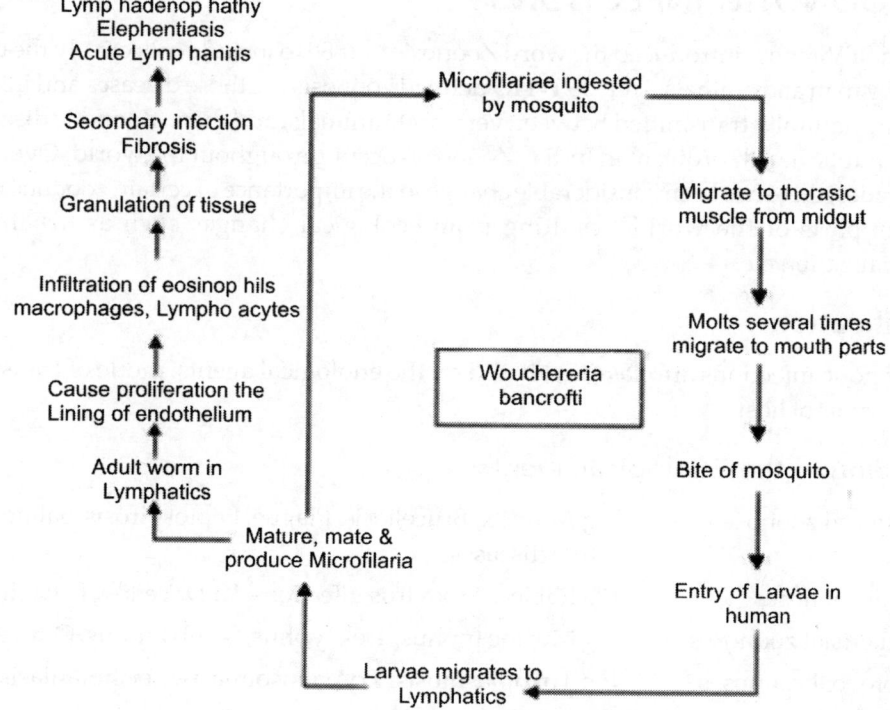

Figure 285   Life cycle of Filarial worm

Microfilariae ingested by a vector mosquito migrate out of the midgut to the thoracic muscles, where they develop, molt several times, and finally migrate to the mouthparts of the mosquito as infective larvae. As the infective mosquito feeds on another host. Larvae quickly migrate to the lymphatics, where they mature, mate, and produce microfilariae in the new host. The time from infection until microfilariae can be detected in the blood varies from 3 to many months.

Infective larvae induce a host inflammatory reaction cause the eventual blockage and oedema characteristic of *W bancrofti* infections. Lymphatic vessels are often partially or completely blocked by dead worms, or by endothelial proliferation, fibrin deposition, and granulomatous reactions. Lymph stasis favours secondary bacterial and mycotic infection.

Early symptoms of lymphatic filariasis consist of intermittent fever and enlarged, tender lymph nodes. The lymphatic vessels that drain into the lymph nodes and that harbor the developing and adult worms also become inflamed and painful and legs of infected human looks like elephants leg.

It is diagnosed by observing Enlarged and tender lymph nodes. Definitive diagnosis is done by identifying microfilariae in thick blood smears.

Mass treatment of this disease is using Diethylcarbamazine citrate. This drug significantly reduces the level of microfilariae in the blood.

AO. ZOONOTIC INFECTIONS

Rudolf Virchow introduced the word Zoonosis in 1880 to include collectively the diseases shared by man and animals. 1n 1959 WHO defined zoonosis are those diseases and infections, which are naturally transmitted between vertebrate animals, and man. Zoonotic diseases are a major public health problem in India. Zoonoses occur throughout the world. Over the last two decades, there has been considerable change in the importance of certain zoonotic diseases in many parts of the world, resulting from ecological changes such as urbanization, industrialization etc.

Classification

Zoonotic infections are classified based on the etiological agents, mode of transmission and reservoir of host

According to the Etiological agents

Bacterial zoonoses	:	E.g Anthrax, Brucellosis, Plague, Leptospirosis, Salmonellosis, Lyme diseases
Viral zoonoses	:	E.g Rabies, Arbovirus infections, KFD, Yellow fever, Influenza
Rickettsial zoonoses	:	E.g Murine typhus, Tick typhus, Scrub typhus, Q-fever
Protozoal zoonoses	:	E.g Toxoplasmosis, Trypanosomiasis, Leishmaniasis
Helminthic zoonoses	:	E.g Echinococcosis, Taeniasis, Schistosomiasis
Fungal zoonoses	:	E.g Deep mycosis – Histoplasmosis, Cruptococcosis, Superficial dermatophytes.
Ectroparasites	:	E.g Scabies, Myasis

According to the mode of transmission

Direct zoonoses

These are transmitted from an infected vertebrate host to a susceptible host (man) by direct contact, by contact with a fomite or by a mechanical vector. e.g Rabies, Anthrax, Brucellosis, Leptospilrosis, Toxoplasmosis.

Cyclozoonoses

These require more than one vertebrate host species, but no invertebrate host for the completion of the life cycle of the agent, e.g Echinococcosis, Taeniasis.

MetaZoonoses

These are transmitted biologically by invertebrate vectors, in which the agent multiplies and or develops host e.g., Plague, Arbovirus infections, Schistosomiasis, Leishmaniasis.

Saprozoonoses

These require a vertebrate host and a non-animal developmental site like soil, plant material, pigeon dropping etc. for the development of the infectius agent e.g., Aspergillosis, Coccidioidomycosis, Histoplasmosis, Zygomycosis.

According to the Reservior Host

Anthropozoonoses

Infections transmitted to Man from lower vertebrate animals E.g., Rabies, Leptospirosis, Plague, Arboviral infections, Brucellosis and Q-fever.

Zooanthroponoses

Infections transmitted from Man to Lower vertebrate animals e.g., Streptococci, Staphylococci, Diphtheria, Enterobacteriaceae, Human tuberculosis in cattle and parrots.

Amphixenoses

Infections maintained in both man and lower vertebrate animals and transmitted in either direction e.g., Salmonellosis, Staphylococcosis.

Factors influencing the prevalence of zoonoses.

Ecological changes in man's environment, Handling animal by – products and wastes (occupational hazards), Increased movements of man, Increased trade in animals products, Increased density of animal population, Transportation of virus infected mosquitoes, Cultural anthropological norms

High risk groups for zoonosis

Laboratory staff who handling infected materials, Veterinarian, Animal handlers and catchers, Wild life officers, Quarantine officers , Naturalists, Slaughter house workers,

Animal researchers

Transmission of zoonotic agents

In human zoonotic agents may transmit and they continuously transfer the etiological agents or they are the final end of disease

Humans are dead end host, Humans may transmit, Organisms enter the human body in varieties of ways.

Penetration of the skin, Inhalation and Ingestion

Control of zoonotic infections

It is controlled by a multidisciplinary way. They are Use human and veterinary medicines, Sanitary engineering, Eradication of infected animal reservoir, Protection of animal, Pasteurization of milk, Proper cooking of food, Eradication of flies and mosquitoes, Mass vaccination

AP. HOSPITAL ACQUIRED INFECTIONS

Introduction

Nosocomial infections are infections acquired in the hospital. The term nosocomial comes from the greak word *nosos* meaning **disease** and *komeion* meaning **hospital.** Nintingale is the first person improves medical care in military and civilian hospital. Semmelweis and Holmes follow Nightingales medial care procedures to reduce the tramission of pathogenic microorganisms within hospital. Pasteur demonstrates that microorganism could be transmitted through air. In this time, Lister realize the significance of Pasteur study and was convinced that microorganism in the air contaminates surgical wounds and He follows antiseptic surgery. Subsequently he also disinfects operating rooms greatly and reduces the mortality of surgical patients.

Rate of Nosocomial Infection

Nosocomial infections occur in approximately (one in 20 patients) 5 % of all patients admitted. Infection rate will vary depending on the type of hospital. Nosocomial infections in acute care institutions are follows. 40% of hospital infections are observed in urinary tract, 20% in surgical wound 10% in respiratory route. 5-10% in primary bacteria and others are due to opportunistic fungal and viral infections (20 – 250). Opportunistic infections are observed in immuno-compromised patients.

Approximately 1% of all nosocomial infections directly cause death and 3% contribute to death.

Important pathogens of nasocomial infections.

Staphylococcus aureus, Enterococcus, E. coli, Group B Streptococci, Coagulase negative Staphylococcus

*Proteus,Pseudomonas aeruginosa,*Antibiotic resistant gram-negative rods,*Candida albicans,Torulopsis*

Aspergillus,Anaerobic bacteria

Sources of microbes that cause nosocomial infection:

There are two categories of infection.

Exogenous – Caused by microbes from an external sources

Endogenous – Caused by microbes that are part of a persons own normal flora.

Hospital personnel, Surgical procedures , Food , Air , Visitors, Medications, Patients Normal Flora, Drains, implants, catherer, Other patients, Water, Fomites, Insects

Factors, which increase the risk of nosocomial infections

Invasive device, Tissue transplant, Extensive skin burn, Sickle cell anemia, Bone marrow failure, Malignant disorder,

HIV infection, Malfunction of spleen, Implants

Methods to avoid Nosocomial infections

Strict isolation

Diphtheria, Smallpox, Conjunctivitis, Lassa fever

Pneumonic plaque

Chicken pox

Contact isolation

Wound infections

Gonococcus conjunctivitis

Herpes simplex

Influenza, Pneumonia

Respiratory isolation

Measles, Meningitis, Mumps

Whooping cough, Pneumonia

Enteric precautions

Amoebic dysentery

Gastroenteritis

Typhoid fever, Cholera

Universal blood and body fluid precautions
AIDS,HBV, Malaria ,Syphilis, Gonorrhoea

Control of nosocomial infections

Because of the seriousness of nosocomial infections, the American Hospital Association (AHA) and the Centers for Disease Control (CDC) recommend that an each hospital develop an infection control programme. One of the most important activities of the control program is surveillance, the systematic observation and recording of cases of transmissible diseases. To accomplish this National Nosocomial Infection Surveillance system (NNIS) was established in 1970 by the CDC.

In 1999 CDC update hospital control program for break cycle of infection. Methodologies are Hand washing, Aseptic technique, Isolation of an infected patient, Proper sanitation, Disinfection, Sterilization.

Components of control programme

1. Hospital policy-making committee

   Develop policies for control of Nosocomial infection

2. Infection control committee

   Develop guidelines for patients care, monitor effectiveness of control program.

3. Microbiology laboratory

   Isolate and identify organism, identify source of infection, monitor disinfection and sterilization process

4. Infection control officer

   Implement infection control practices, surveillance and investigation of infection in patients and in personnel, maintain a continuing education program.

By combining importance of infection and control program Florence Nightingale said that 'the very first requirement in a hospital is that it should do the sick no harm

15

INDUSTRIAL MICROBIOLOGY

Industrial Microbiology deals with the study, utilization, and manipulation of those microorganisms capable of economically producing desirable substances or changes in substances, and the control of undesirable microorganisms. **Fermentation is a** process mediated by microorganisms in which a product of economic value is obtained. In metabolism, fermentation refers to energy generating processes where organic compound acts as both electron donor and acceptor. In context to industrial biotechnology, fermentation is defined as the process by which large quantities of cells are grown under aerobic or anaerobic conditions.

A. STAGES OF FERMENTATION

Industrial fermentation is comprised of two main stages. They are Upstream Processing (USP), and Downstream Processing (DSP).

Upstream Processing involves all factors and processes leading to the product and including the fermentation. It consists of three main areas. They are obtaining a suitable microorganism for the particular product, strain improvement to increase the productivity and yield, maintenance of strain purity, preparation of suitable inoculums and this is followed by Fermentation media and Fermentation Process.

Downstream Processing is the collective term for the processes that follows fermentation. It includes cell harvesting, cell disruption and product purification from cell extracts or the growth medium.

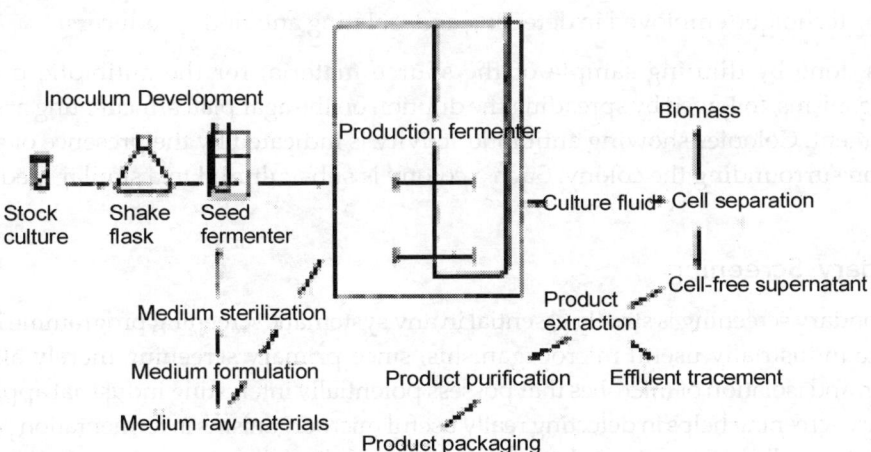

Figure 286 Stages of fermentation

B.   MAJOR CHARACTERS OF SUITABLE INDUSTRIAL MICROORGANISMS

The most important factor for the success of any fermentation industry is the microbial strain. It should have the following characteristics.

- ▲   High yielding strain.
- ▲   Stable biochemical characters.
- ▲   Should not produce undesirable substances.
- ▲   Easily cultivable on a large scale.
- ▲   Should be a stress tolerant strain.
- ▲   Utilization of a wide range of low-cost and readily available carbon sources
- ▲   Amenability to genetic manipulation
- ▲   Safety, non-pathogenicity and should not produce toxic agents, unless there is the target product.
- ▲   Ready harvesting from the fermentation.

C.   SCREENING OF INDUSTRIALLY IMPORTANT MICROORGANISMS

Suitable high yielding strain should be selected from the natural sources like soil. The process of microbial strain from the nature are called screening. There are two types of screening. They are primary screening and secondary screening.

Primary Screening

This consists of simple tests to isolate microbes from the nature with desirable property. It removes worthless microbes using simple fundamental criteria. It is performed by crowded plate technique, auxanography, enrichment culture techniques and use of an indicator dye. Primary screening of antibiotic producer is done by crowded plate technique. It is the simplest screening technique employed in detecting and isolating antibiotic producers.

It is done by diluting sample of the source material for the antibiotic producing microorganisms, followed by spreading the dilution on the agar plates, incubating at specified environment. Colonies showing antibiotic activity is indicated by the presence of a zone of inhibition surrounding the colony. Such a colony is sub- cultured to a similar medium and purified.

Secondary Screening

Secondary screening is strictly essential in any systematic screening programme intended to isolate industrially useful microorganisms, since primary screening merely allows the detection and isolation of microbes that possess potentially interesting industrial applications. Secondary screening helps in detecting really useful microorganisms in fermentation processes. This can be realized by a careful understanding of the following points associated with secondary screening:

1. It is very useful in sorting our microorganisms that have real commercial value from many isolates obtained during primary screening.

2. It provides information whether the product produced by a microorganism is a new one or not.

3. It gives an idea about the economic position of the fermentation process involving the use of a newly discovered culture.

4. It helps in providing information regarding the product yield potentials of different isolates.

5. It determines the optimum conditions for growth or accumulation of a product associated with a particular culture.

6. It provides information pertaining to the effect of different components of a medium.

7. It detects gross genetic instability in microbial cultures.

8. It gives information about the number of products produced in a single fermentation.

D. FERMENTATION MEDIA

Those components that support the growth of microbe and induce the product formation are called fermentation media. Submerged fermentation require liquid media. Solid-substrate fermentations requires solid or semisolid medium. Fermentation media must satisfy all the nutritional requirements of the microorganism and fulfill the technical objectives of the process. The nutrients should be formulated to promote the synthesis of the target product, either cell biomass or a specific metabolite. Media used in the cultivation of microorganisms must contain all elements in a form suitable for the synthesis of cell substance and for the production of metabolic products.

Media Requirements and Media Formulations

Most fermentations, except those involving solid substrates, require large quantities of water in which the medium is formulated. General media requirements include a carbon source, sources of nitrogen, phosphorus and sulphur. Minor and trace elements must also need, and some microorganisms require vitamins, such as biotin and riboflavin. Usually, media incorporate buffers, or the pH is controlled by acid and alkali additions. Antifoam agents also required.

The main factors that affect the final choice of individual raw materials are as follows.

1. Cost and availability: ideally, materials should be inexpensive, and of consistent quality and year round availability.

2. Ease of handling in solid or liquid forms, along with associated transport and storage costs, e.g. require-ments for temperature control.

3. Sterilization requirements and any potential denatu-ration problems.

4. Formulation, mixing, complexing and viscosity characteristics that may influence agitation, aeration and foaming during fermentation and downstream processing stages.

5. The concentration of target product attained, its rate of formation and yield per gram of substrate utilized.

6. The levels and range of impurities, and the potential for generating further undesired products during the process.

7. Overall health and safety implications.

Some of the Frequently used Substrates in Industrial Fermentation

In general, production media is used in the liquid state to falicitate the fermentation process. Each compound production needs individual medium. The production medium must have a suitable chemical composition. It should contains a source of carbon, nitrogen, growth factors and mineral salts. Medium should supply for precursor compound for the production of compound. Medium should have buffereing activity to maintain pH. Antifoaming agents are needed in the medium to avoid the problems of foaming. The raw materials required for designing of the production medium should be freely available in large quantities at reasonable price.

Mostly agricultural products are utilized as a source of raw materials. There are five major categories of raw materials. That are Saccharine materials (Pure glucose or sucrose, cane and beet Molasses, sugarcane juice, cheese); Starchy materials (cereals and root tubers), Cellulosic materials (Sulphite waste liquor, wood molassess, rice straw); Hydrocarbon and vegetable oils (Oleic acid, gasoil); Nitrogenous materials (corn steep liquor, soya bean meal, pharma media, distilled soluble, ammonium salts, urea, or gaseous ammonia, peptone, yeast extract). **Malt extract,** an aqueous extract of malted barley, is an excellent substrate for many fungi, yeasts, and actinomycetes. Dry malt extract consists of about 90-92% carbohydrates, and is composed of hexoses (glucose, fructose), disaccharides (maltose, sucrose), trisaccharides (maltotriose), and dextrins. Nitrogenous substances present in malt extract include proteins, peptides, amino acid, purines, pyrimidines, and vitamins. The amino acids composition of different malt extracts varies according to the grain used, but proline always makes up about 50% of the total amino acids present.

Sterilization of Media

Successful fermentation usually needs the sterilization of equipment, medium, air and subsequent maintanence of sterility of the whole system. Steam, UV, chemical agents are employed in the sterilization process. Fermenter is sterilized by steam. Boiling and steam under pressure are used to sterilize fermentation medium. Air can be sterilized by filteration, UV, heat and chemical agents.

E.   FERMENTOR

Design and Role of Different Parts of Fermentor

In fermentation industries, microbes are to be grown in specially designed vessels loaded with particular type of nutritive media. These vessels are referred to as **Fermentor** or **Bioreactors.** Fermentors are complicated in design, because they must provide for the control and observation of many facts of microbial growth and biosynthesis. The design of fermentor depends upon the purpose for which it is to be utilized. Industrial fermentors are designed to provide the best possible growth and biosynthesis conditions for industrially important microorganisms and allows ease of manipulation for all operations associated with the use of the fermentors. The fermentor used for a particular process should possess following characters:

Characteristics of an Ideal Fermentor or Bioreactor

There cannot be a fermentor ideal for almost all fermentation processes. It should have the following characteristics:

1. Material used in the fabrication of fermentor should be strong enough to withstand the interior pressure due to the fermentation media.

2. It should be resistant to corrosion and free from any toxic effect for the microbial culture.

3. A fermentor should permit easy control of contaminating microbes.

4. It should be provided with the inoculation point for aseptic transfer of inoculum.

5. Should be equipped with the aerating device (Spargers).

6. Should be equipped with a stirring device for uniform distribution of air, nutrients and microbes (Impellers).

7. There should be provision of baffles to avoid vortex formation.

8. Fermentor should be provided with a sampling valve for aseptic withdrawing of sample for different laboratory tests.

9. Fermentor should possess a device for controlling temperature.

10. Fermentor should be provided with pH controlling device for monitoring and maintaining pH.

11. Should be provided with a facility for intermittent addition of antifoam agents for controlling foam formation.

12. There should be provision for feeding certain media components during the progress of fermentation (Precursors).

13. A drain at the bottom is essential for the removal of the completed fermentation broth for further processing.

14. A main hole should be provided at the top of fermentor for access inside the fermentor for different purposes like repairing and thorough cleaning of fermentors between runs.

15. A exit valve should be provided at the top for the exit of metabolic gases produced during fermentation processes.

F. TYPES OF FERMENTOR

Batch Fermentor

Batch reactors are simplest type of reactor. Batch fementors are used to carry out microbiological processes on batch basis. They are available with varying capacities.

The capacity of the fermentor may range from a few hundred to several thousand gallons. They are of four types batch fermentor. They are as follows.

    i.   Small Laboratory fermentors

    ii.   Pilot plant fermentors

    iii.   Large industrial fermentors

    iv.   Horton spheres.

The small laboratory fermentor ranges from 1-2 liters with a maximum up to of 12 –15 liters. Pilot plant fermentors have a total volume of 25 –100 gallons upto 2000 gallons total volume. Larger fermentors range from 5,000 or 10,000 gallons of total volume to approximately 1,00,000 gallons. Horton spheres are rarely employed with a size range of 2,50,000 to 5,00,000 gallons total capacity. Actually the working volume in a fermentor is always less than that of the total volume. In other words, a 'head space' is left at the top of the fermentor above the level of fermentation media. The reason for keeping a head space is to allow aeration, splashing and foaming of the aqueous medium. This head space usually occupies a fifth to a quarter or more of the volume of the fermentor. Batch fermentation is carried out by the reactor is filled with medium and the fermentation is allowed to proceed. When the fermentation has finished the contents are emptied for downstream processing. The reactor is then cleaned, re-filled, re-inoculated and the fermentation process starts again.

Components of Batch Fermenter

pH Control   pH control is achieved by acid or alkali addition, which is controlled by an auto- titrator. The autotitrator in turn is connected to a pH probe.

Temperature control   It is achieved by a water jacket around the vessel. This is often supplemented by the use of internal coils, in order to provide sufficient heat-transfer surface.

Agitation   The agitating device consists of a strong and straight shaft to which impellors are fitted. An impeller, in turn consists of a circular disc to which blades are fitted with bolts. Different types of blades are available and are used according to the requirements.

Aeration   Usually, the aerating device consists of a pipe with minute holes, through which pressurized air escapes into the aqueous medium in the form of tiny air bubbles. This aeration device is called a "SPARGER". The size of the holes in a sparger ranges from 1/64 to 1/32 of an inch or larger.

Foam Control   Aeration and agitation of a liquid medium can cause the production of foam. This is particularly true for the media containing high levels of proteins or peptides. If the foam is not controlled, it will rise in the head space of the tank and be forced from the tank along with the exit valve. This condition often causes contamination of the fermentation from organisms picked up by breaking of some of the foam which then drains back into the tank. Excessive foaming also causes other problems for fermentation.

Types of Fermentation Process

Varieties of processes are followed in fermentation industry depends on the media used and type of operations. Based on the nature of media used the process of fermentation are classified as surface fermentation, submerged fermentation and solid state fermentation.

Surface Fermentations

It is a simplest process by which a culture is allowed to grow on the surface of the medium and allowed the fermentation process. It is a vat process. Eg. Vinegar Production.

Submerged Fermentations

In normal fermentation process the fermentation medium is in liquid condition and is also known as submerged fermentation. Eg. Enzyme Production.

Solid-substrate Fermentations

Solid substrate fermentation involves the growth of micro-organisms on solid, normally organic materials in the absence of free water. The substrates used are cereal grains, bran, legumes and lignocellulosic materials, such as straw, wood chippings, etc. Solid-substrate fermentations lack the sophisticated control mechanisms. Solid-substrate fermentations are the most suitable methods for the production of fungal products. Fungi do not form spores in submerged fermentations. Solid-substrate fermentations are normally involves multistep processing.

1. Pretreatment of a substrate that often requires mechanical, chemical or biological processing
2. Hydrolysis of polymeric substrates, e.g. polysaccharides and proteins -
3. Utilization of hydrolysis products
4. Separation and purification of end products

Based on the type of operation in liquid media, the process of fermentation is of three types they are Batch Fermentation, Continuous Fermentation and Fed batch fermentation.

Batch fermentation is a dynamic process that is never in a steady state. In batch processes the critical parameter is gas exchange or balance between respiration rate and oxygen transfer. In batch operation the sterilized media components are supplied at the beginning of the fermentation with no additional feed after inoculation. When cells are grown in a batch reactor, they go through a series of stages known as Lag phase, Exponential phase, Stationary phase and Death phase.

Continuous fermentations In this method fresh media is continuously added and bioreactor fluid is continuously removed. As a result, cells continuously receive fresh medium and products and waste products and cells are continuously removed for processing. The reactor can thus be operated for long periods of time without having to be shut down.

Fed batch fermentation is the most common type of process used in industry. In this process, fresh media is added continuously or periodically to the bioreactor. There is no continuous removal of product or medium / cells like continueous process. The fermenter is emptied or partially emptied when reactor is full or fermentation is finished.

G. THE MAJOR TYPES OF BIOREACTORS USED IN SUBMERGED PROCESS

Stirred Tank Reactors

In these reactors, mechanical stirrers (using impellers) are used to mix the reactor to distribute heat and materials (such as oxygen and substrates) Bubble column reactors These are tall reactors which use air alone to mix the contents

Air lift Reactors

These reactors are similar to bubble column reactors, but differ by the fact that they contain a draft tube. The draft tube is typically an inner tube which improves circulation and oxygen transfer and equalizes shear forces in the reactor.

Fluidized Bed Rreactors

In fluidized bed reactors, cells are "immobilized" small particles which move with the fluid. The small particles create a large surface area for cells to stick to and enable a high rate of transfer of oxygen and nutrients to the cells

Packed Bed Reactors

In packed bed reactors, cells are immobilized on large particles. These particles do not move with the liquid. Packed bed reactors are simple to construct and operate but can suffer from blockages and from poor oxygen transfer.

Flocculated Cell Reactors

Flocculated cell reactors retain cells by allow them to flocculate. These reactors are used mainly in wastewater treatment.

Stirred Tank Bioreactors (STB)

Stirred tank reactor is the simplest type of reactor. It is composed of a reactor and a mixer such as a stirrer, a turbine wing or a propeller. This reactor is useful for substrate solutions of high viscosity and for immobilized enzymes with relatively low activity.

H. BIOREACTORS USED FOR SOLID-SUBSTRATE FERMENTATIONS

Most solid-substrate fermentations are batch processes.

1. **Rotating drum fermenters,** comprising a cylindrical vessel of around 100 L capacities mounted on its side onto rollers that both support and rotate the vessel. These fermenters

are used in enzyme and micro-bial-biomass production. Their main disadvantage is that the drum is filled to only 30% capacity, otherwise mixing is inefficient.

2. **Tray fermenters**, which are used extensively for the production of fermented oriental foods and enzymes. Their substrates are spread onto each tray to a depth of only a few centimeters and then stacked in a chamber through which humidified air is circulated. These systems require numerous trays and large volume incubation chambers of up to 150-m3 capacities.

3. **Bed systems** consisting of a bed of substrate up to 1 m deep, through which humidified air is continuously forced from below.

4. **Column bioreactors**, consisting of a glass or plastic column, into which the solid substrate is loosely packed, surrounded by a jacket that provides a means of temperature control. These vessels are used to produce organic acids, ethanol and biomass.

5. **Fluidized bed reactors**, which provide continuous agitation with forced air to prevent adhesion and aggregation of substrate particles. These systems have been particularly useful for biomass production for animal feed.

Anaerobic Reactors

Anaerobic reactors are generally used for the production of methane rich biogas from manure (human and animal) and crop residues. They utilise mixed methanogenic bacterial cultures which are characterised by defined optimal temperature ranges for growth. These mixed cultures allow digesters to be operated over a wide temperature range i.e. above 0°C up to 60°C.

I. DOWNSTREAM PROCESSING

The Choice of Recovery Product is based on the Following Criteria

1. The intracellular or extra cellular location of the product
2. The concentration of the product in the fermentation broth
3. The physical and chemical properties of the desired product.
4. The intended use of the product
5. The minimal acceptable standard of purity
6. The magnitude of biohazard of the product or broth
7. The impurities in the fermentation broth
8. The marketable prize for the product Precipitation

The following methods are adopted to carryout the down stream processing of fermentation products.

1. Harvesting of Microbial Cells

Microbial cells are normally separated from the fermentation broth by filtration or centrifugation. Because of the small size of many microbial cells it will be necessary to consider the use of filter aids to improve filtration rates, while heat and flocculation treatments are employed as techniques for increasing sedimentation rates in centrifugation.

Types of Filters used to filter microbial cultures are

1. Plate and Frame Filters
2. Pressure Leaf Filters
3. Rotary Vacuum Filters
4. Cross Flow Filtration (Tangential Filtration)

2. Centrifugation

Microorganisms and other similar sized particles can be removed from a broth by using a centrifuge when filtration is not a satisfactory separation method. Although expensive when compared with a filter. It may be essential when:

1. Filtration is slow and difficult.
2. The cells or other suspended matter must be obtained free of filter aids.
3. Continuous separation to a high standard of hygiene is required.

The Tubular-bowl Centrifuge and The Multichamber Centrifuge are used for centrifugation.

3. Cell Aggregation and Flocculation

In industrial fermentation, the process of fermentation is commonly followed by addition of flocculating agents to the broth to aid de-watering. The use of flocculating agents is widely practiced in the effluent-treatment industries for the removal of microbial cells and suspended colloidal matter. It is well known that aggregates of microbial cells, although they have the same density as the individual cells, will sediment faster because of the increased diameter of the particles (Stokes law). This sedimentation process may be achieved naturally with selected strain of brewing yeasts, particularly if the wort is chilled at the end of fermentation, and leads to a natural clearing of the beer.

4. Cell Disruption

Microorganisms are protected by extremely tough cell walls. In order to release their cellular contents a number of methods for cell disintegration have been developed. Any potential method of disruption must ensure that labile materials are not denatured by the process or hydrolyzed by enzymes present in the cell. Methods available for cell disruption. They are fall into two categories:

A. Physico-mechanical methods
   a. Liquid shear.
   b. Solid shear.
   c. Agitation with abrasives.
   d. Freeze-thawing.
   e. Ultrasonication.

    B.   Chemical methods

        a.   Detergents.

        b.   Osmotic shock.

        c.   Alkali treatment.

        d.   Enzyme treatment.

5.   Precipitation

Precipitation may be conducted at various stages of the product recovery process. It is a useful process in that it allows enrichment and concentration in one step, thereby reducing the volume of material for further processing. It is possible to obtain some products directly from the broth by precipitation, or to use the technique after a crude cell lysate has been obtained.

6.   Liquid-Liquid Extraction

The separation of a component from a liquid mixture by treatment with a solvent in which the desired component is soluble is known as liquid-liquid extraction. The specific requirement is that a high percentage extraction of product must be obtained but concentrated, in a smaller volume of solvent.

7.   Solvent Recovery

A major component in an extraction process is the solvent recovery, which is usually a distillation unit.

Distillation may be achieved in three stages:

1.   Evaporation, the removal of solvent as a vapour from a solution.
2.   Vapour-liquid separation in a column, to sepa-rate the lower boiling more volatile component from other less volatile components.
3.   Condensation of the vapour to recover the more volatile solvent fraction.

8.   Evaporation

A wide range of evaporators is available. Some are operated on a batch basis and others continuously. Most industrial evaporators employ tubular heating surfaces. In batch distillation the vapour from the boiler passes up the column and is condensed. Part of the condensate will be returned as the reflux for counter-current contact with the rising vapour in the column. The distillation is continued until a satisfactory recovery of the lower-boiling (more volatile) component(s) has been accomplished.

9.   Chromatography

Chromatographic techniques are used to isolate and purify relatively low concentrations of products using different solutes. Depending on the mechanism by which the solutes may be differentially held in a column ,the techniques can be grouped as follows:

1. Adsorption chromatography
2. Ion –exchange chromatography
3. Gel-permeation chromatography
4. Affinity chromatography
5. Reverse phase chromatography
6. High performance liquid chromatography

Chromatographic techniques are also used in the final stages of purification of a number of products.

10. Membrane Processes

Membrane processes include two major processes:

    A. Ultrafiltration
    B. Reverse Osmosis

Both processes utilize semi-permeable membranes to separate molecules of different sizes.

11. Drying

The drying of any product is often the last stage of a manufacturing process. It involves the final removal of water from a heat-sensitive material ensuring that there is minimum loss in viability, activity or nutritional value. Drying is undertaken because:

    a. The cost of transport can be reduced.
    b. The material is easier to handle and package.
    c. The material can be stored more conveniently in the dry state.

Contact driers, Spray drier, Freeze drying and Fluidized bed driers are commonly used in fermentation industry.

12. Crystallization

Crystallization is an established method used in the initial recovery of organic acids and amino acids, and more widely used for final purification of a diverse range of compounds. In citric acid production, the filtered broth is treated with $Ca(OH)_2$ so that the relatively insoluble calcium citrate crystals will be precipitated from solution. The calcium citrate is filtered off and treated with sulphuric acid to precipitate the calcium as the insoluble sulphate and release the citric acid. After clarification with active carbon, the aqueous citric acid is evaporated to the point of crystallization. Crystallization is also used in the recovery of amino acids.

J. ETHANOL PRODUCTION

Ethanol is one of the most important organic solvent. It has been used as a solvent, germicide, antifreeze, fuel and chemical raw material.

Microorganisms Involved in Alcohol Production

Yeast is usually used in alcoholic fermentation. *Saccharomyces cerevisae, Saccharomyces ellypriodes, Saccharomyces carlbergensis, Saccharomyces fragillis, Saccharomyces oviformis, Saccharomyces saki* are the yeast used as a alcoholic fermenter. Yeast converts simple sugar into ethanol. Among these *Saccharomyces cerevisae* is the yeast strain used widely in alcoholic fermentation because of having many advantages than others. That are as follows.

1. High ethanol yield.
2. Osmotolerance.
3. High growth & fermentation rate.
4. High temperature fermentation and Low pH fermentation optimum.
5. Ethanol and glucose tolerance.
6. General hardness under physical and chemical stress.

Biochemistry of Ethanol Production

Enzymes in the yeast first convert carbohydrates like maltose or sucrose into glucose and fructose and then convert these in turn into ethanol and carbon dioxide.

$$\text{Sucrose} \xrightarrow{\text{Invertase}} \text{Glucose} \xrightarrow[\text{Pathway}]{\text{EMB}} \text{Pyurvic acid}$$

$$\downarrow \text{ED pathway}$$

$$CO_2 + \text{ethanol} \longleftarrow \text{Acetaldehyde}$$

Raw Materials used for Alcohol Production

Raw materials for fermentative ethanol production can be divided into three types based on the nature of carbohydrate. They are saccharine materials, starchy materials and cellulosic materials.

A. Saccharine Materials

Sugar cane or sugar beet juice, high test molasses, black-strap molasses, fruit pulp and juice wastes, cane sorghum and whey.

B. Starchy Materials

It includes cereal grains, starch root plants. Starch is first hydrated and gelatinized by milling and cooking, and then broken down to fermentable sugars by diastatic enzymes or weak acids. Important cereal grains are corn, wheat, rice, barley and sorghum.

C. Cellulosic Raw Materials

It included the straw components of Corn, Wheat and Rice. Sulfite waste liquor is the residue of partial wood hydrolysis from paper pulp production.

Fermentation Media

Most industries uses cane molasses as raw material. Cane Molasses. It is the waste produced in the sugar mills. It is a dark brown colour viscous syrupy liquid. The molasses contains reducing sugar up to 50%. It may be diluted to obtain optimal concentration (10-18%). Urea is added to fulfill the requirement of nitrogen for good fermentation. Sulphuric acid is added to bring down the pH value to an optimum level (4.5 - 4.9). To prevent the contamination an antibiotic Benzyl Penicillin is used.

Preparation of the Inoculum

Inoculum production is a critical stage in an industrial fermentation process. Inoculum is prepared as a stepwise sequence employing increasing volumes of media. Initial inoculum was prepared in 100ml Erlenmeyer flasks with 50 ml of medium. The medium is inoculated with loopful of conserved stock culture and incubated at 30°C for 30 h, without mechanical agitation. Following to the initial step approximately 0.5 to 5% inoculum by volume is transferred and the volume of inoculums is increased stepwise as 100ml, 250ml, 1000ml, 4000ml (Shaker flask) in the sequence. Microbial cells from the shaker flask can be used as seed culture which are then added to a small fermenter and allowed to grow for 1-2 days. This stimulates conditions that exist in the larger fermenter to be used for production of metabolites. Inoculum media usually are balanced for rapid cell growth and not for product formation. The inoculum level introduced into a production tank also usually is in the 0.5 to 5%, but at times it may be as high as 20%. Finally, the microorganisms are transferred to the main fermentation vessel containing essential media and nutrients.

Fermentation Process

Ethanol is produced by continuous fermentation method using submerged process. 10% inoculums is added to the fermentation media and carried out fermentation for several days. Alcohol production starts within 12 hours of inoculation of seed culture. During fermentation the following conditions are maintained.

Fermentation Conditions

Temperature during fermentation is 28-34⁰C. Settling gravity of yeast vessels 1.045.

Settling pH of yeast vessels 4.5 - 4.9. Duration of fermentation 24 hours. Percentage of alcohol after completion of fermentation is 5-8%. Intermittent agitation is required for maintaining uniform temperature.

Recovery

After the completion of fermentation, fermented liquid is allowed to settle for few hours and the cells are separated to get biomass of yeast. The yeast is used as SCP for animal feed. The supernatant is processed for recovery of ethanol by distilled analysis and rectifier column to obtain rectified spirit. The ethanol is separated from the mixture by fractional distillation to give 96% pure ethanol. It is impossible to remove the last 4% of water by fractional distillation.

K. PENICILLIN PRODUCTION

Penicillin is a group of antibiotics. They include penicillin G, procaine penicillin, benzathine penicillin and penicillin V. Penicillin antibiotics are historically significant because they are the first drugs that were effective against *Treponema pallidum, Staphylococci* and *Streptococci.* All penicillins are β-lactam antibiotics and are used in the treatment of bacterial infections caused by Gram-positive organisms.

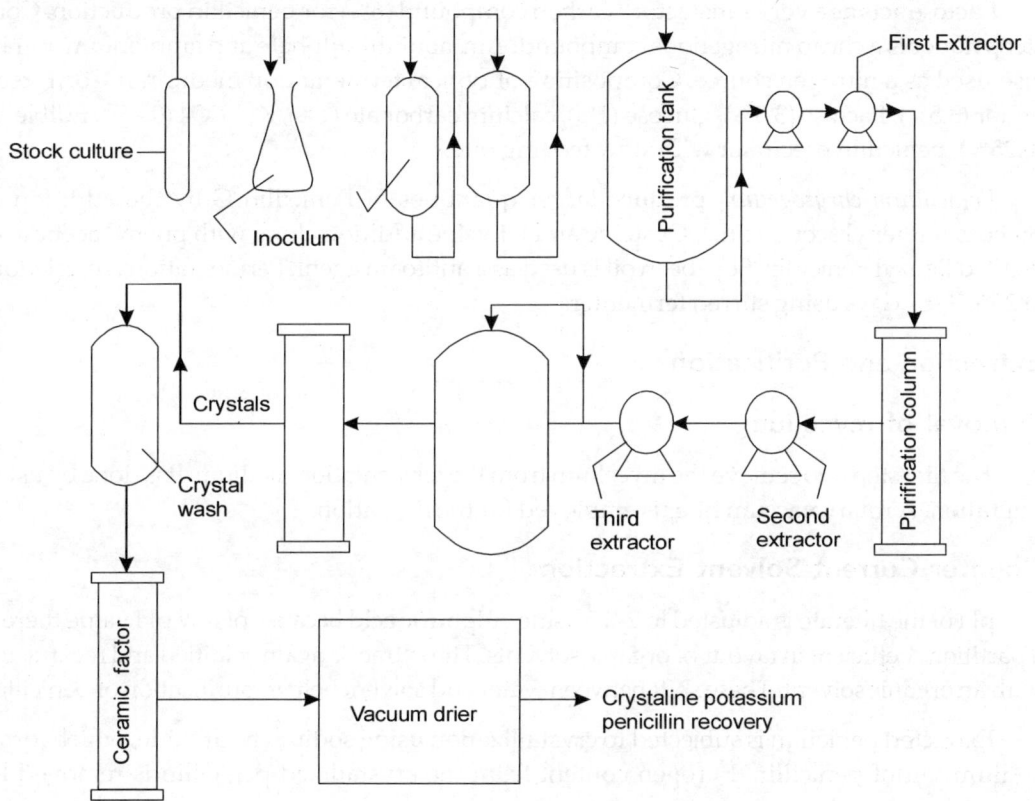

Figure 287   Process of penicillin production

Microorganism

Penicillin is derived from the fungi belongs to the genus Penicillium. *Penicillium chrysogenum* is one of the high yielding strain used for the commercial penicillin production.

Inoculum Preparation

Sporulation media is used for inoculum preparation. This media contains Cane molasses (7.5g/L), glycerol (7.5 g/L), corn steep liquor (2.5 g/L), $MgSO_4 . 7H_2O$ (0.050g/L), KH_2PO_4 (0.060g/L), peptone (5 g/L), NaCl (4g/L), Fe tartarate (0.005g/L), $CuSO_4 . 5H_2O$ (0.004g/L), Agar (2.5 g/L). Multiple stage inoculation is performed to obtain required volume of inoculums. The inoculum is prepared under aerobic condition. Best growth of the fungi is obtained using stirred fermentor.

Inoculation

Fermentation medium is seeded with uniform suspension of nongerminated spore. Uniform suspension of spore is prepared using 1: 10,000 sodium lauryl sulfonate. It is a non toxic wetting agent.

Fermentation Process

Lactose acts as a very satisfactory carbon compound (6%) for penicillin production. Corn steep liquor is a cheap nitrogenous compound. Ammonium sulphate and ammonium acetate also used as a nitrogen source. Composition of typical fermentation medium is Corn steep liquor (3.5%), Lactose (3.5%), glucose (1%), calcium carbonate (1%), KH_2PO4 (0.4%), Edible oil (0.25%), penicillin precursor with slow feeding rate.

Penicillium chrysogenum produce larger quantities of Penicillin G by the addition of precursor Phenyl acetic acid. L Cysteine and L Lysine addition along with phenyl acetic acid yielded Benzyl Penicillin. Soyabean oil is used as a antifoam agent. Fermentation is carriedout at 25°C for 7 days using stirred fermenter.

Extraction and Purification

Removal of mycelium

The first step is to remove the mycelium from the fermentation medium. It is done by using filteration. A rotary vaccum filter is employed for the filteration.

Counter Current Solvent Extraction

pH of the filterate is adjusted to 2-2.5 using sulphuric acid because of low pH value there is a partition coefficient in favour of organic solvents. The extract is again acidified and reextracted with an organic solvent. These shift between water and solvent help in purification of penicillin.

Extracted penicillin is subjected to crystallization using sodium hydroxide, which forms sodium salt of penicillin. Pyrogen content from the crystallized penicillin is removed by charcoal treatment.

L. CITRIC ACID

Citric acid or 2-hydroxy-1, 2, 3-propane tricarboxylic acid was first isolated from lemon juice by Scheele. Citric acid is produced either in the anhydrous form or as the monohydrate. Citric acid is the most versatile and widely used food acidulent. It acts as a chelating agent. It acts as plasticizers in the production of carbonated beverages. Today most of the citric acid is obtained from microbial fermentation process. Citric acid is produced in two processes. They are surface culture process and submerged culture process.

Microorganisms

Aspergillus niger, A. clavatus, P. luteus, Penicillum citrinum, Mucor piriforms are used in industrial citric acid producer. Among these *Aspergillus niger* is used as a best strain due to its higher efficiency and uniform biochemical properties. It is easily cultivated and it produce only negligible amount of oxalic acid.

Surface Culture Process

Culture is inoculated across the surface of the production medium. Culture remains on the surface throughout the fermentation. It is a stationary fermentation process.

Shallow pans are used for inoculums preparation. Spores are inoculated on sporulation medium and incubated at 25°C for 4 to 14 days. Suspension of spore is obtained by suspending the grown spores in a suitable diluent. Spores of inoculums are added to the production medium so as to keep them on the surface.

Medium used for citric acid production should contains Sucrose (140g/l), ammonium nitrate (140g/l), potassium dihydrogen phosphate (140g/l), magnecium sulphate (140g/l), hydrochloric acid to pH 3.4 to 3.5. Optimum temperature is 26-28°C. It is necessary to supply air to the surface of the seeded medium. *Aspergillus niger* produce 60-80gms of citric acid per 100gms of incorporated sugar.

Submerged Fermentation

Submerged fermentation is carried out using beet molasses or glucose syrup as a substrate. Black strap molasses also used as a source of carbon. Remaining components are similar to surface fermentation. *Candida lipolytica* is an active strain used in this method. Larger sterilizable fermenter is used, which is equipped with aerator and agitator.

Recovery

The fermentation medium is drained off to separate the mycelium through filteration. Intracellular citric acid present in the mycelium is obtained by pressing the mycelial mat. Recovered fermented liquid is treated with milk of lime forming the precipitate of calcium citrate. Calcium citrate is treated with an sulphuric acid to liberate citric acid leaving the precipitate of calcium sulphate. Impure solution of citric acid is submitted to discolouration using activated charcoal. Finally pure citric acid solution is evaporated and crystallized.

M. LACTIC ACID

Lactic acid is also known as milk acid. It plays a role in various biochemical processes. It was first isolated in 1780 by Carl Wilhelm Scheele. Lactic acid is employed in pharmaceutical technology to produce water-soluble lactates. It finds further use in topical preparations and cosmetics to adjust acidity and for its disinfectant and keratolytic properties. It is found primarily in sour milk products, such as koumiss, laban, yogurt, kefir and some cottage cheeses. The casein in fermented milk is coagulated (curdled) by lactic acid. It acts as a food additive. Lactic acid is also used in a wide range of food applications such as bakery products, beverages, meat products, confectionery, dairy products, salads, dressings, ready meals, etc. Lactic acid in food products usually serves as either as a pH regulator or as a preservative. It is also used as a flavouring agent. Wooden fermenter are used for lactic acid production.

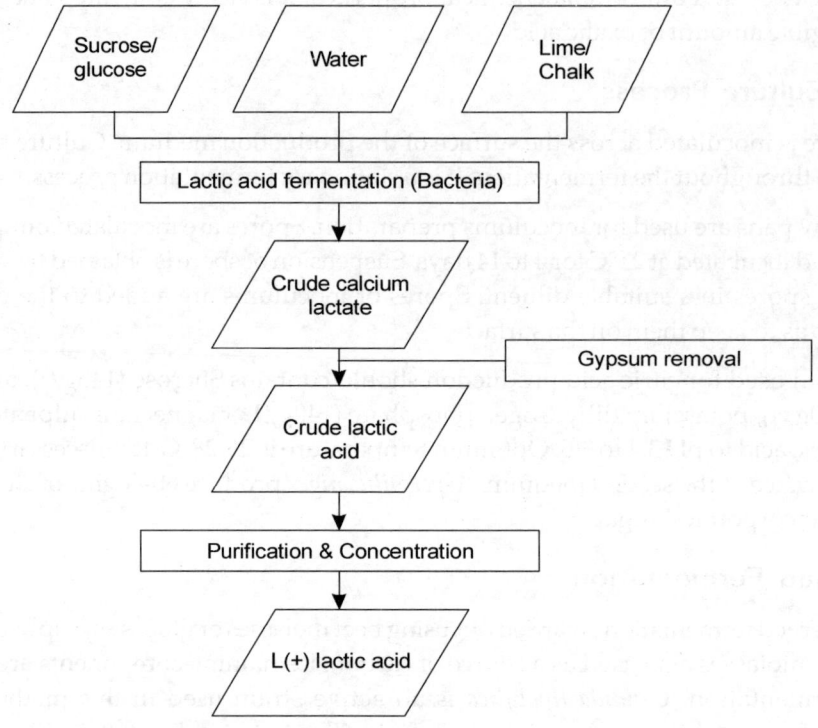

Figure 288 Production and recovery of Lactic acid

Lactobacillus bulgaricus, L. delbrueckii, L. leichmannii, L. casei, Streptococcus lactis, Rhizopus oryzae are the straisns used in lactic acid production. Among these strains *Lactobacillus bulgaricus* is the most preferred industrial strain. Other bacteria which produce lactic acid include *Leuconostoc mesenteroides, Pediococcus cerevisiae, Streptococcus lactis, Bifidobacterium bifidus.*

Lactic acid can be produced naturally or synthetically. Commercial lactic acid is produced naturally by fermentation of carbohydrates such as glucose, sucrose, or lactose. The culture medium contains semi refined sugar (molassess and whey), maltose, lactose, sucrose, calcium carbonate with ammonium hydrogen phosphate.

The natural production process is shown in the figure 288. The raw materials is fermented in a fermenter with the addition of lime and crude calcium lactate is formed. The gypsum is separated from the crude calcium lactate, which results in crude lactic acid. The crude lactic acid is purified and concentrated and L (+) lactic acid.

Inoculum is developed from the stock cultures of *Lactobacillus* and incubate at 45-55⁰C. Each stage inoculums development requires 16-18hours of incubation. 5% of inoculums are added to the fermentation medium. Submerged fermentation process is carried out. Other conditions maintained during fermentation are as follows

   i.   pH – 5.5 – 6.5
   ii.   Temperature – 45-50°C.

iii.   Supply of air is not necessary.

iv.   Medium is stirred to keep the calcium carbonate in suspension.

v.   Duration of fermentation is 5-10days.

vi.   Yield -93-95%

Recovery

Calcium carbonate is added to the fermentation medim and make the pH as 10. Filter the fermentation medium using precoat filters. Filterate is acidified with sulphuric acid for the removal of calcium from the broth. Calcium is precipitated as calcium sulphate.Treat the filtrate with activated carbon to remove organic impurities. Sodium ferrocyanide is used to remove heavy metals by precipitation. Concentration of lactic acid is by evaporation.

N.   ACETIC ACID (Vinegar)

Acetic acid is an organic compound with the chemical formula CH_3COOH. Acetic acid is also called as *vinegar*. Vinegar means sour wine. It is a colourless liquid. When undiluted is also called *glacial acetic acid*. Acetic acid is the main component of vinegar (apart from water). Vinegar is used directly as a condiment, and in the pickling of vegetables and other foods.

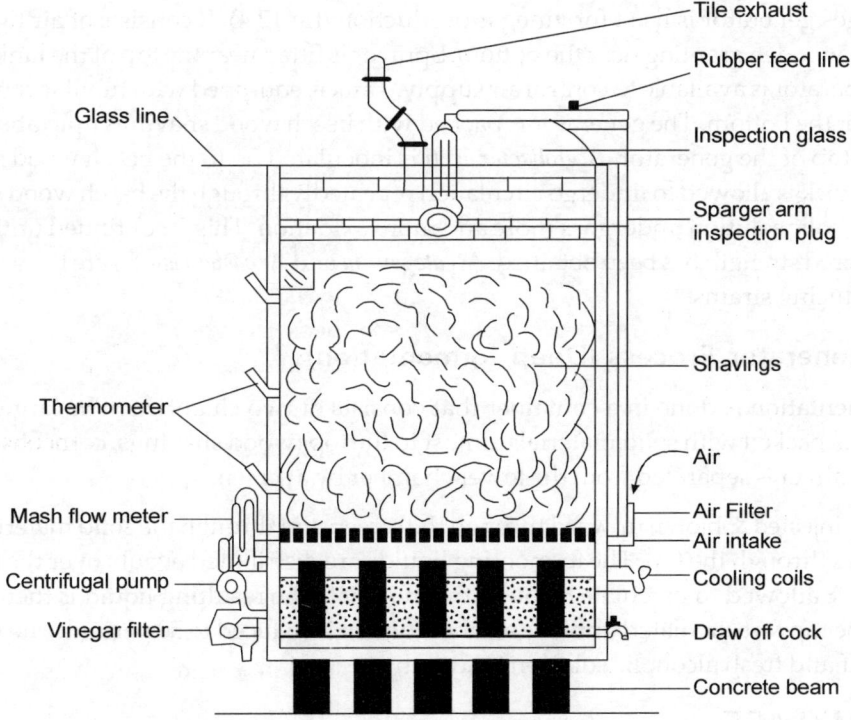

Figure 289   Vinegar production

There are two stages of vinegar production. They are alcoholic fermentation and acetic acid fermemtation

Alcoholic Fermentation

Sugar containing materials are fermented to ethylalcohol using *Saccharomyces cerevisiae*. 1g of glucose yield0.5114g of ethyl alcohol.

Acetic Acid Fermentation

Acetic acid bacteria is a obligatory aerobic bacteria. It is called as *Acetobacter aceti*. These bacteria convert alcohol to acetic acid. There are two types of vinegar production. They are slow and rapid process. Home making vinegar and Orleans process are the examples for slow process. The generator system and submerged fermentation system (bubble method) are good examples for rapid process.

Industrially acetic acid is produced in three methods. They are as follows.

  i.   Orleans process – A bacteria grow very slowly as a film over the surface and produced culinary vinegar.

  ii. Quick vinegar process

  iii. Rapid-generator process

Fringes generator is used for vinegar production (fig 12.4). It consists of air tight tanks. There is a wooden granting near the bottom. Sprayer is fitted near the top of the tank. Vapour liquid separator is available to control air supply. Tank is equipped with tubular cooling coils situated at the bottom. The generator is packed with beech wood shavings upto about 15 feet from the top of the generator. *Acetobacter aceti* is inoculated on to the beech wood shavings. Vinegar stock is allowed to undergo circulation repeatedly through the beech wood shavings. Thus alcoholic solution undergoes more and more oxidation. This is continued until vinegar of the desired strength has been obtained. *A. categenum* and *A. pasteurianum* are the better acetic acid producing strains.

Rapid-generator Process (Deep fermentation)

Fermentation is done in a container that consists of two chambers. The larger (upper) chamber is packed with solid materials almost to the top (wood shavings, corncobs etc.). The upper chamber is separated from the lower chamber by a screen.

Air is injected & blown upward through the screen and through the solid materials, & the air escapes through the top. The fermenting liquids are distributed evenly over the top of the material, & allowed to percolate through the material. The resulting liquid is then pumped back to the top & recirculated until the alcohol content is reduced to 50 percent. The vinegar is drawn off and fresh alcoholic solution is added.

O. AMYLASE

Amylases play a most important role in food technology especially bread making, beer making. Amylases are enzymes, which hydrolyse starch and glycogen by breaking 1, 4 glycosidic linkages. There are two types of amylases. They are α amylases and β amylases. α amylases are endo enzymes. They attack all linkages of starch. β amylases hydrolyze starch and other amylases by splitting off maltose molecules

Production Strains

Fungi - *Aspergillus niger* and *A. oryzae*.

Bacteria - *Bacillus amyloliquefaciens* and *Bacillus licheniformis*.

Medium

1. Fugal amylases are produced by surface culture system using corn starch containg media. Composition of the media are corn starch (24g/L), corn steep liquor (36g/L), KCl (0.2g/L), Na_2HPO_4 (4.7g/L), calcium chloride (1g/L), $MgCl_2$ (0.2g/L).

2. Bacterial amylases are produced by submerged culture using the soyabean meal media. Components of the media are ground soya bean (1.85%). Autolysed brewers yeast (1.5%), distillers dried soluble (0.76%), N-Z amines (0.65%), lactose (4.75%), $MgSO_4$ (0.64%), Antifoam (0.05%). 10% Inoculum is inoculated on fermentation medium and maintain the following fermentation conditions.

   i.   Temperature : 30-40°C

   ii. pH – 7 for bacteria and 4.5-5.5 for fungus.

   iii. Calcium carbonate used as a buffer.

   iv. Required inoculums 10^9 – 10^{10}cells/ml

   v.   Time 100-150hours.

After fermentation amylase containing medium is filtered and subjected to down stream processing.

P.   L GLUTAMIC ACID

Glutamic acid is one of the 20-22 proteinogenic amino acids. It is a non-essential amino acid. Glutamate is an important neurotransmitter that plays a key role in long-term potentiation and is important for learning and memory. It is produced by hydrolysis of wheat gluten, soya bean cake by the cleavage of pyrazolidone carboxylic acid.

Glutamic acid was isolated from wheat gluten hydralyzate by Ritthousen in 1866. Wolff carried out the first chemical synthesis of glutamic acid in 1890. *Corynebacterium glutamicum* was discovered by Kinoshita et.al. in 1957 which produces 30 g/L of L-glutamic acid in glucose medium

The four most widely reported bacteria for glutamic acid production are as follows,

   i.   *Corynebacterium* sp. (*C. glutamicum; C. lilum*)

   ii. *Brevibacterium* sp. (*B. divericartum: B. alanicum*)

   iii. *Microbacterium* sp. (*M. flavum* var. *glutamicum*)

   iv. *Arthrobacter* sp. (*A. globiformis; A. aminofaciens*). Among these preferred organism is *Corynebacterium glutamicum*

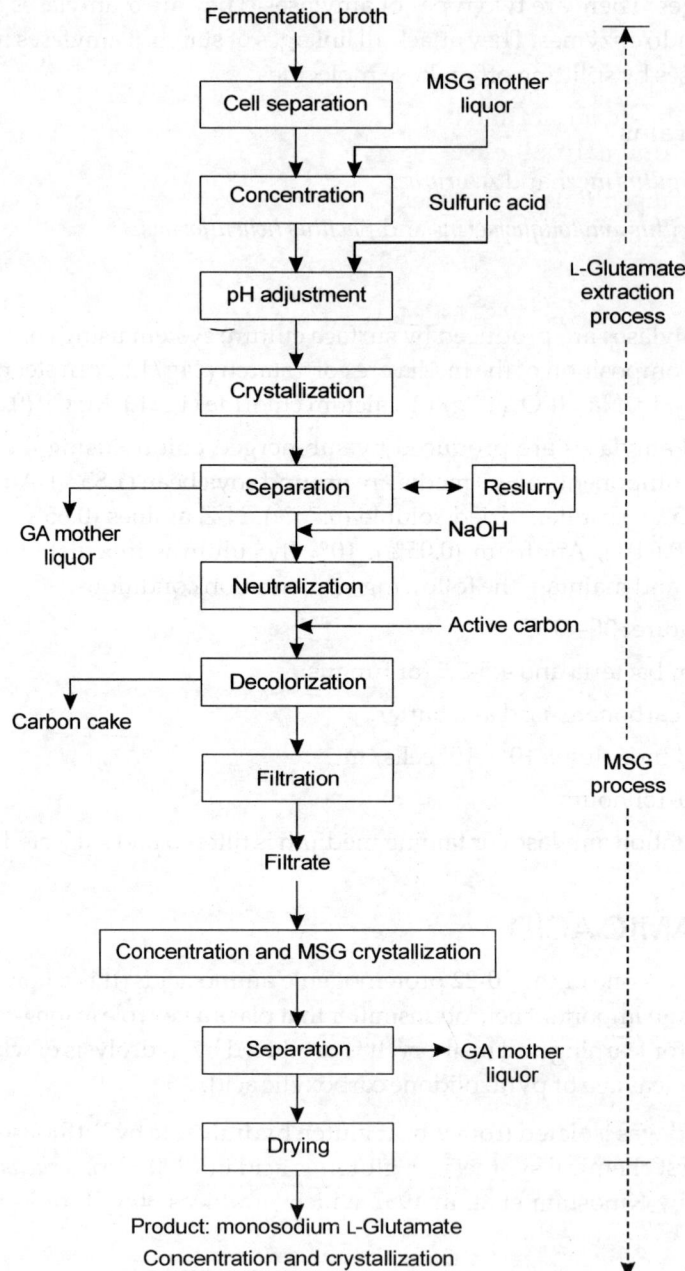

Figure 290 Production and recovery of L-Glutamate

The percentage composition of a medium are Glucose, 10%; Corn steep liquor 0.25%; Enzymatic casein hydrolysate 0.25%; K_2HPO_4 0.1%, $MgSO_4.7H_2O$, 0.25%; Urea, 0.5%. Hydrocarbons have served as carbon sources for glutamic acid production. The optimal temperature is 30° to 35° and a high degree of aeration is necessary.

Cane molasses, beet molasses, and corn syrup are mainly used as the carbon source because of their low cost

Glutamic acid production is greatest when biotin is limited in the medium. When biotin is optimal, growth is luxuriant and lactic acid is excreted. The presence of biotin in the range of 5 to 10 Ug/L is optimal for the excretion of L-glutamic.

Inoculate 6% inoculums for best yield. pH of the medium is set at 8.5 with ammonia and maintained at 7.8 during fermentation. Carryout fermentation upto 35 hours. Glutamic acid yield is 96g/L after 35 hours.

Down stream process is carriedout as per illustration given in Figure 290.

Q. VITAMIN B$_{12}$

Vitamin B$_{12}$ is also called cobalamin. It is a water-soluble vitamin with a key role in the normal functioning of the brain, nervous system and for the formation of blood. It is one of the B vitamin. It is produced industrially only through bacterial fermentation. *Acetobacterium, Aerobacter, Agrobacterium, Alcaligenes, Azotobacter, Bacillus, Clostridium, Corynebacterium, Flavobacterium, Micromonospora, Mycobacterium, Nocardia, Propionibacterium, Protaminobacter, Proteus, Pseudomonas, Rhizobium, Salmonella, Serratia, Streptomyces, Streptococcus* and *Xanthomonas* are able to synthesis vitamin B$_{12}$.

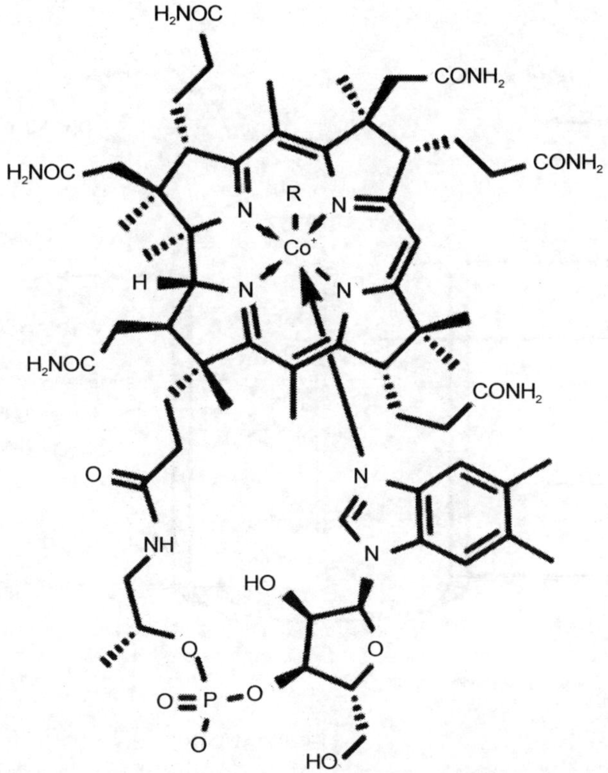

Figure 291   Cyanocobalamine structure

Streptomyces griseus is used to produce vitamin B_{12} in a submerged culture system. Aeration and agitation are essential for vitamin B_{12} production. Time required for the completion of fermentation is 3-5 days. Pure culture is inoculated on 250ml of seed medium. Composition of seed media are Yeast extract (1g/1), beef extract (1g/1), N-Z amine (2g/1), glucose (10g/1), agar (15g/1). 4 liter of inoculums is to be prepared and inoculate 5% inoculums to the **fermentation medium.** Production medium should contain distilled soluble (4%), dextrose (0.5%), calcium carbonate (0.5%), calcium chloride (1-5ppm), pH -7. Cobalt is needed to attain maximum yield. Temperature of incubation is 26.60°C. Optimum rate of aeration is about 0.5volume air/volume media/minutes. Air is sterilized using activated charcoal. Antifoam agent used are soyabean oil, corn oil, lard oil. Yield of vitamin B_{12} fermentation is 1-2mg/L.

Recovery

Most of the cobalamine is associated with mycelium. Heating the mixture to boiling at pH5 or below liberates the cobalamine. Filteration removes mycelium. Filtered broth was treated with cyanide to bring out the conversion of cobalamine to cyanocobalamine. Solution **382** *Microbiology* is passed through a column of adsorbents like activated charcoal, bentonite. Cyanocobalamine is eluted from adsorbent is accomblished by the use of an aqueous solution of materials ranging from organic bases to hydrochloric acid. Finally fermented broths are evaporated to dryness or spray dried.

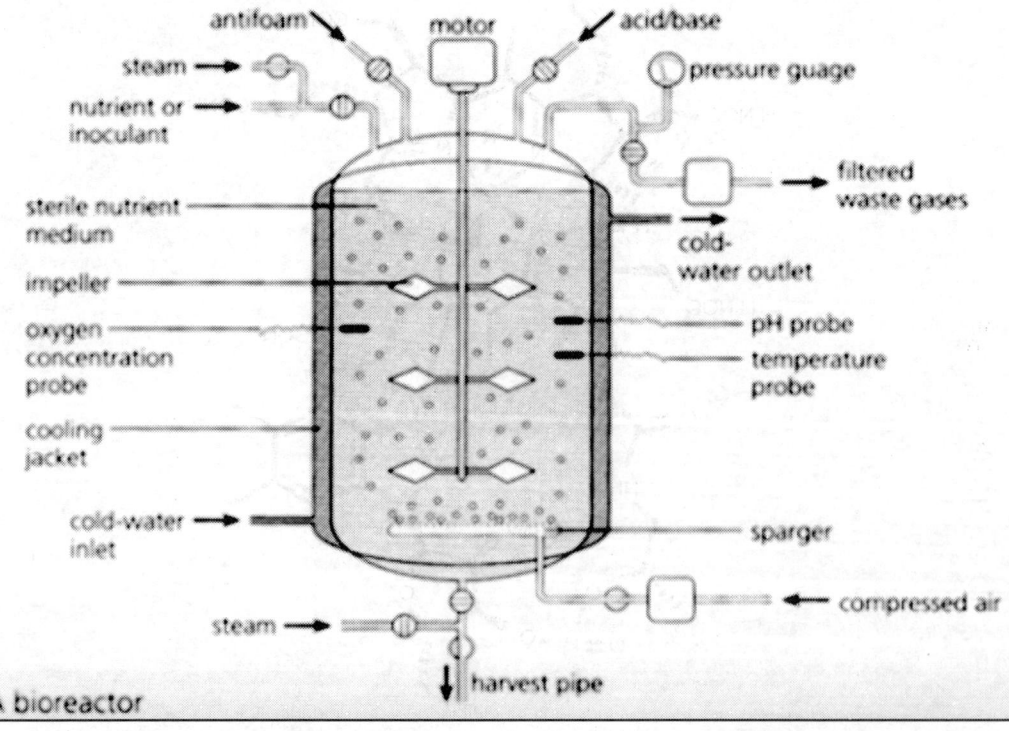

A bioreactor

16

MICROBIAL GENETICS AND MOLECULAR BIOLOGY

A. DNA AS GENETIC MATERIAL

The extensive chemical analysis of chromosomes of different organisms revealed that chromosome contains protein and nucleic acids (DNA or RNA). It was thought that genes that determine heridity might have either protein or nucleic acid. Earlymolecular geneticists had assigned the informational roles of genes to the chromosomal proteins because, they found nucleic acids too simple to carry genetic information. The controversy about the assignment of genetic role either to chromosomal protein or chromosomal DNA existed upto 1949. In 1949 A. Mirsky and H.Ris had found that all cells of an organism appeared to contain the same amount of DNA, whereas different cell types contained quite different amounts and kinds of proteins. It is constancy, therefore, favoured DNA as the genetic material. In 1953, it was universally accepted that DNA is the genetic material of micro organisms (except few viruses) and higher organisms. The following evidences support the DNA as genetic materials.

Griffith's Transformation Experiment

The process of genetic transformation was demonstrated by Frederic Griffith in 1928. This experiment proved DNA as a genetic material and was done by making use of a bacterial strain *Streptococcus pneumoniae*.

Pneumococcus

There are two types of Pneumococcus. Virulent strain and Avirulent strain. Virulent strain possessed a capsule and grown as smooth colonies on nutrient agar. Avirulent strain grown as rough colonies on Nutrient agar. There is no capsule in avirulent strain. Smooth (S) and Rough (R) characters are directly related to the presence or absence of capsule & this trait is known to be genetically determined. In the course of experiment, Griffith injected laboratory mice with Rough Pneumococci (RII) and were found to survive because the injected strain was avirulent. When the mice were injected with SIII strain of pneumococci, the mice suffered from pneumonia and died. When the mice were injected with heat killed S III pneumococci, the mice did not suffer from pneumonia. But when the mice were injected with the mixture of living avirulent R II & heat killed S III virulent, the mice die due to pneumonia. The analysis of dead mice showed that they contain virulent bacteria (SIII). From these results, Griffith concluded that the presence of DNA of the heat killed S III bacteria must have caused a transformation of the living R II bacteria, so as to restore the capacity for capsule formation. This was called Griffith effect or bacterial transformation.

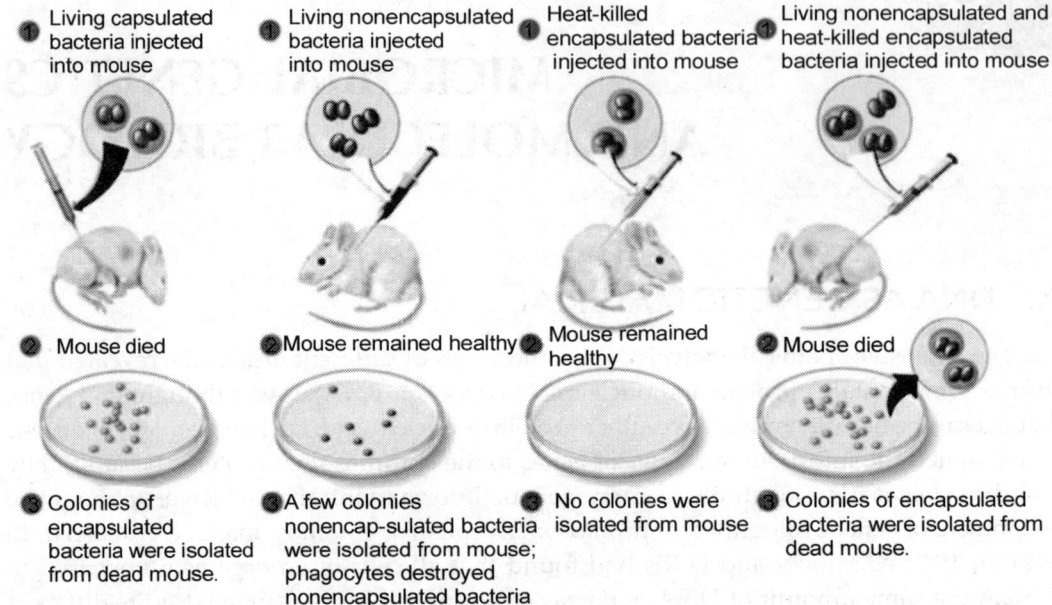

Figure 292 Transformation

Identification of the Transforming Principle

Oswald Avery, Colin Macleod and Maclym McCarty (1944) experimentally proved that the transformation of avirulent form into virulent form is due to the transmission of a chemical substance. They extracted DNA from the SIII bacteria & added to a lilve R culture. After a period of time, the bacterial cells are allowed to grow on solid media. Media contained both virulent and a virulent colonies (S III colonies) again subcultured on fresh medium. The resulting colonies were again S III. Similarly SIII extracted DNA was heated with DNAse enzyme and placed along with R strain. The resulted a colonies are only R strains. When the colonies are treated with protease, and mixed with R strain. The resulting colonies are both R & S type. This clearly evidenced that DNA as a genetic material.

Blender experiment (Hershey & Chase)

It was performed by Alfred Hershey & Martha Chase in 1952. This experiment known as the blender experiment because a kitchen warning blender was used as a major piece of apparatus. The *E. coli* T2 phage was taken as the test agent. In T2 phage, DNA is the only phosphorus containing material. Capsid shell is enriched with sulphur containing Aminoacids like Cysteine & methionine. Sulphur is not available in DNA. Against this information the phage was subjected to radioactive labeling with ^{32}P & ^{35}S. Since phage proteins do not contain phosphorus, only DNA would be labeled with ^{32}P. Phage proteins are labeled with ^{35}S. This labeling would enable one to distinguish between DNA & protein.

Hershey & Chase allowed both kinds of labeled phage particles to infect *E. coli* separately. The phages multiply in *E. coli* and are released as progeny phages after 20 minutes. When progeny phages were studied for radioactivity, it was found that the progeny phage carried labeled only with ^{32}P. The progeny was not labeled with ^{35}S. This clearly indicated that only DNA is being injected into bacterial cells and accordingly prove that DNA is the genetic material.

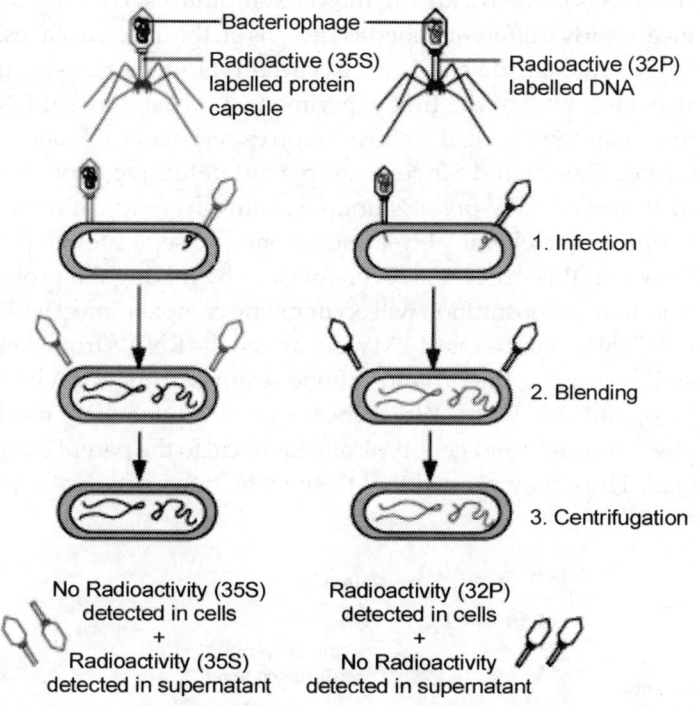

Figure 293   Transduction

Bacterial Conjugation

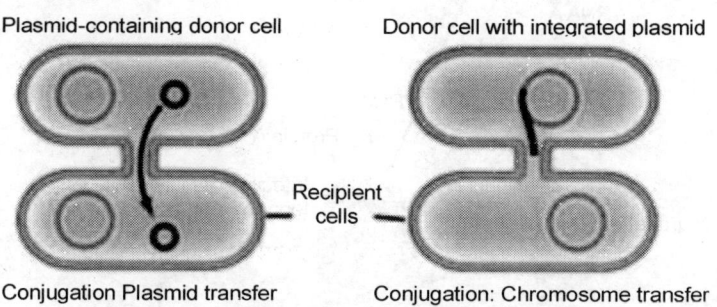

Figure 294   Conjugation

Another evidence for DNA as the genetic material is bacterial conjugation. Conjugation means union of two cells and transfer of DNA from one cell to another through a conjugation canal. It was discovered by Lederberg & Tatum in 1946. They found that there are two types of cells. They are F+ cells (male) conjugated with F- cells (Female). Unidirectional transfer of F+ factor to F+ cell to F– cell took place, so that F– cells are converted F+ cells.

B. RNA AS GENETIC MATERIAL

RNA is the genetic material in some viruses. It was demonstrated by A.Gierer and G.Schramm in 1956 using tobacco plant. Tobacco mosaic virus (TMV) contains RNA as the genetic material. TMV infects tobacco, causing mosaic symptoms. It contains RNA. Different strains of TMV produce clearly different inherited lesions on the infected leaves.The common virus produces a green mosaic disease, but a variant Holmes rib strain (TMV-HR), produces ring spot lesions. One of the first experiments that established RNA as genetic material in RNA viruses was the so called reconstitution experiments of Fraenkel-Conrat and singer published in 1957. Conrat and Singer's simple but definitive experiment was done with TMV. Different strains of TMV possess unique chemical composition of their protein coats. By using the appropriate chemical treatments one can separate the protein coats of TMV from RNA. Moreover, this process is reversible and by mixing the proteins and RNA under appropriate conditions reconstitution will occur giving complete infect ive TMV particles. Conrat and Singer took 2 different strains of TMV, separated the RNA's from their protein coat and then reconstituted by mixing the proteins of one strain withthe RNA of the 2nd strain (Protein A and RNA B) and vice versa. When these mixed viruses were used to infect the tobacco leaves, the phenotypically and genotypically identical to the parent strain from where the RNA was obtained. Thus, they proved that the genetic information of TMV is stored in RNA and not in protein.

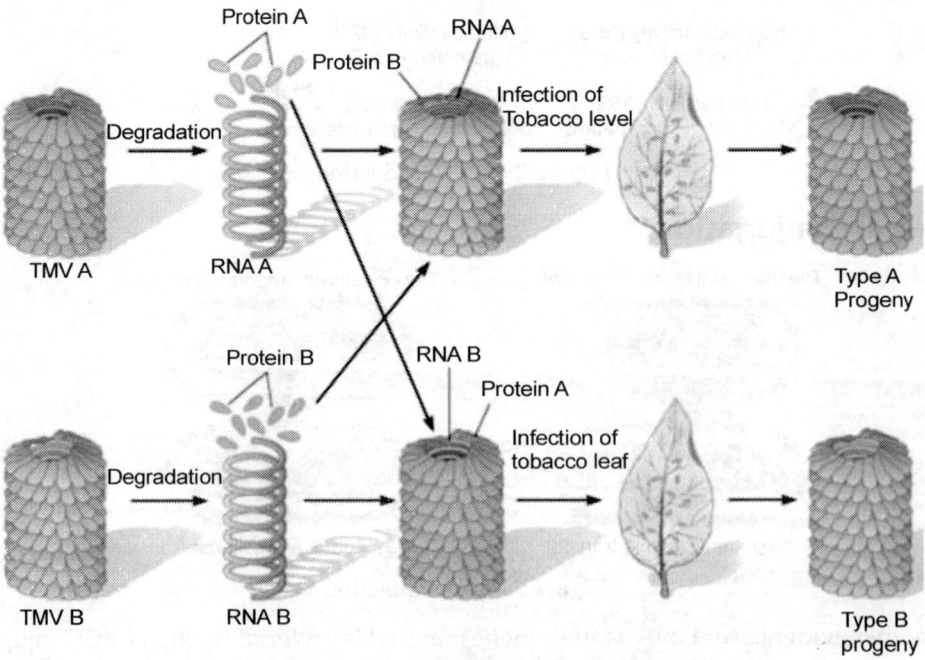

Figure 295 RNA as a genetic material

C. NUCLEIC ACIDS

Introduction

Nucleic acids are the chemical basis of life and heredity. They serve as transmitters of genetic information. Nucleic acids are colourless, complex, amorphous compounds made up of nitrogenous bases, (purine and pyrimidine bases), sugar and phosphoric acid. The nucleic acids are of two types. They are Ribonucleic acid and Deoxyribonucleic acid. As the name implies, their location is mainly the nucleus. However, it is also found to be present in other intracellular organelles (mitochondria and chloroplast). Nucleic acids are present both in the free state as well as conjucated with proteins (Nucleoproteins). In nucleic acids, nucleotides are joined together by phosphodiester linkages. RNA can be seen in the nucleus, risosome and cytoplasm.

Structural Components of Nucleic acids

Components	Ribonucleic acid	Deoxyribonucleic acid
Acid	Phosphoric acid	Phosphoric acid
Pentose Sugar	D-Ribose	Deoxy ribose
Nitrogen Bases		
i. Purines	Adenine	Adenine
	Guanine	Guanine
ii. Pyrimidines	Cytosine	Cytosine
	Uracil	Thymine

Phosphoric acid

The molecular formula of phosphoric acid is H_3PO_4. It contains 3 monovalent hydroxyl groups and a divalent oxygen atom, all linked to a pentavalent phosphorus atom.

Pentose Sugar

The two types of nucleic acids (DNA & RNA) are distinguished primarily on the basis of the 5 carbon sugar pentose which they possess. One possesses deoxyribose, (deoxyribonucleic acid) while the other contains D-ribose (hence called ribonucleic acid). Both these sugars in nucleic acids are present in the furanose form and are of β configuration.

Nitrogenous Bases

Two types of nitrogenous bases are found in all nucleic acids. These are derivatives of purine and pyrimidine.

i. Purine Bases

Purine contains a double ring structure fused to the 5 membered imidazole ring, the purine derivatives found in nucleic acids are adenine and guanine.

ii. Pyrimidine Bases

It is a single ring compound. The common pyrimidine derivatives found in nucleic acids are Uracil, Thymine and Cytosine.

Minor bases in nucleic acid

Apart from the above four bases, certain minor, unusual bases are also found in DNA and RNA. They are 5 methylcytosine, N_4 acetyl cytosine, N_6 methyladenine, N6, N6 dimethyladenine and pseudouracil etc.,

Nucleosides

It is made up of nitrogenase base and a pentose sugar. Two types of pentose sugar are present in nucleoside, they are ribose and deoxy ribose. In the case of purine nucleosides, the sugar is attached to N-9 of the purine ring, whereas in pyrimidine nucleosides, the sugar is attached to N-1 of the pyrimidine ring. The type of linkage is N-glycosidic linkage.

Nucleotides

Nucleotides are the fundamental units of nucleic acids. Each nucleotide is comprising of a

    i.   Phosphate group

    ii. Pentose Sugar and

    iii. Nitrogenous base.

Base Pairing

Base pairing is an essential feature not only to maintain the double helical structure of DNA, but also plays an important role in DNA, RNA and protein biosynthesis.

In DNA

    Adenine (A) pairs with Thymine (T) (A=T)

    Guanine (G) pairs with Cytosine (C) (G°C)

In RNA

    Adenine (A) pairs with uracil (U) (A=U)

    Guanine (G) pairs with cytosine (C) (G°C)

D. DEOXYRIBO NUCLEIC ACID

DNA is the hereditary material. It is found in all cells except few viruses. In prokaryotes, DNA exist as nucleoid in the cytoplasm whereas, In Eucaryotes, it is available within a nucleus. Nucleotide sequence of a nucleic acid is known as its primary structure which confers individuality to the polynucleotide chain. Polynucleotide chain has direction. They are represented in $5' \rightarrow 3'$ and $3' \rightarrow 5'$ directions. Each polynucleotide chain has 2 ends. The 5' end carrys a phosphate group and 3' end carrying an unreacted hydroxyl group. In 1953, J.D. Watson and F.H.C. Crick proposed a precise three dimensional model of DNA structure based on model building studies, base composition and X-ray diffraction studies. This model is popularly known as the DNA double helix.

Salient features of double helix DNA

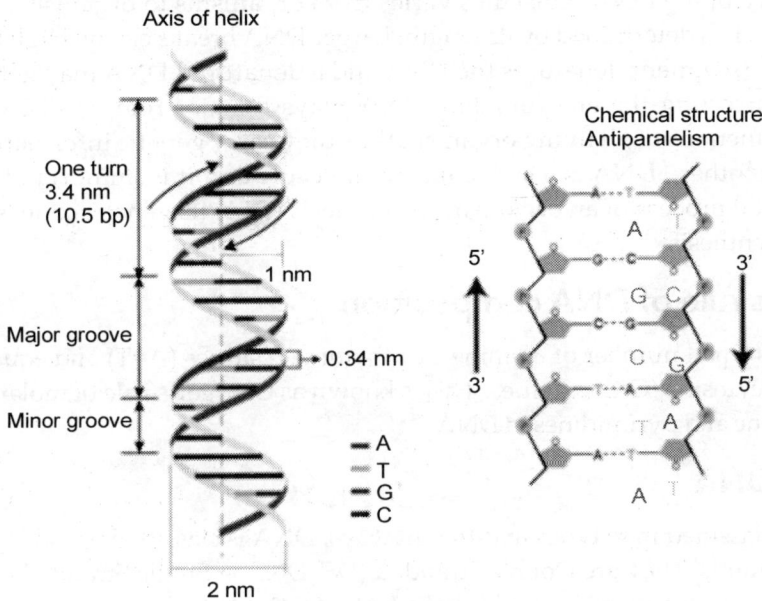

Figure 296 DNA

   DNA is the principal genetic material of all known living organisms. It is chemically called a nucleic acid. DNA contains sugar, phosphoric acid and nitrogenous bases. Watson and Crick designed the structure of DNA in 1953 and won nobel prize in 1962. According to them DNA is in the form of a double helix. DNA is made up of two polynucleotide chains. Each polynucleotide contains three chemical components, they are phosphoric acid, deoxy ribose sugar & a nitrogen base. Nitrogen bases are two ringed purines like adenine (A) & guanine (G) & two or one ringed pyrimidine cytosine (C) and thymine (T). The DNA contains 4 types of nucleotides namely AMP (Adenosine monophosphate), GMP (Guanosine monophosphate) TMP (Thymine monophosphate) & CMP (Cytocine monophosphate). In each nucleotide, the deoxyribose sugar is attached to a phosphoric acid at one side & a nitrogen base at the other side. The nitrogen base is joined to the sugar by a glycosidic bond. Two nucleotides are joined by phosphodiester bond. Many nucleotide are linked together to form a polynucleotide chain. The nucleotide of adjacent chains are linked by hydrogen bond. Adenine always linked with thymine (A – T). Similarly guanine is linked with cytocine (G-C).

The total amount of purine equaled the total amount of pyrimidine (A + G = T + C). Precizely, The amount of adenine is equivalent the amount of thymine & the amount of guamine is equivalent to the amount of cytosine. This is called Chargalf's equivalent rule (1950). The two chains of DNA are complementary to each other.

The two polynucleotide chains of DNA molecules are coiled around each other to form a double helix. The helix has a diameter of 20 A° between the two complementary strands and makes one complete turn for every 34A°: hence each turn contains 10 nucleotides. The DNA helix has two grooves, namely major groove (deep wide) & minor grove (Shallow narrow). The two strands run anti paralley to each other. One strand has phosphodiester linkage in 3' '!5' direction while other strand has phosphodiester 5' '! 3' direction. Size of the DNA molecules varies from organisms to organisms. The DNA is highly fragile. It is determined by its length. Larger DNA breaks easily. High temperature, acid or alkali treatment denatures the DNA and a denatured DNA may gets its original nature by correcting the environment. DNA plays a major role in all biosynthetic & hereditary functions of all living organism thus they carry genetic information from one generation to other. DNA is a stable macro molecule and it is immortal. It controls all developmental process of an organism. It synthesis RNA. Its genetic code is responsible for protein synthesis.

Chargaff's rule of DNA composition

DNA has equal number of adenine and thymine residues (A=T) and equal number of guanine and cytosine (G C) residues. This is known as Chargaffs rule of molar equivalence between purine and pyrimidines in DNA.

Types of DNA

DNA is classified in to types in different ways. DNA is classified into two based on the number of strands. They are Double Stranded (DS) DNA & Single Stranded (SS) DNA. DS DNA : It is a Watson and cricks double helix DNA. SS DNA : Some viruses have SS DNA.

DNA is classified into three on the basis of number of nucleotide residues present in each turn of the helix. They are A DNA, B DNA and Z DNA.

A DNA – It is a double helical DNA have 11residues per turn. It is A° in diameter. It has a right handed helix. It is formed by dehydration of B DNA.

B DNA – It is a biologically important DNA. It is commonly & naturally found in most living system. It has 10 residues per turn of the helix. It is also a right handed helix. It is the Waston & Crick double helix.

Z DNA – It is a left handed double helix having 12 residues per turn.

On the basis of the shape, DNA is classified into 3 types namely circular DNA, Relaxed DNA & super coiled DNA. Circular DNA is circular in shape, eg., φ174. Relaxed DNA is without any helical coiling. Supercoiled DNA is a circular helix and is twisted form of a super helix.

Functions of DNA

1. DNA is the genetic material of living organisms.

2. DNA contains all the information required for the functioning of an individual organism.

3. The genetic information in DNA is converted to characteristic features of living organisms like colour of the skin and eye, height, intelligence, ability to metabolise particular substance, ability to withstand stress, susceptibility to diseases and ability to produce certain substances.

4. DNA is the source of information for the synthesis of all cellular proteins. The segment of DNA that contains information for a protein is known as gene.

5. DNA is transmitted from parents to offsprings and hence transfer genetic information from one generation to another.

E. RNA

RNAs is present in the nucleus, ribosomes and cytoplasm of eukaryolic cells. They are involved in the transfer and expression of genetic information. They act as primer for DNA formation. Some act as enzymes and as coenzymes. RNA also functions as genetic material in some viruses. RNAs are also polynucleotides. In RNA polymer, purine and pyrimidine nucleotides are linked together through phosphodiester linkages. The sugar present is ribose. The nitrogenous bases present in RNA are adenine and guanine (purine bases), uracil and cytosine (pyrimidine bases). The nucleotides present in RNA are adenylic acid, quanidylic acid, cytidylic acid and uridylic acid.

Types of RNA

There are mainly three types of RNA s in all prokaryotic and eukaryotic cells. They are (1) Messenger RNA (mRNA) 2) Transfer RNA (tRNA) 3) Ribosomal RNA (rRNA). They differ from each other by size, formation and stability and importantly with function.

1. Messenger RNA

The term mRNA was coined by Jacob and Monad in 1961. It accounts for 1-5% of cellular RNA. They are single stranded linear molecules and carry genetic message in the form of triplet codon. They consist of 1000-10,000 nucleotides. They have a free or phosphorylated 3' and 5' end. They have different life span ranging from few minutes to days. mRNA molecules are capped at 5' end 5' cap by methylated guanine triphosphate. It is used to bind ribosome. Capping protects mRNA from nuclease attack. The cap is followed by non coding region 1. This is followed by initiation codon, coding region, termination codon and noncoding region 2. At 3'end, a polymer of adenylate (poly A) is found as the tail in eucaryotes. Poly A tail protects mRNA from nuclease attack. Intrastrand base pairing among complementary bases allows folding of the linear molecule. As a result, haripin or loop like secondary structure is formed. Prokaryotic mRNA are polycistronic in nature. The mRNA in nucleus are called hnRNA (heterogenous nuclear RNA). The mRNA is synthesed from DNA by RNA polymerase and the process called transcription. Each mRNA mostly contains the codons for one polypeptide chain although differences do exist Prokaryotic mRNA are called Polycistronic. It contains several sites for initiation and termination of polypeptide due to the presence or sevaral structural genes. Thereforeone mRNA synthesis more than one pdypeptide.

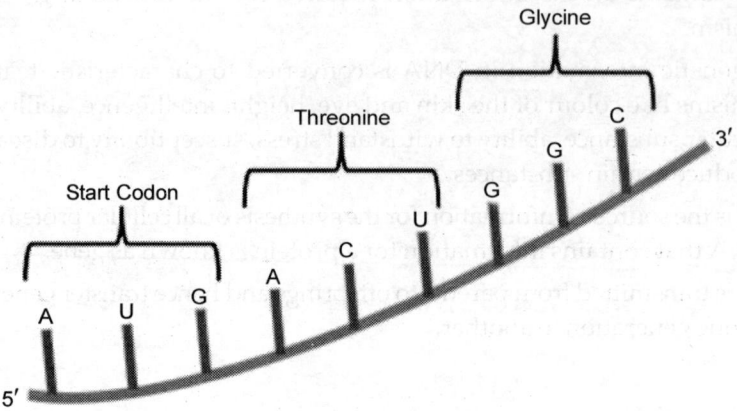

Figure 297   mRNA

Functions

1. mRNA is the direct carrier of genetic information from the DNA in the nucleus to the cytoplasm.

2. It contains information required for the synthesis of protein molecules.

Transfer RNA

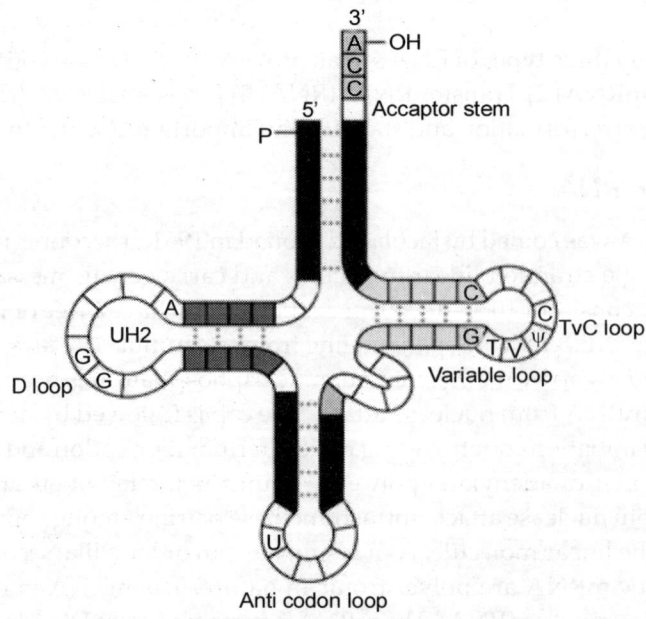

Figure 298   tRNA

It transfers the activated aminoacids to the ribosome. It is also called as soluble RNA, supernatant RNA and adapter RNA. It accounts for 10-15% of total cell RNA. They are the smallest of all the RNAs. Usually they consist of 50-100 nucleotides. They are single standard molecules. They contain unusual bases such as methylated adenine, guanine, cytosine and thymine, dihydrouracil and pseudouridine. These unusual bases are important for binding 6-RNA to intra chain base pairing. Further some bases are not involved in base pairing, resulting in loops and arms formation in tRNA. These folding in the primary structure generates a secondary structure.

Secondary structure of t-RNA is in the form of a clover leaf. The important feature of the clover-leaf structure are,

1. An aminoacid acceptor arm does not end with a loop. It has constant base sequence "CCA" 3'-OH of adenosine moiety of t-RNA.

2. An anticodon arm, stem has 5 paired bases and the loop has 7 unpaired bases. It recognises codon on mRNA.

3. TψC arm which contains unusual base cytosine.

4. D- arm which contains many dihydrouracil residues.

5. Variable arm has miniature stem only present.

Functions

1. It is the carrier of amino acids to the site of protein synthesis.

2. There is at least one t-RNA molecule to each of 20 amino acids required for protein synthesis.

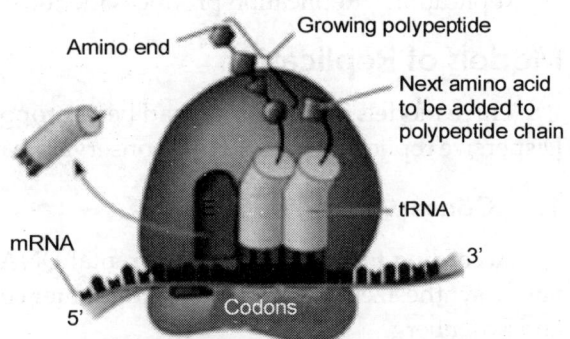

Schematic model with mRNA and tRNA

Figure 299 rRNA

Ribosomal RNA

It is a single stranded RNA. It is a polynucleotide chain. This accounts for 80% of the total cellular RNA. It is present in ribosomes. In ribosomes, r-RNA is found in combination with protein. It is known as ribonucleoprotein. The length of rRNA ranges from 100-600 nucleotides. rRNA molecules have a secondary structure. Intra strand base pairing between complementary bases generate double helical segments or loops. rRNA are classified into 7 types. They are 28S, 18S, 5.8, 5S (Eukaryotic ribosomal RNA), 23S, 16S and 55S (Prokaryotic ribosomal RNA).

Functions

1. They are required for the formation of risosomes

2. They are involved in the initiation of protein synthesis.

Table 16.1   Differences between DNA and RNA

DNA	RNA
1. Sugar moiety is deoxy ribose	Sugar moeity is ribose
2. Uracil is absent	Thymine is usually absent
3. Mostly double stranded molecules	Mostly single stranded molecules
4. Sum of purine and pyrimidine bases are equal G+C = A+T	Sum of purine and pyrimidine bases are not equal.
5. Bases are not modified	Bases are modified
6. Resistant to alkali hydrolysis	Highly susceptible to alkali hydrolysis.
7. No catalytic activity	Some are catalytically active.
8. Mostly DNA is present in nucleolus and nucleus and also in mitochondria chloroplast	Present in the cytoplasm

F.   REPLICATION OF DNA

The process of synthesizing the new double helical DNA from the existing DNA is called as "Replication". Replication produces identical daughter molecules of DNA.

Models of Replication

Three models of replication had been proposed. They are 1.Conservative replication 2. Dispersive replication and 3. Semi conservative replication.

1.   Conservative replication

According to this model, the parental DNA is conserved to one daughter cell and the newly synthesized DNA to another daughter cell. This method was proposed by Cavalieri and Rosenberg.

2.   Dispersive Replication

According to this model, the parental DNA is unequally distributed (randomly) to the daughter cells.

3.   Semiconservative Replication

Semiconseravative model of replication states that the daughter DNA includes one parental strand and one daughter strand. This model was proposed by Watson and Crick.

\

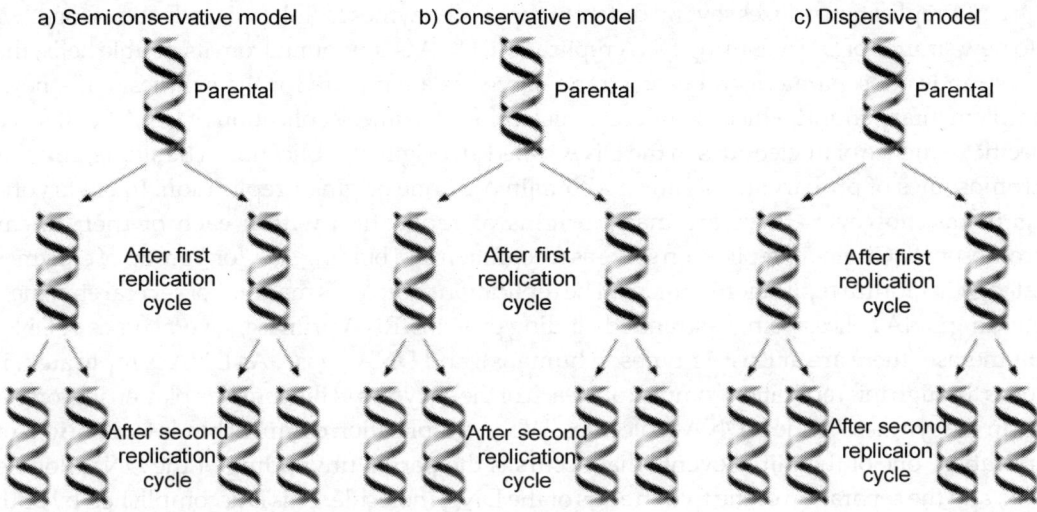

Figure 300 Replication types

DNA replication requires the following components. They are DNA template, DNA unwinding protein, A primer RNA, Superhelix relaxing protein, RNA polymerase, DNA polymerase, Ligase, Mg2+, dATP, dGTP, dTTP, dCTP.

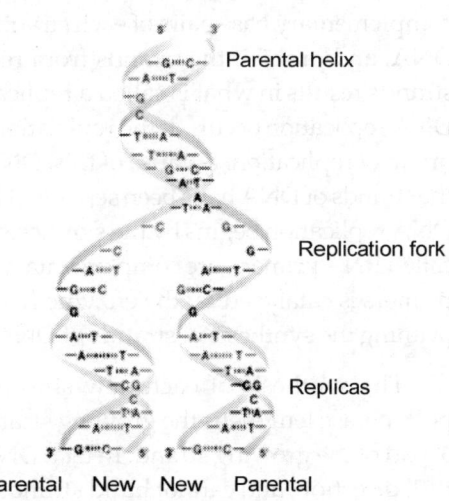

Figure 301 Mode of replication

Initiation of DNA synthesis occurs at a site called origin of replication. In prokaryotes, only one origin is present in the chromosome. In eukaryotes, there are multiple origins with in each of their linear chromosome. The origin consists of a short sequence of A = T base pairs.

The two complementary strands of DNA separate at a site of replication to form a bubble. In eukaryotes many replication bubbles occur.

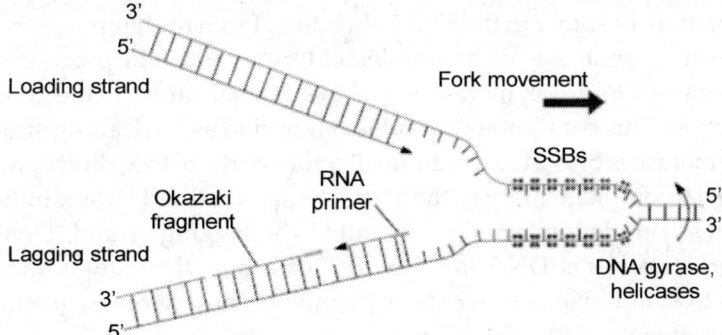

Figure 302 Replication fork

DNA replication utilizes several different types of enzymes to link free nucleotides together into new strands of DNA. During DNA replication, DNA is unwound from its double helix, the two strands are separated, and each strand serves as a template for the synthesis of a new, complementary strand which is built one nucleotide at a time. Replication of DNA begins at a specific sequence of nucleotides in the DNA called an origin of replication. The single, circular chromosomes of prokaryotic organisms contain only one origin of replication. In eukaryotic organisms, however, there are many origins of replication within each of their linear chromosomes. Origins of replication serve as recognition and binding sites for a group of enzymes that together form a replication complex. The replication complex is made up of several enzymes, including DNA helicase, single-stranded binding proteins, RNA primase, several types of DNA polymerase (there are at least 14 types in humans), and DNA ligase. As DNA is replicated, it moves through this replication complex and each of the enzymes in the complex play an important role in the synthesis of new DNA molecules. When a replication complex binds to an origin of replication, one of the initial events that occurs is the partial unwinding of the DNA double helix, and the separation of the two strands of the DNA molecule. This is accomplished by both DNA helicase and single-stranded binding proteins. DNA helicase unwinds the DNA double helix and separates the strands of DNA by breaking the hydrogen bonds between the complementary base pairs of each strand. Single-stranded binding proteins bind each strand of DNA, and prohibit the strands from reforming hydrogen bonds. The separation of the two strands results in what is called a replication bubble, which has a replication fork at each end. DNA replication occurs at the replication forks, and proceeds in both directions away from the origin of replication. Because of this, DNA replication is often referred to as bidirectional. After the strands of DNA have been separated by DNA helicase and single-stranded binding proteins, DNA replication begins by the synthesis of short strands of, surprisingly, RNA. These strands, called RNA primers, are complementary to the template strands of DNA. The synthesis of RNA primers is catalyzed by the enzyme RNA primase. RNA primers act as their name suggests, priming the synthesis of strands of DNA.

The synthesis of each new strand of DNA is catalyzed by DNA polymerase. DNA polymerase lengthens the growing strand of DNA by adding nucleotides, one at a time, to the 3' end of the growing strand. In fact, DNA polymerase is only able to add nucleotides in the 5' to 3' direction, and cannot build strands of nucleotides in the 3' to 5' direction. As replication proceeds at a replication fork, the antiparallel nature of DNA presents a small challenge. As the DNA is threaded through the replication complex, one of the template strands will separate in the 3' to 5' direction, the other in the 5' to 3' direction. From the template strand running in the 3' to 5' direction, synthesis of a complementary strand will proceed smoothly and continuously, because DNA polymerase is able to add nucleotides to the growing strand in the 5' to 3' direction. This continuous strand is referred to as the leading strand. However, since DNA polymerase is only able to add nucleotides in the 5' to 3' direction, synthesis of a new strand of DNA complementary to the 5' to 3' template strand is discontinuous. In order to synthesize this complementary strand, called the lagging strand, DNA polymerase synthesizes short stretches of DNA in the 5' to 3' direction, then jumps ahead toward the replication fork to synthesize another short fragment. These short fragments of DNA are called Okazaki fragments.

As the fragments of the lagging strand are synthesized by DNA polymerase, several other enzymes in the replication complex work together to anneal the lagging strand into one, continuous strand of nucleotides. The complete DNA fragments are then covalently linked through the action of the enzyme DNA ligase.

Through the continuous synthesis of the leading strand, and the discontinuous synthesis of the lagging strand, the entire molecule of DNA is eventually replicated so that two exact copies of the DNA are made. These copies are eventually segregated during cell division and distributed to daughter cells.

DNA replication stops when the polymerase complex reaches a termination site on the DNA in *E. coli*. The Tus protein binds to these *Ter* sites and halts replication. In many procaryotes, replication stops randomly when the forks meet. DNA replication is an extraordinarily complex process. At least 30 proteins are required to replicate the *E. coli* chromosome. Presumably much of the complexity is necessary for accuracy in copying DNA. DNA polymerase III (and DNA polymerase I) also can proofread the newly formed DNA. As polymerase III moves along synthesizing a new DNA strand, it recognizes any errors resulting in improper base pairing and hydrolytically removes the wrong nucleotide through a special 32 to 52 exonuclease activity (which is found in the subunit). The enzyme then backs up and adds the proper nucleotide in its place. Polymerases delete errors by acting much like correcting typewriters. Replication occurs very rapidly. In prokaryotic DNA replication, 750 to 1,000 base pairs per second are added. Eukaryotic replication is much slower, about 50 to 100 base pairs per second.

Deoxyribonucleoside triphosphate The building blocks of DNA replication. A five-carbon, oxygen-containing ribose sugar ring that has three phosphate groups attached to its 5' carbon and either an adenine, cytosine, guanine, or thymine base group attached to its 1' carbon.

Base-pair excision One class of DNA repair system. Recognizes and removes single nucleotide mutations that result from unnatural bases.

Daughter strand Refers to the newly synthesized strand of DNA that is copied via the addition of complementary nucleotides from one strand of pre-existing DNA template during DNA replication.

DNA Helicase The enzyme responsible for separating the two strands of DNA in a helix so that they can be copied during DNA replication.

DNA Ligase The enzyme responsible for sealing together breaks or nicks in a DNA strand. Responsible for patching together Okazaki fragments on the lagging strand during DNA replication.

DNA Polymerase The enzyme responsible for catalyzing the addition of nucleotide substrates to DNA both during and after DNA replication.

Primase The enzyme responsible for initiating synthesis of RNA primers on the lagging strand during DNA replication.

Holoenzyme A term used to describe a collection of different enzymes that work together in a given process such as DNA replication.

HydrolysisThe process in which water is chemically added to a molecule.

Lagging strand In DNA replication, the strand of pre-existing DNA that is oriented in the 5′ to 3′ direction on which synthesis is discontinuous.

Leading strand In DNA replication, the strand of pre-existing DNA that is oriented in the 3′ to 5′ direction with respect to the direction of replication on which replication is continuous.

Mismatch repair One class of DNA repair system. Recognizes and removes mutations that result from base-pairing that is not complementary.

Okazaki fragment Short stretches of newly synthesized DNA found on the lagging strand during DNA replication.

Origin of replication Site of initiation of DNA replication. Short, usually internal stretch in a DNA helix that opens so that each strand is separated for DNA replication.

Parent strand In DNA replication, refers to the pre-existing single strand of DNA that is copied into a new strand of DNA via complementary base pairing.

Replication fork Term used to describe the junction at which nucleotide substrates are being added to a growing DNA chain during DNA replication. Its shape resembles a "Y" where the two branches represent single stranded daughter strands of DNA and the base represents helical DNA.

RNA Primer Short stretches of ribonucleotides (RNA substrates) found on the lagging strand during DNA replication. Helps initiate lagging strand replication and are later removed.

Semi-conservative Refers to the fact that after the replication of one DNA helix each of the two daughter helices that result contain one newly-synthesized and one pre- existing strand of DNA.

Single-stranded binding protein - A protein involved in helping to keep strands of DNA that have been separated by DNA helicase from recoiling in a helix. It works by coating the single strands in such a way as not to cover the bases, allowing them to remain free for base pairing.

Theta model of DNA replication

Replication patterns are some what different in prokaryotes and eucaryotes. For example, when the circular DNA chromosome of *E. coli* is copied, replication begins at a single point, the origin. Synthesis occurs at the **replication fork,** the place at which the DNA helix is unwound and individual strands are replicated. Two replication forks move outward from the origin until they have copied the whole **replicon,** that portion of the genome that contains an origin and is replicated as a unit. When the replication forks move around the circle, a structure shaped like the Greek letter theta (θ) is formed (Figure 303). Finally, since the bacterial chromosome is a single replicon, the forks meet on the other side and two separate chromosomes are released.

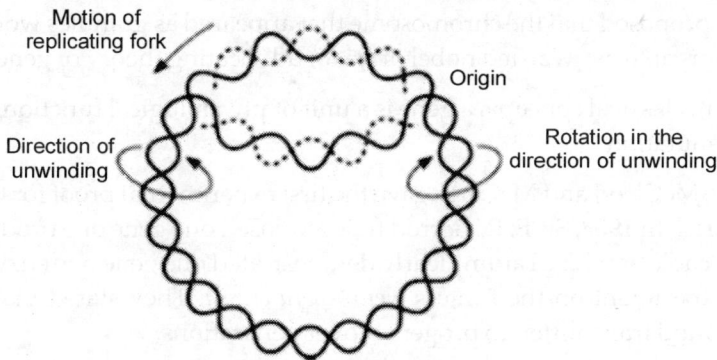

Figure 303   Theta model replication

Rolling circle replication

A different pattern of DNA replication occurs in bacteria (*E. coli* conjugative plasmid) and in viruses, such as phage lambda. In the **rolling-circle mechanism** (Figure 304),

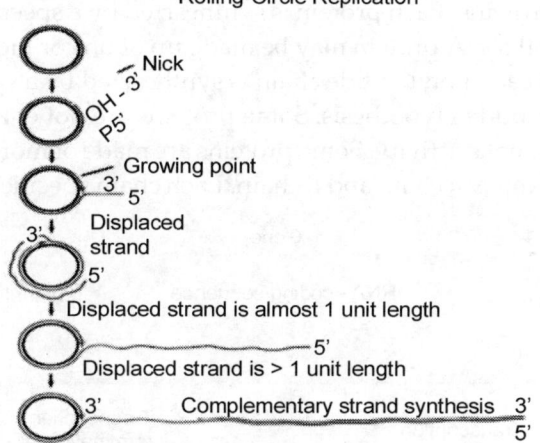

Figure 304   Rolling circle replication

one strand is nicked and the free 3' - hydroxyl end is extended by replication enzymes. As the 3' end is lengthened while the growing point rolls around the circular template, the 5' end of the strand is displaced and forms an everlengthening tail. The single-stranded tail may be converted to the double-stranded form by complementary strand synthesis. This mechanism is particularly useful to viruses because it allows the rapid, continuous production of many genome copies from a single initiation event.

G.   GENE

A gene is a segment of DNA that specifies the sequence of aminoacid in a polypeptide chain. A gene controls a character. Gregor John Mendel stated gene as a factor. In 1909, W. Johanson coined the term gene that act as a hereditary unit. In 1926, T. H. Morgan postulated particulate gene theory. He stated that genes are arranged in a linear order on the chromosome.

In 1928, Belling proposed that the chromosome that appeared as granules would be the gene. In 1933, T.H. Morgan was awarded nobel prize for advocating theory of genes.

According to classical concepts a gene is a unit of physiological function, segregation of characters and mutation.

1944 Avery, McCleod and McCarty gave the first experimental proof for the role of DNA as genetic material. In 1908, Sir E. R. Gerrod first proposed one gene one product hypothesis. In 1941, G. W. Beadle and E.L. Tatum clearly demonstrated one gene one enzyme hypothesis based on the experiment on the fungus *Neurospora crassa.* They stated that genes are the functional units and transmitted to progenies over generations.

The one gene-one enzyme hypothesis is the idea that genes act through the production of enzymes, with each gene responsible for producing a single enzyme that in turn affects a single step in a metabolic pathway. Wild type mold survived in minimal medium agar. The wild type mold could produce all of the enzymes it needs to produce the necessary amino acids to live. Mutants did not survive in minimal medium agar as they could not produce all of the enzymes needed. Each gene dictates the production of one enzyme.

All enzymes are proteins. Each protein is synthesized by a specific gene. It is called one gene one protein hypothsis. A protein may be made up of one or more polypeptide chains. Now, it is proved that each polypeptide chain is synthesized by a specific gene. It is called One Gene, One Polypeptide Hypothesis. Some proteins are not enzymes (ex. insulin) and carryout a specific metabolic activity. Some proteins are made of more than one polypeptide chain (hemoglobin), namely a chain and b chain. Each chain specified by its own gene.

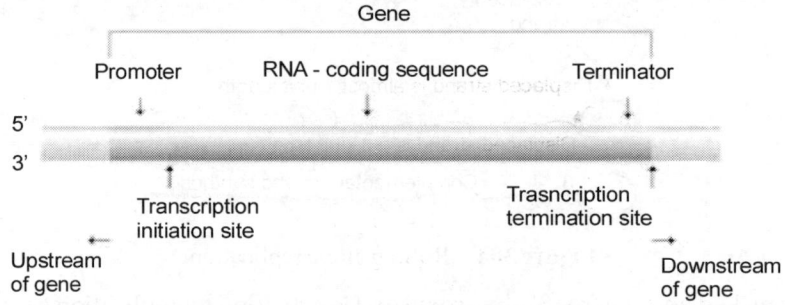

Figure 305   Gene structure

The gene consists of four main regions namely promoter, initiation site, coding sequence, and termination site. Promoter is a region of DNA involved in the initiation of transcription. In the promoter region, the RNA polymerase binds and initiates transcription. It is a first segment of a gene. In almost all promoters, there is a common sequence called consensus sequence. It is also called pribnow box. The consensus sequence is TATAAT. Initiation site is located next to the promoter. The mRNA begins from this site. The coding sequence is the middle segment of the gene. The termination site is at the other end of the gene. The gene coding a single polypeptide chain is called cistron (monocistronic). The operon is said to be polycistronic in nature. Genes are monocistronic in nature.

H. GENETIC CODE

The genetic code is defined as a sequence of three letter "words" in mRNA, called codons. It is sometimes called 'triplets'. Each codon on specifies an amino acid for the synthesis of protein molecule.

Characteristics of Genetic Code

The genetic code has the following general properties :

1. **The code is a triplet codon:** The nucleotides of mRNA are arranged as a linear sequence of codons, each codon consisting of three successive nitrogenous bases. This is called codon, that codes for single aminoacid.

2. **The code is non-overlapping:** In translating mRNA molecules the codons do not *overlap* but are "read' sequentially. Thus, a **non-overlapping code** means that a base in a mRNA is not used for different codons. Each letter of the codon read only once.

3. **The code is commaless:** The genetic code is commaless, which means that no codon is reserved for punctuations. It means that after one amino acid is coded, the second amino acid will be automatically coded by the next three letters.

4. **The code is non-ambiguous:** Non-ambiguous code means that a particular codon will always code for the same amino acid. In case of ambiguous code. , the same codon could have different meanings. Generally, as a rule, the same codon shall never code for two different amino acids. The codons AUG and GUG both may code for *methionine* as initiating or starting codon, although GUG is meant for *valine*. Likewise, GGA codon codes for two amino acids *glycine* and *glutamic acid*.

5. **The code has polarity:** The code is always read in a fixed direction, *i.e.*, in the $5' \rightarrow 3'$ direction. In other words, the codon has a **polarity**. It is apparent that if the code is read in opposite directions, it would specify two different proteins, since the codon would have reversed base sequence.

    Codon: UUG AUC GUC UCG CCA ACA AGG

    Polypeptide: $\rightarrow$ Leu Ile Val Ser Pro Thr Arg

    Val Leu Leu Ala Thr Thr Gly $\leftarrow$

6. **The code is degenerate:** More than one codon may specify the same amino acid; this is called **degeneracy** of the code. **Partial degeneracy** occurs when first two nucleotides are identical but the third (*i.e.*, 3' base) nucleotide of the degenerate codons differs, *e.g.*, CUU and CUC code for leucine. **Complete degeneracy** occurs when any of the four bases can take third position and still code for the same amino acid (*e.g.*, UCU,UCC, UCA and UCG code for serine).

7. **Start codons:** In most organisms, AUG codon is the **start** or **initiation codon**, *i.e.*, the polypeptide chain starts either with **methionine** (eukaryotes) or **N-formylmethionine** (prokaryotes). In rare cases, GUG also serves as the initiation codon, *e.g.*, bacterial protein synthesis.

8. **Stop codons.** Three codons UAG, UAA and UGA are the chain **stop** or **termination** codons. They do not code for any of the amino acids. These codons are also called **nonsense codons**, since they do not specify any amino acid.

The UAG was the first termination codon named as **amber.** other two termination codons were also named as **ochre** for UAA and **opal** or **umber** for UGA.

9. **The genetic code is universal.** Same genetic code is found valid for all organisms ranging from bacteria to man. Such universality of the code was demonstrated by **Marshall, Caskey** and **Nirenberg** (1967) . **Nirenberg** has also stated that the genetic code may have developed 3 billion years ago first with the bacteria.

Anticodon

The codons of mRNA can base pair with complementary codons of tRNA. Codons of tRNA are called **Anticodons.** Three bases of anticodon pair with the mRNA on the ribosomes at the time of aligning the amino acids during protein synthesis.

Wobble Hypothesis: This hypothesis is proposed by **Crick** in 1966. This hypothesis stated that anticodon has the capacity to pair with more than one codon. This is due to non specificity of the third base of the anticodon. The none specific third base is called wobble base. The unusual base pairing of the wobble base is called wobble pairing.

In the cytoplasm, there are 61 codons specifying amino acids. Actually the cell should contain 61 tRNA molecules, each with a different anticodon. But the number of tRNA molecule types discovered is much less than 61. This implies that the anticodons of tRNAs read more than one codon on mRNA.

For example, anticodon UCG recognizes the codon AGC. Here the third base G pairs with C. G-C base pairing is usual. The UCG anticodon can also base pair with UCU. This G-U base pairing is unusual and it is called wobble base pairing.

I.   DECIPHERING THE CODE

Identification of the code for a specific aminoacid are called deciphering the code. It is considered as the codon assignment. It is also called cracking the code. There are three major approaches in deciphering the code. They are 1. theoretical approach, 2. the *in vitro* codon assignment and 3. the *in vivo* Codon Assignment.

A.   Theoretical Approach

The physicist **George Gamow** proposed the **diamond code** (1954) and the **triangle code** (1955) and suggested an exhaustive theoretical framework to the different aspect of the genetic code. He suggested the following properties of the genetic code :

i.   A **triplet codon** corresponding to one amino acid of the polypeptide chain.

ii.   **Direct template translation** by codon-amino acid pairing.

iii. Translation of the code in an **overlapping** manner.

iv.   **Degeneracy** of the code, *i.e.*, an amino acid being coded by more than one codon.

v.   **Colinearity** of nucleic acid and the primary protein synthesized.

vi. **Universality** of the code.

Some of these Gamow's proposals have been contradicted by the molecular biologists. **Brenner** (1957) showed that the overlapping triplet code is an impossibility. Direct template relationship between nucleic acid and polypeptide chain was challenged when **Crick** proposed his **adapter hypothesis**. According to this hypothesis, **adaptor molecules** intervene between nucleic acid and amino acids during translation.

B.   The in vitro codon assignment

1. Discovery and use of polynucleotide phosphorylase enzyme.

Marianne Grunberg-Manago and **Severo Ochoa** isolated an enzyme from the bacteria that catalyzes the breakdown of RNA in bacterial cells. This enzyme is called **polynucleotide phosphorylase. Manago** and **Ochoa** found that outside of the cell (*in vitro*), with high concentrations of ribonucleotides, the reaction could be driven in reverse and an RNA molecule could be made. Incorporation of bases into the molecule is random and does not require a DNA template. Thus, in 1955 **Manago** and **Ochoa** made possible the artificial synthesis of polynucleotides (=mRNA) containing only a single type of nucleotides (U, A, C, or G respectively) repeated many times.

Polynucleotide Configuration

1.   Polyuridylic acid or poly (U) UUUUUU
2.   Polyadenylic acid or poly (A) AAAAAA
3.   Polycytilic acid or poly (C) CCCCCC
4.   Polyguanidylic acid or poly (G) GGGGGG

The deciphering of the genetic code was made possible by the use of synthetic (or artificial) polynucleotides and trinucleotides. The different types of techniques used include the use of polymers containing a single type of nucleotide (called **homopolymers**), the use of mixed polymers (**copolymers**) containing more than one type of nucleotides (**heteropolymers**) in random or defined sequences and the use of trinucleotides (or "**minimessengers**") in ribosome-binding or filter-binding.

2. Codon assignment with unknown sequence.

i. Codon assignment by homopolymer.

The first clue to codon assignment was provided by **Marshall Nirenberg and Heinrich Matthaei** (1961). The first code to be deciphered was UUU. They synthesized a chain of uracil molecules (poly U) as their synthetic mRNA (homopolymer). It codes for phenylalanine. Poly (U) seemed a good choice, because there could be no ambiguity in a message consisting of only one base. It binds well to ribosomes and the product protein was insoluble and easy to isolate.

This discovery was extended in the laboratories of **Nirenberg** and **Ochoa**. The experiment was repeated using synthetic **poly (A)** and **poly (C)** chains, which gave **polylysine** and **polyproline** respectively. Thus, AAA was identified as the code for lysine and CCC as the code for proline. A poly (G) message was found non-functional *in vitro*, since it attains secondary structure and, thus, could not attach to the ribosomes. In this way three of 64 codons were easily accounted for.

ii. Codon assignment by heteropolymers (Copolymers with random sequences).

Khorana synthesized heteropolymers with known sequence using the nucleotides C and U. when combining C and U, CUCUCU... is formed. This polynucleotide containing CUC, UCU codons. CUC codes for leucine and UCU codes for serine.

3. Assignment of codons with known sequences.

i. Use of trinucleotides or minimessengers in filter binding (Ribosome-binding technique).

This was proposed by Nirenberg and Leder (1964). According to them, when a mixture of such small mRNA molecules-ribosomes and amino acid-tRNA complexes are incubated for a short time and then filtered through a nitrocellulose membrane, then the mRNA-ribosome-tRNA-amino acid complex is retained back and rest of the mixture passes through the filter. By using a series of 20 different amino acid mixtures, each containing one radioactive amino acid at a time, it is possible to find out the amino acid corresponding to each triplet by analysing the radioactivity absorbed by the membrane, e.g., the triplet GCC and GUU retain only alanyl-tRNA and valyl-tRNA respectively. All 64 possible triplets have been synthesized and tested in this way. Forty five of them have given clear-cut results. Later on, with the help of longer synthetic messages it has been possible to decipher 61 out of the possible 64 codons.

C. The in vivo codon assignment

Second base of codon

First base of codon	U	C	A	G	Third base of codon
U	UUU Phe	UCU SER	UAU Tyr	UGU Cys	U
	UUC Phe	UCC SER	UAC Tyr	UGC Cys	C
	UUA Leu	UCA SER	UAA	UGA	A
	UUG Leu	UCG SER	UAG	UGG Trp	G
C	CUU Leu	CCU Pro	CAU His	CGU Arg	U
	CUC Leu	CCC Pro	CAC His	CGC Arg	C
	CUA Leu	CCA Pro	CAA Gln	CGA Arg	A
	CUG Leu	CCG Pro	CAG Gln	CGG Arg	G
A	AUU Ile	ACU Thy	AAU Asn	AGU Ser	U
	AUC Ile	ACC Thy	AAC Asn	AGC Ser	C
	AUA Ile	ACA Thy	AAA Lys	AGA Arg	A
	AUG Met	ACG Thy	AAG Lys	AGG Arg	G
G	GUU Val	GCU Ala	GAU Asp	GGU Gly	U
	GUC Val	GCC Ala	GAC Asp	GGC Gly	C
	GUA Val	GCA Ala	GAA Glu	GGA Gly	A
	GUG Val	GCG Ala	CAG Glu	GGG Gly	G

Figure 306 Genetic code

The cell free protein synthetic systems, though have proved of great significance in decipherment of the genetic code, but they could not tell us whether the genetic code so deciphered is used in the living systems of all organisms also. Three kinds of techniques are used by different molecular biologists to determine whether the same code is also used *in vivo* (a) amino acid replacement studies (b) frameshift mutations and (c) comparison of a DNA or mRNA polynucleotide cryptogram with its corresponding polypeptide clear text . Thus, *in vitro* and *in vivo* studies, so far described, gave the way to formulate a code table for twenty amino acids.

GENE TRANSFER METHODS

J.   BACTERIAL CONJUGATION

Conjugation means the transfer of genetic materials from cell to cell by direct contact. It was first demonstrated by Joshua Lederberg and Edward L.Tatum in 1946. They mixed two auxotrophic strains, incubated the culture for several hours in nutrient medium, and then

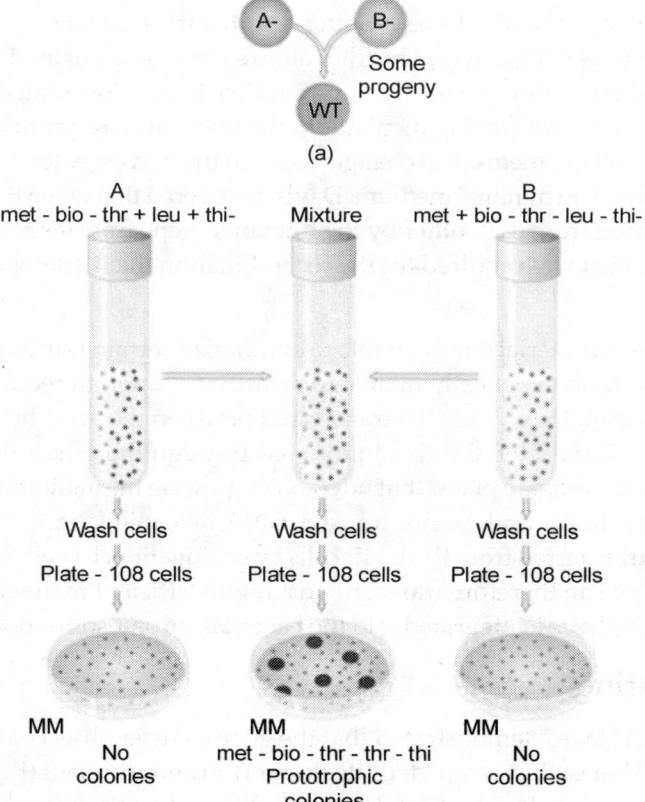

Figure 307 Demonstration by Lederberg and Tatum of genetic recombination between bacterial cells

plated it on minimal medium. Recombinant prototrophic colonies appeared on the minimal medium after incubation. Cells of type A or type B cannot grow on an unsupplemented minimal medium (MM), because A and B each carry mutations that cause the inability to synthesize constituents needed for cell growth. When A and B are mixed for a few hours and then plated, however, a few colonies appear on the agar plate. These colonies derive from single cells in which an exchange of genetic material has occurred; they are therefore capable of synthesizing all the required constituents of metabolism.

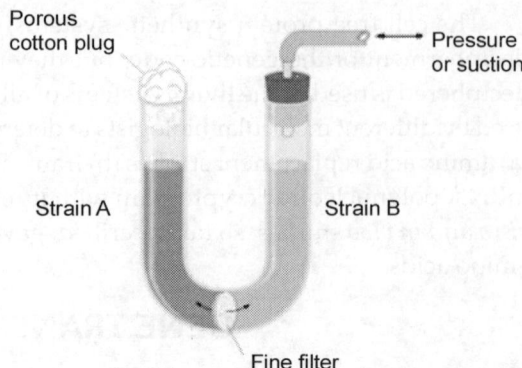

Thus the chromosomes of the two auxotrophs were able to associate and undergo recombination. Lederberg and Tatum did not directly prove that physical contact of the cells was necessary for gene transfer.

Figure 308 Experiment demonstrating that physical contact between bacterial cells is needed for genetic recombination to take place

This evidence was provided by Bernard Davis (1950), who constructed a U tube consisting of two pieces of curved glass tubing fused at the base to form a U shape with a fritted glass filter between the halves. The filter allows the passage of media but not bacteria. The U tube was filled with nutrient medium and each side inoculated with a different auxotrophic strain of *E. coli*. During incubation, the medium was pumped back and forth through the filter to ensure medium exchange between the halves. After a 4 hour incubation, the bacteria were plated on minimal medium. Davis discovered that when the two auxotrophic strains were separated from each other by the fine filter, gene transfer could not take place. Therefore direct contact was required for the recombination that Lederberg and Tatum had observed.

A suspension of a bacterial strain unable to synthesize certain nutrients is placed in one arm of a U-tube. A strain genetically unable to synthesize different required metabolites is placed in the other arm. Liquid may be transferred between the arms by the application of pressure or suction, but bacterial cells cannot pass through the center filter. After several hours of incubation, the cells are plated, but no colonies grow on the minimal medium. F"strains do not contain the F factor and cannot transfer DNA by conjugation. They are, however, recipients of DNA transferred from F+ or Hfr cells by conjugation. F+ cells contain the F factor in the cytoplasm and can therefore transfer F in a highly efficient manner to F"cells during conjugation. Hfr cells have F integrated into the bacterial chromosome, not in the cytoplasm.

F+ and F– Mating

In 1952 William Hayes demonstrated that the gene transfer observed by Lederberg and Tatum was polar. That is, there were definite donor (F+) and recipient (F-) strains, and gene transfer was nonreciprocal. He also found that in F+XF- mating the progeny were only rarely changed with regard to auxotrophy, but F- strains frequently became F+. The F+ strain contains an extrachromosomal F factor carrying the genes for pilus formation and plasmid transfer. During F+ X F- mating or conjugation, the F factor replicates by the rolling-circle mechanism, and a copy copied to produce double-stranded DNA. Because bacterial chromosome genes are rarely transferred with the independent F factor, the recombination frequency is low. The sex pilus or F pilus joins the donor and recipient and may contract to draw them together. The channel for DNA transfer could be either the hollow F pilus or a special conjugation bridge formed upon contact.

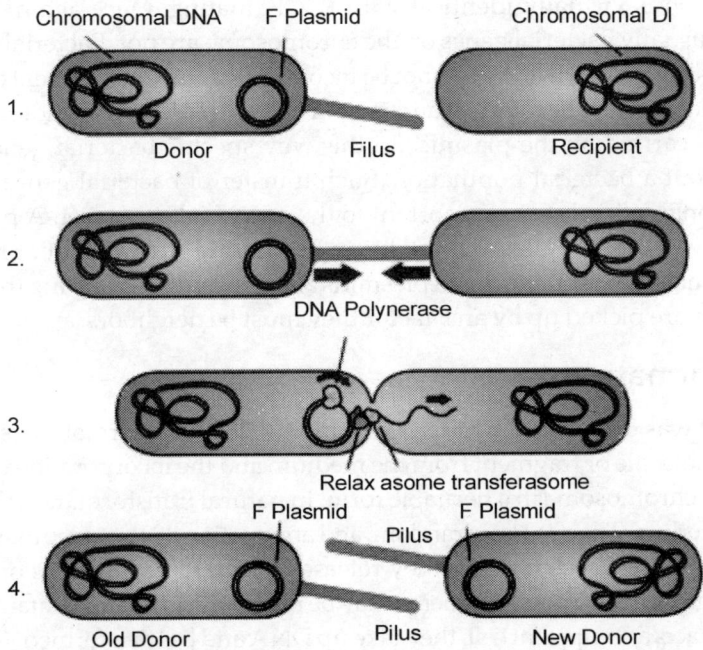

Figure 309 Bacterial conjugation. 1-Donor cell produces pilus. 2-Pilus attaches to recipient. 3-The mobile plasmid is nicked and a single strand of DNA is then transferred 4-Both cells synthesize a complementary strand to produce a double stranded circular plasmid

Hfr Conjugation

Because certain donor strains transfer bacterial genes with great efficiency and do not usually change recipient bacteria to donors, a second type of conjugation must exist. The F factor is an episome and can integrate into the bacterial chromosome at several different locations by recombination between homologous insertion sequences present on both the plasmid and host chromosomes. When integrated, the F plasmid's tra operon is still functional; the plasmid can direct the synthesis of pili, carry out rolling-circle replication, and transfer genetic material to an F- recipient cell. Such a donor is called an Hfr strain (for high frequency of recombination) because it exhibits a very high efficiency of chromosomal gene transfer in comparison with F+ cells. DNA transfer begins when the integrated F factor is nicked at its site of transfer origin. As it is replicated, the chromosome moves through the pilus or conjugation bridge connecting the donor and recipient. Because only part of the F factor is transferred at the start (the initial break is within the F plasmid), the F- recipient does not become F+ unless the whole chromosome is transferred.

F' Conjugation

Because the F plasmid is an episome, it can leave the bacterial chromosome. Sometimes during this process the plasmid makes an error in excision and picks up a portion of the chromosomal material to form an F' plasmid with one or more gene. The F' cell retains all of its genes, although some of them are on the plasmid, and still mates only with an F- recipient. F' X F- conjugation is virtually identical with F+X F- mating. Once again, the plasmid is transferred, but usually bacterial genes on the chromosome are not. Bacterial genes on the F' plasmid are transferred with it and need not be incorporated into the recipient chromosome to be expressed. The recipient becomes F' and is a partially diploid merozygote since it has two sets of the genes carried by the plasmid. In this way specific bacterial genes may spread rapidly throughout a bacterial population. Such transfer of bacterial genes is often called sexduction. F' conjugation is very important to the microbial geneticist. A partial diploid's behavior shows whether the allele carried by an F' plasmid is dominant or recessive to the chromosomal gene. The formation of F' plasmids also is useful in mapping the chromosome since if two genes are picked up by an F factor they must be neighbors.

DNA Transformation

This method was discovered by Fred Griffith in 1928. Transformation is the process of uptaking DNA molecule or fragment from the medium and the incorporation of this molecule into the recipient chromosome in a heritable form. In natural transformation the DNA comes from a donor bacterium. The process is random, and any portion of a genome may be transferred between bacteria. When bacteria lyse, they release considerable amounts of DNA into the surrounding environment. These fragments may be relatively large and contain several genes. If a fragment contacts a competent cell, they take up DNA and be transformed. Component cell posses the ability to uptake a foreign DNA. Natural transformation has been discovered so far only in certain gram-positive and gram-negative genera: *Streptococcus sp, Bacillus sp, Thermoactinomyces sp, Haemophilus sp, Neisseria sp, Moraxella sp, Acinetobacter sp, Azotobacter sp,* and *Pseudomonas sp.*

Transformation in *Streptococcus pneumoniae*

Transformation in Streptococcus pneumonia is mediated by a competence activator protein. Donor DNA attaches to the competence activator protein on the cell membrane. Following attachment, the donor DNA is extensively degraded by the activity of endonuclease I and the degradation process results in transport of polypeptide bound single stranded DNA in to the cell interior. Competetence activator protein facilitates partial autolysis of the recipient cell surface, enabling entry of donor DNA.

Transformation in *Bacillus subtilis*

Transformation in *Bacillus subtilis* is similar to that of *Streptococcus pneumonia* with a few exceptions. DNA attaches to the cell surface and leads to the degradation of one strand. Following degradation, the remaining strand is **transformed in *Bacillus subtilis*** attached to the mesosome, enters the cell and is rapidly converted to mesosome attached double stranded intermediate. Then it is released in the cytoplasm.

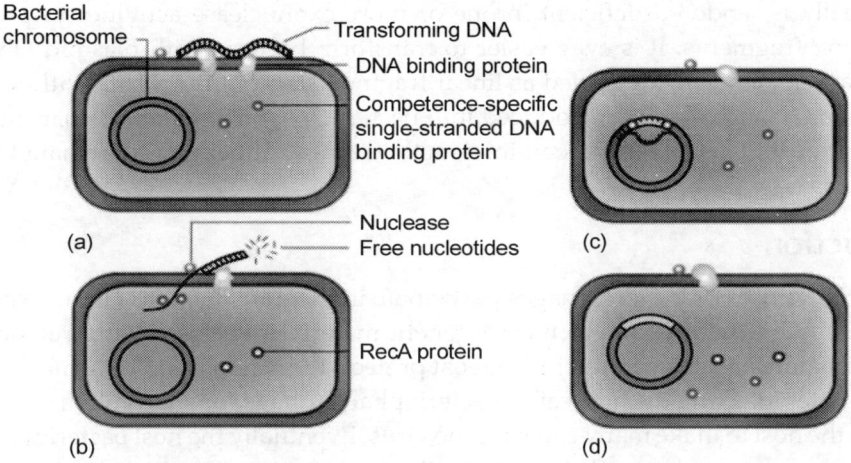

Figure 310 Mechanism of DNA transfer by transformation in a gram-positive bacterium.

    a.   Binding of free doubled-stranded DNA by a membrane-bound DNA binding protein.

    b.   Passage of one of the two strands into the cell while nuclease activity degrades the other strand.

    c.   The single strand in the cell is bound by specific proteins, and recombination with homologous regions of the bacterial chromosome mediated by RecA protein occurs.

    d.   Transformed cell.

Transformation in *Haemophillus influenzae*

Soluble competence factors and nuclease activity are not present in *Haemophilus* transformation. The specificity of *Haemophilus* transformation is due to a special 11 base pair sequence (52 AAGTGCGGTCA32) that is repeated over 1,400 times in *H. influenzae* DNA. DNA must have this sequence to be bound by a competent cell. In *Haemophilus*, during the process of competence, substantial quantities of lipo polysaccharide and three major proteins are formed. Two proteins are found in outer membrane and one protein is found in cell membrane. Double Stranded DNA attached to the cell exterior are transported, intact and encased in membrane vesicles known as transformosome. This vesicles take DNA to the cell interior. Intracellular nuclease activity then produce single stranded DNA and participates in integration with the Chromosome.

Artificial transformation

It is carried out in vitro by a variety of techniques, including treatment of the cells with calcium chloride, which renders their membranes more permeable to DNA. This approach succeeds even with species that are not naturally competent, such as *E. coli*. Relatively high concentrations of DNA, higher than would normally be present in nature, are used to increase transformation frequency. When linear DNA fragments are to be used in transformation,

E. coli usually is rendered deficient in one or more exonuclease activities to protect the transforming fragments. It is even easier to transform bacteria with plasmid DNA since plasmids are not as easily degraded as linear fragments and can replicate within the host. DNA from any source can be introduced into bacteria by splicing it into a plasmid before transformation and hence this is a common method for introducing recombinant DNA into bacterial cells.

Transduction

Bacterial viruses or bacteriophages participate in the transduction. These viruses have relatively simple structures in which virus genetic material is enclosed within an outer coat, composed mainly or solely of protein. The coat protects the genome and transmits it between host cells. After infecting the host cell, a bacteriophage (phage for short) often takes control and forces the host to make many copies of the virus. Eventually the host bacterium bursts or lyses and releases new phages. This reproductive cycle is called a lytic cycle because it ends in lysis of the host. The cycle has four phases. First, the virus particle attaches to a specific receptor site on the bacterial surface. The genetic material DNA, RNA then enters the cell. After adsorption and penetration, the virus chromosome forces the bacterium to make virus nucleic acids and proteins. The third stage begins after the synthesis of virus components. Phages are assembled from these components. The assembly process is complex, but in all cases phage nucleic acid is packed within the virus's protein coat. Finally, the mature viruses are released by cell lysis. Bacterial viruses that reproduce using a lytic cycle often are called virulent bacteriophages because they destroy the host cell. Many DNA phages, such as the lambda phage, are also capable of a different relationship with their host. After adsorption and penetration, the viral genome does not take control of its host and destroy it while producing new phages. Instead the genome remains within the host cell and is reproduced along with the bacterial chromosome. A clone of infected cells arises and may grow for long periods while appearing perfectly normal. Each of these infected bacteria can produce phages and lyse under appropriate environmental conditions. This relationship between the phage and its host is called lysogeny Bacteria that can produce phage particles under some conditions are said to be lysogens or lysogenic, and phages able to establish this relationship are temperate phages. The latent form of the virus genome that remains within the host without destroying it is called the prophage. The prophage usually is integrated into the bacterial genome. Sometimes phage reproduction is triggered in a lysogenized culture by exposure to UV radiation or other factors. The lysogens are then destroyed and new phages released. This phenomenon is called induction. Transduction is the transfer of bacterial genes by viruses. Bacterial genes are incorporated into a phage capsid instead of viral genome because of errors made during the virus life cycle. The virus containing these genes then injects them into another bacterium, completing the transfer. Transduction may be the most common mechanism for gene exchange and recombination in bacteria. There are two very different kinds of transduction:generalized and specialized.

Generalized transduction The transfer of any part of a bacterial genome when the DNA fragment is packaged within a phage capsid by mistake. Specialized transduction - A transduction process in which only a specific set of bacterial gene is carried to another bacterium by a temperate phage; the bacterial genes are acquired because of a mistake in the excision of a prophage during the lysogenic life cycle.

K. TRANSPOSABLE ELEMENTS

The chromosomes of bacteria, viruses, and eukaryotic cells contain pieces of DNA that move around the genome. Movement of gene is also called **transposition.** DNA segments that carry the genes required for this process and consequently move about chromosomes are **transposable elements** or **transposons.** These mobile elements have been variously called **'jumping genes'**, **'mobile elements'**, **'cassettes'**, **'insertion sequences'** and **'transposons'**. The term transposon was coined by **Hedges** and **Jacob** in 1974 for a DNA segment which could move from one DNA molecule (or chromosome) to other and carried resistance for antibiotic ampicilin.

Transposable elements were discovered by **Barbara McClintock** (1965) through an analysis of genetic instability in maize (corn). The instability involved chromosome breakage and was found to occur at sites where transposable elements were located.

Some salient features of the transposable elements are the following:

1. They are DNA sequences that code for enzymes which bring about the insertion of an identical copy of themselves into a new DNA site.

2. Transposition events involve both recombination and replication processes which frequently generate two daughter copies of the original transposable elements. One copy remains at the **parent site** while the other inserted at the **target site** (on the host chromosome).

3. The insertion of transposable elements invariably disrupts the integrity of their target genes. For example, if an IS (= insertion sequence) becomes inserted into an operon, it interrupts the coding sequence and inactivates the expression of the target gene into which it inserts as well as any gene downstream in that same operon. This is because an IS contains transcription and/or translation termination signals that block the expression of other genes downstream in an operon. This "one-way" mutational effect (or polarity) is referred to as a **polar mutation. Ghosal** (1979) reported that DNA sequence of IS1 and IS2 of *E. coli* contains nonsense codons in all reading frames. IS2 appears to have a chain termination codon.

4. Since transposable elements carry signals for the initiation of RNA synthesis, they sometimes activate previously dormant genes.

5. A transposable element is not a replicon (a sequence that contains a site for the origin of replication), thus, it cannot replicate apart from the host chromosome the way that plasmids and phage can.

6. No homology exists between the transposon and the target site for its insertion. Many transposons can insert at virtually any position in the host chromosome or into a plasmid. Some transposons seem to be more likely to insert at certain positions (hot spots), but rarely at base-specific target sites.

Types of transposable elements

Transposable elements are of the following types:

1. Insertion Sequences (IS) or Simple Transposons

The insertion sequences (IS) are shorter (800 to 1500 bp) and do not code for proteins. In fact, IS carry the genetic information necessary for their transposition (*i.e.*, the gene for the enzyme **transposase**).. There are different insertion sequences such as IS1, IS2, IS3, IS4 and so on in *E.coli*. Insertion sequences have been found in bacteriophages, in F plasmids and in many bacteria

2. Transposons (Tn) or Complex Transposons

Transposons (Tn) are several thousand base pair long, and have genes coding for one or more proteins. The hallmark of the transposon is the presence of identical, **inverted terminal repeat (IR) sequences** of 8 to 38 base pairs. Each type of transposon has its own unique inverted repeat. On either side of a transposon is a short (less than 10 bp) **direct repeat**. If a transposon exists in multiple copies, these direct repeats are of different base composition at each site where the transposon exists in the chromosome; the inverted terminal repeats, however, remain the same for a given transposon. The sequence into which a transposable element inserts is called the target sequence. During insertion of a transposon, the singular target sequence becomes duplicated and, thus, appears as direct repeats flanking the inserted transposable element. The direct repeats are not considered part of the transposon. These repeat sequences themselves act like IS or IS-like segments. A single purple flower has appeared in the middle of this double pink African violet as the result of a 'jumping gene'. The wrinkled trait in garden pea is caused by the insertion of transposable element into the structural gene for the starchbranching enzyme.

Examples of Transposons

1. **Tn 3 transposon of *E. coli*.** The molecular structure of transposon Tn 3 of *E.coli* has been worked out. Tn 3 has 4957 bp and contains three genes such as *tnp A*, *tnp R* and *bla*, coding respectively for the following proteins : 1. **transposase** having 1015 amino acids and required for transposition ; 2. a **repressor** (also called **resolvase**) containing 185 amino acids which regulates the transposase ; and 3. β-**lactamase** enzyme which confers resistance to the antibiotic ampicillin. On both sides of the Tn 3 is an inverted repeat of about 38 bp.

2. **Bacteriophage Mu.** The bacteriophage *Mu* (*Mu* = mutator) is a temperate bacteriophage having typical phage properties and could be regarded as a giant transposon. Phage *Mu* was thought to behave in a unique manner since it does not multiply in a lytic way. It always starts its life-cycle by inserting itself into the *E.coli* chromosome at random locations. The resulting *prophage* often mutates the genes into which they become inserted. Thus, phage *Mu* was originally known for its mutagenic properties. During later part of phage life cycle, upon receipt of an appropriate signal, the *Mu* prophage generates its transposition, but also activates the genes encoding for structural proteins which package its DNA. *Mu* DNA phage is unique also in

transposing much often than other transposons, providing nearly a hundred new copies of *Mu* during its hour-long life-cycle. Like other transposons, *Mu* phage has inverted repeats (IR) at or near the ends of its DNA and it makes a short duplication with 5 base pairs of the bacterial DNA into which it is inserted.

3. **Yeast *Ty* elements**. The yeast *Saccharomyces cerevisiae* carries about 35 copies of a transposable element called *Ty* in its haploid genome. These transposons are about 5900 base pairs long and are bounded at each end by a DNA segment called the **delta sequence**, which is approximately 340 base-pairs long. Each ä sequence is oriented in the same direction, forming **acquired character.** The alteration in the morphology or physiology of an organism in response to its ecological factors (environment) is known as acquired character. Acquired characters are usually not heritable.

L.   REGULATION OF GENE EXPRESSION IN PROKARYOTES

In prokaryotes (*e.g.*, *E.coli*), the activities of genes are regulated according to the following mechanisms.

1. Transcriptional Control Mechanisms

In bacteria, there occur several mechanisms of gene regulation at the level of transcription. A notable method depends on whether the enzymes being regulated act in catabolic (degradative) or anabolic (synthetic) metabolic pathways.

A. Negative Control

i. Inducible Operons (Inducible Systems) - Induction

An inducible enzyme is produced only when its substrate (inducer) is present in the environment (*i.e.*, active repressor + inducer = inactive repressor). Most enzymes in this category are catabolic in their activity. This was proposed by **F. Jacob** and **J. Monod** in 1961. They proposed a model. Popularly known as **operon model.** According to this model, an **operon** was defined as a cluster of coordinately regulated genes. The operon is composed of a promoter sequence, followed by an operator gene, followed by one or more structural genes that act as code for proteins. The operon is controlled by a regulator gene found elsewhere on the chromosome.

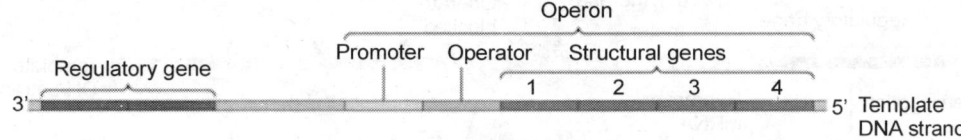

Figure 311   Operon

Operons can be inducible (turned on by a substrate) or repressible (turned off by a product). In the inducible lactose (*lac*) operon, the operator gene is blocked by a repressor protein when the cell is not in the presence of lactose. When lactose is present, it acts as an inducer by binding to the repressor protein, thus preventing it from attaching to the operator. RNA polymerase can then bind to the operator and transcribe the mRNA molecule. Three different proteins are synthesized. When all of the lactose in the cell has been catabolized, the repressor protein binds to the operator and shuts down the operon. One of the good example for operon is **lac operon.** It provides a model system for the study of gene regulation especially in prokaryoles.

Mechanism of lac operon

A *lac* operon contains three structural genes or cistrons, namely *z, y* and *a.* Gene *z* codes for an enzyme b- **galactosidase. It** converts lactose into glucose and galactose. The gene *y* encodes for an enzyme **galactoside permease. It is a** plasma-membrane bound protein and facilitates the entrance of lactose into the cell. Gene *a* specifies an enzyme **thiogalactoside acetylase.** It transfers an acetyl group from acetyl-CoA to β-galactoside. When lactose is

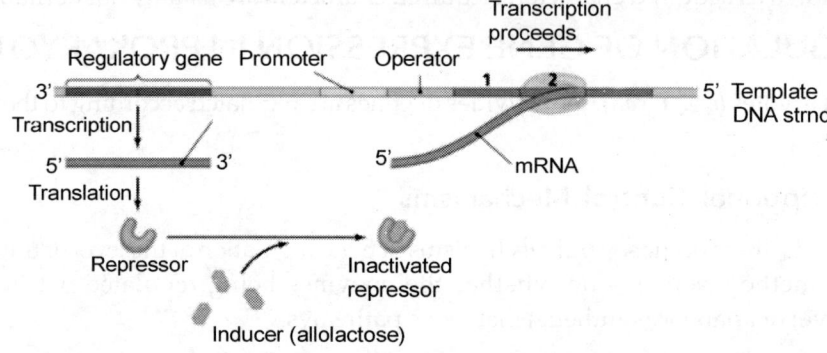

Figure 312   Lac operon

provided in the medium, all three enzymes are synthesized rapidly and simultaneously as co-ordinated response to the presence of this substrate. Thus, lactose acts to **induce** the production of the enzymes needed for its catabolism. Since all three enzymes are synthesized through the translation of a single polycistronic mRNA, it follows that the entire operon is responding as a unit to the presence of inducer. When glucose and lactose are provided in the medium, organisms use glucose first. When glucose is exhausted, a lactose metabolite allolactose induces lactose utilization.

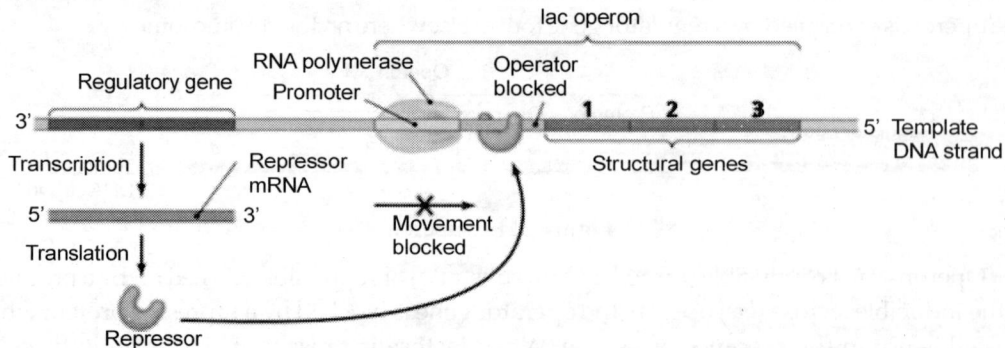

Figure 313   Lacoperon repression and induction

Interaction of allolactose with an active repressor inactivates repressor. Under these condition, the repressor protein cannot occupy the operator. This leads to attachment of DNA dependent RNA polymerase to the promoter region and the enzymes for lactose catabolism like β- galactosidase, and thiogalactoside acetylase are formed. Induction is the negative control mechanism because the presence of an active repressor molecule on the operator site inhibits attachment of DNA dependent RNA polymerase to the promoter region.

ii. Repressible System – Co repression

Some enzymes are normally synthesized when high concentrations of their end product are present. Such enzymes are called **repressible enzymes**; while the end product is called **corepressor**. A gene called **regulator gene** produces a substance called **aporepressor** which unites with corepressor to form a functional **repressor** molecule. This repressor molecule or substance inhibits synthesis of mRNA by all genes specifying enzymes in the synthetic pathway. Most repressible enzymes are found in anabolic pathways.

Example. In *Salmonella ty-phimurium* when the amino acid, histidine occurs in high concentration in the medium, that starts to act as corepressor. As a corepressor, this amino acid terminates the synthesis of 10 enzymes which are required in pathway to histidine. This kind of repression is called **coordinate repression.**

Attenuation

It is a proposed mechanism of control in some bacterial operons which results in premature termination of transcription and which is based on the fact that, in bacteria, transcription and translation proceed simultaneously. Attenuation involves a provisional stop signal (attenuator), located in the DNA segment that corresponds to the leader sequence of mRNA. During attenuation, the ribosome becomes stalled (delayed) in the attenuator region in the mRNA leader. Depending on the metabolic conditions, the attenuator either stops transcription at that point or allows read-through to the structural gene part of the mRNA and synthesis of the appropriate protein.

Attenuation is a regulatory feature found throughout Archaea and Bacteria causing premature termination of transcription. Attenuators are 5'-cis acting regulatory regions which fold into one of two alternative RNA structures which determine the success of transcription. The folding is modulated by a sensing mechanism producing either a

Rho-independent terminator, resulting in interrupted transcription and a non-functional RNA product; or an anti-terminator structure, resulting in a functional RNA transcript.

There are now many equivalent examples where the translation, not transcription, is terminated by sequestering the Shine-Dalgarno sequence (ribosomal binding site) in a hairpin-loop structure. Attenuation is an ancient regulatory system, prevalent in many bacterial species providing fast and sensitive regulation of gene operons and is commonly used to repress genes in the presence of their own product (or a downstream metabolite).

B. Positive Control

Many bacterial genes are under positive control. This mode of gene regulation is attributed to the presence of factors that enhance the attachment of RNA polymerase to the promoters and initiation of mRNA synthesis.

i. Effects of glucose on lac operon (Catabolic repression) or Glucose effect.

The function of b-galactosidase enzyme in lactose metabolism is to form glucose by cleaving lactose. Thus, if both glucose and lactose are present in the growth medium, activity of *lac* operon is not needed, and indeed, no b- galactosidase is formed until virtually all of the glucose in the culture medium is consumed. The lack of synthesis of b-galactosidase is a result of lack of synthesis of *lac* mRNA. No *lac* mRNA is made in the presence of glucose, because in an addition of an inducer to inactivate the *lac i* repressor, another element (*i.e.*, cAMP-CAP) is needed for initiating *lac* mRNA synthesis; the activity of this element is regulated by the concentration of glucose. However, the inhibitory effect of glucose on expression of the *lac* operon is quite indirect. Small molecules, the **cyclic AMP** (**cAMP**), are present in *E.coli* and many other bacteria. Cyclic AMP is synthesized enzymatically by **adenyl cyclase** and its concentration is regulated indirectly by glucose metabolism. When bacteria are growing in culture medium containing glucose, the cAMP concentration in the cells is quite low. *E. coli* contains a protein called the **catabolic activator protein** (**CAP**), which is encoded in a gene called *crp*. Mutants of either *crp* or the adenyl cyclase gene are unable to synthesize *lac* mRNA, indicating that both CAP and cAMP are required for *lac* mRNA synthesis. CAP and cAMP bind to one another to form a unit, called **cAMPCAP** which is an active regulatory element of the *lac* system. The cAMP-CAP complex must be bound to a base sequence in the DNA in the promoter region in order for transcription to occur. Thus, cAMPCAP is a positive regulator, in contrast with the repressor, and *lac* operon is independently regulated both positively and negatively.

ii. Tryptophan operon

The tryptophan (*trp*) operon of *E.coli* is responsible for the synthesis of the amino acid tryptophan. Regulation of this operon occurs in such a way that when tryptophan is present in the growth medium, *Trp* operon is not active. That is, when adequate tryptophan is present, transcription of the operon is inhibited; however, when its supply is insufficient, transcription occurs. The *Trp* operon is quite different from the *lac* operon in that tryptophan acts directly in the repression system rather than as an inducer. Moreover, since the *Trp* operon encodes a set of biosynthetic (or anabolic) rather than catabolic enzymes, neither glucose nor cAMP-CAP has a role in the operon activity. A simple on-off system, as in the *lac* operon, is not ideal for a biosynthetic pathway; a situation may arise in nature in which some tryptophan is available, but not enough to allow normal growth if synthesis of tryptophan was totally shut down. Tryptophan starvation is prevented by a *modulating system in which the amount of transcription in the derepressed state is determined by the concentration of tryptophan*. Such a mechanism is found in many operons responsible for amino acid biosynthesis.

Tryptophan is synthesized in five steps, each requiring a particular enzyme. In the *E.coli* chromosome the genes encoding these enzymes are located adjacent to one another in the same order as they are used in the biosynthetic pathway; they are translated from a single polycistronic mRNA molecule. These genes are called *TrpE, TrpD, TrpC, TrpB* and *TrpA*.

The *TrpE* gene is the first one translated. Adjacent to the *TrpE* gene are the promoter, the operator and two regions called the **leader** and the **attenuator**, which are designated *TrpL* and *Trpa* (not TrpA) respectively. The repressor gene *TrpR* is located quite far from the Trp gene cluster. The regulatory protein of the repression system of the *Trp* operon is the *Trp* R-gene product.

Mutations either in this gene or in the operator cause constitutive initiation of transcription of Trp mRNA as the *lac* operon. This regulatory protein is called *Trp* **aporepressor** and it does not bind to the operator unless tryptophan is present. The aporepressor and the tryptophan molecule join together to form the active *Trp* repressor which binds to the operator.

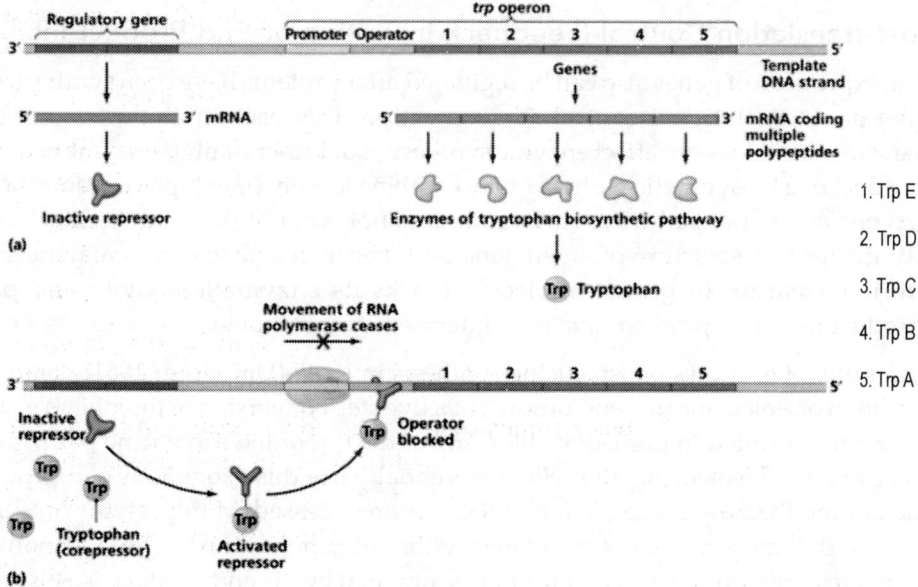

Figure 314   Trp operon

In the repressible tryptophan (trp) operon above, the structural genes involved in tryptophan biosynthesis are transcribed continuously as long as the cell needs the amino acid. When a sufficient concentration of tryptophan has been produced, the amino acid will bind to a repressor protein, causing it to bind to the operator gene and shut the trp operon down.

2. Translational Control

In prokaryotic gene regulation at the translation level, the lifetime of a mRNA molecule may be genetically determined. The average lifetime of many mRNA molecules of *E.coli* is only two minutes at 37°C. The specific nucleotide sequences at the 5′ end may influence its susceptibility to enzymatic digestion. Further, catabolic enzymes are denied access to the mRNA when the ribosome coated at their 5′ ends (*i.e.*, in case of polyribosomes). Hence, the lifetime of mRNAs may also be correlated with the number of free ribosomes available at any given moment to translate mRNA molecules. Bacteria vary their rates of protein synthesis by varying their ribosomal content rather than by varying the translational rate.

Example. In the lactose system of *E.coli*, there are three structural genes under control of a common operator locus determining production of (1) b- galactosidase, (2) galactoside permease and (3) galactoside acetylase. These three proteins are produced in the respective ratios 1:1/2:1/5, reflecting their respective locations relative to the 5′ (operator) end of the polycistronic mRNA in which they are coded (these differences are the examples of translation regulations). Thus, there is a *polarity gradient* within the polycistronic mRNA that reduces the probability of cistron translation as a function of its distance from the 5′ end. It is hypothesized that ribosomes attach to different starting points (ribosome-binding sites) along the polycistronic mRNA at different rates as reflected by the relative amounts of the three proteins synthesized.

3. Post-translation Control (Feedback Inhibition or End Product Inhibition)

The expression of genes also can be regulated after proteins have been synthesized. This is called **post-translational control of gene action**. Feedback inhibition is a regulatory mechanism which does not affect enzyme synthesis, but rather inhibits enzyme activity . The end product of a biosynthetic pathway may combine loosely (if in high concentration) with the first enzyme in the pathway. This union does not occur at the catalytic site, but it does modify the tertiary structure of an enzyme and, hence, inactivates the catalytic site. This **allosteric transition** of protein molecule blocks its enzymatic activity and prevents overproduction of end products and their intermediate metabolites.

Example. The studies on isoleucine synthesis in *E.coli* (**Umbarger** 1961) demonstrated that addition of isoleucine (the end product of a five step conversion of threonine) to a culture of the bacteria resulted in immediate blocking of the threonine′!isoleucine pathway. In the presence of added isoleucine, the cells preferentially use this **exogenous** end product (*i.e.,* isoleucine) and their own isoleucine synthesis becomes ceased. Moreover, the production of each of the five enzymes is not interfered with, but action of an enzyme responsible for deamination of theronine to a-ketobutyrate is inhibited by the end product, isoleucine.

M.   BIOSYNTHESIS OF RNA (TRANSCRIPTION)

The biosynthesis of RNA is very much similar to that of DNA, except that in RNA it is differed by having different types of RNA (mRNA, tRNA, rRNA) and by the nitrogen base uracil instead of thymine. Like DNA, polymerization of 4 nucleoside triphosphates (viz ATP, CTP, GTP and UTP) in the presence of Mg^{2+} (or) Mn^{2+} ion is catalysed by the enzyme RNA polymerase (DNA dependent RNA polymerase) and one strand of DNA serves as template.

Three phases of transcription

1. Initiation

In *E. coli*, all genes are transcribed by a single large RNA polymerase. This complex enzyme, called the holoenzyme is needed to initiate transcription since the s factor is essential for recognition of the promoter. It is common for prokaryotes to have several s factors that recognize different types of promoter (in *E. coli*, the most common s factor is s70).

The holoenzyme binds to a promoter region (about 40-60 bp in size) and then initiates transcription a short distance downstream (i.e. 3′ to the promoter). With in the promoter lie two 6 bp sequences that are particularly important for promoter function and which are therefore highly conserved between species. Using the convention of calling the first nucleotide of a transcribed sequence as +1, there 2 promoter elements lie at position -10 and -35, that is about 10 and 35 bp respectively, downstream of where transcription will begin.

2. Elongation

After transcription initiation, the s factor is released from the transcriptional complex to leave the core enzyme (a2bb'w) which continues elongation of the RNA transcript.

Thus, the core enzyme contains the catalytic site for polymerisation, probably within the subunit. The first nucleotide in the RNA transcript is usually PPPG (or) PPPA. The RNA polymerase then synthesises the RNA in the $5' \rightarrow 3'$ direction, using the 4 ribonucleotide $5'$ triphosphates (ATP, CTP, GTP, UTP), as precursors. The $3'$–OH group at the end of the growing RNA chain attaches to a phosphate group of the incoming ribonucleotide $5'$ triphosphate to form a $3', 5'$ phosphodiester bond. The complex of RNA polymerase, DNA template and new RNA transcript is called a ternary complex (i.e three components) and the region of unwound DNA that is undergoing transcription is called transcription bubble. The RNA transcript forms a transient RNA-DNA hybrid helix with its template strand but then peels away from the DNA as transcription proceeds. The DNA is unwound ahead of the transcription bubble and after the transcription complex has passed, the DNA rewinds.

In each step the incoming ribonucleotide selected is that which can base pair with the next base of the DNA template strand. The incoming nucleotide is UTP to base pair with the A residue of the template DNA. A $3' \rightarrow 5'$ phosphodiester bond is formed, extending the RNA chain by one nucleotide, and pyrophosphate is released. Overall the RNA molecule is grown in a $5'$ to $3'$ direction.

The DNA double helix is unwound and RNA polymerase then synthesizes an RNA copy of the DNA template strand. The nascent RNA transiently forms an RNA-DNA hybrid helix but then peels away from the DNA which is subsequently rewound into a helix once more.

3. Termination

Transcription continues until a termination sequence is reached. The most common termination signal is a G and C rich region is a palindrome, followed by an A = T rich sequence. The RNA made from the DNA palindrome is self complementary and so base pairs internally to form a hairpin structure rich in GC base pairs followed by four or more U residues. However not all termination sites base this hairpin structure. Those that lack such a structure require an additional protein, called rho protein (r) to help recognize the termination site and stop transcription.

N. POST TRANSCRIPTIONAL MODIFICATION

Post-transcriptional modications are not needed for prokaryotic mRNAs. The formation of functionally active RNA molecule continues after transcription is completed in eukoryotes. The product of transcription in eukarryotes are called as primary transcripts and they undergo modification by a process called post transcriptional modification.

Processing of mRNA molecules

mRNA undergoes several modifications before it is being translated. They are

1. Addition of blocks of poly A, containing 200 (or) more AMP residues to the 3' end of messenger RNAs. This addition of poly A tail takes place in the nucleus.

2. A methylated Guanine nucleotide called as "5' cap" is added to the 5' end of mRNA molecule.

3. Methylation occurs in the internal adenylate nucleotides at their N-6 position. It's function is not known. Thus mRNA is processed to get the active mRNA molecule.

Processing of tRNA Molecules

Most cells have 40 to 50 distinct tRNAs. Transfer RNA's are derived from longer RNA precursors of enzymatic removal of extranucleotide units from the 5' and 3' ends.

1. Formation of the 3'-OH terminus

This process involves the action of an endonuclease that recognizes a hairpin loop at the 3' end called RNAse D, which stops two bases at short of CCA terminus, though it later removes these two bases after the 5' end is processed. This enzymatic digestion leaves the molecule called pre-tRNA.

2. Formation of the 5'-P terminus

The 5'-P terminus is formed by an enzyme called RNAse P, which removes excess RNA from the 5' end of a precursor molecule by an endonucleolytic cleavage that generates the correct 5' end.

3. Production of modified bases

The final modification is to produce the altered bases in the tRNA. In tRNA, two uridines are converted to pseudo uridine (y), one guanosine to methyl guanosine (MG), one adenine to isopentyladenine (IPA) etc.

O. MUTATION

Any change in the sequence of DNA is called mutation. In other words, a *mutation* is a permanent change of the nucleotide sequence of the genome of an organism, virus or extrachromosomal DNA or other genetic elements. Mutations result from damage to DNA. A Mutation occurs when a gene is damaged. A Mutagen is responsible for causing mutation. Mutagen is a physical or chemical or biological agent, which cause mutation. The process of causing mautation is called mutagenesis. Ionizing radiations (Gamma rays, alpha particles), UV radiations, hydroxyl and superoxide radicals, deaminating agents (nitrous acid), Polycyclic Aromatic Hydrocarbon, intercalating agents (ethidium bromide, ethyl methane sulphonate), heavy metals (arsenic, cadmium, chromium, nickel) are considered as a mutagenic agents.

Types oF Mutations

Different types of mutations are observed due to alteration in physical, chemical and biological changes indicated in organisms. They are as follows.

a. Somatic mutations: This type of mutation occurs in somatic tissue and are called **somatic mutations**.

b. Gametic mutations: If the mutation occurs in gamete cells like sperms and ova are called **gametic mutations**.

c. Point mutation: This occurs in a small segment of DNA molecule. Changes occurs in a single nucleotide or nucleotide pair. Due to this milder alterations in the single or pair of nucleotides the mutations are called **point mutations**. It is of following types.

i. Deletion mutations: This mutation is due to loss or deletion of single nucleotide pair in a triplet codon of a gene and are called **deletion mutation**.

ii. Insertion or addition mutation: This point mutations is due to addition of one or more extra nucleotides to a gene and are called **insertion mutations**.

iii. Substitution mutation: This point mutation is due to replacement of a nucleotide of a triplet with another nucleotide. They may be of following types:

iiia. Transition: When a purine (*e.g.,* adenine) base of a triplet codon is substituted by another purine base (*e.g.,* guanine) or a pyrimidine (*e.g.,* thymine) is substituted by another pyrimidine base, (*e.g.,* cytosine) then such kind of substitution is called **transition**. The transitional substitution mutations occur due to tautomerization. Uncommon forms of DNA bases are generated by single proton shifts and are called **tautomers**.

iiib. Transversion: These mutations involves the substitution or replacement of a purine with a pyrimidine or vice versa and are called transversion **mutation**.

d. Multiple mutations or gross mutations – This type of mutation alters more than one nucleotide pair or entire **gene**. Then these mutations are called **gross mutations**.

di. Translocation: This type of mutation is due to movement of gene to a non-homologous chromosome and this mutation is known as **translocation**.

dii. Inversion: The movement of a gene within the same chromosome is called **inversion**.

e. Spontaneous mutations: This mutation occur suddenly in the nature and their origin is unknown. They are also called b**ackground mutation**.

f. Induced mutations: This mutation is induced artificially in the living organisms by exposing them to abnormal environment such as radiation, certain physical conditions.

g. Forward mutations: **This** mutation creates a change from wild type to abnormal phenotype. Most mutations are forward type.

h. Reverse or back mutations: The forward mutations are often corrected by error correcting mechanism, so that an abnormal phenotype changes into wild type phenotype.

i. Dominant mutations: This mutation occurs in dominant phenotypic expression are called dominant mutations.

j. Recessive mutations: Most types of mutations are recessive in nature and so they are not expressed phenotypically immediately. The phenotypic effects of mutations of a recessive gene are seen only after one or more generations, when the mutant gene is able to recombine with another similar recessive gene.

k. Lethal mutations: This mutation **results** in the death of the cells or organisms.

l. Subvital mutations: This mutation reduces the chances of survival of the organism.

m. Supervital mutations: This mutation cause the improvement of biological fitness under certain conditions.

n. Missence mutations: This mutation change the meaning of a codon by changing one amino acid into another.

o. Temperature sensitive mutations or Ts mutations: If the gene substitution produces a protein that is active at one temperature (typically 30°C) and inactive at a higher temperature (usually 40–42°C). This mutation is called Ts mutation.

p. Nonsense or chain termination mutations: This mutation results in the stop of protein synthesis. This mutation is due to generation of a termination codon (UAG, UAA or UGA). This also results in the production of a shorter protein.

q. Silent mutations: They change a nucleotide but not the amino acid sequence. This is a silent mutation because it leaves the protein sequence unchanged.

r. Autosomal mutations: This type of mutation occurs in autosomal chromosomes.

s. Sex chromosomal mutations: This type of mutation occurs in sex chromosomes.

Types of Mutagens

The agents that can cause or induce mutations in a gene are known as mutagens and the process of producing a mutation are called mutagenesis. If mutation in nature without the addition of a known mutagen, it is called spontaneous mutagenesis and the resulting mutation is spontaneous mutations. If a mutagen is used, the process is induced mutagenesis and mutation is induced mutation. Mutagens are two types :Physical and Chemical.

Physical Mutagens

These consist of high energy radiations which could penetrate living cells and affect the genetic material. The effect of radiations on living cells and tissues is directly proportional to the degree of penetration of the radiation. Radiations are of 2 types. Electromagnetic radiations and Particulate radiations. Electromagnetic radiations like X-rays and ultraviolet rays can penetrate more in cells and tissues. Penetration power of electromagnetic radiation is inversely proportional to their wave length. Particulate radiations are in the form of sub-atomic particles emitted from the atoms with high energy. Alpha particles, beta particles and neutrons fall in this category. Beta particles have more penetrating power than alpha particles because of its smaller size. However, neutrons without any change are extremely penetrant and can cause severe damage to the living tissues as well as genetic material. The x-rays and particulate radiations provides high energy radiations when they pass through cells and tissues and creates a positively charged ion. The ejected electron cannot remain in free state and therefore, is picked up by another ion creating a negative ion. These ions may combine with oxygen producing highly reactive chemical which may act on genes, chromosomes and other parts of the cells. Peroxides which are mutagenic may be formed in the presence of oxygen following the splitting of water.

Chemical Mutagens

A large number of chemicals are known to cause mutation employing different pathways. These chemical mutagens are classified into 4 major groups on the basis of their specific reaction with DNA. Base analogues are the chemical compounds structurally very similar to the normal nitrogenous bases of DNA. The analogues can get incorporated into replicating DNA in place of a normal base leading to base pair substitution mutations. Like the normal bases they also exist in two alternative forms i.e., keto or enol form or amino and imino form, and change spontaneously from one to another form. Alkylating agents such as ethyl methane sulphonate (EMS), ethylethane sulphonate, nitrogen and sulphur mustards and diethyl sulphate act on DNA by adding alkyl group to all four bases-However these agents show a strong preference for the base guanine. This results either in mispairing of affected base or loss entirely or creating a gap and cause mutation. Acridine dyes are the chemicals that intercalate between the bases of DNA. They include roflavin, acriflavine acridine orange compounds which can mimic base pairs and are able to slip themselves in between the nitrogenous bases. This results in deletion or addition of base pairs during replication. Nitrous acid reacts with the nitrogenous base and deaminates them by removing amino group from adenine, cytosine and guanine by oxidative deamination. This results in mispairing and base pair substitution mutations.

In addition to the above mentioned chemical and physical miutagens, there are a number of chemicals present in the environment that potentially mutagenic. Some major sources of natural mutagenic agents include parasitic fungi of field crops, mushrooms, certain vegetables and medicinal herbs . Other environmental mutagens include air and water pollutants, food additives and preservatives, agricultural chemicals ,cosmetics drugs, pesticides, cigarettes and industrial products such as benzidine ,vinyl chloride ,asbestose . In addition many potentially mutagenic compounds may be carcinogenic or capable of inducing cancer in humans.

Molecular basis of Mutations

Tautomerisation: The purins and pyrimidines in DNA and RNA may exist in several alternate forms called tautomers Tautomerism is a phenomenon where change in chemical forms occurs through rearrangement by electrons and protons in the molecule . As a result some single bonds become double bonds and vice versa .Such change in chemical forms of the base is called tautomeric shift. Although the normal bases possess potentially unstable bonds. They remain chemically stable in one tautomeric form most of the times. This stability is a significant genetic attriute.

Deamination: Deamination of particular bases is a common chemical event that produce spontaneous chemical mutations. Removal of an amino group from a base is called deamination. Example deamintaion of cytosine produces uracil. In case uracil is not repaired back, it will direct the incorporation of adenine in the new DNA strand during replication. This ultimately results in conversion of CG base pair to a TA base pair i.e.,a transition mutations.

Depurination: Depurination involves removal of a purine , from the DNA. This removal occurs due to the breakage of bond between purine and the deoxyribose. If this fault is not repaired, there will be no base to specify a complementary base during DNA replication. In case a randomly choosen base is inserted a mismatched base pair will be produced resulting into a genetic mutation.

Base Analogue: Base Analogue are the chemicals that have molecular structure that are extremely similar to bases of DNA. These chemicals act as mutagens and during DNA replication get incorporated so as to form base pairs with usual bases . One such chemical is 5 bromo uracil (5BU). 5-bromo uracil is a base analogue of thymine and usually atom in 5 BUdR so alter the charge distribution of the molecule that it may tautomerise to a 5 BUdR* form quite frequently. After tautomerisation it possesses the base pair properties of cytosine, i.e., it behaves like cytosine.

Factors effecting Mutations: They may replace a base in the DNA. They may alter the base in such a way that it specifically mispairs with another base. They may damage the base so much that it can do longer pair with any base. They may intercalate themselves in the DNA, paying way for addition or deletion of bases.

P.   DETECTION AND ISOLATION OF MUTANTS

Mutants can be detected and efficiently differentiated from the parent (wild) strain. if one aware of wild strain character, it is easy to detect mutants. Since frequency of mutation is low, it is very important to have very sensitive detection system so that the rare mutant may not be missed from detection. In bacteria, the detection system are straight forward because any new allele should be observed immediately. The following methods are used to detect mutants

Replica plating technique

It was first described by Lederberg and Lederberg in 1952. It is used to detect auxotrophic mutants on the basis of ability to grow in the absence of an aminoacid. It is also used to detect temperature sensitive mutants.

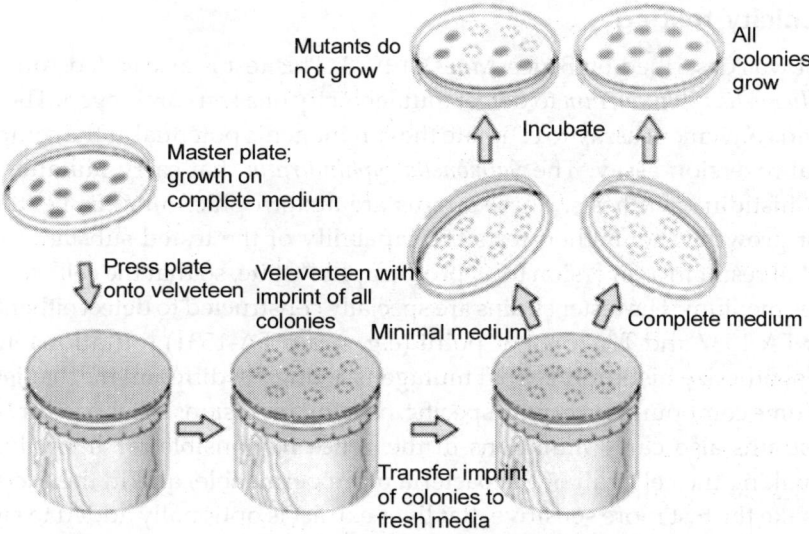

Figure 315   Replica plating

Resistance selection method

Wild type cells are not resistant to antibiotics. Therefore it is possible to grow the bacterium in the presence of the antibiotics.

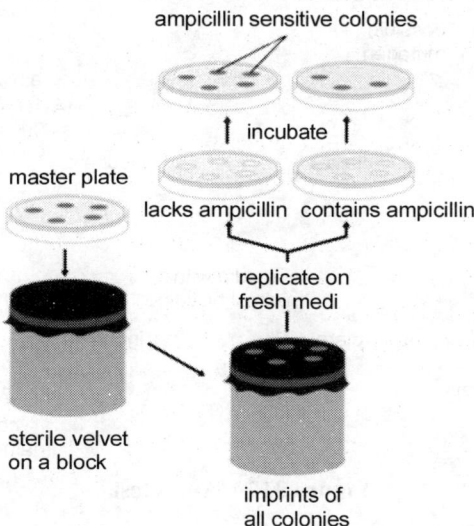

Figure 316   Selection of mutant

Substrate utilization method

Specific substrate utilization of wild or mutant strains are selected using indicator system. Lac- and Lac+ strains are selected using MacConkey agar. Here Lac+ stains are detected as pink colour colonies whereas mutants are detected as yellow colour colonies.

Carcinogenicity testing

This test was described by Bruce Ames in 1973. This test is also called Ames test. This test uses *Salmonella typhimurium* to detect mutagenicity of a test carcinogen. The test serves as a quick and convenient assay to estimate the carcinogenic potential at the compound. It is a mutational reversion assay. The *Salmonella typhimurium* that carry mutations in genes involved in histidine synthesis. These strains are auxotrophic mutants, i.e. they require histidine for growth. The method tests the capability of the tested substance in creating mutations that result in a reversion to a "prototrophic" state, so that the cells can grow on a histidine-free medium. The tester strains are specially constructed to detect either frameshift (e.g. strains TA-1537 and TA-1538) or point (e.g. strain TA-1531) mutations in the genes required to synthesize histidine, so that mutagens acting via different mechanisms may be identified. Some compounds are quite specific, causing reversions in just one or two strains. The tester strains also carry mutations in the genes responsible for lipopolysaccharide synthesis, making the cell wall of the bacteria more permeable, and in the excision repair system to make the test more sensitive. Rat liver extract is optionally added to simulate the effect of metabolism.

The bacteria are spread on an agar plate with a small amount of histidine. This small amount of histidine in the growth medium allows the bacteria to grow for an initial time and have the opportunity to mutate. When the histidine is depleted only bacteria that have mutated to gain the ability to produce its own histidine will survive. The plate is incubated for 48 hours. The mutagenicity of a substance is proportional to the number of colonies observed.

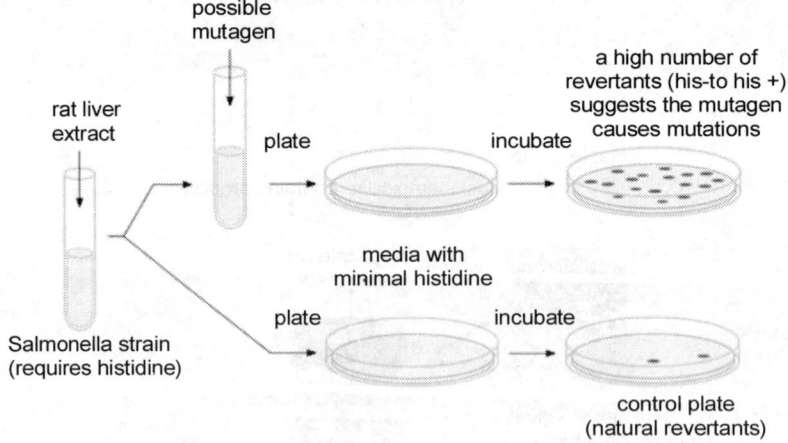

Figure 317   Ames test

Q. DNA REPAIR

There are several types of damage that occur in DNA. Only a few errors accumulate in DNA sequence. The stable error leads to cell death. Unstable or minimal errors are eliminated through repair mechanisms. DNA repair is the collection of processes employed by the cell that recognise and correct damage to the genome (i.e. mutation), either caused by endogenous agents such as metabolic damage or exogenous agents such as ionising radiation.

There are two major types of DNA repair mechanisms. It may depend on light or independent to light. They are photo reactivation (Light repair) repair and light independent repair (dark repair).

A. Light induced repair

Photoreactivation

The damage caused in cells are repaired after exposure of cells to visible light. This is called photoreactivation. In this, an enzyme DNA photolyase cleaves T-T dimer and reverses to monomeric stage. The enzyme photolyase is activated only when the DNA is exposed to visible light.

B. Light independent repair

Exision Repair

It is an enzymatic process. In this process, the damaged portion is removed and replaced by new piece of DNA. Normal DNA acts as a template for the synthesis of new DNA fragment. It is of multiple types.

1.   Base exision repair
2.   Nucleotide excision repair
3.   Recombination repair
4.   Methylation directed very short patch repair
5.   SOS repair

Base excision repair (BER) recognises and corrects minor chemical modifications in DNA bases. Minor modifications are defined as those that do not deley DNA replication. A family of DNA glycosylases recognise bases with a range of modifications, and upon recognition, cleave the glycosidic bond between the sugar and the base to leave an a purine or a pyrimidine site. AP endonucleases recognise AP sites and cleave the phosphodiester backbone at such sites. DNA polymerase I (*E.coli*) or DNA polymerase beta (Eukaryotes) adds the appropriate complementary nucleotide to the exposed 3'OH end. DNA ligase ligates the backbone together.

Nucleotide excision repair (NER) repairs leisons that distort the DNA double-helix (e.g. pyrimidine dimers). The DNA including the leison is removed as an oligonucleotide fragment. This fragment is 12-13 nucleotides long in prokaryotes, and 27-29 nucleotides long in eukaryotes. UvrA recognises and binds to damaged DNA. This recruits UvrB which melts the strands apart, and causes UvrA to dissociate. UvrB recruits UvrC, which cuts the DNA 8 nucleotides 5' of the leison, and 5 nucleotides 3' of the leison. Oligonucleotide fragment is removed with the help of UvrD helicase. DNA polymerase I and DNA ligase refil and seal the gap.

In recombination repair, damaged DNA for which there is no remaining template is restored. This situation arises if both bases of a pair are missing or damaged, or if there is a gap opposite a lesion. In this type of repair the recA protein cuts a piece of template DNA from a sister molecule and puts it into the gap or uses it to replace a damaged strand. Although bacteria are haploid, another copy of the damaged segment often is available because either it has recently been replicated or the cell is growing rapidly and has more than one copy of its chromosome. Once the template is in place, the remaining damage can be corrected by another repair system.

Postreplication repair (Methylation directed repair) is a type of excision repair. Successful postreplication repair depends on the ability of enzymes to distinguish between old and newly replicated DNA strands. This distinction is possible because newly replicated DNA strands lack methyl groups on their bases, whereas older DNA has methyl groups on the bases of both strands. DNA methylation is catalyzed by DNA methyltransferases and results in three different products: N6-methyladenine, 5-methylcytosine, and N4-methylcytosine. After strand synthesis, the *E. coli* DNA adenine methyltransferase (DAM) methylates adenine bases in d(GATC) sequences to form N6-methyladenine. For a short time after the replication fork has passed, the new strand lacks methyl groups while the template strand is methylated. The repair system cuts out the mismatch from the unmethylated strand.

SOS Repair The recA protein also participates in a type of inducible repair known as **SOS repair**. In this instance the DNA damage is so great that synthesis stops completely, leaving many large gaps. RecA will bind to the gaps and initiate strand exchange. Simultaneously it takes on a proteolytic function that destroys the lexA repressor protein, which regulates the function of many genes involved in DNA repair and synthesis. As a result many more copies of these enzymes are produced, accelerating the replication and repair processes. The system can quickly repair extensive damage caused by agents such as UV radiation, but it is error prone and does produce mutations. However, it is certainly better to have a few mutations than no DNA replication at all.

R. SEQUENCING OF GENE

Gene sequencing is a process in which the individual base nucleotides in an organism's DNA are identified. DNA sequencing may be used to determine the sequence of individual genes, larger genetic regions, full chromosomes. Sequencing provides the order of individual nucleotides in DNA or RNA commonly represented as A, C, G, T/U, isolated from cells. This is useful for:

- studying the genome itself, how proteins are made, what proteins are made, identifying new genes and associations with diseases and phenotypes, and identifying potential drug targets
- studying how different organisms are related and how they evolved
- Identifying species present in a body of water, sewage, dirt.
- Helpful in ecology, epidemiology, microbiome research, and other fields.

The following three methods are used for the determination of DNA sequences :

1. Maxam and Gilbert's Chemical Degradation Method

It was developed by Maxam and Gilbert in 1977. It is a chemical degradation method. This technique involves the following steps:

1. The 3' ends of DNA are labelled with ^{32}P.

2. The two strands of this radioactively labelled DNA are separated.

3. The mixture is divided into four samples, each treated with a different reagent having the property of destroying either only G, or only C, or A and G or T and C. The concentration of the

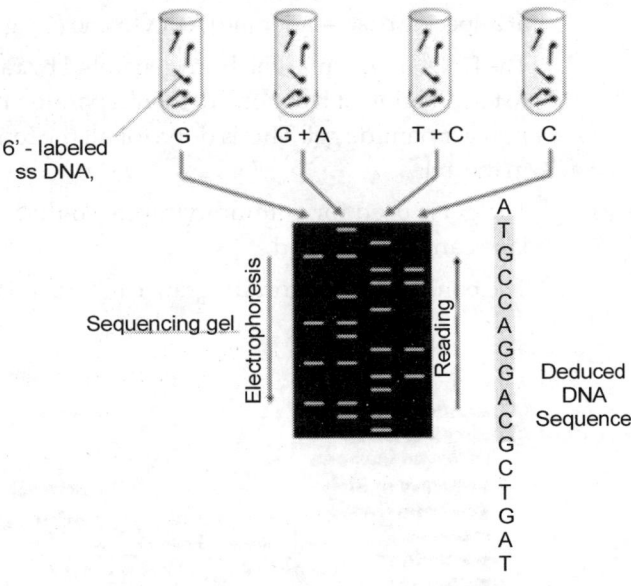

Figure 318 Gilberts-Gene sequences

reagent is adjusted in such a way that 50 per cent of target base is destroyed, so that fragments of different sizes having ^{32}P are produced. Tube 1 is mixed with dimethyl sulphate and piperidine to cut DNA at guanine bases. 2nd tube is treated with dimethyl sulfate, acid and piperidine to cut DNA at adenine bases. DNA in the 3rd tube is treated with hydrazine and piperidine to cut DNA at cytocine or thymine. 4th tube is treated with hydrazine, NaCl and piperidine to break DNA at cytosine residues. NaCl enhances the frequency of cutting.

4. Each of the four samples is electrophoresed in four different lanes of the gel.

5. The gel is autoradiographed to determine the sequence from positions of bands in four lanes.

2. Sanger's Dideoxynucleotide Synthetic Method

Fred Sanger had initially developed a method for DNA sequencing in 1980. It utilized DNA polymerase to extend DNA chain length. This was termed as **plus-minus method**. Later on, **Sanger** (1986) developed a more powerful method, utilizing single-stranded DNA as the template for DNA synthesis, in which **2', 3' dideoxynucleotides** were incorporated leading to termination of DNA synthesis. These dideoxynucleotides are used as triphosphates (ddNTP) and can be incorporated in a growing chain, but they terminate synthesis, since they fail to form a phosphodiester bond with next incoming deoxynucleotide triphosphate (dNTP). Thus, Sanger's dideoxy methods includes the following steps:

1. Four reaction tubes are set up, each containing single stranded DNA sample (cloned in M13 phage) to be sequenced, all four dNTPs (radioactively labelled) and an enzyme for

DNA synthesis (*i.e.*, DNA polymerase I). Each tube also contains a small amount of one of the four ddNTPs, so that each tube has a different ddNTP, bringing about termination at a specific base — adenine (A), cytosine (C), guanine (G) and thymine (T).

2. The DNA fragments which are generated by random incorporation of ddNTP leading to termination of reaction are then separated by electrophoresis on a high resolution polyacrylamide gel. This is done for all the four reaction mixtures on adjoining lanes in the gel.

3. The gel is used for autoradiography so that the position of different bands in each lane can be visualized.

4. The bands on the autoradiogram can be used for getting the DNA sequence.

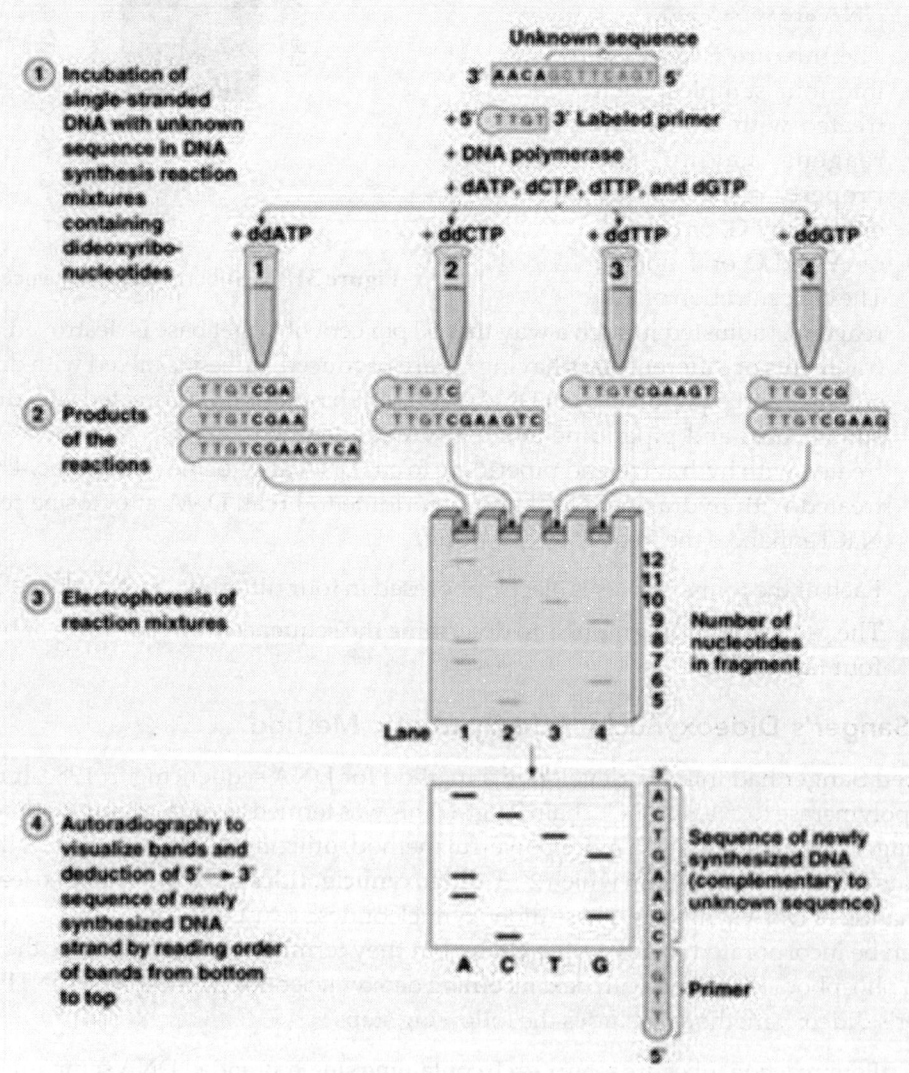

Figure 319 Sangers-Gene sequencing

3.   Automated DNA Sequencing

This method was first developed by Scharf et al., in 1986. This is a modified and advanced version of Sangers method. It is a combined method uses Polymerase chain reaction (PCR) and chain termination method. This method of DNA sequencing is faster and more reliable. It can utilize either the whole genomic DNA or the cloned fragments for sequencing a particular DNA segment.

The DNA sequencing involved the following methods

1.   The DNA is cut into 1000 to 80000bp size using restriction enzyme.

2.   The DNA fragments are separated electrophoretically.

3.   Separated each DNA individually taken in PCR reaction tube.

4.   Each tube is treated with primer, dATP, dCTP, dGTP and dTTP. A small amount of dideoxy nucleotides such as ddATP, ddCTP, ddGTP and ddTTP labeled with different fluorescent dye.

5.   PCR reaction is performed by adding Taq polymerase.

6.   The PCR product obtained in each tube is injected into the capillary tube containing and electrolyte.

7.   Electrodes are connected with the capillary tube and electric current is passed. The DNA fragments move through the capillary tube at different speeds depending on their size. Smallest fragment reaches the other end first.

8.   The capillary is illuminated with lazer beam.

9.   Fluorescence detector measures the extinction coefficient and relative quondam yield to detect the fluorescent dye attached to the specific dideoxynucleotides.

10.   The computer converts the data of quondam yield into the nucleotide or all DNA sequence data as programmed in the computer.

11.   This programme is repeated for all fragments.

12.   Finally, DNA sequence data is aliened to generate the complete sequence data of the DNA.

S.   POLYMERASE CHAIN REACTION (PCR)

This technique a simple and indigeneous method for the amplification of DNA. It was developed by Kary Mullis in 1985. Katy Mullis shared the Nobel prize for chemistry in 1993. Larger quantities of DNA are amplified from a single piece of DNA. For example single fragment of hair or a tiny blood stain left at the site of a crime is enough to amplify the DNA of a particular person.

Steps in PCR

Denaturation (Melting of Target DNA) (95°C): The DNA to be amplified is mixed with deoxyribonucleotides, Taq polymerase (a thermal stable DNA polymerase enzyme) and DNA primers (the DNA primers hydrizide to the end of the gene to be amplified & provide a starting point of Taq polymerase)

Primer Annealing (55°C): The mixture is heated to break down the hydrogen bond in the DNA Forming single-stranded molecules. The mixture cooled to allow the DNA primer to anneal to each end of the segment to be copied. The Taq polymerase then synthesized the complementary stands of DNA, using the primer as starting point.

Primer Extension (Polymerization) : The temperature raised again to separate the DNA stands and then lower the temperature to allow the primers to attached. Taq polymerase now synthesis another set of new complementary stands. This process is repeated until enough DNA has been produced to be identified or used for further research. After 21 cycles, one molecule of DNA can be amplified to over a one million copies. This amount of amplification can be achieved by running the reaction in overnight a thermal cycler, and the instrument that automatically raised and lower the tempemperature. appropriate time interval.

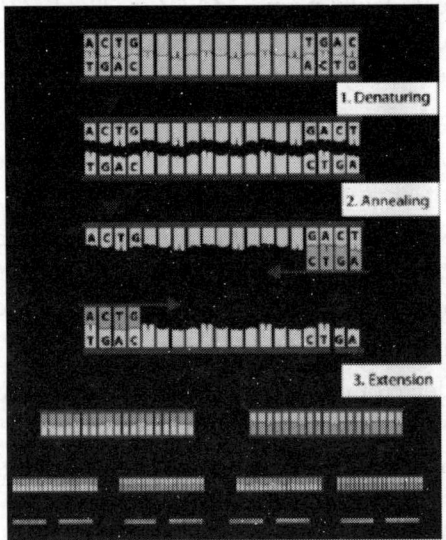

Figure 320   Polymerase Chain Reaction (PCR)

Application of PCR

Diagnosis of Pathogens: There are several pathogens that grow slowly. Therefore, their cells are found less in number in the infected cells/tissues. It is difficult to culture them on artificial medium. Hence, for their diagnosis, PCR-based assays have been developed. These detect the presence of certain specific sequences of the pathogens like bacteria and Virusespresent in the infected cells/tissues.

Diagnosis of Specific Mutation: In humans there are thousands of genetic diseases. Mutations are also related to genetic diseases. Presence of a faulty DNA sequence can be detected before establishment of disease. By using PCR sickle cell anaemia, phenylketonuria and muscular distrophy can also be detected. The other diseases can also be diagnosed by using PCR. For example PCR-based diagnostic tests for AIDS, Chlamydia, tuberculosis, hepatitis, human pappiloma virus, and other infectious agents and diseases are being developed. The tests are rapid, sensitive and specific.

In Prenatal Diagnosis: It is useful in prenatal diagnosis of several genetic diseases. If the genetic diseases are not curable, it is recommended to go for abortion.

DNA Fingerprinting: In recent years DNA fingerprinting is more successfully used in forensic science to search out criminals, rapists, solving disputed parentage and uniting the lost parentage   uniting children to their parents or relatives by confirming their identity. This is done through making link between the DNA recovered from samples of blood, semen, hairs, etc. at the spot of crime and the DNA of suspected individuals or between child and his/her parents/relatives.

In Molecular Archaeology (Palaentology): PCR has been used to clone the DNA fragments from the mummified remains of humans and extinct animals such as the woolly mammoth and dinosaurs from the remains of ancient animals as recently epitomized in Michel Crighton's Jurassic Park. DNA from buried humans has been amplified and used to trace the human migration that occurred in ancient time.

Diagnosis of Plant Pathogens: It is also applied in diagnosis of plant diseases. A large number of plant pathogens in various hosts or environmental samples are detected by using PCR, for example, viroids (associated with hops, apple, pear, grape, citrus, etc), viruses (such as TMV, cauli-flower mosaic virus, bean yellow mosaic-virus, plum pox virus, potyviruses), mycoplasmas, bacteria (Agrobacterium tumifticiens, Rhizobium legunfinosaritin, Xanthomonas compestris, etc), fungi (e.g., collectotrichum gloeosporioides, Glomus spp., Laccaria spp., Phvtophthora spp., Verticillium spp), and nematodes (e.g. Meloido, vne incoginta, M. javanica, etc).

T.   DNA FINGER PRINTING

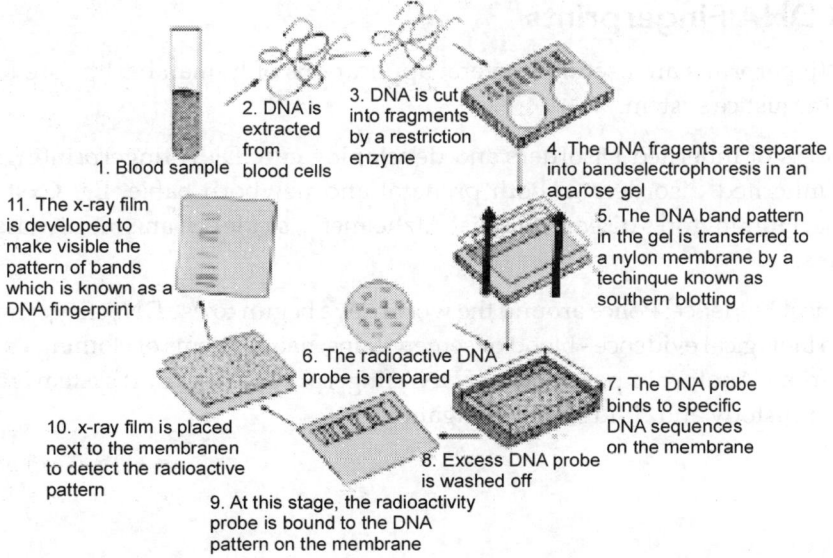

Figure 321   Steps in DNA finger printing technique

DNA fingerprinting is also called as DNA typing, DNA profiling, genetic finger printing, genotyping. The technique was developed in 1984 by British geneticist Alec Jeffreys, after he noticed that certain sequences of highly variable DNA (known as minisatellites), which do not contribute to the functions of genes, are repeated within genes. Jeffreys recognized that each individual has a unique pattern of minisatellites. DNA fingerprinting is rapidly becoming the primary method for identifying and distinguishing among individual genome. DNA fingerprinting is a very quick way to compare the DNA sequences of any two living organisms. DNA fingerprinting is a laboratory procedure that requires the following steps.

Isolation and amplification of DNA: DNA must be recovered from the cells or tissues of the body. Only a small amount of tissue - like blood, hair or skin - is needed. For example, the amount of DNA found at the root of one hair is usually sufficient. Isolated DNA is subjected to amplification by PCR.

Cutting, sizing, and sorting: Special enzymes called restriction enzymes are used to cut the DNA at specific places. For example, an enzyme called EcoR1, found in bacteria, will cut DNA only when the sequence GAATTC occurs. The DNA pieces are sorted according to size by a sieving technique called electrophoresis. The DNA pieces are passed through a agarose gel.

Transfer of DNA to nylon: The distribution of DNA pieces is transferred to a nylon sheet by placing the sheet on the gel and soaking them overnight.

Probing: Adding radioactive or colored probes to the nylon sheet produces a pattern called the DNA fingerprint. Each probe typically sticks in only one or two specific places on the nylon sheet.

DNA fingerprint: The final DNA fingerprint is built by using several probes (5-10 or more) simultaneously. It resembles the bar codes used by grocery store scanners.

Uses of DNA Fingerprints

DNA fingerprints are useful in several applications of human health care research, as well as in the justice system.

Diagnosis of Inherited Disorders and developing cure:DNA fingerprinting is used to diagnose inherited disorders in both prenatal and newborn babie. Eg. Cystic fibrosis, hemophilia, Huntington's disease, familial Alzheimer's, sickle cell anemia, thalassemia, and many others.

Biological Evidence: Police around the world have begun to use DNA fingerprints to link suspects to biological evidence - blood or semen stains, hair, or items of clothing - found at the scene of a crime. Another important use of DNA fingerprints in the court system is to establish paternity in custody and child support litigation.

17

MICROBIAL BIOTECHNOLOGY

A. DEFINITIONS

The term biotechnology was introduced by a Hungarian engineer, Karl Ereky. He defined biotechnology as all lines of work by which products are produced from raw materials with the aid of living things. European Federation of biotechnology defined biotechnology as the integration of natural science and organisms, cells, parts thereof and molecular analogues for products and services. The applications of scientific and engineering principles to the processing of materials by biological agents to provide goods and services.

The integrated use of biochemistry, microbiology and engineering sciences inorder to achieve technological applications of the capabilities of the microorganisms, cultured tissues/cells and part thereof. In general Biotechnology means any modification made on biological system are called Biotechnology.

B. BASIC PRINCIPLES OF BIOTECHNOLOGY

Biotechnology involves the principles of genetic engineering and Molecular cloning.

Genetic Engineering

Genetic engineering is also called genetic modification. It is the direct manipulation of an organism's genome using biotechnology. Genetic engineering alters the genetic makeup of an organism using techniques that remove heritable material or that introduce DNA prepared outside the organism either directly into the host or into a cell that is then fused or hybridized with the host. This involves using recombinant nucleic acid (DNA or RNA) techniques to form new combinations of heritable genetic material followed by the incorporation of that material either indirectly through a vector system or directly through micro-injection, macro-injection and micro-encapsulation techniques.

It involves isolation, manipulation & expression of genetic (engineering) material.

Molecular Cloning

Molecular cloning is a set of experimental methods in molecular biology that are used to assemble recombinant DNA molecules and to direct their replication within host organisms. The use of the word cloning refers to the fact that the method involves the replication of a single DNA molecule starting from a single living cell to generate a large population of cells containing identical DNA molecules. Molecular cloning generally uses DNA sequences from two different organisms: the species that is the source of the DNA to be cloned, and the species that will serve as the living host for replication of the recombinant DNA. Molecular cloning methods are central to many contemporary areas of modern biology and medicine.

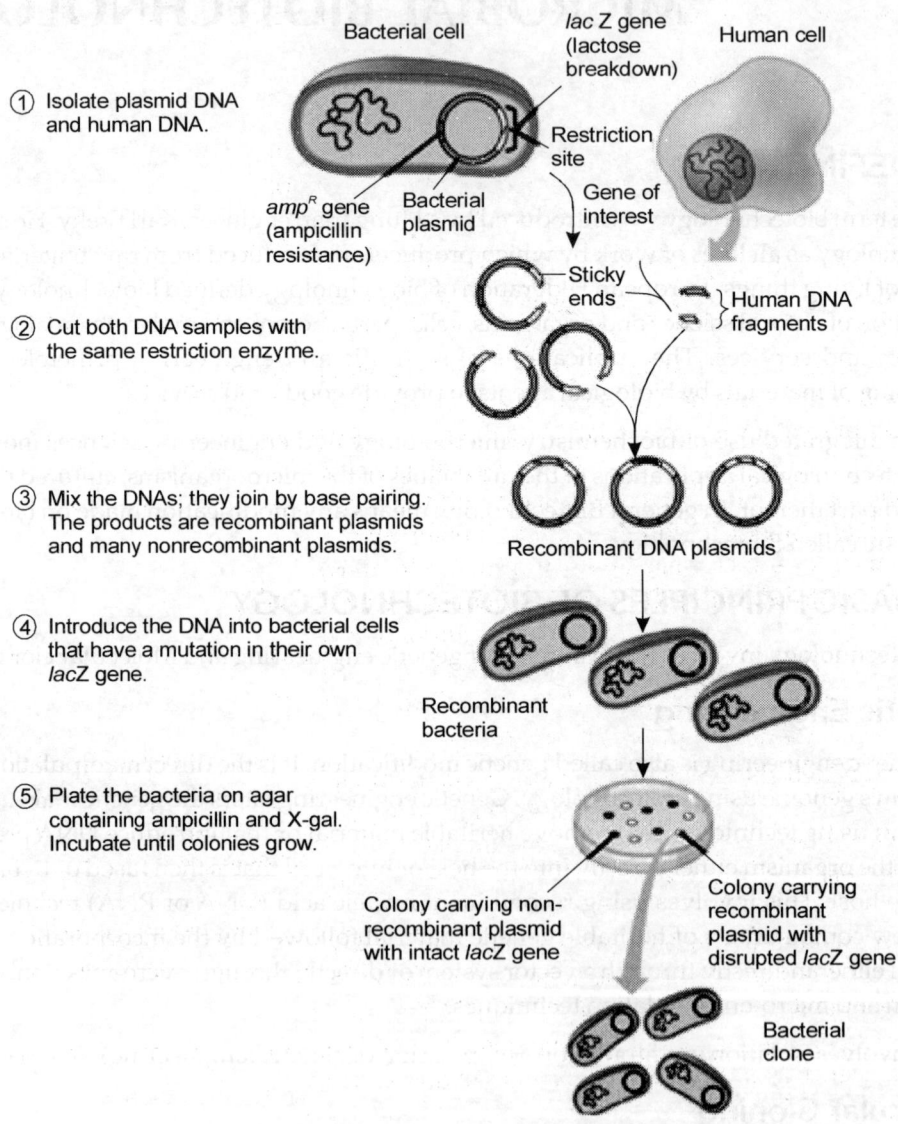

① Isolate plasmid DNA and human DNA.

amp^R gene (ampicillin resistance)

Bacterial plasmid

Bacterial cell

lac Z gene (lactose breakdown)

Restriction site

Gene of interest

Human cell

Sticky ends

Human DNA fragments

② Cut both DNA samples with the same restriction enzyme.

③ Mix the DNAs; they join by base pairing. The products are recombinant plasmids and many nonrecombinant plasmids.

Recombinant DNA plasmids

④ Introduce the DNA into bacterial cells that have a mutation in their own *lacZ* gene.

Recombinant bacteria

⑤ Plate the bacteria on agar containing ampicillin and X-gal. Incubate until colonies grow.

Colony carrying non-recombinant plasmid with intact *lacZ* gene

Colony carrying recombinant plasmid with disrupted *lacZ* gene

Bacterial clone

Figure 322   Basic steps of molecular cloning

Basic steps of molecular cloning

1. Isolation and fragmentation of source DNA (Source DNA is also known as Target DNA (or) Desired DNA (or) FDNA (or) DNA of interest).

2. Insertion of DNA fragments into suitable cloning vector with the help of Restriction enzyme and DNA ligases (Restriction enzymes are also known as molecular scissors or Restriction endo nucleases).

3. The recombinant or Chimeric DNA is introduced into suitable expression system or host cell.

4. Mass production of cell containing desired gene.

C. CONCEPTS IN BIOTECHNOLOGY

Biotechnology is an interdisciplinary pursuit with multidisciplinary applications. Roots of biotechnology include biochemistry, genetics, molecular biology, chemical engineering, physiology, microbiology, cell biology, food sciences and material sciences. A large number of scientists working in this area have contributed to the development of biotechnology.

Several methods, techniques or procedures, which may be collectively called as biotechnological tools. It have been developed for transforming scientific fundamentals (root) into biotechnological applications. These tools include metabolic engineering, immunotechnology, DNA Chip technology, transgenic technology, cell and tissue culture technology, monoclonal antibody technology, bioprocess technology, protein engineering, biosensor technology, pharmaceutical technology, recombinant DNA technology, enzyme technology, agricultural biotechnology, bioinformatics, proteomics, genomics etc., By the aid of these tools of biotechnology, we may consume fruits of biotechnology (Application).

Biotechnology has benefitted medical and health sciences (Diagnostic, vaccines, therapeutics, food), agricultural sciences (improved crop yield, food quality, improved animal health) and environmental sciences (Pollution control, environmental monitoring, bioremediation).

Tissue Culture Technology

It is also known as *in-vitro* propagation of plants, seeds and various parts of the plants. Tissue culture technology is often coupled with rDNA technology, which improve crop yield and quality. Developments in tissue culture technology have helped to produce several pathogen free plants, besides the synthesis of many biologically important compounds including pharmaceuticals.

Genetic Recombination occurs through the insertion of Foreign DNA into suitable plant cell by particle gun method.

Pharmaceutical Technology

Production of biologically important compound which help in the treatment and prevention of severe disease are called Pharmaceutical technology. Through rDNA technology, lots of pharmaceutically important products are produced that are categorized as human protein replacement, therapeutic agents for human diseases and vaccines. Interferons, haemoglobin, relaxin, insulin, tumour necrosis factor, coagulation factor VIII, growth hormone, erythropoietin are important pharmaceutically important products used for the prevention and treatment severe diseases.

Fermentation Technology or Bio Process Technology

Involvement of Microorganisms in the commercial production of industrial products are called Bio process / fermentation technology. It is very widely exploited for industrial applications. Food (cheese, youghurt, vitamin B, B_{12}, aminoacids, baker's yeast, beverages) chemicals (ethanol, butanol, citric acid, lactic acid, enzymes, polymers), Pharmaceuticals (Antibiotics, Vaccines, steroids, diagnostic, enzymes, monoclonal antibodies), Agriculture product (SCP, microbial pesticides, composting process, plant cell tissue culture) are the important products produced through bioprocess/fermentation technology.

Recombinant DNA Technology

This involves the manipulation of genetic materials to achieve the desired goal. rDNA technology involves generation of DNA fragments & selection of the desired piece of DNA, Insertion of the selected DNA into a cloning Vector to create recombinant DNA or chimeric DNA. Introduction of the recombinant Vectors into host cell. Multiplication and selection of clones containing the recombinant molecules Expression of the gene to produce the desired product.

Enzyme Technology

This technology broadly involves production, isolation, purification and use of enzymes in soluble or immobilized form. rDNA technology and protein engineering involved in the production of more efficient and useful enzymes for the benefit of human kind.

Agriculture Biotechnology

Genetic improvement of crop varieties to produce maximum yield and resistance to insects and pathogens. Uses of improved variety of microorganisms are also a part of agriculture biotechnology.

Microbial Mining and Metal Biotechnology

Microorganisms can be successful for the extraction of metal form low grade ores. Mining with microbes is both economical and environmental friendly. Bioleaching and Bio sorption are the process used for the recovery of metal.

Environmental Biotechnology

Environmental pollution detection and monitoring is being done by approaches involving bio systems. Plants, animals and microorganisms are utilized successfully for this purpose. Visual rating, genotoxicity rating, metabolic rating, bio assays, biosensors, biomarkers, DNA probes, reporter genes are employed for the diagnostic, preventive and remedial measures of environmental pollution.

Animal Biotechnology

This involves cultivation of cells for the clinical & diagnostic purpose. Transgenic animal is a part of animal biotechnology

Bioinformatics

It is a combination of biology and Information technology. It broadly involves the computational tools and methods used to manage, analyse and manipulate volumes and volumes of biological data. It is also called computational biology. Genomics, proteomics, DNA micro array are part of bioinformatics. *Mycoplasma genitalium* (small bacterium), *Haemophiles influenza* (Pathogenic human bacteria), *Saccharomyces cerevisiae* (Fungi), Homosapiens (mammals), *Arabidopsis thaliana* (plant) are the first genome sequenced in different groups of life successfully through genomic process.

Genomics

The branch of Bioinformatics that helps in sequencing of complete genomes and their comparative analysis are called genomics.

Table 17.1   Examples of genome sequences

Gene size	Organism	Nature of sequence
5.38 kb	θ × 174	First bacteriophage genome to be sequence
580 Kb	*Mycoplasma genitalium*	Smalled known Gene to be sequenced
1830 Kb	*Haemophillus influenzae*	First human pathogen to be sequenced.
12,050 Kb	*Saccharomyces cerevisiae*	First single cell eukaryote Genome sequenced
97 Mb	*Caenorhabditis elegans*	First multicellular Genome sequenced
3000 Mb	*Homo sapiens*	First mammalian genome to be sequenced
125 Mb	*Arabidopsis thaliana*	First plant genome to be sequenced

D.   PRODUCTS OF BIOTECHNOLOGY

The following products are obtained through Biotechnology.

Animal Biotechnology

Gene therapy, Diagnostic tools, DNA finger printing, Insulin, Relaxin, Interleukin, Interferon, Haemoglobin, erythropoietin, Somatotropin, Growth hormone etc,..

Recombinant Vaccines

Genetic engineering can be applied in many vaccine production for viral and bacterial diseases. Sub unit vaccine for HBV was made cloning Hepatitis B virus DNA in yeast and its expression vectors. Recombinant live vaccines were manufactured using Biotechnology. Eg: Polyvalent vaccine is produced by cloning two different gene in single vector which protect animals from fowl pox and new castle disease. Vaccinia virus is widely employed in the production of Recombinant viral vaccines.

Anti viral vaccines

Rotavirus Vaccines, Influenza A & B Vaccine, chickenpox Vaccines, HSV 2, Dengue, HAV, HBV, Yellow fever, Japanese encephalitis Vaccines for viral diseases.

Antibacterial vaccines

Cholera, Tuberculosis, pneumonia, Tetanus, pertusis, Diphtheria, Typhoid vaccines.

Antiparacitic vaccines

Filariasis, malaria, River blindness, sleeping sickness, Schistosomiasis are the vaccines used for human disease prevention process.

Advantages of Recombinant vaccines are

Safer than attenuated or killed vaccines.

It can be given in high doses without side effects.

Food

Cheese, yogurt, Vitamin B1 & B12, glutamic acid, Lysine, Glucose and Fructose Syrub, mushroom products, Bakers yeast, Antioxidants, colours, flavours, beer, wine, whisky.

Chemicals

Ethanol, butanol, acetone, citric acid, gluconic acid, lactic acid, Enzymes, Polymers like Xanthan, dextran.

Pharmaceuticals

Antibiotics, Vaccines, steroids, diagnostic enzymes, monoclonal antibiodies, enzyme inhibitors.

Agriculture

SCP, microbial pesticides, compost, plant cell & tissue culture, Transgenic plants, Transgenic animals.

Human Genetics & Gene Therapy

Cloning and sequencing of entire human genome is achieved through principles of DNA sequencing or PCR based amplification. It is easy to clone the region containing the genetic defect and compare it with normal sequence.

Eg: Genes for trisomy - 21 and cystic fibrosis.

Gene therapy is the process by which a non functional or dys functional gene is replaced by a functional gene. The first genetic disease treated with genetherapy is severe combined immune Defeciency (SCID) caused by the defective nature of Adenosine deaminase (ADA), the key enzyme involved in purine metabolism.

Table 17.2 Microorganisms in Biotechnology

S.NO	Produce	Micobes involved and Technology	Benefits
1	Corn	*Bacillus thuringiensis* inserted into corn	Insect resistance
2	Potato	Starch production gene from *E.coli* is inserted	Potato with high starch & less fat
3	Soyabean	Improved Strain of *Rhizobium* used to enhance N_2 fixation	Productivity of the Crop improved
4	Tomato	*E.coli* – Antisense RNA technology which delays pectin degradation	Delays ripening of tomatoes
5	Insecticide against Catterpillars	Scorpion tox gene is inserted into insect virus & implanted into plants	Natural, non-polluting cost effective insecticide
6	Bioremediation agent	Citrobacter which absorb nuclear waste used in Biotechnological approach	Pollution management
7	Gene therapy	Functional gene coding for Adenosine deaminase inserted into T Lymphocyte using retrovirus vector.	Therapy for Serve Combined Immuno deficiency (SCID)
8	Recombinant vaccine	Gene for HBV surface antigen cloned into yeast Chromosome	Immunization against Hepatitis
9	Single Cell Protein	Plasmid Containing glutamate dehydrogenase gene for *E.coli*, introduced into *Methylophilus methylothrophs*	5 % increased productivity
10	α amylase	Gene from *Bacillus coagulans* cloned into *E.coli*	Increased enzyme production
11	Insulin	Ins gene Cloned & expressed in *E.coli*.	Recombinant Insulin to treat diabetes
12	Xanthane gene	*Xanthomonas campestris* genetically engineered to grown on whey Lac z, x, y genes inserted	Economically valuable biopolymer from waste.

E. HISTORY OF BIOTECHNOLOGY

Around 7000 BC	Brewing beer, fermenting wine, baking bread with the help of yeast.
8000 BC - 3000 BC	Yogurt and cheese made with lactic acid-producing bacteria by various cultures.
1590	The microscope is invented by Sacharias Jansen.
1675	Micro-organisms discovered by Antonie van Leeuwenhoek.
1856	Gregor Mendel discovers the laws of inheritance.
1862	Louis Pasteur discovers the bacterial origin of fermentation.
1919	Karl Ereky, a Hungarian agricultural engineer, first uses the word biotechnology.
1928	Alexander Fleming notices that a certain mould could stop the duplication of bacteria, leading to the first antibiotic: Penicillin.
1953	James D. Watson and Francis Crick describe the structure of DNA.
1972	The DNA composition of chimpanzees and gorillas discovered. It is 99% similar to that of humans.
1973	Stanley Norman Cohen and Herbert Boyer perform the first successful recombinant DNA experiment.
1975	Method for producing monoclonal antibodies developed by Köhler and César Milstein.
1978	North Carolina scientists Clyde Hutchison and Marshall Edgell show it is possible to introduce specific mutations at specific sites in a DNA molecule.
1980	The U.S. patent for gene cloning is awarded to Cohen and Boyer.
1982	Humulin, produced by genetically engineered bacteria for the treatment of diabetes. It is the first biotech drug approved by the Food and Drug Administration.
1994	The United States Food and Drug Administration approves the first GM food: the "Flavr Savr" tomato.
1997	British scientists, led by Ian Wilmut from the Roslin Institute, report cloning Dolly the sheep using DNA from two adult sheep cells.
2000	Completion of a "rough draft" of the human genome in the Human Genome Project.
2001	The sequence of the human genome was published in Science and Nature.
2004	the first cloned pet-a kitten is deliverd to its owner. She is called Copycat.
2006	a recombinant vaccine against Human papilloma virus got FDA approval.

F. ENZYME TECHNOLOGY

Definition

Enzymes are chemical catalysts that control various biological and chemical reactions. An enzyme is a protein molecule that is a biological catalyst.

Characteristics of Enzymes

All enzymes are globular proteins.

They increase the rate of reactions.

They are very specific in their reaction.

A single enzyme catalyses a single chemical reaction.

They are sensitive to pH, temperature and substrate concentration.

Some enzyme requires co factor for proper functioning.

Classification of Enzymes

The top level classification are

⅄ Oxido reductases catalyse oxidation or reduction reactions E.g., Lactate dehydrogenase.

⅄ Transferases transfer a functional group from a donor to recipient (Methyl or phosphate group) Eg: Phosphosyl transferase.

⅄ Hydrolases catalyse the hydrolysis of various bonds. Eg. Chymotripsin; Peptidases.

⅄ Lyases are enzymes that eleaves C-C, C-O, C-N and other bonds by other means than hydrolysis or oxidation.

⅄ Isomerases catalyse isomerization changes within a single molecule. Eg. Alanine racemases; Cis trans Isomerases.

⅄ Ligases Join two molecules with covalent bonds. Eg.Aminoacid + RNA ligases.

Classification based on the substrate concentration

Constitutive enzyme : This enzyme is always produced in small quantities in the absence or milder quantities of substrate in the medium.

Adaptive enzymes: These enzymes are produced in unusable amount only in response to the particular substrate in the medium.

Classification based on the site of action

Endo Enzymes

Most of the enzymes produced by the cell for the cell function within the cell are called endo enzyme. It is also called as intracellular enzymes. Eg: Helicases.

Exo Enzymes

Some enzymes are liberated by the living cells and catalyse reactions in the cells environment. Such enzymes are called exo exymes. It is also called as extracellular enzyme.

Eg: Amlyase.

Chemical Nature of Enzymes

Many enzymes are pure proteins and are called apoenzyme. Some of the enzyme also have non protein component and are called cofactor. Cofactor is required for catalytic activity. The complete enzyme consist of the apoenzyme and its cofactor. It is called the holo enzyme.

$$\text{Holo enzyme} \rightarrow \text{Apo enzyme} + \text{Co enzyme}$$
$$\text{(Protein)} \qquad \text{(Prosthetic group)}$$

If the co factor is firmly attached to the apoenzyme are called prosthetic group. If the cofactor is loosely attached to the apoenzyme are called co enzyme. NAD is a co enzyme, that carries electrons. Metal ions may bound to apo enzyme and act as a cofactor.

Factors influencing Enzyme Action

Optimum: the environmental state where the enzyme functions the most efficiently

Maximum: the maximum environmental limit in which the enzyme can function

Minimum: the minimum environmental limit in which the enzyme can function

Inhibitors

Competitive inhibitors

Competitive inhibitors are agents that fill the active site of an enzyme

They compete with the **substrate**

Non-competitive inhibitors

These agents do not compete with the substrate for the enzyme's active site, but rather a different region of the enzyme. This is known as **allosteric inhibition**.

Mechanism of enzyme action

Different theories are proposed for explaining mechanism of enzyme action. One of the well accepted model is lock and key theory.

Lock and Key Theory

This theory is first postulated in 1894 by Emil Fischer. This theory postulated that, the lock is the enzyme and the key is the substrate. Only the correctly sized key (substrate) fits into the key hole (active site) of the lock (enzyme).

Smaller keys, larger keys, or incorrectly positioned teeth on keys (incorrectly shaped or sized substrate molecules) do not fit into the lock (enzyme). Only the correctly shaped key opens a particular lock.

Coenzymes

Coenzymes are organic molecules that are required by certain enzymes to carry out catalysis. They bind to the active site of the enzyme and participate in catalysis but are not considered substrates of the reaction. Coenzymes often function as intermediate carriers of electrons, specific atoms or functional groups that are transfered in the overall reaction. The first type of enzyme partner is a group called **cofactors**. Cofactors are not proteins but rather help proteins, such as enzymes, although they can also help non-enzyme proteins as well. Examples of cofactors include metal ions like iron and zinc.

coenzyme	abbreviation	entity transfered
nicotine adenine dinucelotide	NAD - partly composed of niacin	electron (hydrogen atom)
nicotine adenine dinucelotide phosphate	NADP -Partly composed of niacin	electron (hydrogen atom)
flavine adenine dinucelotide	FAD - Partly composed of riboflavin (vit. B2)	electron (hydrogen atom)
coenzyme A	CoA	Acyl groups
coenzymeQ	CoQ	electrons (hydrogen atom)
thiamine pyrophosphate	thiamine (vit. B1)	aldehydes
pyridoxal phosphate	pyridoxine (vit B6)	amino groups
biotin	biotin	carbon dioxide
carbamide coenzymes	vit. B12	alkyl groups

Cofactor

Cofactor is a organic molecules that bind to enzymes and help them function. Cofactors are often classified as inorganic substances that are required for, or increase the rate of, catalysis.

cofactor	enzyme or protein
Zn^{++}	carbonic anhydrase
Zn^{++}	alcohol dehydrogenase
Fe^{+++} or Fe^{++}	cytochromes, hemoglobin
Fe^{+++} or Fe^{++}	ferredoxin
Cu^{++} or Cu^{+}	cytochrome oxidase
K^{+} and Mg^{++}	pyruvate phosphokinase

Micro Organisms in Enzyme Technology

Enzymes have been produced commercially from plants, animals and microbial sources. However, microbial enzymes have the enormous advantage of being able to be produced in large quantities by established fermentation techniques. It is also noted that microbial system is easier to improve the productivity. Animal and plant enzymes are also produced by microorganism through recombinant DNA technology.

Table 17.3 Micro organisms in enzyme production

Enzyme	Activity	Production Strain	Application
Amylase	Starch Conversion	*Bacillus amylolique facience* *Bacillus licheniformis* *Bacillus megaterium* *As per gillus sp.*	Brewing, Baking & Milling industry manufacture of syrups Digestive aids.
Protease	Hydro lyses protein	*Bacillus subtilis* *Aspergillus oryzae* *Aspergillus niger*	Curdling milk Improve dough texture Mashing De hairing Protein hydrolystate
Pectinase	Hydrolyses pectin	*Aspergillus niger* *Aspergillus flavus* *Sclerotina sp* *Bacillus sp*	Coffee bean fermentation Fruit Juice Clarification Soup preparation Pectin degradation
Lactase	Split Lactose to Galactose	*Streptococcus lactis* *Aspergillus oryzae* *Rhizopus sp*	Production of whole milk concentrate Ice cream Frozen desserts
Lipase	Hydrolyses Lipid	*Rhizopus oryzae*	Wax industries
Cellulase	Hydrolyes cellulase	*Trichoderma sp*	Vegetable Soup Cellulose degradation
Collagenase	Hydrolyses collagen	*Clostridium histolyticum*	Medical industry
Glucose Isomerase	Converts glucose to fructose	*Streptomyes sp*	Fructose syrup
Invertase	Hydrolyses Sucrose	*Saccharomyces cerevisiae*	Wine, beer, & alcohol industry
Streptokinase	Digestion of clotted blood	*Streptococcus sp*	Anti blood clot
Cellulose oxidase	Oxidation of Cellulose	*Aspergillus niger*	Food Industry
Glucose oxidase	Removal of glucose	*Bacillus, Aspergillus sp*	Food Industry

- Wide range of biological sources are available to produce commercial enzymes. At present, a great majority (80%) of them are from microbial sources.
- Micro organisms are the most significant and convenient source of commercial enzyme.

The Technology of Enzyme Production - General

The salient features of enzyme production are

- Selection of organisms
- Formulation of medium
- Production Process
- Recovery and purification of enzymes

Selection of Organism

The most important criteria for selecting the microorganisms are as follows

1. The organisms must have the ability to produce enzyme in large amount within a limited time.
2. Strain should have stable biochemical characters.
3. Strain must grown in cheap raw material.
4. Isolation and purification of enzyme should be easy.
5. The strain should be non pathogenic.
6. The strain should capable of genetic manipulation.
7. The organism should adjust to temperature, pH and other condition of the culture medium in the fermenter.

Inoculum can be prepared in a liquid medium using most suitable organism.

Formulation of Medium

Culture medium should contain all the nutrients to support adequate growth of micro organisms. The ingredients of the medium should be readily available at low cost and are nutritionally safe.

Some of the commonly used substrates are starch hydrolysate, molasses, corn steep liquor, yeast extract, whey and soy bean meal. Some cereals such as wheat and pulses (peanut) have also been used.

The pH of the medium should be kept optimal for good microbial growth and enzyme production.

Production Process

Enzymes are mostly produced by submerged liquid conditions and to a lesser extent by solid substrate fermentation.

In submerged culture technique, the yields are more and chances of infection is less.

Solid state fermentation is used for the production of fungal enzymes.

Batch or continuous sterilization techniques are employed for the sterilization of medium.

The growth conditions like pH, temperature, O_2 supply and nutrient addition are maintained at an optimal levels.

The froth formation can be minimized by adding anti foam agent.

Enzyme production is carried by batch fermentation.

The duration of fermentation is around 2-7 days.

Recovery and Purification of Enzymes

Enzymes may be extra cellular or intra cellular. The enzymes may be crude or purified. It also may be solid or liquid form. Recovery of an extra cellular enzyme is easy. Recovery of intra cellular enzymes needs special techniques for cell disruption. Physical, chemical and enzymatic methods are available to break the cell and release their contents.

The recovery and purification steps will be the same for both intracellular and extra cellular enzymes.

Removal of Cell Debris

Filtration or centrifugation can be used for the removal cell debris.

Removal of Nucleic Acid

It interferes with the recovery and purification. It can be precipitated and removed by adding polyamines streptomycin and polyethyleneimine.

Enzyme Precipitation

Enzyme can be precipitated using salts (NH_4SO_4), organic solvents (Iso propanol, ethanol, acetone).

Liquid Liquid Partition

Concentration of desired enzyme is done by liquid liquid extraction using polyethylene glycol.

Separation by Chromatography

Ion exchange, affinity and dye ligand chromatography are used to separate and purify enzymes.

Drying and Packing

Concentrated form of the enzyme can be obtained by drying. This can be done by film evaporation and freeze dryers. The dried enzyme can be packed and marketed. All enzymes used in foods or medical treatments must be of high grade purify.

Regulation of Microbial Enzyme Production

A maximal production of microbial enzymes can be achieved by optimizing the fermentation condition.

Induction

Several enzymes are inducible i.e. they are synthesized only in the presence of inducers such as sucrose, starch. Invertase, Amylase, Lipase, β galactosidase are examples for inducible enzymes. Starch induce production of Amylase.

Feed Back Repression

Feed back regulation by the end product significantly influences the enzyme synthesis.

Nutrient Repression

The inhibition of unwanted enzyme production is done by nutrient repression. The nutrient may be carbon, nitrogen, phosphate or sulfate suppliers in growth medium.

Glucose repression is a good example. If glucose is present in the medium, organism will not produce other enzyme. Hence non glucose carbohydrate may be used.

Genetic engineering for microbial enzyme production

Enzymes are functional products of gene. A enzyme with potential use in industry is identified, the relevant gene can be cloned and inserted into a suitable production host.

Cloning Strategies

The gene responsible for Chymosin (Rennet) isolated from the young calves and inserted into a bacterium. This leads to a production higher qualities of chymosin. Chymosin is a enzyme present in calves which is used for making cheese. In general chymosin supply is very less but requirement is high, through microbes higher quantity in chymosin may be produced.

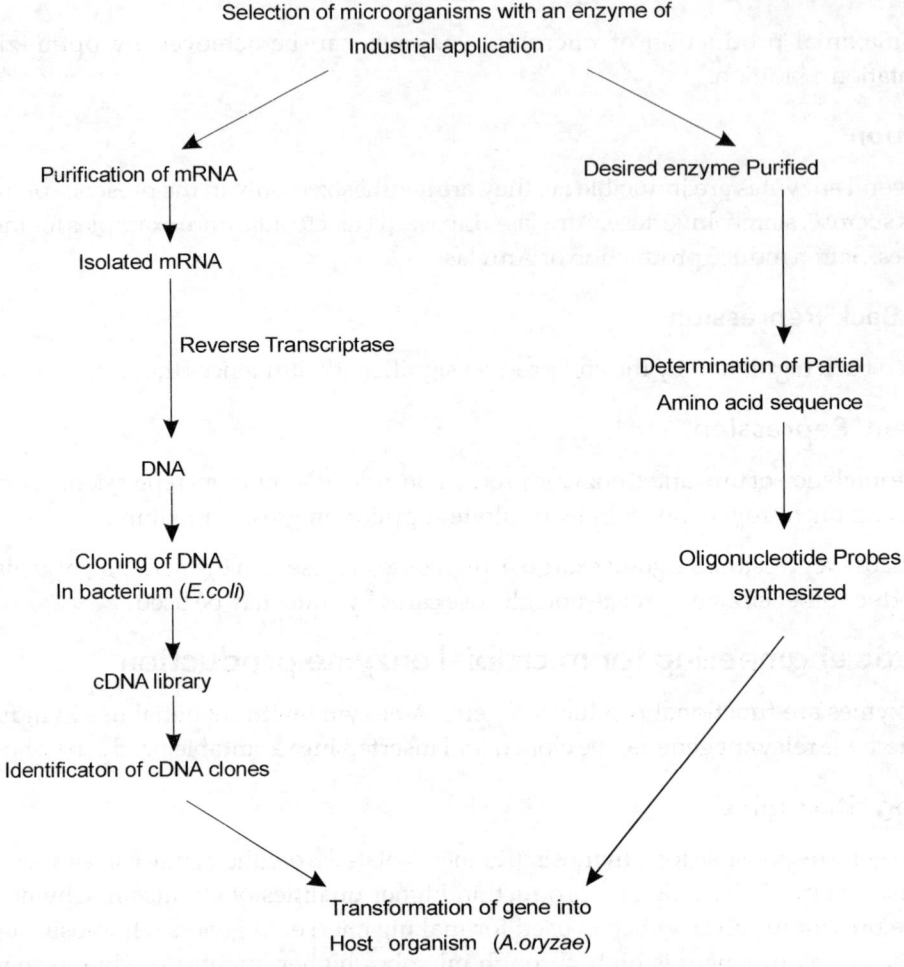

Figure 323   Biotechnology of enzyme production

Protein Engineering for Modification of Industrial Enzymes

Site directed mutagenesis is used for enzyme modification. Selected amino acid at specific position can be changed to produce an enzyme with desired properties. Increases stability, increase catalytic function, resistance to oxidation, changed substrate preference and increased tolerance to alkali and organic solvents are the important properties improved by site directed mutagenis. Protein engineering has been used to structurally modify phospholipase A2 that can resist high concentration of acid and efficiently and used as food emulsifies.

Advantages of Microbial Production of Enzymes

 ⅄   Microbes grow easily in the medium with cheap raw material.

 ⅄   Economically valuable.

▲ Genetic and environmental manipulation possible to increase the yield.

▲ Microbes requires very short generation time.

▲ Microbes can adopt biochemical pathways that help in the production of different enzymes.

▲ Culture and maintenance of microbial cells are easier.

G. APPLICATION OF MICROBIAL ENZYMES

Amylases

▲ Used in beverage industries

▲ Used in the preparation of high maltose syrup

▲ Used to prepare improved detergent

▲ Used in baking industry to increase the rate of fermentation

▲ Used as a drug to cure digestive disorders.

Protease

▲ It is used in beer as chill proofing agent and to control Nitrogen level.

▲ Used to improve the volume of backed products

▲ Used in cheese making.

▲ Used in preparation of rubber sheet from latex

▲ Used in tannery to dehair skin.

▲ Used in detergent industry.

Lactase

▲ Used in ice cream to prevent the formation of lactose crystals.

▲ Used in dairy industry, which removes lactose from milk and whey.

Petinase

▲ Used in coffee industry for fermenting coffee.

▲ Used to extract flavours in cosmetic industry

▲ Used in grape / wine industry to hydrolyse pectin

Cellulases

▲ Used to modify textile fibers.

▲ Used to de size paper pulp.

Glucose Isomerase

▲ Used to remove O_2 from soft drinks.

▲ Acts as stabilizer of citrus flavour in soft drinks.

Catalase

▲ Used in dairy industry for removing water from milk & milk products.

▲ Used in wine to control colour.

▲ Used as stabilizer of citrus flavour in soft drinks.

H. IMMOBILIZATION OF ENZYME

Immobilization is defined as the process in which an enzyme is physically entrapped by an inert insoluble matrix.

Advantages of Immobilization

▲ Stable and more efficient in function.

▲ Can be reused again & again.

▲ Products are enzyme free.

▲ Ideal for multi enzyme reaction systems.

▲ Control of enzyme system is easy.

▲ Suitable for industrial and medical use.

▲ Minimize effluent disposal problem.

Disadvantages

▲ Possibility of loss of biological activity during immobilization.

▲ It is an expensive process.

Methods of Immobilization

Commonly used methods of immobilization of enzymes are adsorption, entrapment, covalent binding and cross linking.

I. Adsorption

It involves the physical binding of enzymes on the surface of an inert support. Support material may be inorganic (Alumina, Silica Gel, Calcium Phosphate Gel, Glass) or organic (Starch, Carboxy methyl cellulose)

II. Entrapment

Enzymes can be immobilized by physical entrapment inside a polymer or a gel matrix. Polymerization is carried out in the

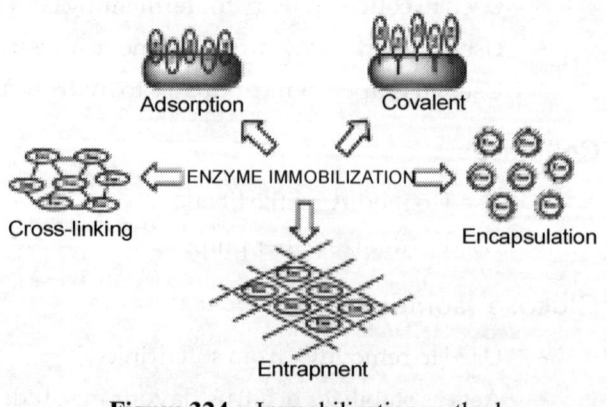

Figure 324 Immobilization methods

presence of the enzyme. After entrapment, enzyme cannot escape by permeation. This technique is commonly referred as lattice entrapment. The materials used for entrapping of enzyme include polyacrylamide gel, collagen, gelatin, starch, cellulose, silicone & Rubber. Enzymes are entrapped by several ways.

Enzyme inclusion in gels : Entrapment of enzymes inside the gel.

Enzyme inclusion in fibres : Enzymes are trapped in a fibre format of the matrix.

Enzyme inclusion in microcapsules:

Enzymes are trapped inside a microcapsule matrix. Hydrophobic & hydrophilic matrix polymerise to form a micro capsule containing enzyme molecules inside.

III. Micro encapsulation

It refers to the process of spherical particle formation where in a liquid or suspension is enclosed in a semipermeable membrane. Pancreatic cells grown in cultures can be immobilized by this method.

IV Covalent Binding

This is achieved by creation of covalent bonds between the chemical groups of enzymes and chemical groups of support. This is a widely used method. Cyanogen bromide activation, diazotation, Peptide bond formation are the methods of covalent binding.

Cellulose, Sepharose, Sephadex, amino alkylated porous glass, albumin are acts as a supporting matrix.

V Cross Linking

Solid support is not used in this method. In this method, enzyme molecules are immobilized by creating cross links between them, through the involvement of poly functional reagents. These reagents create bridges which form the backbone to hold enzyme molecules. Several reagents are used in this method. They include glutaraldehyde, diazobenzidine, hexamethylene di isocyanate etc.

Application of Immobilized Enzymes

Immobilized Aminoacylase used to produce L amino acids from D, L acyl amino acid. L amino acids are used as food, food supplement and medical purpose. Glucose isomerase used to produce high fructose syrup from glucose. β galactosidase used to split lactose to glucose and galactose. They are used for the development of specific analytical assay for the estimation of several biochemical compounds.

They are used in affinity Chromatography, in which it is possible to purify several compounds like antigens, antibodies & cofactor.

Table 17.4 Immobilized enzyme assay

Immobilized Enzyme	Substance assayed
Glucose oxidase	Glucose
Urease	Urea
Cholesterol oxidase	Cholesterol
Lactose dehydrogenase	Lactate
Alcohol oxidase	Alcohol
Hexakinase	ATP
Galactose Oxidase	Galactose
Penicillinase	Penincillin
Ascorbic Acid Oxidase	Ascorbic Acid
L amino acid oxidase	L Amino acid

Immobilized enzyme is used in biosensors. These interact with an analyte and produce physical chemical or electrical signals that can be measured. Eg. Blood glucose bio sensor. pH meter biosensor.

Immobilized antibody to which antigen can directly bind.

Used in pollution control to detect BOD.

Used in improved drug delivery.

Immobilization of Cells

The whole cell or cellular organelles can be immobilized to serve as multi enzyme systems. Immobilized cells have been traditionally used for the treatment of sewage.

Entrapment and surface attachment techniques are commonly used techniques.

The viability of the cells can be preserved by mild immobilization. Immobilized cells are particularly usefully for fermentation.

Advantages

Immobilized cells have higher specific rate of reaction. Enzyme extraction process is not very expensive. They are less susceptible to contamination. They are easy to recover from fermentation medium.

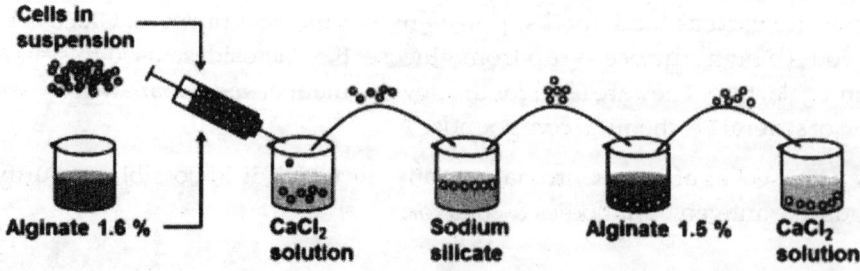

Figure 325 Immobilization of cells

Alginate method

This is a widely used method for the immobilization of microbial cells. Sodium alginate is a polymer of carbohydrate. It is a muco polysaccharide.

Calcium chloride solution (4%) is prepared

Slurry of Sodium alginate in Saline Solution and bacterial / yeast culture as suspension are prepared in a beaker or conical flask.

10 ml of culture is mixed with 50 ml of sodium alginate slurry.

Sodium alginate + culture mixed slurry is taken in a sterile syringe

Sodium alginate slurry is added carefully to the calcium chloride solution and allow to stand for 30 minutes.

Beads of microbial cells are formed, beads contains microbial cells.

These beads are subjected to fermentation of various products.

I. RECOMBINANT DNA TECHNOLOGY (rDNA)

The techniques involved in the construction and use of recombinant DNA molecules are called Recombinant DNA technology. A DNA molecule created in the laboratory by ligating different pieces of DNA which are not normally joined together are called recombinant DNA (rDNA). Genetic engineering primarily involves the manipulation of genetic material to achieve the desired goal in a predetermined way.

The basic steps in gene cloning / rDNA technology are as follows:

1. Generation of DNA fragments and selection of desired piece of DNA. (Human)
2. Insertion of selected DNA into a cloning vector to create a recombinant DNA or chimeric DNA.
3. Introduction of the recombinant vectors into host cell (bacteria).
4. Multiplication and selection of clones containing the recombinant molecules.
5. Expression of the gene to produce the desired products.

Purification of DNA/ Generation of DNA and selection of the Desired DNA

There are three kinds of DNA needed at different times for the successful completion of genetic engineering.

1. Total DNA will often be required as a source of material from which to obtain genes to be cloned. It may be from bacteria, from a plant or from animal.
2. Second type of DNA required is pure plasmid DNA.
3. Phage DNA will be needed if a phage cloning vehicle is to be used.

Preparation of Total cell DNA

Bacterial DNA preparation can be divided into 4 stages

1. Cells are grown and harvested.
2. Breaking of cells and release its contents.
3. Cell contents are treated to remove all unwanted components except DNA.
4. The resulting DNA solution is concentrated.

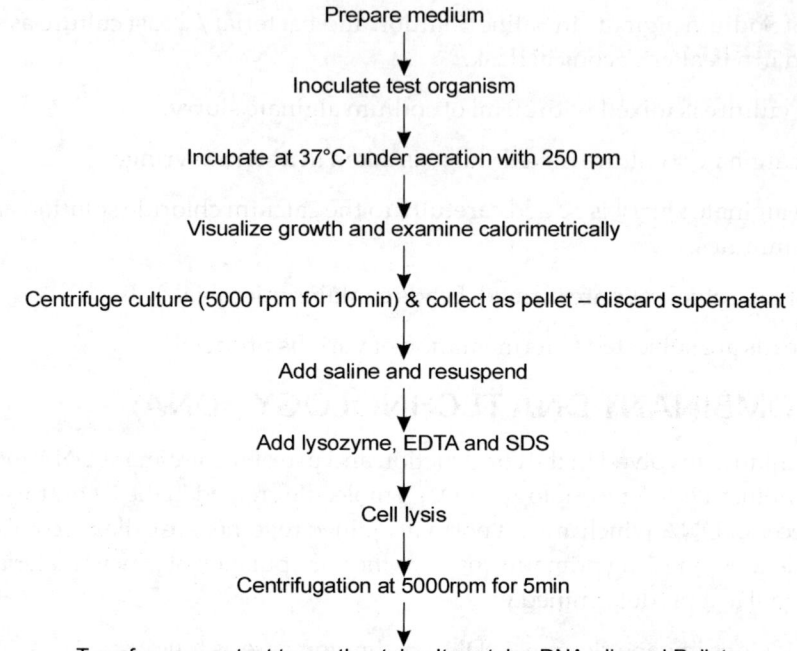

Prepare medium
↓
Inoculate test organism
↓
Incubate at 37°C under aeration with 250 rpm
↓
Visualize growth and examine calorimetrically
↓
Centrifuge culture (5000 rpm for 10min) & collect as pellet – discard supernatant
↓
Add saline and resuspend
↓
Add lysozyme, EDTA and SDS
↓
Cell lysis
↓
Centrifugation at 5000rpm for 5min
↓
Transfer supernatant to another tube. It contains DNA- discard Pellet.

Bacterial cells are grown on M9 medium or Luria Bertani medium. Prepared medium is sterilize properly and inoculated with the bacterial culture. Inoculated medium is incubated at 37°C, aerated by shaking at 150-250rpm. The growth of the culture can be monitored by reading OD at 600nm.

Bacterial cells are harvested by centrifuging grown medium at 5000rpm for 10 minutes.

Physical methods and chemical methods are employed to break open bacterial cell. In physical method, cells are disturbed by mechanical forces. In chemical method, lysozymes, EDTA along with sodium dodecyl sulfate (SDS) are used to disturb the cell wall and cell membrane. Lysozyme digests polymeric compounds of cell wall (eg. PG layer), EDTA removes magnesium ions (that are essential for preserving the overall structure of cell wall), SDS removes lipid molecules and thereby causing cell disruption.

After this process, cell debris is removed by low speed centrifugation at 5000rpm for 5minutes. This results in removal of cell debris as pellet and leaving DNA, RNA and few protein in the supernatant.

Purification of DNA

Bacterial cell extract contains protein and RNA as impurity. Protein will be removed by treating the extract with 1:1 mixture of phenol and chloroform. These mixture precipitate proteins but leave the nucleic acid. RNA is removed by treating the extract with ribonuclease. Precipitated protein and RNA are removed by centrifugation. Proteinase K also used to remove proteins.

Concentration of DNA samples

In the presence of salt (Na+), at a temperature of -20°C, absolute ethanol efficiently precipitates nucleic acid. Precipitated DNA is removed by introducing rod or by high speed centrifugation. Concentrated DNA then redissolved in an appropriate volume of water.

Concentration of DNA in a sample can be measured by Ultraviolet absorbance spectrophotometry. The amount of UV radiation absorbed by a solution of DNA is directly proportional to the amount of DNA in the sample. The absorbance is measured at 260nm. Absorbance (A260) of 1.0 corresponds to 50µg of double stranded DNA per ml.

Preparation of total cell DNA from the organisms other than bacteria

Different methods are adopted to extract DNA from plants, animals and viruses. Purification and concentration techniques are same as bacterial DNA isolation. Breaking of cell and release its contents needs special method for each group of cells, as it needs to digest cellwall / cell membrane. In plants cetyl methyl ammonium bromide (CTAB) is used. It precipitates DNA and leaves carbohydrates, proteins and other contaminations in the supernatant.

Preparation of plasmid DNA

Purification of plasmids from a culture of bacteria involves the same general strategy as preparation of total cell DNA from bacteria.

A culture of cells, containing plasmids is grown in liquid medium, harvested and a cell extract is prepared. The extract is deproteinzed, the RNA removed, and the DNA probably concentrated by ethanol precipitation. This concentrate contains both cell DNA and plasmid. Separating two types of DNA can be difficult, but is essential for the use of plasmid as a cloning vector.

Several methods are available to purify plasmid from the chromosomal DNA. These methods are based on the several physical difference between plasmid DNA and bacterial DNA.

Plasmid DNA is smaller than bacterial DNA. Plasmid is also differ in conformation. Alkaline denaturation, ethidium bromide- Caesium chloride density gradient centrifugation, ultracentrifugation are used to separate plasmid from chromosomal DNA.

Plasmid amplification methods are employed to increase the copy number of a plasmid.

Manipulation of plasmid DNA

Once pure samples of DNA have been prepared, the next step in a gene cloning experiment is construction of the recombinant DNA molecule. To produce this recombinant molecule, the vector, as well as the DNA to be cloned, must be cut at specific points and joined together in a controlled manner. Cutting and joining are the two examples of DNA manipulative techniques.

The cutting and joining manipulations that underlie gene cloning are carriedout by enzymes called restriction endonucleases (For cutting) and Ligases (For Joining).

The Range of DNA Manipulative Enzymes

Enzymes used in gene cloning can be grouped in to five broad classess depending on the type of reaction that they catalyse.

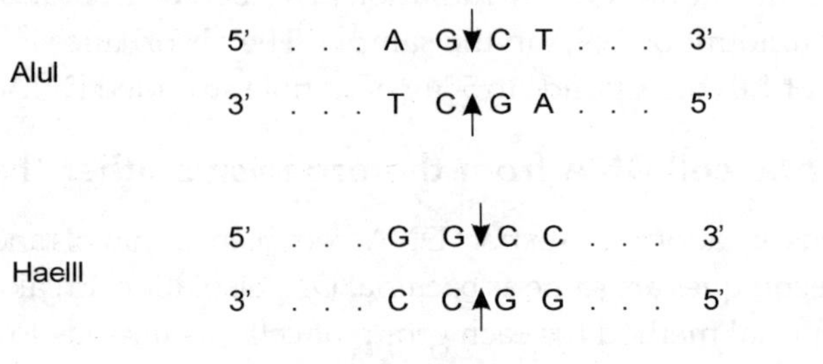

AluI
```
5'  . . . A G▼C T . . . 3'
3'  . . . T C▲G A . . . 5'
```

HaeIII
```
5'  . . . G G▼C C . . . 3'
3'  . . . C C▲G G . . . 5'
```

BamHI
```
5'  . . . G▼G A T C C . . . 3'
3'  . . . C C T A G▲G . . . 5'
```

HindIII
```
5'  . . . A▼A G C T T . . . 3'
3'  . . . T T C G A▲A . . . 5'
```

EcoRI
```
5'  . . . G▼A A T T C . . . 3'
3'  . . . C T T A A▲G . . . 5'
```

AluI and HaeIII produce blunt ends

BamHI HindIII and EcoRI produce "sticky" ends

Figure 326 Restriction sites of RE

1. Nucleases are enzymes that cut, shorten or degrade nucleic acid molecule.
2. Ligases join nucleic acid molecules together.
3. Polymerases make copies of molecules.
4. Modifying enzymes remove or add chemical groups.
5. Topoisomerases introduce or remove supercoils from covalently closed circular DNA.

Nucleases

Nucleases degrade DNA molecules by breaking the phosphodiester bonds that link one nucleotide to the next in a DNA strand. There are two distinct kinds of nucleases.

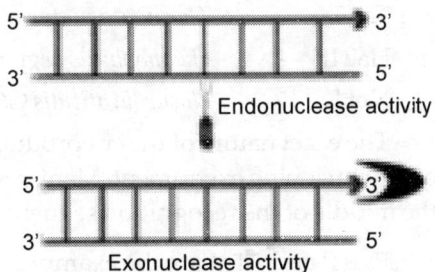

Endonuclease activity

Exonuclease activity

a. Exonucleases - which remove nucleotides one at a time from a end of a DNA molecule.

b. Endonucleases - which are able to break internal phosphodiester bonds within DNA molecule. (Figure 327)

Figure 327 Action of Nucleases

Bal 31 is one of the exonuclease enzyme isolated from *Alteromonas espejiana,* that removes nucleotides from both strands of a double stranded molecule.

Gene cloning requires that DNA molecules be cut in a very precise and reproducible fashion. Each vector molecule must be cleaved at a single position, to open up the circle so that new DNA can be inserted.

There are three kinds of endonucleases. They are type I, II & III. Type I & III are complex and have only a very limited role in genetic engineering. Type II restriction enzymes are the cutting enzymes that are so important in gene cloning.

Type II restriction enzyme (Figure 326)

They are simply called as restriction endonuclease. They have a specific recognition sequence at which it cuts a DNA molecule.

Eg : Pvu I is isolated from *Proteus vulgaris* cuts DNA only at the hexanucleotide CGATCG. At the same time Pvu II which is also from same bacterium, cuts DNA at a different hexanucleotide, CAGCTG.

Many restriction endonucleases (RE) recognize hexanucleotide target sites (6 cutter), but others cut at four, five or even eight nucleotide.

Some examples of Restriction endonucleases are

EcoRI	-	*Escherichia coli* GAATTC Sticky end
BamHI	-	*Bacillus amyloliquefaciens* GGATCC Sticky end
Bgl II	-	*Bacillus globigii* AGATCT Sticky end
Pvu I	-	*Proteus vulgaris* CGATCG Sticky end
Pvu II	-	*Proteus vulgaris* CAGCTG Sticky end
Hind III	-	*Haemophilus influenza* AAGCTT Sticky end
Hinf I	-	*Haemophilus influenza* GANTC Sticky end
Sau 3A	-	*Staphylococcus aureus* GATC Sticky end
Alu I	-	*Arthrobacter luteus* AGCT blunt end

Taq I   -   *Thermus aquaticus* TCGA Sticky end

Hae III -   *Haemophilus aegyptius*   GGCC blunt end

Not I   -   *Nocardia otitidis* GCGGCGC Sticky end

The exact nature of the cut produced by a RE is of considerable importance in the design of a gene cloning experiment. Many restriction enzymes make a simple double stranded cut in the middle of the recognition sequence, resulting in a blunt or flush end.

Pvu II and Alu I are the examples of blunt end cutters.

Large number of Restriction enzymes cut DNA in a slightly different way with these enzymes the two DNA strands are not cut at exactly the same position. Unequal cut is made on the DNA strand which are called sticky or cohesive end.

A restriction digest will result in a number of DNA fragments, the sizes of which depend on the exact positions of the recognition sequence for the endonuclease in the original molecule.

Numbers of DNA fragments are analyzed by separating DNA molecules by gel electrophoresis, as DNA molecules carry an electric change (-). When this DNA placed in a agarose gel along with electric change. Fragments of different size can be separated based on the size.

After separation DNA molecules are visualized by staining (Ethidium bromide), Autoradiography etc.

Ligaton

The final step in construction of a recombinant DNA molecule is the joining together of the vector molecule and the DNA to be cloned. This process is referred to as ligation, and the enzyme that catalyzes the reaction are called DNA ligase.

Sticky ends are easily ligated. Blunt end need to be converted into sticky end. This can be done by making use of linkers and adapters. Linkers are short pieces of double stranded DNA of known nucleotide sequence.

Adapters are short synthetic oligonucleotides. It can be used the DNA fragment that already had one sticky end. Homopolymer tailing also used to produce sticky end.

Introduction of DNA into Living Cells

DNA ligation yields desired recombinant molecule. In addition to rDNA, it also contains unligated mixture, unligated DNA, rDNA with wrong insertion molecules etc.

Most species of bacteria are able to take up DNA molecules from the medium in which they grow. Transformation is a process by which bacterium taken up foreign DNA from the surrounding medium. *Streptococcus, Bacillus and Neisseria* are the good examples for naturally transforming bacteria. *E. coli* needs special treatment to convert it into a competent cells. Competent cells have the capacity to take up foreign DNA naturally or artificially. *E. coli* can be treated with calcium chloride and converted into artificially competent cells.

Cacl$_2$ causes the DNA to precipitate on to the outside of the cells or the salt is responsible for some kind of change in the cell wall that improves DNA binding.

When DNA is added to treat cells, it remains attached to the cell exterior, and is not at this stage transported into the cytoplasm. The actual movement of DNA into competent cell is stimulated by briefly raising the temperature to 42°C.

Transformation of competent cells is an inefficient procedure. 1ng of pUC 8 can yield 1000-10000 transformants.

Marker genes are available on the vector. Ampicillin resistance gene of pBR322 is one of the good example for marker.

Most cloning vectors carry at least one gene that confers antibiotic resistance on the host cell.

Identification of Recombinants

Plating on a selective medium enables transformants to be distinguished from non transformants.

Insertion of foreign DNA into the vector destroys the integrity of one of the genes present on the molecules which are called insertional inactivation.

Screening for pBR322 recombinants is performed in the following way. After transformation the cells are plated on to ampicillin medium and incubated until colonies appear. All of these colonies are transformants. To identify the recombinants the colonies are replica plated onto agar medium that contains tetracycline. After incubation, some of the original colonies will regrow, other will not. Those that do grow consist of cells that carry the normal pBR322 with no inserted DNA. i.e., both amp and tetracycline genes are active. The colonies that do not grow on tetracycline agar are recombinants (Amp^R, Tet^S).

Introduction of Phage DNA into bacterial cells

Transfection and *in vitro* packaging are employed to introduce phage DNA into the bacteria.

Transformation of Non Bacterial Cells

Protoplast can be obtained from fungi, yeast and plant cells, using proper enzyme treatment methods. Protoplasts generally take up DNA quite readily, alternatively transformation can be stimulated by special technique called electrophoration, during which the cells are subjected to a short electrical pulse, through to induce the transient formation of pores in the cell membrane through which DNA molecules can enter the cell.

Microinjection is another method, which makes use of very fine pipette to inject DNA molecules directly into the nucleus of cell to be transformed.

Micro projectile bombardment, by which high velocity micro projectiles, usually particles of gold or tungsten that have been coated with DNA. These micro projectiles are fired at the cells from a particle gun. This unusual technique is termed biolistics and used with a number of different types of cells

J.   CLONING VECTORS

A DNA molecule needs to display several features to be able to act as a vehicle for gene cloning. It must able to replicate within the host cell. Relatively small, ideally less than 10kb in size.

Plasmids are frequently employed as cloning vehicles.

Plasmids

Plasmids are circular molecules of DNA that lead an independent existence in the bacterial cell

Plasmids always carry 1 or more genes and are responsible for a useful characteristic displayed by the host bacterium. Selectable marker gene is present in plasmid. Eg. Amp^R, Tet^S genes.

All plasmids possess at least one DNA sequence that can act as an origin of replication. So they are able to multiply within the host.

Plasmids use host cell replicative enzymes for multiplication.

A few types of plasmid are also able to replicate by inserting themselves into the bacterial chromosome. These integrative plasmids are called as **episomes**.

Two features of plasmids are particularly important as far as cloning is concerned. They are size and copy number.

These are two groups of plasmid, they are conjugative and non conjugative. Conjugative plasmids are characterized by the ability to promote sexual conjugation. Tra genes are controlling conjugation.

Plasmid Classification

The five main types of plasmid are as follows.

Fertility or 'F' plasmids carry only 'tra' genes and promote conjugation. Eg. F plasmid of *E. coli*.

Resistances or 'R' Plasmids carry genes conferring on the host bacterium resistance to one or more antibacterial agents (RP4) of *Pseudomonas*.

Col plasmids code for colicins. Eg. ColE1 of *E. coli*.

Degradative plasmids allow the host bacterium to metabolize unusual molecules such as toluene. Eg. ToL of *Pseudomonas putida*.

Virulence plasmids confer pathogenicity on the host bacterium eg. Ti plasmids of *Agrobacterium tumefaciens*.

The Nomenclature of plasmid cloning vectors.

The name pBR322 confers with the standard rules for vector nomenclature.

P indicates that is indeed a plasmid.

BR - identifies the laboratory in which the vector was originally constructed. Bolivar and Rodriguez - Researches.

322 - Distinguishes this plasmid from others developed in the same laboratory.

Other examples pBR325, 327 328.

Useful Properties of pBR322

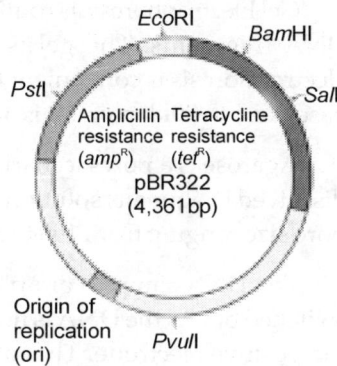

- ▲ It is smaller in size.
- ▲ It's size is about 4361bp.
- ▲ It carries two sets of antibiotic resistance genes - tetracycline, Ampicillin.
- ▲ It is a high copy number plasmid.
- ▲ It has origin of replication site, which initiates replication.
- ▲ EcoR I, Hind III, BamHI, Sal I, Sca I, Pvu I, Pst I restriction sites are available

Figure 328   Structure of plasmid pBR322

Other Vectors of *E. coli*

pBR327, (3273bp), pUC8 (Plasmid University of California - 2750bp, amp^R, LacZ). pUC18 - multiple restriction sites than pUC8. pGEM3Z - 2750bp ampR, LacZ, T7 promotor M13mp1, M13mp7, M13mp2, M13mp8, M13mp9 are some of the M13 based *E. coli* cloning vector.

Phagemid and cosmid vectors also available.

Vectors for Yeast

1. Yeast episomal plasmids or YEps. It is a shuffle vector. Eg. YEPs13.
2. Yeast integrative plasmids (YIps)
3. Yeast replicative plasmids (YRps)
4. Yeast Artificial chromosome (YAC)

Vector for Plant

Ti plasmid

Vector for Animals

Adenovirus

Papillo viruses

SV 40

K.   BASIC TECHNIQUES IN GENETIC ENGINEERING

There are several techniques used in rDNA technology or gene manipulation.

Agarose Gel Electrophoresis

Electrophoresis refers to the movement of charged molecules in an electric field. The negatively charged molecules move towards the positive electrode while the positively charged molecules migrate towards the negative electrode.

Gel Electrophoresis is routinely used technique for the separation or purification of specific DNA fragments. The gel is composed of either polyacrylamide or agarose. Agarose electrophoresis is convenient for the separation of DNA fragments ranging in size from 100 base pairs to 20kb pairs. It is also used for the separation of RNA.

Agarose is a polysaccharide derived from seaweed Gelidium. It forms a solid gel when dissolved in aqueous solution at concentrations between 0.5 and 2%. Agarose forms gel with pore size ranging from 100 to 300nm in diameter.

The DNA samples are placed in the wells of the gel surface and the power supply is switched on. As the DNA is negatively charged, DNA fragments are through the gel towards the positive electrode. The rate of migration of DNA is dependent on the size and shape. Smaller, linear fragments move faster than the larger ones. It is used for the separation of mixture of DNA fragment bases on their size.

The migration of DNA fragments during the course of electrophoresis can be monitored by using dyes with known migration rates.

The bands of the DNA can be detected by soaking the gel in ethidium bromide solution. When activated by UV, DNA base pairs in association with ethidium bromide emit orange fluorescence. Separated DNA can be identified.

Purification of Plasmid DNA

Pure form of plasmid DNA are often required in genetic engineering experiments. The high molecular weight chromosomal DNA can be removed along with the cell debris by high speed centrifugation. This results in the formation of cleared lysate from which pure plasmid DNA can be isolated.

Isopycnic Centrifugation method

The cleared lysate is treated with a solution of Cesium Chloride (CsCl) containing ethidum bromide (Etbr). Etbr can bind with linear DNA molecules and not with circular plasmid DNA. The density of DNA etbr complex much lower than the plasmid DNA. The circular plasmid DNA can be separated by isopyicinic centrifugation in CsCl-etbr gradient.

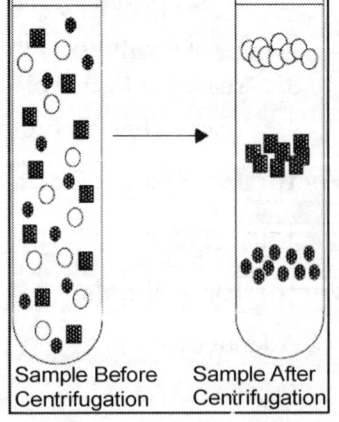

Sample Before Centrifugation   Sample After Centrifugation

Figure 329   Isopycnic centrifugation method

Birnboim and Doly method

This technique developed by Birnboim & Doly (1979). It is widely used in the plasmid DNA purification. This method is based on the principle that there is a narrow range of pH(12-12.5) at which denaturation linear DNA happens and not closed circular DNA.

When the cleared lysate is subjected to carefully monitored alkaline pH(12-12.5), the plasmid DNA remain in solution while all other DNA molecules get denatured & precipitate. This precipitate can be removed by centrifugation. The plasmid DNA in the supernatant can be concentrated by ethanol precipitation.

Purification of mRNA

mRNA is required in a pure form for rDNA techniques.

The purification of mRNA can be achieved by affinity chromatography using Oligo (dT)-cellulose. This method is based on the principle that oligo(olt)cellulose specifically bind to the poly A tails of eucaryotic mRNA. Thus by this approach it is possible to isolate mRNA from DNA, rRNA and tRNA.

As the nucleic acid solution is passed through an affinity chromatographic column, the oligo(dt) binds to poly A tails of mRNA. By washing the column with high salt buffer, DNA, rRNA & tRNA can be eluted while the mRNA is tightly bound. This mRNA can be then eluted by washing with low salt buffer. The mRNA is precipitated with ethanol and collected by centrifugation.

Isolation of Chromosomes

Separation of large chromosomes of eukaryotes is not possible by conventional electrophoresis. The individual chromosomes of eukaryotes can be separated by fluorescence activated cell sorting (FACS) procedure. It is also known as flow cytometry or flow karyotyping.

To carryout FACS, the dividing cells are carefully broken open, and a mixture of intact chromosomes is prepared. These chromosomes are then stained with a fluorescent dye. The quantity of the dye that binds to a chromosome depend on its size. Thus, larger chromosomes bind more dye and fluoresce more brightly than the smaller ones.

The dye mixed chromosomes are diluted and passed through a fine aperture that results in the formation of a steam of droplets. Each droplet contains a single chromosome. The

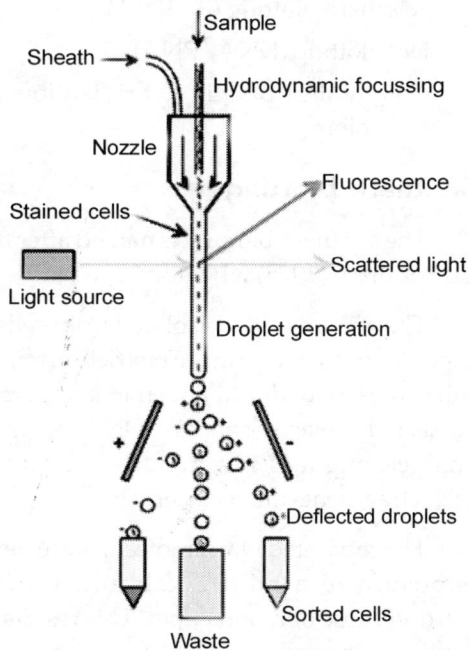

Figure 330 Flow cytometry

fluorescence of the chromosome is detected by a laser. When the fluorescence indicates that the chromosome illuminated by the lased is the one desired, that do not contain the desired chromosome pass through a waste collection vessel.

Collection of Chromosomes with Identical Size

The direct application of FACs is not suitable for the separation of chromosomes with identical size e.g chromosomes 21 and 22 in humans. Collection of such chromosomes can be achieved by use of special dyes (E.G. Hechs + 33258 and chromomycin A3) which bind to AT rich DNA or G rich DNA.

It is convenient to separate two or more chromosomes with identical sizes by different dyes.

Nucleic Acid Blotting Techniques

- ⋏ Blotting techniques are very widely used in analytical tools for the specific identification of desired DNA or RNA fragments.
- ⋏ Blotting refers to the process of immobilization of sample nucleic acids or solid support.
- ⋏ The bottled nucleic acids are then used as targets in the hybridization experiments for their specific detection.

Types of Blotting Techniques

The most commonly used blotting techniques are listed below.

Southern blotting (for DNA)

Northern blotting (for RNA)

Dot blotting (DNA/RNA)

Colony and plaque hybridization (Cell from colony)

Southern blotting

The southern blotting is named after the scientist Ed Southern (1975) who developed it.

The genomic DNA isolated from cells/tissues is digested with one or more restriction enzymes. This mixture is loaded into a well in an agarose gel and subjected to electrophoresis. DNA, being negatively charged migrates towards the anode; the smaller DNA fragments move faster.

The separated DNA molecules are denatured by exposure to a mild alkali and transferred to nitrocellulose or nylon paper. This results in an exact replica of the pattern of DNA fragments on the gel.

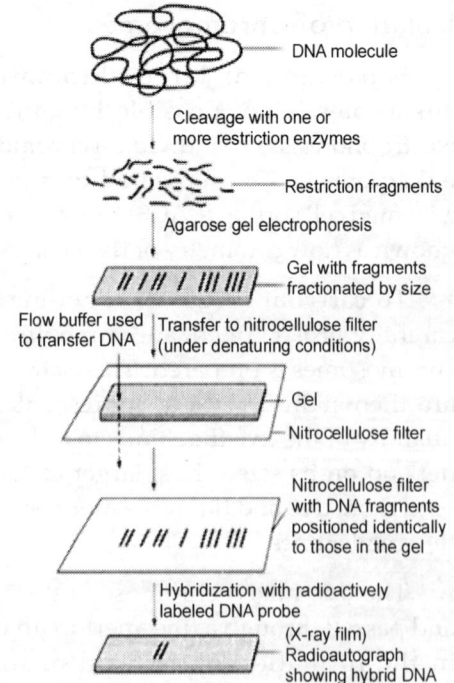

Figure 331 Southern blotting

The DNA can be annealed to the paper on exposure to heat (80°C). The nitrocellulose paper is then exposed to labeled cDNA probes. These probes hybridise with complementary DNA molecules on the paper.

The paper after thorough washing is exposed to x ray film to develope autoradiograph. This reveals specific bands corresponding to the DNA fragments recognized by cDNA probe.

Applications of Southern Blotting

▲ It is an invaluable method in gene analysis. Important for the confirmation of DNA cloning result.

▲ Useful for mapping restriction sites around a single copy gene sequence.

▲ Forensically applied to detect minute quantities of DNA. Highly useful for the determination of restriction fragment length polymorphisms (RFLP) associated with pathological condition.

Northern Blotting

It is used for the specific identification of RNA molecules. This method is similar to southern blotting.

RNA molecules are subjected to electrophoresis, followed by blot transfer, hybridization and auto radiography.

RNA molecules do not easily bind to nitrocellulose paper. Blot transfer of RNA molecules is carried out by using a chemically reactive paper prepared by diazotization of aminobenzyloxymethyl to create diazobenzyloxymethyl paper (DBM). The RNA can covalently bind to DBM paper.

Now a days, nitrocellulose and modified nylon papers are used and are now called as RNA blotting.

The blot transferred RNA molecules hybridize with DNA Probes which can be detected by auto radiography.

Dot Blotting

It is a modification of southern and northern blotting. Nucleic acids are directly spotted on to the filters and not subjected to electrophoresis. The hybridization procedure is the same as in original blotting technique.

It is useful in obtaining quantitative data for the evaluation of gene expression.

Auto Radiography

Auto radiography is the process of localization and recording of a radiolabelled genome within a solid specimen, with the production of an image in a photographic emulsion. These emulsions are composed of silver halide crystals suspended in gelatin.

When a particle or a ray from a radiolabel passes through the emulsions, silver ions are converted to metallic silver atoms. This results in the development of a visible image which can be easily detected.

Direct autoradiography is ideally suited for the detection of weak to medium strength β emitting nucleotides(3H,14C 35S). In this technique, the sample is placed in direct contact with the film. The radioactive emissions results in the development of black areas.

Indirect autoradiography is used for the detection of highly energetic β particles (32P,125I).

Particle energy is first converted to light by a scintillator, which then emits photons on exposure to photographic emulsions.

Applications

It is closely associated with blotting techniques for the detection of DNA, RNA and proteins.

Colony & Plaque Blotting

It is a process of hybridization for the specific identification and purification of colonies.

The desired bacteria are grown as colonies on an agar plate. When a nitrocellulose fitter paper is overlaid on the agar plate, the colonies get transferred. They are permanently fixed on the paper by heat. On alkali treatment, the cells lyse & DNA gets denatured. When these DNA prints are exposed to specific probes, hybridization occurs. This hybrid complex can be localized and detected by auto radiography (Figure 332).

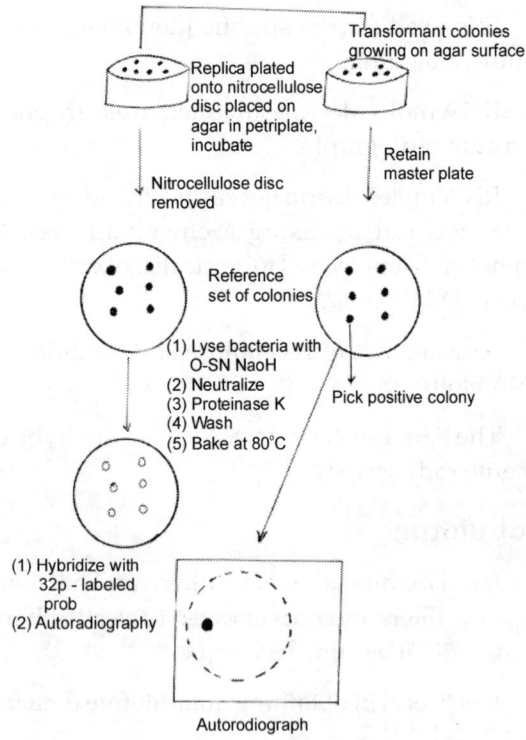

Figure 332 Colony hybridization

L. STRAIN IMPROVEMENT

Improvement of the product producing ability of the microorganisms, plants or animal are called strain improvement.

The bacteria which is isolated from the raw material / environment are called wild strain. Wild strain showed lesser productivity. .To increase the productivity of the particular product, the wild strain should be modified. This procedure is called strain improvement.

The primary goal of an industrial strain improvement / development is the increase in the yield of the desired product.

The improved strain should posses the following characteristics to finally result in high product formation.

Shorter time of fermentation.

Capable of metabolizing low cost substrates.

Reduced O_2 demand.

Decreased foam - formation.

Non production of undesirable compounds.

Tolerance to high concentration of carbon or Nitrogen sources.

Resistance to infections of bacteriophages.

Improved strains can produce only one metabolite.

Methods of Strain Improvement

They are three distinct approaches for improvement of strain that are mutation, recombination, rDNA technology.

Mutation

Any change that occurs in the DNA of a gene is referred to as mutations. Mutation may be spontaneous or induced. The rate of spontaneous mutation is very low. The cause of this mutation include integration and exclusion of transposons, along with errors in the functioning of enzymes such as DNA polymerases, recombinant enzymes and DNA repair enzymes. It is usually not suitable for industrial purposes. Induced mutations frequency is high and is depends on the type of inducer used. Short wavelength UV is one of the effective mutagenic agents. Most important product of UV action are dimers formed between adjacent pyrimidines. Ionizing radiations includes X rays, α rays, β rays, which causes ionization results in the single or double strand breaks. Double strand break results in major structural changes, such as translocation, inversion or similar chromosome mutation. Therefore UV or chemical agents normally preferable for mutagenesis. Nitrous oxide, nitrosoguanidine are examples of chemical mutagen.

In 1943, a strain of *Penicillium chrysogenum* was improved through mutation. It produced 55 fold higher penicillin in aerobic fermentation.

Selection of Mutants

Selection and isolation of the appropriate mutant strains developed is very important for their industrial use. Two techniques are used.

Random Screening

The mutated strains are randomly selected and checked for their ability to produce the desired industrial product. This can be done with model fermentation units. The strain with maximum yield can be selected. It is costly and tedious procedure. But many times, this is the only way to find right strain of mutants developed.

Selective isolation of mutants

There are many ways.

Isolation of Antibiotic Resistant Strains

The mutated strains are grown on a selective medium containing an antibiotics. The wild strains are killed while the mutant strains with antibiotic resistance can grow.

Isolation of Antimetabolite Resistant Strains

Antimetabolites which have structural similarities with metabolites can block the normal metabolic pathways and kill the cells. The mutant strains resistant to anti metabolities can be selected for industrial purposes.

Isolation of Auxotrophic Mutants

An auxotrophic mutant is characterized by a defect in one of the bio synthetic pathways. As a result, it requires specific compound for growth.

Eg. Ampicillin selection of auxotrophs.

Auxotrophs not grown on minimal medium but prototrophs will grow. Those cells grown on Ampicillin medium are selectivity killed finally auxotrophs only remains. This will be selected when cultivating regular medium and selected.

Isolation of Auxotrophs by Replica Plating

Using replica plating auxotrophs are transferred to minimal medium and auxotrophs cannot grow. Finally mutants are picked up from the mother plates.

Isolation of mutant by agar plug method

In which agar cylinders with single colonies are transferred to test plates after incubation in a moist chamber. The diameter of the resulting inhibition zones serves as a measure of the antibiotic production of each strain.

Recombination

The genetic information from two genotypes can be brought together into a new genotype through genetic recombination, which is thus another effective means of increasing the genetic variability of a population.

Natural Gene Transfer Methods

The discovery of natural gene transfer systems in bacteria has greatly facilitated the understanding of the genetics of microbial starter cultures and in some cases has been used for strain improvement. Genetic exchange in bacteria can occur naturally by three different mechanisms: transduction, conjugation, and transformation.

Transduction

Transduction involves genetic exchange mediated by a bacterial virus (bacteriophage). The bacteriophage acquires a portion of the chromosome or plasmid from the host strains and transfers it to a recipient during subsequent viral infection. Although transduction has been exploited for the development of a highly efficient gene transfer system in the gram-negative organism *Escherichia coli*, it has not been used extensively for improving microorganisms used in food fermentations. In general, transduction efficiencies are low and gene transfer is not always possible between unrelated strains, limiting the usefulness of the technique for strain improvement. In addition, bacteriophage have not been isolated and are not well characterized for most strains.

Conjugation

Conjugation, or bacterial mating, is a natural gene transfer system that requires close physical contact between donors and recipients and is responsible for the dissemination of plasmids in nature. Numerous genera of bacteria harbour plasmid DNA. In most cases, these plasmids are cryptic (the functions encoded are not known), but in some cases important metabolic traits are encoded by plasmid DNA. If these plasmids are also self-transmissible or mobilizable, they can be transferred to recipient strains. Once introduced into a new strain, the properties encoded by the plasmid can be expressed in the recipient. The lactic acid bacteria naturally contain one to more than ten distinct plasmids, and metabolically important traits, including lactose-fermenting ability, bacteriophage resistance, and bacteriocin production, have been linked to plasmid DNA. Conjugation has been used to transfer these plasmids into recipient strains for the construction of genetically improved commercial dairy starter cultures.

There are some limitations in the application of conjugation for strain improvement. To exploit the use of conjugative improvement requires an understanding of plasmid biology and, in many cases, few conjugative plasmids encoding genes of interest have been identified or sufficiently characterized. Conjugation efficiencies vary widely and not all strains are able to serve as recipients for conjugation. Moreover, there is no opportunity to expand the gene pool beyond those plasmids already present in the species.

Transformation

Certain microorganisms are able to take up naked DNA present in the surrounding medium. This process is called transformation and this gene transfer process is limited to strains that are naturally competent. Competence-dependent transformation is limited to a few, primarily pathogenic, genera, and has not been used extensively for genetic improvement of microbial starter cultures. For many species of bacteria, the thick peptidoglycan layer present in gram-positive cell walls is considered a potential barrier to DNA uptake. Methods have been developed for enzymatic removal of the cell wall to create protoplasts. In the presence of polyethylene glycol, DNA uptake by protoplasts is facilitated. If maintained under osmotically stabilized conditions, transformed protoplasts regenerate cell walls and express the transformed DNA. Protoplast transformation procedures have been developed for some of the lactic acid bacteria; however, the procedures are tedious and time-consuming, and frequently parameters must be optimized for each strain. Transformation efficiencies are often low and highly variable, limiting the application of the technique for strain improvement.

Electroporation

The above mentioned gene transfer systems have become less popular since the advent of electroporation, a technique involving the application of high-voltage electric pulses of short duration to induce the formation of transient pores in cell walls and membranes. Under appropriate conditions, DNA present in the surrounding medium may enter through the pores. Electroporation is the method of choice for strains that are recalcitrant to other gene transfer techniques; although optimization of several parameters (e.g., cell preparation conditions, voltage and duration of the pulse, regeneration conditions, etc.) is still required.

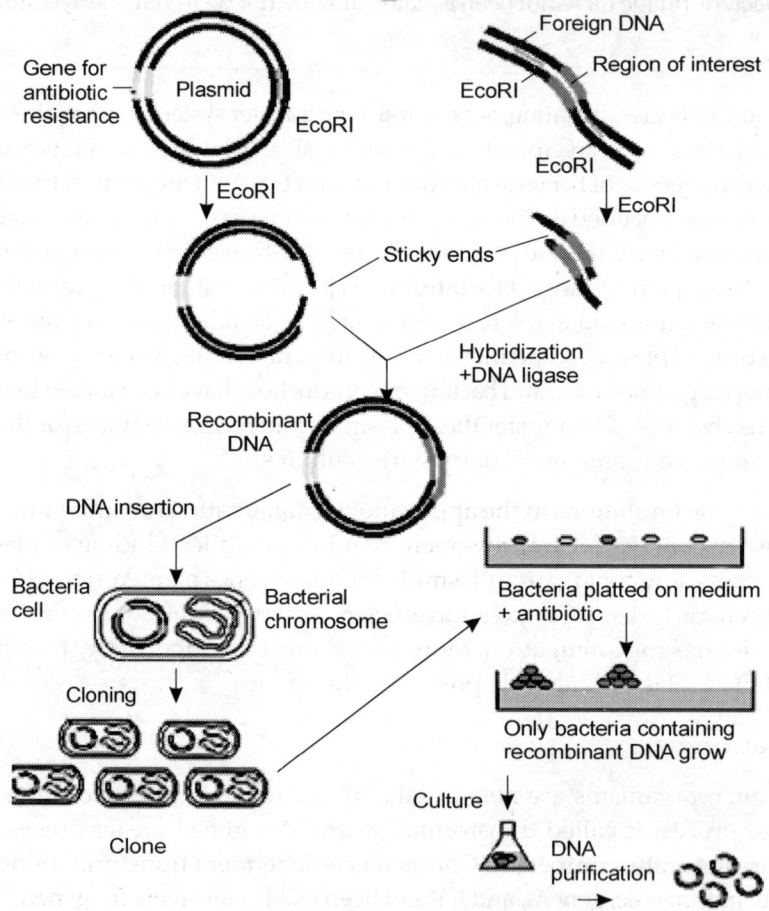

Figure 333   Genetic modification

The advantages of genetic recombination are different alleles of the parent strains with increased metabolite production can be brought together in one strain, so that the cumulative effects of these mutations can be greater than the effect of the single mutation. However, the original hope of attaining a significant yield increase by merely recombining two high-yielding mutants has only been fulfilled in a few cases. In most cases, the productivity of the recombinants usually is intermediate between the values of the parent strains.

In the course of strain development, there is frequently a decline in the increase in yield after each stage of mutation. Besides mutants, which are selectively enriched because of their increased productivity, there is the development of in apparent mutations, which prevent a further increase in the metabolite production through pleiotropic influences. With genetic crosses, these unfavorable mutant alleles may be replaced with alleles of one of the parents in the cross.

High-yielding strains can actually increase the cost of the fermentation because of changed physiological properties (greater foaming, changed requirements for culture medium, etc.). By crossing back to wild-type strains, high-yielding strains with improved fermentation properties may be formed.

Hence an effective strain development approach should involve the use of sister-strain, divergent strain, and ancestral crosses at specific intervals, besides use of carefully mutagenesis to ensure the maintenance of genetic variability.

Gene Technology

Gene manipulation or gene cloning involves separating a specific gene or DNA segment from a larger chromosome, attaching to a small molecule of carrier DNA or simply vector and then replicating this modified DNA thousands or millions of times through both an increase in cell number and the creation of multiple copies of cloned DNA in each cell. The result is selective amplification of a particular gene or DNA segment. A clone is an identical copy. These methods and related tasks are collectively referred to as recombinant DNA technology or more informally genetic engineering.

Cloning of DNA genes from any organism entails five general steps:

- ⋏ Isolation of gene of interests.
- ⋏ Insertion of foreign DNA into a vector.
- ⋏ Introduction of the recombinant vector into host cells.
- ⋏ Selection of transformed host cells.
- ⋏ Cloning or mass culture of transformed host cells. (Figure 333)

Site Directed Mutagenesis

Site-directed mutagenesis has been widely used in the study of protein functions. This method was first developed by Michael Smith who was awarded a Nobel Prize in 1993. Site-directed mutagenesis can also be achieved by using PCR.

1. Cloning the DNA of interest into a plasmid vector.
2. The plasmid DNA is denatured to produce single strands.
3. A synthetic oligonucleotide with desired mutation (point mutation, deletion, or insertion) is annealed to the target region. In figure 334, the T to G point mutation is used as an example.

4. Extending the mutant oligonucleotide using a plasmid DNA strand as the template.

5. The heteroduplex is propagated by transformation in *E. coli*.

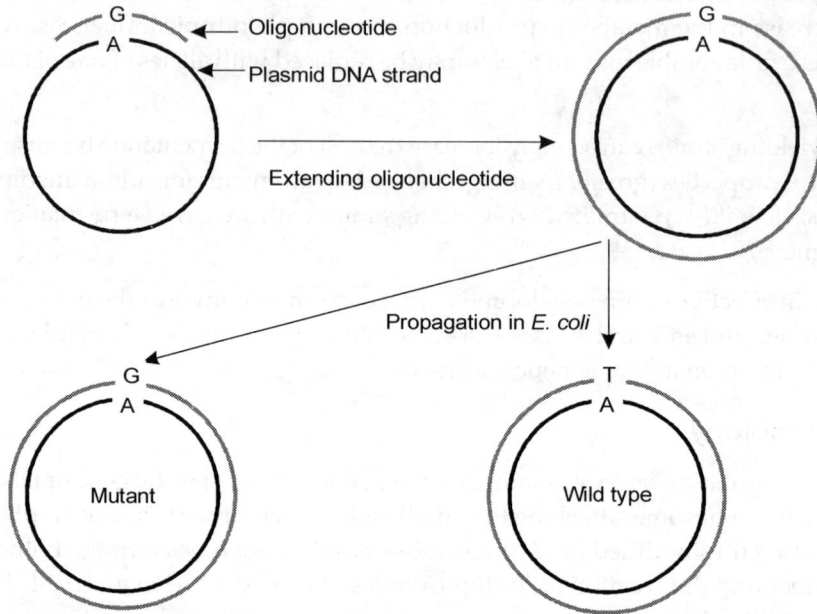

Figure 334 Site directed mutagenesis

After propagation, in theory, about 50% of the produced heteroduplexes will be mutants and the other 50% will be the "wild type" (no mutation).

Parasexual Reproduction

Parasexual reproduction is first seen in Aspergillus. It is known to occur in basidiomycetes, ascomycetes and deuteromycetes. The process involves genetic recombination without the requirement of specific sexual structures. It is a cycle in which plasmogamy, karyogamy and meiosis take place but not at a specified time.

Protoplast Fusion

Somatic fusion, also called protoplast fusion, is a type of genetic modification in plants by which two distinct species of plants are fused together to form a new hybrid plant with the characteristics of both, a somatic hybrid. Hybrids have been produced either between the different varieties of the same species (e.g. between non-flowering potato plants and flowering potato plants) or between two different species (e.g. between wheat triticum and rye secale to produce Triticale).

Uses of somatic fusion include making potato plants resistant to potato leaf roll disease. *Solanum tuberosum* (potato) affected by a viral disease transmitted by the aphid vector. Non-tuber-bearing potato *Solanum brevidens*, which is resistant to the disease. In this process both of these plants were fused and produced disease resistant high yielding potato. The resulting hybrid has the chromosomes of both plants and is thus similar to polyploid plants.

Process for Plant Cells

The somatic fusion process occurs in four steps: The removal of the cell wall of one cell of each type of plant using cellulase enzyme to produce a somatic cell called a protoplast. The cells are then fused using electric shock (electrofusion) or chemical treatment to join the cells and fuse togethor the nuclei. The resulting fused nucleus is called heterokaryon. The somatic hybrid cell then has its cell wall induced to form using hormones. The cells are then grown into calluses which then are further grown to plantlets and finally to a full plant, known as a somatic hybrid.

M. MICRO ALGAL TECHNOLOGY

Introduction

Algae are aquatic plants growing both in fresh water and alkaline aquatic habitats. They are chlorophyllous organisms capable of photosynthesis. About 25,000 species of algae are found worldwide.

Application or Importance of Microalgal Technology

- ⅄ Screening of useful microalgae.
- ⅄ Mass culture and separation of micro algae.
- ⅄ Production of useful compounds from microalgae like agar, alginic acid (Food industry, preservative), β carotene, phycoflours and pharmaceuticals, food, cosmetics, etc.
- ⅄ Genetic improvement of target microalgae by Mutagenesis, gene transfer techniques.

N. CULTIVATION OF SPIRULINA

Introduction

Spirulina is multicellular filamentous blue green algae. Most species of Spirulina colonize fresh waters. *Spirulina* is made up of cylindrical shape cells of about 100μm in diameter and is known as trichomes. Spirulina, now named as Arthospira. It is a microscopic and filamentous cyanobacterium (blue-green algae). It has a long history of use as food. It grows in fresh-water pond, lakes, and in some controlled situations. Its name derived from the spiral or helical nature of its filaments. Spirulina has been produced commercially for the last 20 years for foods and special feeds. It contains very high nutrient compared to any other food. Early interest in Spirulina focused mainly on its rich content of protein, vitamins, essential amino acids, minerals, and essential fatty acids. Spirulina contains over 60% digestible vegetable protein by weight and contains a rich source of vitamins, especially vitamin B_{12} and provitamin A (beta-carotene), and minerals, especially iron. The most widely grown species is *S. platensis* and *S. maxima*. They can colonize extreme environments with high Temparalure. The First commercial spirulina production plant in the world is in Mexico (1970).

Methods of Spirulina Cultivation

Spirulina is cultivated in 2 methods.

1. Out Door Cultivation

Algal ponds are constructed for outdoor cultivation of *Spirulina*. Types of Ponds used for spirulina cultivation are

Circular Ponds with rotating wheels for agitation.

Oblong Ponds with propellers, air injection, air lift etc., for agitation.

Sloped Ponds with gradient flow and propellers.

Closed circular system

Tube System of Farming

Large tubes made of steel or glasses are constructed and the medium is transferred. This system contains inter connected flexible coils. The vertical coils provide a large surface area and also act as solar heat collector.

Photobioreactor Farming

Large transparent tanks are used with an artificial source of light. It is used to produce biochemicals such as phycocyanin, β carotene.

Technological Factors In Spirulina Cultivation

Open Pond Process

⅄ Cement tanks of about 5m are constructed.

⅄ The tanks are coated on the inner surface with algal nutrition.

⅄ The tank is filted with paddle wheels to mix the pond contents for optimum growth of algae.

Nutrients

Commercial fertilizer manufactured by CFTRI / any other company. CFTRI Medium may contain the following ingredients like $NaHCO_3$, K_2HPO_4, $NaNO_3$, K_2SO_4, NaCl, $MgSO_4$, $CaCl_2$, $FeSO_4$.

Carbon Source	The carbon source mainly used is bicarbonate salts.
Light	Shady sunlight is needed for Spirulina.
Agitation	It is done by using stirrer.
pH	pH of the CFTRI medium maintained between 8.5 - 10.5.
Harvesting	Harvesting of algal culture is usually done by gravity Alteration through fine cloth.

Effluent   Effluent after harvesting is still rich in NaHCO$_3$. So effluent is passed through water treatment plant for further purification.

Drying   The algal flakes which are separated and collected is dried by solar drying, air drying.

Indoor Cultivation

Controlled parametes should be provided for best algal production.

⋏   Inoculum is taken from agar slants.

⋏   It is sub cultured in 50ml culture medium followed by 200 ml of culture medium. This process is referred to as scaling up of inoculum.

⋏   The inoculum flask is incubated in a light source consisting of 5 fluroscent lights.

⋏   After a period of growth, the medium is diluted and bottled in glass container.

Factors Affecting indoor Cultivation of Spirulina

Temparature

Lower temperature inhibits algal production.

Sufficient growth of spirulina is observed within a temperature range of 15°-40°C.

Water

Water used for algal cultivation should be clear and never turbid.

Impurities if any should be removed by settling or centrifugation.

Hardness of water should be 7. (1= 18 mg).

pH

⋏ pH range in the medium should 8.5-10.5.

Light

20% incident light energy is required for growth of *Spirulina*.

If light is used directly it results in algal bleaching.

Efficient algal production is observed with 12 hrs light exposure followed by 12 hrs darkness alternatively.

Depth

The depth of Vessel used for spirulina cultivation should be not more than 20 cm.

More depth of vessels interferes with light penetration.

Agitation

Continuous agitation with propellers increase algal production.

O$_2$ Transfer

More O$_2$ transfer occurs during photosynthesis.

During night, availability of O$_2$ should be 75 units / lit.

Harvesting

Harvesting is done by filteration and two types of screens are used for filterations. They are

i.   Stationary Screen - It fiters 10 - 18 m^3 of Spirulina / hrs.

ii. Vibratory Screen - This can harvest 40 m^3 Spirulina / hrs.

Drying

Solar drying is used for end grade Spirulina production.The most widley used method for food grade *Spirulina* preparation is spray drying under vaccum condition for few seconds.

Clean water system cultivation

This system is more expensive because of construction of artificial cultivation farms. These have shallow raceway ponds circulated with paddle wheels and high quality of nutrients. For the fast growth of *Spirulina* in clear water, addition of NaNO$_3$ and NaHCO$_3$ is necessary. pH of the water must be initially maintained to 8.5. It is a self pH adjusting alga which elevates the pH between 10-10.5. At this pH levels there is the least chance of contamination.

The Earthrise Spirulina Farm of California is the world's largest food grade Spirulina farm having the size of about 10 hectares with a capacity of production to about 120 tonnes/ annum. The other big farms are operating in Japan, Thailand, Mexico, Taiwan, Israel, Vietnam, China and India.

In India, food grade Spirulina is cultivated at two main centres, one at Shri A.M.M. Murugappa Chettiar Research Center (MCRC), Madras, and the other at Central Food Technology and Research Institute (CFTRI), Mysore. Madras centre is the biggest food grade Spiulina farm in India. Its annual production capacity is of about 75 tonnes. The products are marketed in India and abroad as health food, baby food and multivitamin tablets.

Waste water system cultivation

This system is applicable in highly populated countries like India where wastes are generated in high quantities and pose environmental problem. In this system, human and animal wastes and sewage are used for growth of Spirulina. The wastes are added into the digester to settle down the solid particles. The liquid effluent is used as a source of the nutrients and added in artificially constructed ponds. As desired $NaNO_3$ and $NaHCO_3$ are also mixed. *S. platensis* is found to grow better in sewage amended with $NaCO_3$ and nutrients in different proportion and also in diluted sewage.

O. APPLICATIONS OF MICRO ALGAE

Algae as food

There are reports that Spirulina was used as food in Mexico during the Aztec civilization some 400 years ago. *S. palmaria palmata* is eaten as 'Dulse' in Canada and contains sugar such as glycerol sorbitol and dulcitol. *Monostroma* is used as a common food called "Anon" in Japan. *Ulva* is used as a dried powder in preparation of sauces. *Nostoc* is sold as "Yuyucho" in China.

Algae as Feed

Algae constitute a source of permenent food for many animals especially in coastal countries. *Laminaria, Saragassum, Fucus, Ascophyllum* are used as fodder. *Spirulina* and corn mixture in ratio off 1:4 is used as the only diet for Tilapia (Fish). Hens which feed on *Fucus* and *Ascophyllum* produces egg with increased iodine content. Sea weed increases the fat content of milk in Cattle. Cyanophyceae and chlorophyceae members are the main food for fishes. Comercial food for sheep are regularly prepared from *Laminaria, Fucus, Ascophyllum*. Major food of many fishes, protozoans and aquatic animals is provided by algae. *Macrocystis* is used for cattle feed because it is rich in Vit A and *Rhodymenia* is used as common food for cattle in France.

Algae as Fertilizers

Seaweeds contain phosphorus, potassium and some trace elements. Hence they are used as fertilizer in coastal regions. The seaweeds are mixed with some organic acid and used as fertilizer. *Lithophylum,* is used in the defeciency of Calcium in the field. *Fucus* is used as common manure. Blue green algae is used as biofertilizers in increasing the rice yield.

Algae as Fuel

Algae during respiration release large amount of CO_2 which servers substrate of CH_4 production. In biomethanogenesis, electron from organic substrates are finally converted to CO_2 which is reduced to methane.

Pharmaceutically And Biologically Active Compound From Algae

Cell extracts and extracts of various unicellular algae are used as a medium for microbial cultivation. *Chlorella vulgaris, Chlamydomonas, Pyrenoids* have been shown to have antibacterial activity against both G+ve and G-ve bacteria. *Gymnodinium* + *Gonyaulax* produce alkyl guanidine compound that is used to treat CNS disorders. *Porphyra* produces arachidonic acid (lipid) which is an essential dietry constituent. Prostaglandins have been isolated from *Gracilaria*. Lyngbya shows antiparasitic activity against *Trichoderma*. *Fucodin* and sodium laminarian sulphate from brown algae is used as an anticoagulant. Chlorellin is an antibiotic frorn *Chlorella*. An chemical effective against *E.coli* obtained from a diatom. Algae are used as antihelminthic, anti hypertensides, anticoagulants, activator, vitamin source. Oinments mixed with 20% of alcoholic extracts of *Scendesmus* cure ulcer, burns. Chlorella tablets treat gastric ulcer, abdominal ulcer and Gastritis. *Scendesmus, Spirullina* are used to lower serum cholesterol level. Creams, oinments, solutions of *Spirulina* are used in treatment of wound. *Spirulina* promote skin metabolism. The blue pigment phycocyanin constitutes 20% of *Spirulina*. It stimulates the immune system and gives protection against a variety of diseases. Phycobilin is used as an antitumor agent as well as for treatment of ulcer and bleeding. *Spirulina* is a natural source of linoleic acid (GLA) which constitutes about 1% of dry weight. GLA is involved in prostaglandin synthesis and metabolism. Prostaglandin is associated with regulation of blood pressure, cholesterol synthesis, cell proliferation. *Asterionella notata* (diatoms) has antiviral and antifungal activity. *Spirulina obliques* is used in post operative care. Extracts of *Phormidium* have anti reverse transcriptase activity.

Algae as SCP

SCP refers to cells of microscopic organism such, as bacteria, yeast, microalgae that are rich in nutrients and grown in large scale.

Algae are rich in Carbohydrates, proteins, minerals, vitamins, fatty acids and hence are used as SCP.

Eg: *Spirulina, Chlorella.*

Nutritive Value of SCP

Carbohydrate	20%
Fat	Lipid present in SCP in the form of triglycerides and fatty acids.
Proteins	45 - 80%
Nucleic Acids	6 - 25%

Spirulina is known as a law-fat, low calorie, cholesterol free protein source.

Pigments

⅄ Microalgae contains pigments such as phycocyanin, phycoerythrin, carotenoids. Carotenoids includes β- carotene, Leutin. etc.

⅄ Pigments are used as colouring agent in food, used as food additives to enhance the flesh colour of salmonoid fish and colour of egg yolk. Phycocyanin made from Spirulina used for cosmetic items. It is insoluble in water and is not removed when it is wetted by water. It does not irritate the skin.

Hydrocarbons

⅄ *Dunaliella salina* contains 30% cyclic hydrocarbons.

⅄ *Botyrococcus sp.* has the hydrocarbon content of 20%.

Steroids

Scendesmus contain chondrilasterol which is used as the starting material for synthesis of hormone cortisone.

Polysaccharides

Porphyridium cruentum and *P.aerugineum* produced extracellular polysaccharide. These polysaccharide are stable at 20° to 90°C and pH 2-9. They have the molecular weight or 4×10^6 dal.

Enzymes

Phosphoglycerate kinase is the enzyme produced from *Spirulina platensis*. It is specific for ATP and it is used for ATP production. Superoxide dismutase is one of the important enzyme produced by *Spirulina* and *Porphyridium*.

P. SINGLE CELL PROTEIN (SCP)

Definition

The dried cells of microorganisms used as food or feed are collectively known as 'microbial protein'. The term 'microbial protein' was replaced by a new term "single cell protein" (SCP) during 1967 at the Massachusetts Institute of Technology (MIT), Cambridge, U.S.A. Eg. algae, bacteria, actinomycetes and fungi.

Historical Importance

- ⋏ Fermented yeast (*Saccharomyces sp.*) was recovered as a leavening agent for bread as early as 2500 B.C.
- ⋏ Fermented milk and cheese produced by lactic acid bacteria (*Lactobacillus* and *Streptococcus*) was used by Egyptians and Greeks during 50-100 B.C. e.g. *Lactobacillus* and *Streptococcus*.
- ⋏ During the first century B.C. the palatability of edible mushrooms was realized in Rome.
- ⋏ In- 16ᵗʰ Century blue-green algae (e.g. *Spirulina*) was consumed as a major source of protein.
- ⋏ The term 'mycoprotein' has been introduced by Ranks Hovis McDougall (RHM) in 1920s for protein produced on glucose or starch substrates.
- ⋏ Importance of mass production of microorganisms as a direct source of microbial protein was realized during World War I in Germany. Baker's yeast (*S. cerevisiae*) was produced in an aerated molasses medium supplemented with ammonium salts.
- ⋏ During World War II (1939-1945) the aerobic yeasts (e.g. *Candida utilis*) were produced for food and feed in Germany.
- ⋏ In the late 1950s, British Petroleum started producing the SCP from hydrocarbons since the crude oil contains 10-25 percent n-alkanes (paraffins).
- ⋏ In India, little attention has been paid on the production of SCP, though mushroom cultivation started in the early 1950s. However, work on mushroom culture at Solan (Himachal Pradesh) from 1970 onward has brought satisfactory results.

Advantages of SCP- (Microbial Protein)

⅄   Microbial cells grow rapidly with shorter generation time. Algal generation time is 2-6h; yeast multiplied within l-3h; bacterial generation time is less than 0.5-2h.

⅄   Microbes and its products are Easily modifiable genetically (e.g. for composition of amino acids);

⅄   Microbial biomass has High protein content of 43-85% in the dry mass;

⅄   Waste materials may used as a raw material.

⅄   Production in continuous cultures.

⅄   Consistent quality.

⅄   It is not dependent on climate.

⅄   Land requirements is low.

⅄   Ecologically beneficial.

⅄   High solar energy conversion efficiency per unit area

⅄   Easy regulation of environmental factors e.g. physical, nutritional, etc

Microorganisms Use as a Single Cell Protein (SCP)

Many groups of microorganisms are used as sources of proteins.

Algae	*Chlorella pyrenoidosa; Scenedesmus acutus; Spirulina maxima*
Bacteria	*Achromobacter delvacvate; Bacillus megaterium; Cellumonas sp.; Methylomonas clara, Pseudomonas sp.*
Actinomycetes	*Nocardia sp.; Thermomonospora fusca*
Yeasts	*Candida lipolytica; C. utilis; Saccharomyces cerevisiae; S. fragilis; Rhodotorula glutinis; Torulopsis sp.*
Mould	*Aspergillus niger; Trichoderma viridae; Paecilomyces varioti*

Substrates used for SCP production

A variety of substrates are used for SCP production. Algae which contain chlorophylls do not require organic wastes. They use free energy from sunlight and carbondioxide from air. Bacteria and fungi require organic wastes. CO_2 (10%), light , combustion gases, bicarbonate, Diesel oil, Bagasse, Methanol, n alkanes, cellulose pulp, potatostarch waste, sulphite liquor, Molasses, milk whey, domestic sewage, straw, starch are the substrate used for SCP production.

Algal SCP

The use of algae as food and feed is known since centuries as they form a part of the diets. Some of the algae like Chlorella, *Scenedesmus, Dunaliella* and *Spirulina* have been found to suitable for mass cultivation and utilization. The advantages in using algae include simple cultivation, effective utilization of solar energy, faster growth and high protein and nutrient content.

Chlorella

Chlorella is a green algae. It belongs to the phylum Chlorophyta. It is spherical in shape, about 2 to 10 nm in diameter. *Chlorella* contains green photosynthetic pigments chlorophyll-a and b. It multiplies rapidly, requiring only carbon dioxide, water, sunlight, and a small amount of minerals. *Chlorella* strains are being used for a variety of applications in biotechnology. It has very high protein contents. They can be used as feed for production of animal protein. In many countries strains of *Chlorella* are utilized for sewage oxidation and waste water treatment. Many people believed *Chlorella* could serve as a potential source of food and energy. It is an attractive potential food source because

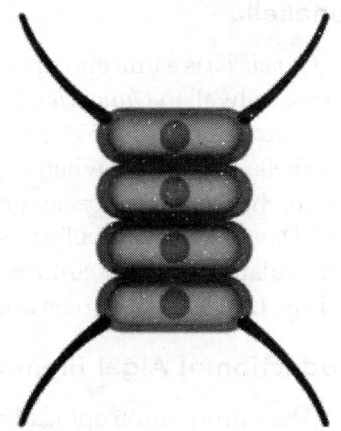

Figure 335   Chlorella

it is high in protein and other essential nutrients; when dried, it is about 45% protein, 20% fat, 20% carbohydrate, 5% fibre, and 10% minerals and vitamins. Mass-production methods are now being used to cultivate it in large artificial circular ponds. It reduce weight of an individual, used in cancer prevention, and immune system support. *Chlorella* yields oils that are high in polyunsaturated fats. *Chlorella* supplementation has a positive effect on the reduction of dioxin levels in breast milk and it may also have beneficial effects on nursing infants by increasing the IgA levels in breast milk. It cures allergy and bronchial asthma.

Scenedesmus

Scenedesmus are small, nonmotile algal cells that are easily recognized by their football shape and their tendency to form orderly looking chains of cells called colonies. They are mostly found in standing water. They can float individually but they are usually found in colonies of 4 to 32, and the two end cells of the chain have spines typically around 200 micrometers in length. They are autotrophic, meaning they get their carbon supply from CO_2 using photosynthesis. *Scenedesmus* can be easily enriched with selenium and fed to chickens. The selenium gives the chicken meat produced better storage qualities. *Scenedesmus* powder added as a dietary supplement could lower cholesterol. It contains 50-56% Protein, 9-14% lipid, 15-17% carbohydrate, 10-13% crude fibres and 10-15% minerals and vitamins a carotenoids are aboundant in *Scenedesmus*.

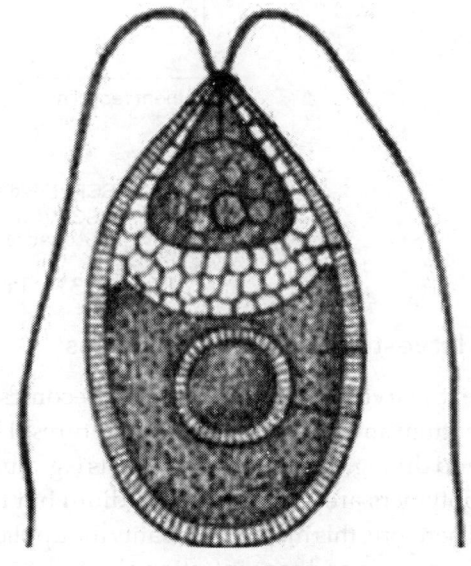

Figure 336   Scenedesmus

Dunaliella

Dunaliella is a unicellular biflagellate green algae. It looks like a *Chlamydomonas*. There is no true cellwall in *Dunaliella*. It occurs in fresh water, saline water and brakish water. It is a rich source of protein (57%). It needs strong light to thrive as it undergoes photosynthesis. Dunaliella grows best when exposed to a light intensity of 60 W/M² for 16 hours per day. Most of the *Dunaliella* species grow best at a pH of 8.2-8.7.They thrive at a temperature of 20-22°C. *Dunaliella* grows well at a salinity of 3,000 - 10,000 ppm (parts per million). They should be inoculated into new cultures at a ratio of at least 1 mL culture to 250 mL medium. When healthy, *Dunaliella* are green and pear shaped.

Production of Algal Biomass

Algae grow autotrophically and synthesize their food by taking energy from sunlight or artificial light, carbon source from carbon dioxide, and nutrients from carbohydrates present in growth medium. It is also cultivated in large trenches i.e. particularly in sewage oxidation ponds by using sunlight or in an artificial illumination conditions.

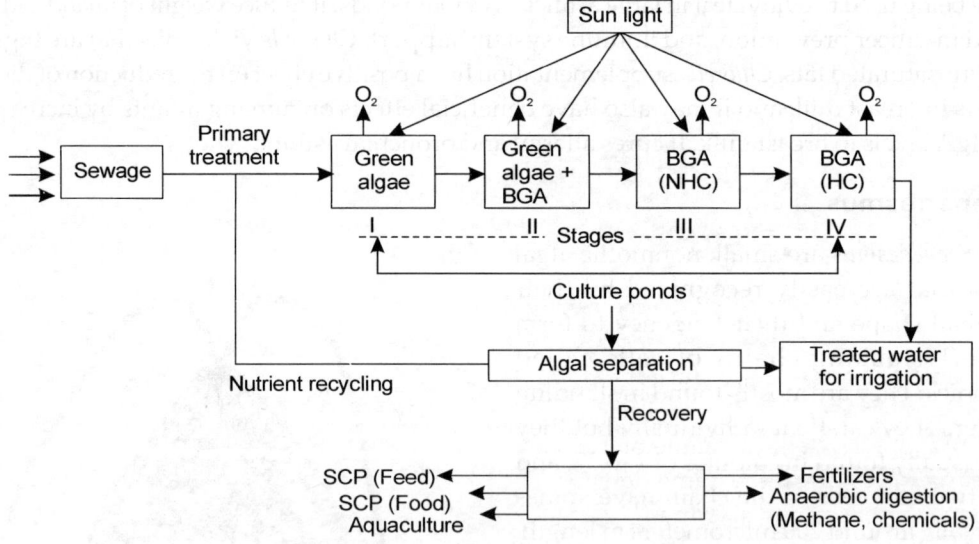

Figure 337 Production of Algal Biomass

Harvesting the Algal Biomass

Harvesting of algal culture becomes problematic because of settling down of cells at bottom and mixing of the algal cultures. The cells are recovered by concentration, dewatering and drying. Sometimes flocculants e.g. aluminium sulfate and calcium hydroxide and cationic polymers are added to the medium but they cannot be separated from the harvested cells. Therefore, this method warrants the application of SCP products in food and feed. Methods of separation and concentration also follow centrifugation, flocculation and centrifugation plus flocculation, but it is not economically feasible so far. Harvesting the cyanobacteria, for example,

Spirulina sp. is less troublesome as their spiral filaments float on the surface of water because of gas filled vacuoles in their cells which result in floating algal mats. Cells are able to fix atmospheric nitrogen. Algal mats are filtered and suspension of *Spirulina* is dried with hot air to get fine powder. Algal yield from stabilization pond is around 70-114 tonnes/ha/year.

Spirulina

Spirulina is multicellular filamentous blue green algae. Most species of *Spirulina* colonize fresh waters. *Spirulina* is made up of cylindrical shape cells about 100μm in diameter and is known as trichomes. *Spirulina*, now named *Arthospira*, is a microscopic and filamentous cyanobacterium (blue-green algae) that has a long history of use as food. It grows in fresh-water pond, lakes, and in some controlled situations. Its name derived from the spiral or helical nature of its filaments. *Spirulina* has been produced commercially for the last 20 years for foods and specialty feeds. It contains very high nutrient compared to any other food, plant, grain, or herb. Early interest in *Spirulina* focused mainly on its rich content of protein, vitamins, essential amino acids, minerals, and essential fatty acids. *Spirulina* contains over 60% digestible vegetable protein by weight and contains a rich source of vitamins, especially vitamin B_{12} and provitamin A (beta-carotene), and minerals, especially iron. The most widely grown species is *S. platensis* and *S. maxima*. They can colonize extreme environments with high Temparalure. The First commercial spirulina production plant in the world is in Mexico (1970).

US federation drug administration recognized spirulina as a supplement of human food and animal feed.

Cultivation of *Spirulina*

In tropical countries *Spirulina* cultured under authotrophic, heterotrphic and mixotrophic conditions. Mass cultivation easier than other algal cultivation because aeration of CO_2 is not necessary for this species since it can maximally utilizes the amounts of carbon. In 1821, *Spirulina maxima* biomass was collected from Texcoco lake and made into biscuits. This biscuits are sold in mexico in the name of Tecuitlatlo. In 1964, Chad biscuits made from *Spirulina platensis* were sole in the name dehi in India. Spirulina is also sole under the trade names nuclena, spirudox and spiruna.

Composition of dried powder of *Spirulina fusiformis* (constituents are in per 100 of powder).

Table 17.6 Nutitive value of SCP

A. Major constituents (%)		C. Minerals (mg)	
Total protein	64.6	Calcium	6.58
Fat	6.7	Phosphorus	977
Crude fiber	9.3	Iron	44.7
Carbohydrates	16.1	Sodium	796
Calories	346	Potassium	1.28
B. Vitamins		D. Essential amino acids (%)	
Beta - carotene	320,000IU	Lysin	2.99
Biotin	0.22 mg	Cystine	0.474
Cyanocobalamin (B$_{12}$)	65.7 mg	Methionine	1.38
Folic acid	17.6 mg	Phenylalanine	2.87
Riboflavin	1.78 mg	Threonine	3.04
Thiamin	0.118 mg		
Tocopherol	0.773 IU		

Uses of Spirulina Single Cell Protein

i. Since *Spirulina* is a rich source of protein (60-72%), vitamins, amino acids, minerals, crude fibers, etc., it is used as supplemented food in diets of under-nourished poor children in developing countries. It has been found that 1 g of *Spirulina* tablets contains as much nutrition as one kg assorted vegetable.

ii. *Spirulina* is very popular as health food. It is the part of the diet of the US Olympic team. Jaggers take *Spirulina* tablets for instant energy. Since it provides all the essential nutrients without excess calories and fats, it is taken by those who want to control obesity. The MCRC (Murugappa Chettiar Research Center) has for the first time launched the project as health and baby food, and multivitamin powder and tablet under trade name 'Multin' and 'Multinal'.

iii. *Spirulina* possesses many medicinal properties. Therefore, it is used as social and preventive medicine also. It has been recommended by medicinal experts for reducing body weight, cholesterol and pre-menstrual stress and for better health. It lowers sugar level in blood of diabetics due to the presence of gamma-linolenic acid and prevents the accumulation of cholesterol in human body. It is a good source of β-carotene and, therefore, helps in monitoring healthy eyes and skin. β-carotene is known as the best anticancer substance. In 1989, UN National Cancer Research Institute announced that substances from blue-green algae are active against AIDS and cancer virus. In Vietnam its tablets are used to increase lactation in nourishing mothers.

iv. *Spirulina* contains high quality of proteins and vitamin A and B. These play a key role in maintaining healthy hair. Many herbal cosmeticians are making efforts to develop a variety of beauty products. Phycocyanin pigment has helped in formulating biolipsics and herbal face cream in Japan. These products can replace the present coaltar-dye based cosmetics which are known as carcinogenic.

Production of Bacterial and Actinomycetous Biomass

Bacteria are widely used as a source of single cell protein because of their short life cycle (20-30 minutes) and capacity to utilize a wide range of organic substrates as a source of energy. Dried biomass of some bacteria are rich in protein, which are easily digestable. Actinomycetes also utilize these renewable sources as they have more or less same generation rate as bacteria. British Petroleum in 1960 produced microbial products by using gaseous and liquid hydrocarbons and chemicals derived from them, for example, methanol, ethanol, etc. The Shell Research Limited, U.K. conducted research on pilot plant scale process for the production of bacterial SCP methane by using *Methylococcus capsulatus* or mixed culture of *Pseudomonas sp.*, *Hyphomicrobium sp.*, *Acinetobacter sp.* and *Flavobacterium*. *Streptomyces sp.* is capable of growing on methanol. *Theromonospora fusca*, a thermophilic species, degrades 60-65 per cent paper mill fines resulting in 30 per cent protein product. Nowadays, cellulose degrading thermophilic actinomycetes offer a great opportunity to yield SCP from cellulosic wastes.

Method of Production

Bacterial biomass is produced in the following ways :

Supply of a nutrient substrate and formulation of a suitable medium

Bacteria needs carbon, nitrogen, Ca, P, K and trace elements for the growth.Diesel oil, bagasse, methanol, n-alkanes and cellulose pulp are used as a carbon source. Ammonia acts as a nitrogen source.

Multiplication of microorganisms

Sterilized medium is pumped into the fermenter and inoculated with the 10% bacterial inoculums. Maintain optimal pH, temperature, aeration, agitation etc., Air needs to be pushed from the bottom.

Separation of cellular substances from the left over medium

Bacterial cells attain 99% of maturation within few days. Harvested broth is subjected to centrifugation, which removes biomass in the form of pellets. One litter broths yields approximately 20g of biomass.

Further treatment to kill and dry the bacterial biomass

Finally biomass is dried and treated with heat, which kills bacterial cells and yields effective killed bacterial SCP.

Bacterial SCP contains 0.3-4% carbohydrate, 45-80% proteins, 16% lipids, 0.5% β caroteine, 6-25% nucleic acids. In nucleic acid, purine and pyrimidine are available in equal percentage. 3-12.5% purine is available in bacterial SCP. Uric acid is produced as a byproduct of purine metabolism. Human body has no mechanism to excrete uric acid. So it gets accumulated in the kidney and other organs. So bacterial SCP is not used for human consumption directly. Birds have the mechanism to excrete uric acid. So poultry is fed with bacterial SCP. Egg and meat produced by such poultry are richer in nutritionally valuable proteins than normal poultry.

Factors Affecting Biomass Production

1. Growth of bacteria and actinomycetes are affected by many factors. These are :
2. Suitable strain of bacterial culture ;
3. Genetic stability of bacterial strain;
4. Absence of bacteriophage;
5. Suitable pH 5-7 of growth medium;
6. Temperature (15-35°C according to strain);
7. Oxygen/agitation to create aerobic condition;
8. Organic substrate and nitrogen concentrations. Optimum C:N ratio which favours high protein contents in cells and inhibits accumulation of lipid is 10:1; and
9. Maintenance of sterile conditions throughout the growth.

Product Recovery

There are many problems related to the recovery of bacterial cells as they are very small and have cell density in the order of 10-20 g/liter. Many pilot plants use flocculants and many have set up decanter type centrifuge. For example, Hoechst (Germany) has developed a device for separation of *Methylomonas clara* from methanol containing culture medium which is based on electrochemical, coagulation and centrifugation. The cells are washed and spent medium is again treated by conventional treatment process as it contains inorganic salts and small amount of cells. Cell biomass is then spray-dried.

Production of Yeast Biomass

Consumption of baker's yeast (*S. cerevisiae*) as food in Germany during World War I increased its importance. Since then, rapid development took place in biotechnological applications of *S. cerevisiae*, as far as culture development, process optimization and scale up of products are concerned. World production of yeast biomass is of the order of 0.4 million metric tonnes per annum including 0.2 million tonnes baker's yeast alone. Many yeasts like *Candida lipolytica*; *C. utilis*; *Saccharomyces cerevisiae*; *S. fragilis*; *Rhodotorula glutinis*; *Torulopsis sp.* are consumed as SCP.

Yeasts get carbon and energy sources from the organic wastes, e.g. molasses, starchy materials, milk whey, fruit pulp, wood pulp and sulphite liquor. *Saccharomyces cerevisiae* prefer molasses, *Candida utilis* prefer starch waste, *Saccharomyces fragilis* prefer milk whey whereas domestic sewage is used for the cultivation of *Rhodotorula glutinis*. pH of the medium is adjusted to 3.5-4.5 to medium is sterilized using autoclave.

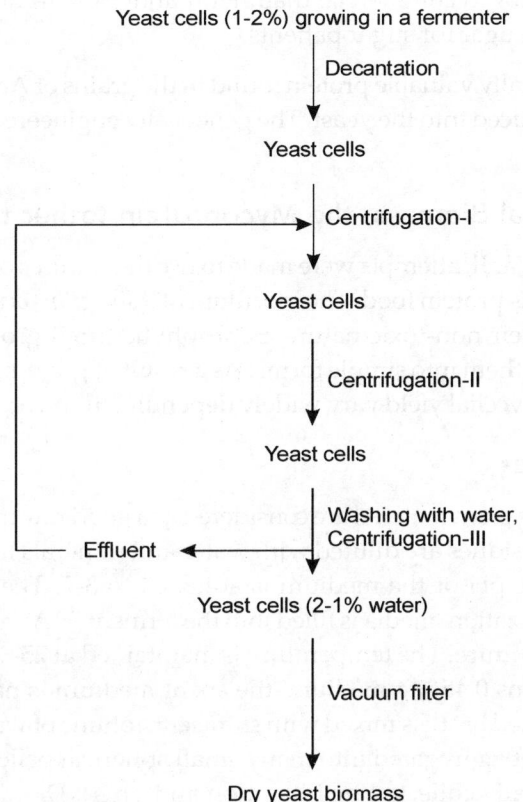

Figure 338 Production of yeast biomass

Factors Affecting the Yield of Yeast Biomass

Growth and yield of yeasts are affected by the following factors. Organic substrate and nitrogen ratio (optimum C : N ratio favouring high protein content should be between 7:1 and 10:1); pH of nutrient medium (pH should be in the range of 3.5 to 4.5 to minimize growth of bacterial contaminates); Temperature (it differs from organism to organism). Most yeast have specific growth rate in the range of 30°C to 34°C. Some strains also grow in the range of 40-45°C; oxygen (for growth on carbohydrates), O_2 required should be 1 g/g of dried cells, and for growth on n-alkanes it should be about 2 g/g dried cells).

Recovery of Yeast Biomass

Yeast cells are small in size (5-8 m), the density of which reaches to 1.1 g/ml. Yeast cells are recovered by centrifugation and washing. Final biomass is obtained through drying.

Genetic engineering of yeast

Thaumatin is a sweet protein found in the fruit of the west African plant, *Thaumatococcus danielli*. It is 5000 times sweeter than 4% sucrose. A DNA encoding for thaumatin was introduced into the Yeast and produced commercial thaumatin under the trade name talin. It is being used as a substitute for sugar for sugar patients.

AMA I is a nutritionally valuable protein found in the grains of Amaranthus. AMA I gene of amaranthus is introduced into the yeast. The genetically engineered yeast produces A MA proteins.

Production of Fungal Biomass, the Mycoprotein (other than Mushrooms)

During the World War II, attempts were made to use the cultures of *Fusarium* and *Rhizopus* grown in fermentation as protein food. The inoculum of *Aspergillus oryzae* or *Rhizopus arrhizus* is chosen because of their non-toxic nature. Saprophytic fungi grow on complex organic compounds and render them into simple forms. As a result of growth, high amount of fungal biomass is produced. Mycelial yield vary widely depending upon organisms and substrates.

Cultivation of fungus

Cellulose containing plant material are considered as a good raw material for mycoprotein production. The crop residues are diluted with water and minerals like ca, Mg, K, P, Zn, Cu, and Co are added to it. pH of the medium is adjusted to 3-7. The media is sterilized by autoclaving. After sterilization, media is filled into the fermenter. A small volume of fungus is inoculated as a starter culture. The temperature is maintained at 25-30°C for 48hours. When the fungal growth attains 0.45g/g medium, the spent medium is harvested to recover the fungal biomass. Harvested broth is mixed with sufficient volume of water and agitates for few hours. Fugal hyphae get aggregated into many small spherical pellets on the surface of the mixture. Biomass is finally collected by filteration and dried. Dried bimass is stored in an aluminium foil or sterile containers.

Advantages of mycoprotein

Chahal (1982) has described the increasing popularity of myco-protein because of the following reasons:

Some of the filamentous fungi grow as fast as most of the single celled organism;

The finished product of filamentous fungi is fibrous in nature and can be easily converted into various textured foods.

Filamentous fungi have a greater retention time in the digestive system than single celled organisms;

Protein content can be as high as 35-50 per cent with comparatively less nucleic acid than single celled organism;

Digestibility and net protein utilization without any pretreatment is higher than single celled organisms;

The overall cost of protein production from filamentous fungi is more economical as compared to that of single celled organism;

Filamentous fungi have greater penetrating power into insoluble substrates and are therefore, more suitable for solid state fermentation of lignocellulosic materials;

Most of filamentous fungi have a faint mushroom like odour and taste which may be more readily acceptable as a new source of food than the yeast odour and green colour associated with yeasts and algae respectively;

The biomass produced by filamentous fungi can be used as such without any further processing because it provides carbohydrates, lipids, minerals, vitamins and proteins. In addition, nucleic acid contents of fungal protein is lower than that of yeast and bacteria

Mushroom

Mushrooms are the members of higher fungi, belonging to the class Ascomycetes (e.g. Morchella, Tuber, etc.) and Basidiomycetes(e.g. Agaricus, Auricularia, Tremella, etc.). They are characterized by having heterotrophic mode of nutrition. According to Chang and Hayes (1978) edible mushroom refers to both epigeous and hypogeous

Figure 339 Mushroom

fruiting bodies of macroscopic fungi. They have a stem (stipe), a cap (pileus), and gills (lamellae) or pores on the underside of the cap. Mushroom may be button like or fan like or umbrella like shaped in nature.

A mushroom develops from a nodule, or pinhead, less than two millimeters in diameter, called a primordium, which is typically found on or near the surface of the substrate. It is formed within the mycelium, the mass of threadlike hyphae that make up the fungus. The primordium enlarges into a roundish structure of interwoven hyphae roughly resembling an egg, called a "button". The button has a cottony roll of mycelium, the universal veil, that surrounds the developing fruit body. As the egg expands, the universal veil ruptures and may remain as a cup, or volva, at the base of the stalk, or as warts or volval patches on the cap.

Figure 340 Oyster mushroom cultivation

Types of mushroom

Presently about a dozen fungi are cultivated in over 100 countries with a production of 2.2 million tons. *Agaricus bisporus* (56%)-white button; *Lentinus edodes* (14%)-shiltake; *Volvariella volvacea* (8%)-paddy straw; *Pleurotus spp* (7%)-oyster; Flammulina (5.5%)-winter mushroom; Tremmula spp (4.6%)-silver ear; Phillota (1.8%)-nameka and Others (1.1%) are the mush room varieties cultivated for human usage. Species of the genus Pleurotus and Agaricus are most commonly cultivated in India.

Nutrients level

Mushrooms are superior to many vegetables. Fresh Nutrient value of mushroom include 89-91%. Moisture, 3.2 % Protein, 4.4% Carbohydrate, 0.3% Fat and 16K Calories. In the dried mushrooms moisture content is very low and the protein content is about 32 - 44%. Dietary mushrooms are a good source of B vitamins, such as riboflavin, niacin and pantothenic acid, and the essential minerals, selenium, copper and potassium.

Uses of mushroom

Crude fibre content is very high and the carbohydrate content is very low. So it can be used for diabetic patients.

Fat, carbohydrate and calorie content are low, with absence of vitamin C and sodium. So it can be used for hypercholesteremic patients.

It contains natural ergosterols, which is a precursor for vitamin D2.

Some mushroom materials, including polysaccharides, glycoproteins and proteoglycans, modulate immune system responses and inhibit tumour growth.

Mushrooms show potential cardiovascular, antiviral, antibacterial, antiparasitic, anti-inflammatory, and antidiabetic properties.

Mushrooms provide a rich addition to the diet in the form of proteins, carbohydrates, minerals and vitamins.

Mushroom can be a substitute for fish, vegetables, meat and egg. Some mushrooms used to reduce blood pressure, gall stones and numbness in hands and foot.

Mushroom cultivation

Mushrooms can be cultivated using bed method, polythene bag method and field method.

Bed method and field methods are more suitable for white button mushroom (*Agaricus bisporus*)., polythene bag method is more suitable for oyster mushrooms.

Oyster mushroom cultivation

Various substrates utilized for the cultivation of Pleurotus are banana pseudostems, wheat straw, paddy straw, ragi straw, compost prepared from straw, saw dust, beech saw dust, sunflower stalks, rice husk and karad hay. However, the highest yields are obtained on rice straw. These can be grown in any container, e.g. earthen pot, cane gasket, polyethylene bags, iron baskets or in wooden trays.

Procedure:

Paddy straw is taken and chopped the straw into 1 to 2 cm bits. Soak the chopped straw into water overnight and Drain off the excess water. Horse gram powder is added at the rate of 8g/kg. Spawn is mixed to the straw at the rate of 30g/kg. This mixture is filled into polyethylene bags with holes. Incubate the filled bags in a room at 21 to 35°C with sufficient light and ventilation for 15-16 days for spawn running.Cut open the polyethylene bags on the sides without disturbing the bed. Spray water over the bags twice a day.

Harvesting

First harvesting is to be done 20-22 days after spawning, 2nd harvesting 27-29 days after spawing and 3rd harvesting 34-36 days after spawing.

Advantages

It can be cultivated in a small space without sophisticated instruments.

It is cultivated by making use of agrowaste.

Simple procedure is enough for the cultivation of the mushrooms

It converts waste into edible fruiting bodies.

White button mushroom cultivation

It grows in areas where the temperature is <25°C. This is not grown in high temperature. Compost is used for the cultivation of white button mushroom. The compost mixture consists of 1000kg straw, 400g chicken manure, 72kg brewers grain, 14.5kg urea and 30kg gypsum. .

Figure 341   Button mushroom cultivation

The compost is filled in small trays or plastic bags for about 15-25cm height. A handful of spawn is spreaded over the compost and a small amount of compost is spread over the spawn. This step is called spawning. After spawning, the compost is sprinkled with water so as to maintain relative humidity between 70 - 80%. The mushroom reaches pinhead stage within 5-15days after spawning. These pinheads take 7-8days to become a button stage

Harvesting

The white button mushrooms are harvested when the cap is still tight with its stalks. The mushroom is slightly pressed to the compost and twisted greatly to harvest it. The soil

Table 17.7   Differentiation features of mushroom grown in India

S. No	Character	Agaricus	Pleurotus
1	Fruiting body	Umbrella shaped	Fan like
2	Pileus	5-10cm diameter	5-10cm diameter
3	Gil colour	Pink or brown	White or grey
4	Stripe	Central & Long	Lateral & indistinguishable
5	Annulus	Present on the stripe	Absent
6	Vulva	absent	absent
7	Other names	White button mushroom; temperate field mushroom	Indian Oyster mushroom
8	Latin name	*Agaricus bisporus*	*Pleurotus sajor caju*

particles adhering on the mushroom maybe removed with a sterile knife. The harvested mushrooms are either cooked immediately or stored at refrigerated temperature.

Q. *BACILLUS THURINGIENSIS - BIOINSECTICIDE*

Bacillus thuringiensis (or Bt) is a common gram positive, spore-forming, soil bacterium. It is used for the control of insect population. It is also called Microbial insecticide or Bio insecticide. During unfavourable condition, vegetative Bt cells undergo sporulation, synthesizing a protein crystal. Proteins in these crystals are called Cry (from Crystal) endotoxins and have been known for decades to display insecticidal activity against specific insect groups.

B. thuringiensis was first discovered in 1901 by Japanese biologist Ishiwata Shigetane. In 1911, *B. thuringiensis* was rediscovered in Germany by Ernst Berliner, who isolated it as the cause of a disease called Schlaffsucht in flour moth caterpillars. In 1976, Robert A. Zakharyan reported the presence of a plasmid in a strain of *B. thuringiensis* and suggested the plasmid's involvement in endospore and crystal formation. Upon sporulation, *B. thuringiensis* forms crystals of proteinaceous insecticidal endotoxins (called crystal proteins or Cry proteins), which are encoded by cry genes. In most strains of *B. thuringiensis,* the cry genes are located on a plasmid. Cry toxins have specific activities against insect species of the orders Lepidoptera (moths and butterflies), Diptera (flies and mosquitoes), Coleoptera (beetles), Hymenoptera (wasps, bees, antsand sawflies) and nematodes.

Bacillus thuringiensis serovar israelensis, is widely used as a larvicide against mosquito larvae. *Bacillus thuringiensis* is only weakly toxic to insects as a vegetative cell, but during sporulation, it produces an intracellular protein toxin crystal, the parasporal body, that can act as a microbial insecticide for specific insect groups.

B. thuringiensis can be grown in fermenters. When the cells lyse, the spores and crystals are released into the medium. The medium is then centrifuged and made up as a dust or wettable powder for application to plants.

Production of BT Toxin by rDNA Technology

Isolate BT toxin gene. Purify it by gradiant centrifugation. 71kb size gene code for BT toxin. Isolated gene is subjected to restriction digestion. Select suitable vector and treat with suitable restriction enzyme. This remove the marker gene without disturbing promoter sequence. Target gene is inserted into the vector. Recombinant plasmid is transferred to the host. Check for production of parasporal crystal toxin. BT toxin containinf plant also developed which also kills insects which act on the plant.

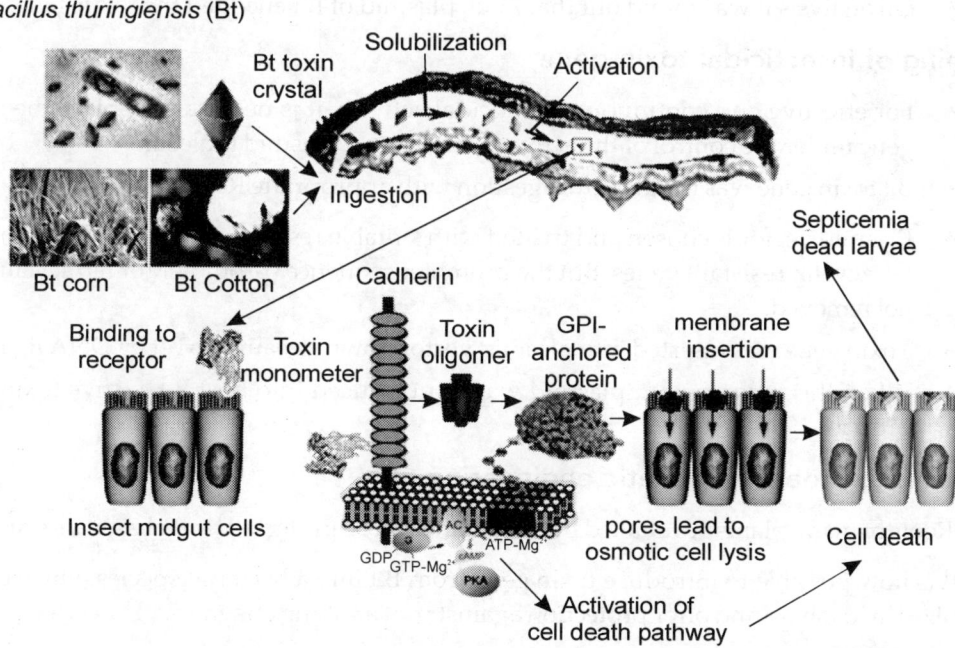

Figure 342 Mode of action of BT toxin

A Microbial insecticide refers to an organism that produces a toxic substance that kills an insect pest.

The most effective and utilized microbial insecticide are the toxin synthesized by *B. thuringiensis*. For eg. *B. thuringiensis* subs kurstaki is toxic to Lepidoptera. (Insect group).

Mode of action of BT Toxin

The insecticidal activity of Bt is present within a very large structure called parasporal crystal which is synthesized during sporulation. Parasporal crystal is a protoxin.

The parasporal crystal, after exposure to alkaline conditions in the hindgut, fragments to release the protoxin. After this, toxin reacts with a protease enzyme, the active toxin is released Six of the active toxin units integrate into the plasma membrane to form a hexagonal-shaped pore through the midgut cell. . This leads to the loss of osmotic balance and ATP, and finally to cell lysis.

Genetic Engineering in BT

- To develop B.t based insecticide that can be effective, it is necessary to isolate and characterise the toxin gene.
- The total DNA of the bacterium is isolated and seperated into plasmid and chromosomal DNA by Gradient centrifugation.
- On analysis it was found out that 71 kb plasmid of B.t encoded the toxin gene.

Cloning of insecticidal toxin gene

- For effective and continuous insecticidal activity, it is necessary to place the toxin gene under the control of the active promoter in a plasmid molecule.
- B.t toxin gene was isolated by digestion with appropriate Restriction Enzyme.
- Plasmid vector is chosen and treated with suitable restriction enzyme that removes tetracyclin resistant genes. But the promoter sequence (promoter for tetracycline) is not removed.
- Toxin gene was inserted into plasmid vector down stream of pTet by DNA ligase.
- When this recombinant plasmid was reintroduced into host B.t., active toxin was produced.

A Novel approach in genetic engineering

Roots of many plants are attaked by insects and can be protected by B.t based insecticides.

It is now possible to introduce toxin gene from B.t into a bacterial species which could colonize rhizosphere and offer protection against root attaking enzymes.

Eg: *P. fluroscence* can be the bacterium of choice.

Tn5 element (Transposon) is modified by removing the right and left border sequences and cloned into a plasmid.

Isolated B.t toxin gene was spliced into the middle of the altered Tn5 element on plasmid.

Wild Tn5 element was inserted into the cromosome of *Pseudomonas fluroscence*.

⅄ Plasmid carrying Tn5 element which inserted toxin gene was introduced into *P. fluroscence*.

⅄ Homologus recombination between altered and wild Tn5 element occurs which leads to the integration of altered Tn5 with B.t toxin gene into the chromosome of *P. fluroscence*.

R. HORMONES

Introduction

Hormones are chemical substances produced by particular cells that function by interaction with a target cell. All hormones act by binding to macromolecular receptors that are located on the cell membrance. Only a little amount of hormone is required to alter cell metabolism. In essence, it is a chemical messenger that transports a signal from one cell to another.

Functions of Some Hormones

Lipotropin   Fatty acid release from adipocites.

Thyrozine   Stimulation of cellular reactions.

Insulin   Glucose metabolism.

S. INSULIN

Insulin is a hormone produced by the β cells of islets of langerhans of Pancreas. It is highly necessary for the utilization of sugar and maintenance of proper sugar level in human blood. Diabetes is a condition due to insufficient secretion of insulin. This condition is also called as hyperglycemia. Insulin facilitates the cellular uptake and utilization of glucose for the release of energy. In the absence of insulin, glucose accumulates in the blood stream and excess will be excreted into urine. The patients of diabetes are weak and tired due to depressed production of energy. Human insulin contains 51 amino acids arranged in the polypeptide chains. The chain A has 21 amino acids and B has 30 amino acids. Both A and B chain are held together by disulfide bond.

Diabetes mellitus affects about 2-3% of population. It is a genetically linked disease characterized by increased blood glucose concentration. This disease is due to insufficient availability of insulin. Insulin is used for the treatment of Diabetes mellitus

Production of Recombinant Insulin

In early years, insulin isolated and purified from the panereas of pigs. It was used for the treatment of diabetes. Attempts to produce insulin by rDNA technology started in late 1970s. The basic technique consisted of inserting human insulin gene and the promoter gene of lac operon on to the plasmids of *E. coli*. Patients of Guy's Hospital London uses first recombinant insulin to their patients in July 1980. Insulin is a first ever recombinant product used for the human applications. In 1986, Eli Lilly company received approval to market human insulin under the trade name Humulin.

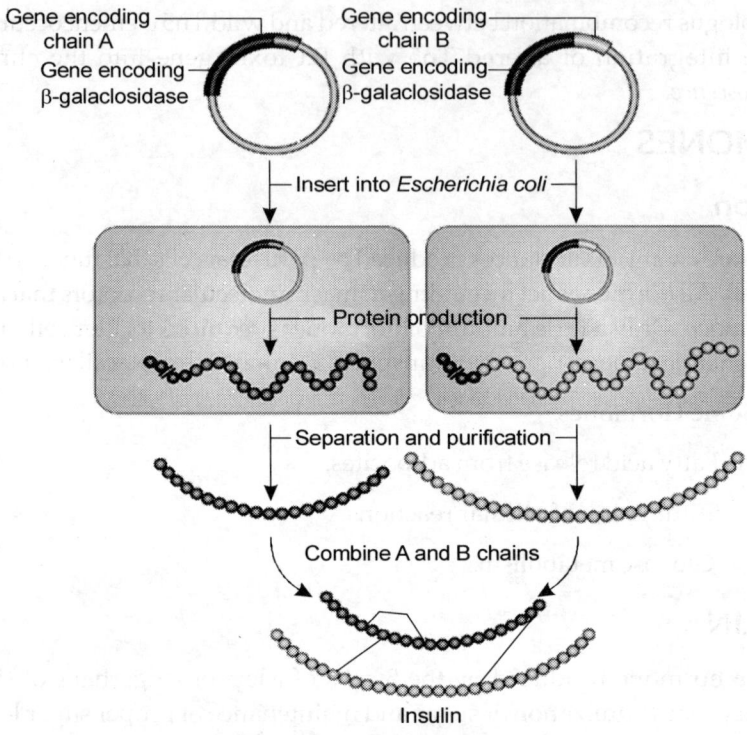

Figure 343   Cloning of insulin gene

Technique for recombinant Insulin production

Insulin molecule is made up of 2 polypeptide chains. A chain is smaller and B chain is larger. Genes for insulin A and B are isolated in the form of mRNA. This is converted to cDNA using reverse transcriptase. Both cDNAs are separately inserted to the plasmids of *E. coli* preferably Lac operon containing plasmid (pUC8 - consisting of inducer, promoter, operator and structured gene Z for β galactosidose). After insertion, the plasmid is transformed into the host *E. coli*. The presence of lactose in the culture medium induces the synthesis of insulin A and B in separate culture. Each clones of *E.coli* produced proinsulin polypeptide. These chains were treated with Cyanogen bromide which cleaves the polypeptide chain at methionine residue. Then A and B chains were chemically combined to produce active insulin.

Second generation recombinant Insulins

In general therapeutic insulin is available as a hexamer. This dissolves slowly to the biologically active dimer or monomer. That's why diabetes patients are instructed to take 30 minutes before meal. Site directed mutagenesis for protein engineering methods are adapted to synthesis second generation Insulin. This is also called as muteins. A large number of insulin muteins have been constructed with an objective of faster dislocation of hexamers to biologically active form.

T.   HUMAN GROWTH HORMONE (hGH)

Growth hormone (GH or HGH), also known as somatotropin or somatropin. It is a peptide hormone. It stimulates growth, cell reproduction and regeneration in humans and other animals. Human Growth hormone is produced by the anterior lobe of pituitary gland. It regulates growth and development. It stimulates overall body growth by increasing cellular uptake of aminoacids, and protein synthesis and promoting the use of fat as body fuel. Insufficient human growth hormone results in retarded growth. It is also called as pitutary dwarfism.

Production of recombinant hGH

Genetic engineering method is adopted to produce hGH. The procedure adopted are

Isolating hGH gene

Inserting hGH gene into *E. coli* plasmid.

Culturing the cells.

Isolation of hGH from the extracellular medium.

The hGH is a protein comprises of 191 amino acids. In natural process, hGH is tagged with signal polypeptide with 26 amino acids and the signal polypeptide removed during secretion to release active hGH for biological function. In rDNA tech, signal polypeptide interrupts hGH production. Biologists have resolved the problem of signal polypeptide interruption by a novel approach. The base sequence in cDNA encoding signal peptide plus the neighbouring 24 amino acid (a total of 50 AA) is cut by restriction Endonuclease Eco RI. A gene for 24 AA sequence is freshly synthesized and ligated to the remaining hGH cDNA. Constructed cDNA is attached to a vector, is transformed to a bacterium such as *E. coli* for culture and production of hGH. A liter of bacterial culture can provide 25-30mg of hGH within 7 hours. Recombinant hGH was approved for human use in 1985. It is marketed as protropin by Genetech Company and Humatrope by Eli Lilly company.

U.   SOMATOSTATIN

Somatostatin is also known as growth hormone-inhibiting hormone (GHIH) or somatotropin release-inhibiting factor (SRIF)) or somatotropin release-inhibiting hormone. It is a peptide hormone that regulates the endocrine system and affects neurotransmission and cell proliferation via interaction with G protein-coupled somatostatin receptors and inhibition of the release of numerous secondary hormones. Somatostatin secreted in hypothalamus of the brain and controls the release of pituitary hormones. Somatostatin is a peptide hormone made up of 14 amino acids.

The first step involves synthesis of the gene coding for the hormone from 8 fragments of oligonucleotides namely A-H. The gene sequence contains 52 bp, of these, 42 code for the hormone and the remainder code for sticky ends that allow the gene to be inserted into the plasmid. They also act as start and stop signals. pBR322 vector, lac control gene and β galactosidase gene sequence was constructed. β galactosidase is an enzyme involved in lactose metabolism. Control region contains regulatory elements for B-gal expression. Somatostatin gene was cloned into the plasmid next to B-gal sequence and transformed into *E.coli*. The hormone was synthesized at the end of the enzyme cleavage with cyanogen bromide released the hormone in active form.

The gene for somatostatin was initially synthesized by chemical methods. Besides the 42 bases coding for somatostatin, the polynucleotide contained a codon for methionine at the 5' end (the N-terminal end of the peptide) and two stop codons at the opposite end. To aid insertion into the plasmid vector, the 5' ends of the synthetic gene were extended to form single-stranded sticky ends complementary to those formed by the EcoRI and BamHI restriction enzymes. A modified pBR322 plasmid

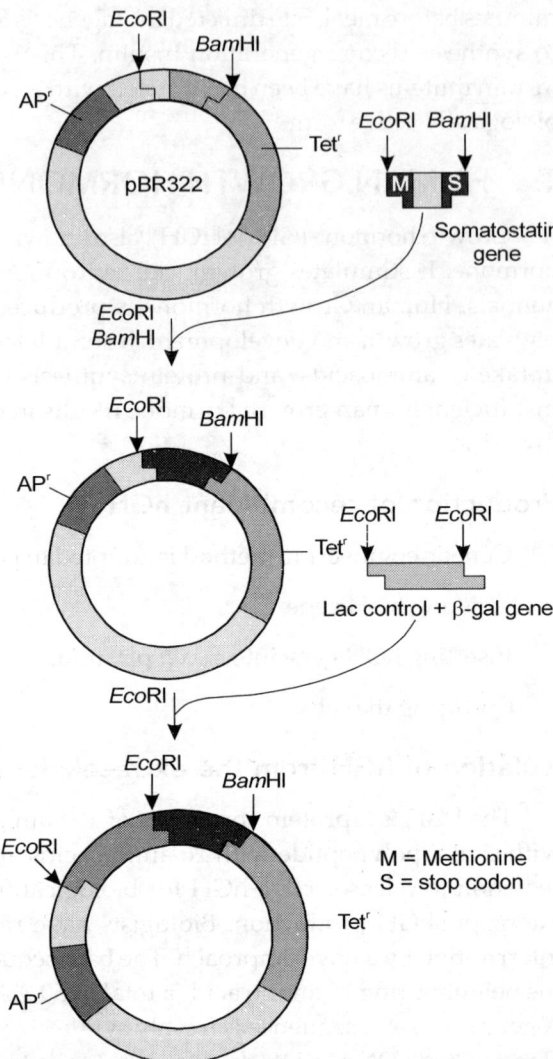

Figure 344   Cloning of somatostatin gene

was cut with both EcoRI and BamHI to remove a part of the plasmid DNA. The synthetic gene was then spliced into the vector by taking advantage of its cohesive ends (figure 344). Finally, a fragment containing the initial part of the lac operon (including the promoter, operator, ribosome binding site, and much of the β galactosidase gene) was inserted next to the somatostatin gene. The plasmid now contained the somatostatin gene fused in the proper orientation to the remaining portion of the β galactosidase gene.

V.   GENE THERAPY

Treatment of genetic disorders by introducing proper or corrected genes into patients cell is called gene therapy. Some diseases of human are due to defects in genetic makeup and are called genetic disorders. Such disorders are treated with gene therapy. Gene therapy is the use of DNA as a pharmaceutical agent to treat disease. It derives its name from the idea that DNA can be used to supplement or alter genes within an individual's cells as a therapy to treat disease. Gene therapy is one of the most important areas of application of genetic engineering. Gene therapy is a valuable tool to cure about 4000 inhereditary disorders.

Hypercholesterolemia, cystic fibrosis, phenylketonuria, Sickle cell anaemia, Thalassaemias, polycystic kidney disease, haemophilia are the examples of genetic disorders.

Genetic disorders are due to mutation in dominant gene in autosomes, recessive genes in autosomes and mutation in X linked recessive genes.

Diagnosis of genetic disorders

Genetic disorders are diagnoses at early stages. They are

1. Parental screening- some genetic disorders are due to defects in parental gene, which have been inherited to children from parents. Such diseases can be detected by testing parents.

2. Antenatal screening - it is a diagnosis of genetic disorders in fetes while it is in mothers womb. The fetal DNA is obtained from the cells of amniotic fluid at 8-16 weeks of pregnancy.

3. Postnatal screening - testing of newborn children for genetic disorder is known as postnasal screening. DNA hybridization with known probes is used for this purpose.

Approaches To Gene Therapy

Addition of Normal gene to replace the function of mutant gene.

Replacement of mutant gene sequence with normal gene sequence.

Establishment of alternative pathways to bypass mutant functions.

The main approaches of gene therapy are Gene augmentation and Gene replacement.

Gene Augmentation

In this process, defective gene remains in the genome. Normal gene (healthy) is randomly inserted in the genome so that - it produces the functional product. It is suitable only for recessive disorders. It is not useful for dominant disorders where defective gene expression interferes with normal gene.

Gene Replacement

It is also known as corrective gene therapy.In this process healthy gene is inserted at specific site in the genome to displace the defective gene.It is useful for treatment of dominant disorders.

Types of Gene Therapy

Somatic gene therapy

In somatic gene therapy, the therapeutic genes are transferred into the somatic cells, or body, of a patient. Any modifications and effects will be restricted to the individual patient only, and will not be inherited by the patient's offspring or later generations. Somatic gene therapy represents the mainstream line of current basic and clinical research, where the therapeutic DNA transgene (either integrated in the genome or as an external episome or plasmid) is used to treat a disease in an individual.

Germ line gene therapy

In germ line gene therapy, germ cells, i.e., sperm or eggs, are modified by the introduction of functional genes, which are integrated into their genomes. This would allow the therapy to be heritable and passed on to later generations. Although this should, in theory, be highly effective in counteracting genetic disorders and hereditary diseases, many jurisdictions prohibit this for application in human beings, at least for the present, for a variety of technical and ethical reasons.

Table 17.8   Difference between somatic and germ line therapy

Somatic Therapy	Germ line Therapy
Genes are introduced into somatic cells (body cells)	Genes are introduced into embryos, egg, sperm.
Genes will not get distributed to germs cell	Genes will get distributed in both germ and somatic
Changes are confined only to the recipient	Changes will be passed on to future generation.
No ethical issues are involved	Ethical issues are involved
Somatic cell manipulation are well developed	Technical difficulties are present in introduction of genes into germ cells.

Somatic gene therapy is of four types. They are embryo therapy, Ex vivo therapy, in vivo therapy and antisense therapy.

Embryo therapy - Treatment of genetic disease by introducing a proper remedial gene into the cells of 2-10 days old embryo is called embryo therapy or fetal gene therapy.

Ex-vivo gene therapy - In this method, some cells are taken from an appropriate organ of patient, remedial gene is introduced into the cell and the cells are transplanted back into that organ. It is also known as transplantation or tissue grafting. Eg. Liver transplantation, kidney transplantation. It is performed by, the following method.

Normal gene is isolated. It is then cloned into retrovirus vector.Target cell (Bonemarrow sample) is taken from patient with genetic defect. Bonemarrow cells are infected with Retrovirus. Transfected cells are reinfused into patient and expression of normal gene occurs.

In vivo gene therapy- the direct delivery of a remedial gene construct into proper organ of a patient to correct gene defect is called in vivo gene therapy. It is less time consuming and easy to perform. Muscular dystrophy, brain tumour are being treated in this way. Adenovirus and herpes virus are used as gene delivery system. Recombinant gene are generated by cloning.It is then introduced into cells by using vectors which are capable of targeting genes to specific sites in the body.

Antisense gene therapy - the treatment of genetic disorders by introducing a remedial gene that prevents the expression of specific defective gene.

Methods for Transport of Recombinant Gene into Target Cell

Micro injection - This method is used to transform embryonic cells.10 to 30 ng of DNA can be transformed.Frequency of stable integration is high.

Electroporation - This method is used to transform pancreatic cells, epidermal cells. Insertion of FDNA occurs under the influence of electric field.Large number of cells can be transformed at a very short time.

Calcium phosphate mediated transfection - DNA to be inserted is mixed with calcium phosphate which is clumped on to recepient cells.Fraction of cells take up the DNA by endocytosis.The frequency of transformation is very low.Only one cell in 10^6 to 10^7 cell get transformed.

Use of Retroviruses - Retroviruses donot kill the cells they infect.They can infect broad spectrum of cell types.Effecient integration and expression occurs in the host.

Liposomes - Liposomes are also used to deliver genes into target cells.Liposomes are tiny spherical molecules with an exterior lipid bilayer that can carry genes across cell membrane.Large DNA sequence can be inserted into the target cells with the help of liposomes.

Cell Target for Somatic Gene Therapy

Normally bonemarrow, hepatic cells, fibroplast, epithelial cells are used for somatic gene therapy.

Table 17.9 Human Genetic Diseases Targeted For Gene Therapy

Disease	Nature of Defect	Gene Replacement
Adenosin deaminase defeciency (ADA)	Destruction of lymphocytes by toxic metabolic product which results in severe combined immuno defeciency	ADA gene inserted into bonemarrow cells using virus vector or liposomes
Cystic fibrosis	Lack of Necessary protein for transport of CL ions across membrane	Cystic fibrosis transfrase gene is placed into adeno rirus or liposomes and
Sickle cell Anaemia	Lack of correct gene to make the globin protein of haemoglobin.	HBA gene is placed in bonemarrow cells by vector.
Muscular distrophy	Lack of gene for producing distrophin Necessary for normal muscle development.	Distrophin gene delivered into muscle tissue
Malignant melanoma	It is the severiest form of Skin cancer. It doesnot respond well to chemotoraphy.	Insertion of B7 genes into tumor cells which increases Natural WBC response

W. MONOCLONAL ANTIBODIES (HYBRIDOMA TECHNOLOGY) -mAbs

Antibodies or immunoglobulins are protein molecules produced by B cells (Plasma cells). In response to an antigen, B cells gear up and produce many types of antibodies. This type of antibodies are called polyclonal antibodies. Monoclonal antibody (MAb) is a single type of antibody that is directed against a specific epitope/antigenic determinant. In 1975 George Kohler and Cesar Milstein achieved large scale production of monoclonal antibodies. They got Nobel Prize in 1984. They could successfully hybridize antibody producing B cells with myeloma cells *in vitro* and create a hybridoma. This results in immortalized B cells, which can produce MAbs. The production of monoclonal antibodies by the hybrid cells is referred to as hybridoma technology.

Principle for creation of Hybridoma Cells

The myeloma cells are not able to produce abs, but the cells are immortal. The selection of hybridoma cells is based on inhibiting the nueleotide synthesizing machinery. The mammalian cells can synthesize nucleotides by two pathways - denovo synthesis and salvage pathway. The denovo synthesis of nucleotides requires tetrahydrofolate which is formed from dihrdrofolate. The formation tetrahydrofolate can be blocked by the inhibitor aminopterin. The salvage pathway involves the direct conversion of purines and pyrimidines into corresponding nucleotides. Hypoxanthine Guanine phosphoribosyl transferase (HGPRT) is the key enzyme in the salvage pathway of purines. It converts hypoxanthine and Guanine respectively to inosine monophosphate and Guanosine monophosphate. Thimidine kinase (TK) enzyme converts thymidine to thymidine monophosphate. Any mutation in any one of the enzymes blocks salvage pathway. When cells deficient in HGPRT are grown in HAT medium (Hypoxanthine aminopterine and Thymidine). They canot survive due to inhibition of denovo synthesis of purine. Thus cells lacking HGPRT, grown in HAT medium die. The hybridoma cells possess the ability of myeloma cells to grow invitro with a functional HGPRT. Thus only the hybridized cells can proliferate in HAT medium.

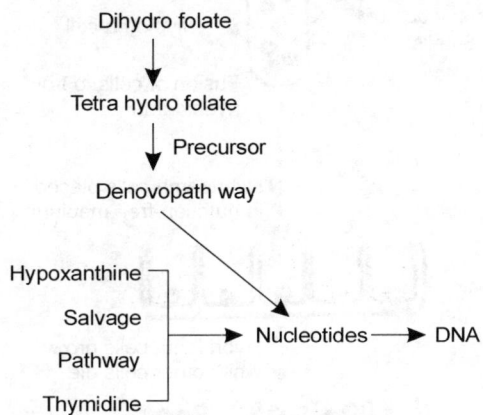

Figure 345   Principle of Myeloma cell growth

Production of monoclonal antibodies

The establishment of hybridomas and production of mAbs involves the following steps.

Immunization,

Cell fusion,

Selection of hybridomas,

Screening of the products,

Cloning and propagation and

Characterization and storage.

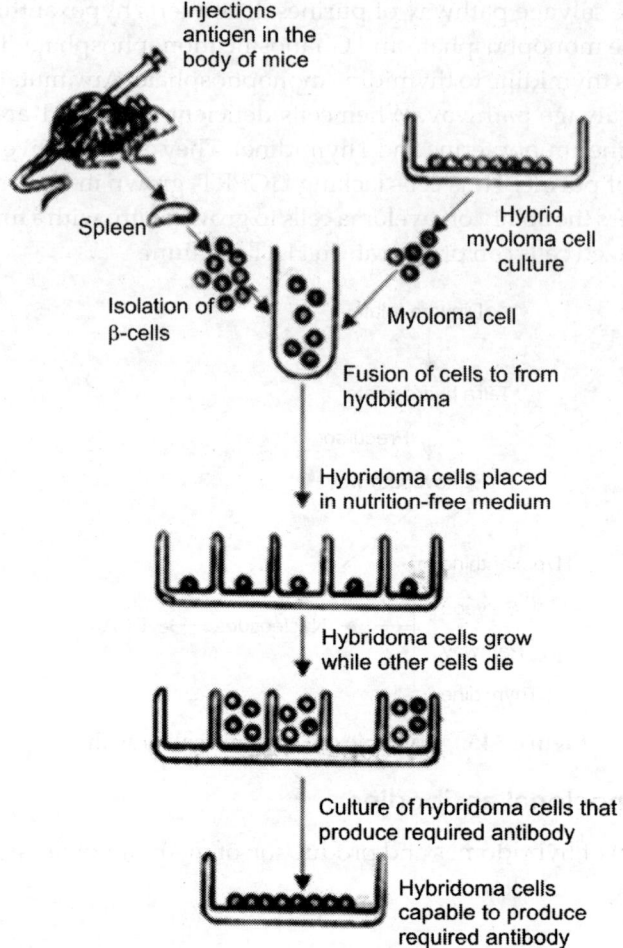

Injections antigen in the body of mice

Spleen

Hybrid myoloma cell culture

Isolation of β-cells

Myoloma cell

Fusion of cells to from hydbidoma

Hybridoma cells placed in nutrition-free medium

Hybridoma cells grow while other cells die

Culture of hybridoma cells that produce required antibody

Hybridoma cells capable to produce required antibody

Figure 346   Hybridoma technology

Immunization

It is the first step in monoclonal antibody technique. Animal is immunized with appropriate ag. The Ag, along with adjuvant like Freunds complete or incomplete adjuvant is injected subcutaneously. Injection of Antigen are repeated several time. This increases stimulation of B cell. Three days prior to killing of animal booster dose is given intravenously. The concentration of desired antibodies are assayed frequently. When the concentration of abs is optimal, the animal is sacrificed. The spleen is aseptically removed and release cells by mechanical or enzymatic methods.

Cell fusion

Throughly washed B cells are mixed with HGPRT - myeloma cells. The mixture of cells is exposed to polyethylene Glycol (PEG) for a short period. Now the mixture contain hybridoma cells, free myelom cells and free lymphocytes.

Selection of Hybridomas

The cells are allowed to grow on HAT medium, only the hybridoma cells grow, while the rest will slowly disappear. This happens in 7 - 10 days of culture. Individual hybridoma is selected.

Screening of Antibodies

The culture medium from each hybridoma culture is periodically tested for the desired antibody specificity. The antibody secreted by the hybrid cells are refered as monoclonal antibody.

Cloning and propagation

The single hybrid cells producing desired antibody are isolated and cloned.

Characterization and storage

The mAbs has to be subjected to biochemical and biophysical characterization for the desired specificity.

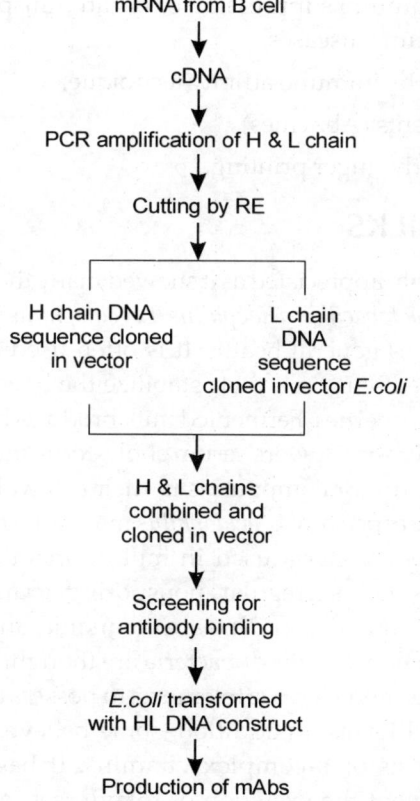

mRNA from B cell
↓
cDNA
↓
PCR amplification of H & L chain
↓
Cutting by RE
↓

H chain DNA sequence cloned in vector   L chain DNA sequence cloned invector *E.coli*

↓
H & L chains combined and cloned in vector
↓
Screening for antibody binding
↓
E.coli transformed with HL DNA construct
↓
Production of mAbs

Figure 347   Monoclonal antibody production

Production of mAbs in *E. coli*

The hybridoma technology is very laborious, expensive and time consuming.

The mRNA isolated from B cells of either human or mouse is converted to cDNA. The H & L sequences of this cDNA are amplified by PCR. The so produced cDNA are then cut by restriction endonuclease. H & L chain are separately cloned in bacteriophage vectors. These sequences are put together and cloned in another phage vector. The combined H & L chains (forming Fr fragment) are screened for antigen binding activity. The specific H & L chains forming plasmid are transformed in *E. coli* and subjected to monoclonal antibody production.

Application of mAbs

1. Diagnostic applications

    Biochemical analysis for the diagnosis of pregnancy, cancer, humoral disorder, infectious diseases. Diagnostic imaging for the detection of mycocardial infarction, deep vein thrombosis, cancer, bacterial infection.

2. Therapeutic applications

    Direct use as therapeutic agents to destroy disease causing organisms, in the treatment of cancers, in the immuno suppression of organ transplantion, in the treatment of AIDS and auto immune diseases.

3. Protein purification by immuno affinity technique.

4. It act as catalytic agents (Abzymes)

5. Used in auto antibody finger printing.

X.   FERMENTED MILKS

Fermented milks are highly appreciated as it showed many therapeutic effects. Acidophilus milk is produced by using *Lactobacillus acidophilus*. *L. acidophilus* is act as a normal flora in the lower intestine and improves general health. It is often used as a dietary adjunct. Many microorganisms in fermented dairy products stabilize the bowel microflora. Some normal flora showed antimicrobial properties. Fermented milk products have multiple health benefits. They minimize lactose intolerance, lowers serum cholesterol and exhibit anticancer activity. Several *Lactobacilli* have antitumor compounds in their cell walls. This indicated that diets including lactic acid bacteria especially *L. acidophilus* may reduce the burden of colon cancer. Another interesting group of bacteria used in milk fermentations are the *Bifidobacteria*. The genus *Bifidobacterium* contains irregular, nonsporing, gram-positive rods that may be club-shaped or forked at the end. *Bifidobacteria* are nonmotile, anaerobic, ferment lactose and other sugars to acetic and lactic acids. Bifidobacteria are thought to help maintain the normal intestinal balance, while improving lactose tolerance; to possess antitumorigenic activity and to reduce serum cholesterol levels. In addition, some believe that they promote calcium absorption and the synthesis of B-complex vitamins. It has also been suggested that *Bifidobacteria* reduce or prevent the excretion of rotaviruses, a cause of diarrhoea among children. Bifidamended fermented milk products are now available in various parts of the world. Intestine of breast fed children showed higher quantities of Bifidobacterium.

Kefir

Kefir is a one of the fermented milk. It is made with kefir "grains". Kefir grains are a combination of lactic acid bacteria and yeasts in a matrix of proteins, lipids and sugars. This components symbiotically forms a grains" that resemble cauliflower. It is a product with an ethanol concentration of up to 2%. This unique fermented milk originated in the Caucasus Mountains and it is produced east into Mongolia. Kefir products tend to be foamy and frothy, due to active carbon dioxide production. This process is based on the use of kefir "grains" as an inoculum. These are coagulated lumps of casein that contain yeasts, lactic acid bacteria, and acetic acid bacteria. In this fermentation, the grains are used to inoculate the fresh milk and then recovered at the end of the fermentation. Kefir grains contain a water-soluble polysaccharide known as kefiran. It imparts a rope-like texture and feeling in the mouth. The grains range in colour from white to yellow. It may grow to the size of walnuts. The composition of kefir depends greatly on the type of milk that was fermented, including the concentration of vitamin B_{12}.

During the fermentation, Lactose is broken down mostly to lactic acid (25%) by the lactic acid bacteria. It results in acidification of the product. Propionibacteria further break down some of the lactic acid into propionic acid. Other substances that contribute to the flavour of kefir are pyruvic acid, acetic acid, diacetyl and acetoin. The slow-acting yeasts, late in the fermentation process, break lactose down into ethanol and carbon dioxide. At the end of the fermentation, very little lactose remains in kefir. People with lactose intolerance are able to tolerate kefir, providing the number of live bacteria present in this beverage consumed is high enough. It has also been shown that fermented milk products have a slower transit time than milk, which may further improve lactose digestion. Several varieties of probiotic bacteria are found in kefir products such as *Lactobacillus acidophilus, Bifidobacterium bifidum, Streptococcus thermophilus, Lactobacillus delbrueckii subsp. bulgaricus, Lactobacillus helveticus, Lactobacillus kefiranofaciens, Lactococcus lactis* and *Leuconostoc* species. *Lactobacilli* in kefir may exist in concentrations varying from approximately 1 million-1 billion colony-forming units per milliliter and are the bacteria responsible for the synthesis of the polysaccharide kefiran. In addition to bacteria, kefir often contains strains of yeast that can metabolize lactose, such as *Kluyveromyces marxianus* and *Kluyveromyces lactis,* as well as strains of yeast that do not metabolize lactose, including *Saccharomyces cerevisiae, Torulaspora delbrueckii* and *Kazachstania unispora.*

Viili

It is also called mold lactic. It is a yogurt-like mesophilic fermented milk. It is originated in the Nordic countries. This cultured milk product is the result of microbial action of lactic acid bacteria (LAB). The bacteria strains used in its production produce exopolysaccharides which gives viili a ropey, gelatinous consistency and a pleasantly mild taste resulting from lactic acid. Viili also has a surface-growing yeast-like fungus *Geotrichum candidum* present in milk, which forms a velvet-like surface. In addition, most traditional viili cultures also contain yeast strains such as *Kluveromyces marxianus* and *Pichia fermentans.* The LAB identified in viili including *Lactococcus lactis* subsp. cremoris, *Lactococcus lactis sub sp.* lactis biovar. diacetylactis, *Leuconostoc mesenteroides sub sp. cremoris.* Among those mesophilic LAB strains, the slime-forming *Lactobacillus latic sub sp*. cremoris produce aphosphate-containing heteropolysaccharide, named viilian. It is prepared by inoculating viili strains inoculated with a mixture of the fungus *Geotrichium candidum* and lactic acid bacteria. The cream rises to the surface, and after incubation at 18 to 20°C for 24 hours, lactic acid reaches a concentration of 0.9%. The fungus forms a velvety layer across the top of the final product, which also can be made with a bottom fruit layer to add extra flavor.

Cheese Production

Cheese is one of the oldest human foods. It is one of the important product of dairy industry. It is thought to have been developed approximately 8,000 years ago. About 2,000 distinct varieties of cheese are produced throughout the world. Cheeses are classified based on texture or hardness as soft cheeses (cottage, cream, Brie), semisoft cheeses (Muenster, Limburger, blue), hard cheeses (cheddar, Colby, Swiss) or very hard cheeses (Parmesan). All cheese results from a lactic acid fermentation of milk, which results in coagulation of milk proteins and formation of a curd. Rennin, an enzyme from calf stomachs, but now produced by genetically engineered microorganisms, can also be used to promote curd formation. After the curd is formed, it is heated and pressed remove the watery part of the milk or whey, salted and then usually ripened. The cheese curd can be packaged for ripening with or without additional microorganisms. Cheese curd inoculation is used in the manufacture of Roquefort and blue cheese. In this case *Penicillium roqueforti* spores are added to the curds just before the final cheese processing. Sometimes the surface of an already formed cheese is inoculated at the start of ripening; for example, Camembert cheese is inoculated with spores of *Penicillium camemberti*. The final hardness of the cheese is partially a function of the length of ripening. Soft cheeses are ripened for only about 1 to 5 months, whereas hard cheeses need 3 to 12 months, and very hard cheeses like Parmesan require 12 to 16 months ripening. The ripening process also is critical for Swiss cheese. Gas production by *Propionibacterium* contributes to final flavor development and hole or eye formation in this cheese. Some cheeses are soaked in brine to stimulate the development of specific fungi and bacteria; Limburger is one such cheese.

Cheese is made from milk : This is carried out by a process of dehydration where in casein and fat are concentrated 5-15 fold. It is a very complicated process and involved 4 statges. They are 1. Acidification of milk, 2. Coagulum formation, 3. Separation curd from whey, 4. Ripening of cheese.

Acidification of milk: it is performed by employing *Streptococcus lactis* and *Lactobacillus lactis*. Sugar of milk is converted to lactic acid. This lowers the pH around 4.6 and thus acidifies milk.

Coagulum formation: when the acidified milk is treated with rennet, casein gets coagulated. Casein consists of three parts namely insoluble α and β casein and κ casein that keep them in insoluble state. By the action of chymosin, κ casein is degraded. Consequently α and β casein and degraded part of κ casein are combined to form coagulum.

Separation curd from whey: when the temperature of coagulum is raised to around 40°C, the curd and whey get separated. The separated curd is cut into blocks , drained and pressed into different shapes.

Ripening of cheese : The cheese curd can be packaged for ripening with or without additional microorganisms. This imparts flavor.

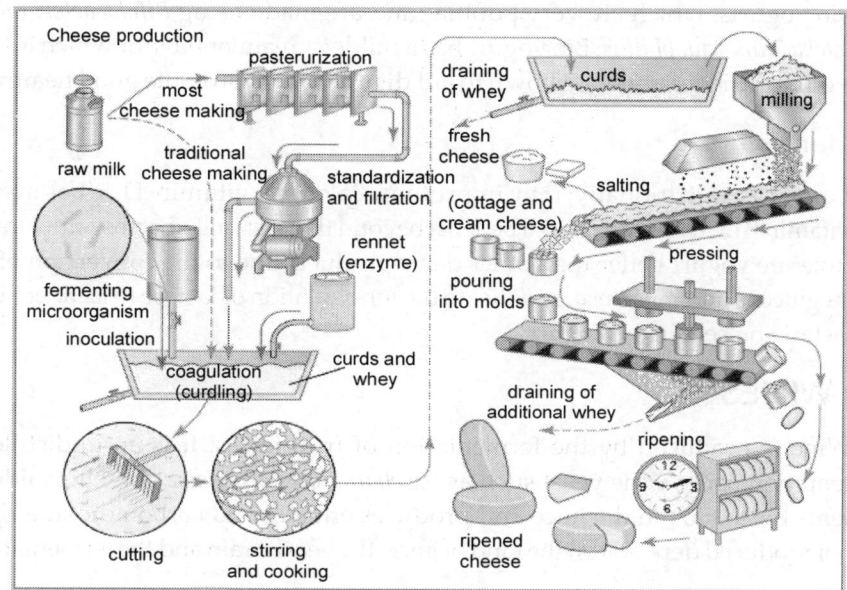

Figure 348　Cheese production

Yogurt

Yogurt is produced through the fermentation of milk by lactic acid bacteria, usually *Lactobacillus bulgarius* and *Streptococcus thermophilus*. The milk is firstly heat treated, homogenised and is then cooled to allow the addition of bacteria or starter culture. Given the right conditions, i.e. correct temperature and moisture, the bacteria are able to ferment the milk sugar (lactose), producing lactic acid. The milk proteins then coagulate and set, to form yogurt. A colourless liquid called acetaldehyde is also produced during fermentation and gives yogurt its distinct flavour. Yogurt can be made from different types of milk, including skimmed, semi-skimmed, whole, evaporated or powdered forms.

Varieties of yogurt

The market now offers a vast array of yogurts to suit all palates and meal occasions. They come in a variety of textures (e.g. liquid, set, smooth), fat contents (e.g. luxury, low-fat, virtually fat-free) and flavours (e.g. natural, fruit, cereal), can be consumed as a snack or part of a meal, as a sweet or savoury food and are available all year-round.

Fermenting milks with different micro-organisms has also provided an opportunity to develop a wide range of products with different flavours, textures, consistencies and, more recently, health attributes. These include:

Live yogurts, which contain harmless bacteria that are added to the milk and are still present and alive.

Probiotic yogurts, which contain live probiotic micro-organisms that are suggested to be beneficial to health.

Bio yogurts, which are very popular and are made using *Bifidobacterium bifidum* and/or *Lactobacillus acidophilius*. Bio yogurt has a milder, creamier flavour which is less acidic than some other varieties and has shown to aid digestion and promote good health.

Nutrients

Yogurt is nutritionally rich in protein, calcium, vitamin D, riboflavin, vitamin B6 and vitamin B12. It has nutritional benefits beyond those of milk. Lactose-intolerant individuals may tolerate yogurt better than other dairy products due to the conversion of lactose to the sugars glucose and galactose, and due to the fermentation of lactose to lactic acid carried out by the bacteria present in the yogurt.

Y. WINES

Wine is produced by the fermentation of fruit juices. It is an undistilled product of fermentation using wine yeast such as *Saccharomyces cerevisiae* var. ellipsoideus. The yeast ferments the sugars in the juice and produces ethanol and carbon dioxide. The amount of alcohol produced depends on the kind of juice, the yeast strain and the fermentation conditions.

Wine production starts with the collection of grapes, continues with their crushing and the separation of the liquid (must) before fermentation, and concludes with a variety of storage and aging steps. All grapes have white juices. To make a red wine from a red grape, the grape skins are allowed to remain in contact with the must before fermentation to release their skin-color components. Wines can be created by using the natural grape skin microorganisms. This natural mixture of bacteria and yeasts gives unpredictable fermentation results. To avoid such problems, one can treat the fresh must with a sulfur dioxide fumigant and add a desired strain of *Saccharomyces cerevisiae* or *S. ellipsoideus*. After inoculation the juice is fermented for 3 to 5 days at temperatures varying between 20 and 28°C. Depending on the alcohol tolerance of the yeast strain, the final product may contain 10 to 18% alcohol. Clearing and development of flavour occur during the aging process.

A critical part of wine making involves the choice of whether to produce a dry (no remaining free sugar) or a sweeter (varying amounts of free sugar) wine. This can be controlled by regulating the initial must sugar concentration. With higher levels of sugar, alcohol will

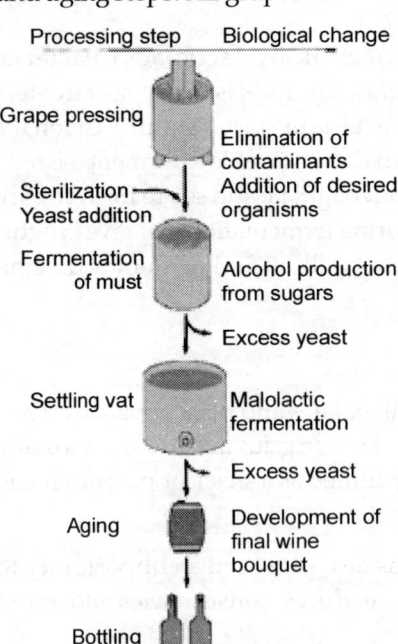

Figure 349   Wine preparation

accumulate and inhibit the fermentation before the sugar can be completely used, thus producing a sweeter wine. During final fermentation in the aging process, flavoring compounds accumulate and influence the bouquet of the wine. Microbial growth during the fermentation process produces sediments, which are removed during racking. Racking can be carried out at the time the fermented wine is transferred to bottles or casks for aging or even after the wine is placed in bottles.

Many processing variations can be used during wine production. The wine can be distilled to make a "burned wine" or brandy. *Acetobacter* and *Gluconobacter* can be allowed to oxidize the ethanol to acetic acid and form a wine vinegar. In the past an acetic acid generator was used to recirculate the wine over a bed of wood chips, where the desired microorganisms developed as a surface growth. Today the process is carried out in large aerobic submerged cultures under much more controlled conditions.

Natural champagnes are produced by continuing the fermentation in bottles to produce a naturally sparkling wine. Sediments that remain are collected in the necks of inverted champagne bottles after the bottles have been carefully turned. The necks of the bottles are then frozen and the corks removed to disgorge the accumulated sediments. The bottles are refilled with clear champagne from another disgorged bottle, and the product is ready for final packaging and labeling.

Grape fruits are crushed to extract the juice. About 0.250 mg of potassium metabisulphate is added per liter of juice. The starter culture yeast is mixed with the juice at the ratio of 1:10. Fermentation is normally carried out at low temperatures (5-6°C) for 7-11 days or longer and after the fermentation is over, the wine is allowed to settle and later clarified and stored for maturation at low temperatures. During maturation, the wine undergoes various chemical changes and these changes are responsible for the production of aroma and bouquet. If appropriate preservatives are not used or the conditions of storage are not adequate, acetic acid bacteria can enter and convert the wine into vinegar (acetic fermentation) and water.

The different types of wines and their percentage of alcohol are

Red wine 11 - 12

White wine 11 - 12

Dessert wine 19 - 21

Appetizer wine 12 - 16

Sparkling wine 11 – 12

Z. BEERS

Beer production uses cereal grains such as barley, wheat, and rice. The complex starches and proteins in these grains must be changed to a more readily usable mixture of simpler carbohydrates and amino acids (Figure 350). It involves germination of the barley grains and activation of their enzymes to produce a malt. The malt is then mixed with water and the desired grains, and the mixture is transferred to the mash tun or cask in order to hydrolyze the starch to usable carbohydrates. Once this process is completed, the mash is heated with hops (dried flowers of the female vine *Humulus lupulis*), which were originally added to the mash to inhibit spoilage microorganisms. The hops also provide flavour and assist in clarification of the wort. In this heating step the hydrolytic enzymes are inactivated and the wort can be pitched — inoculated — with the desired yeast. Most beers are fermented with bottom yeasts, related to *Saccharomyces carlsbergensis*, which settle at the bottom of the fermentation vat. The beer flavor also is influenced by the production of small amounts of glycerol and acetic acid. Bottom yeasts produce beer with a pH of 4.1 to 4.2 and requiring 7 to 12 days of fermentation. With a top yeast, such as *Saccharomyces cerevisiae*, the pH is lowered to 3.8. Freshly fermented (green) beers are aged or lagered, and when they are bottled, CO_2 is usually added. Beer can be pasteurized at 140°F or higher or sterilized by passage through membrane filters to minimize flavour changes. In many places there is increased interest in specialty beers.

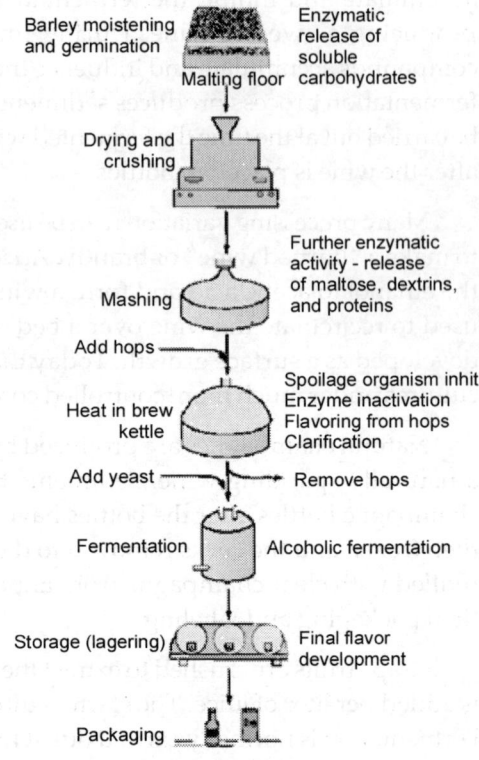

Figure 350   Beer preparation

AA. BREAD

Bread is one of the most ancient of human foods, and is produced with the help of microorganisms. The use of yeasts to leaven bread is carefully depicted in paintings from ancient Egypt. A bakery at the Giza Pyramid area, from the year 2575 B.C., has been excavated. It is estimated that 30,000 people a day were provided with bread from this bakery. Samples of bread from 2100 B.C. are on display in the British Museum. In breadmaking, yeast growth is carried out under aerobic conditions. This results in increased CO_2 production and minimum alcohol accumulation. The fermentation of bread involves several steps: alpha- and beta-amylases present in the moistened dough release maltose and sucrose from starch. Then a baker's strain of the yeast *Saccharomyces cerevisiae*, which contains maltase, invertase, and zymase enzymes, is added. The CO_2 produced by the yeast results in the light texture of many breads, and traces of fermentation products contribute to the final flavour. Usually bakers add sufficient yeast to allow the bread to rise within 2 hours — the longer the rising time, the more additional growth by contaminating bacteria and fungi can occur, making the product less desirable. By using more complex assemblages of microorganisms, bakers can produce special breads such as sour doughs. The yeast *Saccharomyces exiguus*, together with a *Lactobacillus* species, produces the characteristic acidic flavor and aroma of such breads. Bread products can be spoiled by *Bacillus* species that produce ropiness. If the dough is baked after these organisms have grown, stringy and ropy bread will result, leading to decreased consumer acceptance.

AB. SAUERKRAUT

It is also called sour cabbage. It is produced from wilted, shredded cabbage. A concentration of 2.2 to 2.8% sodium chloride restricts the growth of gram-negative bacteria while favoring the development of the lactic acid bacteria. The primary microorganisms contributing to this product are *Leuconostoc mesenteroides* and *Lactobacillus plantarum.* A predictable microbial succession occurs in sauerkraut's development. The activities of the lactic acid-producing cocci usually cease when the acid content reaches 0.7 to 1.0%. At this point *Lactobacillus plantarum* and *Lactobacillus brevis* continue to function. The final acidity is generally 1.6 to 1.8, with lactic acid comprising 1.0 to 1.3% of the total acid in a satisfactory product. Pickles are produced by placing cucumbers and such components as dill seeds in casks filled with a brine. The sodium chloride concentration begins at 5% and rises to about 16% in 6 to 9 weeks. The salt not only inhibits the growth of undesirable bacteria but also extracts water and water-soluble constituents from the cucumbers. These soluble carbohydrates are converted to lactic acid. The fermentation, which can require 10 to 12 days, involves the development of *Leuconostoc mesenteroides, Enterococcus faecalis, Pediococcus cerevisiae, Lactobacillus brevis,* and *L. plantarum. L. plantarum* plays the dominant role in this fermentation process. Sometimes, to achieve more uniform pickle quality, natural microorganisms are first destroyed and the cucumbers are fermented using pure cultures of *P. cerevisiae* and *L. plantarum.* Grass, chopped corn, and other fresh animal feeds, if stored under moist anaerobic conditions, will undergo a lactic-type mixed fermentation that produces pleasant-smelling silages. Trenches or more traditional vertical steel or concrete silos are used to store the silage. The accumulation of organic acids in

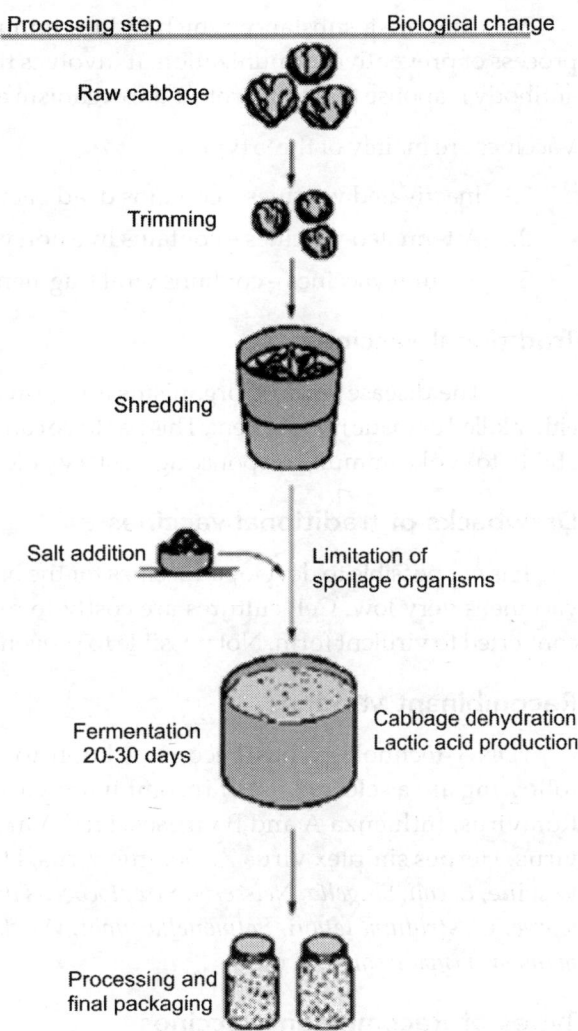

Figure 351 Sauerkraut preparation

silage can cause rapid deterioration of these silos. Older wooden stave silos, if not properly maintained, allow the outer portions of the silage to become aerobic, resulting in spoilage of a large portion of the plant material.

AC. RECOMBINANT VACCINES

Vaccines are a substance which induce immunity of an individual. Vaccination is the process of preventive immunization. It involves the administration of an antigen to elicit an antibody response that will protect the organism against future infection.

Vaccines are mainly of three types.

1. Inactivated vaccines – contains dead bacteria / viruses
2. Attenuated vaccines – contains live non virulent bacteria / viruses
3. Subunit vaccines – contains viral fragments / bacterial molecule.

Traditional vaccines

The disease causing organisms are grown in a culture. They are then purified and either killed or made non virulent. This has to be carefully done without the loss of the organisms ability to evoke immune response against a virulent form of disease causing organism.

Drawbacks of traditional vaccines

It is not possible to develope vaccines for the organisms not grown in culture. The yield of vaccine is very low. Cell cultures are costly to maintain. Some time non virulent cells are converted to virulent form. Not possible to prevent all life threatening infection.

Recombinant vaccines

rDNA technology has become a boon to produce new generation vaccines. The following are a selected list of recombinant vaccines are developed or being developed. Rotavirus, Influenza A and B viruses, HIV, Varicella zoster virus, Japanese encephalitis virus, Herpes simplex virus Z, Dengue virus, HAV, HBV, Yellow fever viruses, Cholera vaccine, *E. coli, Shigella, Neisseria, Streptococcus pneumoniae, Mycobacterium tuberculosis, M. leprae, Clostridium tetani, Salmonella typhi, Wuchereria bancrofti, Plasmodium, Schistosoma manson, Trypanosome sp.* etc.

Types of Recombinant vaccines

1. Subunit recombinant vaccines

These are components of pathogenic microbes. It includes proteins, Peptides and DNA.

2. Attenuated recombinant vaccines

These are genetically modified pathogenic organisms. Virulent forms and made as non virulent lent forms.

3. Vector recombinant vaccines

These are genetically modified viral vectors that can be used as vaccines agains certain pathogens.

Hepatitis B subunit vaccines

HBV causes wide spred Hepatitis infection in Human. HBV causing hepatitis is also called as serum hepatitis. It affect liver and causing chronic hepatitis, cirrhosis and liver cancer. HBV is a 42nm particle called Dane particle. It consist of core containing a viral genome, surrounded by HBc Ag, HBe Ag and HBSAg. HBs Ag is more antigenic and immunogenic. The gene encoding for HBs Ag has been identified. Recombinent HB Vaccine as a subunit vaccine, is produced by cloning HBs Ag gene in yeast cell. The gene for HBs Ag is inserted into PMA56 Vector, which contain alcohol dehydrogenase promoter. This plasmid is transformed into *Saccharomyces cerevisiae* host and cultured.The cells grown on Tryptophan, free medium are selected and cloned.The HBsAg gene expressed and produce 2nm sized particle similar to HBs Ag.This is used for immunization.It is a first synthetic recombinant vaccine (1987) available as the trade names Recombivac and Eugerix B. Individuals must be administered 3 close over a period of 6 months.India is the 4th country in the world to develope an indigenous HB vaccine.

Production of Recombinant HB Vaccine

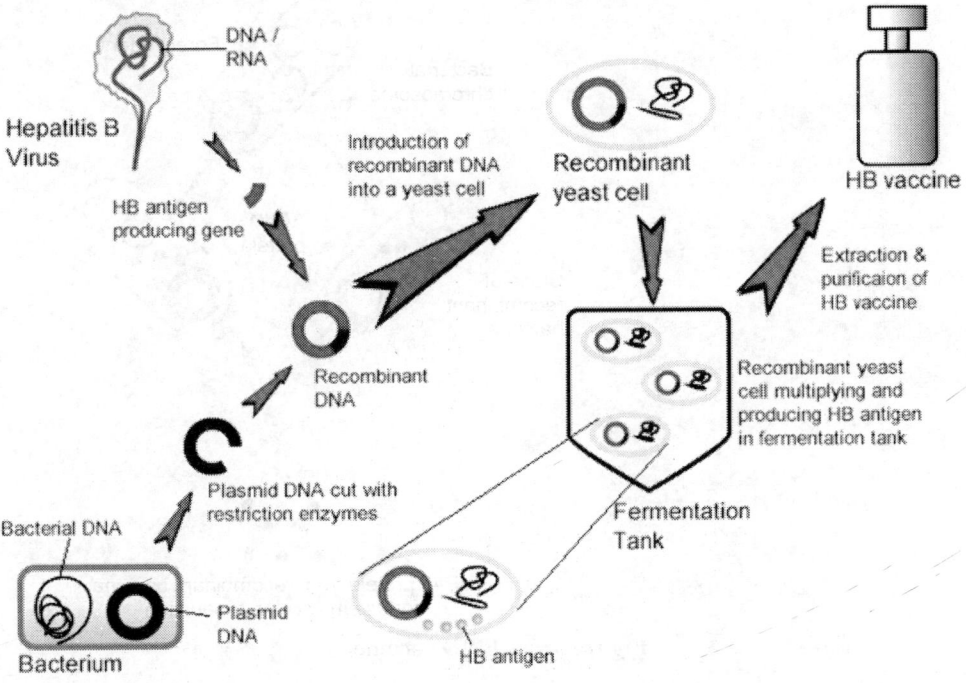

Figure 352 Recombinant Hb vaccine

FMD Vaccine

It is a highly contagious disease affecting cattle and pigs. A formalin killed FMD virus was used previously to vaccinate against this disease. The genome of FMD is SSRNA, covered by 4 viral proteins VP1, VP2, VP3 and VP4. Among these VP1 is more immunogenic. The nucleotide sequence encoding VP1 was identified. A cDNA is developed from the SSRNA. cDNA is cloned in βBR 322 using suitable Restriction enzymes and ligated. cDNA containing vector is transformed into *E. coli*. Cloned *E. coli* produce VP1 protein and used for vaccination.

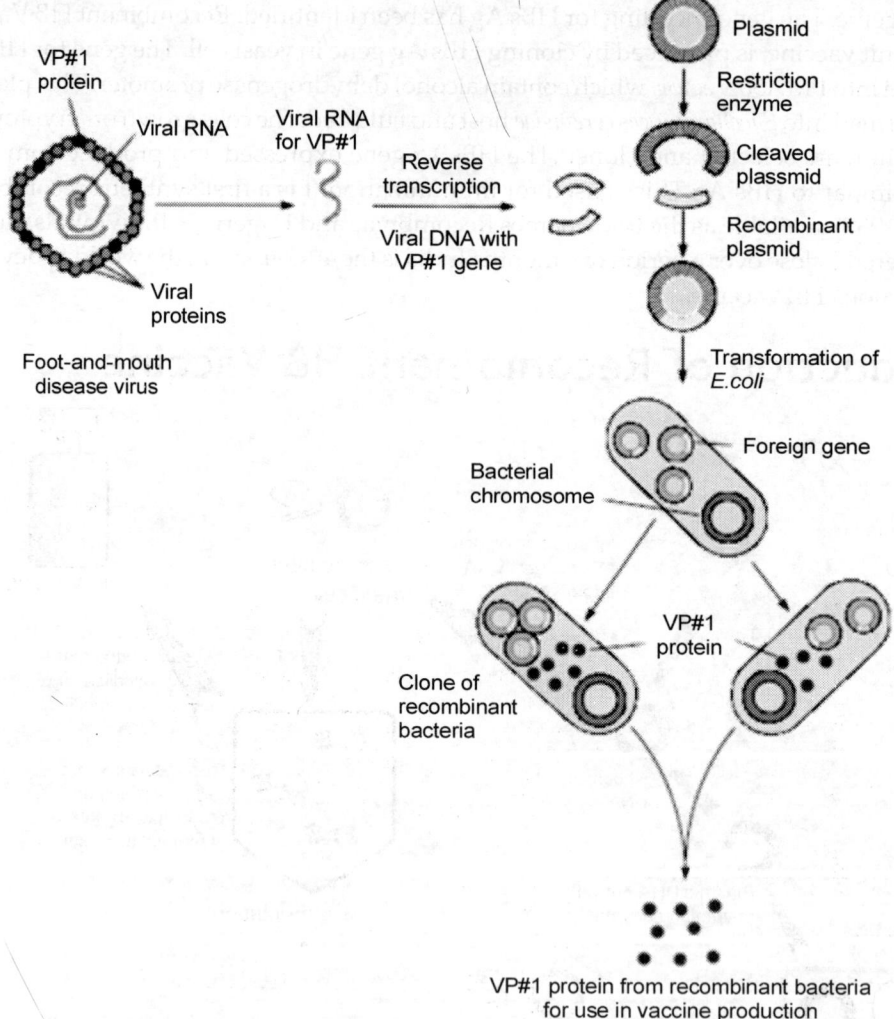

Figure 353   FMD vaccine

Herpes simplex virus

An envelope glycoprotein D (gD) of HSV elicit Antibody production. The gD gene was cloned in a mammalian vector and vaccine is developed.

DNA Vaccines

It is a novel approach started in 1990. The immune response is stimulated by a DNA molecule. It may be administered as nasal spray, intra muscular injection, intravenous injection, intradermal injector or Genegum / bolistic delivery. The plasmid vaccine carrying the Antigenic protein enters the nucleus of the target cell of the host. This DNA produces RNA and inturne the specific antigenic protein. The Antigen can act directly for developing humoral immunity or as fragments in association with MHC molecules for developing cellular imunity.

cDNA also used as DNA vaccine.

Edible Vaccine

Plants serve as a cheap and safe production system for submit vaccines. The edible vaccines can be easily ingested by eating plants. This eliminates the processing of purification. Transgenic tomato, potato have been developed for expressing antigens of animal viruses. First clinical trial was conducted in 1997.

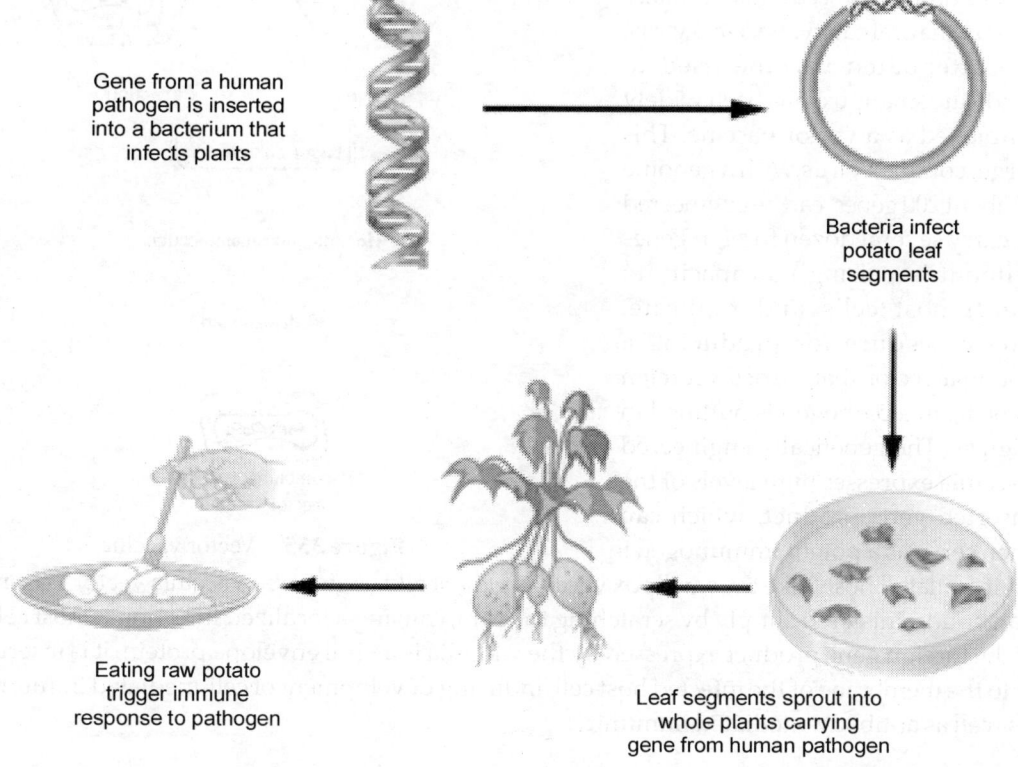

Gene from a human pathogen is inserted into a bacterium that infects plants

Bacteria infect potato leaf segments

Eating raw potato trigger immune response to pathogen

Leaf segments sprout into whole plants carrying gene from human pathogen

Figure 354 Edible vaccine

Attenuated recombinant vaccines

Salmonella, Vibrio, Leishmania sp are used to develop these vaccines.

Genes that encode major antigens of especially virulent pathogens can be introduced into attenuated viruses or bacteria. The attenuated organism serves as a vector, replicating within the host and expressing the gene product of the pathogen. A number of organisms have been used for vector vaccines, including vaccinia virus, the canarypox virus, attenuated poliovirus, adenoviruses, attenuated strains of Salmonella, the BCG strain of *Mycobacterium bovis*, and certain strains of streptococcus that normally exist in the oral cavity. Vaccinia virus, the attenuated vaccine used to eradicate smallpox, has been widely employed as a vector vaccine. This large, complex virus, with a genome of about 200 genes, can be engineered to carry several dozen foreign genes without impairing its capacity to infect host cells and replicate. The procedure for producing a vaccinia vector that carries a foreign gene from a pathogen is outlined in Figure . The genetically engineered vaccinia expresses high levels of the inserted gene product, which can then serve as a potent immunogen in

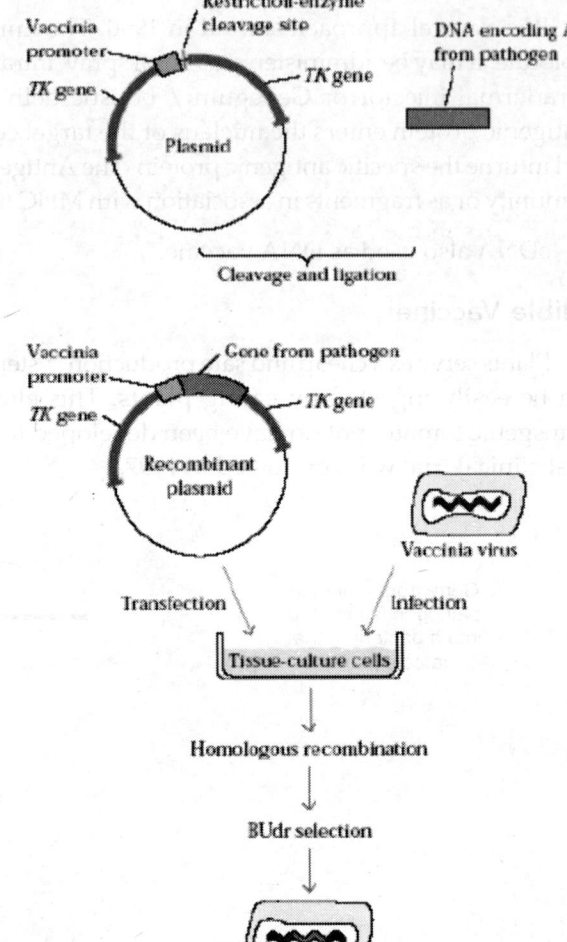

Figure 355   Vector vaccine

an inoculated host. Like the smallpox vaccine, genetically engineered vaccinia vector vaccines can be administered simply by scratching the skin, causing a localized infection in host cells. If the foreign gene product expressed by the vaccinia is a viral envelope protein, it is inserted into the membrane of the infected host cell, inducing development of cell-mediated immunity as well as antibody-mediated immunity.

Recombinant Antigen

Antigen made by making used of rDNA technology are called as Recombinant Antigen.

Refer HBs Ag preparation subunit vaccine.

Index